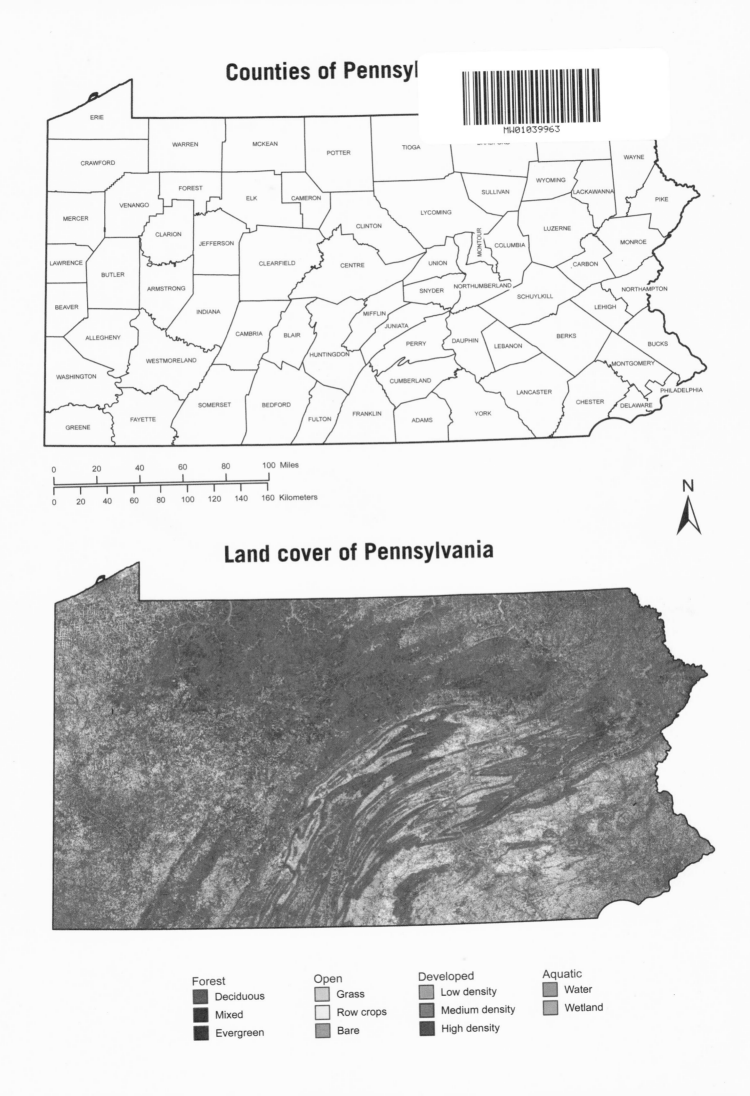

SECOND ATLAS OF

BREEDING

BIRDS *in*

Pennsylvania

SECOND ATLAS OF BREEDING BIRDS *in* Pennsylvania

EDITED BY

Andrew M. Wilson, Daniel W. Brauning,
and Robert S. Mulvihill

Geoff Malosh, *Photo Editor*

Catherine D. Haffner,
Production Assistance and Editing

Andrew Mack, *Bibliographer*

The Pennsylvania State University Press
University Park, Pennsylvania

This project was sponsored and published in association with the Pennsylvania Game Commission,
Carnegie Museum of Natural History, Audubon Pennsylvania, and the Pennsylvania Society for Ornithology.

CARNEGIE MUSEUM OF NATURAL HISTORY
ONE OF THE FOUR CARNEGIE MUSEUMS OF PITTSBURGH

LIBRARY OF CONGRESS CATALOGING-IN-PUBLICATION DATA

Second atlas of breeding birds in Pennsylvania / edited by Andrew M. Wilson, Daniel W. Brauning, and Robert S. Mulvihill ;
Geoff Malosh, photo editor ; Catherine D. Haffner, production assistance and editing ; Andrew Mack, bibliographer.
    p.   cm.
Summary: "Maps the current distribution of all of Pennsylvania's 190 breeding birds and documents
the changes in climate, habitat, and distribution since the first edition of this work.
Includes habitat analyses and color photographs for each species"—Provided by publisher.
Includes bibliographical references and index.
ISBN 978-0-271-05630-2 (cloth : alk. paper)
1. Birds—Pennsylvania—Geographical distribution. 2. Birds—Nests—Pennsylvania.
I. Wilson, Andrew M. II. Brauning, Daniel W. III. Mulvihill, Robert S. IV. Title: 2nd atlas of breeding birds in Pennsylvania.
QL684.P4S43 2013
598.09748—dc23
2012023648

Copyright © 2012 The Pennsylvania State University
All rights reserved
Printed in the United States of America
Published by The Pennsylvania State University Press, University Park, PA 16802-1003

The Pennsylvania State University Press is a member of the Association of American University Presses.

It is the policy of The Pennsylvania State University Press to use acid-free paper.
This book is printed on Fortune matte, which contains 10% post-consumer waste.

Cover artwork by Julie Zickefoose
This painting by Julie Zickefoose, for the cover of the *Second Atlas of Breeding Birds in Pennsylvania*,
was commissioned in memory of Dr. Kenneth C. Parkes (1922–2007), Curator of Birds at Carnegie Museum of Natural
History for more than 33 years. Dr. Parkes was a true aficionado of bird art, and Julie Zickefoose was one of a handful of
contemporary bird artists whose work easily pleased his very critical eye. Julie also painted the cover for the first *Atlas of
Breeding Birds in Pennsylvania*, which depicted a pair of Chestnut-sided Warblers at their nest.

The subject matter for this book's cover also honors Dr. Parkes, who wrote a seminal paper on the genetics of the Blue-
winged X Golden-winged warbler complex in the *Wilson Bulletin* in 1951, and who had a career-long fascination with avian
hybrids. The painting of a first generation male Brewster's Warbler (the offspring of a mixed Blue-winged X Golden-winged
warbler pair) bringing food to its Golden-winged Warbler mate on her nest reflects actual observations made during the
second Pennsylvania breeding bird atlas: On 12 June 2006 in an atlas block in south-central Pennsylvania, and on 18 June
2008 in a block in northeastern PA, Tom Johnson and Dustin Welch, respectively, each confirmed breeding for
Brewster's Warbler by observing a male and its Golden-winged mate carrying food for young.

To my wife, Sue, with thanks for your kindness, encouragement, and endless patience.

—ANDREW WILSON

To Marcia, my wife, for her wise counsel and loving support.

—DANIEL BRAUNING

From among the many hundreds of people acknowledged on the succeeding pages, all of whom I thank sincerely for making this project and this book possible, I would like to dedicate this work to the following individuals whose efforts, I believe, were signally important to the success of this endeavor: first and foremost, at The Cornell Lab of Ornithology, Jeff Gerbracht; at The Pennsylvania State University, Maurie Caitlin Kelly, Ryan Baxter, Doug Miller, and Patryk Soika; at Powdermill Nature Reserve, Cokie Lindsay, Mary Shidel, and my wife, Pam Ferkett; at Carnegie Museum of Natural History, Bill DeWalt, Janet Brombach, and Cathy Klingler.

—ROBERT MULVIHILL

Contents

Foreword *ix*
Acknowledgments *xi*

CHAPTER 1   Aims and Purposes   *1*
Robert S. Mulvihill

CHAPTER 2   The Geography of Pennsylvania   *3*
Andrew M. Wilson and Bernd J. Haupt

CHAPTER 3   Habitats and Habitat Change   *9*
Andrew M. Wilson, Joseph Bishop, and Margaret Brittingham

CHAPTER 4   Atlas Methods   *23*
Daniel W. Brauning and Michael Lanzone

CHAPTER 5   Analytical Methods   *38*
Andrew M. Wilson

CHAPTER 6   Coverage and Results   *48*
Andrew M. Wilson and Daniel W. Brauning

CHAPTER 7   Contributions to Conservation   *70*
Douglas A. Gross, Sarah Sargent, and Catherine D. Haffner

CHAPTER 8   Interpreting Species Accounts   *81*
Daniel W. Brauning and Andrew M. Wilson

CHAPTER 9   Species Accounts   *87*

APPENDIX A   Former Nesting Species   *471*
Daniel W. Brauning

APPENDIX B   Common and Scientific Names of Plants and Animals   *475*

APPENDIX C   Summary of Atlas Results by Physiographic Region (Shaded Gray) and Section   *477*

APPENDIX D   Habitat Associations   *494*

APPENDIX E   Analytical Methods—Statistical Details   *520*

APPENDIX F   Breeding Phenology   *523*

APPENDIX G   Atlas Safe Dates   *527*

Literature Cited   *531*
Index   *569*

Foreword

In 1811, Alexander Wilson—former itinerant Scottish peddler, convicted slanderer, hack poet, and unquestioned ornithological genius—took the first stab at censusing part of Pennsylvania's avian population.

"In Mr. Bartram's Botanic garden, and the adjoining buildings, comprehending an extent of little more than eight acres, the Author has ascertained, during his present summer residence there, that not less than fifty-one pairs of birds took up their abode, and built their nests within that space," Wilson (1808–1814) wrote in the preface to volume four of his *American Ornithology*.

However crude, Wilson's survey of what is now the historic Bartram's Gardens in Philadelphia may well have been the first breeding bird atlas in the Commonwealth. Among the 19 species he recorded were five pairs of House Wrens, a pair of Swamp Sparrows, two each of Baltimore and Orchard orioles, several pairs of "Warbling Flycatchers" and "White-eyed Flycatchers" (Warbling and White-eyed vireos), three pairs of "Summer Yellow Warblers," 10 pairs of Barn Swallows and five pairs of Gray Catbirds— "besides several others whose nests could not be found, but which were frequently observed about; such as the Blue Jay, the Humming-bird, Scarlet Tanager, &c, &c."

Times have changed—and when it comes to bird populations, they are always changing. Almost a century after Wilson conducted his first atlas, an amateur ornithologist named Frank L. Burns (1901, 1907) repeated the exercise a mile and a half from Bartram's Gardens and noted that six of Wilson's species had held their own, one (American Robin) had increased dramatically in abundance, while Swamp Sparrow and Yellow Warbler had disappeared, and the Purple Martin was all but gone.

A breeding bird survey, even a fairly casual attempt like Wilson's, is a snapshot of a fluid, always shifting avian landscape. Play it out on the 46,058-square-mile tapestry that is Pennsylvania, and you needn't wait a century to see the differences. Just the two decades between the first Pennsylvania breeding bird atlas (1983–1989) and the second (2004–2009) brought astounding changes in many bird populations, for good and ill.

That same period also brought a fast-growing sophistication among birders and ornithologists, abetted by amazing technological advances. All of this gave Pennsylvania's second Atlas organizers and volunteers the opportunity to capture incredible amounts of information about what they found in the course of the atlas, from the pinpoint location of individual nests to statewide population estimates. The result is perhaps the most comprehensive landscape-scale survey of any taxa, anywhere, at any time in history.

The undertaking was nothing less than staggering. The planning process incorporated complex land-cover and GAP models, Geographic Information System (GIS) layers, online reporting, and the ability for atlasers to track real-time species totals for their block against goals based on projections drawn from that block's habitat features. More than 2,000 volunteers covered virtually every square kilometer of the state, making more than 77,500 visits to all 4,937 atlas blocks, racking up more than 100,000 hours in the field in the process. They conducted specialized owl and waterbird surveys, while trained crews conducted eight-stop point counts in most blocks—more than 34,000 locations in all. The point counts tabulated not just species but individuals—and allowed, for the first time, estimates of actual statewide populations for more than half of the 190 breeding species detected by the atlas.

Thanks to this monumental effort, we now know that the five most abundant species in the state (Song Sparrow, Chipping Sparrow, Red-eyed Vireo, American Robin, and Gray Catbird) likely have populations ranging from 2.3 million pairs to just under 3 million pairs. In contrast, there may now be more formerly rare Common Ravens (8,000 pairs) than once-common American Kestrels (6,800 pairs).

But as the ravens and kestrels suggest, an atlas— however detailed and ambitious—is still a snapshot of always-dynamic populations. For that reason, one of the most compelling results of the second Atlas is how bird numbers and distributions have changed in the past two decades. This is a period that has seen drastic habitat losses to sprawl development at the same time that forests have continued to mature; decades that saw the decline of critical tree species like Eastern Hemlock to introduced diseases and pests, even as once-rare species of birds made spectacular comebacks.

What changes would a birder, plucked from 1983 and dropped into Penn's Woods today, perhaps notice first? A lot more Canada Geese and Wild Turkeys, for one thing; the goose population rose 393 percent, from an estimated 22,000 to 107,000 pairs, while turkeys jumped 387 percent, from 39,000 to 190,000 pairs. Yellow-bellied Sapsucker populations increased by a whopping 493 percent, from an estimated 8,100 pairs in the first Atlas to 48,000 pairs today, mostly in the state's heavily forested northern tier, where in some places they were more common than all other woodpeckers combined. Detections of another boreal species, the Yellow-rumped Warbler, jumped 278 percent between atlases, while the number of blocks in which Pileated Woodpeckers and Hooded Warblers were found increased by 1,000.

Red-bellied Woodpeckers, Carolina Wrens, Alder Flycatchers, and Common Ravens all were at least three times as widespread in the second Atlas as the first. Carolina Wrens, which ranked 66th in abundance in the first Atlas, jumped 27 places to 39th the second time around. Bald Eagles, found in just 52 blocks during the first Atlas (a time when reintroduction efforts with this species were just gaining traction) have exploded in number; eagles were recorded in more than 10 times as many blocks during the second Atlas.

On the other side of the ledger, Northern Bobwhite, Ring-necked Pheasant, and Wilson's beloved Purple Martin all declined during the same period to levels at barely one-third what they were in the 1980s. Vesper Sparrows, which had been roughly as abundant as Carolina Wrens in the mid-1980s, slipped from being the 69th most common species to 100th, an estimated population decline from 56,000 pairs to just 17,000.

Not surprisingly, the declining species were largely those, like Eastern Meadowlark, and Grasshopper and Field sparrows, dependent on grasslands, or like Yellow-breasted Chat, reliant upon early successional shrubland, two habitats that have suffered serious losses in the intervening years. (A few shrubland specialists, like Chestnut-sided Warbler and Alder Flycatcher, increased significantly between atlases—the warbler likely because of regenerating timber cuts, the flycatcher perhaps thanks to better wetland protection.)

The poster child for Pennsylvania's declining birds was the Golden-winged Warbler, which decreased 82 percent, the steepest drop of any species, from an estimated 35,000 pairs in the first Atlas to just 6,300 in the second, while the number of blocks in which it was found fell by more than 61 percent. The loss of successional habitat, coupled with hybridization with Blue-winged Warblers, continues to drive its steady slide—but the atlas results also suggest that targeted habitat management holds promise for aiding this beleaguered beauty.

All that was then. Literally since the moment when the last atlas report was filed in 2009, Pennsylvania's birdlife has continued to change, in ways that won't be fully evident for years. But the second Atlas has much to say about what we can expect in the decades to come.

For example, the second Atlas documented four new breeding species in the state—Trumpeter Swan, Merlin, Great Black-backed Gull, and Eurasian Collared-Dove. All four are species that have shown explosive population growth in other regions of North America, and with the possible exception of the gull, there is little reason to doubt their numbers will grow rapidly here, too—especially the collared-dove, which has swept across the continent in little more than two decades.

A further example, the number of blocks occupied by Black-throated Blue Warblers increased by more than 55 percent between atlases, suggesting that better management of White-tailed Deer herds has reduced, at least somewhat, the over-browsing of the forest understory where this and other species nest. However, political pressure to increase huntable deer populations remains high, and this tentative progress could quickly and easily be reversed.

By tracking the correlation between the presence of certain breeding birds and Eastern Hemlock, the atlas makes clear that the continuing loss of this vital conifer to woolly adelgid infestations will have a growing impact on many species of forest-nesting birds. Blackburnian Warblers, for instance, showed a striking dependence on hemlocks, especially in eastern Pennsylvania, suggesting that as adelgid losses spread, we may see significant changes in the range and abundance of this and other species.

The coming years carry other major challenges for Pennsylvania's avifauna. As Hemlock Woolly Adelgids move west and north, the invasive Emerald Ash Borer penetrated the state from the northwest, threatening a near-complete loss of native ash trees similar to what is already happening in the upper Midwest.

Marcellus and other deep shale natural gas drilling seems certain to have tectonic effects on Pennsylvania's natural landscape, from aquatic systems to profound levels of fragmentation and noise and light pollution on land. The birds most at risk will likely be those most dependent on very large, intact forests—which, by an unfortunate accident of geology, are exactly the kind of forests most at jeopardy from shale gas extraction.

Finally, the two decades just past already bore the stamp of climate change, with sweeping implications for many species of Pennsylvania birds in the future. While the public and politicians may argue over whether climate change is even real, many of the Commonwealth's birds have already begun responding to it, and in the species accounts that follow, atlas data show time and again that still-subtle climate effects are changing bird distributions in the state.

Nothing ever stays the same, especially not birds. The mountain of data generated by the second Atlas, carefully analyzed and interpreted by the research team and the many authors who contributed to this volume, confirm that—and offers the most extraordinary window ever into the avifauna of this, or any other, place.

Scott Weidensaul

Acknowledgments

Principal funding for this project was provided by:

The Pennsylvania Game Commission, with support from State Wildlife Grant and Wildlife Conservation and Restoration programs, administered by the U.S. Fish and Wildlife Service, Wildlife and Sport Fish Restoration Program.

Additional funding and in-kind support were received from:

Pennsylvania Department of Conservation and Natural Resources
    Pennsylvania Bureau of Forestry
    Pennsylvania Bureau of State Parks
    Wild Resource Conservation Program
Carnegie Museum of Natural History
U.S. Department of the Interior
    National Park Service
Penn State Center for Environmental Informatics
Penn State Institutes of Energy and the Environment
    Pennsylvania Spatial Data Access
    Riparia (formerly the Cooperative Wetlands Center)
Penn State School of Forest Resources
The National Aviary
Gettysburg College

Support for the project was provided by contributions from the following sponsors the 2005 and 2006 "World Series of Birding" teams of atlas point counters (Lewis Grove, Michael Lanzone, Andrew McGann, Cameron Rutt, Paul Sweet, Drew Weber, Andrew Wilson, and John Yerger):

Bill DeWalt, Duane Diefenbach, Jason Dundon, Barb Elliot, Pam Ferkett, Holly Ferkett, Brigid and Ross Ferkett, Michael Fialkovich, Mike and Evelyn Fowles, Theodore Fuller, Roana Fuller, Dan Galbraith, George and Jane Greer, Gregory Grove, Mark Henry, Anna Hoenstine, Tom and Sally Hoover, Dennis and Debra Huber, Katriona Shea, Nick Kerlin, Tom and Janet Kuehl, Alan Maceuchren, F. Arthur McMorris, Theresa Menotti, Dr. David Miller, Jean Miller, Tim Miller, Mead J. Mulvihill, Jr., Thomas Nimick, Jr., Tracy and Tim O'Connell, Joe Pajer, Arthur Scully, Jr., Lula Sheuy, Stephanie Shields, David Smith, Rosemary Spreha, Patty Steffey, Dan Stoltzfus, Katherine Stoner, Dana L. Stuchul, John Thomas, John Thorne, Todd Bird Club, Mike Ward, Nelson and Grace Weber, Leon Weber, Andrew Wilson, Margot Woodwell, Brent Worls, and Lisa Zinn.

Generous financial support to develop and subsidize the publication of this book was provided by:

Pennsylvania Game Commission
Colcom Foundation
Phil Keener
Delaware Valley Ornithology Club
National Audubon chapters (Appalachian, Bartramian,
    Conococheague, Greater Wyoming Valley, Lehigh
    Valley, Presque Isle, York, and Tiadaghton)
Pennsylvania Society for Ornithology
Baird Ornithological Club

The second Pennsylvania breeding bird atlas logo was designed and painted by Larry Barth.

Regional Coordinators

Regional Coordinators played an integral role in all aspects of work with atlas volunteers and data collection. These leaders in bird conservation deserve tremendous recognition for their dedication to recruiting volunteers, coordinating fieldwork, and continually reviewing data for their regions.

26/28	John Tautin	36	Jeff Holbrook
27/29	Chuck Gehringer	37	Robert Fowles
30	Florence McGuire	38	Jerry Skinner
31	Don Watts, Ted Grisez, and Scott Stoleson	39	Nancy Wottrich
32	John Fedak	40	Barbara Leo
33	David Hauber and Keith McKenrick	42	Randy Stringer, Linda Wagner, and Suzanne Butcher
34	Mary and Larry Hirst		
35	Robert Ross and Philip Krajewski	43	Gary Edwards and Russ States

44	Carole Winslow and Michael Leahy
45	Patricia Conway
46	Jocelynn Smrekar
47	Bob Martin
48	Greg Grove and Nick Bolgiano
49	Wayne Laubscher
50	Daniel Brauning
51	Douglas Gross
52	Bill Hintze and Chuck Berthoud
53	Glenn Czulada and Jim Hoyson
54/55	Terry Master
56	Jim Bonner
57	Brian Shema and JoAnn Albert
58	Mark McConnaughy
59	Margaret and Roger Higbee
60	Rory Bower
61	Margaret Brittingham
62	James Dunn and Roana Fuller
63	Mark Henry
64	Allen Schweinsberg
65	Deuane Hoffman
66	Mike Ward and Dave Kruel
67/68	Arlene Koch and Bernie Morris
70	Roy Ickes
71	Mike Fialkovich
72	Dick Byers
73	Janet and Tom Kuehl
74	Neil Woffinden
75	David and Trudy Kyler
76	Mike Lanzone and Dan Ombalski
77	Ramsay Koury
78	Sandy and Gary Lockerman
79	Randy Miller and Rosemary Spreha
80	Steve and Sue Fordyce
81	F. Arthur McMorris
82/83	Bill Etter
84/85	Terry Dayton
86/87	Mark Bowers and Jeff and Retta Payne
88/89	Dan Snell and Regina Reeder
90/91	Dale Gearhart
92/93	Chuck Berthoud and Chris Cunningham
94/95/96	Doris McGovern

Thanks also to these RCs who helped us get started but who were unable to continue as coordinators after the first season: JoAnn Davis, Barbara McGlaughlin, Karen Lippy, and Mark Blauer.

Atlas Volunteers

This project was possible only with the dedication of many birders who volunteered their time and resources to survey Pennsylvania's atlas blocks. We have tried to account for every individual who contributed data to the second Atlas project. We apologize for any omissions or errors in the following listing of volunteer participants in the second Pennsylvania breeding bird atlas.

Travis Abbott
Mary A. Ache
James Acre
Larry Adams
Laura Agee
Roberta Agnew
Cindy Ahern
Theresa Alberici
JoAnn Albert
Kyle Aldinger
Joan Alexander
John Alexander
Kathy Alexandroff
Denise Alleman
Mervin Allgyer
Diane Allison
Daniel Altif
Robert Alwine
Angela Amadio
Robert Amadio
Bill Ambrose
Tom Amico
Julia Amsler
Wendy Andersen
Ann Anderson
Crystal Anderson
Doug Anderson
Merrilee Anderson
Michelle Anderson
Robert Anderson
Dave Andre
Janice Andrews
John Andrews
Scott Angus
Ron Antonucci
Jill Argall
David Argent
David Arrow
Sean Artman
Terry Ashbaugh
Leigh Ashbrook
Mary Assenat
John Atkins
Marianne Atkinson
Jana Atwell
Chad Atwood
Karen Atwood
Audrey Aubrecht
Virginia Aughenbaugh
Dale Aulthouse
Carole Babyak
Lisa Bainey
Sally Bair
Linda Lou Baker
Anna Maria Bakermans
Nancy Baker
Terry Baker
Robert Baldesberger
Sharon Baldridge
Bob Bale
Robert Ballantyne
Colette Ballew
Ken Balliet
Judy Bamburak
Cristine Banaszak
Matt Bango
Dan Barber
David Barber
Patti Barber
Barbara Barchiesi
Tammy Barette
Chelsea Barnes
Joe Barnes
Sherman Barnes
V. Barnes
Jane Barnette

Jason Barnhart
Karen Barnhart
Hannah Barrett
Larry Barth
Jonathan Bastian
William Bastian
Michele Batcheller
Harry Bates
Douglas Bauman
Fern Bauman
Barbara Baxter
Ron Beach
Jon Beam
Scott Bearer
Joe Beatrice
Alice Beatty
Dave Beatty
Dennis Beaver
Carl Beck
Doug Becker
Russell Becker
Terry Becker
Tim Becker
Michele Beckett
Debbie Beer
George Beidler
Eli Beiler
Mary Ann Belin
Stephen Belin
Bud Bell
Ralph Bell
Richard Bell
Mahlon Bender
Charles Bennett
Janis Bennett
Kurt Bennett
Rebecca Bennett
Barbara Benson
Buck Benson
Andy Berchin
George Bercik
Roland Bergner
Henry Berkowitz
Debbie Bernatt
Nancy Bernhardt
Bev Bernoske
Joe Bernoske
Ed Bernot
James Berry
Raymond Berry
Anna Bert
Chuck Berthoud

Tom Betts
William Betts
Gloria Bickle
Paula Bieber
David Bierly
Diane Bierly
Albert Bilheimer
Nancy Bilheimer
Scott Bills
Christina Binder
Lois Bingley
Derry Bird
Barbara Birosik
Ann Bismal
Jim Bissell
Kathy Blaisure
Sidra Blake
Carolyn Blatchley
Joyce Blauch
Mark Blauer
Jenny Bloom Lisak
Linda Blosel
David Blubaugh
Barry Blust
Robert Blye
John Boback
David Bock
Debbie Bodenchatz
Tom Bodenchatz
Ann Bodling
Janice Boehm
Christopher James
    Bohinski
Hendrika Bohlen
Sue Boland
Vaughan Boleky
Nick Bolgiano
Gerry Boltz
Nicole Bond
David Bone
Roseann Bongey
Jim Bonner
Marcia Bonta
Aaron Boone
George Boone
Dorothy Bordner
Carole Borek
Marty Borko
Richard Boshart
Devin Bosler
Gordon Bosler
Howard Bowen

Joan Bowen
Rory Bower
Mark Bowers
Christy Bowersox
JoAnn Bowes
Steve Boyce
Bob Boyd
Mark Boyd
Roy Boyle, III
Jineen Boyle
Deborah Brackbill
Gary Brackbill
Ken Bracken
Sandy Bracken
Tim Bradley
David Brandes
Susan Brandt
Kevin Brant
Emily Brault
Dan Brauning
Marcia Brauning
Sandra Brauning
Stephen Brauning
Becky Brawdy
Keith Breitenstein
Erica Brendel
Mary Brenner
Kas Breslin
Joe Brewer
David Brinker
Jeff Brinker
Margaret Brittingham
Fritz Brock
Judy Brooks
David Broussard
Arlene Brown
Barbara Brown
Larry Brown
Larry Brown, Sr.
Marcia Brown
Mary Brown
Mick Brown
Shoko Brown
Timothy Brown
Walter Brown
Joane Brubaker
Paul Brubaker
Carl Brudin
Jessica Bruland
Debbie Bryant
Janice Buckingham
Melody Buck

Margaret Buckwalter
Ted Buckwalter
Terry Buckwalter
Ron Burkert
Dave Burket
Gregory Burkett
Phil Burkhouse
Sandra Burwell
Joe Busowski
Sue Busowski
Suzanne Butcher
Nan Butkovich
Amy Butler Adams
Stephanie Butler
Terry Butler
Ken Byerly
Dick Byers
Ivan Byler
Marvin Byler
Brian Byrnes
Bill Cadamore
Lois Callahan
Bob Campbell
Elaine Campbell
John Campbell
Karen Campbell
Michael Carey
Bruce Carl
Peter Carlen
Logan Carlton
Marion Carmack
Lee Carnahan
Joseph Carragher
Sarah Carr
Lisa Cass
Mark Catalano
Nancy Chace
Thomas Chaffee
Chuck Chalfant
Lisa Chapman
Ray Charnick
Shirley Chase
Nancy Chludzinski
Lisa Chmil
Larry Choby
David Chronister
Ed Chubb
Frank Chubon
Bill Chupko
Henry Chupp
Joe Church
Sue Cibula

Joshua Clapper
Gary Clark
L. William Clark
Nancy Clark
Ian Clarke
Tom Clarke
Brian Clauser
Tom Clauser
Melanie Claus
Dean Claycomb
Carol Clemens
Dave Clemens
Leslie Clifford
Teresa Clouser
Nancy Clupper
Larry Coble
Ellie Cochran
Lisa Colangelo
Paulette Colantonio
Christopher Colby
Bev Cole
Robert Coley
Larry Collura
Ruth Collura
Debbie Colosi and Family
Flossy Comstock
Ron Comstock
Roger (Skip) Conant
Charles Conaway
Jonathan Confer
Jennifer Conner
Viv Connor
Norm Conrad
Jady Conroy
Richard Conroy
Tom Contreras
Mike Conway
Patricia Conway
Sally Conyne
Deborah Cook
David L. Cooney
Martha Coop
Phillip Cooper
John Corbett
Clay Corbin
John Corcoran
Troy Corman
Anne Cortese
Pasquale Cortese
John Coughenour
Sue Coughenour

Ben Coulter
Susan Courson
Laurie Cowan
Roxann Cox
Mary Craig
Richard T. Cramer
Charles Cravotta
Janet Crawford
Tom Crawford
Dan Creighton
Anita Cressler
Walter Cressler
Kevin Crilley
Robert Criswell
Burt Crowell
Lewis Crowell
Charles Crunkleton
Chris Cunningham
Jean Stull Cunningham
Megan Cunningham
Toby Cunningham
Andrew Curtis
Laura Curtis
Carole Cyphert
Susan Cyr
Glenn Czulada
Alex Dado
Jon Dale
Moira Daly
Barbara Daniels
Bob Daniels
Deborah Danila
Dave Darney
Debbie Darney
Dana Datko
John Datko
Sheree Daugherty
Amy Davis
Brian Davis
Deborah Davis
JoAnn Davis
Josh Davis
Leona Davis
Mary Dawson
Joshua Day
Kimberly Day
Terry Dayton
Barb Dean
George Dean
Jim Dearing
Jonathan Debalko
Anthony De Bonis

Conrad Decker
Susan Dees
Jason Deeter
Karyn Delaney
Jane Delhunty
Martin Dellwo
Colleen DeLong
Joe DeMarco
Jack Deming
Judy Deming
John Dennehy
Nancy Dennis
Dave DeReamus
Robert Derr
Karen DeSantis
Thelma Detar
Scott Detwiler
Gerard Dewaghe
Norman DeWind
Aldolf Deynzer
Gloria Dick
Sally Dick
Thomas Dick
Dallas DiLeo
William Dingman
Kathy Dinsmore
Edward Dix
Donald Dixon
Katharine Dodge
Jim Dolan
Anita Donahue
Rob Donahue
Ned Donaldson
Ed Donley
Denise Donmoyer
Walt Donnellan
James Dowdell
Jack Downs
Michael Drake
Carl Drasher
Carolyn Drasher
Jay Drasher
Jon Dressler
Winifred Dressler
David Drews
Paul Driver
Ted Drozdowski
Joan Duffield
Mary Dundon
James Dunn
John Dunn
Lindsey Duval

Holly Dzemyan
John Dzemyan
Richard Eakin
Jane Earle
Janice Easler
Chris Eberly
David Eberly
Julia Ecklar
Sarah Edge
Robert Edkin
Gary Edwards
Wesley Egli
Brandon Eicher
June Elder
Barb Elliot
Bill Ellis
Bob Elmer
Jennifer Elmer
Catherine Elwell
Adam Erb
Emanuel Erb
Stephen Erb
Robert Erdman
Eric Erickson
Wanetta Escherich
Bill Etter
Bill Evans
Bruce Evans
Sherry Eyster
Richard Faber
David Facey
Joy Fairbanks
PJ Falatek
Lori Falcone
Devich Farbotnik
Iris Fark
Linda Farley
Nancy Faryniak
Will Faux
Bailey Fedak
Danny Fedak
John Fedak
Lisa Fedak
Barbara Feigles
Robert Felton
Shirley Fenstermacher
Rob Fergus
Helen Ferguson
Scott Ferguson
Holly Ferkett
Pam Ferkett
Ross Ferkett

Gary Ferrence
Ruth Ferrier
Dave Ferry
Liam Ferry
Tiarnan Ferry
Betsy Fetterman
Tom Fetterman
Carl Fetzner
Mike Fialkovich
Bert Filemyr
Douglas Filler
Jeff Finch
Charlie Finkbiner
Linda Finley
James Finn
Barbara Fisher
Gideon Fisher
Pamela Fisher
Patience Fisher
Brandon Flaim
Randy Flament
Marilyn Flannery
Phyllis Flasher
Herbert Flavell
Sally Fleming
Jamie Flickinger
Colette Flory
Carol Flowers
Jackie Flynn
Joel Folman
Brad Foltz
Steve Fordyce
Sue Fordyce
Robert Fowles
David Fox
Eric Fox
Peter Fox, Jr.
Donna Foyle
George Franchois
Gloria Francis
Dianne Franco
Linda Frantz
Bill Franz
Deborah Freed
Gary Freed
Patsy Freed
Ron Freed
Linda Freedman
John Freiberg
Ronald French
Don Frew
Marie Frew

Sandy Frey
John Fridman
Bob Friederman
Fred Fries
Rich Fritsky
Nate Fronk
William Frost
Jason Frunzi
Kevin Fryberger
Roana Fuller
Kevinn Fung
Aaron Furgiuele
Walter Fye
Lois Gagermeier
Judy Galbraith
Robin Galebach
Jane Galgoci
Ross Gallardy
Anne Galli
Joan Galli
Todd Garcia-Bish
Dave Gares
Abigail Garfield
Carl Garner
Vernon Gauthier
Logan Gaydos
Dale L. Gearhart
Jay Gebhard
Ronald Gehman
Chuck Gehringer
Dave Gelnett
Don Gensemer
Kevin Georg
Trudy Gerlach
Randi Gerrish
Sarah Gerrish
Janet Getgood
Joe Giacomin
Mario Giazzon
B. Gibson
J. Gibson
Don Gilbert
Jen Gilbert
Wayne Gillespie
Kathleen Gill
Ann Gilmore
Bonnie Ginader
Deborah Gingrich
Lucy Glass
Carolyn Glendening
Shirley Glessner
Gary Glick

Dave Gobert
Dave Gochnauer
Christopher Goguen
Mark Golden
Connie Goldman
Greg Gondella
Candy Gonzalez
Merrill Gonzalez
Kathie Goodblood
Laurie Goodrich
Sandra Goodwin
Gregg Gorton
Steve Gosser
Alain Goulet
Lee Gourley
Edward Gowarty, Jr.
Louise Grace
Shelley Gracey
Steven Graff
Andrew Graham
Robert Grajewski
William Graves
Jim Gray
Maryann Gray
Chris Grecco
Joseph L. Greco, Jr.
Diane Greeley
James Greeley
Ross Greeley
Samantha Greeley
Jo-Ellen Greene
Sayre Greenfield
Thomas Greg
Ian Gregg
Alan Gregory
George Gress
Mary Grey
David Griffin
Debra Grim
Ellen Grim
Kerry Grim
Diane Grimes
Twila Grimm
Janice Grindle
Ted Grisez
Mary Grishaver
William Grohal
Len Groshek
Donald Gross
Douglas A. Gross
David Grove
Deb Grove

Greg Grove
Lewis Grove
Ron Grubb
James Gruber
Trish Gruber
Al Guarente
Carol Guba
Tom Guilfoy
Jason Gulvas
Paul Guris
Robert Gutheinz
Delia Guzman
Barb Haas
Frank Haas
Michael Haas
Nelson Haas
Nikolas Haass
Fred Habegger
Catherine D. Haffner
Aaron Haiman
John Haire
Maryann Haladay-Bierly
Maggie Hallam
Dave Hall
Eric Hall
Eric M. Hall
Kathy Hamilton
Yvonne Haney
Brian Hanna
Jake Hanz
Mary Ann Hanzok
Nate Hardic
Brian Hardiman
Diana Harding
Eleanor Harding
Jim Harkless
Bob Harlan
Elaine Harmon
Earl Harnly
Ralph Harrison
Hank Hartman
Holly Hartshorne
Jeff Hartzell
Joan Hartzell
Mike Harvell
Margret Hatch
David Hauber
Lisa Haught
Cindy Havey
Donna Hawbaker
Rita Hawrot
Douglas Hay

Molly Heath
Jennifer Heckler
Kay Heffner
Frederick Heilman
Allison Heinrichs
Janet Heintz
Annick Helbig
Rolf Helbig
Larry E. Helgerman
Jonathan Heller
Pat Henckler
John Henderson
Raymond Hendrick
Bill Hendrickson
Marsha Hendrickson
Don Henise
Marc Henning
Gary Hennip
Darcy Henry
Mark Henry
Frank Hentschel
Peggy Sue Hentze
Claus Herrmann
Fay Hess
John Hess
Len Hess
Linda Hess
Paul Hess
Stan Hess
Ted Hevener
John Hewett
Todd Hickman
Barbara Hiebsch
Eileen Higbee
Margaret Higbee
Roger V. Higbee
Dusty Hilbert
Sandy Hilbert
Carol Hildebrand
Allen Hill
Amy Hill
James R. Hill, III
Karen Hiller
Carol Hillestad
William Hill
Wayne Hintze
William Hintze
Larry Hirst
Mary Hirst
Anh Ho
David Hochadel
Judy Hochadel

Deuane Hoffman
Hugh Hoffman
Joyce Hoffmann
Judie Hogan
Jeffrey Holbrook
Jennifer Holgate
Nancy Holland
Robert Holman
Jeff Holmes
Michael Holmes
Pam Holmes
Craig Holt
Claire Holzner
Brenda Homan
Kenda Hoovler
Jeff Hopkins
Timothy Hoppe
Janice Horn
Jason Horn
Kathy Horn
Rita Horrell
Kerin Horton
Erika Hough
Donna Housel
Joseph Hovis
Jerry Howard
Marjorie Howard
Rod Howard
Angela Howe
Bill Howe
Daniel Howe
Elaine Howe
Pat Howell
Steve Hower
Joan Howlett
James Hoyson
Matthew Hoyt
Dick Hribar
Linda Huber
Jack Hubley
Tim Hudspath
P. Huff
Thomas Huff
Gabrielle Hughes
Liz Hughes
Carol Hughey
Ron Hughey
Richard Humbert
Joseph Hummer
Kathy Hummon
Cathy Huneke
Ethan Huner

Jeanette Hunkins
Barbara Hunsberger
Deborah Hunsberger
Elizabeth Hunsberger
Peter Hunsberger
Ann Hunt
Connie Hunt
Matthew Hunt
Anne Hurst
Dave Hurst
Katherine Hurst
Corey Husic
Pat Hutcheson
Anna Hutzell
Jane Hyland
Roy Ickes
Bonnie Ingram
Linda Ingram
Ronald Intrieri
Benjamin Israel
Karen Jackson
Laura Jackson
Merle Jackson
Michael Jackson
Dory (Dorothy) Jacobs
Sharon Jacoby
Angelica Jaeger
Bob James
Marie Janoski
Richard Janoski
Tony Jantosik
Mick Jeitner
James Jenkins
Anna Jennings
Ben Jesup
Pat Johner
Denise Johnson
Gail Johnson
Gerri Johnson
Kristina Johnson
Mark Johnson
Ronald Johnson
Sheryl Johnson
Tom Johnson
Virginia Johnson
Connie Johnston
Noel Jones
Anna Jordan
Jane Jordan
Susan Jordan
John Jose
Barb Jucker

Jack Julian
Carl Juris
Nancy Juris
Stephen Kacir
Georgia Kagle
Deborah Kalbfleisch
Brian Kamin
Ezra Kanagy
Mark Kandel
Corey Kanuckel
Brian Karaffa
Stefan Karkuff
Alyssa Karmann
Chad Kauffman
Jeanne Kauffman
Jon Kauffman
Norm Kauffman
Kevin Kearney
Richard Kearns
Peggy Keating-Butler
Bob Keay
Bob Keener
Phil Keener
Robert Keener
William Keim
Lynn Keiser
James Kellam
Jeff Keller
Rudy Keller
Sandra Keller
Wesley Keller
Pamela Kelly
Tim Kelsey
Daniel P. Kelso
Donna Kempt
Dean Kendall
Earl Kenepp
Barb Keough
Kathy Kerber
Nick Kerlin
Nicole Kerlin
Steve Kerlin
Craig Kern
Ed Kern
Joe Kern
Kathy Kern
Margie Kern
Rhoda Kern
Robert Kerner
Rick Keyser
Michael Kiernan
Nancy Ellen Kiernan

Rosemary Kihm
John Kilmer
Jeffrey Kimmel
Ian Kindle
Alison King
Hope King
Jim King
Gary Kinkley
Scott Kinzey
Tina Kirkpatrick
Raymond Kirstein
Tom Kisiel
Stephen Kistler
Timothy Kita
Heather Klees
Daniel Klem, Jr.
Dave Klindienst
Norma Kline
Doris Klint
Terry Kloiber
Daniel Knarr
Dennis Knauss
Danny Knestaut
Cicily Knight
Joan Knight
Joan M. Knight
Kent Knisley
Andrea Knoll
Arlene Koch
Walter Koerber
Meg Kolodick
J. J. Kondrich
Joe Kosack
Stan Kotala
Ramsay Koury
Michael Kovach
Frank Kovaloski
Rick Koval
Helen Kowal
Kristine Kraeuter
Philip Krajewski
Patrick Kramer
Walt Krater
Paul Kreiss
Ann Kriebel
Mary Jane Krotz
Dave Kruel
Nancy Kruel
Dave Kubitsky
Kathryn Kuchwara
Janet Kuehl
Tom Kuehl
Connie Kumer
Dan Kunkle
Michael Kuriga
Thomas Kurtz
David Kyler
Trudy Kyler
Sherri Labar
Carroll Labarthe
Gary La Belle
Patty Lambert
Clayton Lamer
Gloria Lamer
Alex Lamoreaux
Michael Lanzone
Jeff Larkin
John Larocca
Mark Larson
Wayne Laubscher
Todd Lauer
Carole Laughlin
Mary P. Lawrence
Dave Lazor
Jo Leachman
Michael Leahy
Carol Leathem
Amy Leber
Andy Leber
Robert C. Leberman
Ronald F. Leberman
Harold Lebo
Ken Lebo
Jen Lee
Paul Lehman
Dot Leistner
William Lenhart
Joyce Lent
Barbara Leo
Adrienne Leppold
Jen Le Sage
John Leskosky
Jon Levin
Steve Lichvar
David Liebmann
Ann Liebner
Jane Light
Richard Light
Annie Lindsay
Cokie Lindsay
Joe Lipar
Karen Lippy
Barbara Liptak
Emma Lisak
Amanda Livingston
Michael Livingston
E. Loch
Bruce Lockard
Gary Lockerman
Sam Lockerman
Sandy Lockerman
Mark Loewen
James C. Logan
Lonnie Logan
David Long
Joan Lovenbury
Jim Lowe
Gladys Luckenbaugh
Robert Luckenbaugh
Chris Lundberg
Sheila Lunger
Rosemary Lunz
Peter Lusardi
Raymond Lusebrink
Dale Luthringer
Michael Lyman
Dan Lynch
Kevin Lynch
Patrick Lynch
Sherron Lynch
Judith Lynn
Cathy Lyon
Bob Machesney
Donald Mackler
Steve Maczuga
Gerard Madden
Diane Madl
Ryan Magaskie
Mark Major
Carol Majors
Nicholas Mallos
Jim Malone
Regis (Dutch) Maloney
Geoff Malosh
Barbara Malt
Evan Mann
Justin Mann
Charlie Manners
Steve Manns
Audrey Manspeaker
Jody Marcell
James March
Robert March
Jenine Marcus
Tony Marich
Dillon Marino
Andrew Markel
Phoebe Markley
Beth Marshall
Ken Marshall
Matt Marshall
Karlin Marsh
Eva Martin
Jay Martin
Marilyn Martin
Moses Martin
Robert Martin
Russell Martz
Everett Mashburn
Albert Massey
Terry Master
William Mathay
Dennis Mattison
Dale Matuza
Kasia Matyniak
Dave Maust
Bill May
Jane May
Julie Maynard
Daniel Mazlik
Michelle McAllister
Elaine McCann-Bourquin
John McCarthy
Kyle McCarty
Nancy McCaughey
Sam McCaughey
Terry McClelland
Julie McCloskey
Mark McConaughy
Cindy McConnell
Kathleen McConnell
Carol McCullough
Fred McCullough
Molly McDermott
Sally McDermott
Kent McFarland
Andrew McGann
D. L. McGann
Clyde McGinnet
Patty McGinnis (and her 7th grade class)
Barb McGlaughlin
Bill McGlone
Rosemary McGlynn
Doris McGovern
Patrick McGovern
Cindee McGrady

ACKNOWLEDGMENTS xvii

Kelly McGraw
Florence McGuire
Jim McGuire
Marty McKay
Thomas McKenrick
Michael McKiniry
Thomas McKinne
Allen McLaughlin
Jeff McLaughlin
Mark McLaughlin
Wendy McLean
F. Arthur McMorris
Bruce McNaught
David McNaughton
Joel McNeal
Kyla McNulty
Frances McVay
Bob Meacham
Debby Meade
Rick Means
Doris Mearig
Rodney Mee
Robert Megraw
Rick Mellon
John Mercer
Robert Mercer
Holly Merker
Elizabeth Mescavage
Darryl Messinger
Ellwood Meyers
Alfredo Miceli
Jeff Michaels
Bob Michny
Dolly Mignogna
Joseph Mikelonis
Madeline Miles
Ben Miller
Betsy Miller
Char Miller
Emanuel Miller
Eric Miller
Jason Miller
John C. Miller
Kate Miller
Linda Miller
Lisa Miller
Mary Miller
Matthew Miller
Oscar Miller
Randy Miller
Rebecca Miller
Rick Miller

Steven Miller
Tom Miller
Trish Miller
Steve Millere
Daniel Mink
August Mirabella
Janet Mitchell
John Mitchell
Judy Mitchell
Denise Mitcheltree
Marion Mitterer
Betsy Moelk
Peter Moffett
Jake Mohlmann
James Molyneaux
Nadine Molyneux
Colette Monier-
  Coughlin
Judy Montgomery
Maxine Montgomery
Ron Montgomery
Susan Montgomery
Dawn Moore
Lisa Moore
Luis Moore
Marilyn Moore
Owen Moore
Tracey Moore
Greg Morell
Bernadette Morey
Bernie Morris
Amy Morrison
James Morrison
Mary Morrison
Pauline Morris
Eugene Morton
Meg Moses
Joanna Moyer
Mark Moyer
Mary Moyer
Paul Muehlbauer
William Muffley
Lucas Mulfinger
Mary Mulligan-Haines
Robert Mulvihill
Ron Mumme
Dan Mummert
Lisa Mundy
Paul Mundy
Roger Munnell
Bill Murphy
Bob Murray

John Murray
Chuck Musitano
Linda Musser
Andy Myers
Herb Myers
Diane Nace
Anna Marie Nachman
Steve Naugle
Beth Nazelrod
Marge Neel
Susan Neff
Rob Neitz
Beth Nelson
Brad Nelson
Marguerite Nemeth
Kirby Neubert
Harry Neuhard
Felicity Newell
Elizabeth Nicholson
Jeremy Nicholson
Clare Nicolls
Alison Norris
Gary North
Paul Novak
Richard Nugent
Vincent O'Boyle
Michael O'Brien
Gene Odato
Patrick O'Donnell
William Oesterlin
Roy Ogburn
Linda Olczak
Melody Oligane
Fran O'Malley
Christina Ombalski
Dan Ombalski
Katie Ombalski
Linda Ordiway
Rett Oren
Anita Orlow
Claire Orner
Rusty Orner
Linda Orr
Don Orris
Mickey Orris
Stephanie Oss
Lynn Ostrander
Jane Ostroski
Allen Oswald
Jean Oswald
Harold Otto
Sarah Pabian

Martin Page
Annette Paluh
Jeanne Parker
Jason Parkhill
Linda Parlee-Chowns
Edie Parnum
Pamela Parson
Shiela Parsons
Georgette Pascotto
Kathleen Patnode
Audrey Patterson
Eric Patton
Judith Pavelosky
Tom Pawlesh
Chris Payne
Jeff Payne
Retta Payne
David Peachey
Marvin Peachey
Tim Pearce
Alan Pearson
Tom Pearson
Elizabeth Peck
Ernest Peffley
Tony Pegnato
Katie Peight
Louie Peight
Barb Pellam
Carol Pellegrino
Don Pellegrino
Maria Pellegrino
Becky Peplinksi
John Peplinski
Carl Perretta
Jason Perrone
Cal Peterka
Elmer Petersheim
Toby H. Petersheim
Toby J. Petersheim
Ann Pettigrew
Matthew Pettigrew
Doug Phillips
James Phillips
Mathew Phillips
Randy Phillips
Jane Pianovich
Anthony Piccolin
Mark Piekarski
Betsy Piersol
Emily Pifer
Matt Pillars
Alexandra Pinamonti

Steve Pinkerton
Jay Pitocchelli
Linda Pizzela
Wenda Plowman
Tom Pluto
Joann Pochciol
Mike Pochedly
Elizabeth Pokrivka
John Pokrivka
Shelley Pokrivka
Ralph Policichio
Robert Polito II
John Porter
Peggy Porter
William Potter
Rick Povich
Christopher Powell
Joanne Prestia
Gladys Price
Janet Price
Kenneth Price
David Prine
Bill Printz
Mike Pruss
Nick Pulcinella
Dan Pushic
Chase Putnam
Donna Queeney
Gallus Quigley, Jr.
Rosella Radek
Brian Raicich
JoAnn Raine
Jeff Raisch
Catherine Rakow
Paul Rakow
Daniel Ramond
Larry Ramsey
Mary Ramsey
Gail Randall
Sharyn Randig
Steve Rannels
Tom Raub
David Rausher
Chris Rebert
Nancy Redell
Regina Reeder
Jack Reese
Kathy Reeves
Barbara Rehrig
Rich Rehrig
Bill Reid
Carol Reigle
Lee Ann Reiners
Jan Reinhardt
Joe Reinke
Ed Reish
Tink Reish
Donna Reitz
Joan Renninger
Andy Renno
Gideon Renno
Dan Rensel
Eric Rensel
Catherine Renzi
Oren Rett
Theresa Reynolds
Bill Rhone
Sue Ricciardi
Gloria Richardson
Jim Ridolfi
Rich Rieger
Ken Rieker
Dan Riggle
Jerri Rigo
Kathryn Riley
Donna Rinker
Bill Ritchey
Emily Rizzo
Bill Roache
Stone Rob
Chandler Robbins
Richard Robbins
Thomas Roberts
David Robertson
Peter Robinson
Rhonda Robinson
Rick Robinson
David Rockey
Matthew Rockmore
Bruce Rodgers
Gaye Rodgers
Keely Roen
Cindy Rogers
Nancy Romza
Bill Roscher
Emily Rose
Michael Rosengarten
Elizabeth Rosevear Grove
Patricia Rossi
Robert Ross
Brit Roth
Ron Rovansek
Linda Rowan
Carl Rowe
Joan Rowe
Diana Rudloff
Hart Rufe
Linda Ruff
Kyfer Rumburd
Kevin Rung
Jack Runkel
Judith Runkel
Laurel Rush
Keith Russell
Mike Russell
Margie Rutbell
Cameron Rutt
David F. Ryan
Judy Ryder
Adam Sabatine
Margaret Sabo
Frank Sady
Jan Sady
Barbara Saeger
Peter G. Saenger
Robert Sager
Donna Salko
John Salvetti
Kenneth Sander
Jim Sanders
Ron Sanderson
Keith Sanford
Steve Sanford
Sarah Sargent
Greg Sassaman
Julia Saurbaugh
David Saylor
Dave Scamardella
Joyce Schaff
Jeff Schaffer
Michael Schall
LeRoy (Whitey) Schaller
Ronnie Schenkein
Art Schiavo
Phyllis Schiippel
Katie Schill
Peter Schliessman
Isaac Schmelzlen
Mike Schmid
Teresa Schmittroth
Susan Schmoyer
Casey Schneck
Lauren Schneider
Patty Schofield
Jane Schoppe
Wain Schroeder
John Schultz
Tim Schumann
Ruth Schurr
Daniel Schwartz
Devin Schwartz
Sandra Schwartz
Charles Schwarz
Allen Schweinsberg
Joanne Schweinsberg
Lee Schweitzer
Joe Sebastiani
Frederick Sechler
Todd Segner
Jamie Sehrer
Kyle Selepouchin
Edward Seman
Tammy Serata
Alice Sevareid
Win Shafer
Anna May Shaffer
Bob Shaffer
Janet Shaffer
Joni Shaffer
Lauren Shaffer
Lloyd Shaffer
Walter Shaffer
Denny Shaffner
Vicki Shane
Patricia Shank
Matt Sharp
Karen Shaw
Robert Shaw
Skip Shawley
Eliza Sheaffer
Charles Shearer
Larry Sheats
Gordon Shedd
James Sheehan
Anne Sheftic
Daniel Sheldon
Douglas Sheldon
Richard Shelling
Kathie Shelly
Brian Shema
Charlie Sherman
Ethel Sherman
Tom Shervinskie
Gary W. Sherwin
Peg Sherwood
Alex Shidel
Mary Shidel

Debbie Shirey
Rob Shirey
Scott Shirey
Jim Shoemaker
Matthew Shope
Kris Shultz
Theresa Shuman
John Sidelinger
Debra Siefken
Kathy Sieminski
Andrew Sigerson
Linda Signarovitz
Joan Silagy
Brad Silfies
Barbara Silverstein
Priscilla Simmons
Donald Simpson
Lee Simpson
Sam Sinderson
Scott Singer
Ed Sinkler
Sally Sisler
John Skarbek
Jerry Skinner
Logan Slade
Renee Slis
Randy Sliter
John Slotterback
Desere Smart
Nancy Smeltzer
Adam Smith
Andy Smith
Dave Smith
David Smith
Gary Smith
Herbert Smith
James Smith
Janet Smith
Karen Smith
Lisa Smith
Ray Smith
Richard Smith
Robert Smith
Ruth Ann Smith
Sally Smith
Terry Smith
Tom Smith
Vincent Smith
Briana Smrekar
Dan Smrekar
Jocelynn Smrekar
John Smurkoski

Daniel Snell
Patrick Snickles
Cindy Snyder
Donald Snyder
Robert Snyder
Scott Snyder
Gregory Socha
Daria Sockey
Brenda Sollenberger
Jack Solomon
Sue Solomon
Kate Somerville
John Somonick
Joanne Sora
Emily Southerton
Joseph Southerton
Michael Southerton
Darryl Speicher
Matt Spence
Rosemary Spreha
Chris Sral
Bobbi Stack
Kathy Stagl
Sara Stahl
Aidan Stahlman
Tony Stair
Ronald A. Stanley
Olyssa Starry
Russ States
Paul Staudenmeier
Aura Stauffer
Glenn Stauffer
Michael Stauffer
Roberta F. Stauffer
Pat Stawicki
Steve Steele
Terry Steffan
Doug Steigerwalt
Cindy Stephens
William Stephens
Grant Stevenson
Bill Stewart
Bob Stewart
Joanna Stickler
Chris Stieber
Judy Stine
Fred Stiner
Kate St. John
Mary St. John
Linda Stockman-Vines
Grant Stokke
Scott Stoleson

Judith Stoltzfus
Melvin Stoltzfus
Cathy Stone
Derek Stoner
Janet Stoner
John Stoner
Katie Stoner
Roger Stoner
David Storck
Lewis Stout
Bob Strahorn
Jeremy Strait
Joe Strasser
Voni Strasser
John Street
Mark Strittmatter
Mike Strohl
Adam Stuckert
Joyce Stuff
Jennifer Stump
Kacey Sullinger
Jim Sunderland
Mark Swansiger
Sandee Swansiger
Jesse Swarey
Beth Swartzentruber
Bill Sweeney
Edward Sweet
Paul Sweet
Jeff Swope
Dan Syster
Georgette Syster
Chuck Tague
Joan Tague
Nancy Tague
Jeannine Tardiff
John Tautin
Diana Taylor
Fred Taylor
Frederick Taylor
John Taylor
Megan Taylor
Valerie Taylor
Roberta Tedesco
Gerald Teig
Brandon Tepley
Drue Tepper
Carole Terrette
Jeff Territo
Jean Tesauro
Dave Tetlow
Joan Thirion

Annette Thomas
Emily Thomas
Jaime Thomas
Stephen Thomas
Chuck Thompson
Dean Thompson
Judy Thompson
Gabe Tiday
Luke Tiday
Amy Tobin
Daniel Tollini
Lee Tosh
Jess Trainor
John Tramontano
Clark Trauterman
Mariam Trostle
David Troup
Barb Troutman
Adam Troyer
Aden Troyer
Andrew Troyer
David Troyer
Harvey Troyer
Jerry Troyer
Marcus Troyer
Melvin Troyer
Neil Troyer
Steven Troyer
Linda Truman
Norma Trumbull
Samara Trusso
Matthew Tufano
Lisa Tull
Benjamin Turover
Gary Tyson
Bill Uhrich
Janie Ulsh
Rebecca Underwood
Todd Underwood
Gloria Unger
Kevin Upperman
Elissa Uram
Donna Urian
Andy Urquhart
Larry Usselman
Kim Utiss
Joe Valasek
Mary Jo Valasek
James Valimont
John Valko
Doug VanBrunt
Nancy Van Cott

Sara Van Cott
Dennis vanEngelsdorp
Kim Van Fleet
Luise Vankeuren
Roscoe Van Muylwyk
Robert VanNewkirk
Marjorie Van Tassel
Philip Varndell
Ann Vayansky
Tim Vechter
Cyndi Velmer
Kristen Vitkauskas
Anne Vivino-Hintze
George Vivino-Hintze
Francis Velazquez
Caleb Voithofer
Jared Voris
Eizabeth Voytko
Nicholas Voytko
Richard Voytko
Justin Vreeland
Wendy Vreeland
Eric Wachsmuth
Eric Wagner
Linda Wagner
Ronald Wagner
Georgia Wahl
Gerald Walker
Joe Walko
Mark Wall
Christopher Walsh
Larry Waltz
Michael Ward
Robert Wasilewski
Thomas Wasilewski
Todd Watkins
Michael Watko
Don Watts
Harry Watts
Brenda Wayant
Richard Weary
Bruce Weaver
Joan Weaver
Jon Weaver
Marcia Weaver
Drew Weber
Justine Weber
Judy Weglarski
Michael Weible
Scott Weidensaul
Don Weis
Sally Weisacosky
Dustin Welch
Jason Weller
William Wellman
Judy Wenberg
Cory Wentzel
Doug Wentzel
Joanna Wert
Thomas Wescott
Howard West
Brenda Weyant
Jeff Whaling
Susan Wheeler
Bruce Whipple
Curtis Whipple
Ellen Whipple
Corrinne White
David White
Judy White
June White
Linda Whitesel
John Whiting
Susan Whitling
Sue Wicks
Chuck Widmann
Gene Wilhelm
Ann Wilken
Jeff Wilkins
Matt Willen
Carl Williams
Em Williams
Frances Williams
Jack Williams
Lisa Williams
Llew Williams
Mary Williams
Patricia Williams
Richard Williams
Robert Williams
Andy Wilson
James Wilson
Larry Wilson
Rick Wiltraut
Frank Windfelder
Amber Wingert
Carole Winslow
Jean Winslow
Ray Winstead
Doug Wise
Rosanne Wise
Damien Wissolik
Jan Witmer
Matt Wlasniewski
Neil Woffinden
Kimberly Wojnar
Eleanor Wolf
Brett Wolfe
Dawn Wolfe
Ken Wolgemuth
Brian Womer
Patricia Wood
Jack Woolridge
Jill Woshner
Ray Woshner
Nancy Wottrich
Suzanne Wrye
Peter Wulfhorst
Paul Wunz
Jon Wyant
Dan Yagusic
Tom Yannaccone
A. J. Yarborough
Todd Yatsko
David Yeany II
John Yerger
Gideon Yoder
John Yoder
Nancy Yoder
Paul Yoder
Frederick Young
George Young
Joel Young
Ron Young
Susan Young
Joseph Yuhas, Jr.
Eric Zawatski
Elizabeth Zbegner
William Zemaitis
Donna Zemba
Peter Ziebart
Michael Ziegler
Brenda Ziemkiewicz
Fred Zimmerman
Jennifer Zimmerman
Megan Zimmerman
Janine Zinn
Dan Zmoda
Benuel Zook
Jay Zook
Linda Zook
Jim Zoschg
Michael Zrencjak
John Zwierzyna
Amber Zygmunt

Project Personnel

PROJECT DESIGN TEAM
Timothy O'Connell, Joseph Bishop, Robert Mulvihill, Michael Lanzone, Trish Miller, and Robert Brooks.

ADVISORY COMMITTEE

Participant	Affiliation	Participant	Affiliation
Keith Bildstein	Hawk Mountain Sanctuary	Daniel Brauning	Pennsylvania Game Commission
Joseph Bishop	Penn State Cooperative Wetlands Center		

Participant	Affiliation
Margaret Brittingham	Penn State School of Forest Resources
Carole Copeyon	U.S. Fish and Wildlife Service
Greg Czarnecki	The Nature Conservancy
Tony Davis	The Nature Conservancy
Cal DuBrock	Pennsylvania Game Commission
Cindy Adams Dunn	Audubon Society Pennsylvania
Frank Felbaum	Pennsylvania Biological Survey
Douglas A. Gross	Pennsylvania Society for Ornithology
Gregory Grove	Pennsylvania Society for Ornithology
Frank and Barb Haas	Pennsylvania Society for Ornithology
Cliff Jones	First Atlas Advisory Committee
Sally Just	Pennsylvania Department of Conservation and Natural Resources
Maurie Caitlin Kelly	Pennsylvania Spatial Data Access
Robert C. Leberman	First Atlas Advisory Committee
Matt Marshall	National Park Service
Terry Master	First Atlas Advisory Committee
Brad Nelson	Allegheny National Forest
Walter Pomeroy	First Atlas Advisory Committee
William Reid	First Atlas Advisory Committee
Chandler Robbins	Patuxent Wildlife Research Center, USGS
Tim Schaeffer	Audubon Pennsylvania
Matt Sharp	Philadelphia Academy of Natural Sciences
David Smith	Powdermill Nature Reserve
Ron Stanley	Wild Resource Conservation Fund, DCNR
Jim Thorne	Natural Lands Trust
Scott Weidensaul	Author/Naturalist

STEERING/PLANNING COMMITTEE

Steven Balzano, Ryan Baxter, Doug Becker, Joseph Bishop, Carolyn Blatchley, Robert Blye, Daniel Brauning, Margaret Brittingham, Margaret Buckwalter, Ed Chubb, Tom Clauser, Bill DeWalt, Duane Diefenbach, Gary Edwards, George Farnsworth, Ted Floyd, Roanna Fuller, Alan Gregory, Deb Grim, Laurie Goodrich, Doug Gross, Deb Grove, Greg Grove, Lewis Grove, Barb Haas, Frank Haas, Jerry Hassinger, David Hauber, Molly Heath, Debbe Hess, Paul Hess, Margaret Higbee, Roger Higbee, Deuane Hoffman, Steve Hoffman, Roy Ickes, Rudy Keller, Sylvia Keller, Maurie Caitlin Kelly, Doug Kibbe, Nancy Ellen Kiernan, Katrina Knight, Mike Lanzone, Matt Marshall, Terry Master, Mark McConnaughy, Robert Mulvihill, Tim O'Connell, Nick Pulcinella, Bill Reid, Robert Ross, Matt Sharp, Deb Siefken, Bob Snyder, Aura Stauffer, Alana Sucke, and Linda Wagner.

PROJECT STAFF

Robert S. Mulvihill, *Project Coordinator, Newsletter Co-Editor*
Michael Lanzone, *Assistant Project Coordinator*
Trish Miller, *GIS Support*
Pam Ferkett, *Administrative Assistant*
Cokie Lindsay, *Administrative Assistant, Newsletter Co-Editor*
Mary Shidel, *Administrative Assistant, Newsletter Co-Editor*

SUMMER POINT COUNT STAFF

Sidra Blake, Devin Bosler, Ben Coulter, Will Faux, Nate Fronk, Ross Gallardy, Lewis Grove, Ben Israel, Tom Johnson, Alison King, Michael Lanzone, Donald Mackler, Andrew McGann, Jake Mohlmann, Cameron Rutt, Andrew Sigerson, Beth Swartzentruber, Paul Sweet, Bryant Ward, Drew Weber, Andrew Wilson, and John Yerger.

IT SUPPORT, GIS, AND WEB DESIGN

The Internet-based data entry, data display, and mapping site was developed by staff of the Information Science Program at the Cornell Lab of Ornithology: Steve Kelling, Jeff Gerbracht, Paul Allen, Roger Slothower, and Tom Fredericks.

Maurie Caitlin Kelly, Ryan Baxter, and James Spayd from Penn State Institutes of Energy and the Environment's Pennsylvania Spatial Data Access program provided much additional support for custom GIS-based mapping products used by second Atlas volunteers. Cathy Klingler designed and managed an informational website for the second Atlas at Carnegie Museum of Natural History.

Technical assistance and map production for this publication was provided by Joseph Bishop, Doug Miller, Trish Miller, and Patryk Soika at Penn State

University. The following provided valuable additional technical assistance to the second Atlas:

Craig Chapman, *DCNR Bureau of Forestry, GIS data acquisition*
Tamara D. Gagnolet, *The Nature Conservancy, GIS data acquisition*
Fengyou Jia, *DCNR Bureau of Forestry, GIS data acquisition*
Greg McPherson, *DCNR Bureau of Forestry, GIS data acquisition*
Cory Wentzel, *DCNR Bureau of Forestry, GIS data acquisition*
Keith Pardieck, Dave Ziolkowski, and John Sauer (*USGS Patuxent Wildlife Research Center*), assisted with BBS data extraction and analysis

PUBLICATION ADVISORY COMMITTEE
Joseph Bishop, Nick Bolgiano, Daniel Brauning, Margaret Brittingham, Kim Corwin, Greg Czarnecki, Cal DuBrock, Deb Grove, Greg Grove, Doug Gross, Paul Hess, Margaret Higbee, Roger Higbee, Maurie Kelly, Bob Leberman, Andrew Mack, Geoff Malosh, Matt R. Marshall, Terry Master, Doris McGovern, Doug Miller, Robert Mulvihill, David Norman, John Ozard, Sarah Sargent, Scott Weidensaul, and Andrew Wilson.

DATA REVIEW AND PROOFING
Paul Hess, Daniel Brauning, Catherine D. Haffner, Douglas A. Gross, Greg Grove, F. Arthur McMorris, Robert C. Leberman, Geoff Malosh, Robert Mulvihill, and Andrew Wilson.

COPY EDITING
Cassie Blair
Joe Kosack
Catherine D. Haffner

BIBLIOGRAPHY
Andrew Mack
Catherine D. Haffner

PROOFING, EDITING, AND OTHER ASSISTANCE
Patricia Barber
Douglas A. Gross

PHOTOGRAPH SELECTION
Geoff Malosh
Robert Mulvihill

PHOTOGRAPH EDITOR
Geoff Malosh

PHOTOGRAPHERS
Gerard Bailey/VIREO
Glenn Bartley/VIREO
Denny Bingaman
Rick and Nora Bowers/VIREO
John Cancalosi/VIREO
Seth Cassell/DCNR
Rob Criswell
Rob Curtis/VIREO
Gerard Dewaghe

Jacob Dingel
Roger Eriksson/VIREO
Scott Gorring
Steve Gosser
Warren Greene/VIREO
Steve Greer/VIREO
Douglas A. Gross/PGC
Brian Henry/VIREO
Marvin R. Hyett/VIREO
Hal Korber/PGC
Joe Kosack/PGC
Greg Lasley/VIREO
Geoff Malosh
Jeff McDonald
Garth McElroy/VIREO
John McKean/VIREO

F. Arthur McMorris/VIREO
Bob Moul
Chuck Musitano
Claude Nadeau/VIREO
Rolf Nussbaumer/VIREO
Michael Patrikeev/VIREO
Tom Pawlesh
Robert Royce/VIREO
Rob and Ann Simpson/VIREO
Brian E. Small/VIREO
Bob Steele/VIREO
Fred Truslow/VIREO
Tom J. Ulrich/VIREO
Doug Wechsler/VIREO
James M. Wedge/VIREO
Scott Weidensaul
Dustin Welch
Brian K. Wheeler/VIREO
Laura C. Williams/VIREO
Andrew M. Wilson
Rick Wiltraut
Bob Wood and Peter Wood
Christopher Wood
James R. Woodward/VIREO

SPECIES ACCOUNT AUTHORS
Marja Bakermans
Keith L. Bildstein
Nicholas C. Bolgiano
Daniel W. Brauning
David F. Brinker
Margaret C. Brittingham
Brian Byrnes
Michael Carey
Mary Jo Casalena
Seth Cassell
Andrea L. Crary
Robert W. Criswell
Robert L. Curry
John P. Dunn
Christopher J. Farmer
Mike Fialkovich
Jamie Flickinger
Ted Floyd
Laurie J. Goodrich
Ian Gregg
Douglas A. Gross
Greg Grove

Lewis Grove
Paul Hess
Roy A. Ickes
Kevin Jacobs
James S. Kellam
Scott R. Klinger
Arlene Koch
Jeffrey Larkin
Robert C. Leberman
Geoff Malosh
Terry L. Master
F. Arthur McMorris
Eugene Morton
Robert S. Mulvihill
Ronald L. Mumme
Daniel P. Mummert
Michael L. O'Reilly
Robert M. Ross
Cameron Rutt
Jerry Skinner
Jeremy Stempka
Scott H. Stoleson

Bridget J. M. Stutchbury
John Tautin
Kim Van Fleet
Scott Weidensaul
Gene Wilhelm

Andrew M. Wilson
Rick Wiltraut
Matthew A. Young

INDEXING
Andrew M. Wilson

PUBLICATION
Special thanks to Kendra Boileau, Editor-in-Chief, and Jennifer Norton, Design and Production Manager, at Penn State University Press for their enthusiasm, insightful guidance, and endless patience during this massive project. We also thank Melody Negron at Westchester Publishing Services for her patience and attention to detail.

1 Aims and Purposes

Robert S. Mulvihill

Most "first generation" grid-based breeding bird atlases collected presence/absence data for nesting species in order to provide the first-ever detailed "snapshot" of breeding bird distributions for the geographic regions they covered—counties, states, provinces, or entire countries. Pennsylvania's first breeding bird atlas (hereafter "first Atlas") accomplished this goal when it was conducted from 1983 through 1989: just over 2,000 volunteers collectively reported more than 83,000 hours in the field and amassed a total of 318,600 records representing 210 different species, 187 of which were considered breeding birds in the state (Brauning 1992a). Many more individual observations were certainly made by volunteers for the first Atlas, but, ultimately, only one record for each species (representing the highest observed evidence of breeding) was retained in the final database. The comparable total of unique species-by-block observations for the second Pennsylvania breeding bird atlas (hereafter "second Atlas") was just over 350,000, representing 218 species, with 190 considered breeding. Unlike the first Atlas, in the second Atlas many additional records of selected species within the same atlas block (even multiple observations representing the same level of breeding evidence) were collected and retained in the final database for the additional valuable information they could provide regarding breeding phenology and fine-scale distribution patterns (see chap. 6). The collective contribution of the 1,900 volunteers who submitted records for the second Atlas not only added to the solid foundation established by the first Atlas, but will continue to help private and public resource organizations further effective, scientific conservation programs in the years to come.

As groundbreaking as they were, most first atlases did not tell us where—across the length and breadth of their overall distributional ranges—species were more or less common: an observation of a single nesting pair or 100 nesting pairs within a block would be mapped in just the same way. Because geographic differences in abundance, particularly for species of concern, have obvious importance for prioritizing conservation efforts, tackling the challenge of introducing a measure of abundance (i.e., a third dimension of distribution) into the second Atlas was a high priority from the start of its planning in 2001. With advice from a number of experienced field biologists and biostatisticians, we devised a novel, logistically demanding point count protocol, with the express purpose of meeting a principal aim of the second Atlas: not only mapping the distributions of the state's breeding birds but also estimating the true density of as many of these as possible. This customized protocol was executed in every full-sized (i.e., non-border) block in the state over the course of 5 years (see chaps. 4 and 5).

An important lesson learned from the first Atlas provided another set of challenges for the planning and execution of Pennsylvania's second Atlas. Simply put, the specialized natural history of some bird species renders them particularly difficult to find without considerable, targeted effort. Thus, in the first Atlas, the final maps for most nocturnal birds, as well as some habitat specialists (e.g., marsh birds), primarily reflected where the atlas volunteers had made the extra efforts needed to find these species, despite specialized efforts to detect these species. It was impossible to know whether distributional gaps for these species in the first Atlas were real or an artifact of insufficient effort; that is, there was no real way to know whether a species was actually absent from a block or if it was simply missed. Because many of these same species are of particularly high conservation importance and concern in the state, specialized surveys for these species were built into the second Atlas to increase the likelihood that these notoriously difficult-to-detect bird species would, in fact, be detected in blocks in which they likely occurred (see chap. 4). The aim was, therefore, that where these specialized surveys were conducted for the second Atlas and the target species were not detected, we would have greater confidence that they were, in fact, not present.

As another novel feature of the second Atlas, we encouraged volunteers to collect multiple georeferenced records within any given block for selected species of conservation importance (i.e., species in bold or italics on the second Atlas field cards; see chap. 4). Although only one breeding record in one field season was needed to fill in an atlas block for a given species on the final maps in this book, we knew that multiple georeferenced records within the same block could provide better data to support analysis of environmental and habitat relationships for these species, and that these, in turn, could contribute to the work of conservation in the coming years. Taken together, these resources enabled our volunteers to generate a large, precisely georeferenced database of breeding records, particularly for important bird species of conservation concern in Pennsylvania.

Yet another twist in data collection for the second Atlas, compared with the first Atlas, was our attempt to collect information about the co-occurrence of breeding bird species and Pennsylvania's state tree, the Eastern Hemlock. As we faced the rapid spread of the introduced Hemlock Woolly Adelgid from east to west across the state, the second Atlas provided a timely opportunity to find out just how many and which species might be negatively affected by any future losses of the hemlock component of Pennsylvania's forests. Because hemlock cover cannot be distinguished from other "evergreen" cover in satellite land cover data, we encouraged second Atlas volunteers to be our "eyes on the ground," reporting the amount of hemlock present in locations where they observed breeding birds. We requested that volunteers collect one piece of habitat data, especially in connection with confirmed breeding records, for all species. The provisional list of "hemlock-associated" species (see chap. 6) will deserve closer attention and consideration in connection with the conservation impacts of the advancing Hemlock Woolly Adelgid on hemlock stands in Pennsylvania and surrounding states.

Last, but by no means least, we strived to greatly increase participation in the second Atlas by recruiting our volunteers from well outside the conventional birding community circles. Although the state's well-organized birding community clearly contributed the bulk of records to the second Atlas, we used various means, such as backyard bird forms, farmer forms, hunter/fisher surveys, scout projects, and classroom activities promoted through statewide news releases, in an attempt to "popularize" the second Atlas and to increase the number of people who had a stake in the final results. We stressed that "every single breeding bird observation, whether of a common species by a beginning bird watcher or of a real rarity by an expert birder, will put another one of our state birds 'on the map,' in the process adding measurably to our knowledge of the occurrence, status, and distribution of Pennsylvania's birdlife." In an effort to make as many people as possible aware of the aims and purposes of the second Atlas, and to continually recruit new volunteers, we printed newsletters (c. 15,000), *The PennsylAvian Monitor*, once or twice each season and sent it not only to those already enrolled in the second Atlas but also to every public library in the state and many other public distribution points (e.g., nature centers, nature shops, and state park offices). An ulterior motive was to bring more of the state's bird watchers into the fold of local Audubon and other bird clubs, not only for the benefit of those clubs, but also to ensure a ready supply of experienced volunteers for the third Atlas 20 years from now!

Although the hoped-for successes did not follow from all of the aforementioned aims and aspirations of the second Pennsylvania breeding bird atlas, we strongly encourage the organizers of second and subsequent atlases to strive to improve the quantity, quality, and scope of data related to breeding birds, to take full advantage of new and emerging technologies and analytical methods and products, and to reach out to the many people whose interest in (and knowledge about) birds may not yet have led them to membership in a bird organization. The risk of overreaching seems less of a concern than repeating exactly what was done before, especially where there is a clear opportunity to do much more. We hope the second Atlas will, therefore, serve as a source of ideas and inspiration for others embarking on their first or next atlas. There is value in thinking of bird atlases not only as a follow-up to a preceding atlas but also as a step forward for any succeeding atlas.

The welfare of the birds within any political jurisdiction—indeed, the health and vitality of the very environments that we humans share with birds—depends upon practical, up-to-date information that can support sound scientific conclusions that, in turn, can inform and promote effective and efficient on-the-ground conservation efforts. Like most good citizen-science projects, the second Atlas has produced far more data than can be summarized and incorporated in this book, and we look forward to making these data available to conservation biologists and land managers interested in understanding, fostering, and protecting bird diversity in the future. The book you are holding breaks much new ground compared with the first Atlas, but it is just one of what we hope will be a stream of important publications based on the voluminous data collected by thousands of dedicated volunteers and other fieldworkers for the second Atlas.

2 | The Geography of Pennsylvania

Andrew M. Wilson and Bernd J. Haupt

Bird distributions are limited by many factors, among which habitat availability is dominant. Local habitat is, in turn, constrained by physical geography, climate, and anthropogenic land uses. Pennsylvania's northern temperate climate and its topographic features, shaped over geological timescales, leave strong signals on the distributions of its diverse breeding bird populations. Pennsylvania is almost landlocked; the Delaware River, which forms the state's eastern boundary, is tidal for 92 km (57 miles) of its length in the southeast, while the state's northwestern corner includes 82 km (51 miles) of Lake Erie shoreline. Despite the lack of coastal habitats, the state's large size (46,058 mi^2, 119,283 km^2) ensures that it encompasses considerable geographical variation, which in turn results in diverse habitats and a species-rich breeding bird community. This chapter provides maps and brief descriptions of some of the geographic features that either influence bird distributions or otherwise aid the reader's interpretation of the species accounts. For a more thorough treatment, see "Geography of Pennsylvania" in the first Atlas (Brauning 1992a) or *The Atlas of Pennsylvania* (Cuff et al. 1989).

TOPOGRAPHY

The Appalachian Mountains run southwest to northeast through Pennsylvania. They are generally of modest elevation, peaking at 979 m (3,213 feet) at Mount Davis in the Laurel Highlands (fig. 2.1), and sea level at the Delaware Bay shore in Delaware County. Despite the modest elevational range, the signature of elevation on bird distribution in Pennsylvania is pronounced, primarily because it dictates land use: steep slopes and higher elevations are predominantly forested, while flatlands, wide valleys, and lower elevations are dominated by farmland and human development.

Pennsylvania was at the southern limit of glaciation during successive Pleistocene ice ages, the most recent of which, 16,000 years ago, bestowed the northwestern and

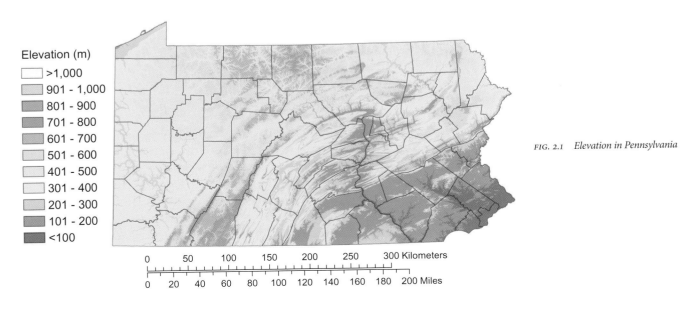

FIG. 2.1 *Elevation in Pennsylvania*

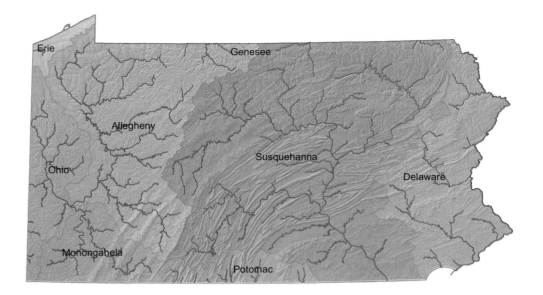

FIG. 2.2 *Drainage basins of Pennsylvania*

northeastern corners of the state with postglacial features. It is in those postglacial landscapes that most of the state's natural lakes and wetlands are found. Farther south, rock folding, due to tectonic shifts, has created a series of anticlines and synclines that form the distinctive "Ridge and Valley" landscape. Dominating central and northern Pennsylvania are the Appalachian Mountains; to the east are the lowlands.

Along the state's eastern border, the Delaware River drains into the Delaware Bay. The state's largest drainage, the Susquehanna River, along with a small component of the Potomac River, drains into the Chesapeake Bay catchment (fig. 2.2). The western third of the state is drained by the Allegheny and Monongahela rivers, which meet in Pittsburgh to form the Ohio River, which flows westward to the Mississippi River. Despite Pennsylvania's proximity to the Great Lakes, only a small portion of the state is within that watershed.

PHYSIOGRAPHIC PROVINCES

Geology and landform characterization provide a framework to delineate Pennsylvania into six geographically homogeneous areas, known as physiographic provinces (fig. 2.3). The Appalachian Plateaus Province covers most of the northern and western area of the state, except for a small sliver of land alongside Lake Erie, which is in the Central Lowlands Province. The Ridge and Valley Province is the second most extensive, forming the eastern flank of the Appalachian Mountains. To the south and east of this, the lowlands are dominated by the Piedmont Province, with a small incursion of the New England Province and Atlantic Coastal Plain Province in the state's southeastern corner. The three largest physiographic provinces are further divided into physiographic sections—regions that exhibit greater geological and topographic homogeneity. Along with counties, physiographic provinces and sections provide the basic geographic units referenced in subsequent chapters and species accounts.

CLIMATE

The data shown in this section are COOP data (Cooperative Observer Program, National Weather Service) provided by the Pennsylvania State Climate Office (http://climate.met.psu.edu/www_prod/data). The climatological data used are daily minimum and maximum temperature and daily precipitation from 123 stations unevenly spread over Pennsylvania (Wisniewski 2011). The daily data are weight-averaged for each weather station separately to bimonthly, 4-month long, or annual datasets, taking into account data gaps. For example, a data gap would occur when new weather stations were either added during the period from 1980 to 2010 or when they did not report continuously. Pennsylvania-wide averages and spatial maps were produced from the gathered time series from each of the 123 weather stations. Data stations that did not encompass 90 percent of data for the considered period of time are omitted because they are not representative data for the analysis.

The averaged mean daily temperatures for the 5-year time interval 2004 to 2009 range from 12°F to 86°F (−11°C to 30°C). The temperatures are strongly affected by Pennsylvania's geologic factors such as soil and elevation. Figure 2.4 shows the averaged mean daily minimum January/February temperatures for the 5-year time interval 2004–2009. In general, the 2004–2009 January/February minimum temperatures show a latitudinal

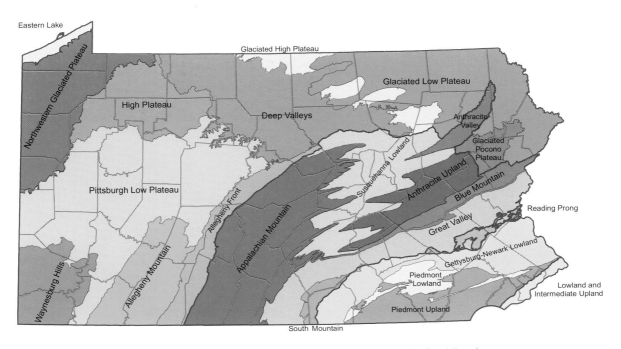

Appalachian Plateaus Province
- Allegheny Front
- Allegheny Mountain
- Deep Valleys
- Glaciated High Plateau
- Glaciated Low Plateau
- Glaciated Pocono Plateau
- High Plateau
- Northwestern Glaciated Plateau
- Pittsburgh Low Plateau
- Waynesburg Hills

Central Lowlands Province
- Eastern Lake

Ridge and Valley Province
- Anthracite Upland
- Anthracite Valley
- Appalachian Mountain
- Blue Mountain
- Great Valley
- South Mountain
- Susquehanna Lowland

New England Province
- Reading Prong

Piedmont Province
- Gettysburg-Newark Lowland
- Piedmont Lowland
- Piedmont Upland

Atlantic Coastal Plain Province
- Lowland and Intermediate Upland

FIG. 2.3 *Physiographic provinces and sections of Pennsylvania*

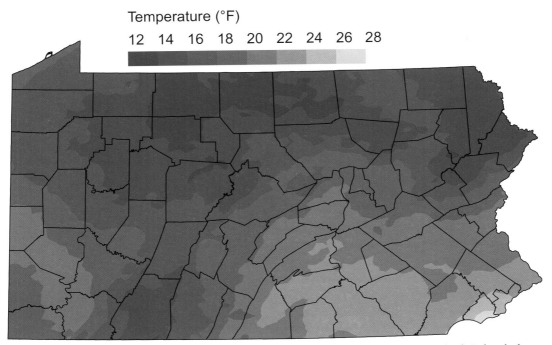

FIG. 2.4 *Mean daily minimum January/February temperatures for the 5-year time interval 2004 to 2009. Analytical method: Cokriging (Spherical) in ArcGIS with an elevation covariate.*

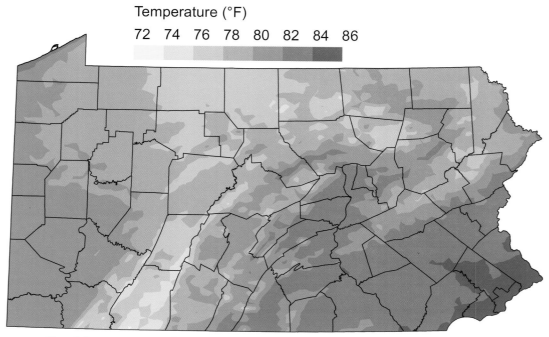

FIG. 2.5 *Mean daily maximum June/July temperatures for the 5-year time interval 2004 to 2009. Analytical method: Cokriging (Spherical) in ArcGIS with an elevation covariate.*

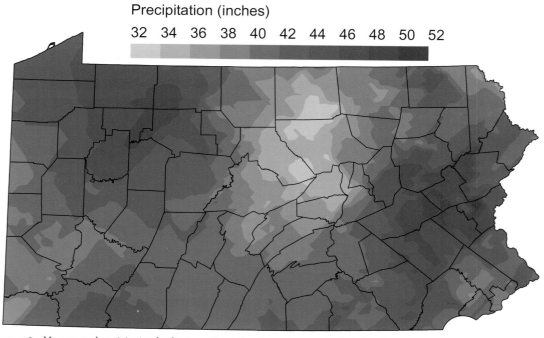

FIG. 2.6 *Mean annual precipitation for the 5-year time interval 2004 to 2009. Analytical method: Kriging (Spherical) in ArcGIS with an elevation covariate.*

gradient; i.e., they show warmer temperatures in southern Pennsylvania. This general trend is modified by Pennsylvania's topography. The temperatures are lower in areas with higher elevations (compare fig. 2.1). The warmest 2004–2009 January/February minimum temperatures (26°F to 28°F, −3.3 to −2.2°C) are located in southeastern Pennsylvania—that is, the Piedmont Lowlands and Uplands as well as the Gettysburg-Newark Lowlands. The coldest 2004–2009 January/February minimum temperatures (12°F to 14°F, −11.1 to −10.0°C) can be found in northern Pennsylvania in the areas of the Deep Valleys and the High Plateau (see fig. 2.4).

Figure 2.5 depicts the averaged mean daily maximum June/July temperatures for the 5-year time interval 2004–2009. In general, this dataset shows trends similar to the averaged mean daily minimum January/

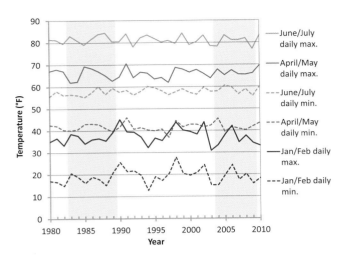

FIG. 2.7 Minimum and maximum temperatures for the time intervals January/February, April/May and June/July in Pennsylvania for the years 1980 to 2010

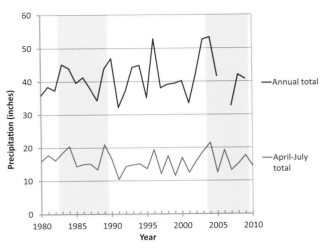

FIG. 2.8 Annual and April–July precipitation totals in Pennsylvania for the years 1980 to 2010. Note: Annual precipitation totals for 2006 and 2010 not displayed due to lower data density (79% and 86%, respectively).

February temperatures. The temperatures tend to be warmer in southern Pennsylvania, especially in southeastern Pennsylvania, and near Lake Erie in the northwest. The temperatures closely follow the topography: they are higher at the lowlands and decrease with elevation. The highest averaged mean daily maximum 2004–2009 June/July temperatures can be found in the area of the Atlantic Coastal Plain Province as well as in the eastern parts of the Gettysburg-Newark Lowland and Piedmont Upland sections (84–86°F, 28.9–30.0°C; compare fig. 2.3).

The averaged mean annual precipitation for the time interval 2004–2009 is shown in figure 2.6. The annual precipitation ranges from 32 to 52 inches (81 to 132 cm). Pennsylvania is exposed to prevailing westerly winds. It is not surprising that the observed precipitation pattern is closely related to elevation (compare fig. 2.1). In general, an increase in precipitation due to orographic lifting can be found west of the Appalachian Mountains and Allegheny Front (compare figs. 2.3 and 2.6), while a precipitation shadow can be found east of the Allegheny Mountains. This precipitation pattern is especially pronounced in the northern part of Pennsylvania. The maximum amount of averaged annual precipitation received in Pennsylvania is located along the Great Valley area, the Blue Mountains, and northeastern part of the Anthracite Upland in eastern Pennsylvania (50–52 inches, 127–132 cm). The term "precipitation" is used because the COOP data provided by the Pennsylvania State Climate Office do not distinguish between the types of precipitation received. The amounts of precipitation shown here include the total amount of rain, snowfall, hail, sleet, ice rain, and other kinds of precipitation.

Figures 2.7 and 2.8 describe the statewide averaged trends of temperature and precipitation during the 1980–2010 period. In general, we do not see clear trends of either decreasing or increasing temperatures for any of the daily minimum and maximum temperature averages for the months January/February, April/May, and June/July. The statewide precipitation shows slight trends; for example, the precipitation for the 4-month interval April to July shows a slight decrease from year 1980 to approximately year 2001, at which time it again starts to increase slightly. The curve for the annual total precipitation does not show such a trend. It is noticeable that the envelope of years with minimum and maximum precipitation increases with time. In other words, we see years with more extremes—that is, low and high precipitation—as time progresses.

It is well established that climate change has resulted in distributional shifts of birds in North America and elsewhere in recent decades (Root et al. 2003; Hitch and Leberg 2007). The potential effects of climate change on bird distributions in Pennsylvania are discussed in chapters 6 and 7. In addition to long-term changes due to climate, individual weather events can have a marked, but often temporary, effect on bird numbers and distributions. Of particular relevance in Pennsylvania are cold winters, which may result in reduced overwinter survival of resident species, and inclement spring and summer weather, which may substantially impact reproductive success (e.g., Jones et al. 2003). Notable weather events during the second Atlas period may have had some effect on both bird numbers and observer activity, but it is impossible to know to what extent this might have affected atlas results. The first year of second Atlas fieldwork followed the winter of 2003/2004, which was the fourth coldest among the 31 winters since 1980 (fig. 2.7). Therefore, it is possible that at the start of the atlas period, populations of some resident species were relatively low; Breeding Bird Survey data for Carolina Wren and Eastern Bluebird certainly support this (Sauer et al. 2011), but it should be noted that populations of both

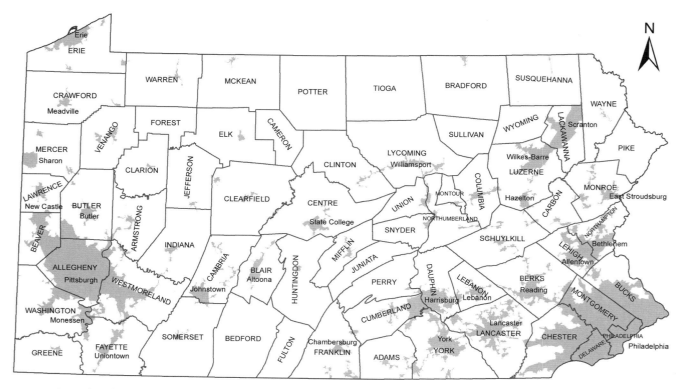

FIG. 2.9 *County boundaries and major urban centers of Pennsylvania*

species rebounded during the 6-year atlas period. The first year of atlas fieldwork was also notably wet; precipitation totals for the spring and summer of 2004 (April through July) were the highest of the 31 years since 1980 (fig. 2.8). This wet weather may have reduced nesting success of species that are particularly vulnerable to wet weather during the breeding season, but again, it is not known whether this was sufficient to impact atlas results.

HUMAN GEOGRAPHY

Pennsylvania's human population grew steadily following the first permanent European settlements in the seventeenth century. The first counties were formed in the southeastern corner of the state in the 1680s, but the current complement of 67 counties (fig. 2.9) was not in place until 1878 (Hottenstein and Welch 1965). During the 1990s, human population generally declined in western counties and increased in eastern counties, with the highest rate of growth in Pike County (RUPRI 2006). In the 2010 Census, the population was 12,702,379 (USBC 2011), more than half of which was concentrated in the Philadelphia and Pittsburgh metropolitan areas. Despite this large and still-growing human population, some areas of Pennsylvania remain sparsely populated; indeed, some of the least populated counties in northern Pennsylvania (Cameron, Elk, and Potter counties) have shrinking populations (USBC 2011).

3 Habitats and Habitat Change

Andrew M. Wilson, Joseph Bishop, and Margaret Brittingham

Pennsylvania is in the heart of the temperate eastern hardwood forest. While evidence grows that Native Americans likely burned and cleared forests for agricultural purposes, or to improve game populations, accounts by early European settlers suggest that the state was predominantly forested in the seventeenth century, when European settlers began to spread across the landscape. European settlement began around Philadelphia with the confirmation of King Charles II's charter to William Penn in the 1680s and spread west and north, eventually to touch the whole state by the 1820s (Miller 1989). Pennsylvania became the energy and timber capital of the country in the late nineteenth century, when coal mines scarred the landscape and timber rafts filled the rivers. Most forests were felled for lumber, tannin, charcoal, or to make way for agriculture. Extensive logging, especially during the nineteenth century, reduced forest extent to only 20 percent of the state's land cover early in the twentieth century (USDA-NASS 2004). Only a few thousand acres remained uncut.

Tumultuous changes during the nineteenth and early twentieth centuries forged Pennsylvania's modern landscape. In the last 80 years, forests have regenerated and are once again the dominant land cover of the Commonwealth (fig. 3.1). Agriculture remains an important industry in Pennsylvania, but it is now concentrated in the fertile lowlands of the Piedmont Province and in valley bottoms elsewhere in the state. Most of the state's intense urban and suburban development is at lower elevations; hence, topography predicts land use, which, in turn, determines the distributions of habitat and, ultimately, birds.

Understanding the changes in these broad habitats, particularly during the period between the first and second Atlases, is critical to appreciating the bird distributions laid out in this atlas. A combination of environmental (abiotic) features, described in chapter 2, and anthropogenic influences determine the vegetation of a given area, which in turn establish the particular conditions suitable for each bird species.

FORESTS

Pennsylvania forest—Pine Creek, Tioga County, Hal Korber/PGC

The amount of forest cover has changed very little since reaching about 60 percent of the state in the 1960s. In 2004, there were 6.7 million hectares (16.6 million acres) of forest in Pennsylvania (McWilliams et al. 2007). Those forests support a multibillion-dollar timber industry and provide significant ecological functions, such as purifying water and providing a home to the majority of the state's wildlife. Pennsylvania contributes significantly to the continental populations of several forest bird species (Stoleson and Larkin 2010) and provides an important link between the southernmost populations of some boreal bird species (in the Appalachian Mountains south of Pennsylvania) and their core populations farther north. Pennsylvania remains a land of forests.

The extent of forest cover varies geographically. Large, contiguous blocks of forest found across much of the Appalachian Plateaus provide greater ecological function

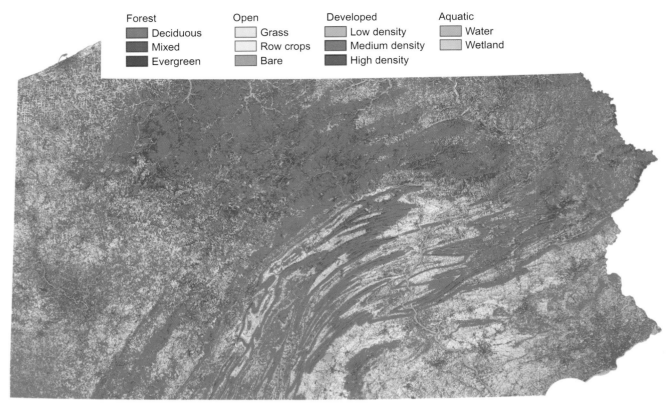

FIG. 3.1 *Land cover in Pennsylvania c. 2005 (Landsat ETM data)*

than do smaller woodlots. Forest fragmentation has been a central theme of ecological research, often focusing on birds, for the past three decades (Whitcomb et al. 1981). The size of forest patches needed by each species varies by the species and the spatial context, but studies have observed that forest within 100 m (300 feet) of a road or permanent edge is fragmented, and many species require at least 100 hectares (247 acres) of contiguous forest cover for populations to be sustainable (Rosenberg et al. 1999). Hence, a smaller subset of Pennsylvania's forested land supports the species of birds most linked to extensive forest (Stoleson and Larkin 2010).

Despite the overall constancy in forest extent over the last 50 years, forest cover remains dynamic at smaller scales, with localized gains and losses counter-balancing each other when aggregated across the state. Landsat ETM land cover data (USGS 2010) shows that between approximately 1994 and 2005 there was a net reduction in forest cover in the eastern half of the state, especially due to exurban sprawl in the Philadelphia metropolitan area (fig. 3.2) and expansion of agriculture in northeastern Pennsylvania (table 3.1). There has been a similar loss of forest within the Pittsburgh commuter belt, but interestingly, there was an increase in forest cover in the city of Pittsburgh itself. Elsewhere in the western side of the state, there has been a net increase of forest cover due to continued farm abandonment and mine reclamation. Forest cover remained practically stable between 1994 and 2005 statewide, but there was considerable variation among the 23 physiographic sections, ranging from a 10.6 percent forest loss to a 7.1 percent gain (table 3.1). The condition of that forest, from the standpoint of species composition, health, and regeneration, also varies dramatically.

Forest types

Two primary forest types dominate Pennsylvania: mixed oak forests cover the Ridge and Valley Province, and northern hardwoods (or maple/beech/birch group) the northern plateaus, together accounting for 87 percent of all forests. Other types, such as elm/ash/cottonwood and pine-dominated stands (sometimes as plantations) are found throughout the state but tend to be small and localized (fig. 3.3). However, Pennsylvania's forests are far from homogenous; tree species diversity is high, and variations in elevation, moisture, slope, and aspect result in a rich pattern of forest habitats (Brittingham and Goodrich 2010).

Red Maple is the most abundant tree in Pennsylvania's forests, estimated to account for 20 percent of all live trees (McWilliams et al. 2007), followed by Black Birch (9.5%), Black Cherry (9%), American Beech (6%), and Sugar Maple (6%) (Forest Inventory and Analysis [FIA] data: McWilliams et al. 2007; Ruefenacht et al. 2008). The sixth-most-abundant species (comprising less than 4% of live trees) is the Eastern Hemlock, Pennsylvania's

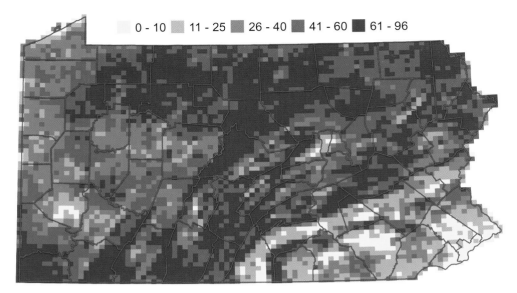

FIG. 3.2 *Percentage forest cover by atlas block c. 2005 (top) and relative change on forest cover by atlas block between 1994 and 2005 (bottom)*

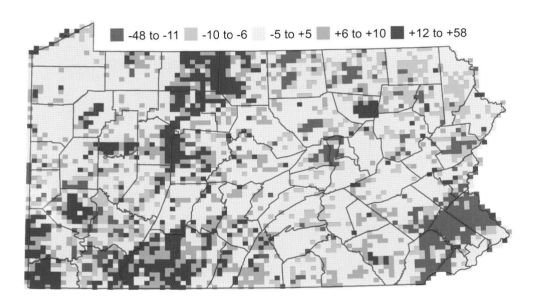

state tree, which has a disproportionate role in shaping ecological communities (Allen and Sheehan 2010). Although once much more abundant than it is now (it was of considerable economic importance during the nineteenth century), around 166,000 hectares (410,195 acres) of hemlock-dominated forest remain, primarily in shady ravines in the northern tier. Even where it is a lesser component of the forest, hemlock is important, partly because it is the most abundant native conifer in much of the state (Allen and Sheehan 2010).

Forest age and maturation

Although Pennsylvania was never completely deforested, most remaining forests are second or third growth (Goodrich et al. 2002). Only a few thousand hectares of unharvested forest remain (Davis 2008); as little as 0.05 percent of the current forest is old growth, and it generally survived in small patches (Goodrich et al. 2002). Mature or old-growth forests support higher densities of birds than in typical mature stands of second-growth forest (Haney 1999). Cavity-nesting species, in particular, benefit from the presence of dead or dying trees, which are characteristic of mature forests. Other species of birds are linked with early successional stages of forest cover, or young forests, so the age of a forest stand is an important factor in determining the suitability of a local site for forest species.

Because most of Pennsylvania's forests are secondary growth, many are still maturing. FIA data show that between 1989 and 2004 there was a significant increase in the proportion of hardwood stands with trees predominantly larger than 28 cm (>11 inches diameter at breast height for hardwoods) to 59 percent of the state's total forestland (McWilliams et al. 2007). Correspondingly,

TABLE 3.1 Land cover in 2005 and changes in land cover (1994 to 2005) by physiographic provinces and sections (fig. 2.3)

Physiographic province (**bold**) and section	Land cover in 2005			% change in land cover 1994 to 2005		
	Forest	Farmland[a]	Developed	Forest	Farmland[a]	Developed
Appalachian Plateaus	**71.7**	**19.6**	**7.2**	**2.4***	**−0.1**	**2.3***
Allegheny Front	77.4	17.6	4.6	7.1*	0.0	2.3*
Allegheny Mountain	70.3	22.1	6.3	6.7*	0.8	3.1*
Deep Valleys	91.9	5.7	1.5	5.8*	−1.0*	0.5*
Glaciated High Plateau	84.3	13.4	1.6	2.7*	1.1	0.9*
Glaciated Low Plateau	76.5	15.2	5.8	−4.8*	3.3*	3.1*
Glaciated Pocono Plateau	81.5	3.6	12.3	−1.3	0.0	3.7*
High Plateau	88.5	7.5	2.9	4.0*	−2.4*	−0.6*
Northwestern Glaciated Plateau	51.5	36.3	10.0	2.0*	−4.0*	4.8*
Pittsburgh Low Plateau	60.6	26.6	11.4	1.9*	0.7*	2.2*
Waynesburg Hills	60.2	26.6	12.1	5.4*	−2.6*	3.0*
Atlantic Coastal Plain	**11.8**	**7.8**	**63.8**	**−0.5**	**3.3**	**−0.2**
Lowland and Intermediate Upland	11.8	7.8	63.8	−0.5	3.3*	−0.2
Central Lowlands	**34.8**	**26.6**	**35.6**	**5.2***	**2.0**	**0.1**
Eastern Lake	34.8	26.6	35.6	5.2*	2.0	0.1
New England	**58.0**	**20.8**	**18.6**	**−8.6***	**−5.5***	**15.7***
Reading Prong	58.0	20.8	18.6	−8.6*	−5.5	15.7*
Piedmont	**32.1**	**37.8**	**28.1**	**−5.7***	**−6.4***	**13.7***
Gettysburg-Newark Lowland	37.2	32.1	28.6	−10.6*	−3.2*	16.8*
Piedmont Lowland	15.3	52.7	30.0	3.8*	−11.1*	7.3*
Piedmont Upland	32.2	39.0	26.9	−3.3*	−8.4*	12.3*
Ridge and Valley	**59.5**	**27.7**	**11.3**	**0.3**	**−2.2***	**5.5***
Anthracite Upland	76.4	12.1	10.0	−1.5*	−1.2*	3.7*
Anthracite Valley	65.4	10.3	21.7	0.6	3.0*	−1.7
Appalachian Mountain	72.1	22.0	5.1	3.4*	−0.7*	3.1*
Blue Mountain	67.3	17.8	12.8	−2.0*	−3.0*	7.0*
Great Valley	26.7	47.2	24.4	0.6	−10.6*	12.4*
South Mountain	81.3	12.0	6.3	−1.4	2.3*	4.4*
Susquehanna Lowland	53.3	35.6	8.9	−3.8*	1.2*	5.1*
PENNSYLVANIA	**63.6**	**23.6**	**11.1**	**1.0***	**−1.3***	**4.3***

*p < 0.05, significant change between 1994 and 2005.
[a] Farmland includes grassland, some of which may not be agricultural grassland.

young age-class forests have declined. Tree stocking densities and canopy cover decreased in the western half of the state between 1989 and 2004; such changes could result from a variety of events and disturbances, such as destructive weather, pests and diseases, and timber harvesting (McWilliams et al. 2007).

Forest changes due to pathogens and pests

In addition to changing age classes and species composition, anthropogenic influences have resulted in more pernicious changes to our forests, notably, through the introduction of pathogens and pests. The virtual loss of once-abundant tree species during the twentieth century, such as American Chestnut and American Elm, due to fungal infestations is well documented and lamented (Khulman 1978). More recently, extensive defoliation of hardwoods due to Gypsy Moth, spanworm, cankerworm, and other insect infestations has resulted in localized timber die-off and, perhaps more importantly for birds, complete loss of canopy cover during the breeding season. Defoliation events reduce populations of birds associated with closed canopy forests during outbreak years (Gale et al. 2001); conversely, severe defoliation, sometimes resulting in tree mortality, creates snags and understory vegetation that, in turn, provides a substrate for cavity-nesting and shrub-associated species (Showalter and Whitmore 2002). During the second Atlas period, extensive defoliation events were documented in many large forest tracts in the eastern half of the state, with up to 300,000 hectares (740,000 acres) affected in each year (data: PA-DCNR Bureau of Forestry; fig. 3.4). Mapped severity of defoliation combines measures of contiguity and extent, where defoliation can be nearly complete in some instances. There is evidence that defoliation events during the last few decades have contributed to reduction in the abundance of oaks and Sugar Maples in Pennsylvania's forests (McWilliams et al. 2007); hence, the effects on the ecology of forests

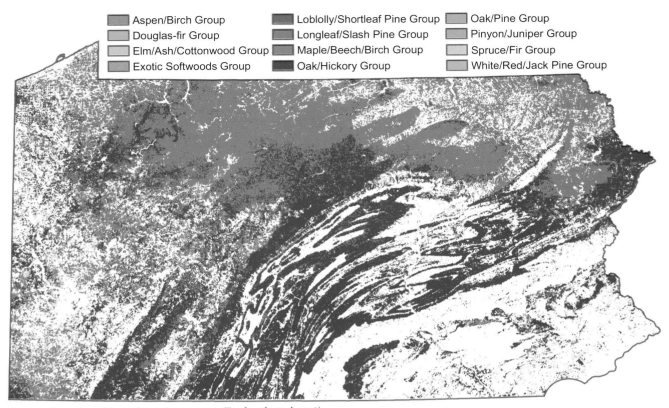

FIG. 3.3 *Forest types of Pennsylvania in 2002–2003 (Ruefenacht et al. 2008)*

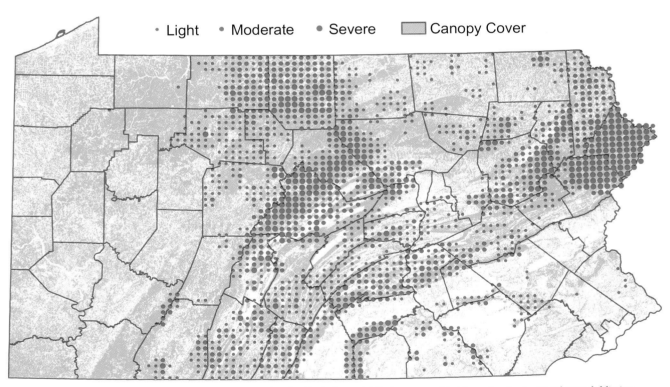

FIG. 3.4 *Severity of defoliation events between 2004 and 2009 by atlas block. Severe defoliation: >20% of deciduous forest experienced some defoliation; moderate is 5 to 20%; light is less than 5%. (Data used with permission of PA-DCNR Bureau of Forestry.)*

HABITATS AND HABITAT CHANGE

Defoliated forest, Centre County, Andrew M. Wilson

can be long term. The worst years for defoliation immediately followed the first Atlas (e.g., 1.8 million hectares, 4.4 million acres, in 1990), and this may have contributed to some of the modest changes in tree species composition observed between atlases. Despite this, there is little evidence that the amount of standing dead wood changed between 1989 and 2004 (McWilliams et al. 2007).

A more recent invasive pest of Pennsylvania's forests is the Hemlock Woolly Adelgid. The adelgid was first detected in Pennsylvania during the 1960s, with defoliation and mortality of natural stands first noted in the southeastern corner of the state during the 1980s (USFS 2010). This nonnative insect has spread north and west across the state, reaching all but 15 counties by 2009 (Allen and Sheehan 2010).

Forest ravine, Sproul State Forest, Andrew M. Wilson

Understory and regeneration

Many forest birds use shade-tolerant vegetation under the forest canopy for nesting or foraging. This vegetation may take the forms of tree seedlings, poised to replace the dominant trees, or shrubs that reach a mature height below the normal canopy. These understory conditions have a dramatic impact on bird communities. There is considerable natural variation in understory types and conditions, largely driven by local topography, soil types, and moisture. For example, evergreen understory plants, such as Mountain Laurel and rhododendron, are

Mountain Laurel understory, Clinton County, Hal Korber/PGC

Overbrowsed forest, Clinton County, Seth Cassell/DCNR

an important nesting substrate for many bird species. They are typically found on cooler north-facing slopes, in shaded ravines, on well-drained acidic soils, or in forested wetlands (Kaeser et al. 2008; Warren 2008). Tree seedlings consisting of shade-tolerant tree species provide similar undergrowth structure for many nesting birds and become the forests of tomorrow. Deer browsing has decreased understory growth and tree regeneration across wide areas (Steiner and Joyce 1999); indeed, the lack of understory plants and trees was described as the "most disturbing finding" of the 2007 FIA report (McWilliams et al. 2007). Over-browsing of understory trees leads to impacts on bird communities, and several birds of high conservation concern are negatively affected by high deer densities (McShea and Rappole 2000). Obvious "browse-lines" (the absence of understory vegetation caused by deer browse) remain a lasting effect on forest structure and future forest composition. However, with White-tailed Deer densities reduced by 25 percent between 2002 and 2007 (WMI 2010), there are many indications that native forest vegetation is rebounding. Ongoing monitoring will assess whether increasing understory regeneration continues.

Areas of early successional forests were historically established through natural processes (wind and fire), and it is likely that Native American actions contributed as well. The extensive timbering of the nineteenth and twentieth centuries (even earlier in the state's southeast) and subsequent abandonment of agricultural lands put large sections of the state into early succession conditions as the exploited forests began to grow. Shrubby habitats may be a climax (or nearly stable) condition when associated with wetlands, and mature scrub oak-pine forests retain similar vegetative structure for many years, but some form of disturbance is generally required to establish, or sustain, young forest conditions. As more of the state came under anthropogenic influence, natural disturbance events were suppressed, and now early successional forest habitats are established predominantly through commercial harvesting of timber. Early successional forest and shrub habitat continues for some years following timbering, but it is influenced by many factors. A poorer representation of shrub habitats may be found along woodland edges and in gardens, tree farms, and older suburban landscapes.

Because early successional and shrub-scrub habitats are of little economic value, there is little interest in tracking changes in their extent. Efforts to reliably identify and monitor changes in early successional habitats remotely are difficult, because these habitats are not reliably defined by remote sensing methods such as the Landsat ETM land cover map. A large decline in successional habitats over the last century, as forests have matured, continued through the period between the atlases (Brittingham and Goodrich 2010). By 2002, successional habitats were estimated to cover 12 percent of the state, including areas that may no longer be suitable for some of the bird species that most depend on these habitats.

Successional and shrub habitats

AGRICULTURAL AND GRASSLAND

Early successional habitat, Schuylkill County, Hal Korber/PGC

Farmland, Fayette County, Geoff Malosh

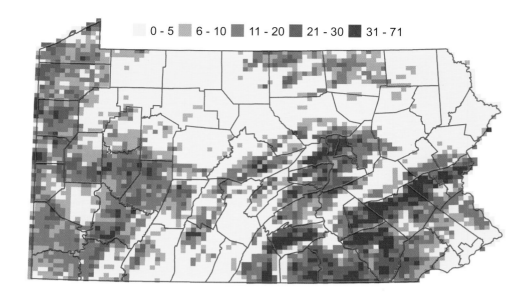
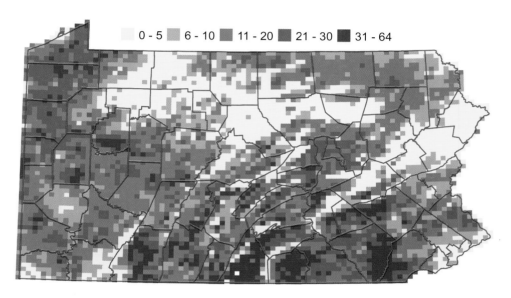
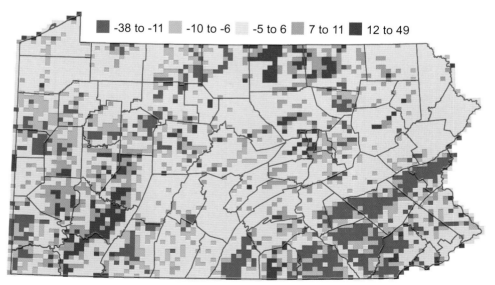

FIG. 3.5 *Percentage agricultural (top) and grassland (middle) cover by atlas block c. 2005, and relative change in agricultural and grassland cover by atlas block between 1994 and 2005 (bottom)*

Agricultural activity in Pennsylvania increased steadily through the nineteenth century as settlers cleared forests for farmland. By the late nineteenth and early twentieth centuries, nearly 70 percent of Pennsylvania was under agricultural production (Miller 1989). During that period, several birds native to the tallgrass prairies colonized agricultural lands in Pennsylvania, while other species once associated with the state's few natural open spaces became common and widespread. Later, during the twentieth century, millions of acres of marginal farmland were abandoned because of their poor condition in favor of more productive lands farther west. By 2005, farmland and grassland had contracted to around 23 percent of the state—approximately 1.1 million hectares (2.7 million acres) of tilled land and 1.7 million hectares (4.1 million acres) of grassland (Landsat ETM land cover data). The largest concentration of farmland remains in the southeast (fig. 3.5); in the Piedmont Lowland and Great Valley sections, 74 percent and 67 percent of all atlas blocks are dominated by farmland, respectively.

Between 1994 and 2005, there was a 1.3 percent decrease in farmland and grassland across the state, with the largest declines in the Pittsburgh Low Plateau Section, the Piedmont Province, and parts of the Ridge and Valley Province (table 3.1; fig. 3.5), likely due to continued exurban sprawl. Conversely, there were modest increases in farmland on the Glaciated Low Plateau of north-central Pennsylvania (table 3.1), where farmland expansion resulted in a loss of forest cover.

Changes in crop types and agricultural practices are among the dynamics that ensure the state's farmed landscape remains in a state of flux. For example, following a long-term decrease (Klinger and Riegner 2008), there was a rapid increase in acreage in permanent pastures between 1997 and 2007 (fig. 3.6), partly due to a move away from the practice of grazing grassland within an agricultural rotation. There was also a substantial increase in conservation cover, largely due to the USDA's Conservation Reserve Enhancement Program, initiated in Pennsylvania in 2000. By the summer of 2007, almost 100,000 hectares (247,000 acres) of conservation cover had been created (USDA-NASS 2009), the majority of it grassland. During the years between the first and second Atlases, there was little change in cropping patterns; hay and corn were dominant crops (fig. 3.7). Soybean cultivation steadily increased, overtaking the combined acreage planted in small grains (barley, oats, and wheat) in 2004.

Even in areas where farmland extent is similar, changes in grassland management practices substantially reduced the availability of fields suitable for nesting grassland-obligate birds. The most critical management changes for grassland birds have been earlier and more frequent cutting. In 1950, the mean first cutting date for timothy and clover hay was 5 July; by 1990, the mean first cutting was a full month earlier (Klinger and Riegner 2008). Alfalfa hay is now cut as early as the first week in May, allowing three

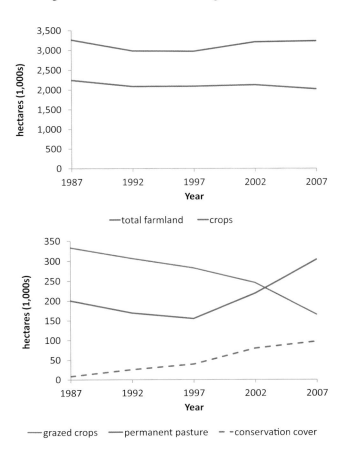

FIG. 3.6 Changes in total agricultural land (top) and certain land uses (bottom) in Pennsylvania between 1987 and 2007. Source: Anon 1987, Anon 1992, USDA-NASS 1999, 2004, 2009.

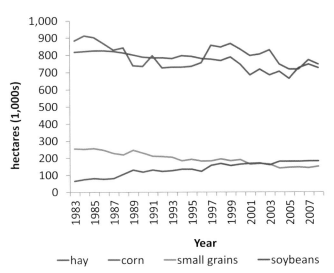

FIG. 3.7 Changes in acreage of major crop types in Pennsylvania between 1983 and 2008. Source: USDA-NASS 2009.

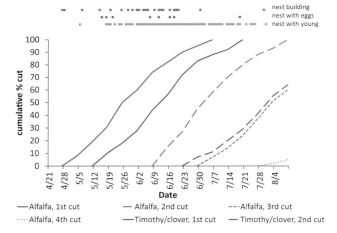

FIG. 3.8 *Average cutting dates of hay fields in Pennsylvania 2004 to 2008 and dates of 2nd PBBA nesting observations of the Eastern Meadowlark (dots). Source of hay cutting dates: USDA-NASS 2008.*

and sometimes four cuts per year, with as little as 5 weeks between harvests. The first and second cuts coincide with the peak nesting season of most grassland-obligate songbirds, such as the Eastern Meadowlark, as indicated by records of confirmed nesting submitted to the second Atlas (fig. 3.8).

Reclaimed surface mine grassland

Reclaimed mine grassland—Imperial grasslands, Allegheny County, Geoff Malosh

Surface mining, principally for bituminous coal in the western half of the state, and to a lesser extent anthracite in the east, has impacted more than 100,000 hectares (247,000 acres) of land in Pennsylvania. Reclamation efforts since the 1960s have mandated reforestation, but grass-seeding areas for erosion control and difficulties of growing trees on degraded land have resulted in grass providing the dominant cover on many reclamation sites. Despite the nonnative grasses used for most reclamation, these reclaimed mines have been found to support significant populations of grassland-obligate birds. Many reclaimed areas have experienced succession at a much slower rate than agricultural areas, resulting in suitable grassland habitat for several decades in many areas of western Pennsylvania. Due to reduced suitability of agricultural grassland for nesting birds, these reclaimed surface mine grasslands have become an increasingly important refuge for a suite of declining species (Stauffer et al. 2010), for which Pennsylvania's reclaimed surface mine grasslands are of global importance. Mattice et al. (2005) estimated 35,000 hectares (86,000 acres) of reclaimed surface mine grassland in a nine-county area of western Pennsylvania (fig. 3.9). Using data on extent of surface mines and land cover data from 2005, we estimate there were 46,000 hectares (114,000 acres) of surface mine grassland available in Pennsylvania during the second Atlas period, the great majority of it in the bituminous coalfield of western Pennsylvania, most notably in Butler, Clarion, Clearfield, and Jefferson counties. Extensive reclaimed surface anthracite mine grassland in Lackawanna and Luzerne counties supports comparatively few grassland birds.

It is not known how much the total amount of reclaimed surface mine grassland changed between the first and second Atlases, but there was likely, at the very least, a geographical redistribution, since some areas were lost through natural succession and others became available due to more recent reclamations. Further, there could have been substantial, but undocumented, net changes in the availability of grasslands at various successional stages, which could explain the loss of species associated with the earliest stages.

WETLAND AND WATER BODIES

Emergent wetlands—Conneaut Marsh, Crawford County, Hal Korber/PGC

Although there are estimated to be 160,000 wetlands in Pennsylvania, most of them are small; in total, they comprise only 2 percent of the state's area, according to

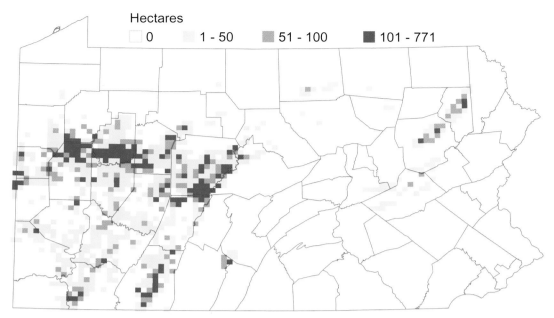

FIG. 3.9 *Estimated area of reclaimed surface mine grasslands of 5 ha or more by atlas block (PA DEP 2009, Abandoned Mine Lines data)*

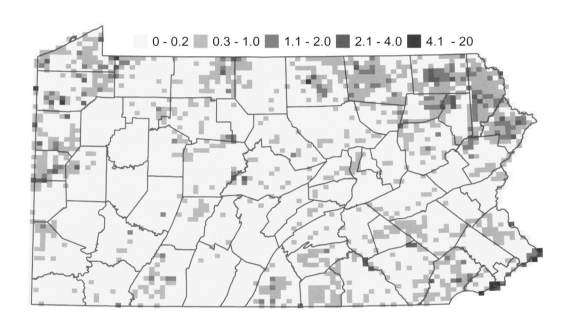

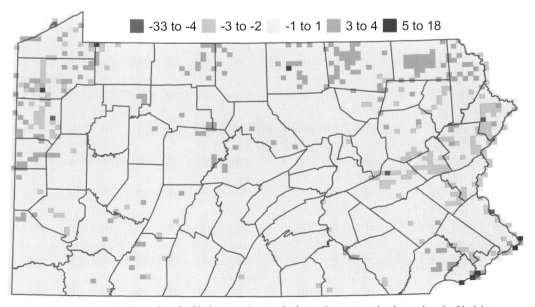

FIG. 3.10 *Percentage wetland cover by atlas block c. 2005 (top) and relative change in wetland cover by atlas block between 1994 and 2005 (bottom)*

the National Wetlands Inventory (Tiner 1990; Brittingham and Goodrich 2010). It is estimated that more than half of the state's original wetlands have been lost to conversion into impoundments for reservoirs and lakes, channelization of rivers, drainage for agriculture, and drainage for development (Goodrich et al. 2002). Extant wetlands are concentrated in the postglacial landscapes in the northwestern and northeastern corners of the state, the lowlying Piedmont Province, and in valleys bottoms. The only tidal wetlands are along the Delaware River in Delaware and Philadelphia counties, but associated wetlands have long been greatly diminished and now extend to only about 260 hectares (650 acres; Gross and Haffner 2010a). Around 20 percent of the state's wetlands are deep-water (lacustrine) wetlands, and 56 percent are swamps, marshes, and bogs (palustrine). Of the palustrine wetlands, around 20 percent are emergent wetlands, typically dominated by cattail and bulrush, and 36 percent are forested wetlands, including shrubby wetlands and forested bogs. Both emergent wetlands and forested wetlands have distinctive bird communities (Brooks and Croonquist 1990) and support some of the state's rarest bird species (Gross and Haffner 2010). The majority of small, isolated wetlands do not support obligate-wetland species but serve as fertile habitat for early successional generalists.

Landsat ETM land cover data show wetland loss continued between the two atlases in some areas, notably in the Pocono region, but also in Crawford and Mercer counties (fig. 3.10). Wetland gain was generally modest; restoration projects under various efforts have resulted in wetland gains in some areas. Natural wetland accrual, due to resurging beaver populations (Lovallo and Hardisky 2009), may also have contributed to gains. As with other habitats, changes in wetland quality between the two atlases is more difficult to quantify, but there are ongoing concerns about runoff of pollutants from industry, agriculture, and residential areas, changes in water levels due to siltation and flood control, and infestation of invasive species (Goodrich et al. 2002; Gross and Haffner 2010).

Natural lakes and ponds can be found throughout the state, but they are most numerous in the northeastern and northwestern corners of the state. Aside from Lake Erie, Conneaut Lake in Crawford County is easily the largest natural lake in the state, at 380 hectares (930 acres). However, this is dwarfed by numerous artificial lakes and reservoirs built during the mid-twentieth century, the largest of which, Pymatuning Reservoir, is 7,070 hectares (17,460 acres). A few species of breeding birds have benefited from the increase in large bodies of open water, but they are outnumbered by the many wetland species displaced by these impoundments (Gross and Haffner 2010).

Rivers and streams

Pine Creek, Lycoming County, Hal Korber/PGC

Ranging from tiny seasonal streams to major rivers, Pennsylvania is home to more than 130,000 km (83,000 miles) of riparian habitat. Headwater streams are found throughout the state, generally decreasing in density from west to east (fig. 3.11). There are very few areas devoid of streams, but the lowest densities are found within big cities and in intensively farmed areas, where they have been lost through channelization, water extraction, and paving. Riparian habitats often have vegetative structure associated with the break in contiguous forest cover that benefits species not strictly because of the water present.

The greatest concern for the health of riparian habitats is pollution, especially runoff of agricultural chemicals (Waters 2010) and acidification due to contamination from mines and acidic precipitation (Mulvihill et al. 2008). As a result, a program of establishing riparian buffers for streams and rivers has been created by setting aside adjacent land in either grassed or forested cover. USDA programs, such as the Conservation Reserve Program (including the Conservation Reserve Enhancement Program), have provided additional nesting cover for a number of bird species in river corridors, but it is not known whether the scale of these plantings has been sufficient to affect bird distributions.

Lake Erie shore and dunes

The shoreline of Lake Erie provides a very small, but unique, component of Pennsylvania's wildlife habitat that attracts many species uncommon to the rest of Pennsylvania. This feature is best preserved along the outer shores of Presque Isle State Park and the sandspit on the

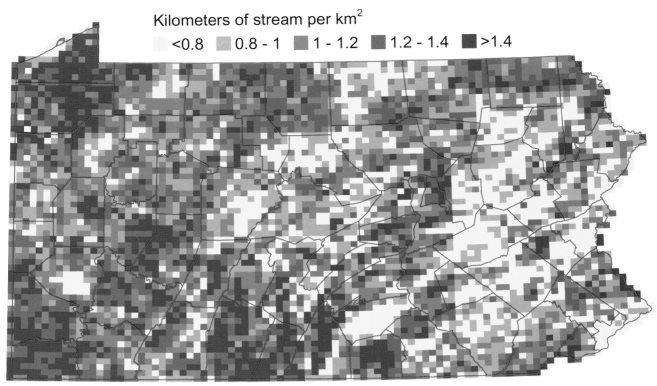

FIG. 3.11 *Densities of streams (Strahler [1952] orders 1–4) by atlas block*

Gull Point—Presque Isle State Park, Erie County, Scott Gorring

tip of the peninsula known as Gull Point. A dynamic area impacted by invasive species, lake-level fluctuations, and other natural and anthropogenic factors, the sand and cobble beaches formerly hosted breeding species (Piping Plover and Common Tern) that nest nowhere else in the state (see appendix A).

DEVELOPED LANDS

Overall, approximately 11 percent of Pennsylvania land cover consists of urban or suburban habitat, collectively known as developed land. The largest concentrations of urban and suburban habitat are in the southeastern and southwestern corners of the state and the city of Erie and its surroundings. In the Ridge and Valley Province, development occurs primarily in the valleys, frequently in areas previously devoted to agriculture. In the years between the two atlases, the amount of developed land increased by approximately 4 percent overall, but the increase was not uniform across the state. Areas of intense development occurred in the New England and Piedmont physiographic provinces and in the Great Valley Section of the Ridge and Valley Province (fig. 3.12). Other parts of the state showed relatively little change in the extent of developed habitat, with a small pocket near Pittsburgh actually showing a decline, likely due to brownfield restoration of former industrial sites.

As an area becomes urbanized, shifts occur in the avian community. Urban and suburban areas tend to favor generalists over species that have very narrow habitat requirements. As a general rule, the diversity of birds is low in urban areas, but the abundance may be very high. Species that can coexist with people often thrive in urban areas, which tend to have a much higher concentration of nonnative species (Rock Pigeon, European Starling, and House Sparrow) than rural areas.

Buildings and associated structures are an important component of the urban environment. Depending on structural design, buildings may provide roost and nest

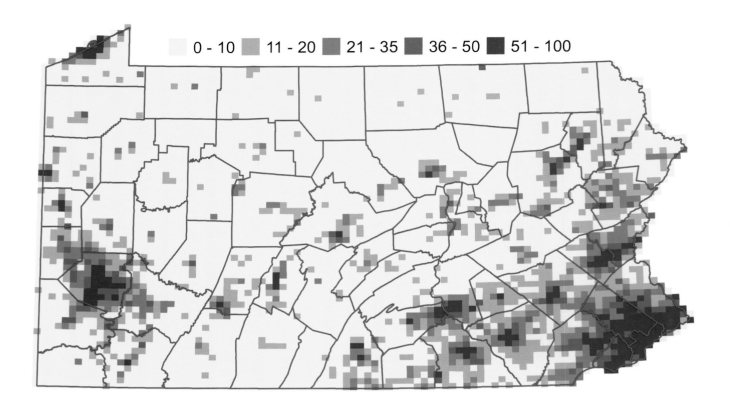

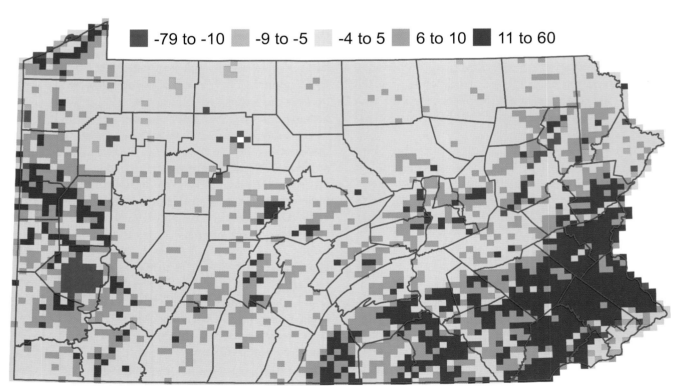

FIG. 3.12 *Percentage of developed land cover by atlas block c. 2005 (top) and relative change in developed land cover by atlas block between 1994 and 2005 (bottom)*

sites. As the density of buildings increases, the number of these species increases at the expense of native species (Gilbert 1989). Windows are another feature of the urban (and rural) environment affecting birds. Klem (1989) reported that 25 percent of bird species that occur in the United States and Canada have been killed by striking windows, suggesting that a large array of species are susceptible to window kills.

4 Atlas Methods

Daniel W. Brauning and Michael Lanzone

PROJECT PLANNING

Since the British initiated broadscale bird atlas projects in the late 1960s (Sharrock 1976), there has been a commitment to repeat atlas projects periodically. The second-generation bird atlas in Britain and Ireland was inaugurated 20 years after the initiation of their first atlas (Gibbons et al. 1993), setting the standard for what is now considered the appropriate interval for repeating an atlas. Upon reaching the 20-year mark, New York and Maryland initiated second atlases in 2000 and 2002, respectively (McGowan and Corwin 2008; Ellison 2010a), with efforts very similar to their first projects. Ontario, Canada, similarly initiated a second atlas project in 2001 and incorporated a number of new design features (Cadman et al. 2007).

With neighboring states repeating breeding bird atlases and the twentieth anniversary approaching for Pennsylvania, discussions among birders suggested that there was support for another round. A group of Pennsylvania's leading birders was convened at a dinner on 18 May 2001, prior to the annual meeting of the Pennsylvania Society for Ornithology (PSO), to consider this initiative. The group became the nucleus of the ad hoc Steering Committee that provided guidance for the project over the following years and represented birder interest in project planning. Despite some concerns about the ability of volunteers to achieve uniform state coverage, strong endorsement of the project was expressed, with a set of conditions for their support. First, a new project had to be directly compatible with Pennsylvania's first breeding bird atlas (1983 to 1989; Brauning 1992a). Second, the renewed effort had to obtain more uniform geographic coverage, improve information gathering for undersampled species (e.g., marsh birds), and incorporate a measure of abundance that could be mapped at high resolution statewide. Last, the committee recognized the novelty of this approach, and concluded that a year of project design and planning should be undertaken before the official start of fieldwork.

In 2002, the Penn State Cooperative Wetlands Center (CWC) was selected to undertake the design phase, entering into a cooperative agreement with the Pennsylvania Game Commission to develop the sampling and design strategy for Pennsylvania's second breeding bird atlas (O'Connell et al. 2004). This design work, initiated during the summer of 2002, was supported with newly available federal funds from the Wildlife Conservation and Restoration Account of U.S. Fish and Wildlife Service's Wildlife and Sport Fish Restoration Program. The CWC formed a core Design Team of five members from the Steering Committee: Tim O'Connell, Joe Bishop, Dan Brauning, Robert Mulvihill, and Michael Lanzone. They received helpful additional input from many sources, including the Cooperative Fish and Wildlife Research Unit at Penn State University, veteran Pennsylvania birders, and representatives of past and concurrent atlas projects in the United States, Canada, and Great Britain.

In addition to the Steering Committee, an Advisory Committee was convened on 17 April 2002 at Fort Hunter, in Dauphin County, to provide a strong organizational base for, and oversight of, Pennsylvania's second breeding bird atlas project (second Atlas). Members of the first Atlas Advisory Committee and representatives of interested organizations and agencies were invited to participate (see acknowledgments). This group provided advice on how to organize the effort, identified potential funding sources, and drafted a position description for a Project Coordinator and sponsoring organization to lead the effort. A subcommittee of the Advisory Committee, including Tim O'Connell (with the Design Team), Cal DuBrock (Pennsylvania Game Commission), Dan Brauning (Project Director), Greg Grove (PSO Board), and Steve Hoffman (Audubon PA), met during the week of 28 October 2002 to interview candidates and make a selection. Mulvihill was selected in December 2002 to serve as

Project Coordinator, and the Carnegie Museum of Natural History agreed that the Powdermill Nature Reserve would continue to employ him in that role. Full-time funding for the position of Project Coordinator was obtained from the new federally funded State Wildlife Grants program starting in the fall of 2003. Michael Lanzone was hired as Assistant Project Coordinator that fall.

With technical planning under way, a larger meeting of the birding community gathered in State College on 10 April 2003, where the Design Team and Project Coordinator were introduced. Additional input was drawn from this group, with the understanding that many would be asked to fill leadership roles or otherwise recruit birders to participate in the project. This expanded Steering Committee agreed upon priorities for the second Atlas: greater attention to underrepresented species and habitats, and implementation of a relative abundance sampling scheme. With this fine-tuned direction and continued enthusiasm, general plans for the project were unveiled at the annual PSO meeting on 17 May 2003 in Indiana, Pennsylvania. This set the stage for solicitation of volunteer participation for the 2004 start of fieldwork. The Advisory Committee met a second time, to review progress, on 27 October 2003 at Olewine Nature Center, Harrisburg.

ATLAS ORGANIZATION

The project was housed at Powdermill Nature Reserve, environmental research center of the Carnegie Museum of Natural History, in Rector, Pennsylvania. A handbook, field cards, and a range of other materials and information were prepared by the Project Coordinators and made available to participants to guide them in conducting all aspects of the project. The Carnegie Museum of Natural History hosted a website to provide instructional materials and data forms to atlas participants, links to relevant Geographic Information System (GIS) spatial data from Pennsylvania Spatial Data Access (PASDA), and access to the interactive atlas database hosted at the Cornell Lab of Ornithology. Second Atlas staff produced a newsletter once or twice a year, mailed it to participants, and made it available through state park offices and other public outlets.

ATLAS BLOCKS AND GRID DESIGN

Like Pennsylvania's first Atlas, and as recommended by the North American Ornithological Atlas Committee, the basic survey unit was the "block," defined as one-sixth of a standard U.S. Geological Survey (USGS) 7.5-minute topographic map (Robbins and Geissler 1990; Brauning 1992a; see fig. 4.1). Blocks ranged in size from 24.8 square kilometers (9.6 square miles) in southern Pennsylvania to 23.9 square kilometers (9.2 square miles) in northern Pennsylvania and were categorized as border, priority, or normal, based on their location. For border blocks on Pennsylvania's eastern, southern, and western borders, where the block boundary extended into a neighboring state, volunteers were instructed to report birds only seen or heard within Pennsylvania. Approximately two-thirds of the area of blocks along the state's southern border fell within the state, while blocks bordered by the Delaware River, to the east, varied in area, including some with very little Pennsylvania territory. The blocks along Pennsylvania's western border included about 1 mile of territory within the state.

Also like the first Atlas, blocks in the southeastern corner of each topographic quadrangle, or block 6, were designated as "priority" blocks for additional coverage. This prioritization was intended to standardize coverage across the state. The second Atlas included 4,937 blocks, of which 787 were priority blocks and 202 were border blocks. This totals nine blocks more than in the first Atlas, in which bird observations from some of the eastern border, Lake Erie shore, and blocks with negligible Pennsylvania territory were combined with neighboring blocks rather than reported separately.

ATLAS REGIONS

Whereas regions in the first Atlas were defined primarily by county boundaries, consisting of one to four counties and ranging from 39 to 306 blocks, region boundaries in the second Atlas were demarcated using a geographic rather than political approach. The DeLorme Pennsylvania Atlas and Gazetteer (DeLorme 2003), hereafter DeLorme Atlas, was used to delineate the regions because it was widely available and organized around topographic quadrangles. Each page of the DeLorme Atlas, or in some cases two pages, comprised a region (fig. 4.1). This regional demarcation was preferred because it established clearly identifiable boundaries based on DeLorme Atlas page numbers and created regions of generally comparable size. Each full DeLorme Atlas map page contained the equivalent of 14 USGS quadrangles, or 84 atlas blocks. For regions where most of a page fell outside of Pennsylvania, two or three pages were joined together to form one region. The smallest region consisted of 55 blocks (DeLorme Atlas, page 40, Region 40) and the

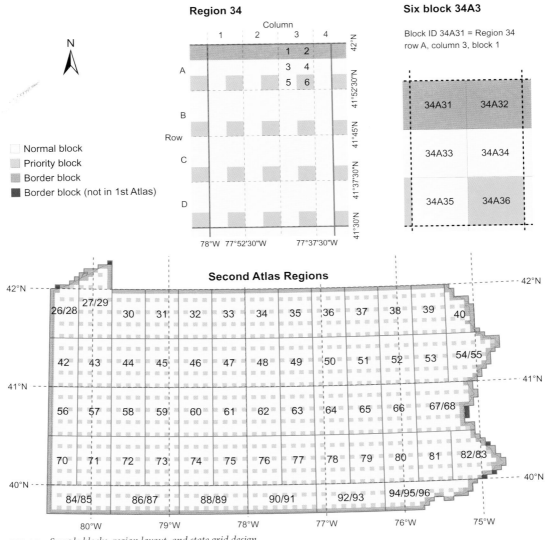

FIG. 4.1 *Sample blocks, region layout, and state grid design*

largest had 125 blocks (Region 67/68). Fifty-seven regions were defined, of which 46 were a full page of the DeLorme Atlas (84 blocks), while the remaining 11 regions consisted of more than one page.

Blocks in the second Atlas were labeled by referencing the DeLorme Atlas page number, in combination with the labels on the vertical axis (rows A–D) and horizontal axis (columns 1–7) and the appropriate sixth of the resulting topographic map. These correspond precisely with the 7.5-minute topographic map boundaries and first Atlas blocks. The DeLorme Atlas pages, therefore, defined region numbers and block labels at a scale of 1:150,000, which correspond exactly with USGS 7.5-minute quadrangle block boundaries. Thus, anyone with a DeLorme Atlas could identify the region in which observations occurred.

The regional boundaries provided the foundation for organizing fieldwork. Volunteer Regional Coordinators, drawn from the local birding community (when possible), worked singly or as co-coordinators to recruit and guide volunteers, manage and oversee coverage, and provide the first round of data review. Regional Coordinators performed the local administrative functions of the project: they recruited birders to participate, disseminated atlas literature, and assigned atlas blocks to volunteers as Block Owners. Block Owners were responsible for achieving the coverage objectives of their block(s), with Regional Coordinators providing backup coverage and oversight. Eventually, 83 individuals served as Regional Coordinators (see acknowledgments for complete list).

BREEDING SPECIES AND OBSERVATION CODES

A master list of species likely to be found breeding in Pennsylvania during the second Atlas was compiled by the Design Team, including species from the first Atlas, species known to have nested within the state since the end of first Atlas fieldwork (1989), and other potential candidates. Species were classified into categories to determine the type of information sought during this atlas effort, including conservation status, rarity, and behavior. Species classified as rare statewide or regionally were targeted for additional attention, including focused surveys, higher levels of evidence of breeding, or precise geographic location (table 4.1). Reports of state threatened or endangered species required validation by regional and state coordinators, specific location information, and narrative comments about breeding behavior or size of species population.

As with other bird atlas projects, a range of breeding behaviors were summarized into a set of codes to record evidence of nesting, ranging from the simplest detection by sight or sound through confirmation of active nests and fledged young. Birds were placed in one of four categories based on breeding evidence—observations outside breeding habitat ("Observed"), or possible, probable, or confirmed breeding—with a two-letter code used for confirmed breeding evidence and single-letter codes for all other categories (Laughlin et al. 1990; table 4.2). The "Observed" category applied to birds seen or heard within the breeding season but not in suitable breeding habitat and not thought to be breeding. "Possible" breeding described birds seen or heard within the breeding season in suitable breeding habitat, but with no discernible breeding behavior exhibited. "Probable" breeding included a range of behaviors associated with breeding. "Confirmed" breeding codes reflected what typically would be considered physical evidence that the species nested nearby. The second Atlas did not recommend undue effort to confirm breeding of common species, because it was believed that the field time could be better spent compiling species lists from multiple blocks. Obtaining confirmed breeding evidence was stressed for species identified in a particular conservation or priority category or for species simply rare or unexpected within the block.

SAFE DATES

Safe dates have been used by breeding bird atlas projects to reflect the period in which a species simply detected by sight or sound, in suitable habitat, could be safely included as a possible breeding species within a block. These dates are narrower than the species' nesting season, but they were used to exclude observations of migrants, nonbreeding individuals, or dispersed post-fledglings from being recorded as breeding records (appendix G). For many species, migratory populations may be found moving through Pennsylvania at the same time the species is nesting locally, and resident birds might not remain within their breeding territory year-round. As a result, both migration periods and nesting dates were used to define safe dates (fig. 4.2). Many data sources were compiled to determine the safe dates, including previous nesting records, bird banding records, personal observations, and several published sources (e.g., Brauning 1992a; McWilliams and Brauning 2000). Bird observations reported as possible breeding codes were not accepted outside safe date periods; probable and confirmed breeding observations could be entered after confirming the record in response to an automated query.

DATA COLLECTION

To be incorporated as a record for Pennsylvania's second Atlas, the following information was required: species name, breeding code, date of observation, block code, and observer's name. Habitat, supplemental comments, and precise location were optional data fields, encouraged for any record but required only for priority species.

A variety of forms were provided to facilitate atlas volunteers' documentation of breeding species. The most basic, the field card, provided blanks to list the required data. A summary field card listed all likely species, with their safe dates, and provided a space to enter breeding codes and dates. Other forms were disseminated, containing targeted lists of species, to focus on groups of species, including Killdeer and American Robin in school yards, several species in agricultural settings, and nocturnal and wetland species. All forms provided space to note time spent looking for birds as well as other effort invested in the atlas project, such as time driving to and from atlas blocks and entering data. These effort data provided both scientific value to assess coverage, and documented administrative time and miles donated to the project. In addition to a variety of data entry forms, other creative means were used to obtain data, such as listing the availability of species occurrences/breeding status in each block from the first Atlas (to indicate where species data were lacking) and the ability for volunteers to print block maps that provided an opportunity to record geographic locations of observations.

Because of the importance of additional information for rare birds, and to provide for more thorough review of such reports, a form was provided to identify precise location, species behavior, and other information. To encourage this documentation, priority species names were printed in boldfaced type on forms and in the handbook. The second Atlas project officially accepted data beginning on 1 January 2004.

TABLE 4.1 Special surveys, data actions, and conservation status for priority species, during the second Atlas, 2004–2009

Species	Conservation status	Special survey type	Actions (**bold** = required)
American Wigeon	Rare	None	**Verify, Point Locate**
Northern Shoveler	Rare	None	**Verify, Point Locate**
Green-winged Teal	Special Concern	None	**Verify, Point Locate**
Hooded Merganser	Rare	None	Point Locate
Pied-billed Grebe	Special Concern	Wetland	**Verify, Point Locate**
Northern Bobwhite	Special Concern	None	**Verify, Point Locate**
American Bittern	Endangered	Wetland	**Verify, Point Locate**
Least Bittern	Endangered	Wetland	**Verify, Point Locate**
Great Blue Heron (colonies)	Conservation Interest	None	Point Locate, Count Nests
Great Egret	Endangered	None	**Verify, Point Locate, Count Nests**
Snowy Egret	Special Concern	None	**Verify, Point Locate, Count Nests**
Black-crowned Night-Heron	Endangered	None	**Verify, Point Locate, Count Nests**
Yellow-crowned Night-Heron	Endangered	None	**Verify, Point Locate, Count Nests**
Osprey	Threatened	None	**Verify, Point Locate**
Bald Eagle	Threatened	None	**Verify, Point Locate**
Northern Harrier	Special Concern	None	**Verify, Point Locate**
Sharp-shinned Hawk	Conservation Interest	None	Point Locate
Northern Goshawk	Special Concern	None	**Verify, Point Locate**
Red-shouldered Hawk	Conservation Interest	None	Point Locate
Peregrine Falcon	Endangered	None	**Verify, Point Locate**
Black Rail	Rare	None	**Verify, Point Locate**
King Rail	Endangered	Wetland	**Verify, Point Locate**
Virginia Rail	Conservation Interest	Wetland	Point Locate
Sora	Special Concern	Wetland	**Verify, Point Locate**
Common Gallinule	Special Concern	Wetland	**Verify, Point Locate**
American Coot	Special Concern	Wetland	**Verify, Point Locate**
Sandhill Crane	Rare	None	**Verify, Point Locate**
Killdeer	Education Focus	Schools	Point Locate, Nesting Dates
Upland Sandpiper	Threatened	Farmer Surveys	**Verify, Point Locate**
Wilson's Snipe	Special Concern	Wetlands	**Verify, Point Locate**
American Woodcock	Conservation Interest	None	Point Locate
Black Tern	Endangered	None	**Verify, Point Locate**
Common Tern	Extirpated	None	**Verify, Point Locate**
Barn Owl	Special Concern	Farmer Surveys	**Verify, Point Locate**
Long-eared Owl	Special Concern	Habitat Model	**Verify, Point Locate**
Short-eared Owl	Endangered	Habitat Model	**Verify, Point Locate**
Northern Saw-whet Owl	Conservation Interest	Nocturnal Survey	Point Locate
Common Nighthawk	Education Focus	Schools	Point Locate, Count Birds
Chuck-will's-widow	Rare	Nocturnal	**Verify, Point Locate**
Eastern Whip-poor-will	Conservation Interest	Nocturnal	Point Locate
Chimney Swift	Education Focus	Schools	Point Locate, Count Birds
Red-headed Woodpecker	Conservation Interest	None	Point Locate
Yellow-bellied Sapsucker	Conservation Interest	None	Point Locate
Olive-sided Flycatcher	Extirpated	None	**Verify, Point Locate**
Yellow-bellied Flycatcher	Endangered	None	**Verify, Point Locate**
Alder Flycatcher	Conservation Interest	None	Point Locate
Loggerhead Shrike	Endangered	None	**Verify, Point Locate**
Bewick's Wren	Extirpated	None	**Verify, Point Locate**
Sedge Wren	Endangered	Wetland	**Verify, Point Locate**
Marsh Wren	Special Concern	Wetland	**Verify, Point Locate**
Golden-crowned Kinglet	Conservation Interest	None	Point Locate
Swainson's Thrush	Special Concern	None	**Verify, Point Locate**
American Robin	Education Focus	Schools	Nesting Dates, Nest Success
Louisiana Waterthrush	Conservation Interest	None	Point Locate, Count Birds
Northern Waterthrush	Conservation Interest	None	Point Locate
Golden-winged Warbler	Conservation Interest	Habitat Model	Point Locate
Prothonotary Warbler	Special Concern	None	**Verify, Point Locate**
Swainson's Warbler	Rare	None	**Verify, Point Locate**
Kentucky Warbler	Conservation Interest	None	Point Locate
Cerulean Warbler	Conservation Interest	None	Point Locate
Blackburnian Warbler	Conservation Interest	None	Point Locate
Blackpoll Warbler	Endangered	None	**Verify, Point Locate**

(continued)

TABLE 4.1 (continued)

Species	Conservation status	Special survey type	Actions (**bold** = required)
Yellow-throated Warbler	Conservation Interest	None	Point Locate
Prairie Warbler	Conservation Interest	None	Point Locate
Black-throated Green Warbler	Conservation Interest	None	Point Locate
Bachman's Sparrow	Extirpated	None	**Verify, Point Locate**
Clay-colored Sparrow	Rare	None	**Verify, Point Locate**
Lark Sparrow	Extirpated	None	**Verify, Point Locate**
Henslow's Sparrow	Conservation Interest	Habitat Model	Point Locate
Summer Tanager	Special Concern	None	**Verify, Point Locate**
Dickcissel	Endangered	None	**Verify, Point Locate**
Red Crossbill	Special Concern	None	**Verify, Point Locate**
Pine Siskin	Special Concern	None	**Verify, Point Locate**
Evening Grosbeak	Rare	None	**Verify, Point Locate**

TABLE 4.2 Breeding codes for second Atlas, 2004–2009

Observed
O — Individual of a species simply observed (seen or heard) within safe dates but not in suitable habitat.

Possible Breeding
X — Individual of a species seen or heard in suitable nesting habitat within safe dates but not exhibiting any of the breeding behaviors described in the following 15 categories.

Probable Breeding
P — Pair (male and female) of a species seen in close proximity to and/or interacting non-aggressively with one another; in sexually monomorphic species, like Song Sparrow, two birds seen feeding or perched close together without displaying aggression.

T — Territorial behavior observed, including counter-singing, drumming in woodpeckers (drumming bird must be seen for species I.D.), aggressive interaction between same-sex individuals, or a singing male in the same location on visits separated by 5 days or more.

C — Ritualized courtship behavior (e.g., aerial displays, courtship feeding) or copulation between two birds observed.

U — Used nest of species found (cannot be used in first year of atlas, 2004); only species with highly distinctive nests may be assigned this code. Because of the difficulty in identifying the old nests of many species, and because used nests, especially in protected settings, can persist for years, "U" is taken only as evidence of probable breeding during the atlas.

A — Agitated behavior or anxiety calls given by adults due to observer or predator presence (does not include agitated reaction to "pishing").

Confirmed Breeding
CN — Adult bird seen carrying nest material. (Note: use of this code for larger species, like crows or herons, which may collect nesting material well outside the block in which they nest, or for species like wrens that build "dummy" nests, will equate to probable breeding; for these species, attempt to upgrade to one of the next nine confirmed breeding codes.)

PE — Physical evidence of breeding condition observed for birds in hand, specifically a highly vascularized (edematous) brood patch and/or visibly gravid condition (supported by appropriately increased body mass); this code limited to use by experienced bird banders.

NB — Nest building observed at nest site. (Note: this code will equate to probable breeding in the case of wrens and excavating species like woodpeckers, kingfishers, and tits, which build "dummy nests" or excavate cavities that may not end up using for nesting; for these species, attempt to upgrade to one of the following seven confirmed breeding codes.)

DD — Distraction display (especially injury feigning), or apparent direct defense of unobserved nest/young.

FL — Recently fledged young observed (downy young of precocial birds, like Ruffed Grouse, or stub-tailed juveniles of altricial species, like American Robin). Use this code very sparingly, because birds truly at this stage should, in fact, be attended/defended by adults—wait for adult to appear as a stronger confirmation (e.g., DD or CF, if possible).

CF — Adult seen carrying either food or fecal sac.

FY — Adult seen feeding fledged young.

NE — Nest of species found containing eggs (nest containing cowbird eggs should count as "NE" both for host species and cowbird).

ON — Occupied nest of species found, but contents unknown because adult is on nest or nest placement prevents examination of contents (also includes observations of hole-nesting species, like woodpeckers, kingfishers, swifts, and tits, seen entering and remaining in cavities).

NY — Nest of species found containing young (nest containing cowbird nestlings count as "NY" both for host species and cowbird).

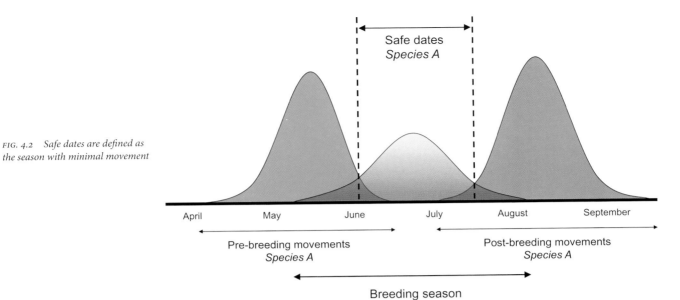

FIG. 4.2 *Safe dates are defined as the season with minimal movement*

OTHER DATA COLLECTED

The only habitat feature incorporated as a standardized data option value was a categorical estimate of abundance of Eastern Hemlock, Pennsylvania's state tree. The Hemlock Woolly Adelgid is an introduced insect pest that threatens hemlocks in eastern North America. Hemlock is an important component of Pennsylvania's forests and its breeding birds (Becker et al. 2008). With the lack of an ecological replacement for hemlocks where they are lost to adelgid infestation, ecologists are concerned this will translate into population declines, or even local extinction, of some of these bird species (Tingley et al. 2002). In an effort to track and understand the relationship between this tree and breeding birds, along with bird observations, hemlock could be recorded as follows: not present (habitat score = H0); one or a scattered few hemlocks, including ornamental trees (score = H1); mixed deciduous-hemlock forest with deciduous trees clearly dominant, i.e., hemlock <50% of trees (score = H2); or hemlock forest or mixed hemlock-deciduous forest with hemlock clearly dominant, i.e., >50% of trees (score = H3).

SPECIAL SURVEYS AND TARGETED EFFORT

In keeping with the goals of the project outlined by the Steering Committee, several resources and survey protocols were developed to help atlas volunteers focus their efforts on finding priority or secretive species within the blocks in which they worked. Species were flagged for additional attention because they were listed by the Pennsylvania Game Commission as Endangered or Threatened, by the Pennsylvania Biological Survey's Ornithological Technical Committee as Special Concern, or because they were simply rare or of some conservation interest (table 4.1). For these species, obtaining confirmed breeding evidence was encouraged, along with precise point locations using the web mapping tool. Central to focused coverage for the second Atlas was an interactive map, housed at PASDA. This web-based tool enabled atlas participants to compile map features (including roads, streams, aerial photography, Important Bird Areas, public lands, three wetland habitat size classes based on the National Wetland Inventory, and other features) into a custom map for any particular block. Among the options were habitat-based predictive maps, using a Gap Analysis approach (Miller 2007) for five species (Barn, Short-eared, and Long-eared owls; Golden-winged Warbler, and Henslow's Sparrow). These maps, indexed to atlas blocks, could be printed from a custom interface and were provided to guide field activities. Habitat-based models can be helpful for directing surveys (Corsi et al. 1999) by directing volunteers unfamiliar with an area into potentially suitable habitats for difficult-to-find species, or for simply suggesting the potential for a species that might otherwise be overlooked. The models were found to be useful in directing surveys of Golden-winged Warblers for the Golden-winged Warbler Atlas Project, run by Cornell University, and during the design phase of the atlas (T. Miller and M. Lanzone, pers. obs.).

Survey protocols were developed for marsh and nocturnal birds to obtain higher uniformity of coverage across Pennsylvania and in recognition that additional efforts were needed to detect these groups of birds. Regional Coordinators were requested to promote these special surveys and resources wherever wetland habitats occurred, but particularly in priority blocks.

Marsh bird survey

Starting in 2005, a marsh bird survey, modeled after the North American Marsh Bird Monitoring Protocol (Conway 2011; Ribic et al. 1999), was promoted through the annual newsletter and by Regional Coordinators. A standardized survey protocol and field form encouraged the reporting of detailed records of marsh bird responses. Audio-lure surveys were to be conducted within 3 hours of twilight or dawn. The selection of target species varied slightly based on wetland size, with wetlands classified as small (0.5–3.0 hectares, 1.2–7.4 acres), medium (>3–10 hectares, 7.4–24 acres), or large (>10 hectares, >24 acres). Compact disks (CDs) provided to volunteers contained three tracks of marsh bird vocalizations, each corresponding to the survey designed for each wetland size. Tracks differed in the number of species incorporated into the playback regime, with smaller wetlands having fewer playback species than larger wetlands. Tracks used for small wetlands included Black Rail, Least Bittern, Sora, Virginia Rail, and King Rail; medium wetlands included the small wetland species and American Bittern; and large wetlands included those species and Common Gallinule, American Coot, and Pied-billed Grebe.

Volunteers were instructed to play the CD, listen for responses, and record all species present. The amount of time spent conducting the survey varied: 10 minutes for small wetlands, 11 minutes for medium wetlands, and 14 minutes for large wetlands. For each, a 5-minute listening period began the survey, followed by 30 seconds of recorded calls for each species and a 30-second listening period. Surveying up to three such wetlands was recommended. Observers were asked to document habitat characteristics within 100 m of the survey location by noting percentage of open water coverage, floating vegetation, cattails, sedges/rushes, Purple Loosestrife, phragmites, and woody vegetation in five distance categories (0 m, 1–25 m, 26–50 m, 51–75 m, 76–100 m; see fig. 4.3 for data sheet).

Within a wetland, observers were asked to complete a survey at points no less than 250 m apart. For smaller wetlands or those with a narrower, linear shape, the center points were placed on the wetland edge to anchor a 180° sampling arc (Ribic et al. 1999; Timmermans and Craigie 2002). Early season surveys were conducted between 15 May and 15 June on warm mornings with light winds and no rain (Ralph et al. 1995). Late-season surveys were completed between 16 June and 15 July. A minimum of 10 days difference between early- and late-season samples, for any one site, was encouraged.

Nocturnal surveys

Three styles of nocturnal surveys were developed to detect owls and other nocturnal species: mini-routes, species-specific efforts, and general owling. For each survey type, a playback CD and data sheets were provided to volunteers. The first type of survey was the mini-route, modeled after the "toot routes" developed by Doug Gross (Gross 2000a), which recommended eight stops at which observers played recorded Northern Saw-whet Owl vocalizations (see fig. 4.4 for data sheet). The full survey cycle was 11 minutes, including quiet listening periods alternating with broadcast calls. Nocturnal mini-routes were initially encouraged in all priority blocks with forested habitat. The second approach targeted Long-eared, Short-eared, and Barn owls with recorded vocalizations of each species. Broadcast of species vocalizations, provided on a CD made available to volunteers, was encouraged on areas targeted as potential habitat for these species (described above). Finally, a general nocturnal survey effort was encouraged through distribution of a CD containing playback of territorial vocalizations to elicit responses from expected owls: Barn, Barred, and Great Horned owls and Eastern Screech-Owl. Vocalizations of other species expected during nocturnal surveys (e.g., nightjars) were provided on the CD for reference, and those species were listed on data sheets to be noted when detected. Each of these tools was promoted to improve detection of secretive species that would otherwise be overlooked without such targeted effort.

Point count surveys

The second Atlas adopted a comprehensive point count sampling protocol designed to support multiple statistical approaches for estimating detection probabilities, breeding bird densities, and population sizes of a wide array of breeding bird species. This effort was conducted by experienced seasonal staff as a parallel project, complementary to the volunteer-based atlas effort. These protocols were developed for the second Atlas project, in part through a grant to Penn State's Cooperative Wetlands Center (O'Connell et al. 2004), and were an integral part of the Planning Team's effort. In addition to the immediate value of these data for assessing bird species

2nd Pennsylvania Breeding Bird Atlas Marshbird Survey

Block ID	Wetland ID	Observer	Date / /200___	Start Time

GPS readings	N		Weather		Species	Silent Period	Playback Period	Total
	W		Weather Code		Black Rail			
Habitat (Check one for each category)			Temp		Least Bittern			
% Cover	0	1-25	26-50	51-75	76-100	Wind Speed		Sora			
Open Water						Wind Direction		Virginia Rail			
Floating veg., e.g., lily pads						**Wetland Size** (Check one)		King Rail			
Cattail						**Small** <3 ha		American Bittern			
Sedge/Rush								Common Moorhen			
Purple Loosestrife						**Medium** 3-10 ha		American Coot			
Phragmites								Pied-billed Grebe			
Woody						**Large** >10 ha					
Other (Explain in comments)											

Comments

☐ Check here if this data **was** entered on the website
☐ Check here if this data **was not** entered on the website

Block ID	Wetland ID	Observer	Date / /200___	Start Time

GPS readings	N		Weather		Species	Silent Period	Playback Period	Total
	W		Weather Code		Black Rail			
Habitat (Check one for each category)			Temp		Least Bittern			
% Cover	0	1-25	26-50	51-75	76-100	Wind Speed		Sora			
Open Water						Wind Direction		Virginia Rail			
Floating veg., e.g., lily pads						**Wetland Size** (Check one)		King Rail			
Cattail						**Small** <3 ha		American Bittern			
Sedge/Rush								Common Moorhen			
Purple Loosestrife						**Medium** 3-10 ha		American Coot			
Phragmites								Pied-billed Grebe			
Woody						**Large** >10 ha					
Other (Explain in comments)											

Comments

☐ Check here if this data **was** entered on the website
☐ Check here if this data **was not** entered on the website

Send completed data sheets by August 1st each year to: Coordinators, 2nd Pennsylvania Breeding Bird Atlas, Powdermill Avian Research Center, 1847 Route 381, Rector, PA 15677

FIG. 4.3 Marshbird survey data sheet

2nd Pennsylvania Breeding Bird Atlas Owl Survey Data Sheet

Block ID	Observer	Date / /200__	Weather Code Start: End:	Temperature Start: End:

Wind Speed Start: End:	Wind Direction Start: End:	Lunar Code (only the postion will change) Start: End:	Version 02.01.2006

Start Time (24 hr) Survey Stop #	Species Surveyed	1 hrs	Species Surveyed	2 hrs	Species Surveyed	3 hrs	Species Surveyed	4 hrs	Species Surveyed	5 hrs	Species Surveyed	6 hrs
Long-eared Owl												
Short-eared Owl												
Barn Owl												
E. Screech Owl	n/a		n/a		n/a		n/a		n/a		n/a	
Barred Owl	n/a		n/a		n/a		n/a		n/a		n/a	
Great-horned Owl	n/a		n/a		n/a		n/a		n/a		n/a	
N. Saw-whet Owl	n/a		n/a		n/a		n/a		n/a		n/a	
Common Nighthawk	n/a		n/a		n/a		n/a		n/a		n/a	
Whip-poor-will	n/a		n/a		n/a		n/a		n/a		n/a	
	n/a		n/a		n/a		n/a		n/a		n/a	
	n/a		n/a		n/a		n/a		n/a		n/a	
Interruptions												
Excessive Noise												
Comments												
Habitat												
GPS Readings (If Possible)	N W		N W		N W		N W		N W		N W	

Start Time (24 hr) Survey Stop #	Species Surveyed	7 hrs	Species Surveyed	8 hrs	Species Surveyed	9 hrs	Species Surveyed	10 hrs	Species Surveyed	11 hrs	Species Surveyed	12 hrs
Long-eared Owl												
Short-eared Owl												
Barn Owl												
E. Screech Owl	n/a		n/a		n/a		n/a		n/a		n/a	
Barred Owl	n/a		n/a		n/a		n/a		n/a		n/a	
Great-horned Owl	n/a		n/a		n/a		n/a		n/a		n/a	
N. Saw-whet Owl	n/a		n/a		n/a		n/a		n/a		n/a	
Common Nighthawk	n/a		n/a		n/a		n/a		n/a		n/a	
Whip-poor-will	n/a		n/a		n/a		n/a		n/a		n/a	
	n/a		n/a		n/a		n/a		n/a		n/a	
	n/a		n/a		n/a		n/a		n/a		n/a	
Interruptions												
Excessive Noise												
Comments												
Habitat												
GPS Readings (If Possible)	N W		N W		N W		N W		N W		N W	

Send completed data sheets by September 1st each year to: Coordinators, 2nd Pennsylvania Breeding Bird Atlas, Powdermill Avian Research Center, 1847 Route 381, Rector, PA 15677

FIG. 4.4 *Owl survey data sheet*

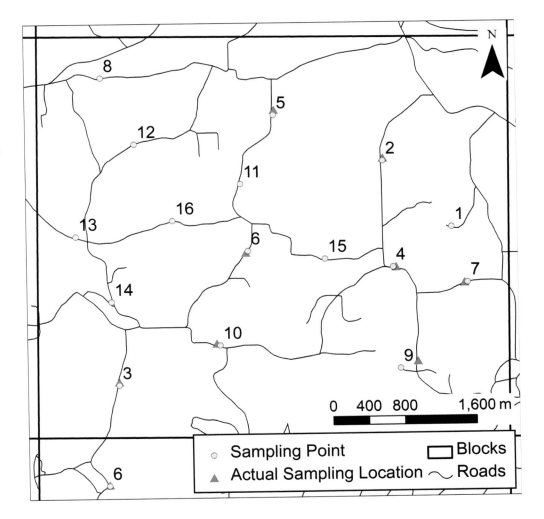

FIG. 4.5 *Example block showing random point count locations*

and associated habitats potentially needing conservation attention, point counts provide essential data for setting quantifiable goals for population maintenance and recovery of select bird species in Pennsylvania.

A road-based random sampling design (O'Connell et al. 2004) was implemented to accommodate coverage in 4,694 atlas blocks fully contained within Pennsylvania. Using a GIS, it was determined that nearly all of the blocks had enough road access to facilitate counts of at least eight randomly selected points, separated by at least 400 m (1,312 feet). Due to safety and noise interference concerns, interstate and major highways were removed from the potential sample points. Sixteen points were randomly selected and numbered within each block and moved to the closest road using an ArcView 3.2 (ESRI, Redlands, CA) extension designed by Ken Corradini of the Penn State Institutes for the Environment (fig. 4.5).

Point counts were conducted at eight locations per block, with up to eight additional point locations provided in each block in case any of the initial eight targeted points did not fall on an actual or accessible road or were for some other reason deemed unsuitable (e.g., excessive noise disturbance, unsafe to stop). In such cases, the preferred point was skipped in favor of an accessible point. In some instances where the road density was very low, fewer than eight points were selected within a block, but in each case observers utilized as many of the selected locations (up to eight) as possible within each block. If more points were needed due to unsuitable locations, observers used additional points, in sequence, until they completed the required eight points per block.

Counts were conducted from 25 May to 4 July each year, beginning at 30 minutes before sunrise and continuing for about 5 hours each morning (approximately 0500 to 1000). Counts were only conducted in suitable weather, and they were not conducted during times of heavy or persistent rain or moderate wind (Beaufort scale 3 or stronger). In addition, if the air temperature was very cold (e.g., <5°C, 41°F) or very hot (e.g., >25°C, 77°F), the observer made a judgment as to whether bird song output was depressed sufficiently to curtail fieldwork (see fig. 4.6).

Prior to the start of each field season, the point count staff was tested for hearing ability and song identification and went through rigorous training pertaining to the point count protocol. Each observer was assigned to

2nd Pennsylvania Breeding Bird Atlas Point Count Data Sheet

Block ID		Observer		Date	/ / 200	Full protocol NOT used	☐
Point #		GPS readings	N	Start Time		Noise interference	☐
			W			Specify (e.g., stream, traffic):	

Habitat/Weather Data:

1 ☐	2 ☐	3 ☐	4 ☐	5 ☐	6 ☐	7 ☐	8 ☐	9 ☐	10 ☐	11 ☐	12 ☐	13 ☐	14 ☐	15 ☐	16 ☐

		<75m >75m	Mark "x" in the time and distance band each ***individual*** of each species is observed within (>1 time band and >1 distance O.K.)					Enter #'s of birds		
Indiv.#	**Species**	**1**	**2**	**3**	**4**	**5**	**Non-song Cue**	**Fly-over**	**Flock**	**Comments**
1										
2										
3										
4										
5										
6										
7										
8										
9										
10										
11										
12										
13										
14										
15										
16										
17										
18										
19										
20										
21										
22										
23										
24										
25										

Notes:

FIG. 4.6 *Point count data sheet*

an area of the state and supplied with a Garmin iQue 3600 WAAS-enabled GPS unit uploaded with the coordinates of each randomly selected point within their survey area. The units provided vocal driving directions to each point. Staff attempted to get as close to the point as possible and recorded the actual latitude and longitude (North American Datum 1983 and decimal degrees) of their locations at the time of the point count. All birds detected were recorded at each point count site; however, for the abundance sampling, only singing males were tracked between time bands, and the point counts were conducted during a 6-minute and 15-second time period, divided into five equal 75-second time bands. Birds were recorded in two distance bands from the observer, within 75 m and greater than 75 m. This survey method allows the use of a "removal model" (Farnsworth et al. 2002) to account for potential bird activity and observer bias (see chap. 5). A stopwatch, set to repeat every 75 seconds, was used to ensure standardized time bands. Each bird detected by song was recorded within or beyond 75 m from the observer during each of five consecutive 75-second time intervals. Significantly, observers also recorded whether an individual bird was redetected by song in subsequent time intervals. All other birds observed (by both sight and sound) during the count were noted as either non-song cue, flyover, or flock, along with the count of individuals.

In addition to recording date, time, location, and noise interference, observers recorded ancillary weather and habitat data at each point. Weather data included exact temperature and categorical data on wind speed, precipitation, and cloud cover. Habitat data were categorical and included the following: hemlock abundance, hemlock health, spruce abundance, spruce qualifier (native vs. planted), extent of ericaceous understory, deer browse impact on understory, recent or active land use change, pastured livestock or working barn, presence of natural nesting cavities, presence of nest boxes, road type, and dominant habitat type.

WEB DATA ENTRY

Working through the Design Team and with input from the Steering Committee, BirdSource, at the Cornell Lab of Ornithology, was subcontracted to develop a user-friendly website to facilitate data entry and exploration. This custom website, adapted from the eBird system already in place, automated many atlas project administrative functions, including registration of volunteers, ownership of blocks, production of block maps, entry and review of records, presentation of results from the first Atlas, and species counts and lists entered by block to track progress during the second Atlas. This system provided the general public access to atlas results and obtained information for registered users in data entry, maintenance, and coordination roles. The website defined administrative roles of permissions and data management. It accepted multiple records of any species within a block, summarized results by highest breeding code per species, and tracked block coverage "completeness."

Filters, incorporated into the data entry routine, prohibited entry of possible breeding records outside the period of safe dates. The web tool also solicited supplemental comments and specific location data for rare species, customized to the atlas region. Additional important website features included the ability to precisely georeference each record through direct entry of latitude-longitude coordinates or through interactive mapping capabilities based on topographic maps or aerial photography at adjustable scales. This system, subsequently adapted by other states, enabled atlas volunteers to enter bird records from an established list of birds in several formats. After entering core data features, observers were prompted to submit supplemental location information and other details, based on the species. This platform was ably supported throughout the atlas by Cornell's Jeff Gerbracht.

Fieldwork, in terms of hours spent in the field (block effort) and time and miles for travel or administrative time (other effort), was requested of all participants upon entry of birds to a block. These effort data were attributed to the participants based on their status as registered observers, block owners, regional coordinators, or administrators. Anyone registered to the website could add time or records to any block.

Volunteers without computer access sent their field cards to their respective Regional Coordinators, who then entered those data on their behalf. Atlas staff at Powdermill Nature Reserve solicited bird sightings through newsletter articles and press releases and also entered records into the database based on data sheets received at the central office.

DATA SCREENING AND REVIEW

With data entry open to anyone willing to register on the website or send in a data sheet, a variety of validation features was needed to ensure data integrity. All observations submitted to the website were automatically classified into three categories using species, seasonal, and geographic filters: "OK," "questionable," and "rare."

Filters were established based on species (statewide rarities and regional rarities required review), seasons (observations outside safe dates were flagged), and on geography (species requiring review were listed for each region). An observation was OK if it fell within the bounds of all filters for the given species, date, and/or geographic limits. If a record did not fall within the bounds of a filter, the user was warned that this species was unusual, and was asked to affirm the observation before the record was accepted into the database. If the user affirmed the record, the observation was advanced for review to the Regional Coordinator, and the disposition of that review was advanced to the Project Coordinators. In addition, observers were asked to submit a verification form for any observations of statewide rarities. In most cases, if the appropriate documentation had been submitted and verification forms were on file, the Project Coordinators generally accepted the recommendations of Regional Coordinators. In the end, 156 species were screened in at least one region, and a total of 34,320 records were reviewed by coordinators through this process during the course of fieldwork. At the completion of fieldwork, a team, which focused on mapped records outside contiguous distributions, was assembled to conduct a final data review.

BLOCK COMPLETION

Standards for adequate coverage of blocks were determined during the project planning phase and incorporated into the data management website. Using Gap Analysis Program (GAP) methodology, a list of potential species was compiled for each block (see chap. 7). The predicted species richness was reduced by 5 percent to exclude very rare species and to account for modeling error. Block coverage goals, as a target species list, were established at 75 percent of the predicted species for non-priority blocks and 90 percent for priority blocks. Instead of a uniform goal of 70 species, established in the first Atlas, this produced a customized target for each block, ranging from 18 to 108 species in regular blocks, and up to 127 in priority blocks. These species targets were incorporated into the block status reports on the data management website, along with a computation of percentage of the total submitted.

In addition to these species targets, a minimum of 25 hours of effort was encouraged for each block. This effort was intended to be spread over the course of the seasons in which all species may be breeding, with the most time spent during the peak of most species' breeding seasons, and spread over the course of the day, to target crepuscular and nocturnal species. For a block to be considered complete by the Regional Coordinator, all of these criteria were carefully taken into account. The Regional Coordinators had the discretion to set a block status as "complete" even if the species target was not reached, in cases in which they felt that an observer had adequately covered all habitats, if 25 hours of effort were adequately spread over the season, and the resulting species list accurately reflected what was anticipated in that block. Once marked as "complete," the block was shaded on the website regional maps and observers were encouraged to work in other blocks.

Upon completion of the fifth scheduled field season, Regional Coordinators recommended that an additional year of fieldwork should be undertaken to achieve the objective of coverage in every block. Targeted blocks were identified in three categories, based on percentage of target species reported to the block by 2008: 236 blocks that had between 40 and 50 percent, 89 with 30 to 40 percent, and the highest priority—20 blocks with less than 30 percent. Data were accepted in any block until 1 July 2009, but Regional Coordinators and volunteers were encouraged to focus on the target block list, starting with the highest priorities.

DATA MANAGEMENT

In order to comprehensively report the breeding birds in Pennsylvania between 2004 and 2009, additional data sources were reviewed and incorporated into the second Atlas project. The most significant source was the point count project, described above and conducted as a parallel project. In anticipation of the final full field season, a record of each species reported by point counters for each block was uploaded into the second Atlas database at the Cornell website during spring 2008. This enabled participants to assess coverage status based on the most complete data available. Point counts completed in 2008 and 2009 were added at the end of the project. Following completion of fieldwork in 2009, data from several other sources were incorporated into the database, including records from USGS Breeding Bird Survey routes with stop locations and ongoing monitoring for CREP (Conservation Reserve Enhancement Program; Wilson 2009). These data were cross-referenced to atlas blocks using georeferenced stop locations and added as possible breeding (X) records. In addition, rare or unusual breeding birds published in the journal *Pennsylvania Birds*, notable species records submitted to the eBird web-based database (CLO 2011), and state-listed bird records compiled by Pennsylvania Game Commission were reviewed and incorporated into the atlas database. Game Commission surveys for American Woodcock, Bald Eagle, Barn Owl, Mute Swan, Osprey, colonial waterbirds, and Peregrine Falcon were added if they were supplemental to second Atlas records.

CERTIFICATION OF FIELD EFFORT

The importance of field effort went beyond scientific assessment of coverage goals. For the second Atlas, the hours spent by volunteers collecting and processing bird data, as well as organizing this effort, had financial value toward federal grants. These grants required non-federal matching funds to enable payment of project expenses. Volunteer time, entered into the atlas website, was summarized, reviewed for reasonableness, and approved by Regional Coordinators and the Project Director as authentic expenses to match the federal grants. A routine, developed by the Cornell Lab, facilitated this accounting procedure and was approved by the USFWS Wildlife and Sport Fish Restoration Program as sufficient to approve or certify volunteer time as matching expenses.

PUBLICATION PLANNING

Planning for this publication started with a meeting of an ad hoc Publication Advisory Committee on 8 November 2008 in State College, Pennsylvania. Kim Corwin, co-editor of New York's second breeding bird atlas (McGowan and Corwin 2008), and John Ozard, also with the New York second atlas, joined this meeting to provide advice and direction on the publication process. This group confirmed that, during the course of the supplemental field season, planning should begin for a full-color, hardback book to be published with the results of the second Atlas. They further concurred that species accounts should be illustrated with photographs. This group met again, formally, on 1 August 2009 at Olewine Nature Center in Harrisburg, Pennsylvania, to assess progress, guide book design decisions, and encourage fundraising to cover publication costs. With the close of fieldwork in 2009, Andrew Wilson was contracted by Carnegie Museum to assist in data analysis and preparation of the manuscript. Doug Miller oversaw Ryan Baxter and Patryk Soika, of the Penn State Institutes of Energy and the Environment, who developed a web-based data management tool that enabled registered users to submit photographs and draft species accounts for review by the editors.

5 Analytical Methods

Andrew M. Wilson

CHANGES IN BLOCK OCCUPANCY CORRECTING FOR SURVEY EFFORT

During the second Atlas, over 110,000 hours were spent in the field, 30 percent more effort than the first Atlas. However, there was not a uniform increase in effort in all blocks; indeed, there was a decrease in hours spent in the field in 37 percent of blocks. Further, the changes in effort were not evenly distributed across the state (see chap. 6).

Although there was considerable variation in effort among blocks, the general pattern in both first and second Atlases was that more field hours resulted in more species being located (see chap. 6). Hence, without accounting for the overall increase in effort, comparison of second Atlas results with the first Atlas would likely result in overly optimistic findings, either over-estimating range expansions or under-estimating range contractions. Therefore, to provide a more reliable assessment of range changes between atlases, we modeled distributions of each species in both atlas periods, correcting for spatial variation in survey effort. The aim of our models was to predict the probability of occupancy in each block, for each species, given a standardized rate of survey effort. We used WinBUGS (Lunn et al. 2000) to fit logistic regression models, with the following form:

$$p = \text{landscape characteristics} + \text{spatial effects} + \text{effort} + \text{noise}$$

where p is the probability of block occupancy (0 to 1); landscape characteristics are the percentage of the block in major land cover types (e.g., farmland and grassland; fig. 5.1) and mean block elevation; spatial effects are occupancy in surrounding blocks (up to eight adjacent blocks); and effort is the total number of hours of fieldwork (fig. 5.1).

Up to three broad groupings of land cover types were selected for each species based on prior knowledge of that species' habitat associations in Pennsylvania and further informed by analysis of atlas point count data (see below; see also appendix D). The spatial effect was an important component of these models. The models assumed that, all other things being equal, a species is more likely to occur in a block if it is found in adjacent blocks. This spatial autocorrelation encapsulates local variation in habitat that may not be reflected in broad land cover types, in addition to larger-scale bird distribution effects; that is, an isolated patch of suitable habitat within the species' range is more likely to be occupied than an identical patch outside of the species' range. Effort was modeled as a function of two estimated parameters (after Link and Sauer 2007), which allowed for a variety of relationships between field survey hours expended and the probability of detecting a species in each block. Hence, the modeled relationship was not necessarily linear—it could reflect diminishing returns with increasing effort. In the final models, the predicted probability of occurrence was based on a correction factor, which assumed 30 hours of fieldwork for all species, in all blocks, in both atlases. Thirty hours was chosen for this standardization, because it was found that most species were likely to be detected in a block given 30 hours of fieldwork.

All of the parameters in the models (including effort) were estimated independently for each species and for both atlas periods. There were sufficient data to estimate models for 132 of the most widespread species; hence, we fitted a total of 264 occupancy models. For each of these 132 species, the results of the models provided a predicted probability of presence for each atlas block, in each of the two atlas periods. We summed the total of these probabilities across all 4,937 blocks to provide an estimate of the number of blocks occupied, corrected for survey effort, in each atlas period (fig. 5.1). A further

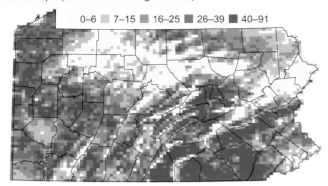
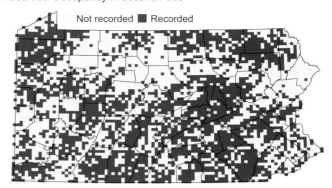
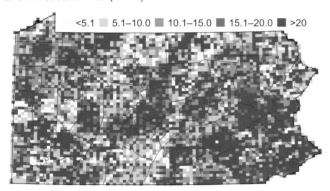
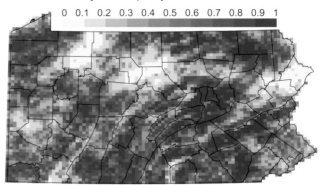

FIG. 5.1 *Map of some of the covariates used in the model of Predicted Occupancy for American Kestrel during the second Atlas (Landscape, Recorded Occupancy, and Effort), and the resulting map of Predicted Probability of Occupancy*

example of the resulting maps of predicted occupancy can be found in chapter 6 (fig. 6.6).

Technical details of these models can be found in appendix E.

ABUNDANCE ESTIMATES AND DENSITY MAPS

Although the atlas point count data used for deriving abundance estimates and density maps are unprecedented in terms of the number and density of data points, there are inherent biases in data that cannot be masked by data volume. Analysis of these data required careful consideration of these potential sources of bias so that we could attempt to correct them to achieve what we believe to be robust population estimates. It should be noted that, due to the need to repeat this time-consuming analysis for the 115 species with sufficient data (>50 detections, but note that the models worked best for species with hundreds or thousands of detections), it was necessary to find analytical techniques that were efficient and, in some cases, heuristic. More robust analysis, which would require the testing and comparison of many potential models for each species, was not logistically possible with available resources. That said, our analysis of these data is thorough, and we have adopted novel techniques to arrive at population and density estimates.

Our analysis aims to quantify and correct for five well-documented biases that occur when counting breeding birds using point count techniques:

1. Not all birds are available for detection.
2. There are differences in detection among observers.
3. Variation in detection by time of season.
4. Variation in detection by time of day.
5. Habitat sampling bias (due to roadside sampling).

Our population estimate model is a generalized linear model, incorporating terms that account for each of the above biases. The equations and comprehensive statistical details can be found in appendix E. In the following sections, we describe the rationale behind the models and illustrate some of the ecological relationships that we incorporated.

For 73 of the 115 species for which we estimated population size, our unit of measurement was "singing males," because the point count protocols were primarily devised to count songbirds. However, for 42 non-songbirds, our estimates are of individual birds. Most of these are non-passerines, crows, and swallows, but this list also includes some passerines with indistinct songs (see table

6.8 in chap. 6). For these species, our population estimates were not directly comparable with those for songbirds, for two reasons. First, we were not able to make corrections based on estimates of birds missed during the count period (see below), and second, we do not know what proportion of birds detected were males. Hence, our estimate of the number of "pairs" for these species (individuals/2) is likely to be a considerable underestimate.

Bird detection

The point count protocol was devised so that the "Removal Method" (Farnsworth et al. 2002) could be used to estimate the proportion of birds missed during a time-limited (6¼ minute) count period. This analytical method was further adapted so that a "detection radius" could be estimated (Farnsworth et al. 2006), thereby allowing the inclusion of detections of birds outside the fixed 75 m distance band (see chap. 4). This is particularly advantageous for scarce species, because it increases sample sizes, thereby increasing the resolution of density maps derived from these data. For some species that are audible over large distances (e.g., Hermit Thrush, Field Sparrow, Eastern Meadowlark), more than 80 percent of observations were outside of the 75 m fixed-distance band.

The detection radius was largely determined by the proportion of birds detected within 75 m. If most detections of a species were <75 m (e.g., White-eyed Vireo; fig. 5.2), the modeled detection radius was small, but if most detections were beyond 75 m (e.g., Eastern Meadowlark; fig. 5.2), the modeled detection radius was large. Estimates of the proportion of birds missed were determined by the rate with which additional singing males were detected in successive time bands (fig. 5.2). If, for example, new detections declined quickly over the five time periods, the model assumed that most individuals were detected (e.g., Red-eyed Vireo; fig. 5.2). In contrast, if significant numbers of additional detections were made during the later time periods, this indicates that more birds were likely to be missed, because they sing infrequently, quietly, or were otherwise easier to miss over a short (6¼ minute) count period.

Estimates of the detection radius ranged from 92 m for Golden-crowned Kinglet to 308 m for Black-billed Cuckoo. Note that these estimates do not represent the absolute maximum detection distance for each species—they represent the average detection limit across a range of conditions, including variations in background noise (traffic, other bird song, wind) and song penetration (which varies with habitat). Estimates of undetected birds (the proportion of birds missed at 0 m distance)

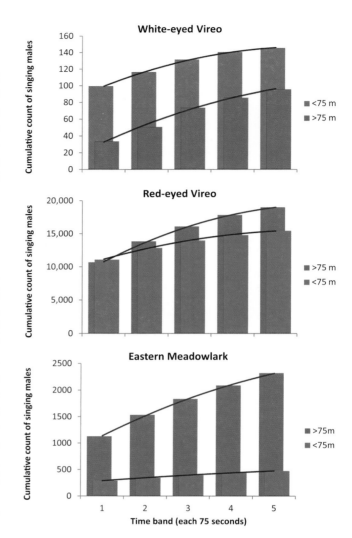

FIG. 5.2 Cumulative atlas point counts of singing White-eyed Vireos, Red-eyed Vireos, and Eastern Meadowlarks in five time bands and two distance bands. Data are from 33,851 point counts.

ranged from 0 (several species) to 0.32 for White-eyed Vireo, with a mean of 0.075. In combination, the estimates of birds missed and the detection radius provide a model for the proportion of birds detected at a range of radial distances from the observer, with detectability described by a spherical function (fig. 5.3). Here we can see, for example, that the White-eyed Vireo was less likely to be detected during the 6¼-minute count period than the Red-eyed Vireo. Also, the Red-eyed Vireo was detectable over larger distances, while the Eastern Meadowlark was detectable over still larger distances.

The direct relationship between the proportions of birds within the fixed 75 m distance band and the estimated detection radius allowed us to estimate the detection radius of non-songbirds, based on the equation shown in figure 5.4. These heuristic estimates of detection radii for non-songbirds were used to estimate population densities for those species.

More details of the methods used to estimate detection rates and detection radii can be found in appendix E.

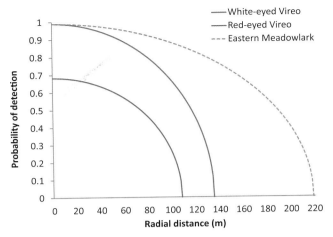

FIG. 5.3 *Modeled probability of detection of White-eyed Vireo, Red-eyed Vireo, and Eastern Meadowlark from second Atlas point counts*

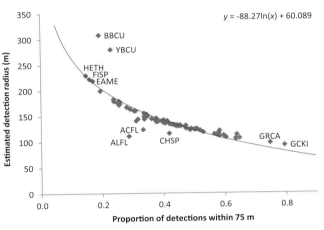

FIG. 5.4 *Estimated detection radius is largely determined by the proportion of detection within 75 m (see appendix E for more details). Extreme values and outliers are labeled: ACFL = Acadian Flycatcher, ALFL = Alder Flycatcher, BBCU = Black-billed Cuckoo, CHSP = Chipping Sparrow, EAME = Eastern Meadowlark, FISP = Field Sparrow, GCKI = Golden-crowned Kinglet, GRCA = Gray Catbird, HETH = Hermit Thrush, YBCU = Yellow-billed Cuckoo.*

Time of season

The seasonal peak in song output of songbirds varies greatly among species, depending on migratory strategy, breeding strategy (e.g., number of broods), and population density (Wilson and Bart 1985; Alldredge et al. 2007; Brewster 2007). For some species, peak output may be a relatively narrow period, while for other species, song output extends for many weeks through the breeding season. Because atlas point count surveys extended over a period of 7 weeks, it is likely that a great many point counts were completed at times that were either before or after the peak of song output for many species; hence, only a proportion of males present were likely to be detected. These seasonal variations in detectability were not only apparent in songbirds; we found that detections of many non-song birds also varied with time of season.

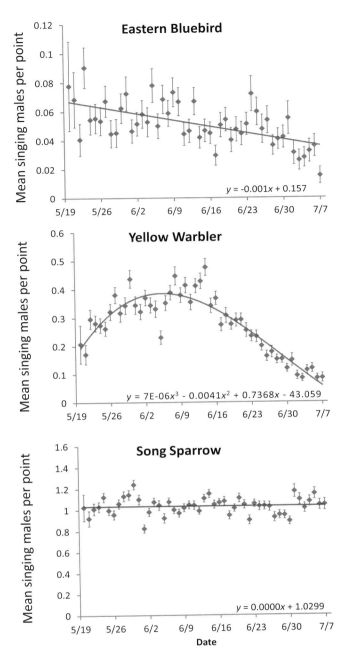

FIG. 5.5 *Mean number of singing male Eastern Bluebird, Yellow Warbler, and Song Sparrow detection per point for each day of the point count survey season (with standard errors). Fitted regression curves (red) show the trend in song output through the season.*

We estimated the effects of seasonality by fitting regression equations to the mean count of singing males per point for each date of the survey season (fig. 5.5). For non-songbirds, the total number of birds detected was the dependent variable. So, for example, the data show that Eastern Bluebird song output declines linearly through the season, Yellow Warbler song output peaks in late May/early June and then declines rapidly, while Song Sparrow song output is relatively constant through the survey season. We fitted several regression lines for each species—linear, logarithmic, exponential, power, or polynomial (2nd through 4th order)—and selected

ANALYTICAL METHODS   41

the regression equation with the highest R^2. Using the resulting regression equations, we were then able to estimate song output for each date, relative to peak output, and apply a correction factor, which estimated the proportion of birds not singing on any given date in the survey season. The same method was also applied to the detection of non-song birds, based on all detections of those species. Of course, for some resident species, peak detectability could have occurred early in the season before the point counts were under way (e.g., Eastern Bluebird; see fig. 5.5); for these species, our estimates may be slightly conservative.

Time of day

Just as song output varies through the season, it also varies through the morning survey period (e.g., Hochachka et al. 2009); again, the pattern of detectability varies among species. Additionally, detections of non-songbirds varied during the morning survey period. As with seasonality in song output, diel variation in song output and general detectability were estimated using regression models and correction factors applied to estimate the proportion of birds missed. For example, Northern Flicker detections were low early in the morning, peaked around 2 hours after sunrise, and then remained high through the morning, Hermit Thrush detections were highest before sunrise and decreased rapidly thereafter, and Song Sparrow detections showed little variation during the morning survey period (fig. 5.6).

Habitat sampling

Use of roadside survey locations is known to introduce significant bias into bird survey samples (Bart et al. 1995; Keller and Scallan 1999). The bias in atlas point count locations was especially marked in certain regions of the state, such as the Ridge and Valley Province (see chaps. 2 and 3). In such areas, most roads are in the valley bottoms, which are dominated by farmland and development, resulting in an oversampling of those habitats and an undersampling of forested habitats, which are found primarily on ridges. Analysis of land cover types within 100 m of point count locations shows that developed lands were oversampled by 110 percent, grasslands by 26 percent, and row crops by 2 percent, while forests were undersampled by 21 percent, when compared with actual land cover statistics for the state (fig. 5.7).

The point count sample has similar biases in elevation—oversampling of lower elevation, undersam-

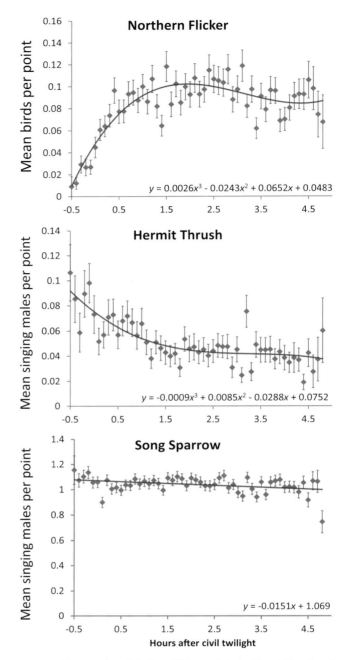

FIG. 5.6 Mean number of Northern Flicker and singing Hermit Thrush and Song Sparrow detections per point for each 6-minute period of the point count survey period (with standard errors). Fitted regression curves (red) show the trend in song output through the morning.

pling of high elevation, and bias in coverage of topographic types (land forms; see appendix E). Without correcting for these biases, population estimates would not be comparable among species and population densities maps would have regional biases.

To correct for sampling biases, we developed predictive habitat models that used remote-sensed data (land cover, elevation, and land form) to provide predictions at an unbiased regular grid of survey points. The prediction grid was chosen to be similar in density to the actual sampling grid, which is approximately one point every 1.9 km (fig. 5.8). The necessity of this analytical

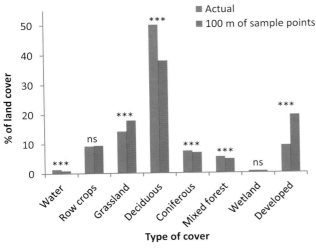

FIG. 5.7 *Bias in land cover (Warner 2007) due to the roadside sampling protocol of atlas point counts. Blue columns are the actual percentages of Pennsylvania in main land cover types, red columns are land cover in a 100 m radius of point count locations. Significant bias is indicated as follows: *P < 0.05, **P < 0.01, **P < 0.001 from two-proportion z-tests.*

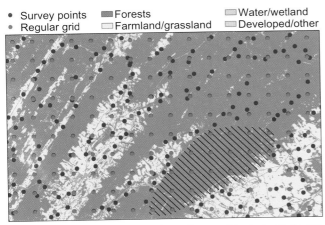

FIG. 5.8 *Actual survey locations and a regular spaced (1.9 km) grid of points in the Ridge and Valley Province. Hatch signifies a forested area where no point count surveys were completed due to a lack of vehicular access.*

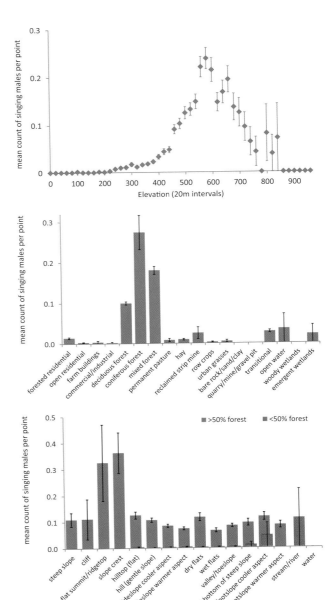

FIG. 5.9 *Mean number of singing male Hermit Thrush detections on atlas point counts, by elevation (top), main habitat type (as assessed by point counters) within 75 m of the point count location (center), and land form (bottom), with standard errors*

step is illustrated by the area of forest highlighted (hatch) in figure 5.8. This area was not accessed during the point count surveys, due to a dearth of accessible roads. Producing a density map based on data from actual survey points, using standard interpolation techniques, would result in density predictions for the highlighted forested area that were based on the nearest sampled points. Close inspection of the map shows that most of the sampled points surrounding that un-surveyed forest were in farmland. Standard interpolation techniques would use data from those sites to infer the presence of birds of open country in that large forest block, while it would predict an absence of forest birds. Of course, the implications of this depend on the resolution of density maps produced; for very coarse density maps, such relatively localized sampling biases would not be apparent. However, multiplied across the state, these biases could have considerable implications for regional patterns of abundance and also for overall abundance estimates. For example, without correcting for this bias, we would be underestimating populations of forest birds by around 21 percent, while overestimating populations of birds associated with human development by 100 percent or more.

Components of the predictive habitat models were chosen by examination of relationships between point count data and remote-sensed data. For example, Hermit Thrush was found mainly in forests, especially coniferous and mixed forests, at higher elevations and on "flat summit/ridgetop" and "slope crest" land forms (fig. 5.9). Because there is correlation between land cover, land

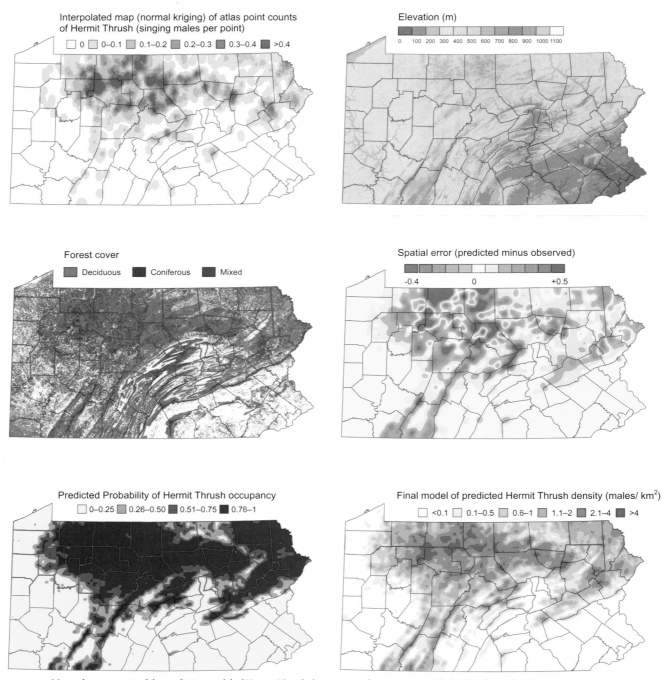

FIG. 5.10 *Mapped components of the predictive model of Hermit Thrush density. Actual counts are modeled with relevant habitat covariates (e.g., elevation, deciduous forest, coniferous/mixed forest). Error of habitat model is mapped so that spatial error can be included in predictive models. Predicted occupancy (modeled on range from second Atlas block data) is also included in the model to refine the predictions. See text and appendix E for more details.*

form, and elevation, we also examined relationships between bird abundance and land form and elevation, just within preferred land cover types. For example, for Hermit Thrush, selecting only those data points in areas dominated by forest, mean counts were still much higher at two land form types (fig. 5.9), while the elevational relationship shown for Hermit Thrush holds true whether or not non-forested points are included. Charts showing these habitat relationships for other species (only those with sample sizes >50) can be found in appendix D.

Final density model

We constructed a generalized linear model to predict bird densities at each point count location, based on the bird detection models and habitat models detailed above. These models made corrections for estimated proportions of birds missed due to differences between observers and deviations from the optimal time of day and time of season for each count location. Habitat covariates were used to model counts (e.g., elevation and forest; fig. 5.10).

The models took the following form:

$$count = (\text{landscape characteristics} + \text{time of year} \\ + \text{time of day} + \text{observer} + \text{spatial error} \\ + \text{noise}) * \text{distribution}$$

It should be noted that predictive models of bird occurrence based on a few key habitat and landscape covariates often have low predictive power. This is especially the case when remote-sensed data are used, because these data sources are coarse and do not provide important information about more subtle habitat requirements. Therefore, the final step of our predictive model was to estimate spatial divergence between the modeled counts and actual counts. If densities of birds could be accurately predicted by land cover, land form, and elevation, any remaining error in predictions would be small and random; a map of the error in predictions at each point (predicted count minus the actual count) would show no pattern. In reality, there are likely to be underlying drivers of density that models do not capture, perhaps because we have no data on that driver (e.g., subtle variations in habitat quality) or because there are variations in underlying drivers of density across the state. For example, it is known that, in some areas, Acadian Flycatchers are almost always associated with hemlocks, while in other areas, they are found in a wider range of forest types. Hence, the error in our models was not random; it contained important information about spatial variation in abundance that our models were unable to capture. The difference between the predictions based on habitat covariates and the actual counts are then mapped (normal kriging, $n = 60$ points in ArcGIS Spatial Analyst) to produce a map of spatial error (fig. 5.10).

Finally, because we have good prior knowledge of the distribution of each species from atlas block data, we included probability of presence as a term in the model. This was from a smoothed (normal kriging, $n = 8$ blocks in ArcGIS Spatial Analyst) version of the predicted range map (corrected for effort; see the beginning of this chapter), which equated to the probability of presence in the area equivalent to one atlas block (c. 15 km^2) around each survey point (fig. 5.10). This ensured that we did not predict abundances of greater than zero in areas in which there is no evidence that the species occurs.

The parameter estimates for the habitat components from the above model were then used to make estimates of densities at each of the c. 36,000 regular grid locations (fig. 5.8), which provide the basis for a density map and overall abundance estimate for the state. Time of day and time of season parameters were also included in this model, but they were set to the optimal times for each species (e.g., for Hermit Thrush, the model assumes that all surveys were conducted 30 minutes before sunrise and on 10 June). For more details about this important, but complex, part of the modeling process, see appendix E.

LATITUDINAL SHIFTS

TABLE 5.1 Interpretation of between atlas changes in mean latitude of occupied blocks. Zero indicates no change, positive indicates northward shift, and negative indicates southward change.

	Mean shift in block latitude		
Description of change	Southernmost block	Median block	Northernmost block
Northward expansion	0	0/+	+
Northward retraction	+	0/+	0
Northward shift	+	+	+
Southward expansion	−	0/−	0
Southward retraction	0	0/−	−
Southward shift	−	−	−
No change	0	0	0

To identify northward or southward shifts in species distribution between atlases, we calculated the latitudes of the block centers of the northernmost, median, and southernmost occupied block across the (up to 94) columns of blocks across the state (see fig. 4.1). We then tested for significant differences (km) in the three measures of latitude using t-tests. Significant changes were then interpreted based on all three measures of change (table 5.1).

BLOCK TURNOVER

Although change in the number of occupied blocks between atlas periods provides a useful metric for assessing change in distribution at the state scale, this metric could mask large distributional changes within the state. It is feasible for substantial shifts in range to occur but the overall number of occupied blocks to remain constant—for example, when species of ephemeral habitats move in accordance with habitat availability. To assess block fidelity between atlas periods, we calculated a *Relative Block Turnover* metric. This metric takes into account (a) that species showing large range changes will have high block turnover, and (b) that block turnover is limited for the most widespread species (turnover tends toward zero as the percentage of occupied blocks approaches 100). To correct for the first of these, we used the following formula:

$$\text{Relative Block Turnover} = \left[1 - \left(\frac{n_{both}}{n_{either}}\right)\right] - \left[\frac{|n_{1st} - n_{2nd}|}{max(n_{1st}, n_{2nd})}\right]$$

where n_{both} is number of blocks in which a species was found in both atlases, n_{either} is the total number of blocks occupied during either atlas period, n_{1st} is number of blocks occupied in first Atlas, and n_{2nd} is number of blocks occupied in second Atlas. The first term in the equation calculates relative turnover, which ranges from 0 (no turnover, present in exactly the same blocks in the two atlases) to 1 (complete turnover, present in completely different blocks in the two atlases). The second part calculates the relative change in the number of blocks occupied, which ranges from 0 (no change) to 1 (colonization or extirpation). *Relative Block Turnover* is, therefore, the difference between the change in the number of occupied blocks and the relative proportional turnover. A *Relative Block Turnover* of 0 would indicate that all block changes are in one direction (either all colonizations or all extirpations), while a *Relative Block Turnover* of 1 would indicate that there were no blocks in which a species was found in both atlas periods.

To provide a way of assessing *Relative Block Turnover* for species across the gradient from rare (where high turnover is likely) to widespread (where high turnover is unlikely or impossible), we regressed (4th-order polynomial) *Relative Block Turnover* against the number of occupied blocks (n_{either}) to provide an estimate of "normal" turnover for a given number of occupied blocks. We then calculated the variation across the gradient by calculating the standard deviation in *Relative Block Turnover* (in windows of $n = 31$ species) and fitting a 4th-order polynomial to the resulting estimates of standard deviation. Species were designated to have high turnover if *Relative Block Turnover* was more than one standard deviation higher than the "normal" turnover curve and low turnover if *Relative Block Turnover* was more than one standard deviation lower than the "normal" turnover curve. Results of this analysis are in figure 6.9 (chap. 6).

SPECIES GUILDS

Atlas point count data were used to assign common species ($n = 122$) to species guilds. Guilds were first established by Cluster Analysis, using "pvclust," a free statistical software package written in R language (Suzuki and Shimodaira 2005). The procedure grouped the 122 species into five clusters. To aid interpretation of these clusters, we used Canonical Correspondence Analysis (CCA) in

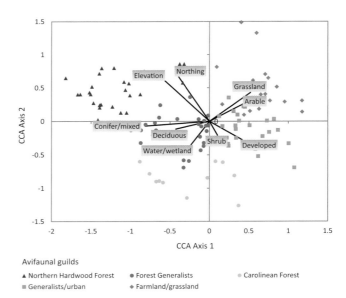

FIG. 5.11 *Bi-plot of species scores and environmental gradients derived from Canonical Correspondence Analysis of atlas point count data. Species assigned to five avifaunal guilds using Cluster Analysis. Each marker represents one bird species; habitat types are land cover within 150 m buffers of point count locations.*

the Vegan package of R (Oksanen et al. 2011) to show the relationships between species scores and environmental gradients. Environmental data included broad land cover types, elevation, and northing. The first (horizontal) axis derived from the CCA shows a gradient from natural/forested habitats (left) to disturbed/open habitats (right), while the second (vertical) axis is correlated with latitude and elevation (fig. 5.11). Because these guilds are defined partly by geography (latitude and elevation), they are not habitat guilds per se, so we refer to them as avifaunal guilds. These five guilds were named Northern Hardwoods, Forest Generalists, Carolinean Forest, Generalists/Urban, and Farmland/Grassland.

For the 68 scarce species not included in the multivariate analysis (due to sparse data), we used expert opinion to assign them to one of the five previously defined avifaunal guilds, where appropriate. However, many of these scarce species are waterbirds, either riparian or found in either freshwater wetlands/lentic habitats. Hence, we define two additional guilds: riparian and wetland.

For common birds (mainly songbirds) in the five terrestrial avifaunal guilds defined by multivariate methods, we derived indices of abundance/species richness to enable mapping of the areas of the state that are most important for each guild. For each species, we divided the predicted bird densities at grid points (fig. 5.8) into deciles (each decile was 1/10 of points with predicted densities of >0) and awarded scores between 1 (bottom decile) and 10 (top decile). We then summed these scores across all species in each guild. Hence, each species received weighting in the index based on the importance of a location for that species, rather than the absolute predicted density.

The theoretical maximum index value is therefore 10 times the number of species in each guild, which would be achieved if the predicted density of every species in that guild was in the top 10 percent statewide for every species.

BREEDING BIRD SURVEY TREND GRAPHS

USGS Breeding Bird Survey (BBS) data were analyzed using the methods of Sauer and Link (2011). Raw BBS data for Pennsylvania were made available by USGS and were modeled in WinBUGS using hierarchical log-linear models, which include observer and route effects. Data were analyzed for all 44 years of the BBS period (1966–2009), and for a total of 136 routes in Pennsylvania, of which a mean of 76 routes were surveyed in any one year. For species for which the second Atlas provided robust population estimates, we modified the models so that the population indices were scaled to approximate population sizes, in effect "back-casting" population estimates from the second Atlas period. To do this, we fitted index values (1966 = 100) for each year and then set the average of index values for the last 6 years—the second Atlas period—equal to our atlas population estimate, with its accompanying 95 percent credible intervals. The results of these models were trend lines identical in shape to those published by USGS (Sauer et al. 2011) but with a different scaling (population as opposed to indices) and wider credible intervals, to incorporate uncertainty around second Atlas population estimates. For seven non-passerine species for which the second Atlas did not provide population estimates, we have presented BBS charts identical to Sauer et al. (2011). For two species for which we have reasonably comprehensive annual population monitoring (Bald Eagle and Peregrine Falcon), we provided actual population trend graphs for 1983–2009.

6 Coverage and Results

Andrew M. Wilson and Daniel W. Brauning

During the years 2004 through 2009, second Atlas volunteers completed surveys in all of the 4,937 atlas blocks defined to have land area in Pennsylvania. In achieving this feat, 854,773 bird records were submitted. This massive undertaking documented 218 species (plus two hybrids), of which 190 were considered breeding during this period. Compiled at the block level, the result is an average of 71.1 species per block (table 6.1), or about 74 percent of targeted species.

SOURCES OF DATA

The records making up the second Atlas include multiple observations of the same species within a block, from multiple sources of data, and reflect the progressive documentation of increased breeding evidence during the course of a field season or across years. Many "lower" breeding codes would not have been entered into the atlas system, since observers "upgraded" breeding evidence on field cards or in the course of fieldwork within a block and entered only the highest code per block. Many preliminary bird observations clearly were made by volunteers during the course of fieldwork. The true number of bird observations in the field is undoubtedly much greater than the impressive number that was ultimately documented.

The vast majority of records, 656,723 (77%), came through the atlas website, entered into the online database; the core of the atlas project, primarily by atlas volunteers overseen locally by volunteer Regional Coordinators. Secondary data sources also contributed significantly to project results and were included to provide the most comprehensive coverage of breeding species during the second Atlas period. For some analyses, these secondary sources were distinguished from the core atlas effort. The second largest source of breeding bird data came from the atlas's parallel effort, the point-count-based abundance sampling, conducted by seasonal staff (see acknowledgments). These data were distilled into records for each reported species and block, totaling 137,372 records, and were incorporated into the atlas block data. The U.S. Geological Survey's ongoing Breeding Bird Survey (BBS) route data during this period were cross-walked with atlas blocks, resulting in 44,344 records incorporated into block tallies. Similarly, georeferenced point count data, collated by an evaluation of the Conservation Reserve Enhancement Program (CREP), were assigned to atlas blocks and entered. Bird reports were also added from the Game Commission's bird-monitoring databases and from published *Pennsylvania Birds* reports, which had not been incorporated by volunteers (the latter are, however, included in the volunteer totals in table 6.2). While these latter sources included a comparable smaller number of records (less than 1% combined), they provided significant sources of coverage for some species, particularly rare or state-listed birds that were well documented outside the atlas project but for various reasons were not noted by volunteers covering those blocks (table 6.2). Noteworthy among these supplemental sources was Maryland's second breeding

TABLE 6.1 Summary of second Atlas coverage, 2004–2009

Atlas effort	All blocks	Priority blocks
Number of blocks	4,937	816
Blocks owned	3,660	641
Blocks with data	4,937	816
Blocks completed	4,402	718
Block effort hours	106,953	22,584
Total submissions	854,773	157,963
Registered participants	1,896	—
Species reported	218	200
Species confirmed	187	166
Mean species per block	71.1	75.5

TABLE 6.2 Number of records submitted, by evidence code and source of data

	Atlas volunteers	Atlas point counts†	BBS	CREP	PGC*	Grand total
Observed (O)	4,851	13	4	0	0	4,868
Possible (X)	352,904	137,359	44,340	16,083	9	550,695
Territorial behavior (T)	95,094	0	0	0	0	95,094
Pair (P)	62,339	0	0	0	5	62,344
Courtship behavior (C)	5,378	0	0	0	0	5,378
Used nest (U)	585	0	0	0	0	585
Agitated behavior (A)	9,757	0	0	0	0	9,757
Carrying nesting material (CN)	6,698	0	0	0	0	6,698
Physiological evidence (PE)	297	0	0	0	0	297
Nest building (NB)	5,800	0	0	0	6	5,806
Distraction display (DD)	1,483	0	0	0	0	1,483
Recently fledged young (FL)	40,297	0	0	0	72	40,369
Adult carrying food (CF)	26,415	0	0	0	0	26,415
Feeding young (FY)	18,367	0	0	0	0	18,367
Nest containing eggs (NE)	3,982	0	0	0	4	3,986
Occupied nest (ON)	14,759	0	0	0	82	14,841
Nest containing young (NY)	7,717	0	0	0	73	7,790
Grand total	656,723	137,372	44,344	16,083	251	854,773

*PA Game Commission Great Blue Heron, Bald Eagle, and Barn Owl monitoring data.
†137,372 individual block records from 501,885 point count records total.

bird atlas (Ellison 2010a), which documented Pennsylvania's only confirmed nesting Loggerhead Shrike record during our second Atlas.

EFFORT

The amount of effort contributed to this project was nothing short of extraordinary. About 1,900 atlas volunteers contributed data to the project (see acknowledgments), tallying 106,953 volunteer field hours and another 34,192 administrative hours. An additional approximately 1,000 people who had registered with the website did not submit data. Undoubtedly, many hours (for time spent in the field, at meetings, traveling, or processing data) were never recorded. Including point count effort, a total of well over 110,000 hours of field effort were documented. Miles spent in fieldwork also were tabulated, totaling 890,000. This volunteer contribution was valued at more than $2 million at standard rates and provided required matching funds for grants supporting the project. In addition to this effort, and not tabulated for each block, point count data contributed from 1.5 to 2 hours per block, and CREP and BBS contributions added a small amount of time to those blocks. More importantly, this block effort (field time) serves as a measure of coverage and was part of the standard goal established for blocks.

Blocks received a median of 14.8 hours of field effort. Slightly over one-quarter of all blocks received (from atlas volunteers) the recommended 25 or more hours. Block effort contributed by block owners totaled 53,020 hours, or about 55 percent of all effort. The balance was by registered volunteers not assigned to the block. About one-quarter of blocks received block effort only from their block owners, and there was no recorded effort by block owners in 45 percent of blocks. Although 3,678 blocks (74%) were owned, 835 (almost one-quarter) of those owners reported no hours within their owned block. The median survey effort in unowned blocks was 9 hours. Therefore, there was clearly a coverage advantage in ownership, but ownership was not a guarantee of acceptable effort in a block.

As with most volunteer efforts, a small proportion of the total participants contributed a large proportion of the effort. The top 252 participants contributed over 50 percent of the block effort. This bias is greater when looking at records of birds submitted, because not only were some observers more active, some of those were also highly efficient.

Block effort varied geographically, by time of day and season. After the planning effort in 2003, second Atlas participation began very strongly with the first field season and continued through the full 5-year planned program. Over 20,000 hours of block effort were documented each year, reaching a high of 22,651 in the third year, 2006. After 5 solid years of effort, a small number of blocks remained without data or had less than 50 percent of the coverage target, so an additional field season was undertaken. Those blocks were targeted for fieldwork during 2009, resulting in 2,039 hours of effort (table 6.3).

The seasonal distribution of block effort reflected the number of species within safe dates through the first

TABLE 6.3 Survey effort by year

Year	Main volunteer effort		Nocturnal surveys (stops)	Point counts (points)	Total hours
	Block visits	Hours			
2004	14,703	20,213	—	3,146	20,541
2005	14,803	20,660	88	5,657	21,256
2006	16,071	22,651	601	6,133	23,339
2007	15,021	20,368	327	7,938	21,222
2008	15,698	21,022	348	10,437	22,138
2009	1,280	2,039	—	1,408	2,185
Total	77,576	106,953	—	34,719	110,682

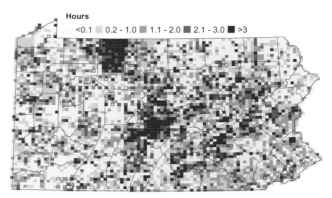

FIG. 6.2 *Total number of reported dusk or nocturnal hours spent surveying each block for the 2004–2009 atlas. Red squares are blocks in which nocturnal surveys were conducted.*

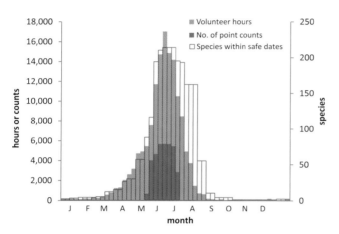

FIG. 6.1 *Total 2004–2009 atlas volunteer and point count effort by weekly period*

half of the year, but effort declined rapidly through July, when many species were still within safe dates. Early-season efforts for owls and other species with early safe dates prompted a little over 1,000 hours per week by the end of March (totaled across all 6 years). Effort rose to almost 5,000 hours per week by early May, peaking at over 10,000 hours per week from the end of May through the second week in July (fig. 6.1).

To the extent to which observers coded their time to the time of day, about two-thirds of fieldwork (not surprisingly) was attributed to the morning hours. Because observers were less likely to record effort for which they didn't detect a species, the documentation of nocturnal fieldwork in particular is likely to be underreported. But nocturnal effort (reported effort during "dark" hours) totaled about 3 percent of effort. Starting in 2005, when the nocturnal surveys were officially unveiled, a total of 1,364 owl points were surveyed. Blocks in which nocturnal surveys were conducted were not uniformly distributed across the state (fig. 6.2).

The combined sources of data provided a smoothing effect on coverage, although in a few cases this may have introduced biases. While point count data predominantly detected songbirds, the large number of stops and many hours of fieldwork resulted in encounters with a wide range of species, including some of the state's rarest, such as a King Rail, three Upland Sandpipers, a Yellow-bellied Flycatcher, and two Red Crossbills, comprising as many as 18 percent of the block records for these species. This does not include the many atlas breeding birds recorded by point counters, between point count surveys or during survey route scouting. The inclusion of point count data resulted in an increase in block totals, over volunteer-based results, of 5 percent or more for 88 species, and 10 percent or more for 22 of these—enough to significantly affect comparisons with the first Atlas. Georeferenced data were available for 52 BBS routes, which generated records for 593 blocks (9% of the total) distributed across the state. CREP records were all from a 20-county area of south-central Pennsylvania and were concentrated in agricultural areas, contributing to 396 blocks. In some instances, this additional coverage of selected species had a notable influence on the total number of records. Only 78 percent of the 781 block detections of Horned Larks were from atlas volunteers; an additional 17 percent were from point count surveys, 3.6 percent from CREP surveys, and 0.6 percent from BBS routes. Data from Bald Eagle monitoring by the Game Commission also contributed 66 blocks (29 percent) of the total with confirmed breeding records.

Population estimates resulting from second Atlas point count data are discussed below, but this standardized effort added, on average, four species to block totals across the state. Species tallies were increased by point counts in 70 percent of blocks, and eight blocks recorded point count data alone. This contributed measurably to the increased coverage, compared with the first Atlas, something frequently mentioned in species accounts. Significantly, point count data resulted in smoothed coverage across regions. In several regions (e.g., 34, 54,

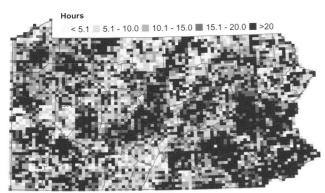

FIG. 6.3 *Total number of recorded block hours spent surveying each block for the 2004–2009 atlas*

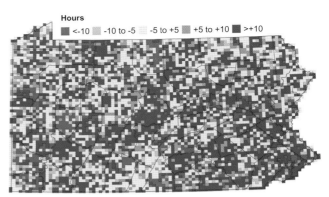

FIG. 6.4 *Change between first Atlas and second Atlas in the total number of hours spent surveying blocks*

and 55), point counts increased the average species count per block by more than 15 percent.

The number of hours reported per block in the second Atlas fits a slightly skewed distribution, centered on the goal of 75 percent of the target, with more blocks falling below the target than above. Few blocks fell below 50 percent of target, and over 300 blocks reached more than 100 percent of the goal. The vast majority fell between 65 and 85 percent of target effort. That equates to a median of 72 species per block, with a range of 24 to 126, excluding "border" blocks. A median of 77 species per block was reported from priority blocks. Border blocks received less coverage, reporting an average of 60 species per block.

Block effort was not as uniformly distributed as was desired (fig. 6.3). The effects of increased coverage are discussed in chapter 5. Block effort was higher in the southeastern corner of the state, resulting in higher detection rates of rare species, and sufficiently lower in regions 58 and 76 to have reduced species detections there.

CHANGE IN EFFORT

The 110,000 hours spent in the field is a 30 percent increase in effort over what had been reported during the first Atlas. However, there was not a uniform increase in effort in all blocks; indeed, there was a decrease in hours spent in the field in 37 percent of blocks. Further, the changes in effort were not evenly distributed across the state (fig. 6.4; table 6.4). In general, effort increased most in the eastern and central parts of the state, but areas with a decline in coverage were widespread and clustered.

Although there was considerable variation among blocks, the general pattern in both the first Atlas and second Atlas was that more field hours resulted in more species being located, up to about 50 hours. The effects of observer effort were especially pronounced in blocks in which less than 10 field hours were expended. Extra effort beyond 50 hours appeared to offer rapidly diminishing returns (fig. 6.5), whereby very few extra species were located no matter how much extra effort was expended in the field. For blocks in which atlas volunteers contributed more than 50 field hours, there was an average of 80 species in the first Atlas and 81.5 in the second Atlas. The mean number of species found in blocks with the median expended hours (9 in first Atlas, 14 in second Atlas) was 70.3 and 69.8, respectively. This suggests that species were overlooked in the majority of blocks in both atlases, although it is likely that blocks received greater effort because they had greater diversity of habitats and the predicted species list was also higher.

The net result of changes in field hours is that a direct comparison of recorded block occupancy between the two atlases is biased in several ways. First, the increased effort between the two atlases results in an overestimate of range expansion for increasing species and an underestimate of range contraction for declining species. Combined, these two effects result in an overly optimistic picture when assessing changes in species richness at the block level. Additionally, the patchiness of changes in field effort may result in apparent local range changes that may merely have been artifacts of variation in observer effort. To address these biases, we used predictive models to estimate species distribution in both atlas periods.

For each species that had sufficient data, we modeled the predicted probability of presence for each atlas block, in each of the two atlas periods (see chap. 5 for methods). We calculated the total of these probabilities across all 4,937 blocks to provide an estimate of the number of blocks occupied, corrected for survey effort. For most species, the "corrected" occupancy maps show infilling within the "recorded" range rather than predicting occupancy in areas in which the species was not recorded (e.g., Wild Turkey; see fig. 6.6).

For most species, the estimated number of occupied blocks, when corrected for survey effort, was appreciably higher than the recorded number of occupied blocks (fig. 6.7). The effect of correcting for effort was more

TABLE 6.4 Block effort in first and second Atlases by physiographic region

Physiographic province	Mean effort hours/block			% change in mean effort hours/block	
	first Atlas	second Atlas			
Physiographic section		volunteers	vols + point counts	volunteers	vols + point counts
Appalachian Plateaus	13.3	17.3	18.0	30	36
Allegheny Front	10.6	19.8	20.5	87	93
Allegheny Mountain	15.2	20.0	20.7	31	35
Deep Valleys	11.8	17.1	17.8	44	50
Glaciated High Plateau	12.8	14.3	15.0	11	17
Glaciated Low Plateau	8.8	16.6	17.3	87	96
Glaciated Pocono Plateau	11.7	16.9	17.6	44	50
High Plateau	14.4	18.8	19.4	30	35
Northwestern Glaciated Plateau	17.2	18.6	19.3	8	12
Pittsburgh Low Plateau	13.3	16.7	17.5	25	30
Waynesburg Hills	18.5	15.1	15.8	−18	−14
Atlantic Coastal Plain	12.2	27.7	28.0	126	128
Lowland and Intermediate Upland	12.2	27.7	28.0	126	128
Central Lowlands	14.8	14.7	15.3	−1	3
Eastern Lake	14.8	14.7	15.3	−1	3
New England	17.1	20.6	21.4	20	24
Reading Prong	17.1	20.6	21.4	20	24
Piedmont	21.0	24.3	25.0	15	19
Gettysburg-Newark Lowland	22.2	23.8	24.6	7	10
Piedmont Lowland	17.7	23.3	24.1	31	36
Piedmont Upland	20.8	25.2	25.9	21	24
Ridge and Valley	14.1	19.8	20.5	40	45
Anthracite Upland	19.8	22.7	23.5	14	18
Anthracite Valley	12.8	15.4	16.2	20	26
Appalachian Mountain	10.4	15.5	16.2	48	55
Blue Mountain	20.5	29.0	29.7	41	44
Great Valley	17.1	26.2	27.0	53	58
South Mountain	15.0	22.4	23.1	48	53
Susquehanna Lowland	12.9	17.5	18.2	35	41

marked on first Atlas results, increasing estimated block occupancy by an average of 16.4 percent (standard error 1.1), compared with 11.2 percent (standard error 0.8) for second Atlas results. The results were more pronounced for scarcer species: the model estimated under-recording of more than 50 percent for five species in the first Atlas period—Hooded Merganser, Red-breasted Nuthatch, Northern Waterthrush, Nashville Warbler and White-throated Sparrow—all locally distributed species found primarily in the northern half of the state, where survey effort was lower. Inevitably, for common species, the net effect of the model on estimated occupancy was limited, partly because common species are less likely to be missed, and partly because block occupancy was closer to the maximum possible, and hence the relationship shown in figure 6.7 always tends to zero for species observed in most blocks.

The net effect of these models was that for species showing an increase in occupancy, increases were tempered,

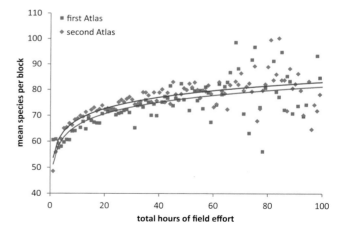

FIG. 6.5 Relationship between hours of survey effort and number of species found in block (possible, probable, and confirmed) in first and second Atlases. Solid lines show logarithmic trend (first Atlas species = (6.49 * log hours) + 51.0, R^2 = 0.48; second Atlas species = (6.54 * log hours) + 53.0, R^2 = 0.57). X-axis is right-truncated at 100 field hours for clarity.

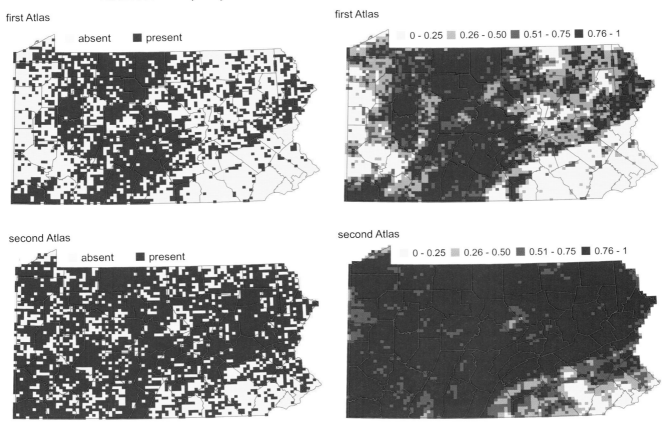

FIG. 6.6 Recorded and predicted occupancy of Wild Turkeys in the first and second Atlases

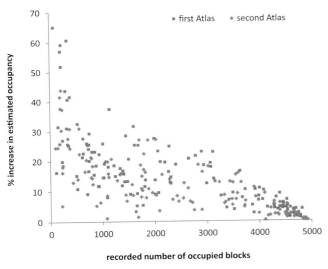

FIG. 6.7 Percentage increase in estimated number of occupied blocks due to modeling efforts effects, plotted against number of occupied blocks, for both first and second Atlases

but there was little overall effect for species showing a range contraction (fig. 6.8). We believe this suggests that many of the species that have increased or expanded their range between atlases were underreported in the first Atlas. In contrast, during the second Atlas, observers may have been particularly aware of species known to be in decline and expended targeted effort in finding them. In any case, second Atlas fieldworkers had much more collective knowledge of bird distributions at their disposal than did volunteers commencing fieldwork during the first Atlas period, largely as a result of the groundbreaking work of the first Atlas and new tools and communications available to volunteers during the second Atlas. Hence, not only did survey effort increase, but survey efficiency, in terms of knowing which habitats to target, almost certainly improved between atlases.

Without correcting for increased effort, the overall results of the second Atlas look positive; recorded block occupancy increased by 15 percent, or more, for 72 species, while occupancy declined by 13 percent, or more for 40 species (table 6.5). Correcting for increased effort shifts these totals slightly, to 67 species showing an expansion in occupancy and 48 showing a decrease. However, we could only make corrections for the 132 most widespread diurnal species. For rare and scarce species, accumulated knowledge of their distributions and habitats in the two decades between atlases may have resulted in increased efficiency in locating them. Hence, the observed changes in block occupancy between the two atlas periods may provide a somewhat optimistic snapshot of changes for species of greatest conservation concern.

BREEDING CODES

Atlas volunteers were advised that obtaining confirmed breeding evidence was less important for common birds; therefore, data were highly skewed toward lower breeding evidence for these species. The most frequently used breeding code used was an "X," possible breeding, totaling 43 percent of codes used, and just 27 percent of records provided confirmed breeding evidence. Even with increased effort, this represents a decline from 31 percent confirmed breeding records in the first Atlas.

SPECIES

A total of 218 species and two hybrids were reported to the second Atlas (of which 190 species were considered breeders during the second Atlas) and given full species accounts herein (table 6.6), 187 of which were confirmed to have bred. Included among these confirmed breeders are three that had never been documented to nest in Pennsylvania prior to the second Atlas: Merlin, Great Black-backed Gull, and Eurasian Collared-Dove. All three species had been undergoing range expansions in recent years. Of these, the Merlin is part of a broad pattern of unaided expansion, which has continued since the completion of this project. Increased Great Black-backed Gull breeding populations in the mid-Atlantic may be due to increased food resources there, but the colonization of the Eurasian Collared-Dove has been more directly aided by humans, having spread rapidly across North America since the 1970s. The Trumpeter Swan, a probable breeder in the second Atlas, was also aided by humans, having been reintroduced into eastern parts of its former range, including neighboring Ontario and Ohio. Additionally, the Double-crested Cormorant, Sandhill Crane, and Blackpoll Warbler had not nested in Pennsylvania at the time of the first Atlas, so they represent new species for atlas projects in Pennsylvania. Other historic nesters, confirmed breeding in the second Atlas but not during the first, were Ring-billed Gull, Herring Gull, and Red Crossbill.

The second Atlas documented major range expansions in a number of species. Notable among these is the Clay-colored Sparrow, not confirmed breeding in the first Atlas and increasing over ninefold, from 3 block records in the first Atlas to 29 in the second. The Bald Eagle expansion, by over tenfold, and the fivefold increase in Peregrine Falcons have been documented by ongoing monitoring by the Pennsylvania Game Commission. Impressive, by sheer numbers, is the doubling of blocks documenting Canada Goose, the largest increase in number of block records (1,816) of any breeding

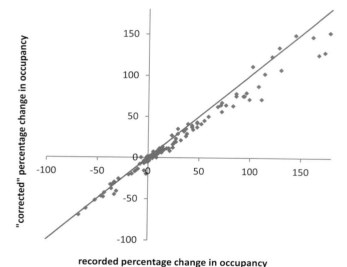

FIG. 6.8 *Recorded percentage change in block occupancy between the first and second Atlases versus estimate change correcting for changes in survey effort, for 132 bird species (blue symbols). The red line represents the line of equality.*

TABLE 6.5 Recorded and modeled (corrected for survey effort) changes in block occupancy between the first and second Atlases

	All widespread species (n = 132)		All species nesting in either atlas (n = 198)	
	Recorded change	Corrected change	Recorded change	Corrected change
Extirpated	—	—	8	(8)
Strong contraction (>33%)	10	11	24	25
Modest contraction (13% to 33%)	5	9	19	23
No change (−13% to +15%)	64	64	71	71
Modest expansion (15% to 50%)	25	23	30	28
Strong expansion (>50%)	28	25	42	39
Colonized	—	—	4	(4)

TABLE 6.6 Summary of atlas results from the first and second Atlases for 198 species either confirmed or probably breeding during either atlas period

Species	Guild	Blocks first Atlas	Blocks second Atlas	Recorded % change All blocks		Priority blocks		Corrected % change	
Canada Goose	Generalist	1,682	3,498	108	***	108	***	87	***
Mute Swan	Generalist†	31	137	342	***	189	**		
Trumpeter Swan	Wetland†	0	1	∞	na				
Wood Duck	Wetland†	1,598	2,027	27	***	26	**	23	***
American Wigeon	Wetland†	3	0	−∞	na	−67	***		
American Black Duck	Wetland†	297	111	−63	***	−67	***		
Mallard	Generalist	2,898	3,021	4	ns	7	ns	0	ns
Blue-winged Teal	Wetland†	100	32	−68	***	−71	*		
Northern Shoveler	Wetland†	7	0	−∞	na				
Green-winged Teal	Wetland†	23	13	−43	ns	−60	ns		
Hooded Merganser	Wetland†	75	211	181	***	150	***	145	***
Common Merganser	Riparian†	227	554	144	***	149	***	149	***
Northern Bobwhite	Farm/grass†	714	231	−68	***	−64	***	−70	***
Ring-necked Pheasant	Farm/grass	2,196	1,057	−52	***	−52	***	−52	***
Ruffed Grouse	Forest gen.†	2,782	1,870	−33	***	−32	***	−30	***
Wild Turkey	Forest gen.	2,402	3,549	48	***	52	***	43	***
Pied-billed Grebe	Wetland†	86	64	−26	ns	22	ns		
Double-crested Cormorant	Riparian†	9	111	1,133	***		na		
American Bittern	Wetland†	53	31	−42	*	−50	ns		
Least Bittern	Wetland†	31	28	−10	ns	0	ns		
Great Blue Heron	Generalist†	2,279	2,908	27	***	26	***	19	***
Great Egret	Riparian†	36	117	225	**	198	**		
Snowy Egret	Riparian†	11	0	−∞	na		na		
Cattle Egret	Riparian†	31	0	−∞	na		na		
Green Heron	Wetland†	1,991	1,792	−10	***	−13	ns	−14	***
Black-crowned Night-Heron	Wetland†	152	77	−49	***	−59	***		
Yellow-crowned Night-Heron	Riparian†	21	9	−57	*	−40	ns		
Black Vulture	Generalist†	352	811	130	***	137	***	107	***
Turkey Vulture	Generalist	3,920	4,390	12	***	9	ns	8	**
Osprey	Wetland†	142	269	89	***	150	***		
Bald Eagle	Riparian†	51	535	949	***	1483	***		
Northern Harrier	Farm/grass†	334	190	−43	***	−40	*	−43	***
Sharp-shinned Hawk	N. hardwood†	1,051	840	−20	***	0	ns		
Cooper's Hawk	Generalist†	1,048	1,512	44	***	59	***		
Northern Goshawk	N. hardwood†	120	86	−28	*	−9	ns		
Red-shouldered Hawk	Forest gen.†	750	1,167	55	***	58	***	45	***
Broad-winged Hawk	Forest gen.†	2,063	1,725	−16	***	−17	*	−20	***
Red-tailed Hawk	Farm/grass	3,633	4,104	13	***	13	*	6	**
American Kestrel	Farm/grass	2,938	2,558	−13	***	−10	ns	−17	***
Merlin	Generalist†	0	13	∞	na		na		
Peregrine Falcon	Generalist†	6	31	417	***	200	ns		
Black Rail	Wetland†	2	0	−∞	na		na		
King Rail	Wetland†	5	5	0	ns	0	ns		
Virginia Rail	Wetland†	120	153	28	**	27	ns		
Sora	Wetland†	88	100	14	**	0	ns		
Common Gallinule	Wetland†	76	48	−37	***	−57	ns		
American Coot	Wetland†	25	18	−28	ns	150	ns		
Sandhill Crane	Wetland†	0	26	∞	na		na		
Killdeer	Farm/grass	3,729	3,721	0	ns	0	ns	−5	*
Black-necked Stilt	Wetland†	1	0	−∞	na		na		
Spotted Sandpiper	Riparian	722	478	−34	***	−13	ns		
Upland Sandpiper	Farm/grass†	54	23	−57	***	−38	ns		
Wilson's Snipe	Wetland†	38	32	−16	ns	−25	ns		
American Woodcock	Forest gen†.	1,469	1,370	−7	ns	1	ns		
Ring-billed Gull	Generalist†	12	171	1,325	na		na		
Herring Gull	Generalist†	2	60	2,900	***	450	**		
Great Black-backed Gull	Generalist†	1	10	900	na		na		
Black Tern	Wetland†	11	3	−73	*		na		
Rock Pigeon	Generalist	3,374	3,356	−1	ns	−2	ns	1	ns

(continued)

TABLE 6.6 (continued)

Species	Guild	Blocks first Atlas	Blocks second Atlas	Recorded % change All blocks		Priority blocks		Corrected % change	
Eurasian Collared-Dove	Generalist	0	15	∞	na		na		
Mourning Dove	Generalist	4,518	4,850	7	***	7	ns	3	ns
Yellow-billed Cuckoo	Forest gen.	2,265	3,238	43	***	41	***	39	***
Black-billed Cuckoo	Forest gen.	1,613	1,703	5	ns	5	ns	7	*
Barn Owl	Farm/grass†	251	117	−53	***	−40	*		
Eastern Screech-Owl	Forest gen†	1,913	1,539	−20	***	−4	ns		
Great Horned Owl	Forest gen†	2,418	1,734	−28	***	−20	**		
Barred Owl	N. hardwood†	1,076	1,262	17	***	36	***		
Long-eared Owl	N. hardwood†	18	14	−22	ns	67	ns		
Short-eared Owl	Farm/grass†	6	5	−17	ns	−50	ns		
Northern Saw-whet Owl	N. hardwood†	96	283	195	***	413	***		
Common Nighthawk	Generalist†	754	219	−71	***	−72	***		
Chuck-will's-widow	Carolinean†	3	2	−33	ns		na		
Eastern Whip-poor-will	Forest gen†	862	496	−42	***	−25	*		
Chimney Swift	Generalist	3,416	3,334	−2	ns	−4	ns	−3	ns
Ruby-throated Hummingbird	Forest gen.	3,518	3,904	11	***	9	ns	11	***
Belted Kingfisher	Generalist	3,063	2,830	−8	**	−6	ns	−8	***
Red-headed Woodpecker	Farm/grass	698	374	−46	***	−42	***	−47	***
Red-bellied Woodpecker	Carolinean	2,006	3,736	86	***	88	***	75	***
Yellow-bellied Sapsucker	N. hardwood	733	1,459	99	***	80	***	71	***
Downy Woodpecker	Forest gen.	4,388	4,531	3	ns	3	ns	0	ns
Hairy Woodpecker	Forest gen.	3,115	3,597	15	***	11	ns	9	**
Northern Flicker	Forest gen.	4,527	4,730	4	*	1	ns	0	ns
Pileated Woodpecker	Forest gen.	2,650	3,653	38	***	32	***	29	***
Olive-sided Flycatcher	N. hardwood†	6	0	−∞	na		na		
Eastern Wood-Pewee	Forest gen.	4,353	4,388	1	ns	−1	ns	0	ns
Yellow-bellied Flycatcher	N. hardwood†	13	13	0	ns	−50	ns		
Acadian Flycatcher	Carolinean	1,766	2,454	39	***	27	***	41	***
Alder Flycatcher	N. hardwood	332	868	161	***	137	***	147	***
Willow Flycatcher	Farm/grass	1,652	2,026	23	***	21	*	27	***
Least Flycatcher	N. hardwood	1,816	1,739	−4	ns	−2	ns	−7	*
Eastern Phoebe	Forest gen.	4,446	4,665	5	*	4	ns	4	ns
Great Crested Flycatcher	Forest gen.	3,493	3,582	3	ns	2	ns	2	ns
Eastern Kingbird	Farm/grass	3,441	3,476	0.8	ns	−0.3	ns	−4	ns
Loggerhead Shrike	Farm/grass†	6	2	−67	ns	0	ns		
White-eyed Vireo	Carolinean	1,022	1,005	−2	ns	0	ns	−2	ns
Yellow-throated Vireo	Forest gen.	1,218	1,760	44	***	39	***	34	***
Blue-headed Vireo	N. hardwood	1,474	2,340	59	***	41	***	50	***
Warbling Vireo	Forest gen.	1,106	1,951	76	***	71	***	64	***
Red-eyed Vireo	Forest gen.	4,586	4,836	5	*	3	ns	4	*
Blue Jay	Forest gen.	4,699	4,704	0	ns	1	ns	−2	ns
Eurasian Jackdaw	Generalist†	1	0	−∞	na		na		
American Crow	Generalist	4,819	4,900	2	ns	2	ns	1	ns
Fish Crow	Generalist	514	956	86	***	62	***	78	***
Common Raven	N. hardwood	850	1,815	114	***	117	***	102	***
Horned Lark	Farm/grass	628	781	24	***	12	ns	12	***
Purple Martin	Generalist	935	520	−44	***	−39	***	−48	***
Tree Swallow	Generalist	3,127	3,929	26	***	28	***	19	***
Northern Rough-winged Swallow	Generalist	1,498	2,138	43	***	47	***	37	***
Bank Swallow	Riparian	521	337	−36	***	−32	*	−38	***
Cliff Swallow	Generalist	883	764	−14	**	−3	ns		
Barn Swallow	Farm/grass	4,537	4,352	−4	*	−4	ns	−5	**
Carolina Chickadee	Carolinean	707	1,075	52	***	55	***	41	***
Black-capped Chickadee	Forest gen.	4,151	4,114	−1	ns	3	ns	−5	*
Tufted Titmouse	Forest gen.	3,989	4,692	18	***	14	*	11	***
Red-breasted Nuthatch	N. hardwood	203	554	173	***	237	***	128	***
White-breasted Nuthatch	Forest gen.	4,082	4,591	12	***	7	ns	7	**
Brown Creeper	N. hardwood	1,139	1,094	−4	ns	−8	ns	−16	***
Carolina Wren	Carolinean	2,070	3,487	68	***	68	***	62	***
House Wren	Generalist	4,540	4,531	0	ns	−2	ns	−4	*

TABLE 6.6 (continued)

Species	Guild	Blocks first Atlas	Blocks second Atlas	Recorded % change All blocks		Priority blocks		Corrected % change	
Winter Wren	N. hardwood	378	799	111	***	69	***	71	***
Sedge Wren	Farm/grass†	13	20	54	ns	−57	ns		
Marsh Wren	Wetland†	77	53	−32	*	−50	ns		
Blue-gray Gnatcatcher	Forest gen.	2,307	2,636	14	***	13	ns	11	***
Golden-crowned Kinglet	N. hardwood	182	370	102	***	68	**	111	***
Eastern Bluebird	Farm/grass	3,866	4,219	9	***	7	ns	8	***
Veery	N. hardwood	2,308	2,543	10	***	7	ns	5	ns
Swainson's Thrush	N. hardwood	43	98	128	***	73	ns	134	***
Hermit Thrush	N. hardwood	1,372	1,877	37	***	31	**	35	***
Wood Thrush	Forest gen.	4,493	4,603	2	ns	−1	ns	0	ns
American Robin	Generalist	4,882	4,923	1	ns	1	ns	0	ns
Gray Catbird	Generalist	4,748	4,841	2	ns	1	ns	0	ns
Northern Mockingbird	Generalist	2,291	2,748	20	***	18	*	11	***
Brown Thrasher	Farm/grass	3,073	3,331	8	**	9	ns	0	ns
European Starling	Generalist	4,389	4,505	3	ns	1	ns	0	ns
Cedar Waxwing	Forest gen.	4,062	4,283	5	*	4	ns	5	*
Ovenbird	Forest gen.	3,674	4,168	13	***	10	ns	7	**
Worm-eating Warbler	Carolinean	735	1,086	48	***	49	***	37	***
Louisiana Waterthrush	Carolinean	1,386	1,784	29	***	26	**	29	***
Northern Waterthrush	N. hardwood	331	229	−31	***	−22	ns	−41	***
Golden-winged Warbler	Forest gen.	615	240	−61	***	−50	***	−62	***
Blue-winged Warbler	Forest gen.	1,403	1,345	−4	ns	−4	ns	−4	ns
Black-and-white Warbler	Forest gen.	2,108	2,281	8	*	10	ns	3	ns
Prothonotary Warbler	Riparian†	43	47	9	ns	−10	ns		
Swainson's Warbler	Carolinean†	2	4	100	ns	200	ns		
Nashville Warbler	N. hardwood	217	213	−2	ns	−2	ns	−20	**
Mourning Warbler	N. hardwood	235	431	83	***	49	*	63	***
Kentucky Warbler	Carolinean	929	662	−29	***	−33	***	−26	***
Common Yellowthroat	Forest gen.	4,733	4,846	2	ns	1	ns	2	ns
Hooded Warbler	Forest gen.	1,418	2,423	71	***	64	***	64	***
American Redstart	Forest gen.	2,992	3,716	24	***	15	*	16	***
Cerulean Warbler	Carolinean	836	776	−7	ns	−11	ns	0	ns
Northern Parula	Forest gen.	499	1,108	121	***	127	***	123	***
Magnolia Warbler	N. hardwood	744	1,439	93	***	70	***	75	***
Blackburnian Warbler	N. hardwood	917	1,580	72	***	56	***	67	***
Yellow Warbler	Generalist	3,984	4,300	8	***	4	ns	3	ns
Chestnut-sided Warbler	N. hardwood	2,008	2,718	35	***	26	**	32	***
Blackpoll Warbler	N. hardwood†	2	3	50	na		na		
Black-throated Blue Warbler	N. hardwood	741	1,278	72	***	64	***	56	***
Pine Warbler	Forest gen.	295	788	167	***	149	***	125	***
Yellow-rumped Warbler	N. hardwood	315	877	178	***	127	***	152	***
Yellow-throated Warbler	Carolinean	152	196	29	*	54	ns	35	**
Prairie Warbler	Forest gen.	1,035	1,117	8	ns	1	ns	4	ns
Black-throated Green Warbler	N. hardwood	1,803	2,675	48	***	36	***	42	***
Canada Warbler	N. hardwood	634	691	9	ns	−7	ns	7	ns
Yellow-breasted Chat	Carolinean	1,242	805	−35	***	−22	*	−33	***
Eastern Towhee	Forest gen.	4,453	4,677	5	*	3	ns	1	ns
Chipping Sparrow	Generalist	4,786	4,888	2	ns	1	ns	1	
Clay-colored Sparrow	Farm/grass†	3	29	867	***		na		
Field Sparrow	Farm/grass	4,291	4,372	2	ns	−2	ns	0	ns
Vesper Sparrow	Farm/grass	1,087	878	−19	***	−20	*	−21	***
Savannah Sparrow	Farm/grass	1,661	2,085	26	***	22	*	19	***
Grasshopper Sparrow	Farm/grass	1,629	1,674	3	ns	4	ns	1	ns
Henslow's Sparrow	Farm/grass	364	229	−37	***	−42	**	−33	***
Song Sparrow	Generalist	4,789	4,867	2	ns	1	ns	1	ns
Swamp Sparrow	Wetland†	1,008	1,332	32	***	31	**	21	***
White-throated Sparrow	N. hardwood	213	142	−33	***	−39	ns	−45	***
Dark-eyed Junco	N. hardwood	1,334	1,831	37	***	38	***	26	***
Summer Tanager	Carolinean†	47	4	−91	***	−75	ns		
Scarlet Tanager	Forest gen.	4,276	4,543	6	**	4	ns	3	ns

(continued)

TABLE 6.6 (continued)

Species	Guild	Blocks first Atlas	Blocks second Atlas	All blocks		Priority blocks		Corrected % change	
				\multicolumn{6}{c}{Recorded % change}			
Northern Cardinal	Generalist	4,442	4,678	5	*	4	ns	2	ns
Rose-breasted Grosbeak	Forest gen.	3,151	3,557	13	***	10	ns	7	**
Blue Grosbeak	Farm/grass	113	219	94	***	85	*	75	***
Indigo Bunting	Forest gen.	4,661	4,851	4	ns	3	ns	2	ns
Dickcissel	Farm/grass†	45	45	0	ns	−25	ns		
Bobolink	Farm/grass	1,527	1,739	14	***	4	ns	13	***
Red-winged Blackbird	Farm/grass	4,537	4,581	1	ns	0	ns	0	ns
Eastern Meadowlark	Farm/grass	3,268	2,896	−11	***	−14	*	−15	***
Western Meadowlark	Farm/grass†	2	2	0	ns		na		
Common Grackle	Generalist	4,461	4,493	1	ns	1	ns	−2	ns
Brown-headed Cowbird	Generalist	4,321	4,510	4	ns	1	ns	−1	ns
Orchard Oriole	Generalist	797	1,561	96	***	90	***	79	***
Baltimore Oriole	Generalist	4,004	4,286	7	**	5	ns	2	ns
Purple Finch	N. hardwood	1,448	1,748	21	***	15	ns	8	*
House Finch	Generalist	3,894	3,995	3	ns	2	ns	2	ns
Red Crossbill	N. hardwood†	6	11	83	na		na		
Pine Siskin	N. hardwood†	71	93	31	ns	11	ns		
American Goldfinch	Generalist	4,650	4,758	2	ns	2	ns	0	ns
House Sparrow	Generalist	4,283	4,209	−2	ns	−5	ns	−5	*

†Species guild assigned by expert opinion. Others designated by multivariate analysis (see chap. 5).
Statistical significance of changes (z-test for differences in proportions):
ns = not significant
*$p < 0.05$
**$p < 0.01$
***$p < 0.001$
na = not available—small sample size or infinity change

bird. Other species increasing more than 1,000 blocks were Wild Turkey, Red-bellied Woodpecker, Pileated Woodpecker, Carolina Wren, and Hooded Warbler.

The number of possible-nesting records of Double-crested Cormorant, Great Egret, and Herring Gull resulted in a large percentage increase in total blocks reported, although only a small number of nest sites is known in the state. After the Clay-colored Sparrow, the songbird with the largest percentage increase (278%) was the Yellow-rumped Warbler. Among the list of species with confirmed nesting during the second Atlas, the largest increases include an eclectic group, including raptors, waterbirds, conifer specialists, and even an early successional species (table 6.6).

Eight of the most widespread species were found in at least 95 percent of blocks during both atlas projects: Blue Jay, American Crow, American Robin, Gray Catbird, Common Yellowthroat, Indigo Bunting, Chipping Sparrow, and Song Sparrow. These species retain such ubiquity that they likely occur in every block.

Many species have previously nested in Pennsylvania but were not documented during this effort. Most of these have been either extralimital or ephemeral, with no established pattern of nesting. Notable among these are the Cattle Egret, Snowy Egret, Black-necked Stilt, and several species of waterfowl. A very brief historical account is provided for these species in appendix A, along with a table listing other species with nesting history in Pennsylvania. Because of historic breeding or their inclusion in the first Atlas, species given full species accounts within this volume (even though nesting was not confirmed during the second Atlas) include Chuck-will's-widow, King Rail, Swainson's Warbler, and Western Meadowlark.

A number of species with established nesting histories during or since the first Atlas showed dramatic declines in block occupancy. Most notable among these are Summer Tanager (−91%), Black Tern (−73%), Common Nighthawk (−71%), Blue-winged Teal (−68%), and Northern Bobwhite (−68%). The nighthawk and bobwhite had previously been widespread. Other striking declines include the Ring-necked Pheasant (−52%) and Ruffed Grouse (−33%), each declining by approximately 1,000 blocks.

Atlas volunteers plotted or entered 84,477 point locations, or about 12 percent of all data entered through the web portal. For state-listed species, or those reported by fewer than 100 records, nearly all records were accompanied with point locations and narrative comments, as required. Several other statewide priority species, with widespread distributions, each received over 2,000 volunteer point locations: Acadian Flycatcher, Hooded

Warbler, Scarlet Tanager, and Wood Thrush. Together, these four species contributed over 14 percent of all volunteer-provided georeferenced records.

RANGE SHIFTS AND BLOCK TURNOVER

Species distributions are constantly in flux. Some species, although not increasing or decreasing dramatically, were predominantly found in different blocks between the two atlas periods. These species show high turnover at the block level and potentially low site fidelity at nest sites. Analysis of relative turnover rates (see chap. 5 for methods) identified 32 species with higher-than-expected turnover of blocks between atlas periods. A preponderance of wetland and early successional species are among this group (fig. 6.9), not surprising for species whose habitat is ephemeral. Other species in this group include some that are widespread but never abundant and perhaps easily overlooked (e.g., Sharp-shinned Hawk, Hairy Woodpecker, Ruby-throated Hummingbird), as well as the Cliff Swallow, which showed a suprisingly high block turnover between atlases. Hence, for some species, high block turnover could be an artifact of uneven block coverage. Twenty-two species were noted to show low block turnover. These included colonial waterbirds (Double-crested Cormorant, Great Egret, Ring-billed Gull), raptors that show site fidelity (Bald Eagle, Peregrine Falcon), and a suite of forest obligates that increased between atlas periods. However, Common Nighthawk is also among this group, due to its large decline in block records but also low turnover within a small number of key areas.

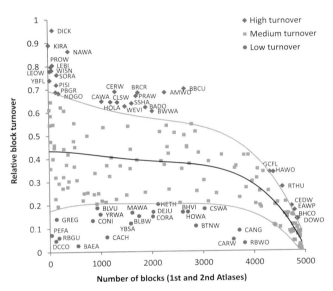

FIG. 6.9 *Relative block turnover between atlases of all species present in five or more blocks in each atlas period (n = 176). Black line is third order polynomial regression line, gray lines are ± standard deviation. Species showing a higher-than-expected block turnover (>1 std. dev. above regression line; red diamonds) are: Pied-billed Grebe (PBGR), Least Bittern (LEBI), Sharp-shinned Hawk (SSHA), Northern Goshawk (NOGO), King Rail (KIRA), Sora (SORA), Wilson's Snipe (WISN), American Woodcock (AMWO), Black-billed Cuckoo (BBCU), Barred Owl (BADO), Long-eared Owl (LEOW), Ruby-throated Hummingbird (RTHU), Downy Woodpecker (DOWO), Hairy Woodpecker (HAWO), Eastern Wood-Pewee (EAWP), Yellow-bellied Flycatcher (YBFL), Great Crested Flycatcher (GCFL), White-eyed Vireo (WEVI), Horned Lark (HOLA), Cliff Swallow (CLSW), Brown Creeper (BRCR), Cedar Waxwing (CEDW), Blue-winged Warbler (BWWA), Nashville Warbler (NAWA), Prairie Warbler (PRAW), Cerulean Warbler (CERW), Prothonotary Warbler (PROW), Canada Warbler (CAWA), Dickcissel (DICK), Brown-headed Cowbird (BHCO), and Pine Siskin (PISI). Species showing lower than expected turnover (>1 std. dev below regression line, blue circles) are: Canada Goose (CANG), Double-crested Cormorant (DCCO), Great Egret (GREG), Black Vulture (BLVU), Bald Eagle (BAEA), Peregrine Falcon (PEFA), Ring-billed Gull (RBGU), Common Nighthawk (CONI), Red-bellied Woodpecker (RBWO), Yellow-bellied Sapsucker (YBSA), Blue-headed Vireo (BHVI), Common Raven (CORA), Carolina Chickadee (CACH), Carolina Wren (CARW), Hermit Thrush (HETH), Chestnut-sided Warbler (CSWA), Magnolia Warbler (MAWA), Yellow-rumped Warbler (YRWA), Black-throated Green Warbler (BTNW), Blackburnian Warbler (BLBW), Hooded Warbler (HOWA), and Dark-eyed Junco (DEJU).*

LATITUDINAL SHIFTS

There is growing evidence that poleward shifts in bird distributions are occurring (Thomas and Lennon 1999; Root et al. 2003; Hitch and Leberg 2007), which is consistent with predicted effects of climate change (Matthews et al. 2011). In neighboring New York, Zuckerberg (2008) found evidence supporting a general northward expansion of southern species, and a northward retraction of northern species, but results were by no means consistent. Analysis of latitudinal shifts within individual states is complicated by the fact that species distributions tend to be much larger than states; hence, shifts are only likely to be detected for species whose southern or northern limits pass through the state. Categorizing these species is not straightforward. In Pennsylvania, for example, many species considered of northern affinity are found at high elevations in the Appalachians, south to the Maryland border and beyond. Further, some species that are found primarily in the southern half of Pennsylvania (e.g., Golden-winged Warbler and Northern Parula) are widespread further north; hence, their southern affinity in Pennsylvania does not reflect their wider distribution.

Our analysis of latitudinal distribution shifts between atlases is on an individual species level (see chap. 5 for methods). In table 6.7, we highlight the species for which there is evidence of latitudinal changes in distribution. These changes include northward and southward expansions, where the leading edge of the range advanced but the trailing edge (which is often well beyond the state

TABLE 6.7 Latitudinal changes in distributions between the first and second Atlases

Species and range shift type	Change in latitude of occupied blocks (km)						Agree with Matthews et al. 2007?
	Southern edge		Median		Northern edge		
Northward expansion							
Great Egret	11	ns	22	ns	34	*	Yes
Black Vulture	0	ns	12	ns	26	*	Yes
Yellow-billed Cuckoo	0	ns	22	***	15	***	Yes
Red-bellied Woodpecker	−3	ns	30	***	45	***	Yes
Acadian Flycatcher	−2	ns	10	*	8	*	Yes
Warbling Vireo	−7	ns	1	ns	14	**	Yes
Northern Rough-winged Swallow	−2	ns	13	***	21	***	—
Carolina Chickadee	1	ns	10	**	16	*	—
Tufted Titmouse	0	ns	13	***	7	**	Yes
Carolina Wren	−3	ns	22	***	35	***	Yes
Hooded Warbler	−9	ns	11	*	18	*	Yes
Pine Warbler	2	ns	19	ns	40	***	Yes
Blue Grosbeak	4	ns	10	ns	21	*	Yes
Orchard Oriole	−7	**	5	ns	19	*	Yes
Northward retraction							
American Black Duck	56	**	21	ns	−11	ns	—
Ruffed Grouse	22	*	15	***	−4	ns	—
Broad-winged Hawk	17	***	15	***	−2	ns	Yes
American Woodcock	19	***	23	***	1	ns	No
Eastern Screech-Owl	10	**	17	***	2	ns	—
Brown Creeper	16	ns	15	**	−1	ns	Yes
Nashville Warbler	29	*	23	**	13	ns	Yes
Northward shift							
Black-billed Cuckoo	8	ns	16	***	5	ns	Yes
Yellow-bellied Sapsucker	3	ns	3	ns	8	**	Yes
Cerulean Warbler	13	ns	16	*	7	ns	Yes
Altitudinal shift							
Northern Waterthrush	1	ns	−8	ns	−21	***	Yes
White-throated Sparrow	−8	ns	−13	***	−17	***	Yes
Southward expansion							
Common Merganser	−55	***	−13	**	17	*	—
Wild Turkey	−16	***	−4	ns	5	ns	—
Great Blue Heron	−9	***	−19	***	0	ns	No
Cooper's Hawk	−28	***	−31	***	−1	ns	—
Blue-headed Vireo	−24	***	−6	ns	4	ns	No
Tree Swallow	−8	***	−15	***	0	ns	No
Hermit Thrush	−22	***	−6	***	2	ns	No
Dark-eyed Junco	−16	*	−3	ns	2	ns	No
Southward retraction							
Northern Harrier	2	ns	−13	**	−26	***	No
American Kestrel	1	ns	2	ns	−5	**	No
Red-headed Woodpecker	8	ns	−20	***	−48	***	No
Horned Lark	−1	ns	−21	***	−26	***	Equivocal
Purple Martin	14	ns	−13	**	−58	***	No
Golden-winged Warbler	4	ns	−17	**	−40	***	No
Yellow-breasted Chat	6	ns	−8	***	−21	***	No
Vesper Sparrow	13	ns	−5	ns	−14	**	No
Southward shift							
Black-crowned Night-Heron	−29	**	−36	***	−41	***	—
Long-eared Owl	−67	***	−61	**	−54	**	—
Yellow-throated Vireo	−11	**	−9	ns	−3	ns	No
Prothonotary Warbler	−40	**	−49	***	−56	***	No

Statistical significance of changes (z-test for differences in proportions):
ns = not significant
*$p < 0.05$
**$p < 0.01$
***$p < 0.001$

border) did not change; northward and southward retractions, where the leading edge receded; and range shifts, where both the leading and trailing edges shifted. Of the 14 species with significant northward range expansions, 12 are predicted to undergo northward expansion in the *Climate Change Bird Atlas* (conservative "PCM Lo" scenario; Matthews et al. 2010); the other two species were not included in the latter study. Of the seven species for which we found evidence of a northward contraction, three show patterns consistent with the *Climate Change Bird Atlas,* while one, Broad-winged Hawk, was not predicted to undergo a significant northward contraction. The three species showing northward shifts are predicted to undergo northward retraction in the *Climate Change Bird Atlas*. Two additional species, Northern Waterthrush and White-throated Sparrow, showed evidence of contractions to high elevations between atlases, which manifests itself as a retraction for northernmost parts of Pennsylvania, which are at lower elevation than the state's highest mountains. Hence, for the great majority of species for which we found evidence of northward changes in distribution in Pennsylvania, our results correspond with projected distribution changes in the *Climate Change Bird Atlas*.

As was found in New York (Zuckerberg 2008), latitudinal distributions in several species are counterintuitive—notably, for a group of species of southern affinity whose ranges are retracting and a group of species of northern affinity whose ranges are expanding southward. We found evidence of such trends for 20 species (table 6.7), which is almost as many as species for which northward shifts are in agreement with predictions. Superficially, this even balance between trends that agree with predictions and those that counter the predictions is puzzling. However, closer scrutiny of these examples reveals some interesting patterns. For example, the eight species showing southward expansion include three that are still rebounding from historical losses due to PCB- or pesticide-induced losses (Common Merganser, Great Blue Heron, Cooper's Hawk) and three forest songbirds that have continued to recover from historical forest loss (Blue-headed Vireo, Hermit Thrush, Dark-eyed Junco). Similarly, there are commonalities among the eight species that underwent southward range contraction between atlases—all are either grassland or early successional species currently undergoing regionwide declines (Sauer et al. 2011).

Other species found primarily in the north or at higher elevations, such as Winter Wren, Black-throated Blue Warbler, and Canada Warbler, respond positively to increased vegetative structure that sometimes results from disturbance. Others, including Golden-crowned Kinglet and Pine Siskin, respond to conifer plantings that imitate natural conifer forests that otherwise may have diminished. Change maps in the species accounts indicate that some species have expanded strongly in the High Plateau Section and adjoining regions (e.g., Winter Wren, Magnolia Warbler), while others are more evenly distributed across the northern third of the state (Alder Flycatcher). Overall, the range changes observed in this suite of species since the first Atlas are not consistent with the simple expectation that northern species ranges will contract northward under a warming-climate scenario. Clearly, other forces are influencing these patterns besides climate change, and we cannot easily separate the effects of climate change from simultaneous changes in land cover, forest age, forest composition and structure, and tree species composition. Increasing average age of forests and resultant vegetation structure, disturbances within these larger-scale forests, and a slowly increasing conifer component in Pennsylvania forests over the last 50+ years may be contributing to the increase in some northern bird species.

Another group of breeding birds that is likely to be affected by climate change includes birds associated with wetlands, especially marsh-nesting birds. Marsh vegetation is strongly tied to hydrological cycles, and extensive periods of either too much or too little precipitation may cause changes in water levels and shifts in vegetation that make the habitat less suitable for this suite of birds. In addition, extreme weather events during the nesting season that result in either flooding or drawing down of the marsh may directly cause nest failure. Despite extra sampling of marsh-nesting birds in the second Atlas, all species in this group, except Sora (+14%) and Virginia Rail (+28%), showed decreases between atlases. The two that increased are the ones that are least area sensitive and are able to make use of small pockets of marsh habitat. Their own increases may be an artifact of increased coverage of this habitat. The steady declines of those wetland species dependent on larger and more complex wetlands has been observed with alarm for decades (Brauning 1992a; Brauning et al. 1994; Gross and Haffner 2010). Climate change may be exacerbating the loss of quality and quantity of wetlands capable of supporting our rare wetland breeding birds.

In general, there is a clear pattern of species with southern affinity (in first Atlas) showing a northward shift in atlas block occupancy between atlases: 18 of 22 species whose center of distribution lay in the southern third of the state in the first Atlas showed northward latitudinal shifts in the second Atlas (fig. 6.10). Exceptions were Black-crowned Night-Heron, Loggerhead Shrike, Yellow-breasted Chat, and Dickcissel. Species of

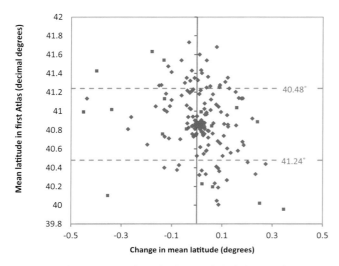

FIG. 6.10 *Changes in mean latitude of occupied blocks between first and second Atlases plotted against mean latitude in first Atlas. Rare species (figure 30 blocks in one or both atlases) in red. Dotted lines indicate lines of latitude approximately 1/3 and 2/3 of the distance between the Maryland border and the New York border.*

northern affinity in the first Atlas were equally divided, with 14 showing a southward shift and 14 showing a northward shift in the second Atlas. Of the other 132 species, almost two-thirds (85) showed a northward shift. Overall, there is ample evidence from the two Pennsylvania atlases that several species have undergone distributional shifts that agree with predictions based on climate change models. However, we caution that patterns are not definitive, and attempts to explain the patterns among species whose patterns are counterintuitive are based on preconceived notions, rather than on a rigorous analysis of cause and effect. We recommend a more detailed analysis of these data, possibly in tandem with atlas data from surrounding states, in order to elucidate larger-scale patterns.

POPULATION ESTIMATES

There were sufficient data from point count surveys to derive population estimates for 115 species (table 6.8). It should be noted that confidence in these estimates varies, and estimates for non-songbirds in particular should be treated with caution (see chap. 5 and species accounts). Populations of species marked with "#" in table 6.8 are likely to be underestimated; hence, comparisons between population estimates of those species and songbirds may be problematic. However, for the majority of common songbirds, the field methods (chap. 4) and analytical methods (chap. 5) allowed for a sophisticated approach to estimating population sizes and density, and, in our opinion, comparisons among these species are robust. Our analysis suggests that no less than 23 species achieve total population sizes of more than 1 million adult birds (~500,000 pairs) in Pennsylvania (table 6.8). The three most abundant species, Red-eyed Vireo, Chipping Sparrow, and Song Sparrow, all have populations that are close to 3 million pairs (and no doubt exceed that number in some years), closely followed by American Robin and Gray Catbird. The population estimates across the 115 species total more than 42 million pairs of birds. Accounting for the previously stated assertion that some of these are underestimates, and the combined total of the other 75 (generally scarce) species, it seems likely that approximately 100 million individual birds attempt to nest in Pennsylvania each year.

Backcasting of population estimates over the 44-year history of the Breeding Bird Survey provides a useful insight into the scale of bird population changes in recent decades. While there are many caveats associated with such backcasting (see chap. 5), the results are remarkable. For example, we estimate the Grasshopper Sparrow population decreased from 270,000 to 92,000 males between the first Atlas and the second Atlas, but projecting backward even further, BBS trends suggest the population in the mid-1960s was between 700,000 and 800,000 pairs. In contrast, the increases shown by other species are equally dramatic. Several species that had very small populations, likely in the low thousands of pairs in the 1960s, now number in the high tens or hundreds of thousands, among them Wild Turkey, Red-bellied Woodpecker, Yellow-bellied Sapsucker, and Carolina Wren. Atlas results across all species show that expanding species outnumber contracting species (table 6.5), which is supported by BBS data. Forty-nine of the 115 species in table 6.8 increased by 15 percent or more, while 34 decreased by 13 percent or more. Interestingly, the sum of all estimates of breeding pairs across these 115 species was similar for both atlas periods (41.3 million in first Atlas, 42.5 million in second Atlas).

SPECIES GUILDS

Multivariate analysis of point count data (see methods in chap. 5) suggest that common breeding birds (primarily songbirds) show five groupings, which we call avifaunal guilds, into the following habitat categories: Northern Hardwood, Forest Generalist, Carolinean Forest, Generalist/Urban, and Farmland (table 6.6).

Northern Hardwood species are most abundant in northern tier counties, especially McKean and Potter, and more locally in Lycoming, Sullivan, and Wyoming counties, as well as in the Pocono mountains of Luzerne and Lackawanna counties (fig. 6.11). Farther south, birds in the Northern Hardwood guild were mainly found at high

TABLE 6.8 Bird population estimates during 1983–1989 and 2004–2009 atlas periods, ranked by second Atlas estimate. Population estimates are provided for 115 most numerous diurnal species. Number of "pairs" assumed equivalent to the number of singing males or half the number of individuals.

		Rank 1st	Rank 2nd	Change in rank	First Atlas "pairs"	Second Atlas "pairs"
	Song Sparrow	2	1	1	3,300,000	2,990,000
	Chipping Sparrow	1	2	−1	3,500,000	2,980,000
	Red-eyed Vireo	4	3	1	2,300,000	2,830,000
	American Robin	3	4	−1	2,400,000	2,450,000
	Gray Catbird	5	5	—	2,100,000	2,380,000
	Ovenbird	13	6	7	1,100,000	1,600,000
	Mourning Dove	11	7	4	1,100,000	1,550,000
	Indigo Bunting	6	8	−2	1,700,000	1,520,000
	Northern Cardinal	9	9	—	1,200,000	1,400,000
	Common Yellowthroat	8	10	−2	1,300,000	1,240,000
	Red-winged Blackbird	7	11	−4	1,400,000	1,160,000
#	European Starling	10	12	−2	1,200,000	1,000,000
	House Wren	12	13	−1	1,100,000	900,000
	Tufted Titmouse	22	14	8	570,000	850,000
	Black-capped Chickadee	18	15	3	730,000	810,000
#	House Sparrow	14	16	−2	980,000	765,000
	Yellow Warbler	16	17	−1	850,000	765,000
#	Common Grackle	15	18	−3	950,000	760,000
	American Redstart	19	19	—	630,000	730,000
	Wood Thrush	17	20	−3	800,000	660,000
	Eastern Towhee	20	21	−1	600,000	610,000
	Scarlet Tanager	21	22	−1	600,000	575,000
	Chestnut-sided Warbler	35	23	12	280,000	520,000
	House Finch	24	24	—	460,000	420,000
#	Cedar Waxwing	26	25	1	410,000	380,000
	Dark-eyed Junco	32	26	6	320,000	380,000
#	American Crow	29	27	2	360,000	372,500
	Blackburnian Warbler	40	28	12	250,000	360,000
	Black-throated Green Warbler	45	29	16	210,000	355,000
	Baltimore Oriole	31	30	1	320,000	330,000
	Eastern Wood-Pewee	23	31	−8	490,000	315,000
	Eastern Phoebe	33	32	1	310,000	315,000
	American Goldfinch	34	33	1	290,000	315,000
#	Barn Swallow	25	34	−9	430,000	295,000
#	Blue Jay	38	35	3	260,000	295,000
#	White-breasted Nuthatch	50	36	14	170,000	280,000
	Veery	42	37	5	240,000	270,000
	Blue-headed Vireo	57	38	19	110,000	270,000
	Carolina Wren	66	39	27	62,000	270,000
	Hooded Warbler	56	40	16	130,000	264,000
#	Brown-headed Cowbird	28	41	−13	370,000	260,000
	Black-and-white Warbler	30	42	−12	330,000	252,000
	Magnolia Warbler	58	43	15	99,000	240,000
#	Downy Woodpecker	43	44	−1	220,000	225,000
	Field Sparrow	27	45	−18	380,000	210,000
	Rose-breasted Grosbeak	44	46	−2	210,000	210,000
#	Rock Pigeon	36	47	−11	270,000	207,500
#	Chimney Swift	39	48	−9	260,000	200,000
	Northern Mockingbird	48	49	−1	180,000	200,000
#	Wild Turkey	77	50	27	39,000	190,000
#	Red-bellied Woodpecker	78	51	27	39,000	177,500
	Eastern Bluebird	60	52	8	94,000	150,000
	Black-throated Blue Warbler	62	53	9	83,000	150,000
	Acadian Flycatcher	51	54	−3	170,000	149,000
	Savannah Sparrow	41	55	−14	240,000	145,000
#	Blue-gray Gnatcatcher	52	56	−4	150,000	140,000
	Least Flycatcher	46	57	−11	200,000	138,000
	Willow Flycatcher	55	58	−3	130,000	135,000

(continued)

TABLE 6.8 (continued)

		Rank 1st	Rank 2nd	Change in rank	First Atlas "pairs"	Second Atlas "pairs"
#	Ruby-throated Hummingbird	64	59	5	72,000	125,000
	Bobolink	54	60	−6	140,000	110,000
#	Canada Goose	95	61	34	22,000	107,500
	Hermit Thrush	68	62	6	57,000	105,000
	Carolina Chickadee	73	63	10	47,000	105,000
#	Great Crested Flycatcher	53	64	−11	150,000	98,000
#	Northern Flicker	59	65	−6	96,000	93,500
	Grasshopper Sparrow	37	66	−29	270,000	92,000
	Eastern Meadowlark	47	67	−20	190,000	89,000
	Northern Parula	75	68	7	41,000	82,000
	Warbling Vireo	71	69	2	52,000	80,000
	Brown Thrasher	63	70	−7	77,000	80,000
#	Mallard	70	71	−1	56,000	70,000
#	Killdeer	65	72	−7	70,000	60,000
#	Yellow-billed Cuckoo	74	73	1	46,000	60,000
#	Wood Duck	86	74	12	34,000	60,000
#	Tree Swallow	90	75	15	29,000	60,000
#	Eastern Kingbird	61	76	−15	86,000	55,000
	Blue-winged Warbler	67	77	−10	61,000	52,000
	Orchard Oriole	96	78	18	20,000	51,000
	Purple Finch	84	79	5	36,000	50,000
#	Hairy Woodpecker	79	80	−1	38,000	48,500
	Yellow-throated Vireo	76	81	−5	40,000	48,000
#	Yellow-bellied Sapsucker	109	82	27	8,100	48,000
	Yellow-rumped Warbler	97	83	14	19,000	45,000
	Worm-eating Warbler	72	84	−12	47,000	44,000
	Swamp Sparrow	87	85	2	33,000	43,500
	Pine Warbler	100	86	14	17,000	42,000
	Horned Lark	81	87	−6	38,000	40,500
#	Ring-necked Pheasant	49	88	−39	170,000	40,000
	Louisiana Waterthrush	91	89	2	29,000	35,000
	White-eyed Vireo	92	90	2	29,000	34,000
	Winter Wren	101	91	10	15,000	32,000
	Alder Flycatcher	108	92	16	9,000	28,000
	Canada Warbler	89	93	−4	30,000	27,000
#	Pileated Woodpecker	104	94	10	13,000	26,500
	Prairie Warbler	83	95	−12	37,000	26,000
	Cerulean Warbler	80	96	−16	38,000	24,000
#	Cliff Swallow	93	97	−4	26,000	21,500
#	Broad-winged Hawk	99	98	1	18,000	18,000
	Kentucky Warbler	82	99	−17	37,000	17,700
	Vesper Sparrow	69	100	−31	56,000	17,000
	Brown Creeper	98	101	−3	18,000	17,000
#	Northern Rough-winged Swallow	106	102	4	11,000	15,000
#	Fish Crow	107	103	4	9,300	15,000
#	Belted Kingfisher	102	104	−2	14,000	12,000
#	Red-tailed Hawk	113	105	8	5,400	11,500
	Yellow-breasted Chat	88	106	−18	32,000	11,200
#	Black-billed Cuckoo	103	107	−4	14,000	11,000
#	Red-breasted Nuthatch	112	108	4	5,500	11,000
	Mourning Warbler	111	109	2	6,400	10,500
	Henslow's Sparrow	105	110	−5	12,000	8,000
#	Common Raven	115	111	4	2,600	8,000
#	Purple Martin	94	112	−18	25,000	7,500
#	Turkey Vulture	114	113	1	5,000	7,500
#	American Kestrel	110	114	−4	7,600	6,800
	Golden-winged Warbler	85	115	−30	35,000	6,300

Population estimate derived from non-song detections, may be conservative (see chapter 5).

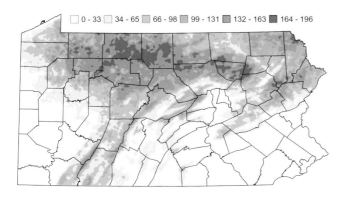

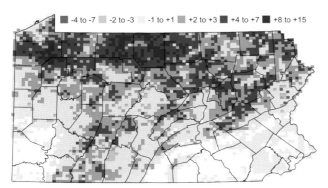

FIG. 6.11 *Northern Hardwood Forest Bird Index (top) and estimated change in species richness between atlases (bottom)*

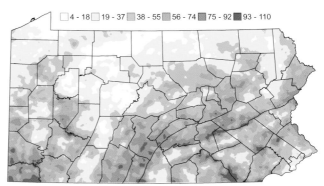

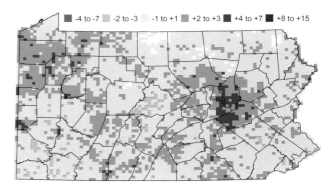

FIG. 6.13 *Carolinean Forest Bird Index (top) and estimated change in species richness between atlases (bottom)*

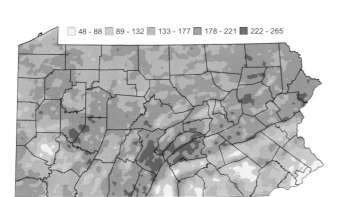

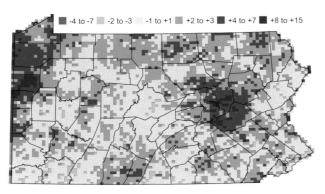

FIG. 6.12 *Forest Generalist Bird Index (top) and estimated change in species richness between atlases (bottom)*

elevations, such as in the Allegheny Mountain Section in Somerset County. Consistent with the results for latitudinal changes (fig. 6.10), species richness of these species of northern affinity increased through most of their range in Pennsylvania (fig. 6.11), but the few indications of localized decline tended to be toward the southern edges of the guild's geographic ranges. The abundance/richness of the Forest Generalist guild was more evenly distributed across the state, and lower in areas with little forest—for example, the intensively farmed Piedmont, and in big cities (fig. 6.12). Highest abundances/species richness of Forest Generalist was in ridgetop forests in the Ridge and Valley Province. Increases in species richness of Forest Generalist species were most pronounced in the northwestern area of the state and in the forests east of the Susquehanna River in Columbia, Northumberland, and Schuylkill counties. Interestingly, these increases do not appear to correlate with changes in forest cover (chap. 3) or increases in survey effort (fig. 6.4). However, unmeasured changes in habitat quality or survey efficiency cannot be ruled out. The Carolinean Forest guild is found in highest abundance/richness at low- to mid-elevations, especially in the southern half of the state (fig. 6.13). Changes in species richness between atlases were modest overall, but it is interesting to note that the pattern of increases and decreases is similar to that of the Forest Generalist guild (fig. 6.12).

COVERAGE AND RESULTS   65

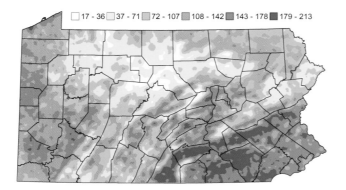
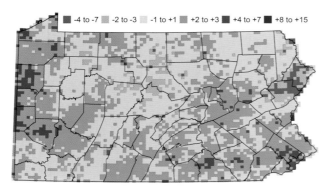

FIG. 6.14 *Generalist and Urban Bird Index (top) and estimated change in species richness between atlases (bottom)*

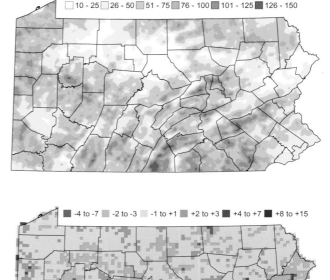

FIG. 6.15 *Farmland Bird Index (top) and estimated change in species richness between atlases (bottom)*

Not surprisingly, Generalist/Urban species were found in highest abundance/richness in the densely populated southeastern corner of the state, and they were generally scarce at higher elevations (fig. 6.14). Species richness increased across the majority of the state between atlases, but there were notable exceptions in north-central Pennsylvania and in Monroe and Pike counties in the east, both of which are densely forested, although there was little net change in forest cover between atlas periods (chap. 3). The distribution of Farmland birds is closely tied to that of open country, including reclaimed surface mines (fig. 6.15). Unlike the aforementioned four avifaunal guilds, the Farmland guild declined in species richness in many areas, but again, these areas do not correspond well with areas in which farmland and grassland declined between atlases (chap. 3).

Species too scarce or underrepresented to be included in the multivariate analysis of avifaunal guilds were assigned to one of the above guilds or to two additional guilds based on prior knowledge: Wetland (fig. 6.16) and Riparian (fig. 6.17) (Brooks and Croonquist 1990). Both of these habitats were underrepresented in point count surveys (chap. 5), and the species associated with them are often localized (table 6.6).

Summarization of changes in distribution between atlases, across these seven avifaunal guilds, encapsulates the broad results of this atlas. For all three forest guilds,

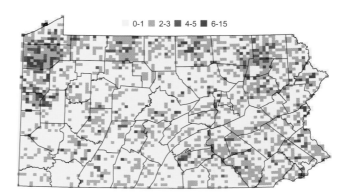
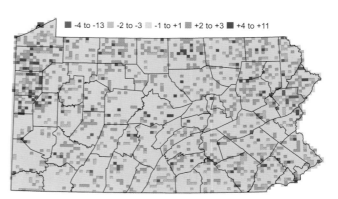

FIG. 6.16 *Species richness of the Wetland avifaunal guild (top) and change in species richness between the first and second Atlases (bottom)*

range expansions greatly outnumbered range contractions, by a ratio of more than 2:1 overall (19 contracting species, 40 expanding species; table 6.9). Of particular interest are the 14 strong expansions (>50%) among the Northern Hardwood guild. Generalist/Urban specialists have fared even better overall, with 15 expansions and only four contractions, of which three are aerial insectivores (Common Nighthawk, Purple Martin, Cliff Swallow). Riparian species have shown a variety of changes, with contractions (five species) balanced by expansions

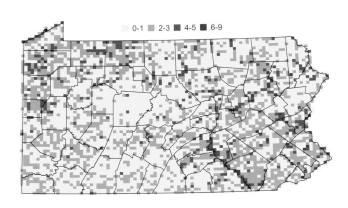

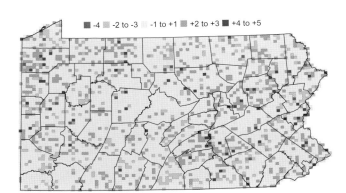

FIG. 6.17 *Species richness of the Riparian avifaunal guild (top) and change in species richness between the first and second Atlases (bottom)*

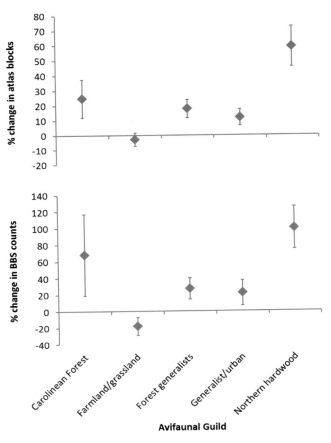

FIG. 6.18 *Mean changes in distribution (corrected change) and BBS counts between the first and second Atlases for 113 common species, across five avifaunal guilds (error bars are standard errors)*

TABLE 6.9   Changes in block occupancy* by avifaunal guild

	Avifaunal Guild						
	Carolinean	Farmland/ grassland	Forest generalists	Generalist/ urban	Northern hardwood	Riparian	Wetland
Extirpated				1	1	2	4
Strong contraction (>33%)	2	7	2	2	2	3	7
Modest contraction (13% to 33%)	2	5	4	1	6		5
No change (−13% to +15%)	2	13	26	22	4	1	3
Modest expansion (15% to 50%)	5	2	6	4	8		3
Strong expansion (>50%)	3	3	4	9	14	4	2
Colonized				2			2
Number of species	14	30	42	41	35	10	26
Ratio of increasing (>15%) to decreasing (>−13%)	2	0.42	1.67	3.75	2.44	0.8	0.44

*Changes corrected for effort for 132 species, not corrected for effort for 66 species. See Table 6.6.

TABLE 6.10 Species (sample size) for which more than 25% of records were in forests with a significant hemlock component

		Hemlock cover category			
	% H0 (none)	% H1 (single or scattered)	% H2 (up to 50% of tree cover)	% H3 (>50% of tree cover)	% H2 + H3
Blackpoll Warbler (5)	0.0	0.0	80.0	20.0	100.0
Swainson's Thrush (83)	7.2	15.7	36.1	41.0	77.1
Yellow-bellied Flycatcher (14)	14.3	14.3	35.7	35.7	71.4
Winter Wren (440)	10.5	23.6	40.5	25.5	65.9
Magnolia Warbler (447)	7.2	27.5	41.2	24.2	65.3
Blackburnian Warbler (691)	12.3	27.6	39.8	20.3	60.1
Northern Waterthrush (139)	12.9	27.3	41.0	18.7	59.7
Northern Goshawk (72)	15.3	29.2	33.3	22.2	55.6
Black-throated Green Warbler (1,368)	20.1	29.9	34.6	15.4	50.0
Canada Warbler (298)	12.8	37.6	38.3	11.4	49.7
Blue-headed Vireo (1,101)	17.6	33.7	33.3	15.3	48.7
Golden-crowned Kinglet (192)	26.6	28.6	25.0	19.8	44.8
Brown Creeper (305)	20.0	37.0	30.5	12.5	43.0
Dark-eyed Junco (614)	17.4	42.3	29.8	10.4	40.2
Red Crossbill (5)	20.0	40.0	40.0	0.0	40.0
Louisiana Waterthrush (909)	28.7	31.9	29.9	9.5	39.4
Hermit Thrush (617)	25.6	37.6	26.4	10.4	36.8
Barred Owl (386)	23.6	41.5	25.1	9.8	35.0
Black-throated Blue Warbler (694)	23.6	42.5	26.5	7.3	33.9
Acadian Flycatcher (1,578)	42.5	24.4	19.0	14.1	33.1
Red-breasted Nuthatch (295)	29.8	37.3	26.4	6.4	32.9
White-throated Sparrow (98)	32.7	36.7	28.6	2.0	30.6
Yellow-rumped Warbler (335)	36.1	37.0	21.2	5.7	26.9
Nashville Warbler (105)	34.3	40.0	22.9	2.9	25.7
Red-shouldered Hawk (597)	37.7	36.7	22.1	3.5	25.6
Northern Saw-whet Owl (293)	31.1	43.3	22.5	3.1	25.6
Ruffed Grouse (576)	31.8	42.9	22.6	2.8	25.3

(four species). In contrast, species in the Farmland guild were much more likely to have shown range contractions than expansions. Similarly, results for wetland species (many of which are scarce and localized in Pennsylvania) also skew strongly negative, with 16 of the 26 species showing modest or strong range contractions or extirpation.

Among 113 of the most common species (primarily songbirds), changes in distribution (number of atlas blocks) and abundance (BBS counts) reflect the patterns shown above. The species in the Carolinean Forest (n = 10) and Northern Hardwood (n = 20) showed marked increases in both distribution and abundance between atlas periods (fig. 6.18). The mean change in abundance of Northern Hardwood species was a remarkable 100 percent. Changes among Forest Generalist (n = 36) and Generalist/Urban (n = 29) guilds were also positive overall, for both distribution and abundance. Changes in both distribution and abundance of species in the Farmland/Grassland guild (n = 18) were, on average, both negative. It is hoped that these results will be of tremendous value in shaping bird conservation planning in Pennsylvania over coming decades, a subject discussed in depth in chapter 7.

BIRD–HEMLOCK ASSOCIATIONS

In all, nearly 130,000 records associated with 200 species were submitted to the second Atlas database with information about hemlock cover, of which 69 percent were H0 (no Hemlocks), 22 percent H1 (single or scattered), 7.5 percent H2 (hemlocks <50% of tree cover), and 1.8 percent H3 (hemlocks >50% of tree cover). More than a quarter of the records of 27 species were associated with forests with a hemlock component (H2+H3; table 6.10), most of which have been previously suggested to be typical of hemlock forests (e.g., Allen and Sheehan, Jr. 2010). Notable among this list are several of Pennsylvania's rarest songbirds are Blackpoll Warbler, Swainson's Thrush, Yellow-belled Flycatcher, Northern Waterthrush, and Red Crossbill. Although these data do not prove a dependency on hemlocks, they do suggest that hemlock is an important component of Penn's Woods for a large suite of species.

OTHER SPECIES

In addition to the 190 breeding species, another 26 species were reported to the second Atlas. Most of these observations represent individuals lingering from the spring, with no association to breeding. Others occurred during the summer months in Pennsylvania (notably shorebirds) but clearly have no breeding association, so they are not considered here. A few were summer observations of species that historically bred or for which their breeding status is uncertain. A brief history of past breeders, including officially Extirpated species, is provided in appendix A. Most intriguing among these was the White-winged Crossbill, which has never been documented breeding in Pennsylvania. With the massive influx of crossbills during the winter of 2008/09, there was great anticipation that another species might be added to Pennsylvania's breeding avifauna. The White-winged Crossbill was seen widely (in nine regions) during May 2009, with some lingering into June. Groups of birds were observed, some offering territorial singing and including pairs present over multiple weeks, but the highest breeding code reported was "C," courtship behavior. Although juveniles were observed by several atlas volunteers in Northampton County, there was insufficient evidence that they had bred locally. Therefore, without indisputable evidence of local confirmed breeding behavior, and given this species' nomadic nature, it was not deemed to be a breeding species in the state.

7 Contributions to Conservation

Douglas A. Gross, Sarah Sargent, and Catherine D. Haffner

BACKGROUND

One of the chief motivating factors for initiating Pennsylvania's second Atlas was the potential for generating data to direct bird conservation. Comparisons with first Atlas results are particularly interesting, because substantial changes in regional and continental-scale bird populations have been documented during the 20-year interval between atlases (NAWMP 2004; USFWS 2008; Sauer et al. 2011). A fresh understanding of bird distributions, fine-tuned with data at multiple scales, raised expectations that the second Atlas would guide conservation for all breeding birds into the new millennium.

This chapter provides an overview of how these results can assist identification and prioritization of bird conservation strategies in Pennsylvania. First, we summarize bird conservation planning efforts. Second, we highlight features of atlases, particularly Pennsylvania's second Atlas, that provide value to bird conservation in Pennsylvania. Third, we identify how the second Atlas complements other breeding bird data sets used for conservation planning. Finally, we discuss how these data are used in status assessment, to target land conservation, and help to shape research and management during climate change.

BIRD CONSERVATION PLANNING

In order to effectively guide bird conservation action, it is critical to understand which species are imperiled, assess possible causes of imperilment, determine knowledge gaps and data needs, and outline goals and objectives to achieve success. Bird conservation planning occurs at global, continental, regional, and state scales by various entities, and there is some consistency in "priority species" (those considered to be most in need of attention), because broader assessments often inform regional and statewide plans.

The International Union for Conservation of Nature (IUCN) and NatureServe are the primary sources for global bird conservation statuses. The IUCN Red List of Threatened Species (Red List) is regarded as the principal resource for global species rankings and trends, for what has been called the "Barometer of Life" (Stewart et al. 2010). Categories in the Red List represent a threat of extinction assessment (IUCN 2001). For the Western Hemisphere, biodiversity is cataloged by NatureServe, a non-profit organization, through partnerships with natural heritage programs and conservation data centers in the United States, Canada, Latin America, and the Caribbean (NatureServe 2006). Species are evaluated and ranked at the global, national, and state/province levels following a standardized system of conservation statuses (table 7.1). Rankings from these organizations inform management decisions, particularly when evaluated in conjunction with continental and regional priorities.

Nationally, the U.S. Fish and Wildlife Service (USFWS) establishes the federal lists of Endangered and Threatened species, under the Endangered Species Act. National bird conservation has advanced considerably since the first Atlas through Partners in Flight (PIF), an international consortium of organizations working toward the goal of "keeping common birds common" (Rich et al. 2004; Panjabi et al. 2005). This landbird conservation initiative followed the model of the North American Waterfowl Management Plan (NAWMP 2004), which set in motion waterfowl conservation and management on a grand continental scale. In conjunction with the North American Bird Conservation Initiative (NABCI), these organizations oversee conservation of all birds, working toward delivering a full spectrum of bird conservation, through government and non-profit collaboration (NABCI-US 2011). These collaboratives have adopted a network of mapped regions, established species priorities and population objectives, and promoted strategic conservation through partnerships across the continent. In addition, the National Audubon Society and the American Bird Conservancy highlighted the most critically imperiled species in the 2007 WatchList for United States Birds (Butcher et al. 2007), which was published as a guide for conservation action.

These national bird conservation plans have further focused at the regional level to identify priority species in specific geographic areas, known as Bird Conservation Regions (BCR). The majority of Pennsylvania is within the Appalachian Mountains BCR, with smaller portions of the state within three other BCRs (fig. 7.1). U.S. Migratory Bird Joint Venture partnerships (NABCI-US 2011) coordinate conservation efforts for the benefit of priority species within BCRs and have developed regional plans, such as the Upper Mississippi Valley/Great Lakes Waterbird Conservation Plan (Wires et al. 2010) and the North Atlantic Regional Shorebird Plan (Clarke and Niles 2000). Other bird conservation priority listings, such as the U.S. Fish and Wildlife Service's Birds of Conservation Concern (USFWS 2008), identify the highest priorities at the regional level, so that proactive measures can be taken to avoid costly federal protection and emergency care under the Endangered Species Act.

Wildlife conservation has advanced significantly at the state level since Congress, in 2001, created the Wildlife Conservation and Restoration Program, which led to the State Wildlife Grants program (AFWA 2011). Coordinated by the USFWS, the program apportions funds annually, based on a state's geographic extent and population, and enables states to address species research and conservation needs before species' populations decline to a critical level and require federal listing. To access these funds, USFWS required each state to develop a comprehensive wildlife conservation strategy, or Wildlife

TABLE 7.1 Standardized conservation status categories and abbreviations (in parentheses) for birds and other organisms, defined by global, national, and state organizations

International Union for Conservation of Nature (IUCN) Red List of Threatened Species [a]
    Extinct (E)
    Extinct in the Wild (EW)
    Critically endangered (CR)
    Endangered (EN)
    Vulnerable (VU)
    Near threatened (NT)
    Least Concern (LC)
    Data deficient (DD)
    Not evaluated (NE)

NatureServe [b]
    Critically imperiled (1)
    Imperiled (2)
    Vulnerable (3)
    Apparently secure (4)
    Secure (5)

Pennsylvania Biological Survey [c]
    Extinct
    Extirpated (X)
    Endangered (E)
    Threatened (T)
    Near Threatened (NT) (formerly Candidate — At Risk)
    Vulnerable (VU) (formerly Candidate — Rare)
    Data Deficient (DD) (formerly Candidate — Undetermined)

[a] See IUCN (2001) for conservation status category definitions and criteria.
[b] Adopted by Pennsylvania Natural Heritage Partnership (PNHP) at the state level.
[c] See Dunstan (1985) for conservation status category definitions.

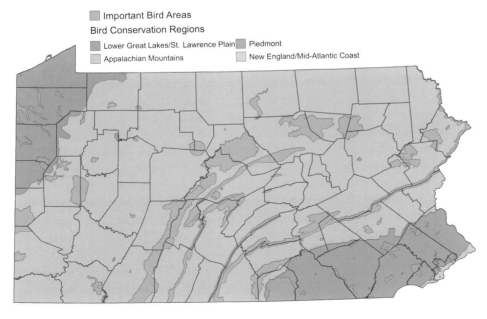

FIG. 7.1 Bird Conservation Regions (BCRs) and Important Bird Areas (IBAs) in Pennsylvania. Source: U.S. Fish and Wildlife Service (BCRs) and Audubon Pennsylvania (IBAs)

Action Plan (WAP), to serve as a blueprint for conservation action.

Pennsylvania's Wildlife Action Plan identified a subset of species of concern in Pennsylvania, or WAP priority species, and outlined habitat-based and species-specific actions for each (PGC-PFBC 2005; Williams 2010). Priority species were categorized into five conservation tiers (immediate concern, high-level concern, responsibility species, Pennsylvania-vulnerable, and maintenance concern) and were further evaluated to reflect Pennsylvania's responsibility to the broader regional, continental, or global population. Species for which Pennsylvania has a high responsibility were defined as Pennsylvania's Species of Greatest Conservation Need—for example, up to 17 percent of the global Scarlet Tanager population breeds in the state (Rosenberg and Wells 1995; PGC-PFBC 2005; table 7.2). Global, regional, and national conservation efforts described above, as well as in Pennsylvania Biological Survey (PABS) and Pennsylvania Game Commission (PGC) assessments, were used to make these determinations. The PABS maintains a list of Species of Special Concern (SSC), which pertains particularly to breeding populations (Dunstan 1985; Brauning et al. 1994; Gross 1998a) and now includes the categories similar to those in the Red List (table 7.1). Changes in species abundance and distribution identified through the second Atlas will be evaluated in the context of broader conservation statuses to update priority lists (table 7.2; see "Conservation Status Assessment," below). Declining species and their habitats will be obvious candidates for inclusion in the revised Pennsylvania WAP (due 2015), to ensure that they receive conservation funding through the State Wildlife Grant program.

ATLASES FOR CONSERVATION

One of the most important aspects of breeding bird atlases is that they are comprehensive, geographically and taxonomically, and therefore provide a good assessment of bird occurrence and distribution to guide conservation. Volunteers are encouraged to generate full lists of species within their blocks, and access to remote and off-road habitats is promoted to provide thorough coverage, which facilitates discovery of rare species. Second atlas projects have added value to conservation planning efforts, because they establish new benchmarks for the status, range, and—in Pennsylvania's case—density of several breeding bird species.

The ability of atlas projects to document range changes has been long recognized (Robbins 1990; Gibbons et al. 1993), and this is amplified with the enhanced features of this second Atlas (chap. 4). For example, Pennsylvania's second Atlas promoted strategies to assist volunteers in comprehensive data collection, including promoting protocols designed to detect night bird surveys and secretive wetland birds (O'Connell et al. 2004). In addition, point count surveys were used to gather bird abundance data. These efforts are critical because, as land use changes, many species that are now common and widely distributed may face new or advancing pressures, such as increasing energy development. The second Atlas website provided background material, such as first Atlas species accounts, maps, and block-specific species lists. These prepared volunteers for more thorough coverage, including that of rare and conservation-priority species, for which there are considerable incentives for inclusion. The coverage afforded by thousands of volunteers and targeted surveys employed during the second Atlas has generated a substantial data set, which will refine conservation actions and monitoring needs, as well as support prioritization and implementation of future bird conservation projects in the state.

These data may also contribute to the revision of state bird population objectives for priority species defined by regional bird conservation partnerships. Population estimates calculated from second Atlas point count data (chap. 6) complement Partners in Flight (PIF) bird population estimates presented in the North American Landbird Conservation Plan (Rich et al. 2004; Blancher et al. 2007), which allocates target population sizes for each state or BCR. Prioritization for monitoring, research, conservation, and management activities is then given to those geographic and political areas in which priority species are most common, are undergoing declines, or have good potential for success. Improving these population estimates with second Atlas data will enhance state contributions to regional conservation objectives.

Another added value of Pennsylvania's second Atlas was the resulting database of georeferenced points, along with descriptive comments by observers, for over 84,400 bird records. Described in chapter 4, this database contributes bird data at a finer scale than is typical of atlas projects. Focused on species of conservation interest, these observations scale down the spatial resolution of atlas results, from the coarse block level of the atlas grid to more precise locations. Confidence in the precision of these locations, based on web-mapping abilities by volunteers, remains to be determined, but the capability is provided to identify the location of a great number of observations within the individual bird's territory. These records complement the grid-based atlas format, which is

TABLE 7.2 Summary of first Atlas and second Atlas results for Pennsylvania Wildlife Action Plan bird species either confirmed or probably breeding during either atlas period and their conservation statuses at various geographical scales

Habitat group / Species	No. blocks first Atlas (1983–89)	No. blocks second Atlas (2004–09)	% change	Conservation status Global[a]	Continental[b]	Regional[c]	State[d]
Agricultural lands and grasslands							
Northern Bobwhite	714	231	−68	G5, NT↓			PBS-NT, PNHP-S1
Northern Harrier	334	190	−43	G5, LC		NE, PIF	PBS-NT*, PNHP-S3
Upland Sandpiper	54	23	−57	G5, LC		FWS, NE, SB	PA-T*, PNHP-S1
Barn Owl	251	117	−53	G5, LC			PBS-NT, PNHP-S3
Short-eared Owl	6	5	−17	G5, LC	WL, PIF	FWS, NE	PA-E, PNHP-S1
Red-headed Woodpecker	698	374	−46	G5, NT↓	WL, PIF	NE, PIF	(PBS-VU)*
Loggerhead Shrike	6	2	−67	G4, LC		FWS, NE, PIF	PA-E, PNHP-S1
Sedge Wren	13	20	54	G5, LC		FWS, NE, PIF	PA-E, PNHP-S1
Grasshopper Sparrow	1,629	1,674	3	G5, LC		PIF	
Henslow's Sparrow	364	229	−37	G4, NT↓	WL, PIF	FWS, NE, PIF	(PBS-VU)*
Dickcissel	45	45	0	G5, LC			PA-E, PNHP-S2
Bobolink	1,527	1,739	14	G5, LC		PIF, PIF-S	
Eastern Meadowlark	3,268	2,896	−11	G5, LC		PIF	
Deciduous and mixed forest							
Sharp-shinned Hawk	1,051	840	−20	G5, LC			
Red-shouldered Hawk	750	1,167	55	G5, LC			
Broad-winged Hawk	2,063	1,725	−16	G5, LC		PIF-S	
Black-billed Cuckoo	1,613	1,703	5	G5, LC		PIF, PIF-S	
Eastern Whip-poor-will	862	496	−42	G5, LC		PIF, PIF-S	(PBS-VU)*
Acadian Flycatcher	1,766	2,454	39	G5, LC		PIF, PIF-S	
Yellow-throated Vireo	1,218	1,760	44	G5, LC		PIF, PIF-S	
Wood Thrush	4,493	4,603	2	G5, LC	WL, PIF	PIF, PIF-S	
Worm-eating Warbler	735	1,086	48	G5, LC	PIF	PIF, PIF-S	
Kentucky Warbler	929	662	−29	G5, LC	WL, PIF	FWS, PIF, PIF-S	
Cerulean Warbler	836	776	−7	G4, LC	WL, PIF	FWS, NE, PIF, PIF-S	
Black-throated Green Warbler	1,803	2,675	48	G5, LC			
Summer Tanager	47	4	−91	G5, LC		PIF	PBS-VU, PNHP-S3
Scarlet Tanager	4,276	4,543	6	G5, LC		PIF-S	
Coniferous and northern forests							
Northern Goshawk	120	86	−28	G5, LC			PBS-VU, PNHP-S2
Long-eared Owl	18	14	−22	G5, LC		NE	PBS-DD*, PNHP-S2
Olive-sided Flycatcher	6	0	−∞	G4, NT↓	WL, PIF	FWS, PIF	PA-X
Yellow-bellied Flycatcher	13	13	0	G5, LC			PA-E, PNHP-S1
Blue-headed Vireo	1,474	2,340	59	G5, LC			
Winter Wren	378	799	111	G5, LC			
Swainson's Thrush	43	98	128	G5, LC			PBS-VU, PNHP-S2
Blackburnian Warbler	917	1,580	72	G5, LC		PIF	
Blackpoll Warbler	2	3	50	G5, LC			PA-E, PNHP-S1
Black-throated Blue Warbler	741	1,278	72	G5, LC			
Canada Warbler	634	691	9	G5, LC	WL, PIF	FWS, NE, PIF	
Pine Siskin	71	93	31	G5, LC			
Ledges, bare soil, and artificial structures							
Peregrine Falcon	6	31	417	G4, LC		FWS, PIF	PA-E, PNHP-S1
Common Nighthawk	754	219	−71	G5, LC			(PA-NT)*
Chimney Swift	3,416	3,334	−2	G5, NT↓		PIF, PIF-S	
Streams, islands, and bottomlands							
Great Blue Heron	2,279	2,908	27	G5, LC		WB	
Great Egret	36	117	225	G5, LC		WB	PA-E, PNHP-S1
Black-crowned Night-Heron	152	77	−49	G5, LC		WB	PA-E, PNHP-S2
Yellow-crowned Night-Heron	21	9	−57	G5, LC			PA-E, PNHP-S1
Bald Eagle	51	535	949	G5, LC		FWS	PA-T, PNHP-S2
Bank Swallow	521	337	−36	G5, LC			(PBS-VU)*
Louisiana Waterthrush	1,386	1,784	29	G5, LC		NE, PIF, PIF-S	
Successional, shrubland, and young forest							
American Woodcock	1,469	1,370	−7	G5, LC		SB	
Alder Flycatcher	332	868	161	G5, LC			
Willow Flycatcher	1,654	2,026	23	G5, LC	WL, PIF		

(continued)

TABLE 7.2 (continued)

Species / Habitat group	No. blocks first Atlas (1983–89)	No. blocks second Atlas (2004–09)	% change	Global[a]	Continental[b]	Regional[c]	State[d]
Brown Thrasher	3,073	3,331	8	G5, LC		PIF, PIF-S	
Golden-winged Warbler	615	240	−61	G4, NT↓	WL, PIF	FWS, NE, PIF	(PBS-NT)*
Blue-winged Warbler	1,403	1,345	−4	G5, LC	WL, PIF	FWS, PIF, PIF-S	
Prairie Warbler	1,035	1,117	8	G5, LC	WL, PIF	FWS, PIF, PIF-S	
Yellow-breasted Chat	1,242	805	−35	G5, LC			
Wetlands, waterbodies, and shores							
American Black Duck	297	111	−63	G5, LC	NAWMP		
Green-winged Teal	23	13	−43	G5, LC			PBS-VU, PNHP-S1S2
Pied-billed Grebe	86	64	−26	G5, LC		FWS, NE, WB	PBS-VU, PNHP-S3
American Bittern	53	31	−42	G4, LC		FWS, NE, WB	PA-E, PNHP-S1
Least Bittern	31	28	−10	G5, LC		WB	PA-E, PNHP-S1
Osprey	142	269	89	G5, LC			PA-T, PNHP-S2
King Rail	5	5	0	G4, LC	WL	WB	PA-E, PNHP-S1
Virginia Rail	120	135	28	G5, LC		WB	
Sora	88	100	14	G5, LC		WB	PBS-VU, PNHP-S3
Common Gallinule	76	48	−37	G5, LC		WB	
American Coot	25	18	−28	G5, LC		WB	PBS-VU, PNHP-S3
Piping Plover	0	0	0	G3, NT↑	WL	SB	PA-X, PNHP-SX
Wilson's Snipe	38	32	−16	G5, LC			PBS-VU, PNHP-S3
Black Tern	11	3	−73	G4, LC		NE, WB	PA-E, PNHP-S1
Common Tern	2	0	−100	G5, LC		NE, WB	PA-E, PNHP-SX
Marsh Wren	77	53	−22	G5, LC			PBS-VU*, PNHP-S2
Prothonotary Warbler	43	47	9	G5, LC	WL, PIF		PBS-VU, PNHP-S2

Bold: Pennsylvania Wildlife Action Plan Species of Greatest Conservation Need (responsibility + imperilment) (PGC-PFBC 2005).
Italics: Pennsylvania Wildlife Action Plan High and Immediate Concern species (PGC-PFBC 2005).
*Species reassessment due to second Atlas data. Pending status category changes are included in parentheses.
[a]**Global Concern**: NatureServe (G; NatureServe Explorer 2011) and IUCN Red List of Threatened Species (IUCN 2011).
[b]**Continental Concern**: Audubon WatchList (WL; Butcher et al. 2007); Partners in Flight Species Assessment 2005 (PIF; www.rmbo.org/pif/scores/scores.html); North American Waterfowl Management Plan (NAWMP 2004).
[c]**Regional Concern**: U.S.Fish and Wildlife Service Birds of Conservation Concern (FWS; USFWS 2008); Northeast Endangered Species and Wildlife Diversity Technical Committee (NE; Therres 1999); Upper Mississippi Valley/Great Lakes (Wires et al. 2010) and Southeast (Hunter et al. 2006) Waterbird Conservation Plans (WB); Upper Mississippi Valley/Great Lakes (de Szalay et al. 2000) and Northern Atlantic (Clark and Niles 2000) Shorebird Conservation Plans (SB), PIF 2005 Species Assessment Regional Concern (PIF) or Regional Stewardship (PIF-S) (www.rmbo.org/pif/scores/scores.html).
[d]**State Concern**: See Table 7.1 for category definitions. Pennsylvania Code (Threatened (PA-T), Endangered (PA-E); 58 Pa. Code § 133.21); Pennsylvania Biological Survey/Ornithological Technical Committee 2005 status assessment (PBS; unpublished); Pennsylvania Natural Heritage Program database (PNHP; www.naturalheritage.state.pa.us/Species.aspx).

essentially species-presence information at the block scale. Combined with the stratified sample of point count locations, and producing a database of at least 585,000 species locations, the second Atlas is a data treasure trove with multiple scales that will take years to mine and potential uses that have not yet been identified.

Beyond the scientific importance of atlas data to conservation, however, is the intrinsic value of volunteers connecting with birds. One of the most powerful aspects of this effort was the intense involvement of the state's birders, who personally experienced many hours of observing birds in their habitats. Like the first Atlas, the second Atlas provided an educational experience for another generation of bird observers and recruited many new bird conservationists.

With new resources and new opportunities, there is great potential for involving this informed and interested public in bird monitoring and stewardship. The first Atlas directly led to the founding of the Pennsylvania Society for Ornithology (PSO) and the inception of the state's Important Bird Areas (IBA) project (fig. 7.1), managed by Audubon Pennsylvania (Audubon Pennsylvania 2011). It also updated the listing of the state's breeding birds of Special Concern and encouraged the establishment of projects for monitoring and conserving those species and their habitats (Brauning et al. 1994). The first Atlas and subsequent projects, such as the PSO Special Areas Project and Audubon Pennsylvania's IBA monitoring, have built up a more knowledgeable volunteer base for the second Atlas; this base has continued to build momentum in enthusiasm and experience. The second Atlas will certainly lead to even more citizen-science projects and even greater advocacy for wildlife and the ecosystems that sustain that resource.

BIRD MONITORING IN PENNSYLVANIA

Breeding bird atlases are intermittent projects of great intensity, and they add valuable elements to the framework of ongoing bird monitoring projects (NABCI 2007; Lambert et al. 2009). While the second Atlas provided statewide data and tens of thousands of spatially explicit data points, many other projects and data sets fill gaps in our knowledge of the state's breeding birds. Principal among these is the Breeding Bird Survey (BBS), coordinated by the U.S. Geological Survey (USGS), which has provided trend data for a majority of North American songbirds for over 40 years, particularly those species that vocalize or are easily seen during a brief roadside survey (Bystrak 1980; Robbins et al. 1989a; Sauer et al. 1994). Because the long, consistent implementation of this survey provides the primary source of trend data for songbirds, extensive use of BBS results has been included in this volume where appropriate. The accessibility of the results and analysis on the USGS website make the BBS particularly valuable as an assessment tool.

As important as it is, the BBS has limitations, however, which are important to consider when using the data to evaluate trends or effects of various environmental changes or stressors on bird abundance and distribution. As a roadside survey, the BBS provides deficient coverage for off-road habitats, including wetlands, remote forest blocks, and scrub barrens (O'Connor et al. 2000). Therefore, it is ineffective in revealing trends of most waterfowl, colonial waterbirds (herons), diurnal raptors, and wetland species—many of which are of conservation concern. It also is less likely to provide adequate coverage of nocturnal and crepuscular species (owls, nightjars, and woodcocks), early-nesters, species with irruptive or nomadic habits, and those that occupy patchy habitats of any kind. BBS results may also be biased toward species associated with disturbed habitats (e.g., American Robin) and underrepresent those found in less anthropogenic habitats (Robbins et al. 1989a; Keller and Fuller 1995; Keller and Scallan 1999). With its focus on long-term trends, the BBS does not typically generate records of rare species suitable for site-specific conservation. The virtues and shortcomings of the BBS monitoring method are discussed by others (Droege 1990; Geissler and Sauer 1990; James et al. 1992; Butcher et al. 1993; Keller and Scallan 1999; O'Connor et al. 2000).

The second Atlas data set uncovers many species not adequately detected by other standardized surveys and, therefore, complements surveys like BBS and Audubon's Christmas Bird Count. The distinctions between data collected by bird atlases and by the BBS are especially relevant to conservation, because often species that fall into conservation categories are too rare or poorly detected to be adequately detected by BBS (Brauning et al. 1994; Gross 1998a). Only 40 percent of Pennsylvania's breeding bird species are surveyed sufficiently by the BBS to assess trends; over 50 species covered by the second Atlas are not included in the BBS summary for the state, due to lack of data (Sauer et al. 2011). Many of the species inhabiting remote locations have unique conservation challenges (Brauning et al. 1994; Gross 1998b, 2010a). Therefore, even occurrence data provided by atlas blocks are valuable for conservation concern species.

The internet-based eBird program is growing by leaps and bounds (CLO 2011). An entirely new approach to documenting bird records developed by Cornell Lab of Ornithology and the National Audubon Society in 2002, this data collection tool provides a platform to collect and continuously update information on bird observations, entered by users. It is emerging as a major source of bird data, and tens of thousands of bird observations are entered by volunteers every month in Pennsylvania (CLO 2011). eBird also incorporates historic data sets and, through the Avian Knowledge Network (AKN), is making vast amounts of bird data more readily available (Iliff et al. 2009). Like atlas data, the visual aspects of eBird and other AKN data can reveal biologically-based geospatial patterns that inspire new hypotheses about bird distributions and trends (Sullivan et al. 2009).

An advantage of eBird is that it enables data collection in all seasons of the year, so it is effective for documenting passage migrants and wintering species as well as breeding species. Although eBird provides another large data set to explore when researching conservation questions, it is not as helpful for providing insight on long-term trends and breeding status questions at this point. Coverage is potentially everywhere, but it is uneven in practice, and breeding status reporting was not included until 2011. Thus, the ability to compare eBird observations with first or second Atlas records is limited. It does, however, have the promise to build greater coverage and, now that breeding information is included, will become more useful in future years, with past atlases providing benchmarks and training volunteers.

Bird occurrence data, collected continuously by Pennsylvania birders, are also referenced when evaluating species distributions, but these reports are not sufficient to effectively evaluate trends. County reports in the quarterly journal *Pennsylvania Birds* provide bird occurrence data, but these tend to emphasize unusual sightings, especially of passage migrants, and are not presented at the fine scale offered by breeding bird atlases. Moreover, some counties are not well covered by resident birders, resulting in significant data gaps. Anecdotal reports on the PABIRDS listserv do not provide the detail necessary to evaluate even the more common species, since "listservs" reports do not quantify effort. They also underrepresent rare nesting species; the locations of and details about rare, listed, or breeding species with a reputation for sensitivity to human disturbance are usually not reported in these forums, as part of the birding code of ethics (ABA 2003). These outlets do, however, provide a diversity of information sources, increase interest in bird distribution and recruit new participants into projects.

Despite certain inadequacies of atlas data that were recognized early in the history of the method (Robbins et al. 1989a), breeding bird atlases are still regarded as the most comprehensive assessment of breeding species, due to the range of birds detected and diversity of habitats covered. The breadth and detail of second Atlas data offer significant insights for conservation planning efforts, including conservation status revisions, defining focal conservation areas, and evaluating effects of emerging issues, such as climate change.

CONSERVATION STATUS ASSESSMENT

The second Atlas provided a significant milestone for evaluating changes in populations of priority species. While other surveys are helpful, Pennsylvania's second Atlas was, like the first Atlas, a cornerstone for conservation prioritization decisions. The Ornithological Technical Committee (OTC) of the Pennsylvania Biological Survey serves as the PGC's advisory committee, with a particular mission to recommend updates to the state list of Endangered and Threatened species (Brauning et al. 1994). The OTC delayed recommending revisions to bird conservation statuses during the second Atlas in order to incorporate these data into review and analysis of status assessment. In Pennsylvania, regulatory listing by the PGC triggers several actions that increase their potential for research, monitoring, protection, and possible conservation initiatives (Brauning and Hassinger 2010). Once listed, species receive protection through the environmental review process and are prioritized for research and other conservation actions. Activities requiring state or federal permits are reviewed through the Pennsylvania Natural Diversity Inventory (PNDI; PNHP 2011). The PNDI database and subsequent environmental review procedures allow protections of nesting sites through spatial and seasonal restrictions specific to each listed species. Hundreds of records of state-listed species generated by the second Atlas were incorporated into PNDI for review of regulatory agencies, their advisors, and species researchers.

Although some experts may question the wisdom of states giving priority to regionally rare species that are not globally threatened (Wells et al. 2010), states do take responsibility for native species within their jurisdictions that face threats to their continued population viability (Brauning and Hassinger 2010). From the state's perspective, such rare species deserve protection as resources and members of the state's biodiversity that future generations will enjoy, and they often represent "umbrella" or "surrogate" species for others with less appeal (Hunter and Hutchinson 1994; Beissinger et al. 1996). These species often reflect diminished ecosystems; for example, over 60 percent of the state's Endangered and Threatened birds are associated with wetlands, a very limited and reduced habitat in the state (Brooks and Croonquist 1990; Gross 1998a; Gross and Haffner 2010). Many of Pennsylvania's Endangered and Threatened species also are listed in neighboring states and are undergoing regional declines due to habitat losses and other causes (see Noss et al. 1995; Steele et al. 2010a).

The second Atlas confirmed that status changes made since the first Atlas were justified and identified more changes that were warranted. Since the first Atlas, the Bald Eagle was upgraded to Threatened from Endangered, the American Bittern and Least Bittern were downgraded from Threatened to Endangered, and the Black-crowned Night-Heron was downgraded from Secure to Endangered (Gross 1998a). Changes in the regulatory listing of several species as a result of second Atlas findings are, at this writing, under consideration. For example, the Upland Sandpiper is a very rare species with restricted range in Pennsylvania and considered Threatened for many years, but the results of the second Atlas lead the OTC to now recommend it be listed as Endangered. Other changes are under consideration as second Atlas results are analyzed.

Changes in the populations of species that were too common and widespread to be considered legally Threatened or Endangered have been recognized by the PABS, resulting in a review of their lists and some changes in the SSC lists (either Near Threatened or Vulnerable) due to their decline and rarity (table 7.1). Labeling these species gives them more priority for monitoring, management, and conservation. The Henslow's Sparrow, a WatchList species, has declined in range size since the first Atlas (table 7.2); however, because its population remains in the thousands of pairs, it is listed as Vulnerable in Pennsylvania, not a higher risk category, despite its continental status of conservation concern. The second

Atlas demonstrated a retraction in its range, making it more vulnerable to extirpation and subject to listing at that level of concern.

Status changes were consistent with continental concerns for Golden-winged Warbler (Confer et al. 2011) and Purple Martin. The addition of Northern Waterthrush to the SSC list signals an additional concern for high-elevation wetlands (Gross and Haffner 2010). The second Atlas results prompted conservation concern for four aerial insectivores, a group of growing concern on a continental scale (McCracken 2008; Nebel et al. 2010): Common Nighthawk, Eastern Whip-poor-will, Bank Swallow, and Purple Martin (table 7.2; Tautin et al. 2009). The second Atlas results also have raised concern for declining populations of two species, Red-headed Woodpecker and Marsh Wren, for which there has long been concern (Gill 1985; Brauning et al. 1994). Finally, the second Atlas showed declines, since the first Atlas, of several species that are difficult to survey and, therefore, problematic for confidently tracking trends and ascribing cause for decreases in range and population.

GEOGRAPHICAL CONSIDERATIONS FOR BIRD CONSERVATION IN PENNSYLVANIA

The population estimates and density maps from second Atlas point count data enable us to focus on those parts of Pennsylvania that support the greatest densities of high-priority conservation species. Analyses could suggest management and conservation actions appropriate in various regions, according to these population densities and apparent trends in populations, as indicated by atlas block coverage and BBS data. Changes in populations could help managers find causes for declines from land use changes, weather events, tree pests and diseases, declines in environmental quality, or human-caused stresses. The increased availability of forest cover, water quality, land use, human population, road density, farming practices, and planned development information can make assessment of bird responses to these variables more accessible, as well as present opportunities to better understand and compensate for observed effects.

Many species of conservation interest are still fairly widespread. Atlases allow us to delineate areas that support strongholds of declining species, or potentially discover new range expansions of others, to determine whether conservation efforts are being directed toward the appropriate places. Spatial information in an atlas can also be used in conservation work from a local perspective. For instance, land managers working at a local scale can refer to an atlas to determine which species occur in or near their area and therefore may merit attention at that site. Forestry professionals, park managers, or state game land managers concerned with maintaining biodiversity within their management areas can turn to the second Atlas to determine whether they are within the range of a particular species or a suite of species.

An example of a declining species for which the second Atlas has been very important in determining the remaining population strongholds is the Golden-winged Warbler. As discussed in the species account, Pennsylvania is part of the global decline of this species (Confer et al. 2011) as its range contracts to higher elevations, larger forest blocks, and higher latitudes. The detailed inventory work that went into this atlas has provided a valuable basis for designating focal areas for the conservation of this species in Pennsylvania (Bakermans et al. 2011). It has future uses in designating focal areas for other conservation priority species such as Cerulean Warbler, Wood Thrush, Louisiana Waterthrush, and other species.

The extensive geographic information afforded through the second Atlas design (chap. 4) has contributed to critical analysis for the conservation of forest interior bird species. As discussed in chapter 3, a majority of Pennsylvania is forested, but ongoing erosion of large forest blocks is likely to continue as exurban development and energy extraction industries encroach upon our most heavily forested areas. As we identify areas with the highest density of representative species, additional conservation efforts can be directed to those locations.

One way to highlight the importance of the remaining large forest blocks for these species would be to designate them as Important Bird Areas (IBAs). Coordinated by National Audubon, the IBA program is a major bird conservation strategy that aims to identify and conserve sites that are vital for supporting bird populations. It uses a set of criteria that a site must meet to qualify as globally or regionally important (see Crossley 1999). Pennsylvania and New York were the first two states to develop IBA programs, in the mid-1990s. Although some forest blocks were designated during the first selection of IBAs, the selection committee lacked sufficient data sets and spatial analysis tools that are now available for designating forest habitat. Abundance data collected in the second Atlas will be particularly helpful here. Currently, there are relatively few IBAs representing high-quality examples of eastern forest bird communities, although these are the most characteristic species of our most widespread habitat. As threats continue to mount in the remaining large forest blocks in Pennsylvania, use of atlas data, especially the density maps, to delineate new forest block IBAs will be tremendously valuable.

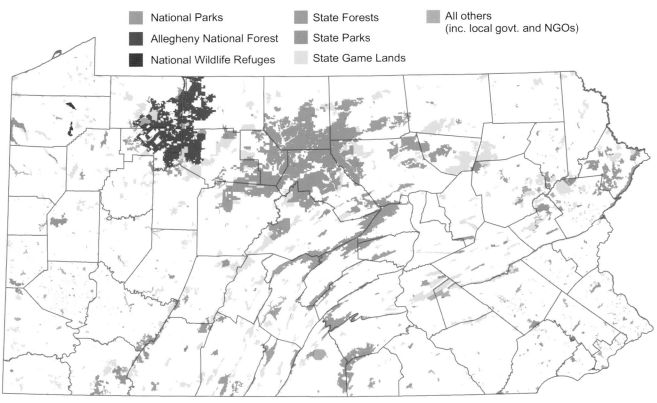

FIG. 7.2 *Protected lands in Pennsylvania. Source: USGS Protected Areas Database (Version 1, 2009)*

The second Atlas also provides valuable information about the range and concentration areas of many species of continental conservation concern. In a concerted effort to provide better information about the importance of any given area to a species' population, estimates of bird populations, based on previously published maps and atlas data, have been made for some species (Wells 1995; Rosenberg and Wells 1995). From these estimates, the percentage of the species' population can be calculated for each geographical region. Early population estimates in the northeastern states gave Pennsylvania high responsibility for neotropical migrant species such as Scarlet Tanager (17%), Worm-eating Warbler (9.7%), Gray Catbird (9.4%), Henslow's Sparrow (8.9%), Golden-winged Warbler (8.7%), and Wood Thrush (8.5%; Rosenberg and Wells 1995). This information is valuable for prioritizing conservation activities regionally, but also as an educational message to the public about the value of maintaining healthy populations of species where they are already common.

Much of the state's core forests and larger wetlands that support important high-priority conservation species also are on public lands (Crossley 1999; Goodrich et al. 2002; Gross and Haffner 2010; fig. 7.2). Almost all of the state's old-growth forest is on public land (Goodrich et al. 2002; Jenkins et al. 2004). These properties are protected, but they also need management to maintain quality habitat for priority species. For example, since some wetland-dependent species use a combination of open water and emergent vegetation, and invasive plant species are detrimental to native wetland plants, management may be needed to maintain habitat for emergent wetland species. Atlas data are helpful to gauge the health of these populations and their habitat. Creating disturbances, to maintain early successional habitat and young forest, is necessary for the continued existence of shrubland species in the Appalachians (table 7.2; Hunter et al. 2001), so it is noteworthy that the second Atlas revealed that 32 percent of the Golden-winged Warbler breeding records were located on state property. The agencies that own this land consequently have responsibility for the continued existence of many species through their own management and stewardship.

Atlas data not only demonstrate the importance of these properties to bird populations but also inform future management and protection. The PGC owns over 566,000 ha (1.4 million ac) in 65 counties and manages these state game lands for the benefit of wildlife. Management plans for these game lands will benefit from information provided by the second Atlas and other sources of bird population data. Through its Bureau of Forestry and Bureau of State Parks, the Department of Conservation and Natural Resources owns over 850,000 ha (2.1 million ac) in the state. The State Parks were particularly well covered by Atlas data because of their accessibility. The State Forests are critical parts of the state's large forest blocks that support substantial populations of forest interior bird species, for which Pennsylvania is so important. The same is true of the Allegheny National Forest, owned by the U.S. Department of Agriculture's Forest Service and serving as the core area of the largest forest block in the northwestern counties. The National Park Service's holdings include Gettysburg National Military Park, Delaware Water Gap National Recreational Area, and the Upper Delaware Scenic and Recreational River. Its Delaware River lands are important areas for Bald Eagles in all seasons and for other riparian species that are not as charismatic. The state's two National Wildlife Refuges, John Heinz NWR and Erie NWR, may not be large (together about 4,250 ha, 10,500 ac), but both were selected as IBAs because they provide wetlands and other important habitats that support rare and imperiled bird populations.

Increasingly, bird species assemblages and guilds are employed as indicators of ecological integrity (Bradford et al. 1998; Canterbury et al. 2000; O'Connell et al. 2000). Species that are at risk, even at the state level, can serve as good measures to ecosystem threats (Hunter and Hutchinson 1994; Beissinger et al. 1996). Acid precipitation has had a negative effect on reproduction, range, and populations of Wood Thrush, Ovenbird, and probably other songbirds, because they reduce soil calcium availability and subsequent food sources (Hames et al. 2002; Pabian and Brittingham 2007, 2011). Acid precipitation and acid mine drainage also affect some bird densities along Pennsylvania's streams, especially that of Louisiana Waterthrush (Mulvihill et al. 2008). Thus, Louisiana Waterthrush may serve as an indicator of headwater stream quality in highlands dominated by public lands, where so many streams originate (Latta and Mulvihill 2010). Forest fragmentation from energy development is a potential threat to our forest bird populations, even those on public lands, because of the large-scale effects these industries have on landscapes, the lack of ownership of mineral rights on many public lands, and other reasons (Brittingham and Goodrich 2010; Johnson 2010). Atlas data, especially the density data on and near our public lands, provide a measure of how development and other stressors may have changed bird populations and ranges in Pennsylvania and also provide benchmarks from which to measure future changes.

Voluntary land protection also can provide valuable bird habitat. The U.S. Department of Agriculture's Conservation Reserve Program (CRP), and the Conservation Reserve Enhancement Program (CREP), administered by the Farm Service Agency and Natural Resource Conservation Service, have increased the grassland component of farmlands (Wilson 2010). By 2007, cumulatively over 100,000 ha (247,000 ac) of private land have been enrolled in CREP, more than 60,700 ha (150,000 ac) of which were either cool- or warm-season grasses (USDA-FSA 2009). Nearly 9,700 ha (24,000 ac) were enrolled as riparian buffer, and about 2,400 ha (6,000 ac) between early successional and filter strips. The second Atlas will help to document the benefits of this and other voluntary assistance programs.

Future conservation design strategies will need a new array of data (Thogmartin et al. 2009). As more stressors enter each landscape, we need to gain more detail, with some geographic precision, about our best options. Even the deepest forests of the most rural counties are not immune to these stressors. Decision-support tools will be necessary to determine the best options (Alexander et al. 2009). The second Atlas results include some pleasant surprises, including the persistence of some forest birds near urban centers, the presence of raptors in some suburbs, and the range expansion of many forest species. The kind of data generated by the second Atlas will certainly support future decision tools needed in Pennsylvania and the region (NABCI 2007). It also will lead managers toward new tools to address those challenges that the second Atlas, or any other bird survey project, has not yet met. Atlas data also have the potential to elucidate research questions such as: Where is it most appropriate to designate areas for early succession and young forests or very mature forests? Which wetlands need modification or amelioration to continue to support our rarest wetland species? What government programs have been most effective at creating habitats for grassland and thicket species that show dramatic regional decline? Which species are most vulnerable to energy extraction or road infrastructure expansion? Which watersheds support species most sensitive to loss of forest cover or water quality and which do not? Are there areas in which tree diseases and pests threaten important bird populations and merit attention and, perhaps, mitigation? The second Atlas provides some of the data needed to answer these questions and will be a valuable tool for the development of sampling strategies for future surveys.

CLIMATE CHANGE AND BIRD CONSERVATION IN PENNSYLVANIA

According to the Intergovernmental Panel on Climate Change (IPCC), the decadal temperature trend on land in the northern hemisphere has increased to over 0.3°C (0.5°F) per decade between 1979 and 2005, compared with less than 0.1°C (0.2°F) per decade when averaged over the 1901 to 2005 time period (Solomon et al. 2007). Climatic changes documented in Pennsylvania between 1970 and 2010 include longer ice-free periods of water bodies, warmer average lows in January, fewer snowy days in winter, more days with heat index over 38°C (100°F) in summer, and longer growing seasons (UCS 2008). Root and Goldsmith (2010) noted that shifts in geographic range are one of the most common ways in which species adapt to changing climatic conditions, often accompanied by shifts in population density within the range. The first and second Atlases represent snapshots of breeding bird population ranges; observed changes between these periods generate questions regarding potential climate change effects on Pennsylvania species' distribution and abundance.

Actual average annual temperature change occurring in the 20-year interval between breeding bird atlases has been negligible (chap. 2; Solomon et al. 2007). Nonetheless, biologically significant changes in climate can occur with only small shifts in average annual temperature (e.g., see Strode 2003), since the interaction of temperature with precipitation and the seasonal patterns of these two factors are what produce life zones and corresponding vegetation types.

Bird species may be most affected with increased temperatures via shifts in Pennsylvania forests' tree species composition, as conditions shift to be less favorable for more northern forest tree species (maples, birches, beech, hemlock) while becoming more favorable for what we now consider to be southern forest species (hickories, southern oaks, pines; Shortle et al. 2009: appendix 6). As tree species composition changes, food and other resources will change as well. For example, large-scale climate patterns have been shown to directly influence breeding bird populations by causing changes in the abundance of moth and butterfly larvae, a key food resource for nesting birds (Sillett et al. 2000; Jones et al. 2003; Anders and Post 2006).

Climate change effects on breeding populations are not limited to changes to vegetation and other resources there. Climate-related changes on wintering grounds also may play a part in the migratory ecology and carryover effects on species that migrate north to the United States. For example, changes in rainfall in its Caribbean wintering grounds may be a factor for the American Redstart's timing of departure from its winter territory and its arrival date and subsequent success on its nesting ground (Studds and Marra 2011; Wilson et al. 2011). Therefore, if climate changes have had subtle effects on breeding populations of migratory birds, the atlases may help reveal them, because of their thorough coverage as well as their quantitative and geographical components.

Many breeding bird ranges in Pennsylvania have shifted noticeably during the 20-year period between atlases (chap. 6). Attributing these changes in distribution unequivocally to climate change is problematic, however. Unfortunately, large-scale phenomena are not amenable to experimentation; thus, we can only infer causation from observed patterns.

Because the rate of climate change is predicted to accelerate over the next century (IPCC 2007), it makes sense to plan for it using an integrated approach, including determining which traits will cause some species to be more vulnerable than others to the negative effects of climate change (Dawson et al. 2011). One trait thought to cause more vulnerability is long-distance migration (Both et al. 2010). A holistic conservation approach of full life-cycle stewardship is a necessary reaction to the international scope of challenges facing long-distance migratory bird species (Greenberg and Marra 2005; Berlanga et al. 2010; Wilson et al. 2011). The conservation of birds into the future will require continued monitoring and documentation of where birds are breeding; therefore, repeated atlases at regular intervals, as well as other monitoring projects, are warranted.

8 Interpreting Species Accounts

Daniel W. Brauning and Andrew M. Wilson

The following two-page accounts present and discuss results for each species in which confirmed breeding was documented during the second Pennsylvania breeding bird atlas, from 2004 through 2009. Species for which at least probable breeding evidence was reported in the second Atlas, and a full species account was provided in the first Atlas (Brauning 1992a), are also included among these accounts. Birds with historic confirmed breeding activity are included in appendix A, where their nesting history in Pennsylvania is briefly summarized.

Each account consists of the following elements: Interpretation of second Atlas results within an historical context, a distribution map of second Atlas results at the "block" scale, a map showing the change in distribution (by block) from that reported in the first Atlas, and a tabulation of the number of blocks reported in three breeding categories along with that tabulation and the percent change from the first Atlas. Map colors were chosen in consultation with the web tool Colorbrewer 2.0 (http://colorbrewer2.org/). Illustrating each species is a color photograph, in most cases taken in Pennsylvania and depicting a breeding behavior. A chart showing the population trend (based on Breeding Bird Survey results) and a density map derived from point count data, along with the statewide population estimate, are included for species with adequate data.

NOMENCLATURE

Common and scientific names for each species identified in each account follow the American Ornithologists' Union (AOU) *Check-list of North American Birds*, 7th edition (AOU 1998) and all supplements through 2011. Accounts (and other tables) list species in the taxonomic sequence defined in the 52nd supplement of the AOU checklist (Chesser et al. 2011). Full names of birds used in this volume follow the AOU checklist standards (e.g., capitalization standards); abbreviated names, such as rough-winged swallow (instead of Northern Rough-winged Swallow), are not capitalized. To maintain consistency, complete common names of all other taxa used herein follow the capitalization pattern adopted for birds (e.g., White-tailed Deer, Hemlock Woolly Adelgid). Scientific names for these taxa used in this volume are provided in appendix B.

ACCOUNTS

Species accounts were developed by 52 authors. Provided to these authors were summaries of results of second Atlas data, including drafts of the distribution, change, and abundance maps; tabular summaries of breeding records with greater detail than provided herein; and the precursors of background habitat analysis provided in the introductory chapters and appendices were made available. In addition, guidance was provided to authors on content, length, and style expected for the accounts. The resulting accounts went through multiple editorial steps to maintain consistency in content and style, although an effort was made to retain the individual styles of the authors. Readers are encouraged to refer to additional species data in appendices C, D and F.

Following is a description of the various components provided in most or all species accounts.

Distribution map 2004 to 2009

The distribution map provided for each species is the primary result of the second Atlas, reflecting all observations at the scale of the atlas block (see chap. 4 for background). Only possible, probable, and confirmed breeding records are used in these maps. The observed

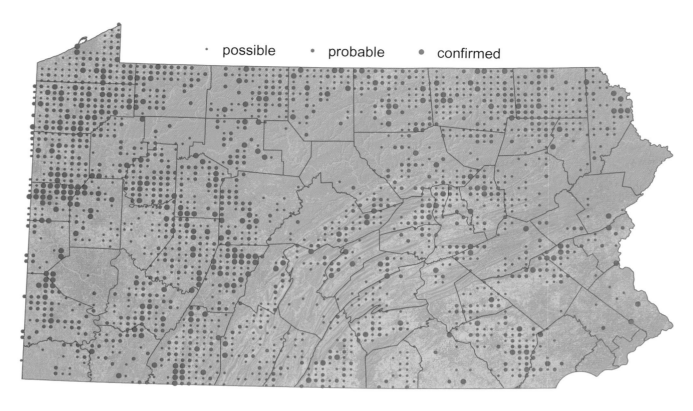

category, not typically considered breeding evidence, is not mapped or otherwise included in atlas summaries, except for the vultures and Great Blue Heron, in which all observed records were lumped with possible breeding evidence to reflect that lowest breeding evidence and to treat uniformly the breeding-season observations of these widely wandering species. This map presents the highest breeding code reported to the project for each block, summarized in the three breeding categories represented by graduated circles. The distribution is overlaid onto a map displaying muted land cover and topography, also presented in more detail in chapter 3. Land cover types are shown as colors in broad categories: urban (gray), agricultural/grassland (yellow), forest (green), and water (blue). All of this is mapped against an outline of county boundaries.

Map of change in distribution

The second map, labeled "Distribution Change," shows the differences in block detection between the first (1983–1989) and the second (2004–2009) Atlases, with county boundaries included for reference. Blocks not uniquely covered in the first Atlas are not displayed as "changed" in these figures. The colors used in the change maps are as follows: yellow, for blocks in which the species was reported only during the first Atlas; green for blocks reporting the species in both projects; and blue for those species detected in the second Atlas but not the first. Care must be used in interpreting the raw changes in block occurrences, as explained in some detail in chapter 6. Changes are best interpreted by noting patterns (increases or decreases) across clusters of blocks, rather than giving undue importance to scattered individual blocks newly reporting, or failing to report, a species in the second Atlas.

Density map

All density maps were derived from atlas point count data (see chap. 4 for field methods). A technical explanation of the methods used to develop the abundance map is provided in chapter 5. The density map, for all species with sufficient data, shows the density of singing males (or all birds) as birds per square kilometer, as noted in the key to each map. Only readily detected songbirds or widespread non-singing birds (e.g., woodpeckers and crows) provide sufficient data for this map. For these, density is particularly valuable in illustrating a third dimension in the distribution, one otherwise not discernible from the presence/absence map of a species with statewide distribution. The resolution of the density maps varies, largely depending on the data density for each species. So, for particularly abundant species (e.g., Red-eyed Vireo), changes in abundances over short distances (<10 km) are mapped with some confidence, while for species with low data density (e.g., Pileated Woodpecker) the maps show more general patterns. For widespread species such as Wood Thrush, which shows more than a tenfold difference across the state, this map presents gra-

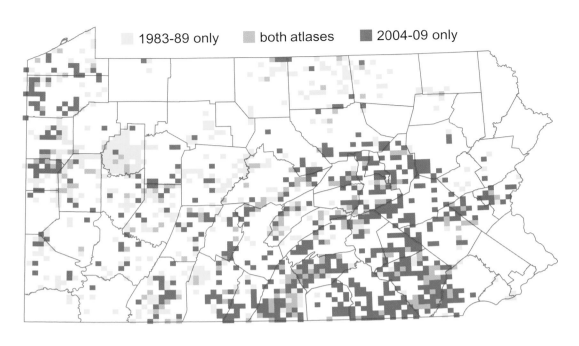

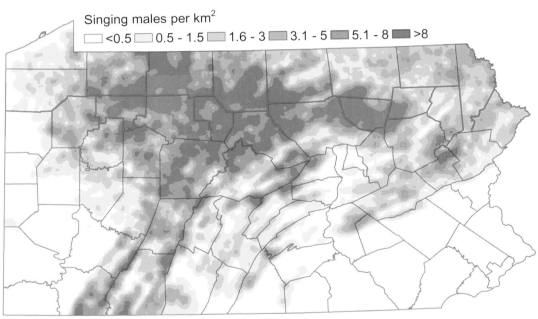

dients of density highlighting where the largest populations reside.

Breeding Bird Survey charts

As in the first Atlas, a chart showing trends from USGS Breeding Bird Survey (BBS) counts in Pennsylvania is presented here over the life of the survey for applicable species. We have used population estimates derived from either the second Atlas (or in a few instances Partners in Flight; Rich et al. 2004) in conjunction with BBS data to "backcast" the population estimates over the BBS period (see chap. 5). Dependent largely on detection of singing males, adequate population trends require a sufficient number of routes, with sufficient detection rates. The trends, from 1966 through 2009, are a key part of interpreting the change in abundance of species between atlases, primarily at a statewide scale. This more than 40-year dataset has been observed to be sensitive to large changes on an annual basis (such as House Finch conjunctivitis outbreaks).

Style

Pennsylvania's second breeding bird atlas project and associated results are referenced throughout as second

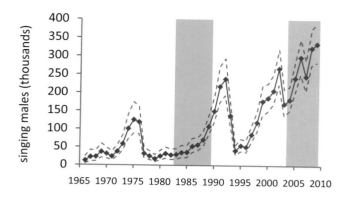

Atlas. Likewise, results of the 1980s project are referenced as the first Atlas. The category of breeding evidence codes (possible, probable, and confirmed) are not capitalized in the accounts and are generally used with the term *breeding evidence*, as in "possible breeding evidence." These refer specifically to the category of the breeding codes used to report nesting evidence (see chap. 4), not the confidence in the record.

Abbreviations and terminology

Table 8.1 is a summary of abbreviations typically used in the text. Metric units are provided, with English units typically given in parentheses.

TABLE 8.1 Acronyms and abbreviations

ac	acre(s)
AOU	American Ornithologists' Union
BBS	Breeding Bird Survey
CBC	Christmas Bird Count
ft	feet
ha	hectare(s)
in	inches
km	kilometers
km^2	square kilometer(s)
m	meters
mi	mile(s)
km^2	square kilometer(s)
NPS	National Park Service
NRCS	Natural Resource Conservation Service
NWR	National Wildlife Refuge
PADEP	Pennsylvania Department of Environmental Protection
PFBC	Pennsylvania Fish and Boat Commission
PGC	Pennsylvania Game Commission
U.S.	United States
USFWS	U.S. Fish and Wildlife Service
USGS	U.S. Geological Survey
WAP	Wildlife Action Plan

Gazetteer

Following is a gazetteer of the major locations mentioned in the following species accounts.

Site names mentioned in the text

Map no.	Site name (county)	Lat., long. (decimal degrees)
1	Algerine Swamp (Lycoming County)	41.542, −77.490
2	Allegheny National Forest (Elk, Forest, McKean, and Warren counties)	41.632, −79.113
3	Allegheny Reservoir (McKean and Warren counties)	41.921, −78.939
4	Allentown (Lehigh County)	40.608, −75.490
5	Atlantic Sand and Gravel ponds (Beaver County)	40.853, −80.404
6	Bald Eagle Mountain (Centre County)	40.809, −78.020
7	Bald Eagle State Forest (Center, Clinton, Mifflin, Snyder, and Union counties)	41.001, −77.234
8	Barrows (Mercer County)	41.443, −80.340
9	Bear Run Nature Reserve (Fayette County)	39.900, −79.464
10	Bellevue Park (Dauphin County)	40.268, −76.852
11	Big Beaver Borough wetland (Beaver County)	40.846, −80.403
12	Bolivar (Westmoreland County)	40.395, −79.151
13	Bradford (McKean County)	41.957, −78.649
14	Broad Mountain (Carbon and Schuylkill counties)	40.877, −75.887
15	Camp Hill (Cumberland County)	40.240, −76.919
16	Carlisle (Cumberland County)	40.202, −77.195
17	Clifford Township (Susquehanna County)	41.666, −75.626
18	Clinton Dam, Allegheny River (Armstrong County)	40.716, −79.579
19	Coalbed Swamp (Wyoming County)	41.471, −76.216
20	Coatesville (Chester County)	39.982, −75.824
21	Conneaut Marsh (Crawford County)	41.558, −80.196
22	Conodoquinet Creek (Cumberland County)	40.265, −76.938
23	Cowanesque Lake (Tioga County)	41.985, −77.159
24	Cumberland Valley (Cumberland and Franklin counties)	40.094, −77.480
25	Custards (Crawford County)	41.540, −80.159

Site names mentioned in the text (continued)

Map no.	Site name (county)	Lat., long. (decimal degrees)
26	Danville (Montour County)	40.963, −76.613
27	Dashields Dam (Allegheny County)	40.551, −80.204
28	Delaware Water Gap National Recreation Area (Monroe and Pike counties)	41.093, −74.993
29	Dutch Mountain (Sullivan and Wyoming counties)	41.435, −76.303
30	Eagles Mere (Sullivan County)	41.411, −76.582
31	Emsworth Dam (Allegheny County)	40.504, −80.090
32	Erie (Erie County)	42.130, −80.086
33	Erie National Wildlife Refuge (Crawford County)	41.606, −79.971
34	Forkston Township (Wyoming County)	41.529, −76.126
35	Glen Morgan/Morgan Lake (Berks County)	40.183, −75.869
36	Greencastle (Franklin County)	39.790, −77.727
37	Gull Point, Presque Isle State Park (Erie County)	42.171, −80.067
38	Harrisburg (Dauphin County)	40.267, −76.884
39	Hartstown Marsh complex (Crawford County)	41.560, −80.368
40	Hawk Mountain (Berks and Schuylkill counties)	40.641, −75.992
41	Hazleton (Luzerne County)	40.959, −75.975
42	Kittanning Dams (Armstrong County)	40.819, −79.529
43	Kiwanis Park (York County)	39.969, −76.742
44	Lackawanna State Forest (Lackawanna and Luzerne counties)	41.229, −75.662
45	Lake Erie (Erie County)	42.110, −80.272
46	Lake Lloyd (Potter County)	41.979, −77.739
47	Lancaster (Lancaster County)	40.038, −76.306
48	Laurel Highlands (Cambria, Fayette, Somerset, and Westmoreland counties)	40.196, −79.151
49	Laurel Hill (Somerset and Westmoreland counties)	40.218, −79.104
50	Lebanon (Lebanon County)	40.340, −76.423
51	Lehigh Valley (Berks, Lehigh, and Northampton counties)	40.583, −75.697
52	Marsh Creek wetlands, aka "The Muck" (Tioga County)	41.791, −77.302
53	Michaux State Forest (Cumberland, Franklin, and Adams counties)	39.834, −77.480
54	Middle Creek Wildlife Management Area (Lancaster County)	40.275, −76.241
55	Moraine State Park (Butler County)	40.944, −80.085
56	Moshannon State Forest (Cameron, Centre, Clearfield, Clinton, and Elk counties	41.175, −78.456
57	Muddy Creek Valley (York County)	39.781, −76.387
58	Natrona Dam, Allegheny River (Allegheny and Westmoreland counties)	40.613, −79.717
59	Niagara Pond, Presque Isle State Park (Erie County)	42.164, −80.087
60	North Mountain (Luzerne County)	41.331, −76.210
61	Pennsy, Black, and Celery Swamps complex (Lawrence and Mercer counties)	41.121, −80.198
62	Philadelphia (Philadelphia County)	39.953, −75.161
63	Philadelphia International Airport (Delaware and Philadelphia counties)	39.869, −75.249
64	Pine Creek (Lycoming and Tioga counties)	41.662, −77.469
65	Pittsburgh (Allegheny County)	40.441, −79.996
66	Plain Grove Township (Lawrence County)	41.055, −80.144
67	Pocono Environmental Education Center (Pike County)	41.171, −74.914
68	Pocono Plateau (Monroe and Pike counties)	41.254, −75.071
69	Pottsville (Schuylkill County)	40.686, −76.195
70	Powdermill Nature Reserve (Westmoreland County)	40.160, −79.272
71	Presque Isle State Park (Erie County)	42.161, −80.119
72	Prince Gallitzin State Park (Cambria County)	40.669, −78.545
73	Promised Land State Park (Pike County)	41.314, −75.209
74	Pymatuning (Crawford County)	41.583, −80.485
75	Pymatuning Reservoir (Crawford County)	41.642, −80.466
76	Pymatuning Swamp (Crawford County)	41.606, −80.393
77	Quakertown Swamp (Bucks County)	40.413, −75.312
78	Raystown Lake (Huntingdon County)	40.373, −78.089
79	Roderick Wildlife Preserve (Erie County)	41.977, −80.515
80	Rookery Island (Lancaster County)	39.987, −76.474
81	Rothrock State Forest (Centre, Huntingdon, and Mifflin counties)	40.711, −77.765
82	Sayre (Bradford County)	41.979, −76.517
83	Scranton (Lackawanna County)	41.408, −75.663
84	Settler's Cabin Park (Allegheny County)	40.421, −80.168
85	Shady Grove (Franklin County)	39.778, −77.669
86	Shamokin (Northumberland County)	40.789, −76.559

(continued)

Site names mentioned in the text   (continued)

Map no.	Site name (county)	Lat., long. (decimal degrees)
87	Shenango River Lake (Mercer County)	41.265, −80.461
88	Sinnemahoning Creek (Cameron and Clinton counties)	41.261, −77.901
89	South Mountain (Adams, Cumberland, and Franklin counties)	39.962, −77.458
90	South Pymatuning Township (Mercer County)	41.319, −80.490
91	Spring Township (Crawford County)	41.805, −80.299
92	Sproul State Forest (Centre and Clinton counties)	41.281, −77.723
93	Spruce Flats Bog (Westmoreland County)	40.123, −79.176
94	State College (Centre County)	40.793, −77.860
95	State Game Lands 57 (Wyoming County)	41.381, −76.193
96	Susquehannock State Forest (Clinton, Potter, and McKean counties)	41.498, −77.695
97	Swamp Run (Butler County)	40.940, −80.034
98	Tamarack Swamp (Clinton County)	41.424, −77.839
99	Tiadaghton State Forest (Clinton, Lycoming, Potter, Sullivan, Tioga, and Union counties)	41.295, −77.382
100	Tinicum/John Heinz NWR (Delaware and Philadelphia counties)	39.888, −75.259
101	Tionesta Natural Area (McKean and Warren counties)	41.644, −78.938
102	Tullytown (Bucks County)	40.138, −74.814
103	Wade Island (Dauphin County)	40.305, −76.908
104	Waggoner's Gap (Cumberland and Perry counties)	40.277, −77.278
105	Warren (Warren County)	41.844, −79.145
106	Washington Boro (Lancaster County)	39.993, −76.468
107	Weiser State Forest (Carbon, Columbia, Dauphin, Lebanon, Montour, Northumberland, and Schuylkill counties)	40.522, −76.745
108	Wildwood Park (Dauphin County)	40.309, −76.885
109	Wilkes-Barre (Luzerne County)	41.245, −75.881
110	Yellow Breeches Creek (Cumberland and York counties)	40.188, −76.927

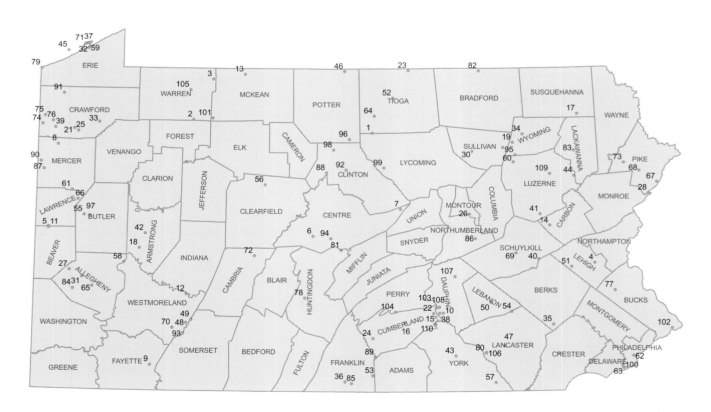

9 | Species Accounts

Jacob Dingel

Canada Goose
Branta canadensis

The Canada Goose is one of the best known and widely distributed waterfowl species of North America; the sight of the familiar "V" formations overhead has long been associated with the return of spring and the arrival of autumn. Due to introductions and translocations, the Canada Goose is now found throughout North America, breeding in every state and Canadian province (Mowbray et al. 2002). Resident or temperate-climate-nesting Canada Geese are defined as geese that were hatched or nest in any state, or in Canada at or below 48°N latitude, excluding Newfoundland. As their name implies, resident geese spend most of the year near their breeding areas, although many in northern latitudes do migrate. Historically, the Canada Goose did not breed in Pennsylvania, but by the 1980s it was found breeding in every county (Hartman 1992a). The present-day population was introduced and established during the early twentieth century and is comprised of various subspecies or races of Canada Geese, including *B. c. maxima*, *B. c. moffitti*, *B. c. interior*, *B. c. canadensis*, and possibly others, reflecting their diverse origins (Dill 1970; Pottie and Heusmann 1979; Dunn 1992). Beginning in the 1970s, the Pennsylvania Game Commission began translocation efforts, in and out of state, to deal with nuisance and crop damage problems, but these efforts were terminated in 1995 because of their high cost and ineffectiveness.

Preferred habitat for the Canada Goose is best characterized as wet areas (lakes, streams, ponds) that provide refuge and roosting areas adjacent to farmland or grazing areas, particularly lawns, golf courses, and parks. With their abundance of lush lawns for grazing, suburban/urban landscapes with areas of water also provide preferred habitat. Since resident Canada Geese live in temperate climates with stable breeding habitat conditions, low predation rates, and an abundance of preferred habitat, they exhibit high annual survival and production rates (Dunn and Jacobs 2000).

During the first Atlas period, the Canada Goose was most prevalent as a breeding bird in the glaciated sections of the Appalachian Plateaus or in the Piedmont Province. By the second Atlas, there had been a significant expansion of its breeding range into the northern, central, and southwestern portions of the state. The number of blocks with Canada Goose records more than doubled between these periods, to 71 percent of all blocks. As well as expansions into previously unoccupied areas, there was infilling in the state's southeastern and northwestern corners, resulting in occupancy rates of more than 90 percent in several physiographic sections (appendix C). Areas of higher elevation in the Appalachian Plateaus Province, where there are few wetlands or ponds for nesting and a paucity of open areas for foraging, remain unoccupied. Elsewhere, there are now few blocks with suitable habitat in which the Canada Goose is not found. The highest densities continue to be in the agriculturally dominated areas of the Piedmont Province.

The rapid expansion of Canada Goose populations between atlases was also documented by a fivefold increase in counts on Breeding Bird Survey routes over that period (Sauer et al. 2011) and a similar increase from the first statewide Atlantic Flyway Breeding Waterfowl Survey, which estimated 11,819 pairs in 1986 to 1989 (Sheaffer and Malecki 1998). Breeding waterfowl surveys conducted by the Pennsylvania Game Commission resulted in an estimate of 88,600 breeding pairs and a total population (pairs plus non-breeders) of 289,900 in 2009 (Jacobs et al. 2009). Following the rapid growth observed during the 1990s, Pennsylvania's Canada Goose spring breeding population appears to have stabilized to between 250,000 and 300,000. Expansion of hunting seasons, and other lethal and non-lethal programs implemented to control overabundant Canada Goose populations, may be having an effect on the population's growth rate. Overabundant populations of resident Canada Geese are often involved in damage to property, agriculture, or natural resources, as well as conflicts with public health and safety (Conover and Chasko 1985).

JOHN P. DUNN

Distribution

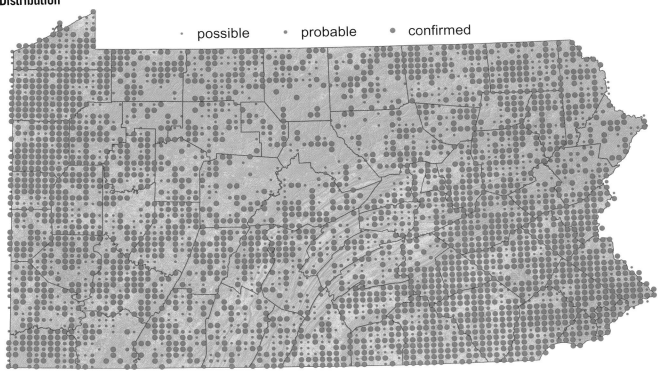

Distribution Change

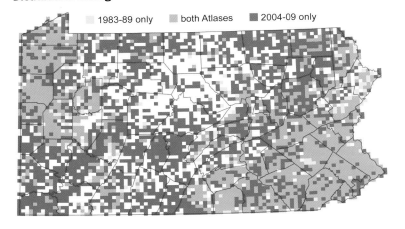

Number of Blocks

	first Atlas 1983–89	second Atlas 2004–09	Change %
Possible	341	738	116
Probable	235	391	66
Confirmed	1,106	2,369	114
Total	1,682	3,498	108

Population estimate, birds (95% CI):
215,000 (205,000–225,000)

Breeding Bird Survey Trend

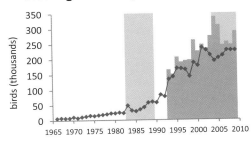

Gray bars are Atlantic Flyway Breeding Waterfowl Survey estimates, 1993–2009 (Gregg et al. 2000; Jacobs et al. 2010)

Joe Kosack/PGC

Mute Swan
Cygnus olor

Native to Eurasia and originally introduced into North America in captivity, the Mute Swan became well established in the wild on this continent during the twentieth century. Self-sustaining populations occur up and down the Atlantic coast (particularly in the Chesapeake Bay), in the Great Lakes region, and in the Pacific Northwest (Ciaranca et al. 1997). Its large size, all-white plumage, and tolerance for human activity have made the Mute Swan a conspicuous and often popular component of the wetland avifauna. However, like many other exotic species, it poses ecological problems, which have grown in severity as populations have increased. Major concerns include overgrazing of aquatic vegetation (Tatu et al. 2007), aggression toward native birds (Therres and Brinker 2004), and nuisance issues (Allin 1981). The Mute Swan is an invasive species and is not protected under state or federal law; the Migratory Bird Treaty Reform Act of 2004 specifically identified the Mute Swan on a list of non-native species to which international migratory bird treaties do not apply.

In the second Atlas, the Mute Swan was observed infrequently overall, occurring in about 3 percent of blocks statewide. However, this represented a significant expansion; block detections quadrupled from the first Atlas. Mute Swan was found in a wide variety of wetland habitats, predominantly in the portions of the Commonwealth dominated by agricultural or developed land. It continued to be most common in southeastern Pennsylvania, but the relatively few, localized clusters in this region during the first Atlas expanded and coalesced to produce larger (though still scattered) concentrations across a broader swath of the southeast. Similarly, the increases and expansions in second Atlas records for western Pennsylvania appear to be centered on a handful of first Atlas locations. The species remained scarce or absent across the more mountainous, forested portions of Pennsylvania, except for a few records near urban areas or major rivers.

The expansion of the Mute Swan's range between atlases likely reflects a combination of reproduction and dispersal of feral swans, both instate and from large populations in the Great Lakes and Chesapeake Bay (the proximity of the latter area to dramatic between-atlas expansions in York, Lancaster, and Chester counties is suggestive), and ongoing escapes and intentional releases of domestic birds. The fact that establishment and maintenance of Mute Swan populations tends to be closely associated with civilization produces a broad spectrum of "wildness," which complicates interpretation of atlas records. Many swans are located in highly altered habitats and receive artificial feeding, yet the species is free-ranging and not necessarily dependent on humans for survival. Decisions to include such birds as wild breeders in the second Atlas were sometimes subjective and may not always have been consistent between observers or atlases.

The general picture of population expansion provided by the atlas data is similar to results from Maryland (Brewer 2010) and New York (McGowan 2008a) and corresponds well with population trends and distributional data from other surveys, including the annual Midwinter Waterfowl Survey and the periodic (normally at 3-year intervals) Atlantic Flyway Mid-Summer Mute Swan Survey (Jacobs et al. 2009). Mute Swan populations in the latter survey peaked around 2002 in both Pennsylvania (348 individuals) and the flyway (14,344), and have since declined in two consecutive surveys. Populations remain well above 1980s levels (120 to 140 in Pennsylvania and 6,000 to 8,000 for the flyway), but population reduction efforts have been effective; by 2010, swan numbers had been reduced from a high of 3,995 birds to 201 in the Maryland portion of the Chesapeake Bay (L. Hindman, pers. comm.).

A Mute Swan management plan is in place for the Atlantic Flyway, with the goals of reducing populations to minimize impacts on wetland habitats and native waterfowl and preventing range expansion into new areas (AFC 2003). Under this plan, Pennsylvania has a dual goal of zero free-ranging Mute Swans and a maximum of 250 in captivity. Educational efforts, possession regulations for captive swans, and active population control activities will be important tools in achieving management goals for this invasive species.

IAN GREGG

Distribution

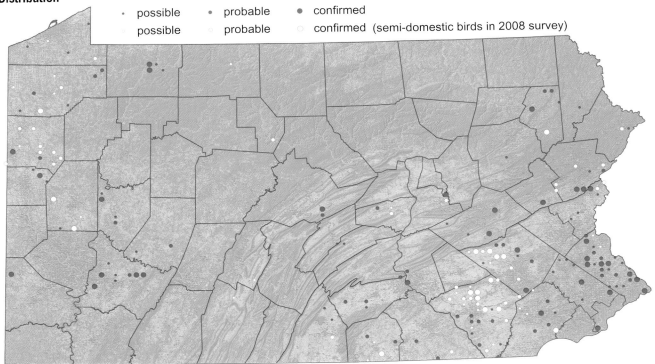

Distribution Change

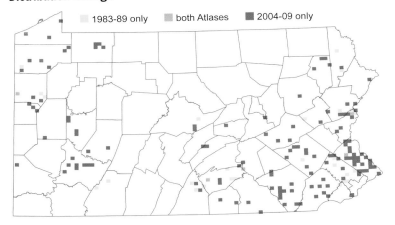

Number of Blocks

	first Atlas 1983–89	second Atlas 2004–09	Change %
Possible	10	56	460
Probable	9	43	378
Confirmed	12	38	217
Total	31	137	342

MUTE SWAN

Glenn Bartley/VIREO

Trumpeter Swan
Cygnus buccinator

The Trumpeter Swan is the largest species of native North American waterfowl. Prior to the arrival of European settlers and fur traders, its breeding range extended along a wide band from the Bering Sea, east through almost all of Canada, and south to Missouri, Illinois, and Indiana, but it subsequently contracted due to market hunting for feathers and skins in the seventeenth to nineteenth centuries (Banko 1960; Mitchell 1994). By 1900, a century of unregulated killing had nearly wiped out the entire continent's population of Trumpeter Swans from all but several isolated breeding areas. With protection from the 1918 Migratory Bird Treaty Act, a remnant population of fewer than 70 birds, discovered in 1932 near Yellowstone National Park, represented the only wild population in the lower 48 states and became the core population for the eventual recovery of the species (Mitchell 1994).

Today, estimates show about 16,000 Trumpeter Swans residing in North America. The exact historic range limits and distribution of Trumpeter Swan remain uncertain because the species disappeared so early in the period of European colonization; evidence is limited to fossil records, a few museum collections, and early historical accounts. Ornithological publications generally have not included the Trumpeter Swan as a species native to Pennsylvania, not even as an extirpated species (e.g., McWilliams and Brauning 2000), although skeletal remains, dating from around 1600, were found at Native American archaeological sites in what are now Huntingdon and Lancaster counties (Mitchell 1994).

Early twenty-first century observations of trumpeters in Pennsylvania, although still uncommon, are likely the result of successful reintroduction programs in Ontario (Lumsden and Drever 2002) and Ohio (Shea et al. 2002; OBBA 2011). In 1999, the Ohio Division of Natural Resources, in conjunction with the Cleveland Zoo, released 15 birds (from eggs acquired from Alaska) within the Mosquito Creek Wildlife Area in Trumbull County, Ohio, resulting in several nesting pairs there. The second Ohio atlas confirmed breeding of Trumpeter Swans in 41 blocks, including along the Shenango River drainage (OBBA 2011). Ontario confirmed breeding north and west of Lake Ontario (Lumsden 2007). New York's second atlas project confirmed the species in six blocks, predominantly in the Lake Ontario Plain (McGowan 2008b). As a result, there is a small and growing population of Trumpeter Swans established around the eastern Great Lakes.

Most trumpeters observed in Pennsylvania are marked with coded neckbands, wing tags, or leg bands from these reintroduction efforts. Birds from the above-mentioned population in Trumbull County, Ohio, have dispersed into the Pennsylvania portions of Shenango River Lake, predominantly within South Pymatuning Township, Mercer County, although no nesting sites have been located there (C. Brudowsky, pers. comm.). An unbanded pair of Trumpeter Swans was reported to the second Atlas during 2004 at a small lake in Lawrence County (off Edwards Road), where the species had been seen since 1999 (B. Dean, pers. comm.). Two birds, acting as a pair, were observed at this small lake on and off from January through the summer, and at a nearby sand and gravel excavation pond. A pair of Trumpeter Swans, possibly the same birds, was observed at the nearby Atlantic Sand and Gravel ponds, in Beaver County, on 2 July and 12 August that year (M. Vass, pers. comm.). A similarly unmarked pair was seen at nearby Big Beaver Borough wetland, in Beaver County, during June 2006. None of these reports included physical evidence of nesting, such as a nest or fledged young, but the behavior observed in 2004 provided the strongest suggestion that a nesting attempt had been made.

As observations of this striking bird continue to increase around the Great Lakes and nesting is documented very close across the border in Ohio and at scattered locations in New York, the day may be near when the Trumpeter Swan is confirmed breeding within Pennsylvania.

JOHN P. DUNN

Distribution

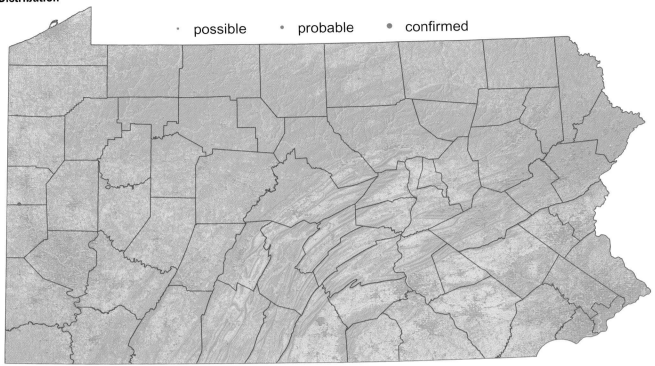

Distribution Change

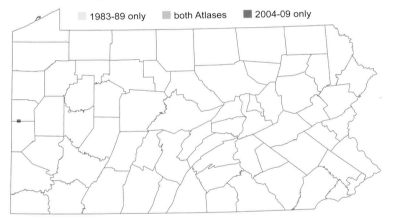

Number of Blocks

	first Atlas 1983–89	second Atlas 2004–09	Change %
Possible	0	0	0
Probable	0	1	∞
Confirmed	0	0	0
Total	0	1	∞

Geoff Malosh

Wood Duck
Aix sponsa

The Wood Duck can instill passion for wild creatures and places by all who cherish Penn's Woods. "Woodies" are the second most harvested duck (after Mallard) in the state, with recent annual harvests ranging from 20,000 to nearly 40,000 (Raftovitch et al. 2009). Local breeding Wood Ducks and their offspring are very important in supporting these seasons, comprising 83 percent or more of the state's harvest annually, based upon band recoveries (Jacobs et al. 2009).

Wood Ducks have not always been taken by hunters in robust numbers. In the early twentieth century, they were in significant decline from the loss of wetlands and nesting cavities and underregulated harvest. Closed-season protection and the reintroduction of the beaver, a habitat provider, bolstered the woody population. Additionally, the Pennsylvania Game Commission and conservation organizations in other states helped Wood Ducks by constructing and placing nest boxes afield, a practice continued to this day.

The Wood Duck has experienced a significant population expansion in Pennsylvania between the first and second Atlas periods, having been observed in over 27 percent more blocks. The species was observed in more blocks during the second Atlas period than in the first Atlas in 21 of 23 physiographic sections. Significant increases in occupied blocks were observed for the Appalachian Plateaus, Piedmont, and Ridge and Valley provinces. No significant declines in block occupancy were observed between atlas periods. The highest block occupancy rates were in the southeastern provinces (Piedmont 56%, New England 52%), followed by the Central Lowlands (46%), Ridge and Valley (43%), and Appalachian Plateaus (38%) provinces.

Significant Wood Duck population growth has been documented by the Breeding Bird Survey, with a 3.9 percent annual growth from 1966 through 2009 (Sauer et al. 2011), and more than a doubling in counts between atlas periods. However, this index should be used with caution, since the species is detected at low frequencies due to its secretive nature and habitat preferences. Bellrose and Holm (1994) estimated a Pennsylvania breeding population of nearly 35,000 between 1961 and 1970, and 45,000 between 1981 and 1985, using band recovery and harvest statistics. The Atlantic Flyway Breeding Waterfowl Survey (AFBWS), initiated in 1989 (Heusmann and Sauer 2000), was a major step forward in estimating annual Wood Duck populations. Estimated spring breeding population abundance in the state has ranged from a low of nearly 81,500 in 1993 to a high of nearly 133,000 in 2002 (Jacobs et al. 2009). The highest average breeding population densities were in the Northwestern Glaciated Plateau (1.73 birds/km^2) and Glaciated Low Plateau (1.57 birds/km^2) sections. The density in the Piedmont declined 44 percent from 1.3/km^2 in the 1990s to 0.73/km^2 in the 2004 through 2009 period. In contrast, densities in the combined Pittsburgh Low Plateau and Waynesburg Hills sections and Ridge and Valley Province increased by 60 percent and 229 percent, respectively, while more modest increases were noted in most other regions. The Wood Duck was observed in 41 percent of blocks and 47 percent of priority blocks, suggesting that it was often missed in blocks with low effort. Dates of nests with eggs varied from 7 March to 13 July, while broods were observed from 21 April to 5 August.

Wetlands with mature trees that can host natural cavities are important habitat components for Wood Duck nesting. Statistically significant increases in large-class trees (those with a diameter greater than 11 inches [~28 cm]) have occurred statewide between 1989 and 2004, with increases in the 16-inch (~41 cm) and larger classes exceeding 20 percent (McWilliams et al. 2007). However, forest cover declined notably between survey periods in the Piedmont (chap. 3), which may partly explain declines in AFBWS counts there. Current status of beaver populations appears strong, with 76 percent of Pennsylvania Game Commission Wildlife Conservation Officer districts reporting increasing or stable populations (Lovallo and Hardisky 2009). Continued conservation, along with proper forestry management and wetlands protection, will ensure Wood Duck populations are maintained or increased.

KEVIN JACOBS

Distribution

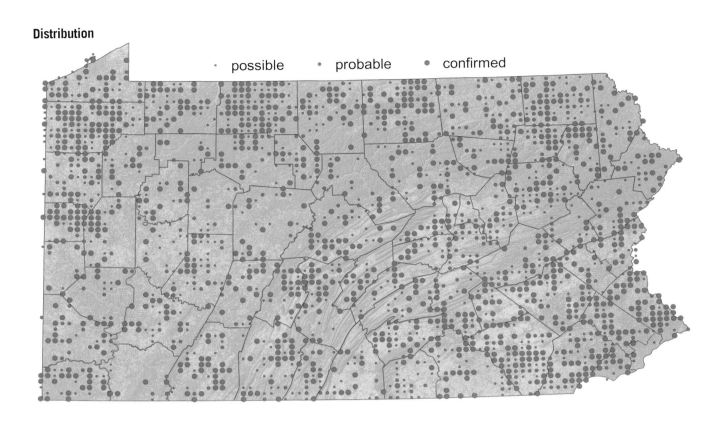

Distribution Change

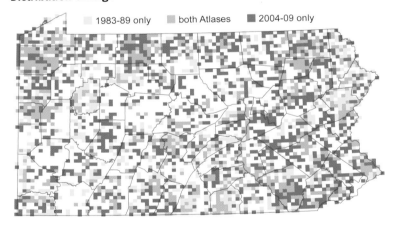

Number of Blocks

	first Atlas 1983–89	second Atlas 2004–09	Change %
Possible	425	576	36
Probable	346	402	16
Confirmed	827	1,049	27
Total	1,598	2,027	27

Breeding Bird Survey Trend

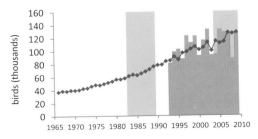

Gray bars are Atlantic Flyway Breeding Waterfowl Survey estimates, 1993–2009 (Gregg et al. 2000; Jacobs et al. 2010)

Andrew M. Wilson

American Black Duck
Anas rubripes

The American Black Duck, as its name implies, has a dark plumage, and the sexes are similar in appearance. Unlike other dabbling duck species, the male black duck does not exhibit a colorful alternate plumage. Similar in size and appearance to the female Mallard, the species is often found with Mallards in similar breeding habitat, and the two species will frequently hybridize (Longcore et al. 2000).

The American Black Duck is found primarily east of the Mississippi River, breeding from eastern Manitoba to the Canadian Maritimes, and southward through the mid-Atlantic states to coastal North Carolina (Longcore et al. 2000). Historically, it was more or less confined to forested and shrub-scrub wetlands and was, therefore, mostly isolated from the Mallard. It once rivaled the Mallard in abundance in parts of Pennsylvania (Todd 1940; Grimm 1952). In fact, numbers of resident black ducks banded at Pymatuning in the late 1930s exceeded Mallards by more than two to one (Warren 1950).

The first Atlas revealed the breeding American Black Duck to be widely distributed, but a locally uncommon breeding bird in Pennsylvania. It occurred in only 6 percent of blocks statewide, with the highest concentration in the Glaciated Pocono Plateau Section (39% of blocks; Hartman 1992b). By the second Atlas, the species had decreased in statewide occurrence by 63 percent, being found in only 111 blocks, compared with 297 during the first Atlas. Declines of similar magnitude were observed during the recent second Atlas efforts conducted in New York (Swift 2008) and Ontario (Ross 2007). Declines were noted in almost all physiographic sections where the American Black Duck was observed in the first Atlas (appendix C), but perhaps of greatest concern is that in the former stronghold of the Pocono region, the number of occupied blocks declined by 68 percent. Simply put, the black duck is becoming a very uncommon and localized breeding species in the state, with the only remaining concentrations occurring in the glaciated sections of the Appalachian Plateaus and northern portion of the Ridge and Valley Province.

Loss and degradation of emergent and shrub-scrub wetlands statewide, and high losses in the Pocono region where black duck nesting densities are highest, have been especially detrimental (PGC-PFBC 2005). Expansion of Mallard populations into former black duck habitat, with resulting competition and hybridization (Merendino et al. 1993; Kirby et al. 2004; Mank et al. 2004), may also be responsible for limiting black duck populations in otherwise suitable habitat.

Pennsylvania's breeding American Black Duck population was estimated at about 300 breeding pairs by Pennsylvania Game Commission surveys over the 1993 through 2009 period, about 1 black duck pair for every 150 Mallard pairs (Jacobs et al. 2009). However, both the BBS and PGC waterfowl surveys record too few black ducks to draw meaningful conclusions about breeding population trends.

The range contraction of breeding American Black Ducks in Pennsylvania coincided with a long-term decline in winter numbers in the Atlantic Flyway, as measured by the Mid-Winter Waterfowl Survey; counts declined by about 50 percent, from an average of 400,000 in the late 1950s to about 204,000 from 2006 through 2009 (Klimstra and Padding 2009). Winter counts of black ducks in Pennsylvania during the years 2006 through 2009 averaged 1,432 birds, only 10 percent of the 1955 through 1960 average of 13,555 (Klimstra and Padding 2009). Paralleling the northward shift in breeding range in Pennsylvania, there is some evidence for a northward shift in the winter range of black duck, which may bias traditional winter count data (AFC 2006).

The decline of the American Black Duck has received much attention from government agencies and nongovernmental organizations concerned about its conservation status (Tautin 2010). The PGC lists the black duck as a species of Conservation Concern, and the U.S. Fish and Wildlife Service formally lists it as a "Game Bird Below Desired Condition" (USFWS 2007). Numerous conservation plans address the species' conservation in both breeding and wintering areas. It remains uncertain whether or not these conservation actions, if implemented, can ensure that breeding populations of American Black Duck in Pennsylvania remain viable.

JOHN P. DUNN

Distribution

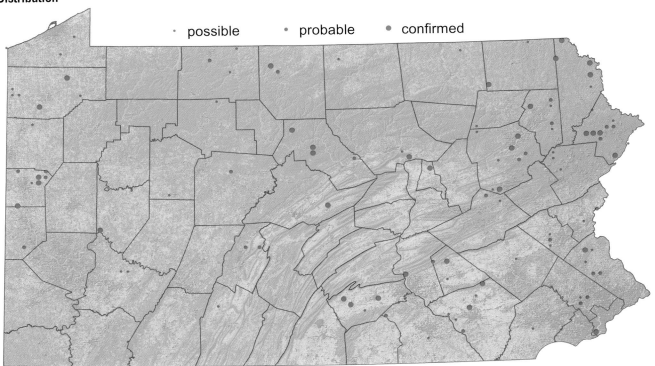

Distribution Change

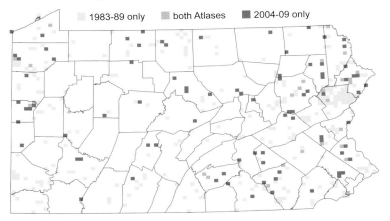

Number of Blocks

	first Atlas 1983–89	second Atlas 2004–09	Change %
Possible	117	49	−58
Probable	81	27	−67
Confirmed	99	35	−65
Total	297	111	−63

AMERICAN BLACK DUCK

Chuck Musitano

Mallard
Anas platyrynchos

The Mallard has long been Pennsylvania's most widespread, abundant, and economically important duck (Conklin 1938; Hartman 1992c; McWilliams and Brauning 2000). Mallards are the most prized duck pursued by waterfowlers in the state, with recent annual harvests ranging from 60,000 to over 90,000, easily comprising half or more of the annual state duck harvest (Raftovitch et al. 2009). Pennsylvania's breeding Mallard population is very important in supporting these seasons; these birds and their offspring comprise 80 percent or more of the state's harvest annually (Sheaffer and Malecki 1996; Jacobs et al. 2009). With such levels of harvest, it is imperative to maintain robust monitoring programs to ensure Mallard conservation. Nesting typically occurs from mid-March to mid-June, with the majority of nest initiation occurring from 24 March to 7 April (Stempka 2009). Mallards prefer to nest in grassland, hayfields, and shrubby vegetation, but they are also found in forested and wetland areas (Hoekman et al. 2006). Nests are most commonly constructed on the ground and lined with down after the last egg is laid.

In the second Atlas, the highest rates of Mallard detections were in the southeastern Pennsylvania provinces (Piedmont, New England, and Atlantic Coastal Plain), where Mallards were found in 83 to 89 percent of blocks (appendix C). Occupancy rates were lower in the Ridge and Valley Province and lower still in the densely forested sections of the Appalachian Plateaus Province. Although over 500 Mallards were detected on atlas point counts, these birds were highly aggregated and do not provide a sufficient sample to produce a map of density estimates.

The distribution of the Mallard in Pennsylvania appears to have changed little between atlas periods, and no significant changes in the number of occupied atlas blocks were observed either statewide or by physiographic province (appendix C). Increases in the numbers of blocks with Mallards in the Northwestern Glaciated Plateau and Anthracite Upland sections could have been, at least in part, due to increased observer effort (chap. 6). Some of the significant decline in blocks with Mallard records in the Appalachian Mountain Section of the Ridge and Valley could be due to reduced observer effort, notably in southern Huntingdon County and western Juniata County.

Breeding Bird Survey data show a 2 percent annual increase in Pennsylvania Mallards during the period 1966 through 2009 (Sauer et al. 2011). Extrapolating from BBS counts, we estimate that Mallard numbers were 19 percent higher in the second Atlas period than in the first. In contrast, there has been a decline in the number of Pennsylvania Mallards banded in late summer, as shown by the Pennsylvania portion of the Atlantic Flyway Breeding Waterfowl Survey (AFBWS; Jacobs et al. 2009). Estimated spring Mallard breeding population abundance in the state has ranged from a high of nearly 271,000 in 1994 to a low of over 131,000 in 2008.

Long-term AFBWS counts show patterns similar to atlas block occupancy rates: breeding Mallard densities are highest in the combined southeastern provinces (3.0 birds/km^2), followed by the combined Pittsburgh and Waynesburg Hills sections (2.9 birds/km^2), the glaciated northwest (1.6 birds/km^2), and glaciated northeast (1.6 birds/km^2) sections. There was evidence of a declining trend in average Mallard density between the periods 1993–1998, 1999–2004, and 2005–2009 for the southeastern provinces (−26%), Pittsburgh Low Plateau, Waynesburg Hills (−34%), and Northwestern Glaciated Plateau sections (−18%), whereas densities in the other provinces and sections appeared stable over the same periods. Habitat change is likely the culprit; the areas with declining Mallard populations have the greatest losses of agriculture and grassland habitats, as well as highest increases in the amount of developed land between atlas periods (chap. 3).

Conservation of this species will be driven by the continuation of landscape-level programs such as the Conservation Reserve Enhancement Program, as well as myriad wetlands and associated uplands habitat partnerships among government, non-governmental organizations, and private landowners across the Commonwealth.

KEVIN JACOBS AND JEREMY STEMPKA

Distribution

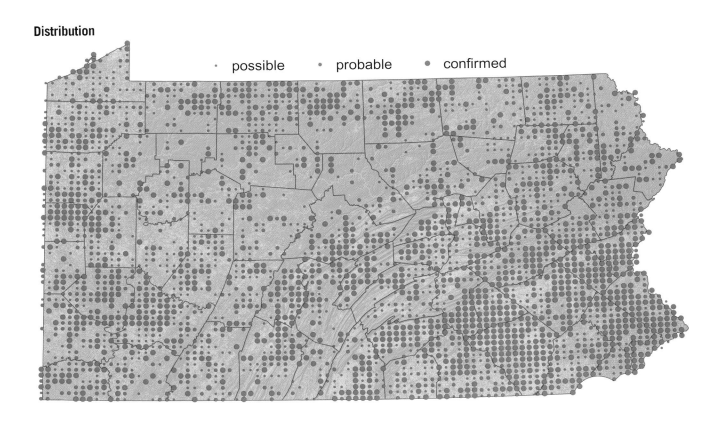

Distribution Change

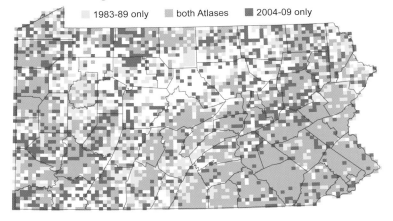

Number of Blocks

	first Atlas 1983–89	second Atlas 2004–09	Change %
Possible	629	903	44
Probable	711	774	9
Confirmed	1,558	1,344	–14
Total	2,898	3,021	4

Population estimate, birds (95% CI): 107,000 (78,000–130,000)

Breeding Bird Survey Trend

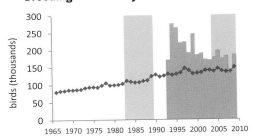

Gray bars are Atlantic Flyway Breeding Waterfowl Survey estimates, 1993–2009 (Gregg et al. 2000; Jacobs et al. 2010)

Geoff Malosh

Blue-winged Teal
Anas discors

The Blue-winged Teal is among the most abundant waterfowl in North America; unfortunately, it has become significantly rarer as a breeding species in Pennsylvania since the first Atlas. It is an especially opportunistic migratory waterfowl, known for very low rates of natal fidelity (Lokemoen et al. 1990). Spring settling patterns depend upon an abundance of temporary wetlands within healthy grassland landscapes and can vary annually, depending on the amount of vernal wetland habitat present during migration.

Historical accounts of breeding Blue-winged Teal in Pennsylvania began in the 1930s at Presque Isle (Todd 1940) and in the Pymatuning region (Conklin 1938; Grimm 1952). The first Atlas documented the Blue-winged Teal in 100 blocks, a third of which reported confirmed breeding, and considered the species to be "more widely distributed in Pennsylvania than previously suspected" (Hartman 1992d). Most records were in distinct clusters, notably in the Northwestern Glaciated Plateau, but also in some southern counties, including traditional sites such as Tinicum (John Heinz National Wildlife Refuge) along the Delaware River.

By the second Atlas, the species' presence was reduced to only 32 blocks, including just 11 breeding confirmations. Fewer were noted in all physiographic provinces, even in the northwestern strongholds. There were no breeding records from some areas that had been apparent strongholds in the first Atlas, including Tinicum (four blocks in first Atlas) and the Pennsy, Black, and Celery Swamp complexes in Lawrence and Mercer counties (seven blocks in first Atlas). Blue-winged Teal were found during both atlas periods in only 10 blocks, 7 of them in the Conneaut Marsh complex in southern Crawford County, which now is undoubtedly the breeding focal point for this species in Pennsylvania. Elsewhere, there were widely scattered records, with breeding confirmed in six counties in all.

Because it nests in wetlands that are sometimes large and inaccessible, it is likely that the Blue-winged Teal was underreported during the second Atlas. Of the 11 second Atlas records of broods, 1 was on 26 May, but the rest were between 10 June and 28 July. With so few breeding confirmations, and none at all during two atlas years (2006 and 2008), the current breeding population must be considered very small, likely in the tens of pairs at the most—a far cry from the optimistic estimate of 1,425 pairs at the end of the first Atlas period (Hartman 1989).

Migratory Blue-winged Teal pass through Pennsylvania later than most other waterfowl. As a result, the Atlantic Flyway Breeding Waterfowl Survey, initiated in 1989 (Heusmann and Sauer 2000), is more likely to detect migrants in Pennsylvania, since the survey period (15 April to 5 May) is well before the blue-wing's normal nesting season.

The Blue-winged Teal should receive recognition as a breeding Species of Special Concern within the Commonwealth, in order to enhance protection of Pennsylvania's habitat types upon which they are dependent: healthy grasslands associated with emergent and vernal wetlands. Other species that rely on these landscapes have experienced declines between atlas periods (e.g., Spotted Sandpiper and Northern Harrier). Conservation focus should be directed toward identification and protection of remaining habitats, as well as continued emphasis on delivery of landscape-level habitat conservation such as USDA-NRCS conservation practices (i.e., Wetlands Reserve Program, Conservation Reserve Enhancement Program, Wildlife Habitat Incentive Program, etc.), USFWS Partners for Wildlife, and other governmental and nongovernmental initiatives.

KEVIN JACOBS

Distribution

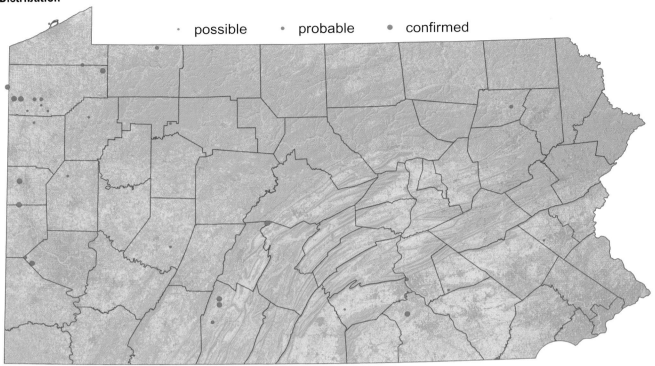

Distribution Change

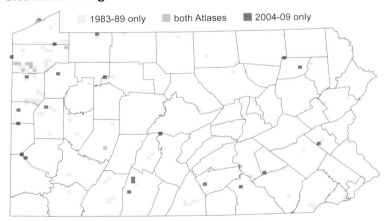

Number of Blocks

	first Atlas 1983–89	second Atlas 2004–09	Change %
Possible	34	13	−62
Probable	37	8	−78
Confirmed	29	11	−62
Total	100	32	−68

Gerard Dewaghe

Green-winged Teal
Anas crecca

The Green-winged Teal, the smallest native North American duck, is a relatively abundant breeding waterfowl on this continent. However, Pennsylvania lies at the southeastern edge of its breeding range, and there are no indications that it was ever anything other than a rare breeding species, at best (McWilliams and Brauning 2000).

The secretive Green-winged Teal uses somewhat different habitat types than the Blue-winged Teal, preferring dense emergent marshes and shrubby swamps more so than grasslands and vernal wetlands, resulting in difficult detection of this species (Belrose 1976). Hartman (1992e) contended that the Commonwealth's habitats could support enough nesting pairs of Green-winged Teal for it to be considered more than incidental. However, neither of the state's breeding bird atlases has detected much breeding evidence. It appears that the Green-winged Teal was even rarer in the second Atlas period than in the first, with a near halving of records and a drop from confirmation in five blocks to just one. The smattering of possible and probable breeding records in both atlas periods, primarily in late June or July, is difficult to interpret, but they could reflect wandering non-breeders. Conneaut Marsh, in Crawford County, has the best credentials for supporting a sustaining population of Green-winged Teal. It was the site of a possible breeding in the first Atlas and of the only confirmed breeding in the second Atlas: a female with one small young on 29 June 2004. Pairs were noted in midsummer at nearby Pymatuning in two atlas years; this is also an area with prior breeding records (Hartman 1992e), including the state's first documented nesting attempt in 1936 (McWilliams and Brauning 2000).

Monitoring programs for breeding birds do a poor job of detecting this and other rare waterfowl. Even the Atlantic Flyway Breeding Waterfowl Survey (AFBWS), initiated in 1989 (Heusmann and Sauer 2000) and designed to provide annual estimates for duck populations in the northeastern United States, appears to detect primarily migratory individuals in Pennsylvania. Estimates are not indicative of true breeding populations of teal in Pennsylvania, since many migrating teal are encountered during the survey period (15 April through 5 May). Like several other common migratory waterfowl, the rare individuals that remain to nest in Pennsylvania are not easily detected, tucked away in impassable swamps and wetlands. The Green-winged Teal has a long history of nesting within the state, but the marginal population contributes almost nothing to the species' extensive national populations and can be viewed primarily as accidental to the state's breeding avifauna. As such, no special conservation status is designated within the state.

Continental Green-winged Teal populations are healthy. As Pennsylvania is on the edge of its breeding range, the status of habitat types in Pennsylvania upon which they are dependent should receive conservation focus largely because they provide habitat to a host of other sensitive wetlands-dependent species (rails, Marsh Wrens, bitterns, etc.). Conservation focus should be directed toward identification, protection, and management of remaining habitats. In addition, there should be continued emphasis on delivery of landscape-level habitat conservation such as USDA-NRCS conservation practices (i.e., Wetlands Reserve Program, Wildlife Habitat Incentive Program, etc.), USFWS Partners for Wildlife, and other governmental and non-government wetlands conservation initiatives.

KEVIN JACOBS

Distribution

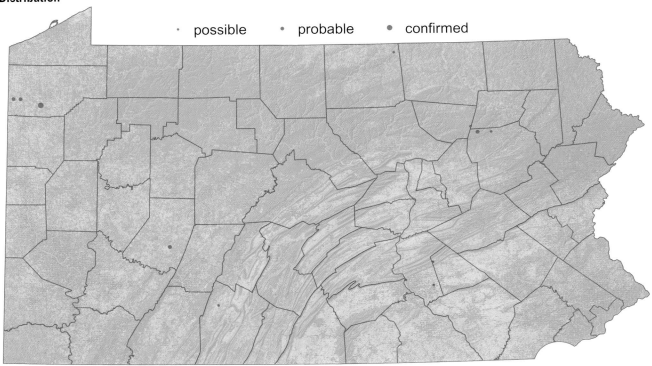

Distribution Change

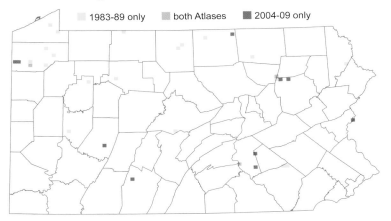

Number of Blocks

	first Atlas 1983–89	second Atlas 2004–09	Change %
Possible	14	8	−43
Probable	4	4	0
Confirmed	5	1	−80
Total	23	13	−43

Rick and Nora Bowers/VIREO

Hooded Merganser
Lophodytes cucullatus

The handsome Hooded Merganser is a common spring and fall migrant in Pennsylvania, but it is comparatively uncommon as a breeding bird. Pennsylvania is toward the edge of the species' breeding range in eastern North America, where it is found from the Upper Mississippi, north through the Great Lakes and northeastern United States, and into southern Canada. It nests in tree cavities; hence, it is typically found on tree-fringed lakes and wetlands, notably beaver impoundments.

Although summering Hooded Mergansers were noted in Pennsylvania's earliest avifaunal accounts, with nesting confirmed in the late nineteenth century (Todd 1940), there are few indications that it was ever common, and it was found in only 75 atlas blocks in the first Atlas, concentrated in the northwestern corner of the state (Gross 1992a). Since then, it has experienced a significant range expansion, with records in 211 blocks in the second Atlas, a 181 percent increase. The Northwestern Glaciated Plateau Section is still a stronghold for this species, where it was found in 22 percent of the blocks (appendix C). However, records increased across the northern tier, especially in the Northeastern Glaciated Low Plateau, where the Hooded Merganser was rare in the first Atlas. In addition to this growth in the north, there was a notable southward expansion: the number of occupied blocks in the southern half of the state (south of 40.9°N) increased from 10 in the first Atlas to 54 in the second. But the species remains scarce and localized, observed in just 4.3 percent of all blocks statewide.

The Hooded Merganser has a protracted breeding season, with atlas volunteers finding nests with eggs from 25 April to 9 July, and broods from 19 April to 9 August. This long season could result in the species being missed during atlas fieldwork if visits were concentrated during the middle of the survey season. Additionally, the species often nests in isolated and sometimes remote forested wetlands, some of which would have gone unchecked by atlas volunteers. Hence, there is reason to suggest that the species may have gone underreported.

Breeding Bird Survey counts have shown an increase across the species' breeding range in recent decades, and Pennsylvania is no exception, with a mean annual increase of 13.2 percent (Sauer et al. 2011). However, this index should be used with caution, since Hooded Merganser detections are few, due to its secretive nature and habitat preferences. The Atlantic Flyway Breeding Waterfowl Survey (AFBWS), initiated in 1989 (Heusmann and Sauer 2000), estimated that the breeding population of Hooded Merganser in the state averaged 3,586 from 1993 through 2009, ranging from a low of 283 in 1994 to a high of 9,625 in 2005 (Jacobs et al. 2009). Average breeding densities were 0.12 birds/km^2 in the Northwestern Glaciated Plateau Section, 0.08 birds/km^2 in the Glaciated Low Plateau Section, and much lower elsewhere.

Important elements of Hooded Merganser breeding habitat in Pennsylvania are available wetland habitats, especially with clear waters for foraging, and mature trees that can host natural cavities. Statistically significant increases in mature trees (>33 cm [11 inches] diameter at breast height) have occurred statewide between 1989 and 2004 (McWilliams et al. 2007), although a more important metric may be the number of decaying or dead trees in the large-size class. In addition to increases in large trees, the resurgent beaver population in Pennsylvania has played a key role in provision of suitable wetlands for the Hooded Merganser. The current status of beaver in Pennsylvania is favorable, with 76 percent of Pennsylvania Game Commission Wildlife Conservation Officer districts reporting increasing or stable beaver populations (Lovallo and Hardisky 2009). Additionally, the Hooded Merganser readily takes to Wood Duck nest boxes in Pennsylvania, and the provision of large numbers of boxes, especially on state game lands, has undoubtedly facilitated the increase and spread of the species. Continued conservation of the beaver, provision of Wood Duck boxes, retention of large trees, and protection of wetlands will ensure that mergansers maintain or improve their status.

KEVIN JACOBS

Distribution

Distribution Change

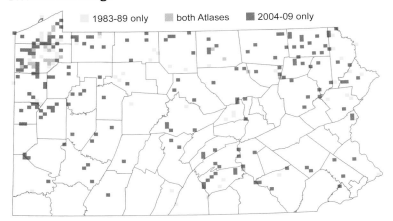

Number of Blocks

	first Atlas 1983–89	second Atlas 2004–09	Change %
Possible	25	65	160
Probable	8	19	138
Confirmed	42	127	202
Total	75	211	181

John Cancalosi/VIREO

Common Merganser
Mergus merganser

One of our largest native ducks, the Common Merganser is an increasingly familiar species on larger waterways, especially forested rivers, during the breeding season. The handsome male is easily spotted due to its eye-catching white flanks and breast, but the more subdued plumage of the female causes her to be more easily overlooked. Males often leave females soon after the start of incubation (Mallory and Metz 1999) and may depart the breeding ground completely (McWilliams and Brauning 2000).

Pennsylvania is at the southern edge of the large breeding range of the Common Merganser, which stretches from the northern United States through much of Canada to Alaska. Following European settlement, persecution and pollution drove the Common Merganser to extirpation as a breeding bird in the state, only for it to reclaim its former range during the second half of the twentieth century (Reid 1992a), when more enlightened attitudes paved the way for the resurgence in populations of this and other piscivores.

The comeback of the Common Merganser has been nicely documented by the two atlases. By the time of the first Atlas it was already well established, with records in 227 blocks, almost all in the northern tier (Reid 1992a). In the second Atlas period it was found in 554 blocks, a 144 percent increase. It overtook the American Black Duck to become the third most widespread breeding duck in the state (after Mallard and Wood Duck), and it was observed in more than twice as many blocks as the Hooded Merganser statewide. Increases were noted almost everywhere, but they were especially apparent in the southern half of the state, where the Common Merganser has gone from being rare during the first Atlas period to increasingly widespread in the second, especially along the Delaware River. It is now scarce only across the southwestern sections.

The net result of these changes was that the mean latitude of the Common Merganser's southern range limit moved south by 55 km (33 miles). Despite this, the core of the merganser's range is still in the northern tier, especially in McKean County and surrounding areas, where high levels of atlasing effort revealed this species to be ubiquitous on sufficiently large waterways. This suggests that, despite the huge increase in block occupancy, the Common Merganser may still have been underreported elsewhere, partly because of limited river access points in some areas. Additionally, the species has a protracted breeding season (young were noted 21 May to 22 August); hence, repeated visits may be necessary to detect this species, which can easily elude detection, especially during incubation, when males may have already departed.

The resurgence of the Common Merganser population in Pennsylvania is estimated at 11 percent per year by the Breeding Bird Survey (Sauer et al. 2011), but the BBS index should be interpreted with caution due to low detection rates and habitat biases. The Atlantic Flyway Breeding Waterfowl Survey (AFBWS) has ensured much more robust monitoring of breeding duck numbers since 1989 (Heusmann and Sauer 2000). Estimated spring Common Merganser populations in the state were between 8,000 and over 18,000 in 15 of the 17 years between 1993 and 2009 (Jacobs et al. 2009). AFBWS data show that the highest breeding densities (average of 0.4 birds/km^2) and plot occupancy are in the northeastern corner of the state, followed by the Northwestern Glaciated Plateau and the remainder of the Appalachian Plateaus.

As with the Hooded Merganser, forest maturation in Pennsylvania and the resulting increase in large trees (McWilliams et al. 2007) may have facilitated the spread of the Common Merganser by ensuring a plentiful supply of suitable tree cavities. Perhaps of greater importance, however, is the availability of clean, biologically productive rivers and streams. The resurgence of the breeding Common Merganser population in Pennsylvania will, therefore, depend on sympathetic forestry management and rigorous protection of wetlands and water quality. With these in place, there is the potential for this species to spread to the many as yet unoccupied rivers, particularly in the south of the state.

KEVIN JACOBS

Distribution

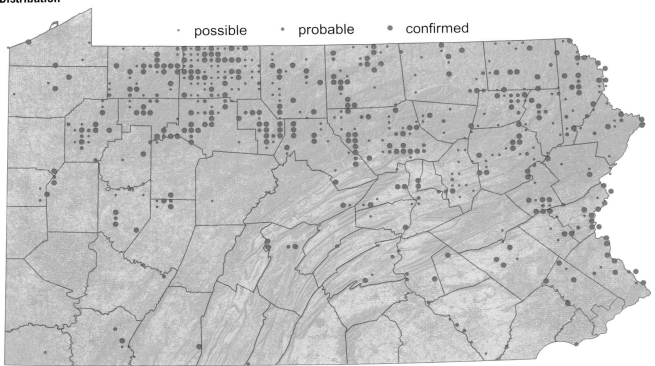

Distribution Change

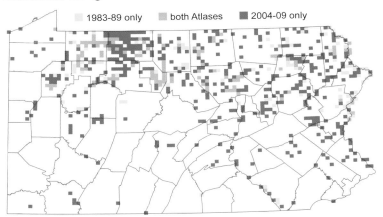

Number of Blocks

	first Atlas 1983–89	second Atlas 2004–09	Change %
Possible	77	228	196
Probable	31	85	174
Confirmed	119	241	103
Total	227	554	144

COMMON MERGANSER

Jacob Dingel

Northern Bobwhite
Colinus virginianus

The Northern Bobwhite is one of the most popular game birds in North America. The male's distinctive *bob-white* call was once commonly heard across farmland in much of southern Pennsylvania, but since the mid-1960s the bobwhite's range and populations have declined dramatically. Its native range includes most of the eastern United States, west to the Rockies, and south into Mexico.

Bobwhites were probably present in most of Pennsylvania counties prior to 1850 (Bolgiano 2000) and likely reached their greatest abundance and distribution between 1820 and 1860. Their range contracted to southern Pennsylvania and the border with Ohio during the first half of the twentieth century. The population core became Fulton and Chester counties and the Lower Susquehanna Valley (Harlow 1913; Jenkins 1942; Ickes 1992a). Pennsylvania's historical abundance of bobwhites followed a common pattern shared with other birds of open country. Before European settlement, a small number of bobwhites lived around natural forest openings. Then, with the expansion of agriculture, bobwhite populations reached their zenith. During the latter part of the twentieth century, farmland lost to development in primary bobwhite counties exceeded one-half million ha (1.4 million ac; Klinger et al. 1998; Goodrich et al. 2002). As farming practices and use of herbicides and pesticides intensified on remaining agricultural lands, bobwhite populations declined further in response to diminished food and cover, although they sometimes temporarily increased as farm abandonment once again created favorable habitat (Leopold 1931; Schorger 1944). Populations declined rapidly between 1945 and 1955 but made a brief recovery in the early 1960s due to set-aside programs, such as the Soil Bank Program (Klinger et al. 1998).

Since 1966, the Northern Bobwhite has shown a dramatic decline on Breeding Bird Survey routes in Pennsylvania (Sauer et al. 2011), with average annual declines of 17 percent between 1970 and 1980 (Sauer et al. 2004), after which they have been too scarce to monitor. Concurrent with the population decline, range contractions resulted in a statistically significant 68 percent reduction in occupied blocks between the first and second Atlas periods and a 76 percent reduction in the number of blocks with confirmed breeding.

Records in the second Atlas were widely scattered, and in some northern counties there were just as many occupied blocks as in the first Atlas. In the former core range in the south, however, the loss of this species is stark; for example, it was found in 81 percent fewer blocks in the Piedmont, exactly the same decline between atlases as was recorded in Maryland's Piedmont (Ellison 2010b). The status of wild Northern Bobwhite populations in Pennsylvania is greatly confounded by the annual release of 60,000 pen-reared bobwhites by game bird breeders, sportsmen's clubs, and private individuals (Dunn et al. 2008). It is not known how many of the second Atlas records were of genuinely wild birds, but most were almost certainly survivors of sporting releases. The fact that 15 confirmed atlas reports involved young birds suggests that the bobwhite still breeds sparingly in the wild, but it is questionable whether there are now any truly self-sustaining wild populations left in Pennsylvania.

Restoring Northern Bobwhite populations in Pennsylvania will require the restoration of farmland ecosystems on a landscape scale. The intensity of agricultural activity today leaves little habitat for Northern Bobwhites or other shrubland-dependent species. The Northern Bobwhite was included within the Wildlife Action Plan by the Pennsylvania Game Commission (PGC), due to long-term population declines (PGC-PFBC 2005). The PGC recently completed the draft Pennsylvania Northern Bobwhite Quail Recovery Plan 2011–2020 (Klinger 2011), which outlines the efforts needed over the next 10 years to restore and maintain the bobwhite as a breeding species in the state.

SCOTT R. KLINGER

Distribution

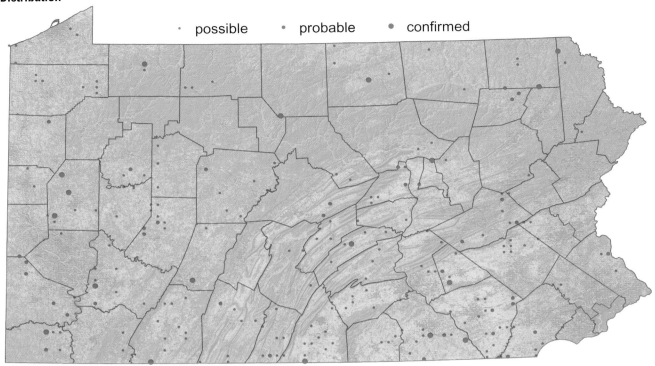

Distribution Change

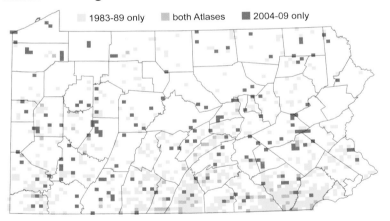

Number of Blocks

	first Atlas 1983–89	second Atlas 2004–09	Change %
Possible	406	177	−56
Probable	233	36	−85
Confirmed	75	18	−76
Total	714	231	−68

NORTHERN BOBWHITE

Chuck Musitano

Ring-necked Pheasant
Phasianus colchicus

One of the most popular game birds among hunters in Pennsylvania, the Ring-necked Pheasant has become a well-known member of the state's avifauna since the first successful introductions from Asia in the 1890s (Gerstell 1935). Native to Asia, it was introduced into Europe as long as 1,000 years ago. A resurgent enthusiasm for pheasant hunting in Europe in the early nineteenth century was soon exported to North America by European colonists and their descendants. The male's beautiful plumage, long tail, and far-carrying territorial crowing call have ensured that this species has become as culturally entrenched in parts of North America as it has in Europe. However, in many parts of North America, pheasant populations are now much lower than they were just a few decades ago (Sauer et al. 2011).

Following initiation of propagation programs by the Pennsylvania Game Commission (PGC), the Ring-necked Pheasant was described as "very common" in Pennsylvania by the 1920s (Sutton 1928a) and remained so during the middle decades of the twentieth century. Since then, populations have declined more than 95 percent, according to Breeding Bird Survey (BBS) data (Sauer et al. 2011). By the time of the first Atlas, the pheasant had undergone a "dramatic decline" (Brauning 1992b), but it was still widespread in agricultural areas such as the Piedmont.

In the time between the two atlases, the pheasant continued to decline to half the number of blocks with records. The most spectacular range contraction was in the Piedmont, where the number of block records declined by 76 percent, similar to the decline in the Maryland Piedmont (Ellison 2010c). Statistically significant declines in occupancy were recorded in most physiographic sections (appendix C). Perhaps even more telling, breeding was confirmed in only 100 blocks in the second Atlas, compared with 697 in the first Atlas effort.

Despite the overwhelming statewide decline of pheasants, some localized areas sustained a stable or even increased number of blocks, such as Crawford County and parts of Somerset and Fulton counties. However, it is difficult to ascertain whether such localized gains are genuine or due to increased atlas coverage or increased releases of captive-bred birds (R. Boyd, pers. comm.). Areas of highest density on atlas point count surveys coincide with areas in which pheasants were imported from wild populations in the Midwest and areas in which there is habitat and where habitat management for this species is ongoing (e.g., Somerset and Montour counties); hence, it is difficult to know to what extent atlas records of Ring-necked Pheasants reflect naturalized populations. Regardless, it is clear from both BBS and atlas results that this species' populations across the state are no longer self-sustaining.

The decline in Ring-necked Pheasant numbers has been linked to changes in agricultural practices, including earlier hay mowing, a reduction in the acreage of small grains, and reduced winter cover (Klinger and Riegner 2008). Perhaps of most significance was the loss of USDA set-aside programs, which provided hundreds of thousands of acres of idle fields from the mid-1950s to the mid-1970s (Bolgiano 1999). With the end of these programs, it was estimated that secure pheasant nesting habitat was reduced by 86 percent between 1966 and 1992 (Klinger and Hardisky 1998).

With the aim of restoring "self-sustaining and huntable" Ring-necked Pheasant populations to Pennsylvania, the PGC has recommended the establishment of several Wild Pheasant Recovery Areas of at least 4,050 ha (10,000 ac) in extent (Klinger and Riegner 2008). Conservation actions within those areas would include increasing secure nesting/brood habitat and winter cover through state and federal conservation programs, such as the Conservation Reserve Enhancement Program and Wildlife Habitat Incentives Program. Given the precipitous decline of "wild" pheasant numbers, the recovery plan is undoubtedly ambitious; success is not guaranteed, but the leverage provided by this popular game bird could benefit a suite of declining grassland and shrub-scrub species that nest in conservation grasslands (Wilson 2009; Wentworth et al. 2010).

ANDREW M. WILSON

Distribution

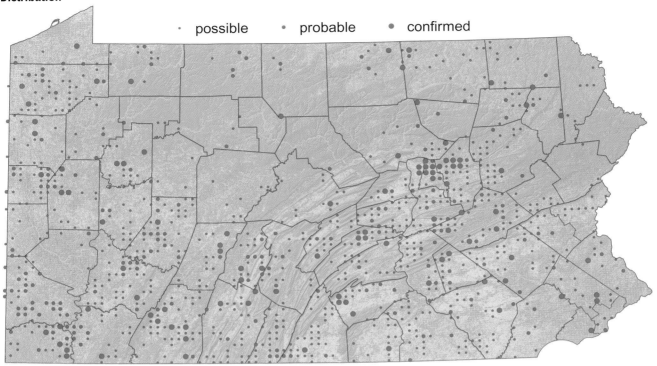

Distribution Change

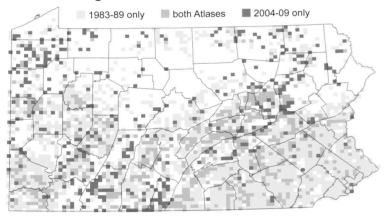

Density

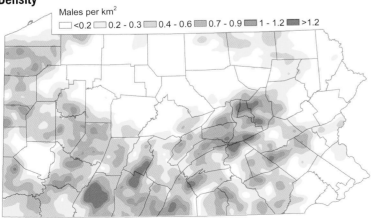

Number of Blocks

	first Atlas 1983–89	second Atlas 2004–09	Change %
Possible	860	770	−10
Probable	639	187	−71
Confirmed	697	100	−86
Total	2,196	1,057	−52

Population estimate, males (95% CI):
40,000 (34,000–48,000)

Breeding Bird Survey Trend

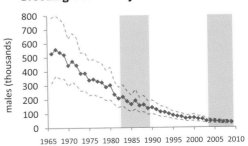

RING-NECKED PHEASANT 111

Jacob Dingel

Ruffed Grouse
Bonasa umbellus

The quintessential species of upland forests, the Ruffed Grouse is a fitting state bird for Penn's Woods. The male's drumming display and the thunderous takeoff of a bird flushed from cover are unmistakable sounds of the outdoors, and "ol' ruff" remains prized by Pennsylvania hunters, who harvested about 90,000 annually during the second Atlas period (Weaver and Boyd 2008).

The Ruffed Grouse is the most widely distributed nonmigratory North American game bird. Its range coincides closely with that of aspen, extending from Alaska across southern Canada and the northern United States and southward along the Rocky and Appalachian mountains. Although northeastern forests do not support grouse densities as high as those of aspen-dominated forests in the upper Midwest, they have historically contained healthy populations of grouse wherever they provide substantial proportions of early successional growth. Regenerating timber harvest areas, shrub thickets, and reverting farmland provide the high stem densities that are an important determinant of grouse habitat quality, particularly for breeding and brood rearing (Dessecker et al. 2006).

During the second Atlas period, volunteers most frequently located Ruffed Grouse in the interior portions of the Appalachian Plateaus and Ridge and Valley provinces, where they were generally found in one-third to two-thirds of the blocks, with over 70 percent occupancy in the Anthracite Upland and Deep Valleys sections. Breeding grouse were much more scattered in the periphery of these two provinces (western tier of counties and the Great Valley Section) and almost entirely absent from the other four physiographic provinces. Overall, about 38 percent of blocks were occupied in the second Atlas, down significantly from 56 percent in the first Atlas; declines occurred statewide, but they were especially dramatic south and east of Blue Mountain and in the Waynesburg Hills in the southwest.

This overall trend is linked to the maturation of Pennsylvania's forests; statewide acreage of small-diameter forest decreased 26 percent between 1989 and 2004 (McWilliams et al. 2007), even while total forest cover remained essentially stable. Southeastern Pennsylvania lost over half of its young forest during this period. With grouse and their habitat uncommon there during the first Atlas, and immigration potential limited for a nonmigratory species with a small home range and short dispersal distance, the virtual disappearance of grouse from the isolated forest fragments in this portion of the Commonwealth is not surprising. In the western counties, the loss of early successional forest between atlases was not markedly more rapid than the state as a whole. However, western Pennsylvania had the highest proportion of early successional habitat in the state during the first Atlas period (McWilliams et al. 2007), probably due in large part to extensive farm abandonment in earlier decades, and grouse may have been near historic highs that could not be sustained as these forests (which are less contiguous than those in the northern tier) matured.

Habitat-driven grouse population declines are not unique to Pennsylvania. Similar trends were evident in New York (Post 2008a) and Maryland (Ellison 2010d), and Ruffed Grouse are a Priority Bird Species for the Appalachian Mountain and Piedmont Bird Conservation regions (Rosenberg 2004). Still, the long-term outlook for Ruffed Grouse in Pennsylvania is not necessarily bleak. Outside of the southeastern portion of the state, where forests have probably become too patchy to support grouse populations at historic levels, potential grouse habitat in Pennsylvania remains widespread, but suboptimal, due to increased forest age. Factors such as economic maturity of timber resources, increased interest in biofuel production, and improved awareness of early successional habitat issues may favor increased levels of active forest management and a resurgence in the proportion of young forest on the landscape in coming decades. Should this occur, many "good old days" may still lie ahead for grouse enthusiasts in the Keystone State.

IAN GREGG

Distribution

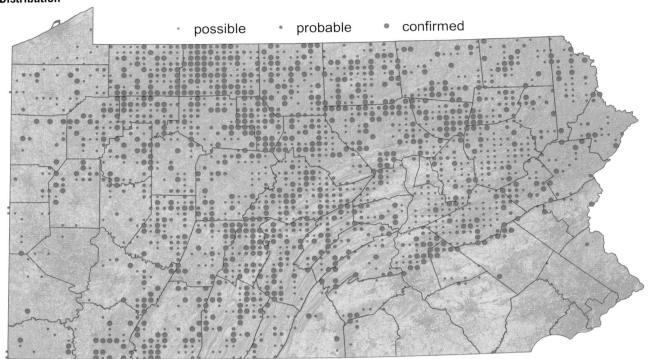

Distribution Change

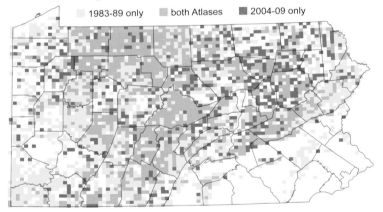

Number of Blocks

	first Atlas 1983–89	second Atlas 2004–09	Change %
Possible	927	810	−13
Probable	439	398	−9
Confirmed	1,416	662	−53
Total	2,782	1,870	−33

Jacob Dingel

Wild Turkey
Meleagris gallopavo

The Wild Turkey is native to North America and found in every mainland state but Alaska (Eaton 1992). Of six subspecies, the Eastern Wild Turkey (*M. g. silvestris*) is the largest and the only one to occur in Pennsylvania. One of the state's largest birds, gobblers can exceed 11 kg (25 lb), but they average less; hens average 3.5 to 5 kg (8 to 11 lb; Pelham and Dickson 1992). Predominantly vegetarians, feeding on grass, seeds, soft and hard mast, roots and tubers, young turkeys and brooding hens also consume insects, amphibians, small reptiles, grubs, crustaceans, and other animal matter.

The Wild Turkey was abundant and widely distributed throughout most of North America during European settlement, although it was originally scarce in Pennsylvania's north-central counties, due to prolonged deep snow and understory shading from vast expanses of forests containing mature White Pine and Eastern Hemlock (Wunz and Hayden 1981). Native Americans and early settlers depended on turkeys as a reliable and important source of food. By the 1930s, Wild Turkey populations were extirpated or reduced to very low levels, due to loss of forests by logging, agricultural development, and unregulated hunting, except in the remote and rugged portions of their range (Kennamer et al. 1992). Populations across New England and most Mid-Atlantic states were extirpated, but south-central Pennsylvania's Ridge and Valley Province maintained about 10 percent of the nation's estimated 30,000 remaining turkeys; this area proved essential for Wild Turkey restoration. Following farm abandonment and reforestation, Wild Turkey populations began to expand northward. By the early 1950s, the secondary growth of broadleaf forests in that area provided suitable habitat for turkeys to expand naturally (Wunz 1978). The Pennsylvania Game Commission (PGC) released almost 200,000 turkeys raised in game farms, without much success, between 1915 and 1981 (Wunz and Hayden 1981). But several trap and transfer programs initiated by the PGC, beginning in the 1950s, restored and introduced Wild Turkey throughout Pennsylvania; similar programs occurred in nine other states (Tapley et al. 2007; Casalena 2010a).

The significant increase in number of blocks occupied between the first and second Atlases reflects the fact that populations were still expanding since the PGC relocated more than 2,000 Wild Turkey from 1958 through 1983. However, the range also expanded into western and southeastern Pennsylvania since the first Atlas, due to transfers of 719 turkeys, from 1984 through 2003, into suitable-but-unoccupied habitats in western and southeastern Pennsylvania, as well as closing the fall turkey hunting season in that area. Since the Wild Turkey is a permanent resident of Pennsylvania, probable and possible breeding records most likely represent breeding birds. The densest turkey populations in Pennsylvania now occur where there is optimal turkey habitat of 60 percent wooded and 40 percent agricultural or reverting field/shrub habitat—in the extreme northwestern, southwestern, and northeastern counties, as well as counties on the east side of the Susquehanna River basin. These areas contain a diversity of habitats within the normal annual range of about 5 km^2 (~2 miles^2; Casalena 2006, 2010b).

Spring Wild Turkey populations in Pennsylvania increased from an estimated 139,000 by the end of the first Atlas period to an estimated 410,000 by 2001, but they decreased considerably to an estimated 272,000 by 2005, due to the natural lower recruitment populations exhibited when nearing habitat carrying capacity, coupled with liberalized hunting seasons and several consecutive years of harsh winters and inclement spring weather (Casalena 2006). With hunting season restrictions beginning in 2004, and favorable spring weather, turkey populations quickly rose again to an estimated 360,000 in 2010 (Casalena 2010b).

Recovery of Wild Turkeys from near extinction to present levels is a success story attributable to the efforts of wildlife agencies and private conservation organizations in applying sound habitat management, resulting in abundant populations for the enjoyment of all.

MARY JO CASALENA

Distribution

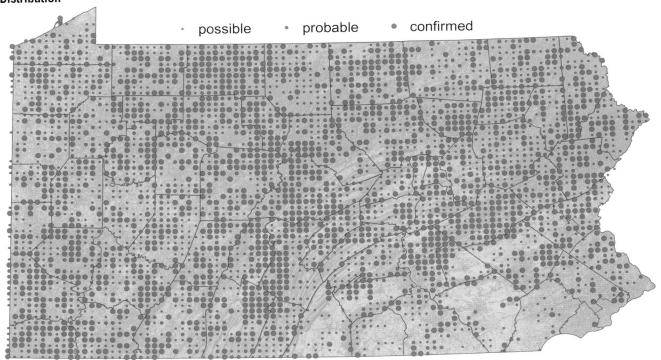

Distribution Change

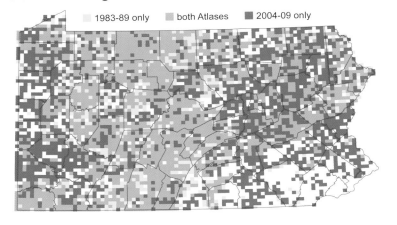

Number of Blocks

	first Atlas 1983–89	second Atlas 2004–09	Change %
Possible	748	1,429	91
Probable	336	545	62
Confirmed	1,318	1,575	19
Total	2,402	3,549	48

Rolf Nussbaumer/VIREO

Pied-billed Grebe
Podilymbus podiceps

The only member of its family to breed in Pennsylvania, the Pied-billed Grebe may be found in a variety of wetland habitats where open water and emergent vegetation are present. Although it is not particularly shy most of the year, during the nesting season it becomes secretive and tends to sink out of sight when approached or disturbed. The species breeds across southern Canada and throughout the United States. However, due to the decline in populations and habitats in the state, it is listed as a Maintenance Concern species in Pennsylvania's Wildlife Action Plan (PGC-PFBC 2005).

The Pied-billed Grebe has never been a common breeder in the state, but it was historically most regular in extreme southeastern and northwestern Pennsylvania (Poole 1964). The first Atlas indicated that the species was found most frequently in northwestern blocks and had become extremely rare in southeastern Pennsylvania (Ickes 1992b).

During the second Atlas, the Pied-billed Grebe was reported in only 64 blocks, a decrease of 26 percent compared with the first Atlas. However, this difference was due mainly to fewer possible breeding records; the number of blocks with confirmed records was the same in both atlases. The majority of Pied-billed Grebe reports were again in northwestern Pennsylvania, with about 30 percent recorded from six counties. But, unlike the decrease observed in the southeastern counties during the first Atlas (just two possible breeding records in the five-county Philadelphia area; Ickes 1992b), pied-bills were found in seven blocks there in the second Atlas, including confirmed nesting in two.

When the 26 confirmed grebe records of the second Atlas are examined, there appears to be a relationship with the amount of wetland present: 20 of the confirmations were located in 16 counties having blocks with at least 3 percent emergent wetland cover and as much as 11 to 20 percent wetland cover (chap. 3). The population size of the Pied-billed Grebe during the second Atlas was estimated using all of the probable and confirmed breeding records. The possible breeding records were not considered, since many of those may represent non-breeding summer visitors that failed to migrate north. Assuming records for all atlas years represented different birds, the maximum number of breeding pairs in Pennsylvania during the second Atlas period was 45. However, the average number of probable and confirmed breeding records per year during the atlas period was only nine, suggesting that the statewide population could be in the low double-figures.

Because the number of Pied-billed Grebe blocks in the state with confirmed records was the same for both first and second Atlas periods, and given that the species' population has remained stable in Maryland (Ellison 2010e) and increased 47 percent in New York between first and second atlas periods (McGowan 2008c), there is reason to be cautiously optimistic about the conservation outlook for this species in Pennsylvania and surrounding states. The keys to the grebe's success appear to be the preservation and creation of wetlands having certain characteristics. For example, recent modeling studies (Osnas 2003; Lor and Malecki 2006) concluded that wetland habitat characteristics such as amount of emergent vegetation and water depth are important in determining whether Pied-billed Grebe nests are present; notably, grebe nests were located in sites with at least 70 percent emergent vegetation, with a mean water depth of 24 to 56 cm (9 to 22 inches), and an average vegetation height that ranged from 69 to 133 cm (27 to 52 inches). However, a possible limiting factor for the Pied-billed is human disturbance, since non-motorized boat traffic entering water lily beds and waves from boats towing water skiers can result in grebe nest destruction (Muller and Storer 1999).

ROY A. ICKES

Distribution

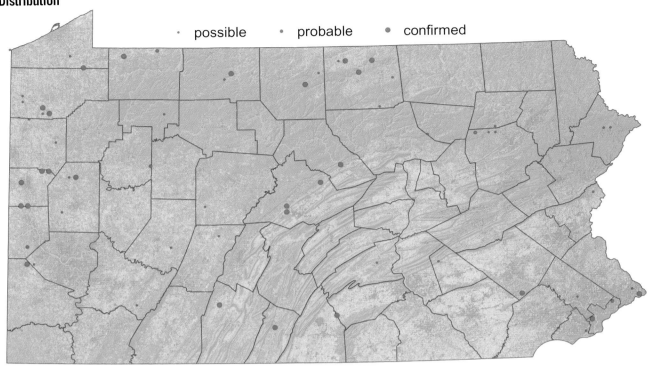

Distribution Change

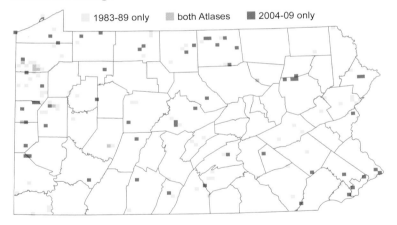

Number of Blocks

	first Atlas 1983–89	second Atlas 2004–09	Change %
Possible	48	33	−31
Probable	12	5	−58
Confirmed	26	26	0
Total	86	64	−26

PIED-BILLED GREBE

Rob Criswell

Double-crested Cormorant
Phalacrocorax auritus

Among a handful of currently breeding species without confirmed breeding evidence during the first Atlas effort, the Double-crested Cormorant is Pennsylvania's newest colonial waterbird nester. The "sea crow," as its scientific name implies, is a large, black, almost exclusively piscivorous bird that prefers open water (lakes, reservoirs, and major rivers) as foraging habitat. It ranges from Alaska to Baja California, across parts of the North American interior, to the Atlantic seaboard, Gulf of Mexico, and Cuba.

The recent expansion of breeding populations into the Great Lakes, where there were no breeding records prior to the twentieth century, and Pennsylvania, where there are no historical breeding records, has been attributed to a variety of causes. These include release from chlorinated hydrocarbon pesticide burdens, protection as a migratory species (1972 amendment to the Migratory Bird Treaty Act), reduced over-winter mortality (utilization of southern catfish ponds), and greater food sources in the Great Lakes and large rivers (primarily the invasive Alewife; Hatch and Weseloh 1999). Double-digit breeding population increases of this piscivorous bird in the Great Lakes and many eastern states since the 1970s have caused concern in the commercial and sport fish fishery industries about potentially negative impacts on fish populations and fisheries (Ross and Johnson 1999). Resource agencies are also concerned about competition for nesting trees and sites where the Double-crested Cormorant has invaded breeding colonies of waterbird species at risk or listed as Endangered or Threatened, such as Wade Island in Harrisburg.

Though scattered, possible and probable breeding observations occurred throughout the state during the second Atlas effort, but only four confirmed breeding records were obtained, two from the lower Susquehanna River, one from the Delaware River, and one from nearby Peace Valley Reservoir (Bucks County). Three of those records involved nest building or courtship behavior, while only one, from the established breeding colony at Wade Island in the Susquehanna River, documented young birds at the nest. This was the site of the first Double-crested Cormorant nest recorded in Pennsylvania in 1996; the number of nests has increased rapidly since then, to 120 in 2009 (Haffner and Gross 2009; Ross 2010a).

The number of blocks in which breeding evidence was observed increased from 9 to 111 between the first and second Atlases, mostly from the same geographic locations (lower Susquehanna and Delaware rivers). These sites represent optimum habitat and are logical extensions of existing breeding colonies in New York state (Lakes Ontario and Champlain as well as Oneida Lake), the Chesapeake Bay, and the Mid-Atlantic coast. Future expansion of Pennsylvania's breeding population into large reservoirs scattered throughout the state seems likely, though at a slower rate in western Pennsylvania, due to fewer (Lake Erie) or absent (Ohio River) breeding colonies in those drainages. A lack of suitable islands (preferred breeding sites) in Lake Erie may prevent nesting in Pennsylvania's portion of Lake Erie.

The Breeding Bird Survey trend for Double-crested Cormorant in the eastern region of North America from 1999 to 2009 was an increase of 13 percent per year (Sauer et al. 2011). Although this significantly positive trend will likely slow in coming years, continued expansion of the breeding population may well occur, especially in previously unpopulated states such as Pennsylvania. Because cormorants tend to invade existing waterbird colonies, eventually displacing other species, the Pennsylvania Game Commission, charged with protecting listed species, recently took steps to manage nesting cormorant numbers. Culling of breeding birds took place in 2006 at Wade Island, which houses the state's only nesting Great Egrets and the largest Black-crowned Night-Heron colony (both species are listed as Endangered in Pennsylvania). Concern about state fisheries is presently premature but in need of continued monitoring (Ross 2010a). The U.S. Fish and Wildlife Service continues to allow limited take of this species where fisheries or local economic impacts are significant (USFWS 2003a).

ROBERT M. ROSS

Distribution

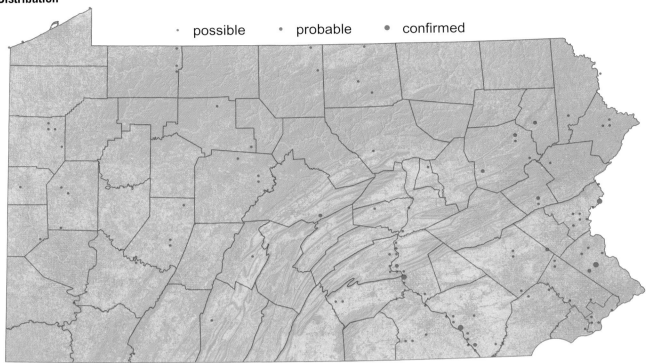

Distribution Change

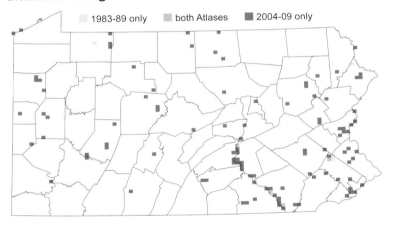

Number of Blocks

	first Atlas 1983–89	second Atlas 2004–09	Change %
Possible	9	98	989
Probable	0	9	∞
Confirmed	0	4	∞
Total	9	111	1,133

Bob Moul

American Bittern
Botaurus lentiginosus

The American Bittern, often solitary and difficult to see, is a cryptically plumaged heron that breeds in freshwater emergent marshes from the mid–United States almost to the tree line in northern Canada. Known locally in Pennsylvania as "Bum Chuck," "Stake-driver," or "Thunder-pumper" for its peculiar territorial vocalization, the American Bittern often stands motionless, with bill pointed skyward, its vertical stripes blending perfectly with its cattail milieu.

Surprisingly, little is known about the American Bittern's basic ecology (Gibbs et al. 1992). It is a migratory bird that typically arrives in Pennsylvania from early April to mid-May, with the majority of its population to our north. Area-sensitive, it breeds predominantly in extensive freshwater marshes, especially ones with dense stands of cattails, spatterdock, bulrushes, and sedges interspersed with open water (Gibbs et al. 1992; Leberman 1992a).

The American Bittern was formerly much more numerous in Pennsylvania (McWilliams and Brauning 2000) and was described as common in Pymatuning Swamp (Sutton 1928b), Swamp Run (Geibel 1975), and Conneaut Marsh (Hartman 1989) in western Pennsylvania. By the time of the first Atlas, it was a rare breeding bird in most of the state (Leberman 1992a). It was found in 53 atlas blocks and confirmed in only 5; of these, 17 of the occupied blocks and 2 of those with confirmed breeding were in the Northwestern Glaciated Plateau. The second Atlas period documented a further statewide decline, with observations in only 31 blocks and confirmed breeding in only 2. The Northwestern Glaciated Plateau Section did not escape the decline; in fact, a 71 percent drop in block occupancy there greatly exceeded the 28 percent drop across the remainder of the state. The stronghold for this species in the second Atlas shifted to Tioga County, notably the extensive Marsh Creek wetlands, where the only record of successful breeding in the second Atlas was obtained: a recently fledged juvenile on 4 July 2007. The only other confirmed breeding record was of one in distraction display in Centre County on 31 May 2006. Even allowing for some underrecording of this secretive species, it seems improbable that its Pennsylvania population now exceeds the low double digits of breeding pairs in any given year.

Declines in American Bittern numbers in Pennsylvania are part of rangewide decreases (Gibbs et al. 1992). A targeted marshbird-monitoring program, conducted throughout the Great Lakes Basin, showed an 8.8 percent yearly decline from 1995 through 2003 (Crewe et al. 2005). Although the American Bittern is not as rare in New York as it is in Pennsylvania, block occupancy declined by 10 percent in New York between atlases (McGowan 2008d), and there were significant declines between atlases in the southernmost region of Ontario (Timmermans 2007a). The species was listed as Threatened in Pennsylvania from 1979 through 1997 and then downgraded to Endangered (Gill 1985; McWilliams and Brauning 2000). In Pennsylvania, over 50 percent of historic wetlands have been lost, and many of the remaining areas are degraded (Goodrich et al. 2002). Net increases in wetland acreage in Pennsylvania since 1990—due to the successful Partners for Wildlife Program, managed by U.S. Fish and Wildlife Service and the Wetland Reserve Program—were apparently not sufficient to sustain bittern populations.

Conservation efforts need to minimize future loss and degradation of extant emergent wetlands as well as increase their area and contiguity, through habitat restoration, where possible. Because the American Bittern is difficult to monitor, we do not have good information about where most birds are breeding or about how successful they are. The second Atlas has identified a few key sites, and site-specific conservation plans should now be developed. Restoring and sustaining a viable population of American Bitterns and other wetland species in Pennsylvania will require restoring and sustaining the ecosystems upon which they depend.

GENE WILHELM

Distribution

Distribution Change

Number of Blocks

	first Atlas 1983–89	second Atlas 2004–09	Change %
Possible	40	21	−48
Probable	8	8	0
Confirmed	5	2	−60
Total	53	31	−42

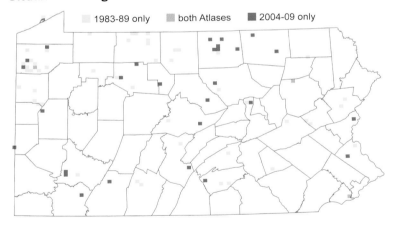

AMERICAN BITTERN

Geoff Malosh

Least Bittern
Ixobrychus exilis

The Least Bittern, the smallest North American heron, is quite secretive in its movements through the dense emergent vegetation that characterizes its preferred marsh habitats. It is a striking bird with a tan neck and sides, additional tan streaking on its white undersides that contrasts with a glossy greenish-black back, head, and crown, and chestnut wings with large buff-colored patches. It has an array of vocalizations; the cuckoo-like *cu-cu-cu* and rail-like ticking are the most common and easily recognized (Horn 2009; Gibbs et al. 2009). The Least Bittern occurs throughout the western hemisphere from southern Canada southward to Central and South America (Horn 2009). The majority of the North American population breeds in the eastern half of the United States, with scattered pockets occurring in several western states (Gibbs et al. 2009).

More restricted in its habitat preferences than other marsh birds, the Least Bittern prefers larger wetland complexes containing dense stands of tall emergent vegetation, primarily cattails, and small open areas of water interspersed throughout the habitat (Brown and Dinsmore 1986; Moore et al. 2009; NYNHP 2009). Roughly equal amounts of emergent vegetation and open water, and fairly stable water regimes and water depths of 10 to 50 cm (4 to 20 in), are key habitat requirements (Gibbs et al. 2009; Horn 2009). Its preference for specific vegetation types could preclude any tolerance for invasive plant species such as Purple Loosestrife and Common Reed.

Although the Least Bittern is not federally listed, its numbers have declined throughout much of its breeding range, primarily because of continuing loss of suitable habitat from wetland draining, impoundments, land development, and agricultural practices (Moore et al. 2009). The most notable population declines in the United States have occurred in the northeastern and north-central states. As a consequence, it is listed as either Threatened or Endangered in 16 states, including Pennsylvania, where it is listed as Endangered (Brauning 2010a). It was formerly more numerous in Pennsylvania, especially in the lower Delaware River marshes, where counts of as many as 27 nests were made at Tinicum (John Heinz National Wildlife Refuge) in the 1950s (Brauning 2010a).

There is evidence of a modest decline in Least Bittern numbers between atlases, with records from only 28 blocks in the second Atlas (confirmed in only 6), compared with 31 and 9 respectively in the first Atlas. The expectation would be for increased detections during the second Atlas period, relative to the first, if populations were stable. Therefore, the results pertaining to this species, as well as other secretive marsh birds, should be interpreted with caution.

Only four blocks were occupied in both atlas periods, but there was little evidence of a change in distribution overall, with the Northwestern Glaciated Plateau providing roughly one-third of records in both periods. The extensive marshes of Crawford County have long been noted as a stronghold (Brauning 2010a). In the southeast, Tinicum remains important; five nests were found there in the first year of the second Atlas. Elsewhere, records were very thinly scattered, reflecting the paucity of suitable habitat for this species across most of the state. Due to its elusive nature, little is known about the Least Bittern's population size. Most records submitted to the atlas were of single birds, pairs, or their offspring, but there was evidence of more than one pair at a handful of sites. Even so, it seems doubtful that there were more than 100 breeding pairs in the state, and possibly far fewer.

Habitat loss and degradation are the greatest threats to Least Bittern. Areas in which this species regularly occurs should be protected and managed to maintain appropriate native vegetative composition and structure. In addition, water should be monitored for quality and levels properly maintained at stable levels. Consideration should also be given to expanding emergent wetlands where possible to create additional large-scale complexes conducive to increasing potential Least Bittern breeding habitat.

KIM VAN FLEET

Distribution

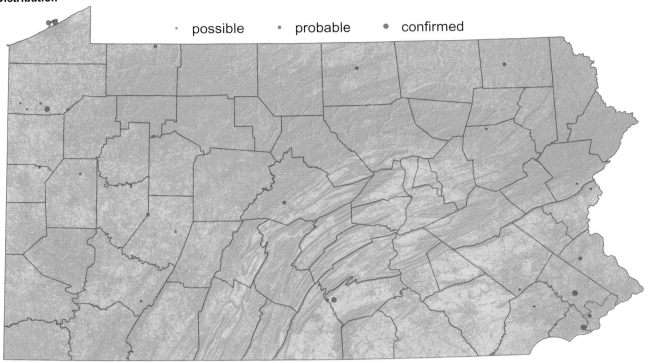

Distribution Change

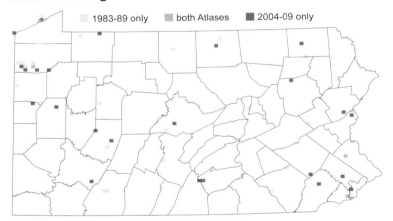

Number of Blocks

	first Atlas 1983–89	second Atlas 2004–09	Change %
Possible	15	14	−7
Probable	7	8	14
Confirmed	9	6	−33
Total	31	28	−10

Jacob Dingel

Great Blue Heron
Ardea herodias

At 1.5 m (5 ft) in height, the Great Blue Heron is Pennsylvania's largest piscivorous wader, exceeded in size, among all the state's breeding birds, only by Wild Turkey, Bald Eagle, Sandhill Crane, and Mute Swan. The species breeds across North America from Alaska to Guatemala, east to Cuba, and north to Nova Scotia. Primarily a colonial nester, its greatest densities are found along coasts and on Great Lakes islands (Butler 1992). However, Pennsylvania harbors a significant inland breeding population, with numerous small- to medium-sized colonies scattered throughout the state (Ross 2010b). The Great Blue Heron is listed neither globally nor at the state level, and its state breeding status is characterized as secure.

Historically, the Great Blue Heron was a widespread nesting bird in Pennsylvania, but direct persecution and tree felling reduced numbers such that, by the mid-twentieth century, it was restricted to the northwestern corner of the state (McWilliams and Brauning 2000). A population recovery was well under way by the first Atlas period, but in colonies away from the northern tier, they remained few in number. Between atlases, there was a significant 28 percent increase in the number of atlas blocks with Great Blue Heron records, and the number of blocks with confirmed breeding increased from 114 to 203. The greatest increases in block records occurred in the Piedmont (88%) and Ridge and Valley (53%) physiographic provinces, resulting in a significant southward range expansion for this species. The breeding distribution of Great Blue Heron in Pennsylvania is now nearly uniform across the state, with apparent gaps only along the Appalachian Plateaus Province, coincident with the West Branch Susquehanna River and perhaps a few additional drainages southwest to the Maryland border. Many of these waters continue to be impaired by mine drainage or acid precipitation, with absent or reduced fish populations (Cooper and Wagner 1973).

As in the first Atlas, only observations of nests were included as confirmed breeding records, since this species wanders widely in search of food, and fledged young disperse considerable distances from nest sites (Schwalbe and Ross 1992). All other records were listed as possible breeding, except for a few reports of birds carrying nesting material, which were classified as probable breeding. The Great Blue Heron nests early; atlas volunteers observed occupied Great Blue Heron nests as early as 28 February, so there is a chance that nests of this species were overlooked in blocks that were not visited before leaf out. Nests with eggs were noted from early May to mid-June, nestlings as late as mid-July, and fledged young from mid-June to late July.

The Great Blue Heron has increased steadily in abundance and distribution in Pennsylvania since the mid-1960s, according to Breeding Bird Survey data (Sauer et al. 2011), which suggests that the population doubled between atlas periods. The first thorough statewide survey of colonies was in 1993, when 1,654 nests were counted in 36 counties (Ross 2010b). In a 2007/08 survey, 2,217 nests were counted in 116 active colonies in 52 counties (Haffner and Gross 2008), representing a 35 percent increase since 1993. Records submitted to the second Atlas show, by comparison, a total of 2,734 nests throughout the state, although this spanned a 6-year period. The state's largest colony, at Barrows in Mercer County, held 225 nests in fall 2009, fewer than the peak of 441 counted in 1999 (Haffner and Gross 2010a). The average colony size of atlas records was 13.8 nests.

The conservation outlook for this migratory species looks favorable in Pennsylvania, with certain cautionary conditions. Nesting habitat is abundant statewide, and the species seems to be highly adaptable to human activities, extending its range into increasingly urbanized and developed landscapes. However, logging, mineral extraction, and energy development, particularly gas drilling activity and wind farm development, constitute an increasing threat for colonies in both private and public forests. Further loss of wetlands and contaminants in foraging habitats poses risks both in Pennsylvania and on wintering grounds of southern coasts and the Caribbean (Butler 1992).

ROBERT M. ROSS

Distribution

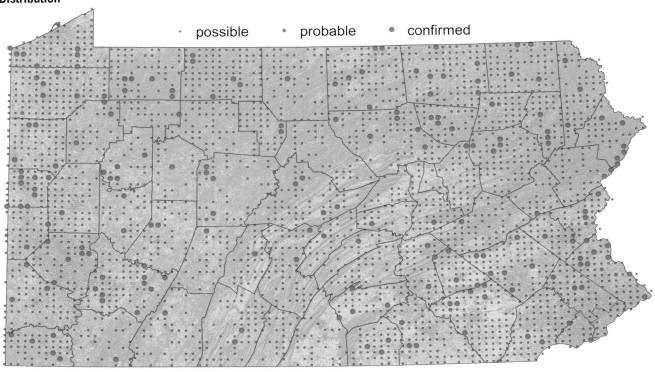

Distribution Change

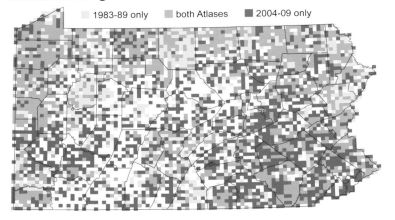

Number of Blocks

	first Atlas 1983–89	second Atlas 2004–09	Change %
Possible	2,148	2,695	25
Probable	17	10	−41
Confirmed	114	203	78
Total	2,279	2,908	28

Breeding Bird Survey Trend

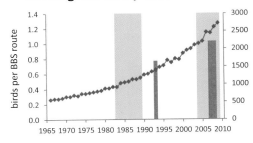

Gray bars are active nest in 1993 and 2007/08 surveys (Haffner et al. 2008) - right axis.

Joe Kosack/PGC

Great Egret
Ardea alba

The elegant and graceful lines of a Great Egret quietly stalking a shoreline contrast markedly with the raucous nature of a nesting colony. We are fortunate that this species still nests and forages in Pennsylvania, since rangewide extinction was imminent at the turn of the twentieth century due to plume hunting for the millinery trade. Unlike the smaller Snowy Egret, whose head is adorned with plumes in the breeding season, Great Egrets develop only scapular plumes that extend beyond the tail. Otherwise, breeding plumage is entirely white, contrasting with a yellowish-orange beak, bright green lores and eye-ring, and glossy black legs. Juveniles are also white, but scapular plumes are lacking, lores and eye-ring are black, and the beak is shorter and stocky in appearance with a blackish tip (McCrimmon et al. 2001).

Probably known by more names than any other heron, this species occurs on every continent except Antarctica. Its range extends across North America, although most of the population is found in coastal areas. Inland, the species prefers river valleys in the East and Midwest, and well-watered valleys in the West (McCrimmon et al. 2001). The Great Egret is seen across Pennsylvania as a post-breeding wanderer, but it nests only in the Lower Susquehanna River Valley at two sites: Wade Island in Harrisburg and Kiwanis Park in the city of York. It is listed as Endangered in Pennsylvania due to its low population and limited number of colony sites (Master 2010a). Rangewide, the population is stable to slightly increasing (Kushlan et al. 2002).

Wade Island is the state's largest mixed-species wading bird colony. It averaged 171 active Great Egret nests over 6 years of monitoring during the second Atlas period (2004–2009), compared with 57.5 over 5 years during the first Atlas period (Haffner and Gross 2009; PGC, unpublished data). The second colony, at Kiwanis Park in York, predominantly hosts Black-crowned Night-Herons, but it included three to eight Great Egret nests per year during the second Atlas (Haffner and Gross 2009). An interesting single probable-breeding record came from Bedford County, where a single pair was observed mating, but there was no subsequent proof that nesting occurred there. Individuals observed in other blocks likely represent birds either foraging away from established colony sites or non-breeding individuals.

The first Atlas confirmed nesting in Philadelphia along the lower Delaware River and at Rookery Island in Lancaster County. These sites have long been abandoned (Schutsky 1992a; Ross 2010a). Thus, the center of distribution in the lower Susquehanna River Valley remains the same: the number of colonies has declined but the number of nesting individuals has doubled since the first Atlas.

Great Egrets arrive at Wade Island on a remarkably consistent schedule, usually appearing from 13–17 March each year (S. Lockerman, pers. comm.). Nesting begins about a month later, following courtship and nest repair activities. Small fish compose 48 percent of their diet, rusty crayfish 30 percent, and tadpoles 18 percent. They prefer feeding in water willow shallows surrounding river islands and on the nearby Conodoquinet Creek (Romano 2008). Fifty-eight percent of the Great Egrets observed during an aerial survey of the Susquehanna River were found foraging within 5 km of Wade Island, with a few birds ranging as far as 28 km north of Wade Island. When mean depth reaches approximately 2 m (6 ft) at the USGS Harrisburg gauge, individuals leave the Susquehanna River and forage on ponds in the surrounding countryside, including Wildwood Park in Harrisburg (Romano 2008).

The increase in records of wandering, non-breeding birds during the second Atlas offers hope that this species may expand its range in Pennsylvania. In both Maryland (Therres 2010a) and New York (McCrimmon 2008a), small populations are now established well away from traditional coastal nesting sites. Continued erosion and subsequent loss of nesting trees on Wade Island, as well as increasing numbers of Double-crested Cormorant that compete for nesting sites and are known to kill nesting trees as a result of their more acidic droppings (Master 2010a; Romano 2008), are of great concern regarding the continued existence of the Great Egret as a breeding bird in Pennsylvania.

TERRY L. MASTER

Distribution

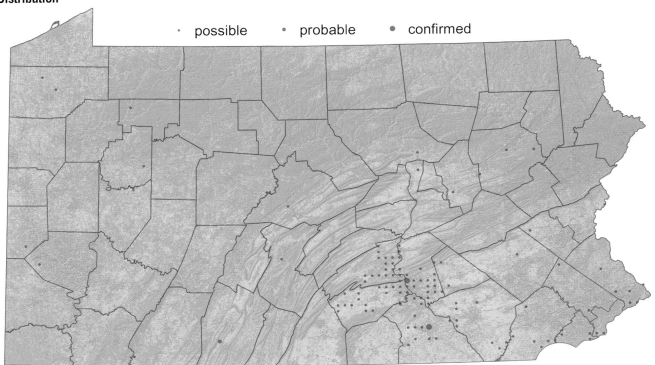

Distribution Change

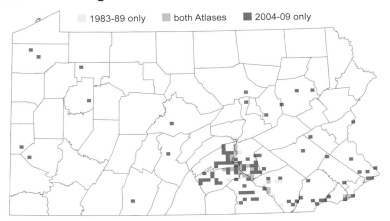

Number of Blocks

	first Atlas 1983–89	second Atlas 2004–09	Change %
Possible	29	114	293
Probable	3	1	−67
Confirmed	4	2	−50
Total	36	117	225

Geoff Malosh

Green Heron
Butorides virescens

Less than half a meter in height, this small wader is primarily a solitary nester throughout its wide range, which includes the entire eastern United States, Caribbean region, Central America, and western American coastal areas from Baja California to Puget Sound. Primarily a fish predator, but also feeding on a broad range of small vertebrates and invertebrates, both aquatic and terrestrial (Davis and Kushlan 1994), the Green Heron's foraging habitat in Pennsylvania overlaps that of Great Blue Heron (open wetlands and rivers). However, the Green Heron seems to prefer nesting in swampy thickets associated with smaller streams and wetlands, where it often forages amid dense vegetation or under closed riparian canopies. Nests may be found near or some distance from streams or wetlands, but these usually are built in small or mid-level trees, such as wild crabapples and hawthorns, rather than in tall woodland trees. Because of this heron's largely solitary habit, population numbers anywhere are difficult to estimate (Davis and Kushlan 1994). However, the Green Heron is widespread and adaptable to human habitats, so it is not considered at risk in Pennsylvania or regionally.

The Green Heron was found in just over one-third of atlas blocks during the second Atlas, being somewhat concentrated in the northwestern and southeastern corners of the state. Its block occupancy was lowest (30%) in the Appalachian Plateaus Province and highest in the Piedmont Province (60% of blocks). This distribution is similar to that found during the first Atlas. However, comparing the second Atlas with the first Atlas, block occupancy in the Appalachian Plateaus Province was significantly lower (13% decrease), in particular within the Waynesburg Hills and Glaciated Pocono Plateau sections (down 55% and 60%, respectively). Changes in the other provinces and sections, although predominantly negative, were not statistically significant.

Highest rates of block occupancy during the second Atlas were observed in the Piedmont Province (60%), the Blue Mountain (61%) and Great Valley (54%) sections of Ridge and Valley Province, and the Northwestern Glaciated Plateau Section (58%) of Appalachian Plateaus Province. Only slightly different occupancy rates occurred in priority blocks within these same sections, suggesting little or no coverage bias. Additional breeding "hotspots" were found in scattered areas across the northern tier. A common habitat feature of all of these areas is an abundance of wetlands and low-gradient streams with intact riparian zones.

Atlas volunteers confirmed Green Heron breeding as early as 1 May and as late as 21 August, with an average date of 27 June. Nests with eggs were observed between 19 May and 8 July, nests with young between 1 June and 28 July, and fledged young as late as 21 August. These egg dates extend those previously reported in Pennsylvania by nearly 2 weeks (latest reported as 26 June by Master [1992a]).

Although Master (1992a) reported no discernible population trend for Green Heron through the first Atlas period, the trajectory for this species has since changed. The 1966–2009 Breeding Bird Survey trend was downward, with an estimated 15 percent decline between atlas periods (Sauer et al. 2011). BBS trends across North America showed a significant annual decline of 1.5 percent between 1966 and 2009 (Sauer et al. 2011).

Although Pennsylvania still has a substantial breeding population of Green Heron, its principal habitats may have diminished in quality, if not quantity, since the first Atlas. Emergent wetland acreage may have increased modestly, statewide, in the past two decades (PADEP 2010), but loss of small wetlands and degradation of streams and their associated riparian margins continue, due to a variety of conflicting land uses such as exurban development, longwall coal mining, mine drainage, and intensive agriculture. These factors undoubtedly contribute to the current slow decline of this species. Though a potential problem at aquaculture facilities and fish hatcheries, this solitary species does not typically cause as much damage as some colonial waterbirds and thus is not at risk from such behavior (Parkhurst et al. 1992; Glahn 1997).

ROBERT M. ROSS

Distribution

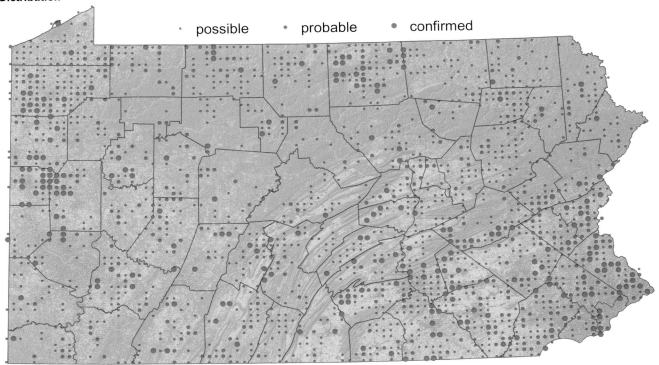

Distribution Change

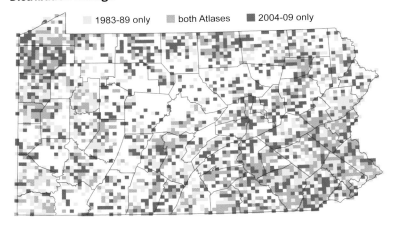

Number of Blocks

	first Atlas 1983–89	second Atlas 2004–09	Change %
Possible	1,278	1,199	−6
Probable	480	366	−24
Confirmed	233	227	−3
Total	1,991	1,792	−10

Breeding Bird Survey Trend

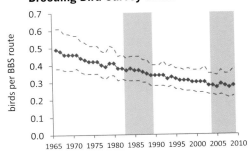

GREEN HERON

Rob Curtis/VIREO

Black-crowned Night-Heron
Nycticorax nycticorax

An overall grayish color, enhanced by a black back and crown contrasting with two pure white head plumes, and brilliant red eyes characterize the appearance of the Black-crowned Night-Heron. This heron is found in wetland and riparian habitats on every continent except Australia and Antarctica. In North America, it ranges southward from central Canada throughout the entire continent but is most common in coastal areas. In Pennsylvania, major colonies are well known but ephemeral. The existence of scattered, smaller colonies and individuals contributes to infrequent sightings and detection difficulty during atlas efforts. The species is listed as Endangered in Pennsylvania and Delaware and Threatened in New Jersey and Ohio (Hothem et al. 2010). The population is considered stable in North America and stable to slightly increasing globally (Kushlan et al. 2002; Hothem et al. 2010).

The Black-crowned Night-Heron was detected in 77 blocks, almost exclusively in the Piedmont Province. A few blocks were occupied along the North Branch of the Susquehanna River in northeastern Pennsylvania and in the lower Delaware River Valley. Confirmed breeding records came from only 16 blocks, half of which included colonies in Berks, Dauphin, Lancaster, and York counties. Wade Island in Dauphin County is the largest and most reliable nesting site for this species, with an average of 91 active nests per year during the second Atlas period (Gross and Haffner 2011). A colony at Kiwanis Lake, York County, is second with 82 nests annually on average. The remaining six colonies averaged 12 nests per year. The other eight block confirmations were all of fledged young in blocks in which no colonies or active nests were known.

There has been a 49 percent decline in the number of blocks occupied since the first Atlas, but only two fewer blocks with confirmed breeding (Schutsky 1992b). The most notable colony loss has occurred along the lower Susquehanna River at Rookery Island, Lancaster County, which was abandoned in 1989 after reaching a high of 456 nests in 1985 (McWilliams and Brauning 2000; Ross 2010a). This, the largest-known wading bird colony in Pennsylvania at the time, was abandoned for unknown reasons. Historic colonies in the Philadelphia area were also abandoned during the early 1990s, with only possible-breeding records noted on the lower Delaware River during the second Atlas. No individuals were observed across central and western Pennsylvania, as was the case during the first Atlas.

Black-crowned Night-Herons arrive at the state's largest remaining colony on Wade Island in early April and begin nesting approximately 2 weeks later. They nest in this mixed-species colony with Great Egret and Double-crested Cormorant. On average, pairs hatch three eggs, of which 80 percent survive to fledging (Detwiler 2008). Adults and juveniles forage along river shorelines and on the Conodoquinet Creek, generally within 1 km (~0.6 mi) of Wade Island. They consume small fish and Rusty Crayfish, likely an increasingly important dietary component (Detwiler 2008).

Concern exists for this bird as a breeding species in Pennsylvania. There has been a long-term decline in nests in the state, and on Wade Island in particular, where only 42 nests were counted in 2010, the lowest number since surveys began in 1988 (Gross and Haffner 2011). A similar trend is reported for other Mid-Atlantic states (Walsh et al. 1999; McCrimmon 2008b). Historically, depredation at fish hatcheries was a major source of mortality, but covering these facilities has largely eliminated the problem. An exact explanation for the current decline remains elusive. Competition with increasing numbers of Double-crested Cormorant has been suspected on Wade Island as cormorant numbers increase. However, evidence for direct negative interactions between the two is lacking. In fact, night-herons readily consume fish dropped by cormorants (Detwiler 2008). A new colony was recently found in Lancaster County that contained 54 nests in 2010. This is a promising development that emphasizes the importance of continued population monitoring and colony protection for survival of this species in Pennsylvania.

TERRY L. MASTER

Distribution

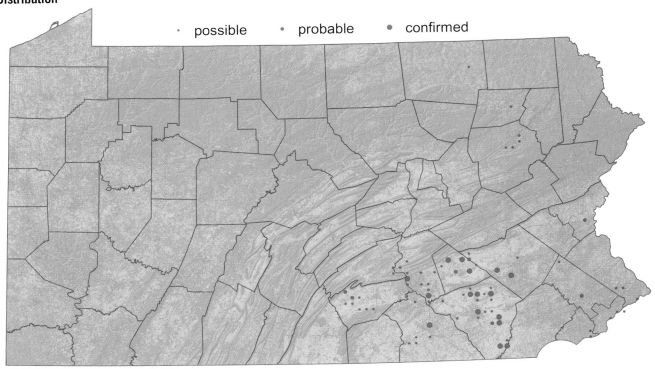

Distribution Change

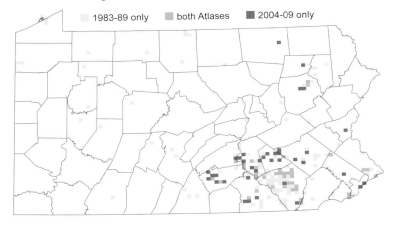

Number of Blocks

	first Atlas 1983–89	second Atlas 2004–09	Change %
Possible	112	54	−52
Probable	22	7	−68
Confirmed	18	16	−11
Total	152	77	−49

Bob Moul

Yellow-crowned Night-Heron
Nyctanassa violacea

The Yellow-crowned Night-Heron is one of Pennsylvania's rarest breeding birds. This most distinctively patterned heron has an instantly recognizable breeding plumage, with its black-and-white facial pattern, cream-colored forehead, and relatively long neck. It breeds from southern New England along the Atlantic Coast to southeastern Brazil, west to central Oklahoma, and south along the Pacific Coast, from Baja California to Peru (Watts 1995). Inland colonies are scattered and typically found along rivers and streams. Overall distribution is, in part, determined by availability of crustacean prey (Watts 1995). A northward expansion of its breeding range occurred from 1925 through the 1960s and included Pennsylvania, where breeding activity has been confined to several sites in the lower Susquehanna River Valley for the past 35 years (Schutsky 1992c; Ross 2010a). The species is listed as Endangered in Pennsylvania and neighboring Delaware and Threatened in New Jersey and Ohio. Its rangewide population is considered to be stable (Kushlan et al. 2002).

The Yellow-crowned Night-Heron was found in only nine blocks during the second Atlas period—six in the lower Susquehanna River Valley, near Harrisburg, and three in the lower Delaware River drainage near Philadelphia. Nesting was confirmed in only four of these blocks, three near Harrisburg, including one encompassing the city's Bellevue Park neighborhood in Dauphin County, and two on the Conodoguinet Creek in Camp Hill, Cumberland County. The fourth block was occupied by one or two nesting pairs within the mixed-species colony at Kiwanis Park, York County, between 2004 and 2006 (Gross and Haffner 2011). The current distribution represents a 57 percent drop from 21 blocks in the first Atlas period, primarily due to abandonment of former nesting sites on the Conestoga and Little Conestoga creeks in Lancaster County (Schutsky 1992c). The number of nests observed annually ranged from 1 to 10 during the first Atlas period and 3 to 6 during the second Atlas period (Ross 2010a; Gross and Haffner 2011). Thus, there has been a range contraction since the first Atlas period, but the average number of nests has remained similar.

This species has typically nested solitarily or in loose, single-species colonies, often in American Sycamores lining the banks of streams such as Conodoguinet Creek and Yellow Breeches Creek (historically) in Cumberland County. However, the largest colony of Yellow-crowned Night-Herons during the first Atlas period was located on an island in the Susquehanna River, across from the Governor's mansion in Harrisburg, where 10 nests were observed in 1987 and 1988 (Schutsky 1992c). The largest nesting concentration during the second Atlas period was in Bellevue Park, Harrisburg, where six nests were located in 2009. In 2010, as many as 14 birds were seen feeding together on insects and earthworms in a vacant lot adjacent to the neighborhood, and individuals were observed grabbing insects disturbed by lawn mowers (C. Dunn, pers. comm.). It is likely that some nests located on remote, sycamore-lined stream reaches in south-central Pennsylvania remain undetected in most years (Master 2010b).

This night-heron maintains a consistent, though tenuous, breeding presence in and around Harrisburg. A species with such a small number of nests located in just a few sites is always vulnerable to extirpation. However, the species' tolerance of human activity (Watts 1995), and the abundant supply of tall nesting trees and food resources along the Susquehanna River and its tributaries, bodes well for its continued existence as long as nests are not disturbed and food supplies remain adequate. The burgeoning, invasive Rusty Crayfish population, although problematic in most respects, may represent a new and abundant source of food for this crustacean specialist (pers. obs.). Residents of Bellevue Park are very interested in and protective of nests in their neighborhood, in large part due to educational efforts by the Pennsylvania Game Commission. Hopefully, these encouraging developments will ensure the continued existence of this beautiful night-heron as a breeding species in Pennsylvania.

TERRY L. MASTER

Distribution

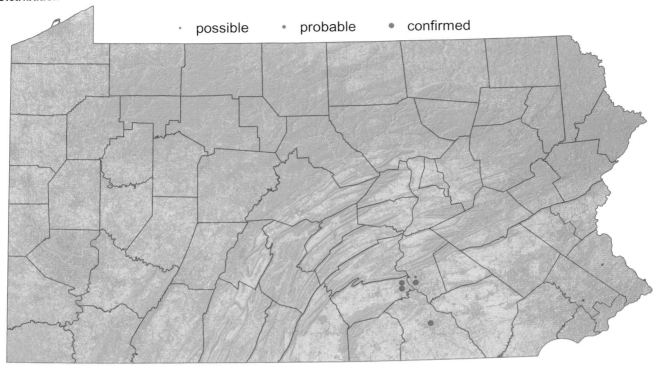

Distribution Change

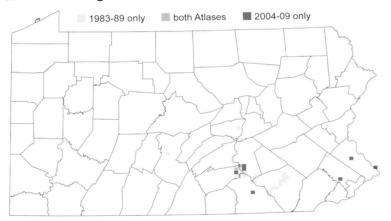

Number of Blocks

	first Atlas 1983–89	second Atlas 2004–09	Change %
Possible	12	5	−58
Probable	3	0	∞
Confirmed	6	4	−33
Total	21	9	−57

Geoff Malosh

Black Vulture
Coragyps atratus

Black Vultures are an increasingly common sight in the skies of southeastern Pennsylvania, where they frequently soar, roost, and feed together with the far more abundant Turkey Vulture. More social than Turkey Vultures, Black Vultures, which search for food exclusively by sight, often follow Turkey Vultures to carrion, the latter having located it with their keen sense of smell. Found in both open and forested habitats, Black Vultures often associate with and depend heavily upon human garbage. Like Turkey Vultures, they also feed regularly on deer gut piles and roadkills. Although less migratory, overall, than Turkey Vultures, the majority of the Black Vulture population evacuates the Commonwealth in winter, although Winter Raptor Surveys from 2001 through 2007 found this species to be widely but thinly distributed throughout the southeastern corner of the state (Grove 2010). Distances traveled south during their migrations remain unknown, although it is suspected that they are less extensive than those of Turkey Vultures in this region.

The Black Vulture breeds in North America as far north as southern New York and New England and in South America as far south as south-central Chile. However, the bulk of its population occurs in the tropics, unlike the Turkey Vulture, which can be equally abundant in the temperate zone.

Like the Turkey Vulture, the Black Vulture nests on the floors of abandoned and semi-abandoned buildings, as well as in crevices in fallen logs and rocky outcrops and in caves (Buckley 1999). During the second Atlas, the Black Vulture was commonly observed in the southeastern quarter of the state, including both forested and open rural and suburban areas and exurban areas around cities. This cryptic nester was observed in 811 blocks, but breeding was confirmed in only 16 of those blocks. Active nests described in the atlas generally were found in human structures, although these are more easily discovered than in the rock piles also described. Overall, Black Vultures increased their range in Pennsylvania by 130 percent since the first Atlas. The species' range remains largely confined to the southeast, but there was a notable northward shift in its range limit, by an average of 26 km (16 miles). Coupled with this was a consolidation within previously occupied areas; for example, the percentage of blocks with Black Vultures in the Piedmont increased from 30 to 71 percent (appendix C).

The expansion of this species' range in Pennsylvania reflects region-wide increases; during the last 20 years it has colonized southeastern New York state (McGowan 2008e) and continues to increase south of Pennsylvania in Maryland (Ellison 2010f). Trends in occurrence detected in the second Atlas concur with the results of other recent surveys. Breeding Bird Survey counts in Pennsylvania increased by more than 10 percent per year from 1966 through 2009 (Sauer et al. 2011), making this one of the most rapidly increasing species in the Commonwealth. Farmer et al. (2008a) reported significant increases of migrants counted at all three migration watch sites in Pennsylvania.

As is true of the Turkey Vultures, historic and current increases in Black Vulture populations in eastern North America are likely due to increasing wildlife populations, including those of White-tailed Deer and other game and nongame wildlife (Kirk and Hyslop 1998), increasing road traffic resulting in a growing availability of roadkills, and decreased direct persecution. Although there appears to be little immediate cause for concern about this species globally, populations of scavenging raptors are disproportionately threatened compared with other diurnal birds of prey (Bildstein 2006). Given the ease with which regional populations of vultures can be surveyed (Bildstein et al. 2007), it behooves the conservation community to continue to monitor Black Vulture numbers.

KEITH L. BILDSTEIN

Distribution

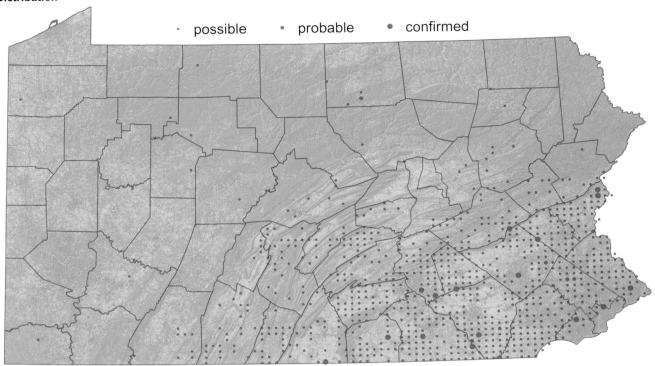

Distribution Change

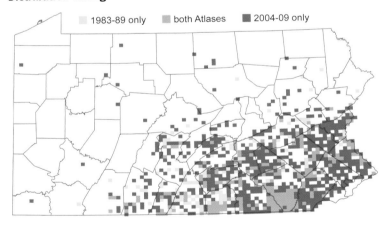

Number of Blocks

	first Atlas 1983–89	second Atlas 2004–09	Change %
Possible	330	788	139
Probable	5	7	40
Confirmed	17	16	−6
Total	352	811	130

Breeding Bird Survey Trend

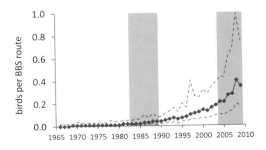

BLACK VULTURE 135

Geoff Malosh

Turkey Vulture
Cathartes aura

The most abundant and widespread avian scavenger in the state, the Turkey Vulture is a common sight in the skies of both open and forested habitats throughout most rural and suburban areas in the Commonwealth. The species also is a regular inhabitant along the edges of many urban areas. In summer, the Turkey Vulture sometimes seems to be as common as clouds in the skies above Pennsylvania. It breeds throughout most of the Americas and is largely migratory at high latitudes in both the Northern and Southern hemispheres. Much of the Pennsylvania breeding population evacuates the Commonwealth in winter for areas as far south as southern Florida (Kirk and Mossman 1998; Mandel et al. 2008).

Turkey Vultures are eminently capable soaring birds that search for carrion both by sight and smell. Although the species is visible when searching for food and while feeding, its nests, which are often on the floors of abandoned and semi-abandoned buildings, as well as in crevices in fallen logs, rocky outcrops, and caves (Kirk and Mossman 1998), can be inaccessible and difficult to find. Further, up to 70 percent of Turkey Vultures in a given area may not be breeding birds (Kirk and Mossman 1998), clouding the picture of the species' breeding status as documented by breeding bird atlases.

During the second Atlas, Turkey Vultures were commonly observed in all parts of the state, including both the heavily forested areas of the north-central counties and the exurban areas around the state's largest cities. Although it was reported in 89 percent of all blocks, breeding was confirmed in just 54 blocks, reflecting the difficulty of locating nests. The number of blocks with Turkey Vultures increased significantly in Pennsylvania between atlases, with a 13 percent increase in occupied blocks in the Appalachian Plateaus driving most of this change. The Turkey Vulture's expansion from its former range in southern Pennsylvania (McWilliams and Brauning 2000) to the rest of the state is now almost complete. This expansion into unoccupied areas was even more pronounced in neighboring New York, where blocks with Turkey Vultures increased by 71 percent between atlases (McGowan 2008f), reflecting a considerable range expansion in the northeastern United States.

Trends in occurrence detected between atlases concur with the results of other surveys. The Breeding Bird Survey reported a significant annual increase of 4.6 percent in Pennsylvania between 1966 and 2009 (Sauer et al. 2011), or approximately a 50 percent increase between the two atlas periods. On a regional scale, Farmer et al. (2008a) reported significant, "strong and steady" increases of migrants counted at six of seven hawk watch sites, including both Waggoner's Gap and Hawk Mountain, where the species increased by 29.8 percent and 5.3 percent annually, respectively, from 1990 to 2000. The regional increases in the Northeast are largely in line with similar increases at migration watch sites in the western United States (Smith et al. 2008).

Historic and current increases in Turkey Vulture populations in eastern North America are likely due to increasing populations of wildlife (including those of White-tailed Deer; Kirk and Mossman 1998), increasing road traffic resulting in a growing availability of roadkills, and decreased direct persecution, especially in the southeastern United States. Although there appears to be little immediate cause for concern for this species, populations of scavenging raptors are disproportionately threatened globally compared with other diurnal birds of prey (Bildstein 2006). For example, the reason the Turkey Vulture and the Black Vulture now rank as the world's two most common species of vultures is because the two previous most common species (both Old World species in the genus *Gyps*) declined catastrophically during the past 30 years in response to unintended poisoning. Currently, the Turkey Vulture is not considered a species of concern within the Commonwealth, and breeding populations in Pennsylvania appear to be secure. However, given the problems other vultures have encountered in recent times, it is obviously important to continue to monitor Turkey Vulture.

KEITH L. BILDSTEIN

Distribution

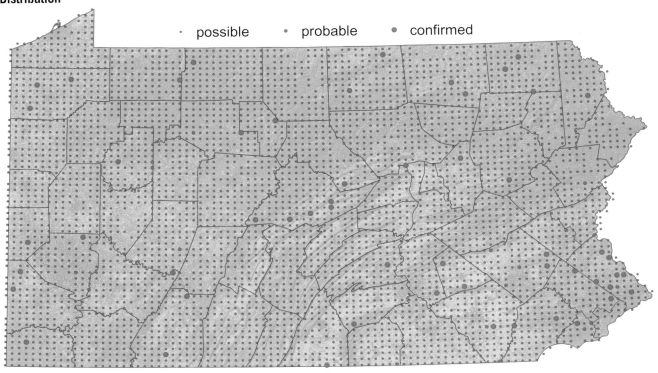

Distribution Change

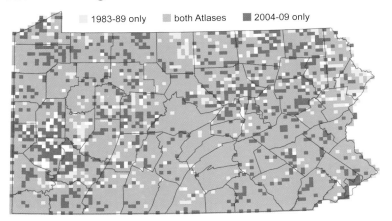

Number of Blocks

	first Atlas 1983–89	second Atlas 2004–09	Change %
Possible	3,831	4,320	13
Probable	45	16	−64
Confirmed	44	54	23
Total	3,920	4,390	12

Breeding Bird Survey Trend

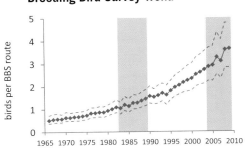

TURKEY VULTURE

Rob Criswell

Osprey
Pandion haliaetus

Over the past three decades, Osprey populations have undergone a surging recovery like only two other birds of prey, Bald Eagle and Peregrine Falcon, with which they share much in common. One of few Pennsylvania breeders found throughout the globe, the Osprey benefited from substantial environmental protections and Pennsylvania's five reintroduction projects between 1980 and 2007. The then-innovative strategy known as "hacking," pioneered successfully for the Osprey in Pennsylvania and replicated elsewhere, fueled this recovery (Schaadt and Rymon 1983; Rymon 1989). The resulting breeding distribution of the second Atlas is greater than at any time previously documented in Pennsylvania.

Because the Osprey typically does not disperse widely from its natal site (Houghton and Rymon 1997), blocks with confirmed breeding in the second Atlas can be seen as clusters that are linked to those historic reintroductions. Nests in several northeastern counties have roots in those earliest reintroductions and the state's first recovered nest in the Poconos during the first Atlas (Rymon 1992). Nests in Tioga, Butler, and Mercer counties are linked to the U.S. Army Corps of Engineers' reintroductions at Cowanesque Lake (Tioga County) and Moraine State Park (Butler County). The group of nests along the Allegheny Reservoir and in the southwestern counties likely drew from birds released, respectively, in New York (Nye 2008) and West Virginia (WVDNR 2010) in the late 1980s. More recent reintroductions at Raystown Lake (2003–2005; Juniata College 2009), and Prince Gallitzin State Park (2007; S. Rorer, pers. comm.) have not yet resulted in local nesting, although hope remains for those areas.

The strings of nests found along the Delaware River in Philadelphia and the Susquehanna River in Lancaster County provide an exception to the historical link to historic reintroductions. These nest sites are believed to represent natural expansions from the robust populations in the Delaware and Chesapeake bays (Watts and Paxton 2007). The link to the Osprey's reintroduction is beginning to fade as the expanding populations blend into each other and natural expansion fills in distribution gaps.

The maximum number of active nests in any given year is hard to determine, because annual Osprey surveys have not been conducted since about 2004 (Brauning and Siefken 2005). Sites plotted by atlas volunteers include at least 82 different nests, most of which were occupied for multiple years. A statewide breeding population estimate of at least 115 pairs was compiled in 2010 (Haffner and Gross 2010b). The number of blocks hosting confirmed Osprey breeding records increased tenfold, from 9 in the first Atlas to 90 in the second.

Recovery had just begun by 1989, resulting in a limited distribution at the time of the first Atlas. Many observations during both projects were not associated with established nests, resulting in an apparent random scatter of non-confirmed observations across the state. These may represent transient birds rather than potential breeders: late migrants, sub-adults, and other non-breeding Ospreys have a long history of overlapping the breeding season in Pennsylvania in a way that obscures the nesting population. As a result, the very few blocks reporting Ospreys in both first and second Atlases reflect only the haphazard detection of summering Ospreys along suitable water bodies, rather than changes in distribution between the two studies. The many new confirmed locations in the second Atlas comprise the primary story.

Osprey habitat and nest sites reveal a strong association with human activity. Since Pennsylvania has few natural lakes, most of its Osprey nests are associated with impounded waterways, lakes, reservoirs, or rivers. Moreover, approximately 97 percent of known nests are placed on human structures, either poles or towers provided specifically for them, or cell towers, power lines, or other parts of the human infrastructure. Future conservation of this species is interwoven with cell phones and power lines, and their associated conflicts. The security of such nesting activity will be a consideration when reviewing the recovery potential of this state-threatened bird.

DANIEL W. BRAUNING

Distribution

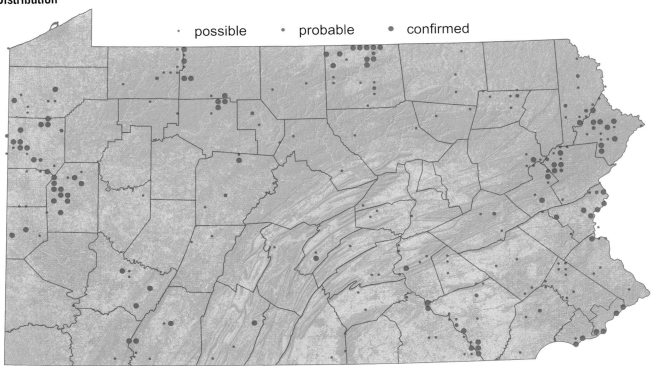

Distribution Change

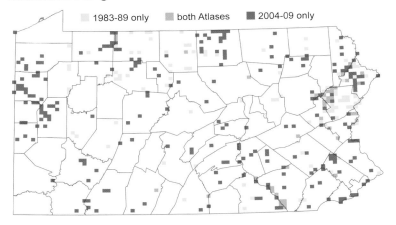

Number of Blocks

	first Atlas 1983–89	second Atlas 2004–09	Change %
Possible	111	158	42
Probable	22	21	−5
Confirmed	9	90	900
Total	142	269	89

OSPREY 139

Jacob Dingel

Bald Eagle
Haliaeetus leucocephalus

As our national symbol and, for many people, a symbol of wildness, the Bald Eagle has attracted the attention, concern, and support that it required to elicit a remarkable recovery from virtual extirpation in the lower 48 states. The Bald Eagle is a uniquely American species, confined to the North American continent (Buehler 2000). Pennsylvania is in the heart of its range in the East, but not as important to its population as the upper Great Lakes region, Maine, Florida, or the Chesapeake Bay. It is even more common in Alaska and the Pacific Northwest (Otto and Sauer 2007).

The remarkable recovery of the Bald Eagle in Pennsylvania, and elsewhere, is one of the great success stories of wildlife conservation. From a very precarious status of only one or two nesting pairs in Crawford County in the early 1980s (Leberman 1992b), the population has expanded through reintroduction and protection to 48 counties by the end of the second Atlas period (Gross and Brauning 2010). A 15 percent annual population growth has been sustained throughout the recovery period (Gross 2009). Expansion has mainly been through the gradual spread of eagle territories along the major waterways, but observations of banded birds show that this recovery is also fueled by immigration from neighboring states, which also have successful eagle recovery programs.

Although it can be an opportunistic scavenger and general predator, the Bald Eagle is primarily piscivorous during the nesting season (Buehler 2000). As a result, nests are strongly associated with streams, lakes, wetlands, and reservoirs with adequate fish populations. Bald Eagles now occupy the major tributaries of the Susquehanna, Delaware, and Allegheny rivers, while some larger waterways of southwestern Pennsylvania (the Monongehela, Youghiogheny, Beaver, Ohio, and Casselman) remain mostly unoccupied and show potential for further colonization by eagles. Unlike Ospreys, Bald Eagles rarely use artificial structures for nest support, despite the ready availability of such structures in eagle habitat (Ryman 2006).

Not only has the Bald Eagle increased since the first Atlas period, it also increased during the second Atlas period. During this period, the number of active Bald Eagle nests more than doubled, from 77 to 174, which is itself an underestimate of the actual population size (Gross 2009). Many more blocks were recorded (535) than known to hold nests, because eagles can range far out from their nests into other blocks. Although Bald Eagles have the deserved reputation of nest site fidelity, there is also some turnover of nest sites on a year-to-year basis (Watts and Duerr 2010), so some eagle pairs occupy different blocks from year to year. The overall number of territories occupied by eagles during the atlas period exceeded the total of 174 in 2009 (Gross and Brauning 2010).

To supplement the volunteer contributions to the second Atlas, Pennsylvania Game Commission (PGC) staff contributed data for 91 eagle nests from their monitoring data to ensure complete coverage. The high percentage of nests not registered by second Atlas volunteers may represent a lack of coverage of the remote riverside hills, wetlands, and mountains where the species nests, or it may suggest a reliance by volunteers on data being submitted by the PGC, which conducts annual nest surveys and issues reports.

The centers of Pennsylvania's breeding eagle population, in the northwestern wetlands, the Lower Susquehanna River, and the Upper Delaware River Watershed/Pocono Mountains, still correspond with the original remnant population and locations where recovery was targeted in the 1980s. Eagles have since expanded up the rivers and tributaries to fill more of the habitat where water quality and human disturbance are not limiting. In 2006, they even recolonized the Philadelphia urban landscape, with new nests in Bucks, Chester, Delaware, and Philadelphia counties. Other pairs are now nesting successfully near human habitations, but usually where human disturbance is light during the early part of the nesting season. The future of Bald Eagles looks extremely bright; as a result, conservation focus may change from reactive protection to proactive management of the riparian forest habitat, wetlands, and high-quality watersheds required by this flagship species, currently listed as Threatened in the state (Gross and Brauning 2010).

DOUGLAS A. GROSS

Distribution

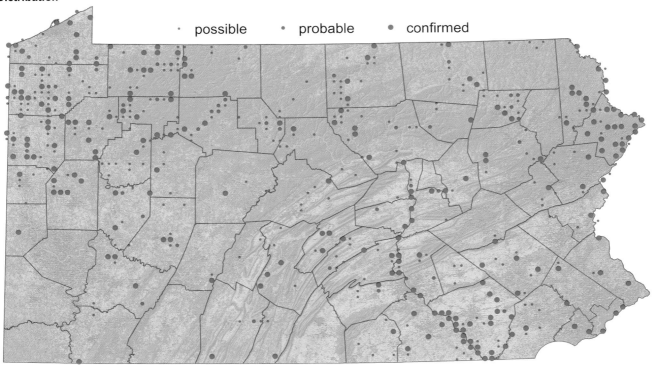

Distribution Change

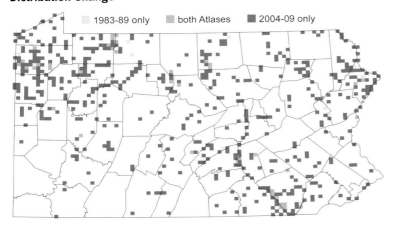

Number of Blocks

	first Atlas 1983–89	second Atlas 2004–09	Change %
Possible	33	267	709
Probable	7	42	500
Confirmed	11	226	1,955
Total	51	535	949

Fred Truslow/VIREO

Northern Harrier
Circus cyaneus

A slim, white-rumped hawk with long tail and legs and an owl-like face (Terres 1980), the Northern Harrier breeds throughout North America from Alaska, south across the United States as far south as southern California, northern Texas, Kentucky, and Maryland, and winters throughout this range. The degree of sexual dimorphism in its plumage (gray male, brown female), and its propensity for polygyny, are exceptional among birds of prey (MacWhirter and Bildstein 1996).

Harriers prefer extensive, open, grassy fields, marshes, scrub-wetlands, and old fields for feeding and nesting. In Pennsylvania, reclaimed surface mines provide a high percentage of suitable habitat due to the loss of wetlands and grassy fields elsewhere (Palmer 1988). Suburban sprawl, reforestation, conversion of hay crops to row crops, and intensive farming have contributed to the continuing widespread trend in reducing Northern Harrier nesting habitat (Goodrich 1992a). Moreover, a steady decline in Pennsylvania's harrier population has been noted for decades (Poole 1964).

During the second Atlas period, breeding Northern Harriers were scarce, reported in only 190 blocks compared with 334 blocks in the first Atlas, a 43 percent decline in block occupancy. In both atlas periods, the harrier was found statewide, with a few concentrated clusters of blocks with confirmed breeding and a wide scatter of possible and probable breeding records. During both periods, the harrier was rarest in the Piedmont Province. Blocks with confirmed breeding were few; this is explained, at least in part, by the species' extensive foraging areas, which can make it difficult to "pin-down" nest locations. In the first Atlas, three areas were especially important for this species: the reclaimed surface mines of Clarion County, grasslands in the northern tier (especially Potter County), and the marshes along the lower Delaware River near Tinicum (John Heinz National Wildlife Refuge). The latter area was apparently deserted by breeding harriers by the second Atlas period, and the number of occupied blocks in the two other former strongholds was much diminished. Localized block occupancy increases in Somerset County (reclaimed surface mines) and agricultural areas in the Cumberland Valley and Bradford County partly compensated for losses elsewhere, but the overall picture was rather bleak. However, there were second Atlas records from 51 out of the 67 counties of Pennsylvania, demonstrating that this species is still widely, but now very thinly, distributed across the state. Due to long-term declines, the Northern Harrier is listed as a species of High Conservation Concern in Pennsylvania's Wildlife Action Plan (PGC-PFBC 2005).

The apparent decline in Pennsylvania's Northern Harrier population was not noted in neighboring Ontario (Sandilands 2007), New York (Post 2008b), or Maryland (Ellison 2010g) between atlases. Surveywide, the species declined on Breeding Bird Survey routes during the 1960s through the 1980s, but trends have been rather stable since then (Sauer et al. 2011). Further, counts of harriers on Winter Raptor Survey routes in Pennsylvania increased by 20 percent per year from 2001 through 2008, partly due to the provision of new foraging opportunities through the Conservation Reserve Enhancement Program (CREP; Wilson et al. 2010). However, winter counts may be largely made up of birds from farther north, and there is, as yet, no evidence that CREP fields have been used as new nesting habitat.

Because the Northern Harrier uses a range of open habitats during the breeding season, it could benefit from general conservation and management measures suggested for wetland (Gross and Haffner 2010), farmland (Wilson 2010), and reclaimed surface mine bird populations (Stauffer et al. 2010). Maintaining mosaics of interspersed upland and wetland habitats and avoiding disturbance during the breeding season also could be crucial for this species (Sechler 2010).

GENE WILHELM

Distribution

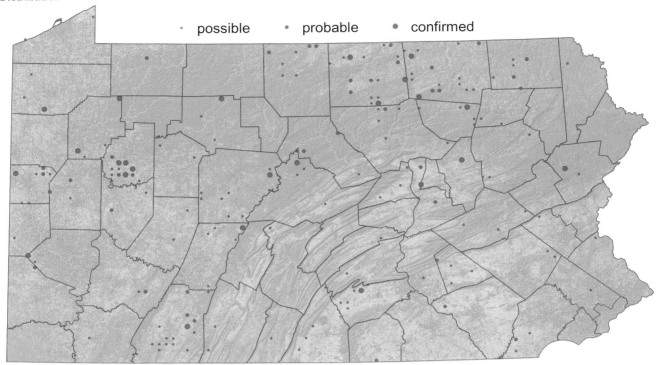

Distribution Change

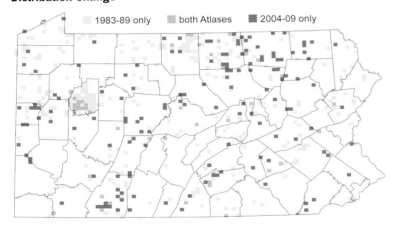

Number of Blocks

	first Atlas 1983–89	second Atlas 2004–09	Change %
Possible	236	128	−46
Probable	79	38	−52
Confirmed	19	24	26
Total	334	190	−43

Brian K. Wheeler/VIREO

Sharp-shinned Hawk
Accipiter striatus

The smallest forest raptor nesting in Pennsylvania, the Sharp-shinned Hawk is often seen at feeders or in suburban areas during winter. It usually nests in large expanses of forest, often placing its nest in conifers or dense deciduous stands (Bildstein and Meyer 2000). However, some studies suggest that this species may be adapting to nesting in smaller forests, and closer to humans, than in the past (Coleman et al. 2002; Grimm and Yahner 1986; Sebastiani 2008). The Sharp-shinned Hawk breeds across Canada in boreal forests and in the eastern United States, south through New England and the Appalachian Mountains. Only a few breeding records occur from Pennsylvania's Piedmont Province, southern New Jersey, and the Coastal Plain (Goodrich 1992b; Walsh et al. 1999). Pennsylvania is south of the main breeding area for this species but is within the southward range extension into the Appalachian Mountains.

In the second Atlas period, the Sharp-shinned Hawk was sighted in 840 atlas blocks, with 100 confirmed nesting records. Its statewide distribution appears similar in the first and second atlas periods, although there was a 20 percent decline in block occupancy between the two atlases. Most of the change was in the Appalachian Plateaus and Ridge and Valley provinces, where block detections dropped by 19 and 29 percent respectively. The recorded decline in block occupancy is likely to reflect a genuine decline, since atlasing effort was higher overall during the second Atlas.

Despite the species' penchant for conifers (Grimm and Yahner 1986; Bildstein and Meyer 2000), 82 percent of Sharp-shinned Hawk nests found by atlas volunteers were located in areas with few or no hemlocks. Pennsylvania's Sharp-shinned Hawks initiated courtship during the second Atlas as early as 10 March (an average of 19 May) and nest building was observed as early as 22 April (an average date of 12 May). The earliest fledged young were observed on 18 June, with an average date of 16 July. Occupied nests were observed as late as 5 July, with food carrying observed as late as 5 August. The protracted breeding season increases the chance of detection by atlas volunteers; however, estimates of breeding season inter-nest distance range from 1 to 5 km (Bildstein and Meyer 2000), so overall nesting densities, even in Pennsylvania's most forested counties, are likely to be low. Hence, it is likely that some Sharp-shinned Hawks were missed in some blocks, especially in well-forested areas, where the species can easily elude detection.

The Sharp-shinned Hawk is considered a species of Maintenance Concern, due to its dependence on conifers and contiguous forests (PGC-PFBC 2005). Migrating populations declined by a significant 3.7 percent annually at Hawk Mountain Sanctuary between 1994 and 2004 (Farmer et al. 2008a). Declines in migration numbers were also observed in New Jersey and Connecticut, whereas Waggoner's Gap counts showed a non-significant trend (Farmer et al. 2008a). Declines at migration watch sites have been attributed to migratory short-stopping (Viverette et al. 1996). Increases in wintering Sharp-shinned Hawks in the Coastal Plain and Piedmont provinces may support this hypothesis (Bolgiano 1997). In contrast to the reduction in Sharp-shinned Hawk detections in Pennsylvania between atlas periods, there was a 68 percent increase in block occupancy in New York over a similar period (Hames and Lowe 2008a), suggesting that possible declines in Pennsylvania may not be part of a wider phenomenon.

Environmental factors that may impact this species include Spruce Budworm cycles, acid rain (Kirk and Hyslop 1998), and organochlorine contaminants (Wood et al. 1996). Contaminants are a lingering issue for the Sharp-shinned Hawk, since it preys on songbirds that can be exposed to them on their wintering grounds (Kirk and Hyslop 1998). The effects of organophosphates used for forest pest control are poorly known. Window and vehicle collisions are a threat during the non-breeding season (Hager 2009). Breeding season surveys for forest raptors could improve our understanding and ability to conserve this species in the future.

LAURIE J. GOODRICH

Distribution

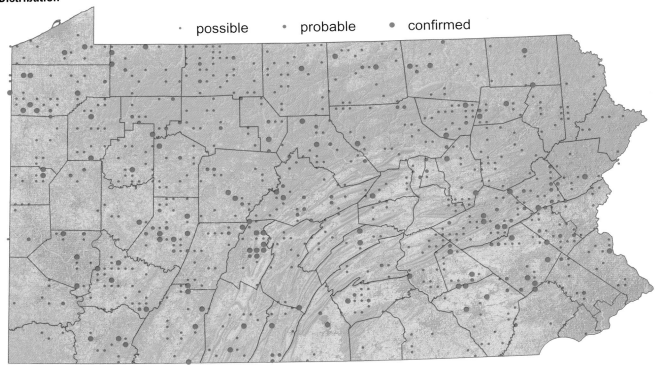

Distribution Change

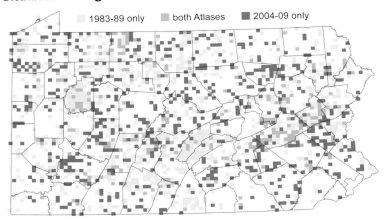

Number of Blocks

	first Atlas 1983–89	second Atlas 2004–09	Change %
Possible	751	652	−3
Probable	202	88	−56
Confirmed	98	100	2
Total	1,051	840	−20

SHARP-SHINNED HAWK

Bob Wood and Peter Wood

Cooper's Hawk
Accipiter cooperii

Although once a bird of large forests and rural woodlots, the Cooper's Hawk is fast adapting to urban and suburban landscapes, taking advantage of their abundant prey and nest sites (Rosenfield et al. 1996; McConnell 2003). As a result, the Cooper's Hawk has shown one of the most dramatic changes in distribution between atlas periods of any nesting bird within the state. During the late nineteenth century, W. E. Hughes suggested that the Cooper's Hawk was one of the most common breeding hawks in southeastern Pennsylvania (Hughes 1898). However, the species was scarce throughout the Piedmont during the first Atlas. Today, this region of the state represents a key stronghold for the species.

The Cooper's Hawk breeds throughout the lower 48 states and southern Canada, and it winters south through Central America (Curtis et al. 2006). It is considered secure throughout its range, although dramatic declines occurred in Pennsylvania and other states from 1900 through the 1950s, due to persecution and pesticides (Goodrich 1992c; Curtis et al. 2006). Since the 1960s, the Cooper's Hawk has been increasing and, more recently, expanding its range into urban and suburban areas (McWilliams and Brauning 2000; McConnell 2003).

In the second Atlas, the Cooper's Hawk was found nesting regularly in southern counties and more sparsely in the more forested north-central and northeastern regions, a pronounced switch since the first Atlas. There was a 10 percent reduction in block occupancy on the Appalachian Plateaus, which went from hosting 73 percent of the species' occupied blocks in the first Atlas to 45 percent in the second. In contrast, block occupancy increased tenfold in the Piedmont and more than doubled in the Ridge and Valley Province (appendix C). The net result of these changes was a 44 percent increase in blocks with Cooper's Hawks between atlases.

The Cooper's Hawk uses a variety of woodland types for nesting, from single trees in urban yards to large contiguous forests (Curtis et al. 2006). Large, more mature trees are often used for nesting (Titus and Mosher 1981). Hemlocks were not a strong component of the nesting habitat for Cooper's Hawk during the second Atlas period, with 69 percent of occupied blocks showing no hemlock cover. However, the Cooper's Hawk often uses pine stands or trees, particularly White Pine, for nesting in suburban Pennsylvania (McConnell 2003). This species may benefit from the increased forest fragmentation occurring throughout southern counties of the state (Brittingham and Goodrich 2010). In urban areas, the Cooper's Hawk may occur at high densities, with inter-nest distances as low as 0.5 km (0.3 miles), and high productivity (Rosenfield et al. 1996). Prey availability undoubtedly influences this pattern. This species appears to have adapted well to human-modified landscapes, and it is now found in highest numbers in the state's most developed counties.

During the second Atlas period, the 316 confirmed breeding records spanned from 22 March to 6 August, with young in the nest detected from 16 May through 16 July (appendix F). The early nesting of this secretive nesting species may have led to some being missed where atlas volunteers and atlasing visits were limited, particularly in northern regions.

Migration counts from Pennsylvania watch sites show a significant annual increase of 4 to 5 percent per year from 1994 to 2004 (Farmer et al. 2008a). Similar increases were seen on Breeding Bird Survey routes within the northeastern states, with a 5.3 percent annual increase in Pennsylvania (Sauer et al. 2011). The increase observed between atlas periods appears to reflect patterns observed across the northeast region (Curtis et al. 2006; Hames and Lowe 2008b). Although increasing, the Cooper's Hawk is susceptible to significant mortality from window-strikes, vehicles, electrocution, and predators (Hager 2009). These risks are even higher in urban environments. Other possible conservation threats include West Nile virus (Wünschmann et al. 2004) and contaminants (Bildstein 2006). However, the Cooper's Hawk is clearly doing very well in Pennsylvania at this time, and it has shown a remarkable ability to adapt to radically altered landscapes.

LAURIE J. GOODRICH

Distribution

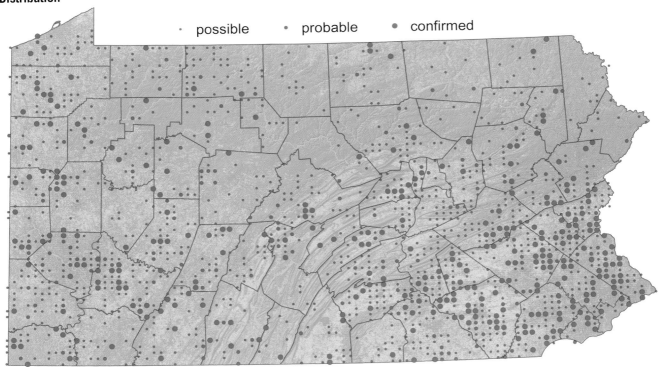

Distribution Change

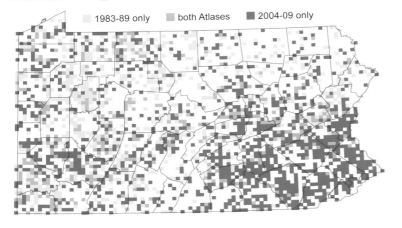

Number of Blocks

	first Atlas	second Atlas	Change
	1983–89	2004–09	%
Possible	715	1,033	44
Probable	163	163	0
Confirmed	170	316	86
Total	1,048	1,512	44

Breeding Bird Survey Trend

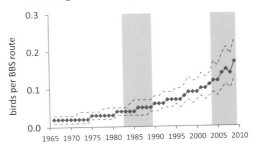

COOPER'S HAWK 147

David Brinker

Northern Goshawk
Accipiter gentilis

The ultimate forest raptor, Northern Goshawk is the avian monarch of the expansive mixed hardwood/conifer forest in central and northern Pennsylvania. Perfectly adapted, with relatively short wings and long rudder-like tail, goshawks expertly navigate the forest, executing surprise attacks on prey such as Red Squirrel, Ruffed Grouse, and Snowshoe Hare. The Northern Goshawk is a fierce defender of its nest, and many a hiker, turkey hunter, or trout fisherman in Penn's Woods has hastily retreated when pursued by an aggressive goshawk defending its nest and chicks. An encounter with a breeding goshawk is a memory seldom forgotten. Outside the breeding season, goshawks are secretive permanent residents seldom observed in their forest home.

Northern Goshawks have a Holarctic distribution, with breeding populations spread throughout the cold and cool forested regions of the northern hemisphere. In eastern North America, Pennsylvania is the southern limit of the primary breeding range (Squires and Reynolds 1997), with a narrow extension of its breeding range historically, at least, south along the highest elevations of the Appalachian Mountains. In Pennsylvania, as a result of its statewide rarity, goshawks are listed as Vulnerable in the state's Wildlife Action Plan (PGC-PFBC 2005). Dispersal from a healthy Pennsylvania goshawk population is essential to sustainability of breeding to the south in Maryland and West Virginia.

In the second Atlas, the Northern Goshawk was found in 86 blocks, a significant decline (28%) from the 120 blocks documented during the first Atlas. Goshawks were documented only in eight blocks in common between both atlas periods. In both atlases, most records were in the Appalachian Plateaus Province (79% first Atlas, 92% second Atlas), with the remainder in the Ridge and Valley Province. There was only one confirmed breeding Goshawk record south of 41°N during the second Atlas, and the number of occupied blocks below that latitude dropped from 22 to 4.

Northern Goshawks are now almost exclusively restricted to the Allegheny High Plateau Section of the Appalachian Plateaus Province, where they are most frequently recorded in the Allegheny National Forest and Pennsylvania's extensive system of state forests, game lands, and parks. A greater proportion of observations in the first Atlas occurred in the Appalachian Mountain and Glaciated Low Plateau sections than during the second Atlas. Although it is difficult to demonstrate statistically with the relatively small sample of atlas records, the breeding population of goshawks in Pennsylvania has declined and retracted into the core forests of central and northern Pennsylvania. This decline parallels New York's second atlas, which documented a 20 percent decline in blocks with goshawks (Crocoll 2008a), while to the south, since 2006, Northern Goshawks have nearly disappeared as a breeding species in Maryland and West Virginia (pers. obs.).

The reasons behind this contraction of the Northern Goshawk distribution are subtle and currently unproven. In the Ridge and Valley, conifers are an important habitat component of goshawk nesting areas (Kimmel and Yahner 1994). As Hemlock Woolly Adelgid damage to hemlock stands spreads northwestward across Pennsylvania, the loss of hemlock could be reducing Northern Goshawk habitat quality. Monitoring in the Allegheny National Forest from 2001 through 2010 documented a steep decline in goshawk nest success rates and the occasional loss of adult females to nest predators, most likely Fishers. The increase in nest failures coincides with a dramatic increase in indices of Fisher abundance in Pennsylvania (Lovallo 2008). More generally, Northern Goshawks have been shown to be susceptible to West Nile virus–induced mortality (Wünschmann et al. 2005), and the establishment of West Nile virus may have played a role in the observed decline.

The future of Northern Goshawks in Pennsylvania is clouded by a lack of understanding of the ultimate causes of population change in this charismatic forest raptor. The root cause of the change occurring in central Appalachian goshawk populations is not simple habitat loss, as it often is with so many wildlife species. Targeted research will be needed to identify the factors responsible for the observed distributional change, so that conservation actions can be taken to maintain a healthy goshawk population in Pennsylvania.

DAVID F. BRINKER

Distribution

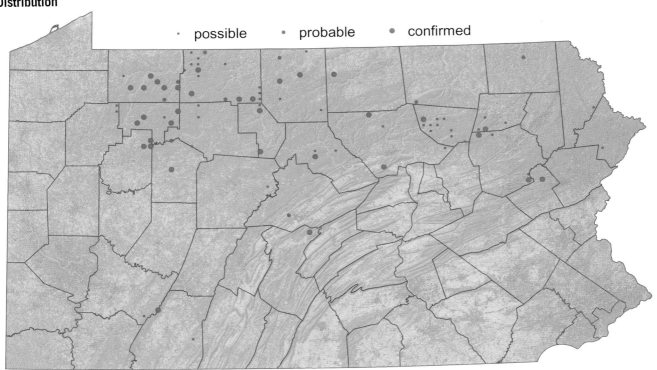

Distribution Change

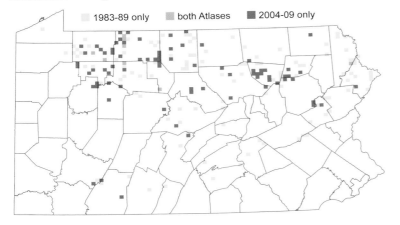

Number of Blocks

	first Atlas 1983–89	second Atlas 2004–09	Change %
Possible	65	44	−32
Probable	12	9	−25
Confirmed	43	33	−23
Total	120	86	−28

NORTHERN GOSHAWK

Jacob Dingel

Red-shouldered Hawk
Buteo lineatus

The colorful and vocally conspicuous Red-shouldered Hawk was once a common component of eastern lowland, deciduous, and mixed forests (Stewart 1949; Titus and Mosher 1981; Crocoll 1994). Its breeding range extends from New Brunswick, west to Minnesota, and southward into Texas and Florida. A separate breeding population occurs in coastal Oregon, California, and Baja California (Crocoll 1994). Pennsylvania is located near the northeastern corner of the breeding range, and individuals from the northern half of the range, including Pennsylvania, generally migrate to the southern United States for the winter (Goodrich and Smith 2008).

The Red-shouldered Hawk is a species of Greatest Conservation Need in nine Northeastern states (A. Haskel, pers. comm.). It is a species of Maintenance Concern in Pennsylvania's Wildlife Action Plan, because here, its reliance on riparian forests makes it an indicator of high-quality and large-scale contiguous forests (PGC-PFBC 2005). Due to its restricted geographic range, breeding populations in Pennsylvania probably are important to the overall conservation of this species.

Red-shouldered Hawks were detected in approximately 24 percent of all second Atlas atlas blocks. These detections represented a highly significant increase of 55 percent compared with the first Atlas. Increases in block occupancy between the atlases were recorded in 19 of the state's 23 physiographic sections, with 10 of these changes significant. No significant decreases were recorded. Statewide, Red-shouldered Hawks were found in 416 more blocks than in the first Atlas. The vast majority of these gains occurred in the Appalachian Plateaus (314 blocks) and the Ridge and Valley provinces (83 blocks).

From 1999 through 2009, the Breeding Bird Survey (BBS) estimated a significant increase in Red-shouldered Hawks of 3.6 percent per year in Pennsylvania, but this was based on a very small number of routes (Sauer et al. 2011). On a continental scale, the surveywide trend estimate from the BBS from 1966 to 2009 was a statistically significant increase of 3.1 percent per year (Sauer et al. 2011). Counts of migrating birds at Hawk Mountain Sanctuary and Audubon's Hawk Watch at Waggoner's Gap showed no change during the period 1974 to 2004 (Farmer et al. 2008a). Bednarz et al. (1990) reported a significant long-term decline in counts at Hawk Mountain Sanctuary from 1949 to 1986—that is, the period leading up to the first Atlas. As with BBS estimates, relatively low detection rates for this species at migration watch sites limit the value of these trend estimates (Farmer et al. 2008a).

Previous authors have noted that the larger Red-tailed Hawk may outcompete the Red-shouldered Hawk for nest sites in some locations (Bryant 1986; Titus et al. 1989). With Red-tailed Hawk populations increasing in Pennsylvania according to BBS data (+4.3% per year, 1966–2009; Sauer et al. 2011), a concomitant decline in Red-shouldered Hawks might be expected, but thus far, the species appears to be stable or increasing. While these two species certainly can be found nesting in proximity to one another, the Red-shouldered Hawk is primarily a bird of extensive bottomland forests or forested valleys with wetlands, whereas the Red-tailed Hawk generally selects more fragmented upland forest habitats for nesting. It is likely, therefore, that these two species rarely are in direct competition for nest sites.

Roughly 60 percent of the Commonwealth is estimated to have land cover characteristics identified as primary or secondary habitat for nesting Red-shouldered Hawks (Myers et al. 2000); Pennsylvania is considered important to the overall conservation of this species (Reitz and Nelson 2010). Maintenance and restoration of large tracts of mature bottomland forest on public and private lands, especially in the Appalachian Plateaus and Ridge and Valley provinces, may help ensure a bright future for the Red-shouldered Hawk and many other species of animals and plants that share this habitat in Pennsylvania.

CHRISTOPHER J. FARMER

Distribution

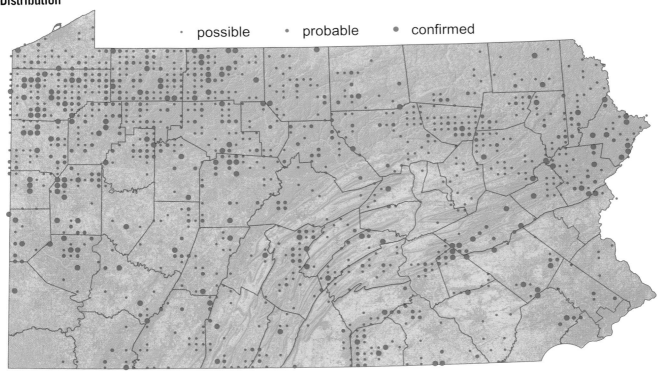

Distribution Change

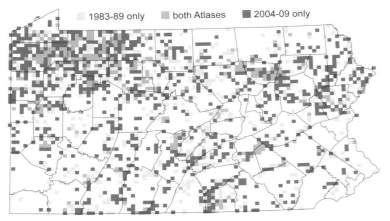

Number of Blocks

	first Atlas 1983–89	second Atlas 2004–09	Change %
Possible	440	760	73
Probable	175	253	45
Confirmed	135	154	13
Total	750	1,167	55

Breeding Bird Survey Trend

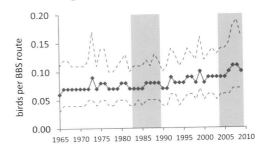

RED-SHOULDERED HAWK 151

Jacob Dingel

Broad-winged Hawk
Buteo platypterus

The Broad-winged Hawk is one of Pennsylvania's most conspicuous raptors during migration, swirling overhead by the hundreds during spring and fall. In contrast, although it is a common breeding bird of Pennsylvania forests, its secretive nature makes it much less apparent to birdwatchers during nesting.

Restricted to continuous deciduous or mixed forests, Broad-winged Hawks often move quietly beneath the canopy during the breeding season (Matray 1974). They range from central Alberta east across Canada, and from the western Great Lakes south through eastern Texas and northern Florida (Goodrich et al. 1996). Higher densities correspond to the northern hardwood forest and the mixed conifer-deciduous forests of northern Pennsylvania, north through New England (Grimm and Yahner 1986; Goodrich et al. 1996).

The Broad-winged Hawk, although "secure" within the United States and Canada (Farmer et al. 2008b), is considered a species of high responsibility for Pennsylvania, due to its association with large, contiguous forests and the importance of the state in its breeding range (PGC-PFBC 2005). Because a large proportion of its diet consists of amphibians, it may be sensitive to contaminant-related prey declines as well as habitat loss (Matray 1974; Goodrich et al. 1996).

The Broad-winged Hawk nests throughout the state's larger forests. Atlas records show that its current range corresponds well with the distribution of core forest. In southern counties, nesting Broad-winged Hawks are found where contiguous forest remains, such as in the South Mountain Section of the Ridge and Valley Province.

Broad-winged Hawks return to Pennsylvania during mid-April (McWilliams and Brauning 2000). The earliest confirmed nest building during the second Atlas occurred on 15 April. Nests were most often confirmed with observations of food carrying ($n = 44$), fledged young ($n = 73$), and adult observed in a nest ($n = 38$). Young were observed as fledglings as early as 7 June, with an average date of 11 July and a late date of 10 August.

Breeding season records were submitted for 35 percent of atlas blocks, a significant 16 percent decline since the first Atlas. Notable declines occurred in the Ridge and Valley Province (−21%), Piedmont (−67%), and the Pittsburgh Low Plateau and Waynesburg Hills sections of the Appalachian Plateaus (−27% and −58%, respectively). These areas have all seen increases in human development in recent decades (Goodrich et al. 2002; Brittingham and Goodrich 2010). In northern and central counties, lower numbers of atlas volunteers may have resulted in reduced numbers of records of this reclusive species, as was likely in the first Atlas (Senner and Goodrich 1992). Localized increases in atlas records for the second Atlas resulted from increased effort by this atlas as well as localized increase in forest cover in north-central and south-central counties.

Migrant Broad-winged Hawks declined at Hawk Mountain by 3.4 percent annually from 1994 to 2004, while at Waggoner's Gap, near Carlisle, counts increased by 7.8 percent (Farmer et al. 2008a). To the north of Pennsylvania, the Broad-winged Hawk appears to be holding its own in New York state (Crocoll 2008b), but to the south, the species has been lost as a breeding bird from large portions of the Maryland Piedmont (Ellison 2010h). These changes echo regional changes in Broad-winged Hawk occupancy in Pennsylvania, and they highlight regional declines, which may be due to forest loss and fragmentation.

Although still a common nesting bird, the Broad-winged Hawk's recent declines serve notice that Penn's Woods may warrant attention. The range declines between atlas periods closely follow regional and state trends in forest cover and quality (Bishop 2008; Brittingham and Goodrich 2010). Conservation of contiguous forest habitat throughout Pennsylvania is key to conservation of nesting Broad-winged Hawks as well as other forest-interior birds. As a long-distance migrant, Broad-winged Hawks also are subject to habitat constraints and human harassment during migration and wintering periods (Goodrich et al. 1996; Ruelas Inzunza et al. 2009).

LAURIE J. GOODRICH

Distribution

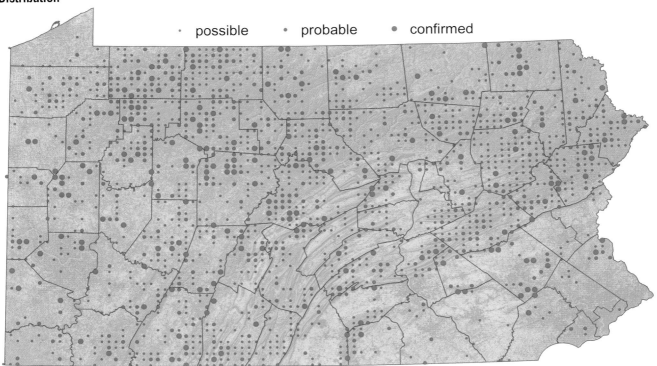

Distribution Change

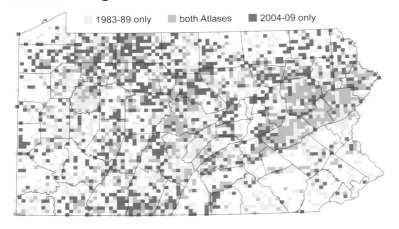

Number of Blocks

	first Atlas 1983–89	second Atlas 2004–09	Change %
Possible	1,216	1,166	−4
Probable	494	370	−25
Confirmed	353	189	−46
Total	2,063	1,725	−16

Breeding Bird Survey Trend

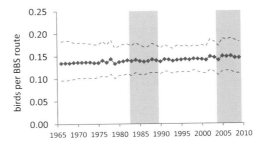

Geoff Malosh

Red-tailed Hawk
Buteo jamaicensis

The Red-tailed Hawk nests from Alaska east, across Canada and south into Mexico, Central America, and the Caribbean (Preston and Beane 2009). Pennsylvania lies at the heart of the eastern range for the *B. j. borealis* subspecies of Red-tailed Hawk (Preston and Beane 2009).

As one of the most widespread breeding hawks found in Pennsylvania, the red-tail's large size and habit of perching in open areas ensure that it is the most commonly observed hawk in much of the state. It nests in open or semi-open habitats wherever large trees or manmade structures provide adequate nest and perch locations (Titus and Mosher 1981; Preston and Beane 2009). Its generalist diet allows it to adapt to a wide variety of habitats, from urban parks to secluded forest, although it is most often seen in open country. Once heavily persecuted, it has recovered from declines during the first half of the twentieth century (McWilliams and Brauning 2000). During the second Atlas period, it was recorded in more blocks (4,104) than any other raptor, or 83 percent of all blocks.

In the second Atlas period, the Red-tailed Hawk was most commonly observed throughout southeastern and western counties, where suburban development and farmland provide abundant habitat. It was absent only in some north-central and northeastern blocks, where forest cover dominates the landscape. The overall distribution was little changed between atlas periods, although the number of blocks with Red-tailed Hawks increased by 13 percent. This change also may reflect the Red-tailed Hawk's increased comfort with urban and suburban environments (Stout et al. 2006).

The Red-tailed Hawk displayed breeding behavior in most months during the second Atlas. The earliest probable evidence of breeding, a pair sitting together, occurred on 27 January, and the latest on 5 August. Occupied nests were observed from 24 February through 15 July, while nests with young were recorded from 9 April through 15 July and eggs were noted as early as 24 March. Almost half of breeding confirmations were of fledged young, which were noted as early as 1 May and as late as 2 October. Active nests were found in almost 500 blocks.

The estimated population of the Red-tailed Hawk within the state was 23,000 individuals, which represents an average of five birds per occupied block. These data suggest that densities were highest in areas with interspersed farmland and woodland, such as Mercer, Franklin, Union, Montour, and Mifflin counties. Currently, Red-tailed Hawks are increasing in Pennsylvania and surrounding states (Sauer et al. 2011), showing a 4.3 percent annual increase on Breeding Bird Survey routes, 1966–2009, in the state. However, wintering birds detected on winter road surveys have not changed substantially since 2001 (Grove 2010), and Pennsylvania migration watch sites show declines in migrant populations (Farmer et al. 2008a). Such contrasts remind us that the composition of the Red-tailed Hawk population in Pennsylvania varies with the season, which suggests that northern populations may be becoming less migratory.

Red-tailed Hawks have adapted well to Pennsylvania's varied landscape. Their habit of hunting along roads and living in agricultural environments makes them highly susceptible to vehicle collisions and pesticide poisoning. Electrocutions and illegal shooting and trapping also remain current threats throughout their range (Hager 2009; Preston and Beane 2009). Despite these threats, the Red-tailed Hawk continues to thrive, and it remains Pennsylvania's most numerous diurnal raptor.

LAURIE J. GOODRICH

Distribution

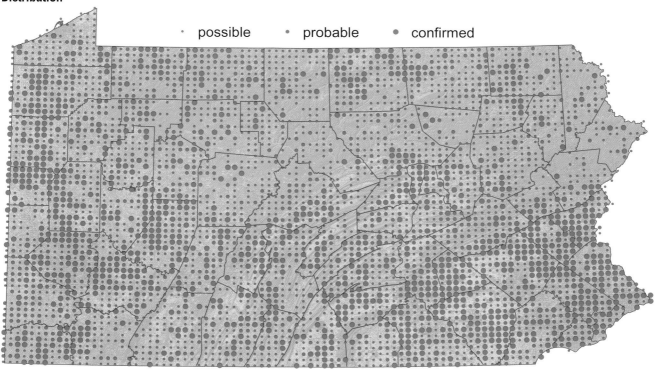

possible · probable · confirmed

Distribution Change

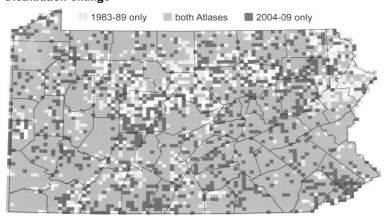

1983–89 only | both Atlases | 2004–09 only

Density

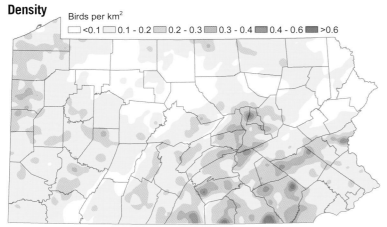

Birds per km²: <0.1 | 0.1 - 0.2 | 0.2 - 0.3 | 0.3 - 0.4 | 0.4 - 0.6 | >0.6

Number of Blocks

	first Atlas 1983–89	second Atlas 2004–09	Change %
Possible	1,690	1,987	18
Probable	1,301	1,031	2
Confirmed	930	1,086	17
Total	3,633	4,104	13

Population estimate, birds (95% CI):
23,000 (21,000–25,000)

Breeding Bird Survey Trend

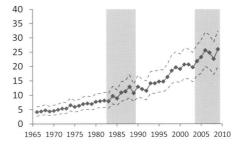

RED-TAILED HAWK 155

Jacob Dingel

American Kestrel
Falco sparverius

The American Kestrel, North America's smallest falcon, is commonly seen near open fields in rural areas and sometimes in urban centers. Although hover-hunting is its most conspicuous behavior, the kestrel is often seen perched on trees or utility wires near grassland habitats. These colorful little raptors are obligate secondary cavity nesters that use existing natural and man-made cavities, including Northern Flicker holes or other tree cavities, abandoned buildings, and nest boxes (Smallwood and Bird 2002).

The kestrel breeds in eastern and western North America, north to the tree line and south into most of Central and South America. Pennsylvania is centrally located within the eastern portion of this breeding range. The kestrel is not considered a species of Special Concern within the Commonwealth, but breeding populations in Pennsylvania are likely to be important to the conservation of this species.

American Kestrels were found across most parts of the state in the second Atlas, except in the heavily forested areas of the north-central counties. There were records in 52 percent of blocks, a significant 13 percent decline since the first Atlas. The decline in the Piedmont (28%) was most notable and pervasive, with significant losses in the greater Philadelphia metropolitan area. Blocks in which kestrels were found in the second Atlas, but not in the first, had a higher-than-average increase in atlas effort (11.8 vs. 8.1 hours). When corrected for increased effort, the estimated decrease in the number of blocks with kestrels between the first and second Atlases was 17 percent.

The decline detected between atlases concurred with other recent surveys of kestrel populations. Counts of kestrels on Breeding Bird Survey routes in Pennsylvania have shown a steady although non-significant downward trend since 1966 (Sauer et al. 2011). On a regional scale, Farmer and Smith (2009) reported significant declines of 4.5 to 1.7 percent per year in migration counts in eastern North America, 1974–2004, and significant declines of up to 8.5 percent per year, 1994–2004. These declines reflect a continental-scale decline in kestrels over the last decade (Farmer and Smith 2009). Smallwood et al. (2009a) reported significant declines in occupancy rates in six nestbox programs throughout the eastern United States from the mid-1980s to 2007.

American Kestrels were most frequently reported on atlas point counts in the Northwestern Glaciated Plateau Section, farmed valleys in the Ridge and Valley Province, and the western portion of the Piedmont Province. Only 194 kestrels were recorded on point count surveys, resulting in a population estimate of 13,600 birds, the equivalent of 2.7 pairs per occupied atlas block. This is much lower than the Partners in Flight estimate of 36,000 birds (Rich et al. 2004), but due to methodological caveats (chap. 5), these results should be interpreted with caution.

Declines in kestrel populations in eastern North America may be due to a variety of causes. Populations of the larger Cooper's Hawk increased throughout the region in the last 30 years (Farmer et al. 2008a), and studies at Hawk Mountain and elsewhere have suggested that this species preys upon American Kestrels. Since 1999, West Nile virus also has impacted numerous bird populations in the region, including kestrels (Medica et al. 2007). Smallwood et al. (2009a) examined these competing hypotheses but did not find evidence supporting them on the breeding grounds. They suggested that the cause of kestrel declines may lie along migration routes or on winter range. Recent research on breeding kestrel populations in New Jersey suggests that changes in the distribution and abundance of large (>1,000 ha, ~2,500 ac) patches of open habitats could drive changes in kestrel abundance (Smallwood et al. 2009b). The apparent losses of kestrels in the Philadelphia and Pittsburgh metropolitan areas between atlases suggest that land use change, specifically the loss of farmland to exurban development (chap. 3), could be contributing to declines in the American Kestrel's range in Pennsylvania.

CHRISTOPHER J. FARMER AND KEITH L. BILDSTEIN

Distribution

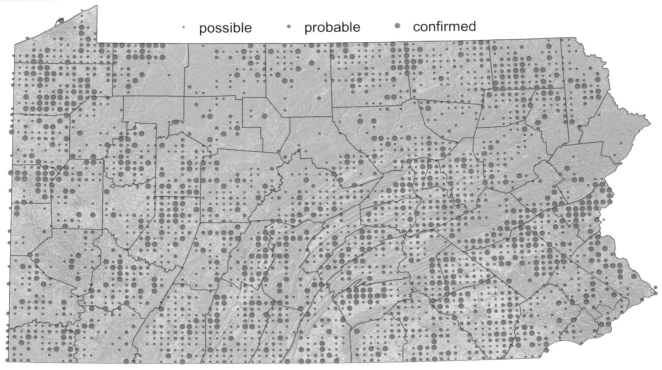

possible · probable · confirmed

Distribution Change

1983–89 only | both Atlases | 2004–09 only

Density

Birds per km²
<0.1 | 0.1 - 0.2 | 0.2 - 0.3 | 0.3 - 0.4 | 0.4 - 0.6 | >0.6

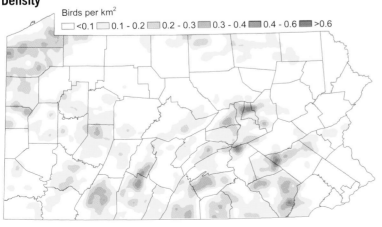

Number of Blocks

	first Atlas 1983–89	second Atlas 2004–09	Change %
Possible	1,290	1,336	3
Probable	910	621	−32
Confirmed	738	601	−19
Total	2,938	2,558	−13

Population estimate, birds (95% CI):
13,600 (11,600–15,600)

Breeding Bird Survey Trend

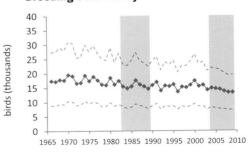

AMERICAN KESTREL 157

Rick Wiltraut

Merlin
Falco columbarius

Without a doubt, one of the most exciting discoveries of the second Atlas was of nesting Merlins. Formerly known as "Pigeon Hawk," the Merlin is a small but powerful falcon that nests mostly at high latitudes (>50°N) throughout Canada, Europe, and Asia. In eastern North America, it had never before been observed nesting as far south as Pennsylvania.

The Merlin's occurrence as a breeding bird during the second Atlas follows earlier expansions into the Maritime Provinces and Maine in the 1980s (Erskine 1992) and Vermont, New Hampshire, and New York in the 1990s (Warkentin et al. 2005). Currently, northern Pennsylvania represents the southern limit of its breeding distribution in the East; its range limit in the western United States is southern Oregon, which is about the same latitude as northern Pennsylvania.

During the second Atlas, volunteers recorded Merlins in 13 blocks and confirmed nesting in 6. The first-ever Pennsylvania nest was discovered by John Fedak on 18 May 2006 in Bradford, McKean County, following an observation of food transfer between two birds by Lynn Ostrander. In addition to the Bradford site, nests or recently fledged young were observed in Eagles Mere (Sullivan County), Promised Land State Park (Pike County), Sayre (Bradford County), and Warren (Warren County).

Through most of their range in eastern North America, the Merlin prefers to nest in the old stick nests of crows as well as those of other hawk species, with conifers their preferred nesting trees (Warkentin et al. 2005). Two of the Merlin nests in Pennsylvania were well hidden near the tops of White Pines and were believed to be old Fish Crow nests (pers. obs.). White Pine was also used at another location (D. Gross, pers. comm.), and Norway Spruce was used at two other locations (J. Fedak, R. Wiltraut, and B. Fowles, pers. comm.).

Incubation lasts 28 to 32 days, and the young remain in the nest from 25 to 30 days (Baicich and Harrison 2005). At the Promised Land site, Merlin courtship behavior was seen as early as 18 April, and young fledged as early as 8 July (pers. obs.); elsewhere in Pennsylvania, nests still containing young were observed as late as 24 July. Fledglings remained in the vicinity of their nests for about 5 weeks, dispersing from breeding sites in mid-August (Warkentin et al. 2005).

The Merlin is rather tolerant of people around the nest site and, increasingly often, will nest in close proximity to towns where ornamental plantings of pines and spruces interspersed with open areas such as parks and cemeteries provide the necessary mix of good cover and open areas for hunting. All six Merlin breeding sites found during the second Atlas were in residential areas, and all successfully raised young. In Pennsylvania, as elsewhere (Warkentin et al. 2005), the Merlin also seems to prefer nesting near water; two Pennsylvania sites bordered lakes, and two sites were next to major rivers. At the Promised Land site, Merlins consistently hunted prey near a lake adjoining a regenerating clearing, the result of a F-2 tornado in May 1998 (pers. obs.). Although tolerant of people, the male Merlin vigorously defends its territory from larger birds, such as crows, ravens, and other raptors. The species is very vocal around its nesting areas, and its loud, ear-catching *Ki-ki-ki-kek-kek-kek* calls often were the first things that alerted atlas volunteers to the bird's presence.

The Merlin preys mainly on birds, often focusing on the most locally abundant species (Warkentin et al. 2005). At the Promised Land site, at least 15 species of songbirds were identified as prey items over a 3-year period, with Barn Swallow, Tree Swallow, and Cedar Waxwing being particular favorites (pers. obs.). The Merlin also occasionally eats small mammals and reptiles, and the recently fledged young will take insects such as dragonflies (Warkentin et al. 2005). The remains of an unidentified small bat were found under a tree regularly used by fledglings at Promised Land (pers. obs.); at the Bradford site, food transfer of a small mammal was noted (J. Fedak, pers. comm.).

RICK WILTRAUT

Distribution

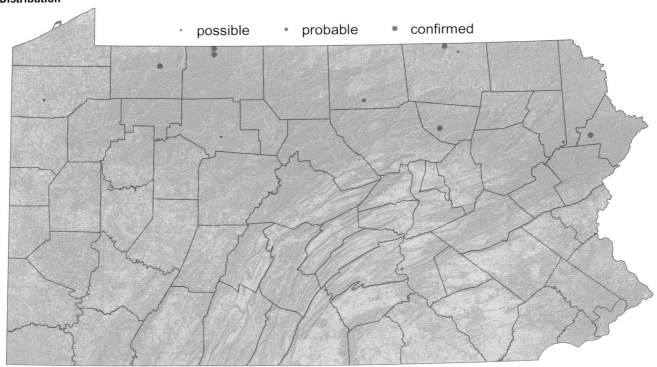

Distribution Change

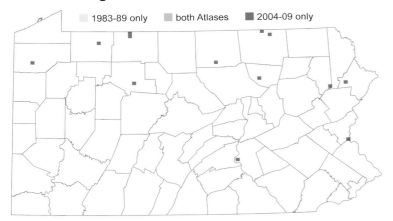

Number of Blocks

	first Atlas 1983–89	second Atlas 2004–09	Change %
Possible	0	6	∞
Probable	0	1	∞
Confirmed	0	6	∞
Total	0	13	∞

Joe Kosack/PGC

Peregrine Falcon
Falco peregrinus

The Peregrine Falcon is a charismatic species, a brilliant aerialist, and one of the fastest animals on earth. It also experienced one of the most dramatic conservation successes of the twentieth century. The species was extirpated from eastern North America in the 1940s through the 1960s by the use of dichloro-diphenyl-trichloroethane (DDT) and other chlorinated hydrocarbon insecticides (Hickey 1969). Since then, it has been reintroduced, its population has expanded, and the increasing presence of this dramatic raptor in urban areas is a local phenomenon.

A cosmopolitan species, the Peregrine Falcon inhabits all continents except Antarctica and lives in diverse habitats where avian prey, open areas for hunting, and suitable nest sites are available. Within North America, the majority of the species' historical population was in the West and in the Arctic, where the terrain provided an abundance of cliff nest sites. A smaller population was found in the East, primarily in the Appalachian Mountains. Pennsylvania's historic population prior to the 1950s was estimated at 44 pairs, all but 1 of which was found on cliff ledges overlooking the Delaware and Susquehanna rivers and their major tributaries (Hickey 1969). As a result of the population crash, the Peregrine Falcon was listed as Endangered nationally in 1973 and subsequently by all northeastern states, including Pennsylvania. The ban of DDT use, the passage of the U.S. Endangered Species Act, and the reintroduction of thousands of captive-bred peregrines launched the recovery of this species across its historic, U.S. range (USFWS 1999).

The first Atlas documented the very early stage of that recovery, with nesting confirmed on three bridges in the Philadelphia area beginning in 1986 (Brauning 1992c). In 1999, the Peregrine Falcon was removed from the federal Endangered Species list, due largely to its healthy population recovery in the West (USFWS 1999), but it remained on the state Endangered Species lists in Pennsylvania and many other eastern states, where its recovery has been more gradual. Pennsylvania's breeding population doubled, from 11 to 26 nesting pairs, during the second Atlas period (McMorris and Brauning 2010). Fifty percent of the Peregrine Falcon's 2009 nests were on bridges, 23 percent on tall buildings, and 8 percent (two nests) on smokestacks. The remainder, 19 percent (five nests), were on natural cliffs along the Delaware River in Northampton County, the Susquehanna River West Branch in Lycoming and Union counties, and the Susquehanna River North Branch in Luzerne County (McMorris and Brauning 2010).

Although the Peregrine Falcon is still listed as Endangered in Pennsylvania, the conservation outlook is positive. Productivity averaged 2.5 fledged young per nesting pair during the second Atlas period. The species has adapted well to the urban environment in spite of the unique hazards associated with nesting on man-made structures, recently including structures less than 15 m (50 ft) tall. However, the recovery and health of its population are highly dependent on active management. As in most eastern states, the Pennsylvania Game Commission monitors nesting at all known nest sites and bands young at all accessible nests, resulting in a population data set unsurpassed by any species in eastern North America. Most pairs that nest on man-made structures use nest boxes placed specifically for their use. Coordination with bridge, building, and power plant managers provides protection, often by restricting or rescheduling construction and maintenance activities to minimize disturbance. Fledglings at urban nests are frequently rescued from accidents unique to the location (e.g., falling into the river or onto the roadway, colliding with buildings), significantly improving nest productivity. Cliff-nesting peregrines are less dependent on these management measures, but the slow recolonization of cliffs is delaying the restoration of this dramatic species to its historic habitat.

DANIEL W. BRAUNING AND F. ARTHUR MCMORRIS

Distribution

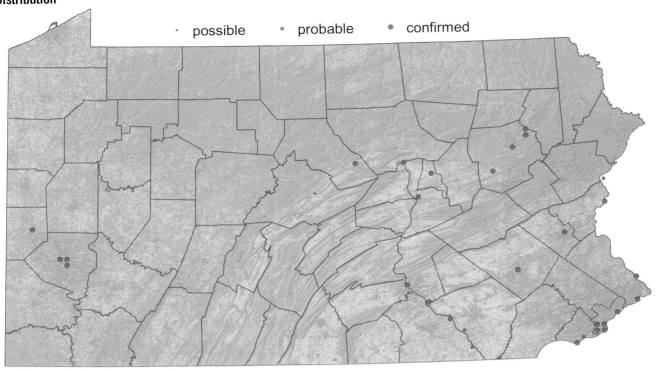

Distribution Change

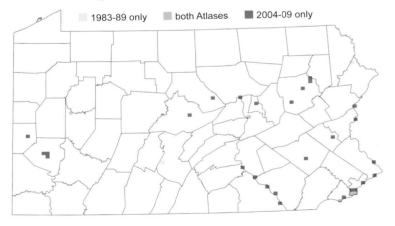

Number of Blocks

	first Atlas 1983–89	second Atlas 2004–09	Change %
Possible	3	1	−67
Probable	0	4	∞
Confirmed	3	26	767
Total	6	31	417

Rick and Nora Bowers/VIREO

King Rail
Rallus elegans

The largest of all North American rails, the King Rail is also one of the rarest breeding birds in Pennsylvania. It was designated a state Endangered species in 1985 because of its scarcity and has retained that precarious status ever since (Gill 1985). It is known by its plump, chicken-like body, long bill, striking cinnamon-brown appearance, secretive behavior, distinctive courtship call (heard day or night), and primary association with fresh and brackish water marshes in the eastern United States. Pennsylvania is at the northern periphery of its breeding range. Historical records indicate that the King Rail was once more common than it is today, though never abundant (Gill 1985).

The King Rail prefers extensive emergent cattail marshes in Pennsylvania, although a diversity of sites has been described across its range (Poole et al. 2005). Historically, there were numerous nesting records from eastern Pennsylvania, with most reports coming from the freshwater tidal marshes of the lower Delaware River Valley (Brauning 1992d), but there have been no confirmed breeding reports there since the early 1990s (McWilliams and Brauning 2000). Confirmed breeding records for western Pennsylvania are far fewer, and some historical nesting sites have been destroyed, notably by flooding to create reservoirs (Geibel 1975; Wilhelm 1993a). However, sporadic King Rail records have persisted in the northwestern counties through the first Atlas period (Brauning 1999d) and more recently (McWilliams and Brauning 2000; Criswell 2010).

As in the first Atlas, there were records of King Rails in five atlas blocks in the second Atlas period. However, there was not a single instance of confirmed breeding in the second Atlas. All but one of the atlas records were of birds heard during targeted surveys. One of the sites in which breeding was confirmed in the first Atlas (Butler County) has subsequently been destroyed by an adjoining surface mining operation, and there were no records during the second Atlas at the other confirmed breeding site (Tioga County), where development there may have affected habitat quality. Although the status of the King Rail is dire, the few records in the second Atlas give us a glimmer of hope for the species' future. Two King Rails were found in Quakertown Swamp, Bucks County, a historic nesting site for the species prior to the first Atlas. Additionally, three single King Rails responded to taped recordings in the Pennsy, Black, and Celery Swamps complex in Lawrence-Mercer counties during Audubon Pennsylvania point count surveys in early June 2007 (Van Fleet 2008). What is so encouraging about these three records is that the King Rail was confirmed nesting there in the 1980s (McWilliams and Brauning 2000), and King Rails have consistently been heard there by volunteers of the Bartramian Audubon Society, who monitor the area almost annually (pers. comm.).

The status of the King Rail in Pennsylvania during the second Atlas paralleled findings in New York, where it was also reported in five blocks, but breeding was confirmed in none (Medler 2008a). In Maryland, where this species is still found widely in Chesapeake Bay marshes, the number of occupied blocks halved between atlas periods (Therres 2010b). Clearly, in Pennsylvania and surrounding states, the King Rail is in trouble. It is listed as Threatened in New York and Massachusetts, and is considered Endangered in Ontario and Ohio, where it was once the most common rail in some of the state's marshes (Peterjohn and Rice 1991).

In summary, the previous loss of emergent cattail wetlands in Pennsylvania was the single most critical threat to the King Rail's future (Gill 1985). Every effort must be made to preserve the species' few known breeding sites and expand other cattail marshes throughout the state. Annual nocturnal monitoring, especially in May and early June, is essential. Yet, with a scarcity of recent records everywhere in eastern North America, there may also be some unknown factors contributing to the species' precipitous decline. The future for the King Rail in Pennsylvania looks uncertain, and its status as a breeding bird in the state is very tenuous.

GENE WILHELM

Distribution

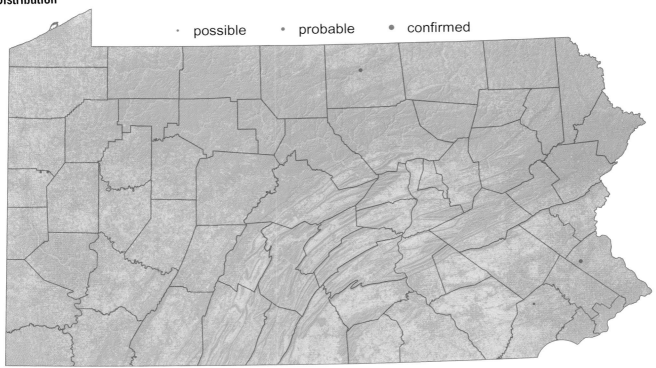

Distribution Change

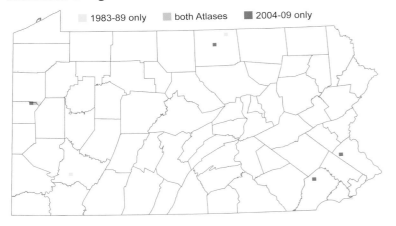

Number of Blocks

	first Atlas 1983–89	second Atlas 2004–09	Change %
Possible	2	3	50
Probable	1	2	100
Confirmed	2	0	−∞
Total	5	5	0

Bob Wood and Peter Wood

Virginia Rail
Rallus limicola

The Virginia Rail is a secretive bird associated with emergent wetland habitats of Pennsylvania. Its chicken-shaped, laterally compressed body allows for ease of movement through dense vegetation. The grunting call that often occurs as a duet in breeding pairs, and the *kid-dik, kid-dik, kid-dik* call are the two most common vocalizations heard during this species' breeding season (Conway 1995).

The global breeding range of the Virginia Rail spans the North American continent from the lower Canadian provinces, south to central California, and eastward to North Carolina (Conway 1995). The overall continental status of the Virginia Rail is uncertain, but it is commonly agreed that populations declined throughout the twentieth century. Breeding Bird Survey trends are inconclusive and sometimes contradictory, due to inadequate coverage, since generally few roads traverse small or large wetland complexes. In Pennsylvania's Wildlife Action Plan, the Virginia Rail is designated as a High Level Concern, primarily due to continued loss of wetlands throughout its range (PGC-PFBC 2005).

The Virginia Rail is probably more common than any other rail species in Pennsylvania, but it is rather scarce and localized in distribution. In both atlases, it was reported most frequently from the northwestern and northeastern corners of the state, where postglacial landscapes harbor plentiful natural wetlands. Elsewhere, there were scattered records from the rest of the state, with rather few from southern border counties and the Pittsburgh Low Plateau. Throughout its range, the Virginia Rail prefers marshlands containing adequate cover in the form of tall emergent vegetation, most often cattails, shallow to intermediate water levels (0–20 cm, 0–8 inches), mud-flats, and an abundance of macroinvertebrates (Eddleman et. al 1998; Craig 2004; Zimmerman et al. 2002a). It can occupy small wetlands, but it is more readily found in the larger marsh complexes with a near equal mix of water-filled openings and vegetation. The Virginia Rail arrives on Pennsylvania breeding grounds from early April to late May. From mid-May to late May, behaviors associated with territorial aggression and pair formation are widely noted. Although the earliest documentation of confirmed breeding during the second Atlas was of fledged young on 19 May, most confirmed breeding reports were later, and fledged young were typically reported from late May to early August.

During the second Atlas, atlas volunteers frequently used playback calls, which are proven to elicit responses from Virginia Rails, thus increasing detection rates (Conway 1995; Conway and Gibbs 2005). The Virginia Rail was found in 153 blocks, a 28 percent increase over the first Atlas. However, it is unclear whether this represents an actual increase in Pennsylvania's breeding population or is partially an artifact of increased use of playback calls. The number of blocks with breeding confirmation was slightly lower in the second Atlas (31) than in the first Atlas (34). Although the Virginia Rail population size in Pennsylvania is difficult to ascertain, it is likely to be in the low hundreds of pairs. Interestingly, only 39 blocks were occupied in both atlas periods, which suggests either that this species was significantly underreported or that suitable habitat is somewhat ephemeral. Certainly, marsh birds present detection challenges, such as the frequently remote and inaccessible nature of wetlands and yearly variation in water levels due to overabundant rainfall or drought. The latter could result in atlas volunteers missing Virginia Rails and other marsh birds, unless wetlands are surveyed during years in which suitable conditions prevail.

Habitat loss from development, draining, filling, or invasive plant species, along with succession to less suitable habitats, continue to affect Virginia Rail breeding populations throughout their range. However, rail populations are known to respond positively to habitat creation of adequately sized mitigation projects in appropriate areas (Porej 2003). Therefore, it is important to better protect and manage remaining marshlands across the state, while at the same time create new habitats to ensure the long-term stability and survival of the Virginia Rail in Pennsylvania.

KIM VAN FLEET

Distribution

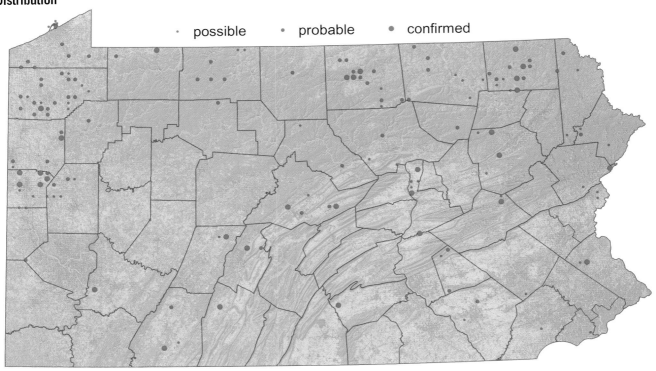

Distribution Change

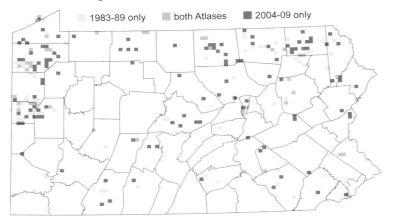

Number of Blocks

	first Atlas 1983–89	second Atlas 2004–09	Change %
Possible	42	55	31
Probable	44	67	52
Confirmed	34	31	−9
Total	120	153	28

VIRGINIA RAIL

Bob Wood and Peter Wood

Sora
Porzana carolina

The Sora is a small, chunky, gray-brown rail primarily associated with emergent marshes and wet meadows. Like the Virginia Rail, it is hard to monitor because of its secretive habits. However, when it vocalizes, its characteristic descending whinny cannot be mistaken for the call of any other bird.

The Sora's breeding distribution extends over most of the United States, except the southeastern states, and much of temperate Canada, although it is local over much of its range. Pennsylvania is situated at the southernmost extent of its eastern range. Its wintering grounds include the northern portions of South America, through Central America to southern California, and coastal regions of the southeastern United States (Melvin and Gibbs 1996; Zimmerman et al. 2002b; Meyer 2006).

The Sora inhabits emergent wetlands and is often found in association with the Virginia Rail (Conway et al. 1994; Eddleman et al. 1988). However, competition for food resources between it and the Virginia Rail is limited, since the Sora's primary diet consists of seeds of wetland plants, with invertebrates playing a secondary role (Melvin and Gibbs 1996). The Sora avoids wetlands with large expanses of open water, preferring complexes of permanent or semi-permanent wetlands characterized by extensive stands of emergent vegetation interspersed with mudflats and floating or submerged residual vegetation and some open water (Melvin and Gibbs 1996; Zimmerman et al. 2002b; Meyer 2006).

The Sora arrives on its Pennsylvania breeding grounds from April through early May, but the presence of migrants during that time clouds the species' breeding status (McWilliams and Brauning 2000). Hence, only records after 15 May were included in the second Atlas, unless accompanied with at least probable breeding evidence. The earliest confirmed breeding was of two separate pairs carrying nesting material on 7 May, and the latest was of fledged young on 22 August.

Over the last two centuries, a considerable amount of Sora habitat has been lost in Pennsylvania (Tiner 1990); as a result, populations are likely much lower than before European settlement. Declines continued through the first Atlas period (Reid 1992b), when the species was found to be very local, present in only 88 blocks scattered across 33 counties. There was little evidence of any change in status by the second Atlas, when it was found in 100 blocks. As in the first Atlas, the Northwestern Glaciated Plateau Section was a notable stronghold for this species. Occupied blocks were otherwise scattered sparsely across the state, with four contiguous blocks containing the Marsh Creek wetlands ("The Muck"), in Tioga County, the only notable concentration elsewhere. Although the Sora's status appears little changed between the two atlases, this result should be viewed with caution, since it is widely known that the use of playback calls increases the likelihood of detection (Conway and Gibbs 2005), and playback was employed more extensively in the second Atlas.

The Sora is currently considered to be a species of Maintenance Concern in Pennsylvania for several reasons, including a general decline in numbers throughout its range, loss of wetlands (Tiner 1990) on breeding and wintering grounds, and the fact that it is generally a poorly studied species. Since little data is available about nesting productivity, annual survival rates, or even harvest level impacts on this species (Conway et al. 1994), it is best to assume that population numbers are declining, since current information across its range suggests a population decline. Efforts should be made to regularly monitor known populations as well as to locate additional ones. Since larger wetland complexes offer more benefits to rail species (Porej 2003), statewide efforts should be focused on prevention of further loss and degradation of remaining wetlands through conservation easements or direct acquisition, maintaining and enhancing extensive wetland complexes across the state, and removal of non-beneficial invasive plant species. Finally, the employment of wetland management practices, such as adjusting water levels to compensate for seasonal variation in rainfall and germination times for certain plants species, would also benefit this species (Zimmerman et al. 2002b).

KIM VAN FLEET

Distribution

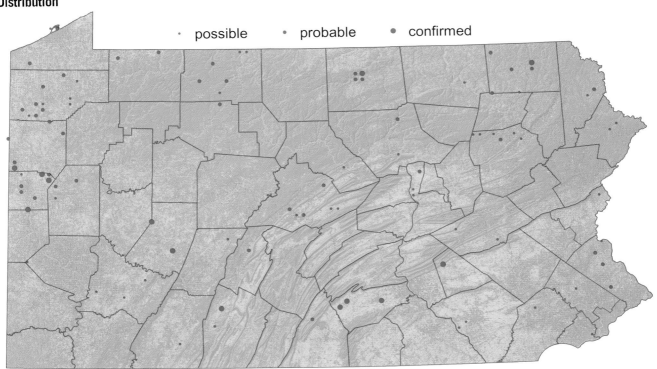

Distribution Change

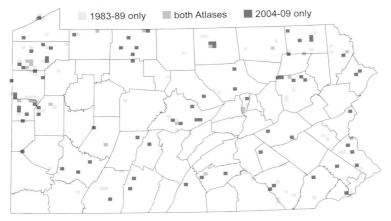

Number of Blocks

	first Atlas 1983–89	second Atlas 2004–09	Change %
Possible	35	44	26
Probable	31	43	39
Confirmed	22	13	−41
Total	88	100	14

Geoff Malosh

Common Gallinule
Gallinula galeata

At first glance, the Common Gallinule, when seen skulking gracefully through a marshy wetland, looks something like an American Coot. However, the red bill, a white line along the sides, and the white under-tail coverts are distinctive. In fact, as noted by Bannor and Kiviat (2002), the Common Gallinule also is both ecologically and behaviorally intermediate between the coot and most of the rails. In 2011, the Common Gallinule was split from the Common Moorhen, which is found throughout the Old World (Chesser et al. 2011).

During the breeding season, the Common Gallinule can be unusually noisy, uttering harsh explosive chicken-like notes: usually four or five loud *squawks* are followed by a series of *checks* or abrupt frog-like *kups* (Pough 1951). These calls, along with the fact that the species is not particularly shy while moving along in open water near clumps of aquatic vegetation, made it easier for atlas volunteers to find nesting Common Gallinules than other species of rails in Pennsylvania's wetlands. Even so, the Common Gallinule sometimes occurs on small ponds and wetlands, where it may have been overlooked.

In eastern North America, the Common Gallinule is most common in southern states, but it occurs locally north to Maine, southern Ontario, and Minnesota. In Pennsylvania, where it had been listed as a species of Maintenance Concern due to its perceived population declines (PGC-PFBC 2005), the species has always been uncommon because of limited wetlands habitat availability. Historically, most breeding records in Pennsylvania have come from the Northwestern Glaciated Plateau or the Coastal Plain. Statewide, there was a 37 percent decline in the number of blocks where Common Gallinules were found between atlases, and the number of blocks with confirmed breeding fell even more steeply from 32 to just 13. In the species' traditional northwestern stronghold, Common Gallinules were found in 28 percent fewer blocks than in the first Atlas. Elsewhere, there were records in 21 widely scattered blocks, most relating to possible or probable breeders, but adults with chicks were noted in Tioga and Bucks counties.

Although studies have not enumerated the breeding population size in the important Conneaut Marsh and Pymatuning region of Crawford County (many sections of these wetlands have limited access available to birders), the actual number of birds nesting at least within these traditional areas seems fairly stable over the long term, although numbers vary from year to year (pers. obs.). Although the species was formerly common in the marshes of the lower Delaware River at Tinicum (John Heinz National Wildlife Refuge) and surrounding wetlands, this area was virtually abandoned during the 1990s with increasing development adjacent to the wetlands (McWilliams and Brauning 2000), and there was no evidence of breeding in that area during the second Atlas period. Surveys in the Lake Erie Basin between 1995 and 2003 showed a significant decline in Common Gallinule numbers (Crewe et al. 2005), and there were declines between atlases of 35 percent in Ontario (Timmermans 2007b) and 33 percent in New York (Medler 2008b). Hence, declines noted between atlases in Pennsylvania may be part of a wider decline in northern Common Gallinule populations.

Studies have shown that breeding densities of Common Gallinules are highest in marshes that are interspersed, with approximately half open water and half emergent vegetation (Bannor and Kiviat 2002). Improving management of marshes to ensure they retain these interspersed characteristics might not only increase the number of nesting Common Gallinules in Pennsylvania but should also benefit an entire guild of wetland bird species, including a variety of ducks, Virginia Rails, Sora, Least and American bitterns, and Marsh Wrens, as well as, perhaps, aid in the recovery of Black Terns in the state. The Common Gallinule is listed as a game species in Pennsylvania, but little is known of what, if any, effect hunting may have on this incidentally taken species. Local studies of this and other rail species could be useful in the formation of effective management programs.

ROBERT C. LEBERMAN

Distribution

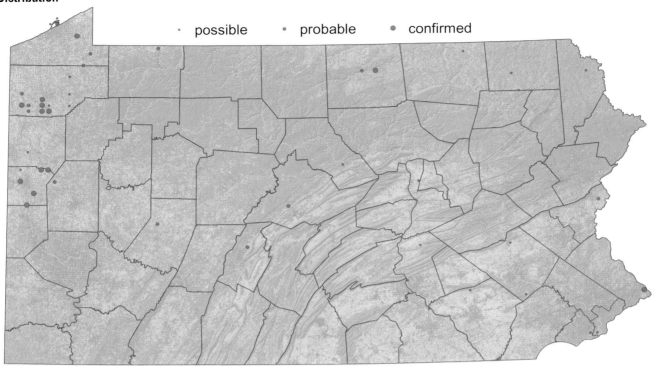

Distribution Change

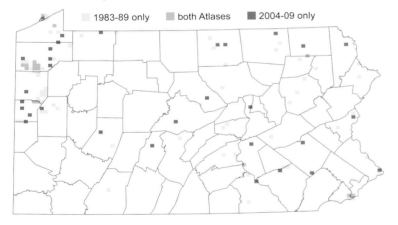

Number of Blocks

	first Atlas 1983–89	second Atlas 2004–09	Change %
Possible	34	22	−35
Probable	10	13	30
Confirmed	32	13	−59
Total	76	48	−37

Jacob Dingel

American Coot
Fulica americana

The most aquatic member of the rail family, the American Coot is a rather awkward-appearing, almost chicken-like bird, with a mostly blackish-gray body and a contrasting short, white bill. Coots are fairly common as migrants in Pennsylvania in both spring and fall; they often gather in large numbers, particularly in the autumn, in areas like Presque Isle Bay at Erie, at Pymatuning Reservoir and Conneaut Lake in Crawford County, and at favorable feeding areas along the Susquehanna and Delaware rivers in southeastern Pennsylvania. Almost every year, a few non-breeding coots are found summering either as single birds or in small groups on ponds and in marshy areas across the state. Nesting American Coots, however, are a rarity anywhere in Pennsylvania, reflecting the scarcity of suitable large freshwater marshes that have an abundance of floating aquatic vegetation and are interspersed with areas of open water.

There have been scattered instances of breeding across the state since the late nineteenth century, but the American Coot has never been common in Pennsylvania (Leberman 1992c). For a period of a few years in the 1930s, the flooding of the Pymatuning Reservoir temporarily created habitat for at least a dozen nesting pairs of coots, but numbers declined thereafter (Grimm 1952). During the first Atlas, the American Coot was confirmed breeding in 11 blocks, all but 1 from marshes located within the Northwestern Glaciated Plateau Section. In the second Atlas, volunteers established breeding through the observation of adults with their recently fledged young in just two widely separated areas. In the northwestern area, an adult with young chicks was observed at Conneaut Marsh, near Custards, Crawford County, on 24 June 2006 (at a location about 2 km [ca. 1 mile] southwest of an area of the same marsh complex where coots were confirmed in the first Atlas) and at Morgan Lake in Berks County, where breeding has been documented regularly since 1995 (McWilliams and Brauning 2000). All but one of the four probable breeding records for coot also came from northwestern Pennsylvania; three of these were clustered in the ancient wetlands near the 11,000-year-old terminal moraine of the Wisconsin Glacial Age (Jennings and Avinoff 1953). Details submitted by atlas volunteers for the 12 possible records scattered across the state suggest that most of these coots were probably late migrants or other summering but non-breeding birds.

American Coot populations declined rangewide in the late nineteenth century due to habitat loss and overhunting (Brisbin et al. 2002). More recently, populations appear to be stable (Sauer et al. 2011). Coots are a little more numerous in neighboring New York and Ontario than in Pennsylvania, and populations there appear to be stable between atlases (Medler 2008c; Tozer 2007). Hence, there is nothing to suggest that the apparent decline in nesting coot populations in Pennsylvania between atlases is due to wider population scale changes.

The American Coot is listed as a species of Maintenance Concern in Pennsylvania's Wildlife Action Plan (PGC-PFBC 2005) because of the decline in the quality and quantity of suitable wetlands in the state. Considering their less than high priority as a game species in Pennsylvania (like the late comedian Rodney Dangerfield, coots "get no respect"), habitat management specifically for the American Coot seems unlikely, but the species undoubtedly would benefit from those management practices suitable for other wetland species such as the Least Bittern, American Bittern, Common Gallinule, Sora, Virginia Rail, teal, and a variety of other duck species. Any habitat restoration on the breeding grounds that focuses on establishment of wetlands with a mosaic of open water and emergent vegetation should result in increased breeding densities of the American Coot (Brisbin et al. 2002), as well as the above-mentioned guild of marshland species, several of which are rare and declining in the state.

ROBERT C. LEBERMAN

Distribution

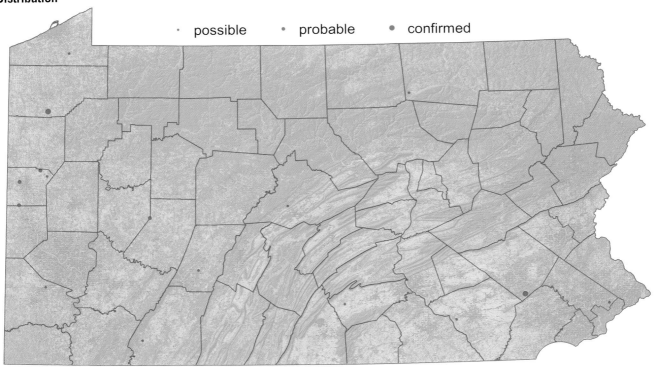

Distribution Change

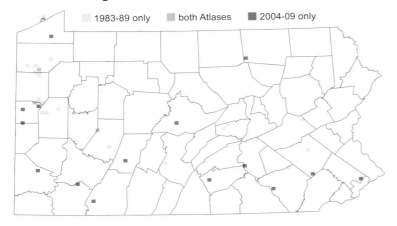

Number of Blocks

	first Atlas 1983–89	second Atlas 2004–09	Change %
Possible	10	12	20
Probable	4	4	0
Confirmed	11	2	−82
Total	25	18	−28

AMERICAN COOT

Bob Moul

Sandhill Crane
Grus canadensis

The Sandhill Crane is a long-necked, long-billed, long-legged, and heavy-bodied gray bird of grasslands, cultivated fields, and freshwater marshes. It superficially resembles the slimmer, more widespread Great Blue Heron, which is called a "crane" in some parts of Pennsylvania.

Unlike tree-nesting herons, the Sandhill Crane is a ground-nesting bird noted for its distinctive trumpeting vocalizations and elaborate courtship dancing displays. The species has an extensive breeding range from Siberia to Cuba and the Isle of Pines. The migratory North American subspecies (from which the Pennsylvania breeding birds undoubtedly are drawn) nests from Alaska south and east through central Canada and the upper Great Lakes (Tacha et al. 1992).

The Sandhill Crane was an exciting addition to the list of Pennsylvania breeding birds shortly after the first Atlas was published (Wilhelm 1993b, 1993c). Summering birds also were being found in New York about the same time, and breeding was confirmed for the first time during the state's second atlas, 2000–2005 (McGowan 2008g). It was recorded in just one location during the first Ohio breeding bird atlas (Peterjohn and Rice 1991), but in more than 25 blocks during Ohio's second atlas (2006–2010). Similarly, its range "exploded substantially" in Ontario between that province's first (1981–1985) and second atlases (2001–2005; Sutherland and Crins 2007).

Farmers in Plain Grove Township, Lawrence County, first noticed two adult Sandhill Cranes in June 1991. What may have been the same pair reappeared on 27 March 1992 in the same area, but this time the pair roamed northeastern Lawrence and southeastern Mercer counties (Wilhelm 1992). The birds left on 18 October but again returned on 28 March 1993 to the same locality, to reappear on 3 August in the company of a juvenile crane. This marked the first documented breeding record of the species for the state (Wilhelm 1993b). A second juvenile was confirmed in the same area on 15 August 1994 with its parents and another adult pair of cranes (Rodgers 1994). Concurrent with the first breeding in 1993, a flock of 30 birds overwintered in northwestern Butler, northeastern Lawrence, and southeastern Mercer counties for the first time. The flock has overwintered every year since and had increased to at least 66 individuals by January 2009 (Malosh 2009).

Sandhill Crane nesting expanded into Crawford County by 1997, with breeding confirmed near Erie National Wildlife Refuge with the presence of juveniles (McWilliams and Brauning 2000). A summering pair was found in Sullivan County as early as 2001 (Kerlin 2001), and the nearby Bradford County sites were documented in 2004 (Gerlach 2005). Will Faux reported the state's first nest in Bradford County in 2004 (Brauning 2004), but the first photographed nest was not found until Bonnie Dersham unexpectedly encountered a recently hatched chick and an unhatched egg on 5 May 2009 in State Game Lands 294, Mercer County, while conducting a survey of Massasauga rattlesnakes.

Most observers reported that birds were active each year at established sites. Nesting was confirmed in Bradford (since 2004), Butler (since 2004), Columbia (in 2007), Crawford (since 1997), Lawrence (since 1993), and Sullivan (since 2001) counties. The precocial young were reported from 6 May through 20 July. Most of these sites included brushy wetlands within 2 km (1.2 miles) of fallow grasslands or agricultural fields, indicating a lack of shyness on the part of cranes for nearby human activities. The observations in the southern counties did not provide evidence of nesting.

Maintaining a substantial and viable population of this new breeding species, however, will depend not only on protecting currently known nest sites but also on identifying, prioritizing, and securing other suitable breeding and overwintering habitats and adjoining agricultural lands. Conservation easements with local farmers may be one feasible way of establishing an adjoining buffer zone and helping to preserve these essential wetland ecosystems.

GENE WILHELM

Distribution

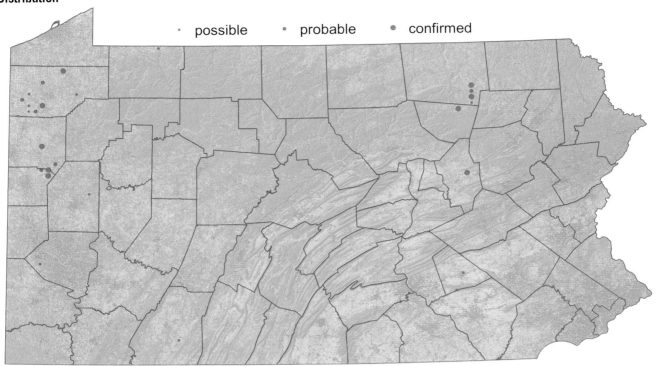

Distribution Change

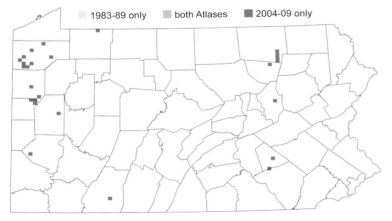

Number of Blocks

	first Atlas 1983–89	second Atlas 2004–09	Change %
Possible	0	11	∞
Probable	0	6	∞
Confirmed	0	9	∞
Total	0	26	∞

Joe Kosack/PGC

Killdeer
Charadrius vociferus

Widespread and vocal, the Killdeer is undoubtedly the most familiar shorebird found in Pennsylvania. Regularly residing in open, developed areas such as parking lots, lawns, golf courses, agricultural fields, and even gravel roofs, the Killdeer is also common in natural environments, including the shores of lakes, ponds, and rivers. Its presence is well known by the general public, particularly due to its attention-drawing calls and animated nest defense displays.

The Killdeer ranges throughout most of North America, breeding in the southern half of Canada, the United States, and Mexico. It winters in the southern part of the United States, southern Mexico, and Central America (Jackson and Jackson 2000). Historically, the Killdeer was considered a game species and was widely hunted in the late nineteenth and early twentieth centuries (Bent 1929). In some parts of North America, hunting pressure was so severe that the species was nearly eliminated—for example, in New England and Quebec (Bent 1929)—but it was described as common in all Pennsylvania avifaunal accounts from the period (Santner 1992a). In Pennsylvania, it has long been associated with settled areas (McWilliams and Brauning 2000) and has no doubt benefited greatly from anthropogenic change.

There is little evidence of any significant change in the status of the Killdeer in Pennsylvania in recent decades. It occupied 3,729 blocks during the first Atlas period (76% of total) and 3,721 blocks during the second Atlas period (75% of total), remarkably similar numbers. In both atlases, it was absent from blocks with little open ground, mainly the densely forested blocks of the Appalachian Plateaus or forested ridgetops elsewhere. There is also no evidence of substantial changes in distribution, with gains and losses in block occupancy largely confined to margins of large forests, where habitat availability may be low and Killdeer populations small and ephemeral. It is interesting to note that the Killdeer was absent in both atlas periods from many blocks in Chester and Delaware counties, in the Philadelphia metropolitan area, despite the fact that it is adept at finding nesting opportunities in the most developed of areas.

Pennsylvania's Killdeer population during the second Atlas was estimated at 120,000 birds (but see caution in chap. 5), which equates to an impressive average of 16 pairs per occupied block. Highest densities occur in the Northwestern Glaciated Plateau, the farmed valleys of the Ridge and Valley Province, and the western half of the Piedmont Province, reflecting areas with either shorelines or extensive agricultural areas, and moderately high human population densities.

Pennsylvania Breeding Bird Survey counts suggest that the Killdeer population has been relatively stable since the 1960s (Sauer et al. 2011), but a modest decline in counts in the last 10 years suggests that the number may be about 16 percent lower than in the first Atlas period. Recent declines have been noted in several surrounding states (Sauer et al. 2011), and the Killdeer is listed as being of Moderate Concern in the U.S. Shorebird Conservation Plan because of such declines (Morrison et al. 2006). The behavior and life history of the Killdeer is reflected in the breeding codes reported by atlas volunteers. Distraction displays and fledged young were the most frequently observed confirmations of breeding. Nesting was observed as early as 26 March, and a nest with eggs was found as late as 15 August; late-season breeding records may involve renesting attempts, since this species often loses clutches and is rarely double-brooded in the north of its range (Jackson and Jackson 2000).

Living in human-altered environments does expose the Killdeer to potential dangers, including pesticides, oil pollution, lawnmowers, and automobiles (Jackson and Jackson 2000). On the other hand, the adaptability of the Killdeer to survive in environments altered by man is undoubtedly responsible for its success as a breeding species in Pennsylvania.

MIKE FIALKOVICH

Distribution

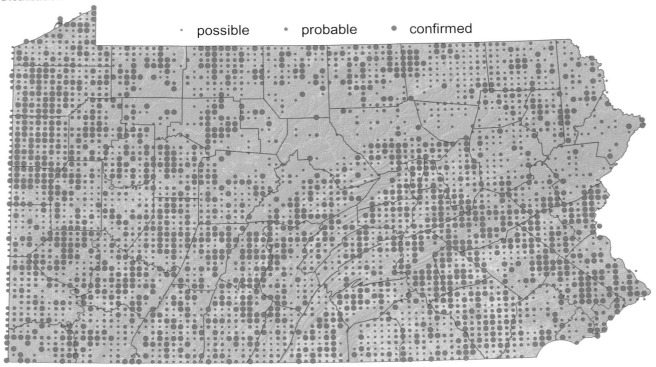

Distribution Change

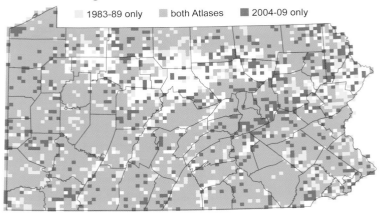

Density

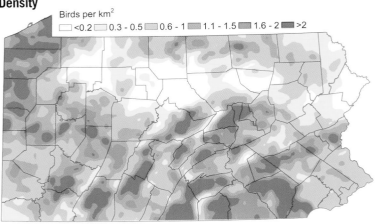

Number of Blocks

	first Atlas 1983–89	second Atlas 2004–09	Change %
Possible	963	1,508	57
Probable	935	867	−7
Confirmed	1,831	1,346	−26
Total	3,729	3,721	0

Population estimate, birds (95% CI):
120,000 (114,000–127,000)

Breeding Bird Survey Trend

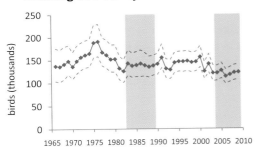

Geoff Malosh

Spotted Sandpiper
Actitis macularius

Of the five shorebird species that breed in Pennsylvania, the Spotted Sandpiper is the one most closely associated with open water. Breeding adults are easily identified by their heavily spotted underparts, which is a characteristic unique among shorebirds. Also characteristic of the species are its continual teetering motion when perched or walking and its habit, when flushed, of flying off low over the water, with pulses of quick, stiff, shallow wing beats. Spotted Sandpipers breed from coast to coast across the upper two-thirds of the lower 48 states as well as in most of Canada and Alaska (Oring et al. 1997). They nest throughout Pennsylvania, which is near the southeast edge of it range, but nesting is uncommon and widely scattered in the state.

The Spotted Sandpiper was recorded in only 478 blocks during the second Atlas, or less than 10 percent of all blocks. Overall block occupancy was 34 percent lower than during the first Atlas. This decrease was fairly uniform across all major physiographic regions; no region saw a decline of less than 29 percent or of more than 44 percent. The decline in priority blocks between atlases was only 13 percent, which suggests that the overall 34 percent decline may be an overly pessimistic assessment. A 12 percent decline occurred between the first and second atlas periods in New York, where the Spotted Sandpiper is a more common breeder than in Pennsylvania (McGowan 2008h).

Always associated with water, Spotted Sandpipers use both small ponds and streams and large lakes and rivers. Near the shorelines where they feed, they require patches of semi-open habitat for nesting and dense vegetation for brood cover (Oring et al. 1997). Just 10 active nests of this species were found during the second Atlas, between the dates of 11 May and 26 June. Atlas reports were most numerous in the eastern one-third of the state and in the western counties along and near the Ohio border. As might be expected, there were few breeding records of this shorebird from heavily forested regions of the state, especially the northern highlands and the Ridge and Valley region south of the Juniata River. Perhaps surprisingly, there also were relatively few records along substantial reaches of major rivers and their principal tributaries, including the West Branch of the Susquehanna, the Juniata, and the upper Allegheny and Monongahela. This same pattern was noted during the first Atlas (Brauning 1992e). There were localized clusters of records on the Susquehanna River, but very few reports in the long stretch between Danville and Harrisburg. Access to shorelines of major rivers is often difficult or prohibited without a boat; thus, low coverage in this area probably accounts for this pattern to a significant extent. However, as noted previously (Brauning 1992e), water pollution is probably the historic reason for the lack of records along the West Branch of the Susquehanna, which drains the former coalfields in Clearfield, Clinton, and Centre counties. The paucity of Spotted Sandpiper records in that area is mirrored by the Belted Kingfisher, another species likely to be adversely affected by decades of acid mine drainage into the freshwater food chain.

The Spotted Sandpiper appears to be slowly declining as a breeder in Pennsylvania. Because of its dependence on aquatic habitats, environmental contaminants such as heavy metals and other agricultural and industrial pollutants may pose direct and indirect threats to its survival and reproduction (Stratford 2010). Atlas results raise, but do not answer, the question of whether the apparent scarcity of breeding Spotted Sandpipers along Pennsylvania's major river corridors is due to adverse environmental conditions or simply a coverage artifact. To address this issue would require targeted surveys and studies along the shores of the large rivers.

Because of the declines observed in the second Atlas, as well as long-term declines on Breeding Bird Survey routes in Pennsylvania (−4.5%/year) and surrounding states (Sauer et al. 2011), the Spotted Sandpiper may warrant consideration as a species of concern in Pennsylvania.

GREG GROVE

Distribution

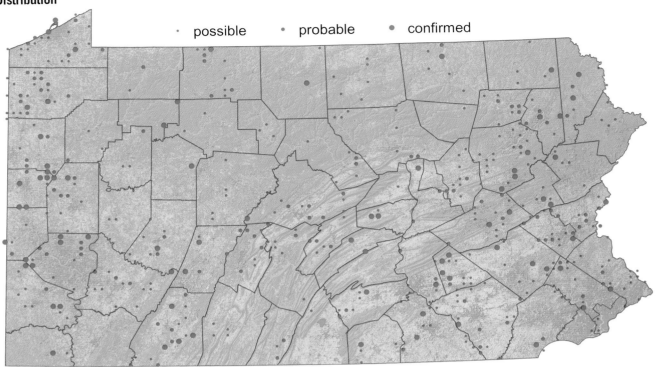

Distribution Change

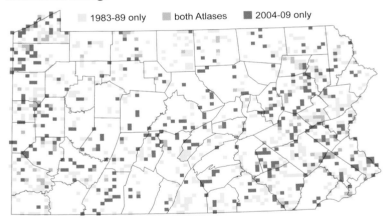

Number of Blocks

	first Atlas 1983–89	second Atlas 2004–09	Change %
Possible	449	279	−38
Probable	184	121	−34
Confirmed	89	78	−12
Total	722	478	−34

Rob Criswell

Upland Sandpiper
Bartramia longicauda

The Upland Sandpiper is characteristic of the Great Plains, from which it presumably originated (Todd 1940). Formerly called Bartramian Sandpiper (for the Pennsylvanian botanist William Bartram), then Upland Plover, its name was finally changed to Upland Sandpiper in 1973 (Houston and Bowen 2001). This unusual sandpiper, larger than the Killdeer, with a small, dove-like head and long neck and tail, is seldom seen near water (Terres 1980). Although it breeds most commonly in the prairies and plains, it ranges as far north as the Yukon and Alaska and as far east as the Canadian Maritime Provinces. It spends nearly 8 months of the year on grasslands in southern South America (Houston and Bowen 2001).

The Upland Sandpiper has experienced a dramatic population fluctuation in Pennsylvania. It first increased as Native Americans opened the forest, continued to expand during European settlement (peaking before the end of the nineteenth century, when agricultural land was at its height), and then declined as abandoning farmland, returning forest, and intensified mono-agriculture became the norm. Overhunting in the Great Plains in the late nineteenth century persecuted the species so harshly that it has never fully recovered (Houston and Bowen 2001). More recently, continued loss of habitat, as well as heavy use of pesticides in South American winter grounds, only exacerbated the species' precipitous decline. By 1960, the Upland Sandpiper was considered uncommon in Pennsylvania (Poole 1964), and it was listed as Threatened in Pennsylvania in 1985 (Gill 1985).

In western Pennsylvania, the Upland Sandpiper exhibits a sequential, seasonal use of habitats, possibly due to its historic association with the roaming ancient bison in the Great Plains. The species seems to be drawn to Amish farms, which employ a similar rotational field system (Wilhelm 1995). Amish properties adjoining extensive reclaimed surface mine grasslands provide ideal conditions; one such area in northwest Pennsylvania has supported a colony for several years (Brauning 1998), and 12 birds (10 adults, 2 juveniles) were noted there in June 2009 (pers. obs.). The Upland Sandpiper rarely nests alone (McWilliams and Brauning 2000), but since Amish farms are limited in size, each one can only support one or two sandpiper families.

In the second Atlas, the Upland Sandpiper was noted in only 23 atlas blocks statewide, less than half the total of 54 in the first Atlas; further, breeding was confirmed in just 2 blocks, compared with 21 in the first Atlas. Moreover, the Upland Sandpiper's range is becoming increasingly restricted, with no records in the second Atlas east of Adams County and a virtual desertion of former strongholds in Erie and Crawford counties. Thus, in Pennsylvania, the Upland Sandpiper is now not only a rare breeder but also largely confined to a few Amish farms in northwestern Pennsylvania and reclaimed surface mines. The coal-bearing counties stretching from Clarion westward to the Ohio border are now very much the stronghold for this species, along with mosaics of reclaimed surface mines and agricultural areas in Somerset County.

Breeding Bird Survey (BBS) data show a significant negative trend for the Upland Sandpiper in the eastern states throughout the period from 1966 through 2009, contrasting with a slight positive trend in the species' core range in the Central BBS Region (Sauer et al. 2011). The U.S. Shorebird Conservation Plan listed it as a High Conservation Concern (Morrison et al. 2006). Currently, the species is listed as Endangered in most states surrounding Pennsylvania (Houston and Bowen 2001).

Urgent measures are needed to prevent further Upland Sandpiper population decline, and possibly extirpation, in Pennsylvania. Preservation of suitable grassland mosaics, delaying of mowing until early July to help preserve eggs and flightless young, and use of moderate grazing in rotation (Dechant et al. 1999) could be encouraged to maintain the relict populations of this species. Such measures would require the direct help of farmers and surface mine owners.

GENE WILHELM

Distribution

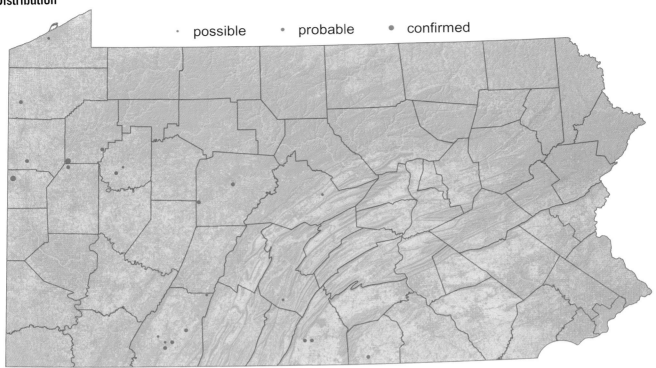

Distribution Change

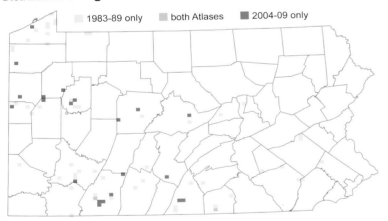

Number of Blocks

	first Atlas 1983–89	second Atlas 2004–09	Change %
Possible	16	8	−50
Probable	17	13	−24
Confirmed	21	2	−90
Total	54	23	−57

Bob Wood and Peter Wood

Wilson's Snipe
Gallinago delicata

The Wilson's Snipe (called "Common Snipe" in the first Atlas) breeds widely across North America, from Alaska to Newfoundland, throughout most of subarctic Canada, to northern California, Colorado, the Great Lakes area, and northern New England (Mueller 1999). Thus, Pennsylvania is on the southeastern margin of its large breeding range.

The continental breeding population of this unusual "shorebird" (and migratory game bird) is approximately 2 million (Brown et al. 2001; Trevor and Andres 2006); the U.S. Fish and Wildlife Service estimates about 120,000 were harvested by hunters in the United States in 2007 (Case and McCool 2009). Wilson's Snipe is not officially listed at the federal or state level, but the Pennsylvania Wildlife Action Plan identifies it as a species of Maintenance Concern (PGC-PFBC 2005). This ranking is based on the state's breeding population, not migrants from farther north.

Similar in size, shape, and aerial courtship behavior to the American Woodcock, the snipe is primarily found in wet meadows, low-lying pastures, and more open marshes and swamps, whereas the woodcock inhabits forested uplands and wetlands. Its quick zig-zag flight and startling, raspy *scaipe* call, given when it flushes, as well as its habit of dropping abruptly back down into the marsh vegetation, help to identify and differentiate it from the American Woodcock, which bursts up and flies straight away on twittering wings when flushed. The snipe is also readily identified by its distinctive courtship flight sounds, especially the "winnowing" or "booming" sounds produced by the air rushing through its spread lateral tail feathers during the aerial dives that are part of its crepuscular courtship flight displays. As with the American Woodcock, it is not unusual to see and hear several snipe displaying over a breeding marsh at one time.

The second Atlas shows a sparse breeding distribution for Wilson's Snipe, found primarily across the northern glaciated provinces of Pennsylvania, with scattered breeding records in the Allegheny Mountains (i.e., Allegheny Front) southeast of Pittsburgh. This distribution remains essentially unchanged from the first Atlas, except for the loss of possible breeding sites in the Ridge and Valley Province. Nearly all confirmed or probable breeding sites are on the glaciated plateaus of northwestern and northeastern Pennsylvania or at high-elevation wetlands in the mountain regions to the south. Nesting snipe tend to be found rather irregularly in Pennsylvania, even at known breeding sites; annual weather is a factor, especially the extent of dry or wet conditions in winter and early spring (Leberman 1992d; McWilliams and Brauning 2000).

Atlas volunteers reported Wilson's Snipe pairs as early as 3 April, courtship behavior from 21 April to 3 June, and territorial behavior as late as 3 July. Confirmed breeding evidence (one each of distraction display and nest with young) was found only in mid-June. The author has personally heard "winnowing" displays or territorial vocalizations on the large wetlands of Tioga County (known locally as "The Muck") as early as 25 March and as late as 16 June.

Good estimates of the Wilson's Snipe breeding population size in Pennsylvania are lacking, but given the number of atlas records, it certainly must be on the order of hundreds of birds, at the most. Since it is a cryptic and crepuscular wetland species, the Breeding Bird Survey barely detects snipe in the Appalachian or Piedmont regions that include Pennsylvania (Sauer et al. 2011); hence, regional trends in breeding populations are uncertain. Because Pennsylvania's small breeding population is already at the margin of its North American range, the conservation outlook for this species in Pennsylvania is far from secure; the continued presence of this unique breeding bird in Pennsylvania will depend entirely on the effective protection and management of suitable wetland habitats in the state.

ROBERT M. ROSS

Distribution

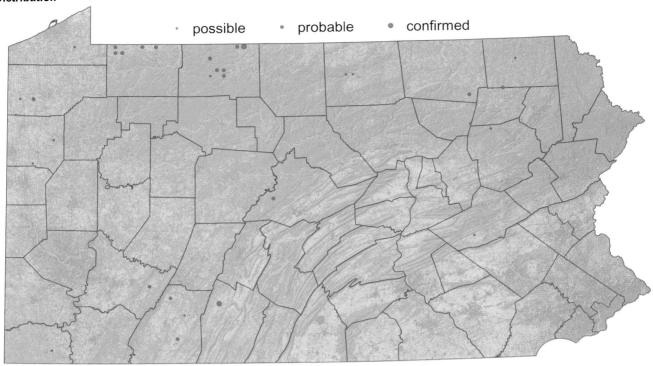

Distribution Change

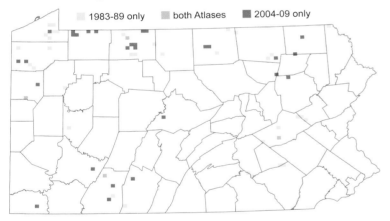

Number of Blocks

	first Atlas 1983–89	second Atlas 2004–09	Change %
Possible	24	13	−46
Probable	12	17	42
Confirmed	2	2	0
Total	38	32	−16

Bob Wood and Peter Wood

American Woodcock
Scolopax minor

The American Woodcock is an unusual upland shorebird, popular with both birders and hunters. Birders like it for its odd appearance and behaviors. Upland game bird hunters like the woodcock because it relies on its cryptic coloration for protection and because it holds tightly in front of pointing dogs before flushing at the last second. The woodcock requires young forest habitats with small openings for spring courtship; brushy, woody vegetation for cover; and rich, moist soils that support an abundance of its primary food, earthworms. In prehistoric times, this habitat would have been associated with beaver flowages and other wetland areas, and with forests disturbed by fires, storms, and other natural phenomena. Following the cutting and subsequent regeneration of much of Pennsylvania's forests during the nineteenth century, woodcock habitat and populations likely increased. Today, much of the woodcock's habitat in Pennsylvania consists of old farm fields reverting to forest, with aspen, alder, hawthorns, dogwoods, and viburnums dominating early stages of succession. Recently cut forests also provide good habitat.

Historic accounts of woodcock in Pennsylvania vary with respect to its distribution and abundance. Warren (1890) stated that "this species is very generally dispersed throughout the swampy districts of state," and he mentioned several counties, mostly in northwestern Pennsylvania, favored by sportsmen. Sutton (1928a) wrote that the woodcock was an "uncommon and somewhat irregular migrant and summer resident," and Todd (1940) reported that "there are no parts of this area [western Pennsylvania] where the Woodcock may not be found at one season or another." Liscinsky's (1965) classic work on Pennsylvania's woodcock does not comprehensively define its range in Pennsylvania, but it mentions alder in northeastern and north-central Pennsylvania and aspen in northwestern Pennsylvania as being predominant habitats. McWilliams and Brauning (2000) described woodcock as uncommon breeders with a statewide distribution.

Results of the second Atlas show that the woodcock remains widely distributed in Pennsylvania and, consistent with historic accounts, is still concentrated in the northwest. Since the first Atlas period, however, the distribution appears to have shifted slightly northward, with the many southern areas not detecting them this time. However, caution is advised in drawing inferences from these results, because during the second Atlas it appears that concerted efforts to find Woodcock were made in some second Atlas regions (e.g., 28, 32 and 43). In early spring, a well-organized observer could find woodcock in two or more atlas blocks in a short time in a single evening.

The apparent concerted effort to find woodcock in some regions should also be taken into account in interpreting the non-significant, 7 percent reduction in block occupancy between the atlases. The long-running U.S. Fish and Wildlife Service Woodcock Singing Ground Survey (Cooper and Parker 2010), data from which was included in the second Atlas, documented a significant long-term decline (−0.94% per year, 1968–2010) in Pennsylvania's woodcock population. Since 1968, woodcock have declined throughout North America (Cooper and Parker 2010). In neighboring Maryland, there was 48 percent decline in block occupancy between atlas periods (Ellison 2010i), but long-term declines may have slowed further north (Cooper and Parker 2010).

Although the American Woodcock remains fairly common and widespread in Pennsylvania today, that should not be taken for granted. The early successional habitat on which the woodcock and many other species depend is being lost in Pennsylvania as forests mature, development consumes open space, and farmland abandonment slows. That abandonment may actually cease as demand for energy from renewable resources increases and abandoned fields go back into production. Overhunting was once seen as a cause of woodcock population decline (Todd 1940; Sutton 1928a), but woodcock hunting today is well regulated (USFWS 2010) and, with hunting interest declining (Boyd and Cegelski 2008), is not a threat. Habitat loss is the long-term threat to American Woodcock in Pennsylvania.

JOHN TAUTIN

Distribution

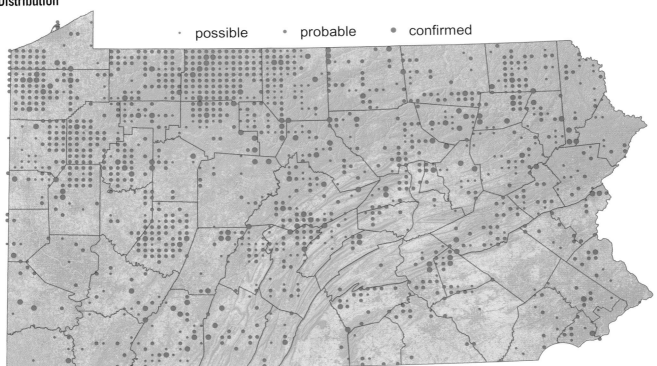

· possible • probable • confirmed

Distribution Change

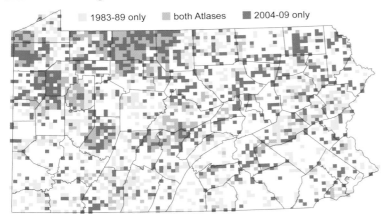

1983–89 only | both Atlases | 2004–09 only

Number of Blocks

	first Atlas 1983–89	second Atlas 2004–09	Change %
Possible	543	390	−28
Probable	716	840	17
Confirmed	210	140	−33
Total	1,469	1,370	−7

AMERICAN WOODCOCK

Geoff Malosh

Ring-billed Gull
Larus delawarensis

The Ring-billed Gull is Pennsylvania's most commonly and widely observed gull—the "seagull" that catches even casual observers' attention while it forages for scraps in parking lots, at shopping malls, and fast food restaurants. When breeding, however, this species is not primarily associated with the sea. It nests throughout much of northern interior North America, reaching the ocean as a breeder only at the far eastern end of its range in Newfoundland.

The Ring-billed Gull breeds colonially, usually nesting on the ground along shores of lakes and rivers, on grassy and sandy islands, and occasionally in marshes. Roofs, dredge spoils, piles of rubble, and other human-made habitats are also used widely (Ryder 1993). In fact, a roof was used at Pennsylvania's single active colony during the second Atlas.

Pennsylvania lies at the southeastern fringe of this gull's breeding range, and predators and human interference have combined to prevent the species from attaining a viable and permanently undisturbed breeding colony in the state. Erie County has hosted Pennsylvania's only two historical nesting locations, one at aptly named Gull Point in Presque Isle State Park and the other in nearby urban Erie.

Nesting occurred before and since the first Atlas period, but not during. The first (unsuccessful) breeding attempt came in 1983 at Gull Point, where 20 nest scrapes, three of them containing eggs, were discovered (Stull et al. 1985; Hill 1986). Predators destroyed the nests, no young were produced, and the remaining nests were abandoned. There were several more attempts there in subsequent years, including larger colonies of approximately 50 nests in 1995 and 1997 (McWilliams 1995; McWilliams and Brauning 2000), but again, no young were produced.

The state's first successful colony—a huge one, estimated to contain 700–800 adults and 120 chicks—was discovered a few miles away from Presque Isle in 1999, at an industrial plant in the city of Erie (McWilliams 1999). The gulls first nested in a vacant lot at the plant and then expanded to the plant roof. Because of property damage and danger to health and safety, the U.S. Department of Agriculture Wildlife Services was asked to disperse the colony. Nesting attempts and control measures continued until 2009. During the second Atlas period, 962 Ring-billed Gull nests were removed (H. Glass, pers. comm.).

Prospects for breeding expansion of the Ring-billed Gull in Pennsylvania are difficult to predict. Prominent clusters of blocks reporting ring-bills during the second Atlas were at Lake Erie and Pymatuning Lake, in the northwestern area of the state, Allegheny Reservoir and the upper reaches of the North Branch Susquehanna River in the northern tier, an extensive stretch of the Lower Susquehanna, and along the lower Delaware River. All of those areas could probably furnish the species an adequate diet of fish and appropriate nest substrates. The Ring-billed Gull's breeding distribution has expanded in New York, where confirmed breeding blocks doubled to 28 blocks during the second atlas from 2000 through 2005 (Richmond 2008a). Breeding was confirmed in three blocks in Chemung County along the Chemung River, adjacent to Pennsylvania's northern border. Provisional results for the second Ohio breeding bird atlas show breeding in at least four blocks in the northeastern portion of the state (OBBA 2011), where none was found in the first Ohio atlas (Peterjohn and Rice 1991). Judging by the New York and Ohio occurrences, a colony could be established anywhere in northern or western Pennsylvania with a substantial lake or waterway.

As with the Herring Gull, population control rather than conservation could face the Ring-billed Gull wherever it might attempt to establish colonies in Pennsylvania, particularly in urban areas and industrial sites, or in other places where predation on nesting wading birds and terns might be a concern. In the 1980s, extensive control programs were conducted at industrial locations in southern Ontario, where negative impacts on tern colonies were reported (Blokpoel and Tessier 1986, 1987). As the opportunistic Ring-billed Gull becomes more interconnected with humans, wildlife managers will watch closely.

PAUL HESS

Distribution

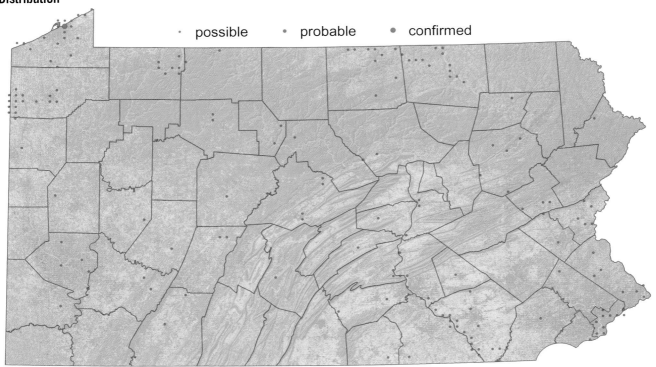

Distribution Change

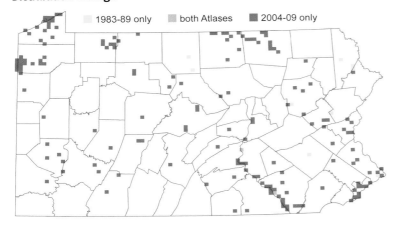

Number of Blocks

	first Atlas 1983–89	second Atlas 2004–09	Change %
Possible	11	170	1,445
Probable	1	0	−∞
Confirmed	0	1	∞
Total	12	171	1,325

Geoff Malosh

Herring Gull
Larus argentatus

The Herring Gull breeds colonially or as isolated pairs in Arctic and northern temperate regions worldwide. In North America, it nests from Alaska across Canada, around the Great Lakes, along the Atlantic coast south to North Carolina, and locally on the Gulf coast. It has been expanding its range, often in conflict with humans and wildlife (Pierotti and Good 1994).

Pennsylvania's first colony of Herring Gulls was discovered in Pittsburgh in 1994, not long after the first Atlas was completed (Floyd 1994). According to a range map published at that time (Pierotti and Good 1994), Pittsburgh was the species' southernmost breeding locality in interior North America. During the second Atlas period, nesting was confirmed in seven blocks. Six blocks in the southwest appear on the map as a "V," with the Allegheny River as its eastern arm and the Ohio River its western arm. The single block in the northwest represents a single large colony at Erie.

Proximity to water and protection from predators seem to be key factors in nest site selection (Pierotti and Good 1994), and the Pennsylvania sites provided both features. Although the species nests on diverse natural sites such as islands, lakes, rivers, beaches, marshes, cliffs, tundra, and occasionally even in trees, it is known around the world for adapting to human habitats. This adaptation is clearly important in Pennsylvania, where every nest confirmed in the second Atlas was on a man-made structure. The two initial nests at Pittsburgh were built on concrete warning pylons at an Allegheny River dam (Floyd 1994). By the second Atlas period, as many as 13 nests were active on the pylons and the concrete piers, as well as on the steel structure of a high bridge over the dam (Fialkovich 2003; S. Kinzey, pers. comm.). Isolated pairs were confirmed nesting northward on the Allegheny River at the Natrona, Clinton, and Kittanning dams. One to three nests were found on the Ohio River at the Emsworth and Dashields dams. All of these were on warning pylons or flat-topped mooring pilings. Meanwhile, the presence of pairs indicated probable breeding at the Kelly Station Dam on the Allegheny River and at the Montgomery Dam on the Ohio River.

The species' history at Erie is very different. Four unsuccessful nests were discovered in 1995 on the beach at Gull Point, Presque Isle State Park (McWilliams 1995), but none in subsequent years. Next, several nests were located in 1999 within a large Ring-billed Gull colony at an industrial plant (McWilliams 1999). The gulls first nested in the plant parking lot and later moved to the plant roof, where the colony's Herring Gull component grew quickly. Because of property damage and danger to health and safety, the U.S. Department of Agriculture's (USDA) Wildlife Services was asked to disperse the entire colony. During the second Atlas period, 562 Herring Gull nests were removed from this site (H. Glass, pers. comm.). The eradication program continued until no nesting gulls remained.

Observed and possible breeding record reports came from 53 blocks, concentrated mainly at Pymatuning Reservoir in the northwest and along the Susquehanna and Delaware rivers in eastern Pennsylvania. Southern Bucks County, on the Delaware River, has hosted a Christmas Bird Count record of more than 140,000 wintering Herring Gulls (Mirabella 2005; Ortego 2005), and suitable man-made habitat is available in that area for pairs to remain and breed. Consequently, it is reasonable to expect further nesting in the state in the years following the second Atlas.

The nesting ecology of the species in southwestern Pennsylvania—isolated pairs nesting on structures in and over water—has not yet raised management concerns. However, a return of the colony at Erie, in addition to raising health concerns by the USDA, might not bode well for the hoped-for recovery at nearby Presque Isle of two of the state's recently extirpated breeding species, the Piping Plover and Common Tern.

PAUL HESS

Distribution

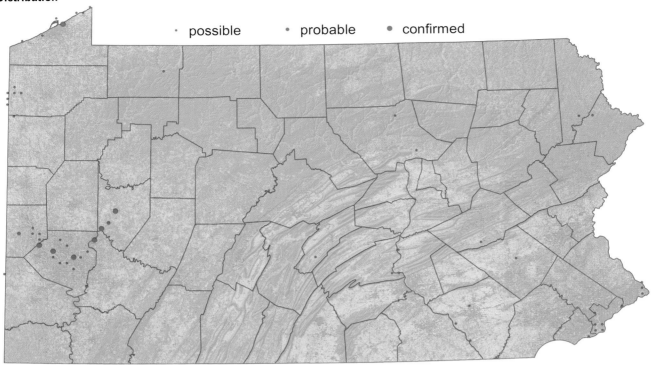

Distribution Change

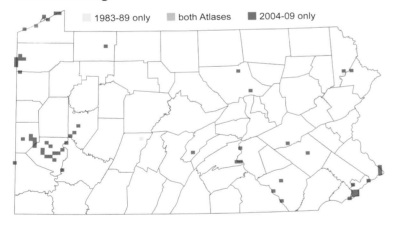

Number of Blocks

	first Atlas 1983–89	second Atlas 2004–09	Change %
Possible	1	51	5,000
Probable	1	2	100
Confirmed	0	7	∞
Total	2	60	2,900

Geoff Malosh

Great Black-backed Gull
Larus marinus

The Great Black-backed Gull is a denizen of the North Atlantic. Its breeding range includes northeastern North America, Greenland, Iceland, and much of northern Europe. In North America, its coastal breeding range stretches from Labrador to the Chesapeake Bay, with small numbers south to coastal North Carolina (Good 1998). Away from the coast, breeding pairs and small colonies are found locally along the St. Lawrence River and the Great Lakes, as well as very rarely on smaller inland lakes and rivers.

Historically, the Great Black-backed Gull was very rare in Pennsylvania until the middle of the twentieth century. The species was thoroughly persecuted by feather hunters and egg collectors in the nineteenth century, which resulted in a contraction of its breeding range to the far north. Only after it became a protected species under the Migratory Bird Treaty Act of 1918 did it begin to recover in the United States, and by the 1960s it may have actually exceeded its former range in North America (Good 1998). Most summer records of Great Black-backed Gulls in Pennsylvania are of immature birds, primarily on the southern stretches of the Delaware and Susquehanna rivers and on Lake Erie. In summer, adults are uncommon, but regular, in these same areas.

One of the most notable discoveries of the second Atlas was a nesting pair of Great Black-backed Gulls on the Delaware River in Delaware County, the first in Pennsylvania's recorded history. In May 2006, an employee at the Sunoco fuel terminal, near the Philadelphia International Airport, discovered a Great Black-backed Gull nest on a pier at the facility, over what was described as a "quiet portion" of the Delaware. The nest was constructed primarily of grasses and contained three eggs on 26 May. Three chicks hatched on 16 June, but one soon disappeared. The remaining two were banded on 30 June and fledged in early July (McGovern 2006).

At the outset of the second Atlas, the Great Black-backed Gull was not expected to be among the state's breeding species. Safe dates had not been set, but with the confirmation in 2006, the safe dates were established as 15 May to 31 July, and records of adults on the Delaware River during the entire atlas period were upgraded from observed to possible, as appropriate.

Despite the unprecedented nature of the discovery, the fact that the Great Black-backed Gull has now bred in Pennsylvania is not such an unexpected surprise. Good (1998) speculated that their expansion may be bolstered by the abundant availability of human refuse. Indeed, some of the greatest attractors for huge aggregations of large white-headed gulls in winter, anywhere in the Mid-Atlantic, are the landfills in the Tullytown area of Bucks County. The highest recorded total of Great Black-backed Gulls on the Southern Bucks County Christmas Bird Counts, which includes Tullytown, is a staggering 15,867 in 2004 (Mirabella 2005; Bolgiano 2005). The reason for the gulls' specific affinity for Tullytown as a wintering site, as opposed to other landfills in the region, is uncertain, although the consequences of this are clear: with the species already expanding its range southward, and with thousands finding a winter home on the Delaware River, the number of Great Black-backed Gulls lingering into the summer there will probably continue to increase. Whether they will be able to consistently find sufficiently undisturbed nest sites along this urbanized stretch of river, and thus become a more regular breeder in the area, is another matter.

On Lake Erie, the prospects for future nesting remain unclear. Great Black-backed Gulls remain very local on the Great Lakes, with breeding documented only on Lake Huron and Lake Ontario. In Pennsylvania, adults are uncommon but fairly regular on the Lake Erie shore in summer, yet evidence of possible nesting has not materialized. As with the difficulty they face on the Delaware River, the lack of consistently undisturbed and predator-free sites along the well-developed Pennsylvania lakeshore is likely the primary factor that keeps the species from establishing a breeding population in the region.

GEOFF MALOSH

Distribution

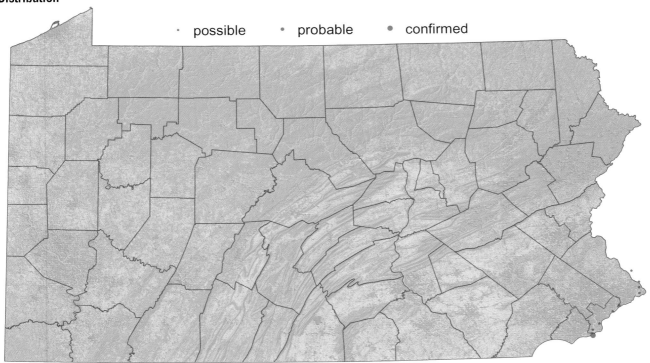

Distribution Change

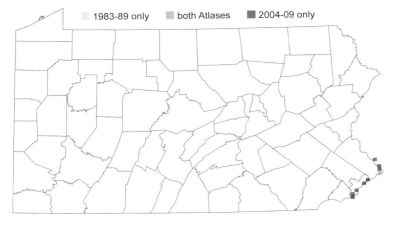

Number of Blocks

	first Atlas 1983–89	second Atlas 2004–09	Change %
Possible	1	9	800
Probable	0	0	0
Confirmed	0	1	∞
Total	1	10	900

Geoff Malosh

Black Tern
Chlidonias niger

The flight of Black Terns feeding over an inland marsh or shallow lake, with their erratic, buoyant nighthawk-like zigzags, is a charming and inspiring sight—but it is a sight that is at risk of being lost on breeding grounds in Pennsylvania. The North American breeding range of the Black Tern stretches across much of southern Canada and east to northern New England, New York, northwestern Pennsylvania, and the Lake Erie marshes in northwestern Ohio. These eastern populations have been on the decline in recent decades: the second Ontario atlas found a 25 percent decline in the number of blocks with Black Terns (Weseloh 2007), New York listed the species Endangered since 1999 (Mazzocchi and Muller 2008), and preliminary data from Ohio (OBBA 2011) suggest significant declines in the Lake Erie marshes—now the only breeding sites in that state. In Pennsylvania, where the Black Tern has only ever been documented as nesting within Crawford and Erie counties (McWilliams and Brauning 2000), the number of blocks from which terns were reported fell from 11 in the first Atlas to just 3 in the second Atlas.

The Black Tern has been listed as an Endangered species in Pennsylvania because of the very small and erratic nature of the breeding population, its limited distribution in the state, its susceptibility to human disturbance, as well as a decline in the size and quality of the wetlands available to it. The species prefers large, relatively shallow, undisturbed wetlands with plenty of emergent vegetation for nest sites and plenty of adjacent open water for foraging.

During the second Atlas, the Black Tern was confirmed nesting only at Presque Isle State Park, Erie County, where a breeding pair was discovered at its nest on Niagara Pond on 4 June 2004. This is the same area where they apparently also had attempted nesting during the previous year. Breeding was strongly suggested within two adjacent blocks in the Hartstown Marsh complex, Crawford County, when two adult birds were seen flying low together over the open marsh and repeatedly landing together in the emergent vegetation on 19 June 2004. Apparently, however, there was no follow up on these Hartstown area observations, the last site in Crawford County where Black Terns had nested (PGC unpublished data). Elsewhere in Crawford County, there was no evidence of breeding terns at the other historic nesting sites, such as the Upper Reservoir at Pymatuning Lake or at the nearby Conneaut Outlet Marsh or at Smith's Marsh (Leberman 1992e).

Apparently, the Black Tern hovered on the brink of extirpation as a breeding bird in Pennsylvania during the second Atlas, a situation deserving of increased management attention by state wildlife biologists. A few of the terns' former state nesting sites, like Conneaut Marsh, appear to have substantially declined in habitat suitability for the terns, due to changes in water level and the encroachment of woody vegetation, such as the water-willow, buttonbush, and the various shrubby dogwoods. Other areas, such as the Upper Reservoir at Pymatuning, appear to be rather stable and remain suitable to the needs of the Black Tern. The species is well known for its low site tenacity, due to changes in water level and other wetlands conditions (Heath et al. 2009), giving some hope for the species' recovery in Pennsylvania. As suggested in the account for the first Atlas (Leberman 1992e), limiting human access to important tern breeding sites during the critical nesting season should be considered, since birds are sensitive to disturbance. In managed areas, water levels that are regulated to maintain semi-open tracts of emergent vegetation could benefit the Black Tern and many other freshwater marsh species, such as the Pied-billed Grebe, Least and American bitterns, American Coot, Common Gallinule and Marsh Wren, most of which are also in decline or now found in perilously low numbers in the state.

ROBERT C. LEBERMAN

Distribution

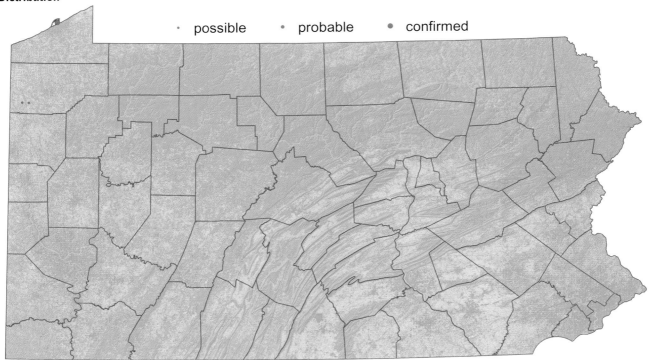

Distribution Change

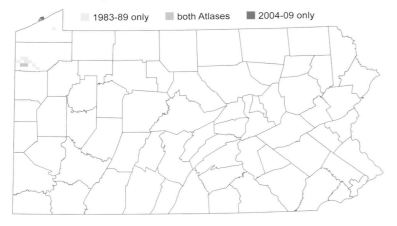

Number of Blocks

	first Atlas 1983–89	second Atlas 2004–09	Change %
Possible	4	2	−50
Probable	3	0	−∞
Confirmed	4	1	−75
Total	11	3	−73

Claude Nadeau/VIREO

Rock Pigeon
Columbia livia

The wild Rock Pigeon is a native of caves, cliffs, and mountains in the Old World, but it has a long history of domestication and feral forms that are now a familiar presence in most areas of dense human settlement across the world. It was deliberately introduced to North America in the early seventeenth century (Schorger 1952), when it was an important food source for European colonists. Artificial selection of specific traits in show birds and racing pigeons has added greatly to the domesticated varieties from which feral populations have descended, such that feral Rock Pigeons now show a wide variety of color forms, although many still show traces of the handsome plumage of their wild ancestors.

Because of its domestic origins, the Rock Pigeon has often been overlooked by ornithologists, and its historical status in Pennsylvania is uncertain. It is likely, however, that feral populations followed quickly in the footsteps of European colonists, and to this day it remains closely tied to human activity. It has long been a naturalized bird, with self-sustaining populations in urban and agricultural areas across North America. It nests predominantly on ledges of buildings, such as city blocks, under bridges, and farm outbuildings, but in some areas it has reverted to natural cliff nest sites (Johnston 1992). Use of bridge overpasses as nest structures, even in highly forested areas, enables this bird to occur far from other human structures. The Rock Pigeon is often considered a pest and is not afforded legal protection.

In the first Atlas, when it was known as the Rock Dove, the Rock Pigeon was widespread in Pennsylvania, absent only from densely forested country. The species' distribution changed little between the first and second Atlas periods. However, there were localized areas of range contraction, especially in the southwestern corner of the state, while in other areas, particularly in the periphery of the High Plateau, there was modest range expansion. The pattern of these changes does not consistently match broad-scale changes in land use (chap. 3). Because this species has a lowly status among the birding community, it is possible that it was sometimes overlooked by atlasers in one or both of the atlases, which could have resulted in an overestimate of changes in block occupancy.

From second Atlas point count data, we estimate a population of 415,000 Rock Pigeons in Pennsylvania, an average of more than 120 birds per occupied block. Densities were highest in agricultural areas in the Ridge and Valley Province, Lancaster County, and in the environs of the largest cities. Estimated densities of 15 birds per km^2 across Philadelphia County are in keeping with densities in other northern metropolitan areas, which generally have the largest numbers on Christmas Bird Counts (Johnston 1992). Both point count numbers and block occupancy were higher in arable farmland than in predominantly pastoral areas (appendix D).

The Rock Pigeon has proven to be a very adaptable species. Its numbers may be limited by nest site availability, but it flies considerable distances to find food (Johnston 1992). Spilled grain, especially corn, forms most of the diet in rural areas, while town- and city-dwelling pigeons are well known for being voracious consumers of all manner of human food scraps (Johnston 1992). It breeds year-round; atlas volunteers found occupied nests in early January, fledged young in early February, and nest building in November. Thus, it is a species that has proven resilient to landscape-level changes, and the only likely cause of population change is deliberate population control or significant changes in food supply. Breeding Bird Survey data show a steady decrease in numbers in Pennsylvania since the 1960s, the cause of which is uncertain. Similar declines are evident in surrounding states (Sauer et al. 2011). Rock Pigeons do not appear to be in direct competition with any native species, and they often form an important part of the diet of Peregrine Falcons and other birds of prey (White et al. 2002). Notwithstanding the sometimes considerable localized damage to property, the Rock Pigeon is a somewhat benign non-native that is now a thoroughly established addition to our avifauna.

ANDREW M. WILSON

Distribution

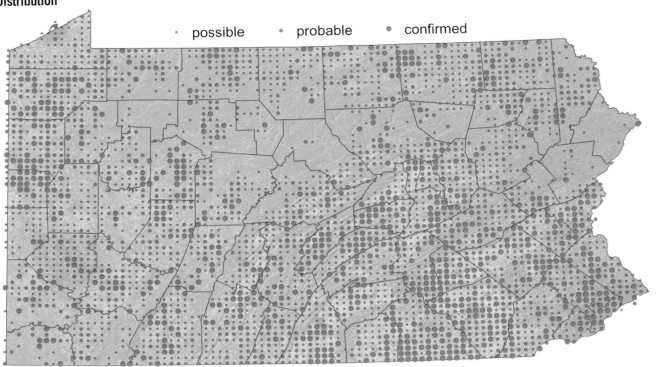

Distribution Change

Density

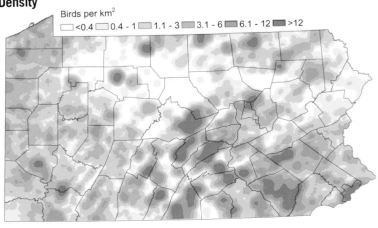

Number of Blocks

	first Atlas 1983–89	second Atlas 2004–09	Change %
Possible	1,168	1,764	51
Probable	888	854	−4
Confirmed	1,318	738	−44
Total	3,374	3,356	−1

Population estimate, birds (95% CI):
415,000 (395,000–435,000)

Breeding Bird Survey Trend

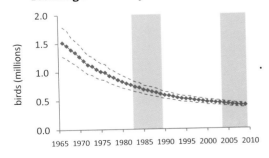

Geoff Malosh

Eurasian Collared-Dove
Streptopelia decaocto

The Eurasian Collared-Dove is a large, pale dove native to the Indian subcontinent, from which it spread rapidly through natural colonization across most of eastern Asia and Europe during the twentieth century. Following an accidental release of captive birds in the Bahamas in the 1970s, the species colonized nearby Florida in the early 1980s and from there spread northward (Romagosa and Labisky 2000). By 2009, it had spread to most states of the United States and into southern Canada, but curiously, it remained relatively rare in the mid-Atlantic states and New England (CLO 2011). It thrives wherever there is human habitation (Fujisake et al. 2010), and it does not appear to be limited by temperate winters; indeed, it is resident within the Arctic Circle in Europe (Hagemeijer and Blair 1997).

The first accepted record of a Eurasian Collared-Dove considered to be wild in Pennsylvania was of one present in Spring Township, Crawford County, from 28 July to 1 August 1996 (Pulcinella 1997). Following a sighting in Schuylkill County in 1998, reports increased in frequency over the next few years, primarily in the southeastern area of the state, culminating in the establishment of a small population in Shady Grove, Franklin County, in 2004. Up to nine birds were seen in the Shady Grove/Greencastle area throughout the second Atlas period. Statewide, there were reports from 15 blocks, one in Westmoreland County being the only report away from the southeast. Breeding (unsuccessful) was attempted in Lebanon in 2004 and suspected in Coatesville, Chester County, in 2006, but most other atlas records were of single birds or transients.

Six years after the establishment of the population in southern Franklin, it appears that this species has yet to gain a toehold in other areas, although clusters of records around Lebanon and northern Franklin/Cumberland/southern Huntingdon counties suggest that there may also be birds persisting in those areas. Further, it is possible that small populations have gone undetected in villages or suburbs little-visited by birders. Even so, all evidence suggests that colonization is progressing slowly, and the statewide population may be fewer than 10 pairs.

As with all non-native invasive species, the potential for the Eurasian Collared-Dove to conflict with native wildlife or damage economic interests should be taken seriously. It has been suggested to be behaviorally dominant over native species at feeding stations (Romagosa and Labisky 2000), but experimental research showed no evidence that the species outcompetes Mourning Doves (Poling and Hayslette 2006). Despite being a non-native bird, the Eurasian Collared-Dove currently is protected in Pennsylvania but may be taken by hunters in other states. Because it tends to be found around human habitation, hunting likely will not be an effective or practical means of population control. Other than allowing dove hunters to take this species where possible, there appears to be little concerted effort to control or eradicate this species.

Given that economic and biological impacts from the invasion of this species seem to be minor (thus far), it seems unlikely that coordinated eradication programs will be attempted, and we can expect a consolidation of this species' tenuous presence in Pennsylvania in coming years. Survey-wide Breeding Bird Survey data show that counts of this species have been doubling every 2 years (Sauer et al. 2011). A similar population increase in Pennsylvania would result in a population in the hundreds of pairs within 10 years and in the thousands by 2025. The paucity of collared-dove records elsewhere in the mid-Atlantic and northeastern United States is perplexing; for example, it has yet to be confirmed breeding in Maryland (Ellison 2010j). Therefore, it is possible that there is an unknown range-limiting factor which will temper this species' spread in Pennsylvania.

ANDREW M. WILSON

Distribution

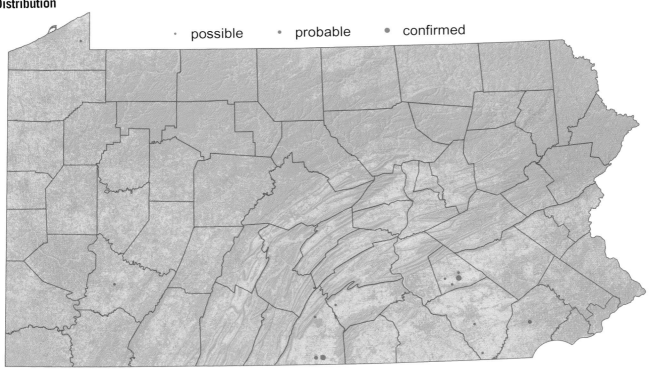

Distribution Change

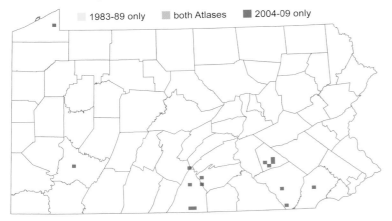

Number of Blocks

	first Atlas 1983–89	second Atlas 2004–09	Change %
Possible	0	11	∞
Probable	0	2	∞
Confirmed	0	2	∞
Total	0	15	∞

Joe Kosack/PGC

Mourning Dove
Zenaida macroura

The most widely distributed of North America's migratory game birds, the Mourning Dove breeds in southern Canada, throughout the lower 48 United States, and in most of Central America and the Caribbean. The species' cooing call, whistling wings, and distinctive silhouette, along with its typically close association with civilization, make it familiar to even casual wildlife observers.

Although long-term population declines in the western United States are of some concern, Mourning Dove populations in the East appear stable or increasing (Sanders and Parker 2010). Relative to most other states, Pennsylvania is not considered a dove hotspot; historic and current harvest levels and estimated breeding densities rank at or below median values of the 27 states in the Eastern Management Unit (Martin and Sauer 1993; Sanders and Parker 2010). Still, it is found in impressive numbers: during the second Atlas period, Pennsylvania hunters bagged approximately 425,000 Mourning Doves annually, the most of any game bird species (Weaver and Boyd 2008), and second Atlas point count data yielded an estimate of over 1.5 million singing males statewide, easily making this the most abundant non-passerine in the state.

The Mourning Dove was among the top 10 species for both occurrence and abundance in the second Atlas, occupying over 98 percent of blocks. The few existing gaps in distribution were widely scattered across the most heavily forested portions of the Commonwealth and generally corresponded in location (but shrank in size) compared with the first Atlas. The small-but-significant range expansion reflected by this infilling of gaps was the only notable change in distribution; for the remainder of the state, occupancy was over 90 percent for all physiographic sections in both atlases.

As suggested by its presence throughout Pennsylvania's diverse landscape, the Mourning Dove can be classified as a habitat generalist, able to obtain needed food, grit, water, and roosting/nesting sites from a wide variety of natural and man-made sources. This is not to say, however, that doves are equally numerous in all habitats; point count data clearly indicate higher abundances in less-forested areas (appendix D). It is noteworthy that this includes suburban settings as well as agricultural areas. Evidently, doves have been able to adjust to the rampant loss of farmland in Pennsylvania (especially the southeastern counties) in recent decades, which has proven so detrimental to many agriculture-linked species.

Pennsylvania's Mourning Dove population is increasing as well as spreading into the last remaining gaps in its distribution. Breeding Bird Survey data for the state show a sustained increase, averaging 1.7 percent annually, 1966–2009 (Sauer et al. 2011), and a 36 percent increase in the population between atlas periods. Mourning Doves have high reproductive potential, with a protracted breeding season (dates of confirmed breeding in the second Atlas ranged from early March to late September; appendix F), along with annual production of multiple broods, offsetting the small size of individual clutches. In light of the fact that annual survival rates appear to have decreased since the 1970s (Otis et al. 2008), reproductive output may be increasing over time.

Coinciding with the second Atlas, wildlife agencies intensified monitoring of dove population demographics to provide a more rigorous basis for harvest management (USFWS 2005). An intriguing topic for future research would be to evaluate quality of various habitats by comparing dove survival and reproductive rates. For example, is annual production from the "Big Woods" areas newly occupied by doves between atlases sufficient to offset mortality, or do such areas represent dove population "sinks" sustained by immigration from elsewhere? Such research should lead to further advances in stewardship, which, combined with the species' own prolific nature, demonstrated ability to benefit from landscape changes brought about by humans, and its popularity with a broad constituency of consumptive and non-consumptive users, suggests that the outlook for the Mourning Dove in Pennsylvania is bright indeed.

IAN GREGG

Distribution

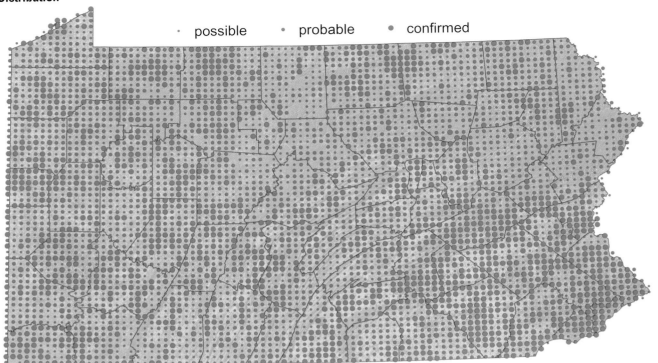

Distribution Change

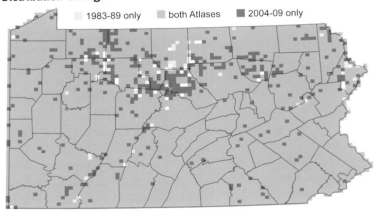

Density

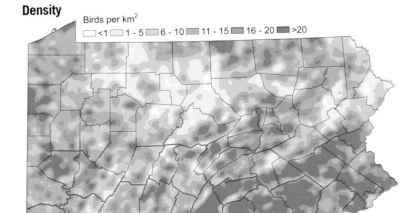

Number of Blocks

	first Atlas 1983–89	second Atlas 2004–09	Change %
Possible	1,021	1,349	32
Probable	1,542	2,141	39
Confirmed	1,955	1,360	−30
Total	4,518	4,850	7

Population estimate, males (95% CI):
1,550,000 (1,500,000–1,600,000)

Breeding Bird Survey Trend

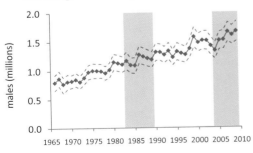

MOURNING DOVE 197

John McKean/VIREO

Yellow-billed Cuckoo
Coccyzus americanus

The shy, elusive Yellow-billed Cuckoo breeds statewide and has a particular fondness for hairy caterpillars. It breeds across most of the United States, except in the extreme north-central and western regions (Hughes 1999). Historically, it has been more common in Pennsylvania than the Black-billed Cuckoo, particularly in the southern counties (Jacobs 1893; Stone 1894). Also, unlike its close relative, the Yellow-billed Cuckoo prefers open woodland over dense mature forest, and it is more likely than its congener to be found in suburban areas (Hall 1983).

According to the first Atlas, the Yellow-billed Cuckoo was more widespread than the black-billed, more often found in fragmented forests, and was much more prevalent in suburban and rural lightly forested counties. Although the yellow-billed was still less common than the black-billed in northern, extensively wooded Pennsylvania, first Atlas data suggest that it has continued its northward expansion that began in the state 60 to 70 years ago (Ickes 1992c).

The second Atlas shows a highly significant 43 percent increase in occupied blocks by the Yellow-billed Cuckoo, with no significant decreases in any of the state's physiographic sections. Rather, the species has continued its northward expansion dramatically and now is at least as common as its close relative in northern extensively wooded Pennsylvania. Significant increases in five of the Appalachian Plateaus sections are at least 140 percent, and those sections are all in the northern third of the state. The tendency for the Yellow-billed Cuckoo to be less common than the black-billed in heavily forested areas has been reversed statewide, not just in northern areas, since every county in Pennsylvania with extensive forest cover has more yellow-billed records than its congener. Furthermore, data from the second New York atlas suggests that the species is moving into more extensively wooded areas there as well. Although only a slight 3 percent increase in Yellow-billed Cuckoo-occupied blocks occurred statewide, the number of Adirondack blocks with records more than doubled (McGowan 2008i).

Second Pennsylvania Atlas data indicate that the Yellow-billed Cuckoo is still fairly common in some suburban areas; at least two-thirds of the blocks are occupied in such counties as Cumberland and Washington. Nevertheless, yellow-bills showed a non-significant increase (lowest of the major provinces) in the highly developed Piedmont Province.

The highly significant increase in the Yellow-billed Cuckoo's population may be at the expense of the Black-billed Cuckoo, since it is reported that, under certain circumstances, the yellow-billed negatively impacts its close relative (Richards 1976; Hughes 1997). Additionally, in all of the second Atlas physiographic sections in which the Yellow-billed Cuckoo increased, in several instances significantly, the Black-billed Cuckoo decreased significantly. As both Pennsylvania cuckoo densities have been reported to increase during gypsy moth defoliation (Gale et al. 2001), and several of the areas with increased Yellow-billed Cuckoo numbers during the second Atlas correspond to locations where severe defoliation occurred in Pennsylvania between 2004 and 2009 (chap. 3), insect outbreaks also might have contributed to the species' increase in the state.

The total Pennsylvania population for the Yellow-billed Cuckoo was estimated to be about 120,000 individuals during the second Atlas period. However, Breeding Bird Survey data show no significant trend in Yellow-billed Cuckoo counts in the period 1966 through 2009; rather, wide population fluctuations were noted, and the second Atlas fieldwork was completed during a particularly strong peak (Sauer et al. 2011).

While it appears that the Yellow-billed Cuckoo population is increasing in Pennsylvania, particularly in the northern regions, this could be an artifact of population cycles. In contrast, populations are declining throughout much of its range, and factors that may impact the species' numbers need to be determined and monitored (Hughes 1999). Additional information about the interspecific relations of the two Pennsylvania cuckoos would be very useful in determining the future of both species in the state.

ROY A. ICKES

Distribution

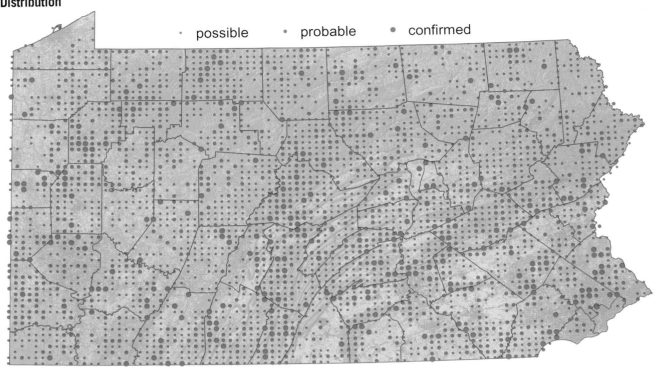

Distribution Change

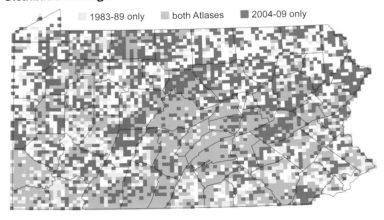

Density

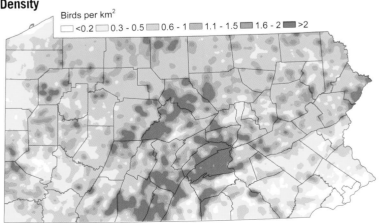

Number of Blocks

	first Atlas 1983–89	second Atlas 2004–09	Change %
Possible	1,307	2,144	64
Probable	764	850	11
Confirmed	194	244	26
Total	2,265	3,238	43

Population estimate, birds (95% CI):
120,000 (114,000–127,000)

Breeding Bird Survey Trend

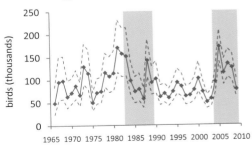

YELLOW-BILLED CUCKOO 199

Bob Wood and Peter Wood

Black-billed Cuckoo
Coccyzus erythropthalmus

Both Pennsylvania cuckoos breed statewide and are voracious consumers of enormous quantities of hairy caterpillars. The Black-billed Cuckoo is less widely distributed nationally than the Yellow-billed Cuckoo, since its breeding range is confined to the northern half of the United States, east of the Rocky Mountains. It is listed as a Maintenance Concern species in Pennsylvania's Wildlife Action Plan (PGC-PFBC 2005), due to the listing by Partners in Flight as a IIA priority species because of significant regional declines in its breeding population (Rich et al. 2004). Historically, the black-billed was reportedly more common than the Yellow-billed Cuckoo in the extensive, mature forests of the northern portions of Pennsylvania (Harlow 1913; Street 1976).

During the first Atlas, the species' distribution pattern appeared to be similar to historical reports. The Black-billed Cuckoo was found in only about one-third of the state's atlas blocks but was still more commonly reported than the Yellow-billed Cuckoo in seven of the eight heavily forested northern tier counties (Ickes 1992d).

The results of the second Atlas suggest that the population of the Black-billed Cuckoo has increased slightly in Pennsylvania and that its distribution has shifted northward marginally. There was no change in the overall number of occupied blocks between the first and second Atlas periods, but during the same time, highly statistically significant increases (from 42 to 95 percent) were noted in four of the Appalachian Plateaus Province sections that make up much of the northern forested third of the state. The only province with a significant decrease in blocks occupied by Black-billed Cuckoos was the Piedmont in southeastern Pennsylvania; however, nonsignificant decreases were reported in six other southern sections of the state. Second Atlas efforts in Maryland and New York also suggest that the Black-billed Cuckoo's distribution may be shifting northward, with a 4 percent increase in New York (McGowan 2008j) and a 5 percent decrease in Maryland (Ellison 2010k). Furthermore, in their study of the effect of climate change on North American birds, Hitch and Leberg (2007) reported a statistically significant northward distributional shift for the Black-billed Cuckoo.

The marginal northern shift of the Black-billed Cuckoo's range may be due in part to displacement in the south by the yellow-billed (Richards 1976; Nolan and Thompson 1975). In two of the four physiographic sections in which Black-billed Cuckoo records decreased significantly between atlas periods, Yellow-billed Cuckoo records increased significantly (Appalachian Mountain and High Plateau). Moreover, yellow-billed records increased in the other two sections (Allegheny Mountain and Piedmont Upland), although not significantly.

The density of both Pennsylvania cuckoos has been reported to increase during gypsy moth defoliation (Gale et al. 2001), and defoliating insect outbreaks might have contributed to the black-billed's distribution shifts between atlas periods. Several of the areas into which the species shifted during the second Atlas correspond to severe defoliation locations in Pennsylvania between 2004 and 2009 (chap. 3). The black-billed increased in the Pocono region and the central and north-central parts of the state, all of which experienced extensive defoliation. The Pennsylvania population of Black-billed Cuckoos was estimated to be 22,000 individuals during the second Atlas period. Breeding Bird Survey (BBS) counts have declined by 2.3 percent annually, 1966–2009, including a decline of approximately 22 percent between atlas periods (Sauer et al. 2011).

Given the somewhat conflicting results of a significant increase in occupied blocks between atlas periods and a significantly decreasing population based on BBS data, the status of the Black-billed Cuckoo in Pennsylvania is uncertain. Information is needed on the effects of pesticides and habitat fragmentation on cuckoo populations and how habitat is partitioned between the Black-billed Cuckoo and its sympatric congener, the Yellow-billed Cuckoo (Hughes 2001; Anders 2010). Second Atlas reports of the Black-billed Cuckoo decreasing and the Yellow-billed Cuckoo increasing in some of the same physiographic areas of Pennsylvania suggest the latter information may be especially important for understanding the Black-billed Cuckoo's future in the state.

ROY A. ICKES

Distribution

Distribution Change

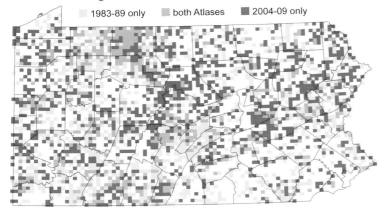

1983–89 only | both Atlases | 2004-09 only

Density

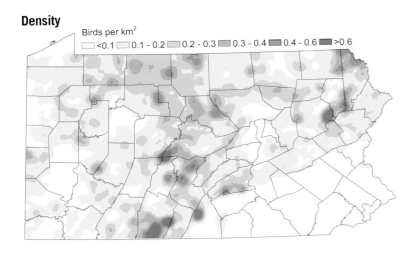

Birds per km² : <0.1, 0.1 - 0.2, 0.2 - 0.3, 0.3 - 0.4, 0.4 - 0.6, >0.6

Number of Blocks

	first Atlas 1983–89	second Atlas 2004–09	Change %
Possible	1,047	1,281	22
Probable	428	302	−29
Confirmed	138	120	−13
Total	1,613	1,703	5

Population estimate, birds (95% CI):
22,000 (19,000–25,000)

Breeding Bird Survey Trend

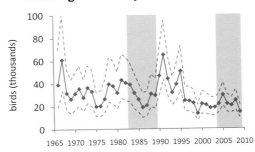

BLACK-BILLED CUCKOO 201

Joe Kosack/PGC

Barn Owl
Tyto alba

Closely tied to agriculture and landscapes with extensive grasslands, the Barn Owl earned its name by frequently nesting in barns, silos, and other similar structures. With 36 subspecies and populations on every continent except Antarctica, the Barn Owl is one of the world's most widely distributed birds (Taylor 1994). Though this species is globally secure, significant population declines have occurred in North America over the past 50 years (Colvin 1985; Marti 1992). The Barn Owl is currently designated a species of concern in Pennsylvania's Wildlife Action Plan (PGC-PFBC 2005).

During the first Atlas period, the Barn Owl was concentrated in the southeastern quarter of the state but had a wide distribution including 56 of the state's 67 counties. By the second Atlas period, the Barn Owl's distribution had greatly contracted, reduced to only 30 counties, representing a 53 percent reduction in occupied blocks between the first and second Atlas periods.

Loss of foraging habitat, due to changes in farming practices and commercial development of farmland, is a primary cause of declining Barn Owl populations (Rosenburg et al. 1992). In Pennsylvania, over 767,000 acres of pasture and more than 420,000 acres of cropland were lost to development between 1982 and 1997 (Goodrich et al. 2002). In addition, the conversion of hayfields and pasture to row crops has further decreased abundance of optimal foraging habitat for Barn Owls. The loss of nest sites is also considered to be a main factor leading to declining Barn Owl numbers in North America (Taylor 1994; Rosenburg et al. 1992). The gradual deterioration and disappearance of barns and silos, in concert with the screening of entrances to prevent access by Rock Pigeons, has eliminated many previously productive and secure nesting sites for Barn Owls (Rosenburg et al. 1992).

Much of the range contraction in Pennsylvania, especially in the southeastern corner of the state, is likely due to loss of farmland to exurban development. Between 1969 and 1992, 27 percent of farmland was lost in the Lehigh Valley, and 37 percent was lost around Philadelphia (Goodrich et al. 2002). During the first Atlas period, breeding was confirmed in 13 atlas blocks in the southeastern corner of the state, including Lehigh, Montgomery, Bucks, Philadelphia, and Delaware counties. During the second Atlas period, breeding was not confirmed in these counties, and there was only a single block where breeding was considered probable in Lehigh County. In addition, during the second Atlas period, Barn Owls had significantly declined in the western part of the state, an 82 percent decline in the Appalachian Plateaus Province.

Data from the first and second Atlases provide evidence that Barn Owls are declining in abundance and distribution throughout the state, in spite of the Pennsylvania Game Commission's active search for Barn Owl breeding activity as part of a statewide Barn Owl Conservation Initiative. Nearly 70 percent of the Barn Owl records for the second Atlas period were submitted by these agency staff. Without this targeted effort, it is likely that an even greater decline in the number of occupied blocks would have been apparent. This is supported by the fact that larger declines were documented between atlases in New York (−78%; McGowan 2008k) and Maryland (−72%; Ellison 2010l). In all, Barn Owls were confirmed in 85 blocks, and even though there were no more than 35 confirmations in any one year, it seems likely that the current Barn Owl population of Pennsylvania is approximately 100 pairs.

The future for Barn Owls in Pennsylvania will depend on the abundance and distribution of grassland-dominated landscapes and suitable nesting locations. Programs that preserve agriculture and encourage farmers to establish and maintain grassland habitats will likely be important for maintaining adequate foraging habitat for this species. In addition, providing nest boxes is a valuable conservation tool, especially in areas where foraging habitat is plentiful but nest sites are scarce (Marti et al. 1979).

JAMIE FLICKINGER AND DANIEL P. MUMMERT

Distribution

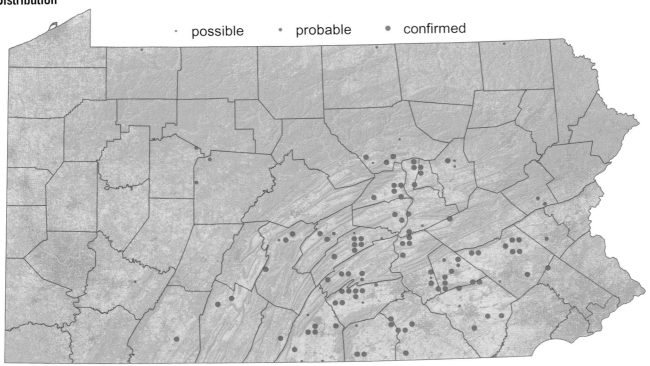

Distribution Change

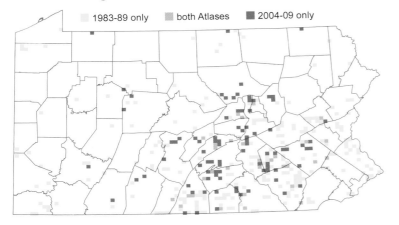

Number of Blocks

	first Atlas 1983–89	second Atlas 2004–09	Change %
Possible	101	21	−79
Probable	43	11	−74
Confirmed	107	85	−21
Total	251	117	−53

Chuck Musitano

Eastern Screech-Owl
Megascops asio

The descending, whinnying call of the Eastern Screech-Owl is one of the most familiar nocturnal sounds of the Pennsylvania woods, as adults begin to actively court in midwinter, their tremolo vocalizations filling the otherwise silent nights. By early May, the parents may be shuttling food (mostly small mammals, birds, large insects, reptiles, and amphibians) to the chicks. The latest an occupied nest was detected during the second Atlas was early August, a time of year when some homeowners may have been lying awake listening to the raspy begging calls of fledged young drifting discordantly through open windows.

Although the Eastern Screech-Owl has "the widest ecological niche of any owl in its range" (Gelbach 1995) and is found from urban parks and suburban backyards to farm woodlots and remote northern-tier forests, it is not uniformly common across Pennsylvania. Although it was detected in all 67 counties and on 31 percent of blocks, targeted owl surveys in a subset of blocks showed that the screech-owl was, like most owls, probably underreported and was likely to have occurred in at least 43 percent of blocks.

What is more, the screech-owl proved to be far more widespread, in the second Atlas, from the Appalachian Mountain Section south and east through the Great Valley and Piedmont than in regions north and west of the Ridge and Valley. This echoes comments made by early ornithologists, who likewise found the Eastern Screech-Owl to be more a creature of smaller woodlots, long ridges, and settled lands rather than large, contiguous blocks of mature forest. Harlow (1918), for example, noted that it was "much rarer in the mountainous districts and not found at all in the primæval forests." Both atlases reflect the species' general scarcity above 600 m (~2,000 ft), which may be a function of denser, more contiguous forest cover at such elevations.

Atlas volunteers, however, generally found fewer Eastern Screech-Owls everywhere in the state during the second Atlas than in the first (Reid 1992c). Looking at all blocks, every physiographic province of the state showed declines, although some, like those in the Atlantic Coastal Plain and Central Lowlands, were not statistically significant. However, some sections showed steep, significant drops from the first Atlas; screech-owls were down more than 50 percent in the Allegheny Front Section, 45 percent in the Waynesburg Hills, and almost 46 percent in the Pittsburgh Low Plateau, for example. Statewide, the species showed a significant decline of almost 20 percent.

When only priority blocks were analyzed, however, the results were far more mixed, and there was no significant trend up or down. What this may suggest is that screech-owls were less common statewide than during the first Atlas, making them harder for atlasers to find in the matter of routine atlasing, but with the additional effort put into priority blocks, at least a few could often be detected. The analysis also showed a northward contraction in the Eastern Screech-Owl's range, although this may have been skewed by the greater nocturnal surveying effort in the second Atlas. Interestingly, there were no significant changes in atlas block occupancy in either New York (Smith 2008a) or Maryland (Ellison 2010m) over similar time periods. One potential explanation for the pervasive decline in the screech-owl population may be the accidental introduction in 1999 of West Nile virus (WNV) into North America, which causes illness and death in a variety of raptors, including Eastern Screech-Owls (LaDeau et al. 2008). Experiments have shown that WNV is a significant disease in screech-owls (Nemeth et al. 2006), although the extent to which it has impacted the species in Pennsylvania is unclear. Christmas Bird Count data corrected for effort show no clear trend through 2006, after which a steep drop in counts occurred from 2006 through 2009 (Bolgiano 2010).

Intriguingly, experiments have shown that female screech-owls that survive a bout of WNV infection are able to pass along protective antibodies to their eggs, providing at least some protection for young chicks (Hahn et al. 2006). It remains to be seen whether Eastern Screech-Owl populations will develop a more general resistance to the disease.

SCOTT WEIDENSAUL

Distribution

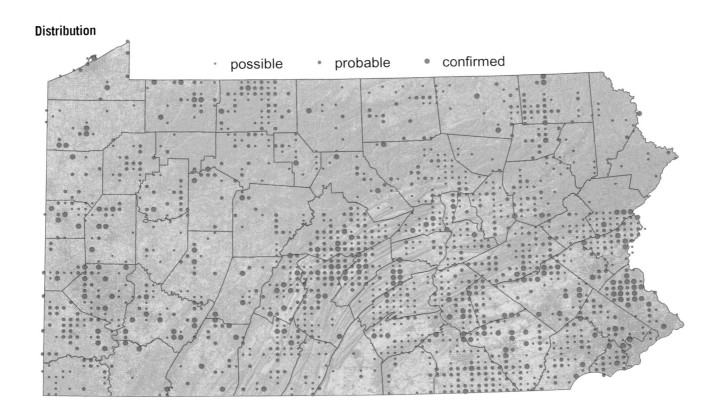

Distribution Change

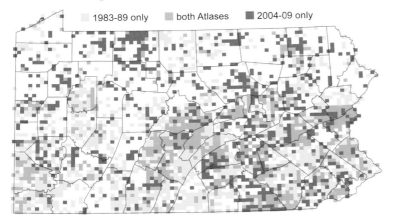

Number of Blocks

	first Atlas 1983–89	second Atlas 2004–09	Change %
Possible	995	782	−21
Probable	650	565	−13
Confirmed	268	192	−28
Total	1,913	1,539	−20

Geoff Malosh

Great Horned Owl
Bubo virginianus

Pennsylvania's largest owl, the Great Horned Owl, occurred in more blocks than any other owl during both Pennsylvania atlases. Remarkably adaptable, it has the most extensive range and varied habitat requirements among North America's owls, occurring throughout the continent below the treeless Arctic, as well as in much of Central and South America (Houston et al. 1998).

The Great Horned Owl prefers open, mature woodlands, particularly edge habitat and woodlots in agricultural areas (Morrell and Yahner 1994); it is sometimes considered the nocturnal counterpart to the Red-tailed Hawk. Pellet analysis revealed that its primary Pennsylvania food sources are opossums, rabbits, and rats (Wink et al. 1987), all common near forest edge or in agricultural areas. The Great Horned Owl is most frequently detected by its calls. While it is most vocal in late fall and winter, before and during its nesting season, it will continue calling into spring and summer, when many atlas detections occurred.

Regions in which the Great Horned Owl was most common included southeastern Pennsylvania, where it was found in 57 percent of Piedmont blocks (61% of priority blocks), and the Ridge and Valley, where it was found in 52 percent of blocks (64% of priority blocks). Wherever substantial night birding was conducted in those regions, the Great Horned Owl was often found. During the atlas period, 15 Christmas Bird Counts sites from southeastern Pennsylvania or the Ridge and Valley tallied 20 or more Great Horned Owls during a single count. Such numbers suggest that wherever there is a sizeable woodlot with significant edge in those areas, there is a high likelihood that it is used by a Great Horned Owl.

In contrast, the Great Horned Owl was much less frequently found in blocks within Pennsylvania's large interior forests, which include much of the north-central counties, parts of the Poconos, and the larger forested areas of the Ridge and Valley and Laurel Highlands. In those areas, it occurred at forest openings, particularly around towns. Although snow-covered roads prevented winter owling in substantial portions of those regions, Great Horned Owl was infrequently detected there during spring/summer night birding, when some hooting could be expected. The common large owl of the big woods is the Barred Owl, which was found primarily where the Great Horned Owl was scarce. Similar contrasting distributions also occur in New York, where Barred Owl was most common in large core forests and Great Horned Owl was most common in fragmented forests (McGowan 2008l). This relationship is more clearly defined in Pennsylvania's second Atlas than it was during the first Atlas (Schwalbe 1992a).

In areas of the Appalachian Plateaus where farmland and woodlots predominated, the Great Horned Owl was regularly encountered where night birding was conducted, being found in 24 percent of plateau blocks (30% of priority blocks). More consistent nocturnal effort would probably reveal this owl to be more common here where there is suitable habitat. This region experienced the largest between-atlas decline in Great Horned Owl detections, at 43 percent. It is possible that owl populations underwent a bigger contraction there than elsewhere in Pennsylvania. The 28 percent decline in the number of block occurrences from first to second Atlas probably reflects a genuine population decline; however, as with all nocturnal species, variations in effort across the state and between atlases reduce our confidence in this conclusion.

The Great Horned Owl is susceptible to West Nile virus (Huffman and Roscoe 2010). A downturn in Christmas Bird Count (CBC) numbers from northeastern states, including Pennsylvania, was consistent with a West Nile virus effect (Caffrey and Peterson 2003); Pennsylvania CBC numbers remain depressed. While it was not severe enough to eliminate the Great Horned Owl over large areas of Pennsylvania, the epidemic probably contributed to the decline in detections between atlases. However, the Great Horned Owl's outlook in Pennsylvania appears to be favorable because of its wide range and adaptability.

NICHOLAS C. BOLGIANO

Distribution

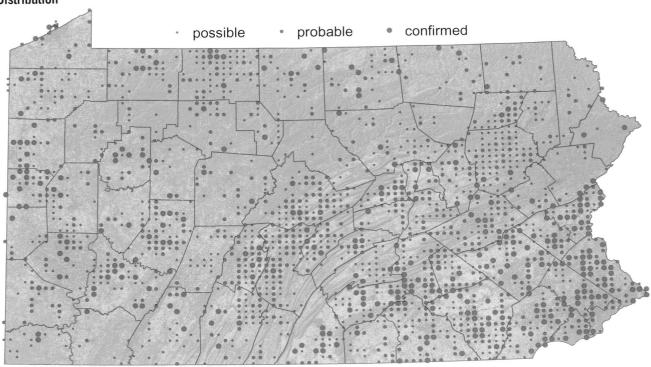

Distribution Change

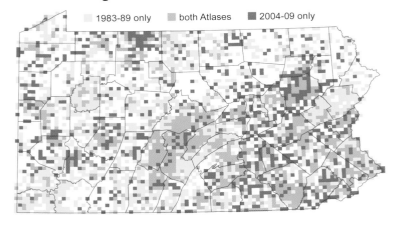

1983–89 only | both Atlases | 2004-09 only

Number of Blocks

	first Atlas 1983–89	second Atlas 2004–09	Change %
Possible	1,096	832	−24
Probable	871	582	−33
Confirmed	451	320	−29
Total	2,418	1,734	−28

Claude Nadeau/VIREO

Barred Owl
Strix varia

Few birds reflect the extent and health of Pennsylvania's forest better than the Barred Owl. Its well-known *who cooks for you, who cooks for you all* call resonates throughout the night from Penn's Woods wherever there are sufficiently sized mature forest stands, thus marking Pennsylvania as a "Keystone State" for the species within its eastern North American range, which extends from Canada to Florida. Almost certainly an abundant bird before the felling of Pennsylvania's forests, the Barred Owl was confined to remote remnant tracts by the end of the nineteenth century; populations have slowly recovered since that time (Leberman 1992f).

The Barred Owl was located in 186 more blocks in the second Atlas than in the first Atlas, a gain of 17 percent in all blocks and 36 percent between priority blocks. Not surprisingly, the greatest concentrations of owls were in the heavily forested parts of the state, with particularly good numbers found within the Allegheny National Forest, along the ridges of Centre and Huntingdon counties, and eastward along the Kittatinny Ridge and in the South Mountain Section. The Barred Owl was found by atlas volunteers to be absent or rare only in areas that are under extremely intensive agriculture, such as much of the Piedmont, and from such highly urbanized areas as Philadelphia and Pittsburgh. The Barred Owl was the most frequently reported owl on second Atlas nocturnal owl surveys, detected at 10.7 percent of all stops. It was also found in 60 percent of atlas blocks in which owl surveys were conducted. Hence, it is safe to assume that the statewide recorded block occupancy rate of 26 percent is a considerable underestimate.

Due to wide variation in effort concentrated on nocturnal birds, both spatially and between atlases, ascertaining changes in the Barred Owl's status and distribution is difficult. The Christmas Bird Count (CBC) provides the best estimate of long-term trends for this species in Pennsylvania. Although numbers recorded in the CBC vary considerably from year to year, there has been an increasing trend—of about 6 percent per year—since the 1960s (NAS 2010). Increasing trends and increases in atlas blocks records were also noted in both Maryland (Ellison 2010n) and New York (McGowan 2008m)

Although the Barred Owl nests in a variety of mixed forest types, including predominantly deciduous tracts, mixed deciduous and conifer woodlands appear to be preferred (Mazur and James 2000). With an eye toward measuring the long-term impact of the Hemlock Woolly Adelgid invasion on birds, atlas volunteers were asked to evaluate the abundance of Eastern Hemlocks as part of their routine reporting of this and many other birds. Of the seven species of owls breeding in the state, the Barred Owl exhibited the strongest association with hemlocks, with 76 percent of the records associated with some degree of Eastern Hemlock. The Barred Owl was also frequently observed in plantings of non-native conifers, such as Scots Pine or Austrian Pine, when they adjoined other more extensive forest types that might provide suitable nesting sites. Increasingly, these owls also appear to be adapting to smaller woodlots and suburban areas, or even to cities like Pittsburgh, where there are parks and other wooded areas offering suitable nesting sites. This is a trend that might be encouraged by providing nesting boxes for the birds (Mazur and James 2000), although this should not substitute for proper management of forest ecosystems.

Like the other owls, the Barred Owl is a frequent victim of traffic kills in wooded areas along Pennsylvania's busy highways. The shooting of these owls as vermin, however, seems to have greatly lessened in the last few decades, and the species' future in Pennsylvania appears to now be linked to the availability of large tracts of forest. Barred Owl populations should be carefully monitored within our forested areas that are under consideration for energy development such as natural gas and wind power, where an increase in forest openings and fragmentation of contiguous woodlands may create habitats more suitable to the more aggressive Great Horned Owl (Mazur and James 2000).

ROBERT C. LEBERMAN

Distribution

possible · probable · confirmed

Distribution Change

1983-89 only · both Atlases · 2004-09 only

Number of Blocks

	first Atlas 1983–89	second Atlas 2004–09	Change %
Possible	541	676	25
Probable	420	474	13
Confirmed	115	112	−3
Total	1,076	1,262	17

BARRED OWL

Warren Greene/VIREO

Long-eared Owl
Asio otus

Mysterious and elusive, the Long-eared Owl's status was uncertain after the first Pennsylvania Atlas (Santner 1992b). The second Atlas also failed to lift the veil of mystery, despite attempts to provide volunteers with additional information about this species' habits and habitat. After two atlas projects, the Long-eared Owl remains one of the state's rarest and most poorly known nesting bird species.

The Long-eared Owl has long been considered a "Candidate—Undetermined" species in Pennsylvania, rare enough to raise conservation concern but not well enough known to confidently declare it either Endangered or Threatened (Gill 1985; Brauning et al. 1994; Gross 1998a). Although it is one of the world's most broadly distributed owls, Pennsylvania is near the southern edge of its range in eastern North America (Kain 1987; Marks et al. 1994; Konig et al. 1999). In much of its range, and especially in the northeastern United States, it is sparsely distributed and difficult to detect (Marks et al. 1994; Sutton and Sutton 1994). It is nocturnal and highly secretive, "freezing" in a slim, cryptic pose against a tree trunk when approached. The male's brief advertising *hoot* is easily missed or mistaken for another owl or a distant canine. Because reports are mixed concerning its response to imitations of its vocalizations (Shepherd 1992; Konig et al. 1999), since it has a variety of vocalizations (Takats et al. 2001), and since at least one study indicates an avoidance by owls to vocalizations (Evans 1997), the Long-eared Owl is one of our most difficult birds to inventory or monitor.

In the second Atlas, the Long-eared Owl was discovered in only 14 blocks and confirmed breeding in only 4 blocks. Although the Long-eared Owl is a species of northern affinity, most records in the second Atlas were in the southern half of the state. Typically, the Long-eared Owl is found in interspersed habitats with dense shrubs, tree lines or woodlots for nesting, and open ground for foraging (Santner 1992b; Konig et al. 1999). Most records in the second Atlas were in landscapes of such mosaics, often including conifer stands.

It has been understood that "because of its shyness, nocturnal habits, and infrequent calling, the Long-eared Owl is easily missed and may be more common than realized" (Gill 1985). After the first Atlas, Santner (1992b) observed: "The Long-eared Owl is so secretive and difficult to find that no accurate assessment of its numbers in Pennsylvania has ever been made." Even with additional decades of bird studies, this statement remains true: the Long-eared Owl remains one of the rarest and most poorly understood breeding species in the state. There is insufficient evidence to suggest that there has been a change in status between atlases. Interestingly, of the 32 atlas blocks with records in either atlas (18 in first, 14 in second), none had records in both periods, affirming the low detection rate, as found in other states and provinces (e.g., Konze 2007; Medler 2008d). The Long-eared Owl is an early nester (Marks et al. 1994), and most atlas reports were before June; however, one intriguing breeding confirmation, of full-grown young in a well-wooded setting in Centre County on 15 July, suggests that other pairs may nest later in the season than suspected and in more densely forested areas than previously thought (M. Brittingham, pers. comm.). The Long-eared Owl does not build its own nest, but rather uses old stick nests built by crows, hawks, herons, or squirrels (Marks et al. 1994).

Consensus among the many ornithological writers is that this species is not only difficult to find but also fairly rare and has declined in number. Even in 1940, Poole stated that it was "formerly much more common" (Poole 1964). The Long-eared Owl is listed as Threatened in New Jersey, Vulnerable in New York, and critically imperiled in several other states (Gross 2010a). More intensive surveys for this species are necessary to better understand both its status and the potential for conservation.

DOUGLAS A. GROSS

Distribution

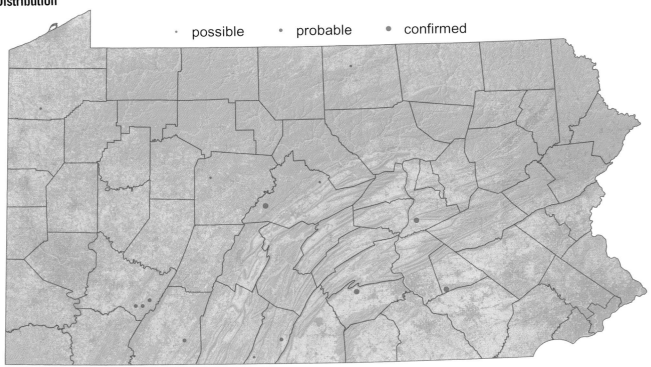

Distribution Change

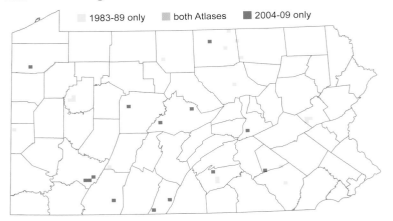

Number of Blocks

	first Atlas 1983–89	second Atlas 2004–09	Change %
Possible	11	5	−55
Probable	1	5	400
Confirmed	6	4	−33
Total	18	14	−22

Rob Criswell

Short-eared Owl
Asio flammeus

Drifting at twilight on silent wings, the Short-eared Owl looks ghostlike as it courses over grassy meadows and fields—one of the rarest breeding birds in Pennsylvania, and one now almost wholly dependent on man-made grasslands.

Associated across its nearly global range with open habitats such as prairies, heathlands, tundra, and grassy wetlands (Wiggins et al. 2006), the Short-eared Owl was likely always a rare breeding bird in the heavily forested Commonwealth, which today marks the southernmost limit of its nesting range in the East. Historical nesting records in Pennsylvania were few and widely scattered, confined to bogs and large marshes, with nineteenth- and early-twentieth-century nests from Lehigh, Berks, and Crawford counties, and (in a still-earlier report from John James Audubon) probably what is now Carbon County (Master 1992b). The total number of documented nests prior to the first Atlas was fewer than 10 (Brauning 2010b), and the Short-eared Owl has been listed as Endangered in Pennsylvania since 1985 (Gill 1985). Threats include habitat loss, modern agricultural techniques, and pesticide and herbicide use (Brauning et al. 1994).

For decades, beginning with the first confirmed nesting in 1966 (Miller 1966a), the only known breeding location in the state was a 20 ha (50 ac) field along the Delaware River at the Philadelphia International Airport, where wintering Short-eared Owls were also commonly observed. Breeding was periodically recorded at this site through at least the late 1990s (Miller 1999), but the field was developed into an airport cargo hub in 2005 (K. Russell, pers. comm.), and habitat loss at this site is "probably irreversible" (Brauning 2010b).

The discovery of a tiny nesting population on reclaimed surface mines in northwestern Pennsylvania was one of the major revelations of the first Atlas. In the years since then, the population there has expanded and contracted, probably in response to cyclical prey availability and immigration from more northerly breeding populations. In 1997, for example, Short-eared Owls' nests were reported from Clarion, Allegheny, Jefferson, Lawrence, and Venango counties (McWilliams and Brauning 2000), but southern Clarion County has remained the core of its range in the state since the mid-1980s (Brauning 2010b).

During the course of the second Atlas, short-eareds were recorded in only 7 of the 4,937 blocks, and at only five locations was breeding considered at least possible. Two sites were in Clarion County and one in Lawrence, but only one breeding pair was confirmed, with an adult carrying food in an extensive area of reclaimed surface mine in Clarion County where the species had been confirmed breeding in the past. No nests or young were found.

Interestingly, a pair of Short-eared Owls was reported flying together over a reclaimed surface mine in Clinton County in late May, while a single individual flew silently in response to a recorded vocalization in northern Centre County, less than 16 km (10 miles) west of the Clinton County site. If these sightings did indicate breeding, they represent a significant range extension for the species.

The creation and management of grassland areas 50 ha (120 ac) or larger would be beneficial to the Short-eared Owl. Reclaimed surface mines managed for grassland habitat are likely to continue to provide the best nesting sites for this species, but only if such sites are maintained to minimize nesting-season disturbance and to provide optimum cover for microtine prey species (Brauning 2010b).

SCOTT WEIDENSAUL

Distribution

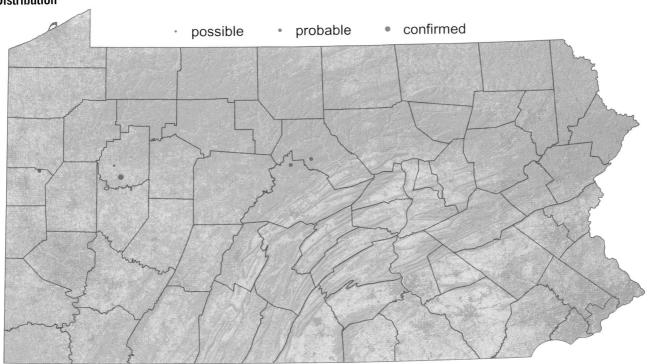

Distribution Change

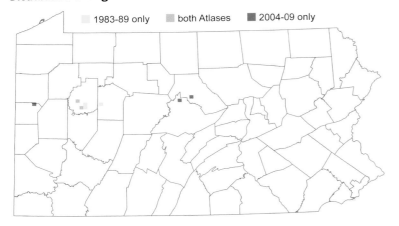

Number of Blocks

	first Atlas 1983–89	second Atlas 2004–09	Change %
Possible	2	1	−50
Probable	1	3	200
Confirmed	3	1	−67
Total	6	5	−17

Scott Weidensaul

Northern Saw-whet Owl
Aegolius acadicus

Few Pennsylvania birds have been cloaked in as much mystery, or elicited such curiosity, as the Northern Saw-whet Owl—the smallest of the Commonwealth's nocturnal raptors and, until recently, the least known. Since the mid-1990s, coordinated migration netting, using a recorded audiolure of the saw-whet's call, has shown that it is a widespread and common migrant throughout Pennsylvania, probably drawing on source populations in Canada and New England (Weidensaul 2010). Its breeding status in the state has been far from clear, however.

Sutton (1928a) concluded that the saw-whet "nests rarely in the northern and mountainous counties," and the first Atlas did little to change that perception; the saw-whet was recorded in just 2 percent of blocks, widely scattered, primarily across the Appalachian Plateaus and Ridge and Valley provinces, but also occurring in South Mountain and along the lower Delaware River. Ornithologists assumed the species was a rare or accidental breeder, perhaps more common following irruption years (McWilliams and Brauning 2000).

Gross (2000b) initiated "Project Toot Route," in which volunteers surveyed randomly selected forested habitat in the mountainous regions of the Commonwealth using tape-playback to elicit vocal responses from territorial male Northern Saw-whet Owls. During the first field season, saw-whets were detected on 44 percent of the 88 routes completed, and in the second year, 57 percent of 79 routes. Although the species was again most reliably detected on the plateaus, it was found at more southerly locations and at lower elevations than expected, including three encounters within 14 km (10 miles) of the Mason-Dixon Line. Surveyors noted the species most often in moist forest with a well-developed understory, such as rhododendron or mountain laurel. Gross concluded that the Northern Saw-whet Owl is "consistently fairly common in forested Pennsylvania highlands" (D. Gross, pers. comm.), regardless of migration irruptions.

Although the Northern Saw-whet Owl remains a much sought-after—if rarely seen—species, it now appears that the bird's cryptic and elusive nature, rather than any real rarity, accounts for the paucity of reports. The second Atlas provides a particularly stark example of this. Thanks to targeted owl surveys, the species was found in 6 percent of all blocks, a nearly 200 percent increase from the first Atlas, and in 10 percent of all priority blocks. Owl survey coverage was still relatively limited statewide, however, and the dramatic results of a concentrated effort in Atlas Region 32 in northwestern Pennsylvania—where the species was found in almost every block—suggests that this species remained badly underreported in the second Atlas. Region 32 may also encompass some of the best potential nesting habitat in the state for this species, since it scores very high in total forest cover, percentage of conifer forest cover, percentage of core forest, and elevation (chap. 3), all presumed to be important for the nesting Northern Saw-whet Owl (Rasmussen et al. 2008).

A second, more diffuse area of concentration appears in central Pennsylvania, but here detections of saw-whet owls did not correlate as closely to intensive nocturnal effort (chap. 6), suggesting that the species was more widely scattered. This is likely due to the more fragmented nature of the woodland in this part of the state, which lacks the high percentages of core forest and conifer component found in the plateaus. The Kittatinny Ridge and South Mountain Section formed the southeastern limit of the saw-whet owl's distribution during the second Atlas, with a number of records from southwestern Pennsylvania, connecting to the small breeding population in high-elevation bogs of western Maryland (Brinker and Dodge 1993).

The Northern Saw-whet Owl's association with higher elevation forests may make it susceptible to the effects of climate change in the decades ahead. Changing forest composition, especially the loss of hemlocks to adelgids and the threat of fragmentation of core forests to gas drilling and other development, likewise pose potential challenges. As this atlas proves, however, there is still much to be learned about even the basic distribution of this retiring species.

SCOTT WEIDENSAUL

Distribution

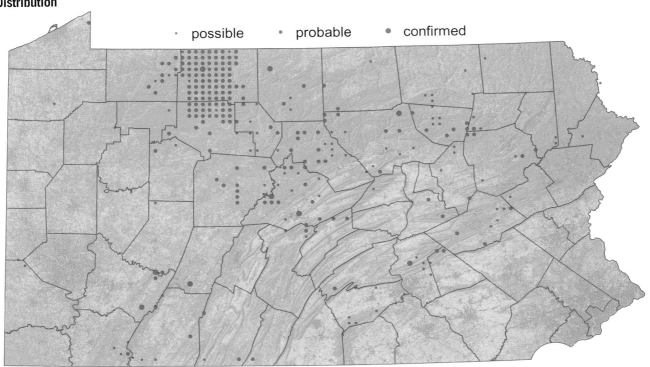

Distribution Change

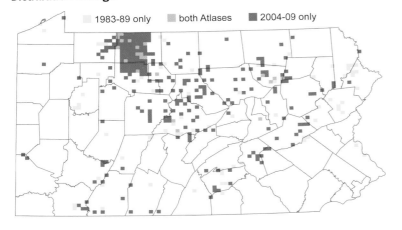

Number of Blocks

	first Atlas 1983–89	second Atlas 2004–09	Change %
Possible	57	95	67
Probable	25	179	616
Confirmed	14	9	−36
Total	96	283	195

Bob Wood and Peter Wood

Common Nighthawk
Chordeiles minor

In Pennsylvania, the Common Nighthawk is an urban bird, uttering its characteristic nasal *peents* while foraging acrobatically for insects above the city rooftops, its long, slender wings interrupted at the bend by bright white patches. However, it is equally at home on the wide-open spaces of the Great Plains, and its breeding range spans most of North America, from southern Canada to Central America (Poulin et al. 1996). Nesting occurs across Pennsylvania, but it has become highly localized and scattered since the first Atlas.

At the time of the first Atlas, the Common Nighthawk was already declining in Pennsylvania (Brauning 1992f) and elsewhere in the United States (Sauer et al. 2011). From the first Atlas to the second Atlas, the number of blocks reporting nighthawks dropped 71 percent, from 754 to 219. The decrease in confirmed records was even more extreme, from 62 to only 10. There were only four confirmed reports of occupied nests with eggs or young, occurring from 12 June to 3 July. The results of the second New York and Maryland atlases tell a similar story, with 71 percent and 83 percent declines in blocks occupied since the first atlases in those states, respectively (Medler 2008e, Ellison 2010o).

Nighthawks were found in a small fraction of blocks in all major physiographic regions of Pennsylvania, their pattern of distribution determined largely by locations of cities and towns, with a few reports from former coalfields. The aforementioned decline was uniform throughout the state. The most extensive clusters of second Atlas reports came from the greater Pittsburgh area, the Scranton/Wilkes-Barre corridor, and an eastern cluster of urban areas including Hazleton, Pottsville, Shamokin, and Allentown. However, in all of these areas, as elsewhere, nighthawks have declined substantially since the first Atlas.

Most atlas reports from urban areas presumably involved birds nesting on rooftops, a characteristic for which the Common Nighthawk has long been known (Warren 1890; Todd 1940). There were three confirmed atlas reports of urban ground nesting (a parking lot, gravel adjacent to a building, and along a railroad bed). However, one confirmed site was on a remote former surface mine in northern Clinton County, on bare, disturbed ground among sparsely scattered, low vegetation, similar to other natural sites described previously (Fowle 1946; Harrison 1975). Additional reports of probable nesting in former surface mines came from Clinton, Centre, and Clearfield counties and from former coalfields in Schuylkill, Lackawanna, and Luzerne counties. A further example of non-urban nesting comes from a site in Carbon County formerly denuded of vegetation by industrial pollution and now revegetated with grass (D. Kunkle, pers. comm.). In New York, clustered reports of ground-nesting nighthawks came from non-urban areas (Medler 2008e).

There was no obvious correlation between nocturnal coverage and the number of nighthawk reports in individual atlas regions; furthermore, nighthawks are often detected during daytime surveying. Therefore, the pattern of atlas records probably reflects the current distribution fairly accurately. However, nesting in some strip mine areas may have been missed, because many of those areas are either remote or privately owned.

The decline of the Common Nighthawk in the eastern United States is believed to be largely due to use of rubberized surfaces unsuitable for nesting, rather than gravel overlay for flat rooftop construction and repair. As a remedy, small gravel patches placed on rubberized rooftops may allow continued nesting by nighthawks (Marzilli 1989; Poulin et al. 1996). Of unknown significance is the effect of pesticide use on this insect specialist, or the impact of predation, notably from growing numbers of crows in the urban environment. The Common Nighthawk winters in South America, where little is known of conditions that may adversely affect its survival (Poulin et al. 1996).

Unfortunately, the prospects for the Common Nighthawk in Pennsylvania are not good, in view of the steep, long-term decline. The discovery of breeding on former strip mines suggests that these lands may become an important refuge. Post-atlas efforts to more thoroughly assess Common Nighthawk usage of former coalfields should be undertaken, especially on public lands where active management for the species could be encouraged.

GREG GROVE

Distribution

Distribution Change

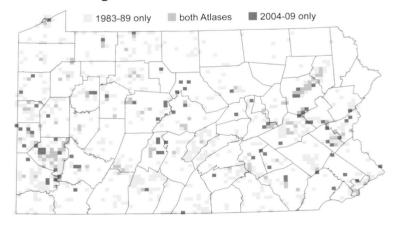

Number of Blocks

	first Atlas 1983–89	second Atlas 2004–09	Change %
Possible	478	144	−70
Probable	214	65	−70
Confirmed	62	10	−84
Total	754	219	−71

Rob and Ann Simpson/VIREO

Chuck-will's-widow
Caprimulgus carolinensis

The Chuck-will's-widow is known in the United States as a bird of the Southeast. It ranges from Florida along the Atlantic coast, north to Long Island, and west to central Nebraska, Oklahoma, and Texas. Pennsylvania lies on the extreme northern periphery of the Chuck-will's-widow's eastern breeding range. Not surprisingly, therefore, it is considered a casual-to-accidental species in the state, with most observations comprised of single birds heard calling in late spring (migrants that have overflown their normal range). Indeed, the majority of Pennsylvania records are of 1-day visitors (McWilliams and Brauning 2000).

Despite this, both the first and second Atlas efforts found probable breeding evidence for the species in the state. In the first Atlas, the most notable records were of multiple males heard calling in Armstrong County during three consecutive summers from 1985 to 1987. Breeding was never confirmed in Armstrong County in those years, but it was strongly suspected (Santner 1992c). These Armstrong County birds also represented Pennsylvania's first record of multiple Chuck-will's-widows heard or seen in the same area.

There were just two records of the species during the second Atlas period; however, both were classified as probable breeding and each was compelling in its own right. The first was of a singing male, likely unmated, found in early July 2004 at a private residence adjacent to Bald Eagle Mountain in southwestern Centre County. It was heard calling for at least 8 consecutive nights and was documented by audio recording. The record became more significant when the bird returned in 2005 and was heard calling regularly from early May until early August. It was the first Chuck-will's-widow to return to the same site in Pennsylvania 2 years in a row since the Armstrong County birds were found during the first Atlas period.

In 2007, two Chuck-will's-widows were found in late May in southeastern Fulton County, marking only the second record of multiple "Chucks" in the same area in Pennsylvania's history (the first being the already mentioned Armstrong birds). One of the Fulton birds, a calling male, was documented by photograph and audio recording. Most interestingly, the second bird was seen foraging while the calling male could be heard in the distance, but it never sang, suggesting it was likely a female. Breeding was not confirmed, but this record of a likely pair should be considered the most compelling evidence of breeding chucks in the state of Pennsylvania to date.

The sites of both records from the second Atlas and the Armstrong County bird in the first Atlas were similar in habitat and geology: rural areas near 300 m (1,000 ft) elevation, characterized by woodlots interspersed among farmland (Santner 1992c). All three records are consistent with the species' suspected habitat preferences. Openings in woodlots, either natural or man-made, appear to be an important habitat component for foraging (Straight and Cooper 2000).

The Fulton record was particularly interesting because south-central counties have been suspected as the most likely area in the state for breeding chucks (McWilliams and Brauning 2000), based largely on the region's habitat similarities and geographic proximity to areas of known breeding pairs in nearby Maryland (Reese 1996). However, during the second Maryland/District of Columbia atlasing effort between 2002 and 2006, there were no records of Chuck-will's-widow from this region (Ellison 2010p).

Both sites from the second Atlas, and the Armstrong County site from the first Atlas, also share notable commonalities with breeding sites in Ohio, where the species is similarly at the northern extreme of its summer range (Peterjohn 2001). In New York, where the species reaches its far northeastern limit, chucks are primarily associated with extensive pine-oak barrens or with coastal barrier beaches (Mitra 2008), as is the case also in New Jersey. Of course, neither of these habitats is particularly abundant in Pennsylvania. Thus, despite the proximity of the New York and New Jersey populations to southeastern Pennsylvania, the southwestern and south-central counties of Pennsylvania remain the most likely areas to host breeding Chuck-will's-widows in the future.

GEOFF MALOSH

Distribution

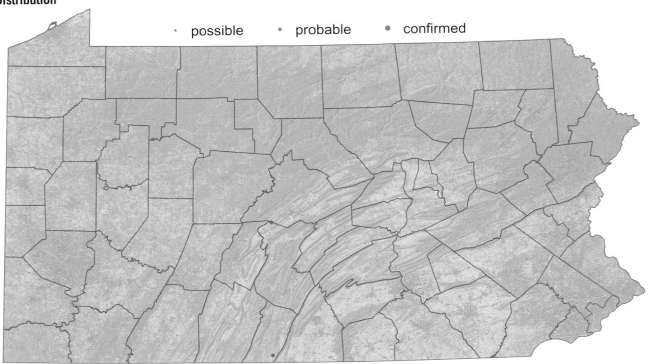

Distribution Change

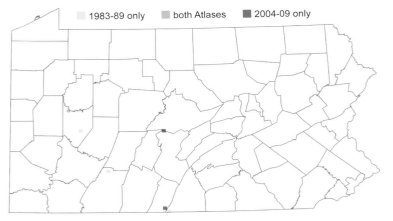

Number of Blocks

	first Atlas 1983–89	second Atlas 2004–09	Change %
Possible	1	0	−∞
Probable	2	2	0
Confirmed	0	0	0
Total	3	2	−33

Warren Greene/VIREO

Eastern Whip-poor-will
Caprimulgus vociferus

The onomatopoeical song of the whip-poor-will, so iconic of its Appalachian setting, is still heard at select Pennsylvania locations, though apparently at many fewer than previously. A nocturnal summer resident of barrens and woodlands with scattered clearings, its Pennsylvania population may have peaked roughly a century ago as forests were regenerating. Warren (1890) described the whip-poor-will as "a rather common summer resident in the wooded and mountainous portions of Pennsylvania." Its range has gradually contracted, as fewer locations retain its required habitat characteristics. It was formerly considered conspecific with the Mexican Whip-poor-will (*C. arizonae*), but the two species were split by the American Ornithological Union in 2010 (Chesser et al. 2010). The Eastern Whip-poor-will's breeding range includes much of the southern United States, extending north to southern Ontario and Quebec (Cink 2002). Partners in Flight has estimated a breeding population of 8,000 whip-poor-wills in Pennsylvania, which is less than 1 percent of the estimated global population (Rich et al. 2004).

During the first Atlas period, the whip-poor-will was found in 17 percent of all blocks, with the highest concentrations in the Ridge and Valley and in a western-central cluster of counties from Indiana to Venango. At that time, numbers had been declining in Pennsylvania for perhaps a half-century, since Todd (1940) reported noticeable decreases in the 1930s. Breeding Bird Survey data show a steep decline of 6 percent per year since 1966 (Sauer et al. 2011). During the second Atlas, whip-poor-will was found in just 10 percent of all blocks, 42 percent fewer than in the first Atlas. Losses are most noticeable in the south-central and northeast Ridge and Valley Province, the Glaciated Low Plateau, and throughout the western sections of the Appalachian Plateaus Province (appendix C). Even larger declines in blocks with whip-poor-wills (−57%) have been noted in neighboring New York and Maryland (Medler 2008f; Ellison 2010q).

The decline in whip-poor-will occupancy of priority blocks in the second Atlas (25%) is less than for all blocks (42%), which suggests that special surveys to enhance detection of nocturnal species conducted in many priority blocks were successful at reducing coverage bias for this species. Hence, apparent changes in distribution between atlases are partly driven by changes in observer effort. Nonetheless, strongholds are still apparent in some large contiguous forests, such as Sproul, Moshannon, and Rothrock state forests in the center of the state, where there was a substantial effort to detect nocturnal birds. Conversely, fewer whip-poor-wills were found in the southern counties of Somerset, Bedford, and Fulton, likely because relatively little nocturnal atlasing was done there during the second Atlas. Losses in western Pennsylvania, even in areas where there was considerable nocturnal atlas effort, were substantial and likely reflect a real range contraction.

Because of the whip-poor-will's requirement for large forest tracts, maintaining large forested core habitats will be essential for retaining healthy whip-poor-will numbers in Pennsylvania. A large percentage of the state's forests are approaching maturation, and the method by which these trees will be harvested will be a key determinant of the future of many forest birds (Stoleson and Larkin 2010), including the whip-poor-will, which probably will benefit from cutting if sufficient intact forest remains nearby. Within these areas, whips need sites with open ground or open understory for foraging, dense growth for nesting, and an abundant supply of large moths, their favorite food (Hess et al. 2000; Cink 2002).

In general, nocturnal birds pose a real challenge for biologists and agencies trying to assess and monitor their populations. The two Pennsylvania breeding bird atlases seem to indicate that both of the state's widespread nightjars may be in trouble. Although this atlas and existing breeding bird surveys are probably not adequate for assessing numbers and trends in whip-poor-wills, a specialized nightjar survey recently begun in Pennsylvania and other eastern states could add to our knowledge about this enigmatic bird.

NICHOLAS C. BOLGIANO

Distribution

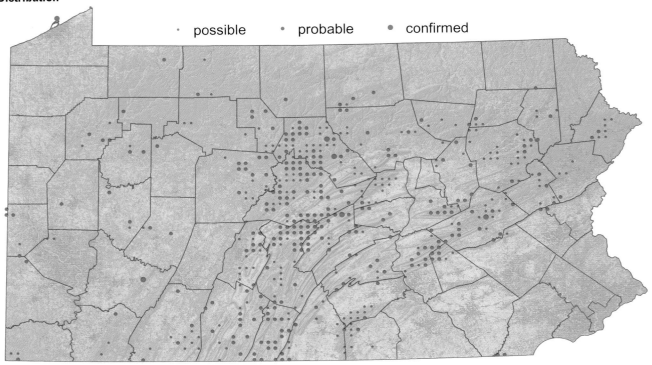

Distribution Change

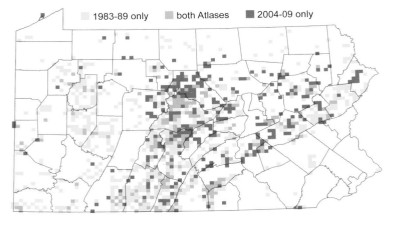

Number of Blocks

	first Atlas 1983–89	second Atlas 2004–09	Change %
Possible	464	228	−51
Probable	375	262	−30
Confirmed	23	6	−74
Total	862	496	−42

Greg Lasley/VIREO

Chimney Swift
Chaetura pelagica

In cities and towns across the Commonwealth, the Chimney Swift is Pennsylvania's most familiar aerial insectivore. It is nearly constantly airborne, using its slender body and rapid wingbeats to prey on a variety of flying insects. Its primary habitat requirements are open air for hunting and sheltered vertical walls on which to build nests. Today, the Chimney Swift is nearly completely dependent on urban and suburban habitats because of the nesting structures they provide, including its namesake "chimney" (Cink and Collins 2002).

The Chimney Swift breeds throughout the eastern United States and southern Canada and west to the Great Plains states. In the fall, it gathers in large roosts at large chimneys before migrating to South America for the nonbreeding season (Cink and Collins 2002). Populations of Chimney Swifts appear to be declining across its North American range, including Pennsylvania. Breeding Bird Survey data reveal a rangewide decline of nearly 2.1 percent per year since 1966, with a more modest 1.2 percent decline over the same period in Pennsylvania (Sauer et al. 2011). These declines contributed to listing as a species of Maintenance Concern in Pennsylvania's Wildlife Action Plan (PGC-PFBC 2005).

The Chimney Swift is widely distributed in Pennsylvania, with breeding observations from every physiographic section and every county. The species is more patchily distributed in the most heavily forested regions of the Commonwealth, including north-central and northeastern Pennsylvania. These areas likely lack the man-made structures generally used for nesting. Only in rare cases has the Chimney Swift been discovered nesting in hollow trees and other naturally occurring cavities that the species used before European colonization (Blodgett and Zammuto 1979; Turner et al. 1984), but the bird's behavior, and its broad distribution in many remote areas, suggests that historic habitat may still be used in Pennsylvania.

From the first Atlas to the second, block occupancy declined nearly 2.5 percent. Most of Pennsylvania's physiographic sections showed a reduction in block counts, but a 36 percent decline in the Glaciated Low Plateau was the only statistically significant change (appendix C). The Chimney Swift's breeding season in Pennsylvania starts shortly after birds arrive in the spring. Second Atlas breeding evidence shows that courtship typically occurs during the months of May and June, with an early date of 6 May. Swifts carrying nesting material were observed between 23 May and 30 June. Most young birds have fledged by early July, although one nest with young was documented on 14 August.

High Chimney Swift densities were associated with virtually every city, with the highest densities occurring in Pennsylvania's most urbanized regions, including concentrations around Allentown, Erie, Harrisburg, Philadelphia, Wilkes-Barre, and, most notably, in the industrial river corridors around Pittsburgh. Occupancy rates were highest in areas with largest atlas effort, but this pattern likely reflects actual differences in swift distribution between developed and undeveloped areas because swifts are easily detectable, are visible, and vocalize throughout the day.

Pennsylvania's Chimney Swift population declined 27 percent between the first and second Atlases (Sauer et al. 2011). Point count data collected during the second Atlas suggest a total state population of 430,000 individuals, but this is approximate because Chimney Swift densities are not reliably determined from point counts (chap. 5).

Due to long-term population declines, Chimney Swift populations are approaching a crossroads. Pennsylvania still harbors a substantial population of these impressive aerialists, but the coming decades will be telling for the long-term sustainability of the species. If current trends continue, the species may become a conservation priority. Threats include decreasing availability of suitable nesting sites and changes in aerial insect populations in urbanized areas. Conservationists could engage building managers, chimney sweeps, and individual landowners to avoid disturbance of nests in established chimneys and also promote placement of swift towers, which function as chimney substitutes (Graham 2011).

BRIAN BYRNES

Distribution

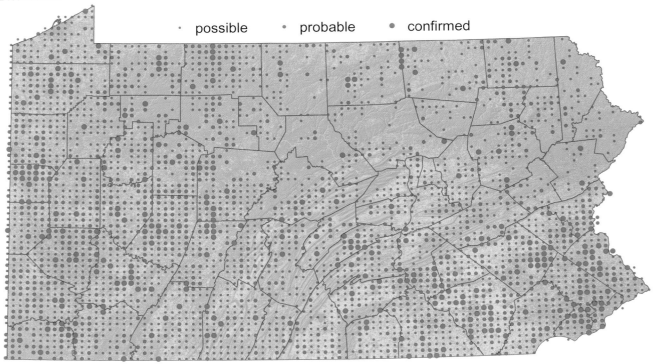

Distribution Change

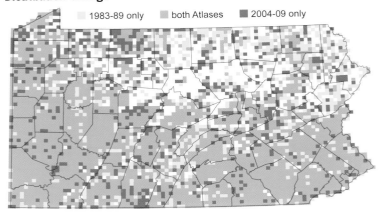

Density

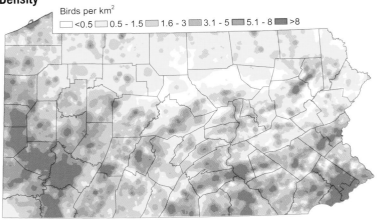

Number of Blocks

	first Atlas 1983–89	second Atlas 2004–09	Change %
Possible	1,624	2,218	37
Probable	1,163	830	−29
Confirmed	629	286	−55
Total	3,416	3,334	−2

Population estimate, birds (95% CI):
430,000 (410,000–450,000)

Breeding Bird Survey Trend

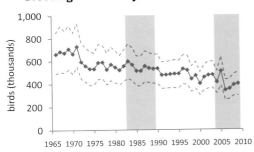

Bob Wood and Peter Wood

Ruby-throated Hummingbird
Archilochus colubris

The Ruby-throated Hummingbird is one of Pennsylvania's most easily recognized and popular birds. No other bird looks or acts like it, zipping back and forth from perch to nectar feeder or hovering at flowers. Because of its small size, it is often detected first by sound—the fast buzz of its wings as it flies past, or the sharp, squeaky chattering that it makes when it chases an interloper away from its territory or favorite feeder—rather than by sight. The male has a daybreak territorial song (Sargent 1999), but it is usually not heard. More often seen and heard is the male's vigorous arcing flight display, which is directed toward potential mates.

Hummingbirds are only found in the Western Hemisphere. Seventeen species have nested in the United States, but the ruby-throat is the only one that breeds east of the Mississippi River. There is little evidence of a change in its status in Pennsylvania through the last century and a half; in the first Atlas, it was found throughout the state, but its distribution was somewhat patchy in the southeast and areas in which coverage was sparse (Mulvihill 1992a).

In the second Atlas, the ruby-throat was found in almost 80 percent of blocks, with reports from every part of the state. It was apparently absent or overlooked only in blocks dominated by dense forest, large agricultural tracts, and the most urban areas. Overall, the block-level occurrence of ruby-throats increased a significant 11 percent compared with the first Atlas. A similar increase was noted between first and second atlases in Maryland (Ellison 2010r), while a larger increase (+21%) was noted between first and second New York atlases (McGowan 2008n).

Its overall distribution in the second Atlas, as in the first, clearly shows that the ruby-throat has adapted to a wide range of human-induced habitat change, as evidenced by population growth even in highly urbanized areas. This is supported by Breeding Bird Survey results, which show a 49 percent increase between atlas periods (Sauer et al. 2011). Point count data collected and analyzed for the second Atlas give a conservative estimated statewide breeding population of 250,000 ruby-throats. The highest densities occurred in the Appalachian Mountain and Allegheny Mountain physiographic sections in the central parts of the state. The lowest densities were in predominantly agricultural areas and extensively urban areas in the southeastern, northeastern, and southwestern portions of the state. The ruby-throat has long been associated with well-forested and riparian areas (Warren 1890); however, it generally does not favor densely forested tracts away from streams, perhaps because its favorite native nectar-producing plants—spotted jewelweed and bee balm—do not grow well in those areas.

The Ruby-throated Hummingbird's walnut-sized, lichen-covered nest usually is built on a small, downward-sloping branch that often overhangs an unimproved road, trail, or stream and is extremely well camouflaged. Thus, only 13 percent of the reports in the second Atlas, similar to results in the first Atlas, were in the confirmed category. Although June through mid-July is the normal expected breeding time, in the second Atlas, nest building was confirmed as early as 1 May, and fledged young were reported as late as early September. The plausibility of the later date for breeding evidence, by which time many hummingbirds are actively migrating or have already migrated through Pennsylvania, is supported by other second Atlas reports of nests with young on 26 August and "occupied nests" on 28 August.

The outlook is good for the Ruby-throated Hummingbird, and it is not currently in need of conservation or management efforts. Recent publications of North American hummingbird guide books (Howell 2002; Williamson 2001), and the expansion of a nationwide hummingbird banding network, have led to an upsurge in the number of people maintaining hummingbird feeders and planting hummingbird-friendly flowers. Thus, observations of the ruby-throat and knowledge of its behavior, longevity, migration times, and migration routes have increased and will no doubt continue to do so.

ARLENE KOCH

Distribution

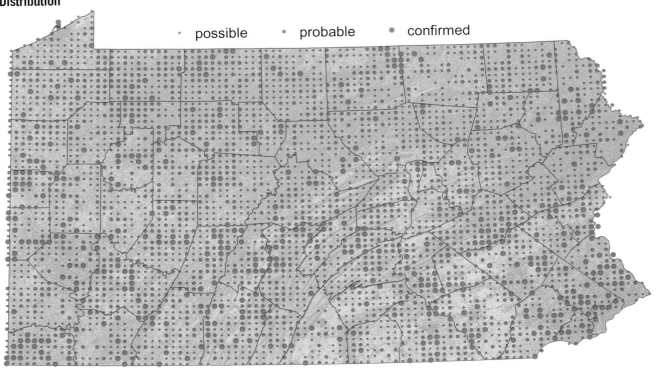

Distribution Change

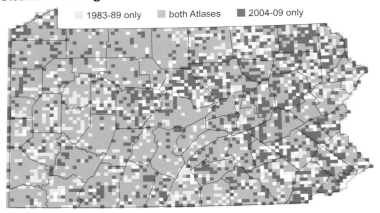

Density

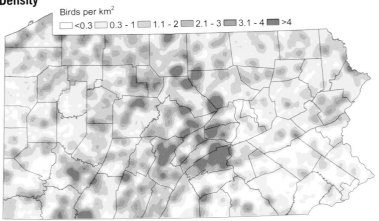

Number of Blocks

	first Atlas 1983–89	second Atlas 2004–09	Change %
Possible	1,772	2,325	31
Probable	1,141	1,125	−1
Confirmed	605	454	−25
Total	3,518	3,904	11

Population estimate, birds (95% CI):
250,000 (235,000–275,000)

Breeding Bird Survey Trend

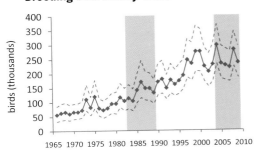

Jacob Dingel

Belted Kingfisher
Megaceryle alcyon

A sighting of a Belted Kingfisher is often preceded by its rattling call, a familiar sound on Pennsylvania's waterways. This is our only breeding species whose female is more colorful than the male, and evidence suggests that females may be more territorial than males (Kelly et al. 2009). The relatively large head and heron-like beak are obvious adaptations for catching fish, either quietly from perches or more dramatically while hovering above streams, ponds, and lakes. This species is widely distributed across the state and is the only kingfisher occurring from Alaska to central Texas and coast to coast. Its distribution is limited primarily by availability of clear water for hunting and suitable substrates for digging nest burrows (Kelly et al. 2009). The overall North American status of the Belted Kingfisher appears secure compared with persistent continental population declines of herons and egrets that share a similar diet (Sauer et al. 2011).

The kingfisher was recorded in 62 percent of blocks during the first Atlas and was relatively evenly distributed across the entire state. During the second Atlas, it was recorded from 57 percent of blocks distributed across all physiographic provinces. The species was confirmed in 433 blocks, representing 15 percent of the total, similar to the first Atlas. Confirmation of breeding within blocks is often not straightforward, because suitable nesting sites are scarce and foraging individuals may be detected relatively far from nest burrows (McGowan 2008o).

Occupancy by the Belted Kingfisher was somewhat reduced in several parts of the state, but with little overall geographic pattern. The noticeable lack of occupancy in Clearfield County is similar to the first Atlas and may be due to continued effects of surface coal mining on streams and their fish populations (Master 1992c). Occupancy was highest in the Ridge and Valley Province and in the Piedmont, but there were some extensive areas in all regions where no kingfishers were found. Areas of high occupancy do not coincide with higher-order stream density, but rather with lower elevations characterized by larger, slower-moving streams that typically have better bank development. However, the Belted Kingfisher is generally found in low densities, so it is an easy species to miss when surveying atlas blocks, especially if block effort is low. Analysis of second Atlas data shows that the Belted Kingfisher was found in only 30 percent of blocks with less than 5 hours of effort, increasing to more than 70 percent of blocks with more than 20 hours of effort. Thus, due to the lower overall effort in the first Atlas, block turnover between atlases may be overestimated, and the actual range decline may be greater than the documented 8 percent decline in occupied blocks.

Nevertheless, the decline in block occupancy is corroborated by the Breeding Bird Survey (BBS). Although the BBS may not be particularly efficient at detecting a water-dependent species, data from 1966 through 2009 show a slight decline in counts within the state (Sauer et al. 2011). Atlas and BBS trends from Pennsylvania are remarkably similar to those from Maryland (Ellison 2010s) and New York (McGowan 2008o). The Green Heron shares habitat and prey preferences with the Belted Kingfisher and also shows a modest decline in numbers on BBS routes. The Bank Swallow, which also nests in burrows in riparian areas, declined precipitously since the first Atlas. Taken together, the evidence warrants conservation concern for all of these species whose food resources and/or nesting habitat are associated with aquatic environments.

For now, the Belted Kingfisher, with its rattling call and undulating flight, remains a familiar sight across our state. The conservation concerns expressed above may be heightened by water-quality impacts from newly emerging shale-gas development in watersheds across Pennsylvania. Continued and consistent environmental enforcement is essential to protect the water resources upon which the Belted Kingfisher depends.

TERRY L. MASTER

Distribution

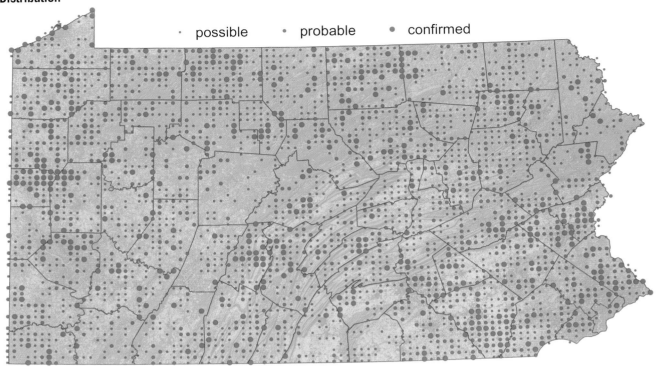

Distribution Change

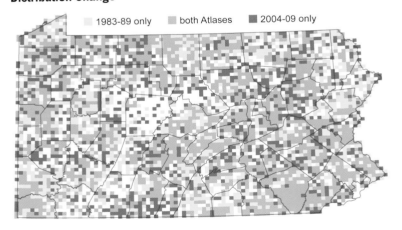

Number of Blocks

	first Atlas 1983–89	second Atlas 2004–09	Change %
Possible	1,530	1,731	13
Probable	1,052	666	−37
Confirmed	481	433	−10
Total	3,063	2,830	−8

Breeding Bird Survey Trend

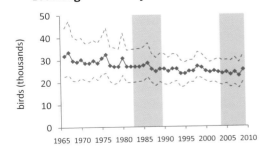

BELTED KINGFISHER

Bob Wood and Peter Wood

Red-headed Woodpecker
Melanerpes erythrocephalus

The strikingly plumaged Red-headed Woodpecker is Pennsylvania's most conspicuous, yet rarest, woodpecker. Its red head, contrasting white underparts and black upperparts, and conspicuous white wing patches and rump make this an unmistakable bird. Pennsylvania and New York represent the northeastern periphery of the species' breeding range, which extends westward across Michigan to southern Manitoba and southward to Florida and Texas. Breeding Bird Survey data from 1966 through 2009 show almost universal declines, especially in northern populations from the Great Plains and Midwest, and east to New York state (Sauer et al. 2011). The species appears on the Maintenance Concern list in Pennsylvania's Wildlife Action Plan (PGC-PFBC 2005).

The number of blocks with Red-headed Woodpecker records in Pennsylvania decreased by 46 percent from the first Atlas to the second Atlas. Only the three southern counties of Franklin, Adams, and York seemingly escaped losses; the woodpecker was found in 101 blocks in these three counties, more than a quarter of the statewide total. A southward retreat from former isolated outposts in the north is apparent, with no fewer than 13 counties in the northern half of the state losing this species between the first and second Atlases. It is no coincidence that a dramatic loss of this species was also documented in New York's southern tier between atlas periods (McGowan 2008p). This species appears to be undergoing a rapid range contraction.

Despite the steep decline in range size over the last decade, the Red-headed Woodpecker is still widely, if thinly, distributed in Pennsylvania. Records came from 51 counties, although there were no confirmed records in 19 of those. In addition to the aforementioned stronghold along the state's southern border, there were clusters of occupied blocks in the northwestern and central counties of the state, but these were greatly diminished since the first Atlas period. Estimates from point count data indicate a total population of 4,500 Red-headed Woodpeckers in Pennsylvania, although it should be noted that this was based on a small sample size (66 birds on point count surveys). Atlas results show that this species was often detected in open habitats and at elevations below 500 m (1,600 ft; appendix D). Typical Red-headed Woodpecker habitat features small woodlots with snags juxtaposed with grazed permanent pasture (Rodewald et al. 2005; A. Wilson, pers. comm.). These observations reflect the importance of an open landscape, within which Red-headed Woodpeckers forage for flying insects (Smith et al. 2000).

Historical fluctuations in Red-headed Woodpecker populations and distribution at a regional level are well documented and attributed to variations in annual mast production by oak and beech trees, grasshopper and cicada population cycles, and tree mortality caused by disease outbreaks (Smith et al. 2000). Interspecific competition for nesting cavities and food sources has been described (i.e., European Starlings and several woodpecker species), but large-scale impacts on Red-headed Woodpecker populations remain to be documented. They are highly dependent on acorn caches during the winter, when other food sources become scarce (Smith et al. 2000). Hence, the long-term decline in oak-hickory forests in Pennsylvania (McWilliams et al. 2007) and elsewhere could be another factor contributing to the loss of this species, especially after years when a greater proportion of woodpeckers has vacated the state in winter in response to winter severity and poor foraging opportunities (McWilliams and Brauning 2000). However, the Red-headed Woodpecker has one of the broadest diets among any North American woodpecker species (Smith et al. 2000). Thus, any single change in food abundance (with the exception of acorns) is unlikely to have an effect on population size. Changes in predator density may be more important. Known predators of eggs or adults include snakes, raccoons, and raptors, specifically the Cooper's Hawk (Smith et al. 2000), a species that has increased markedly in Pennsylvania between atlas periods (Sauer et al. 2011). The causes of the decline in Red-headed Woodpecker numbers are, therefore, complex and manifold. The future prospects for this species, in much of its former range in Pennsylvania, are decidedly bleak.

JAMES S. KELLAM

Distribution

possible   probable   confirmed

Distribution Change

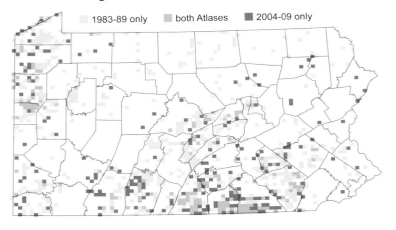

1983-89 only   both Atlases   2004-09 only

Number of Blocks

	first Atlas 1983–89	second Atlas 2004–09	Change %
Possible	321	173	−46
Probable	183	82	−55
Confirmed	194	119	−39
Total	698	374	−46

Population estimate, birds (95% CI):
4,500 (3,300–6,000)

Breeding Bird Survey Trend

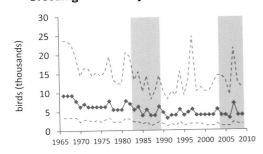

Tom Pawlesh

Red-bellied Woodpecker
Melanerpes carolinus

Identified by its bright red head patch and distinctive call, rather than the inconspicuous pink abdomen for which it is named, the Red-bellied Woodpecker is well known to birders and is a frequent feeder visitor in much of Pennsylvania. It breeds throughout the eastern United States, is the most abundant woodpecker in the southeastern part of the country, and is expanding its range into southern Canada. Nesting in almost any type of large Pennsylvania tree in suburban areas or woodland, the Red-bellied Woodpecker was recorded in slightly more than three-quarters of the blocks during the second Atlas.

Historically, the Red-bellied Woodpecker was considered regular in the southwest (Jacobs 1893), questionable in the southeast (Michener 1863; Stone 1894), and a sporadic breeder in some northern parts of the state (Sutton 1928a). However, by the 1950s it had spread across Pennsylvania and was located in most areas of the state by the 1980s (Street 1976; Stull et al. 1985). First Atlas records indicate that the Red-bellied Woodpecker was widespread in the southwest and had continued its statewide expansion, becoming common in the southeastern physiographic provinces and more widespread in the northern half of the Commonwealth (Ickes 1992e).

The second Atlas demonstrates a continued and dramatic northward expansion of the Red-bellied Woodpecker's range in Pennsylvania, with an 86 percent increase in occupied blocks between atlas periods. Increases in occupied blocks were noted in every physiographic province, with substantial increases (between 198 and 537 percent) in the six northernmost sections (appendix C). Further evidence of the woodpecker's impressive northward expansion comes from the second New York atlas: an increase of 123 percent in occupied blocks between atlas periods was recorded (McGowan 2008q). Various factors have influenced the Red-bellied Woodpecker's Pennsylvania expansion, probably including periods of mild winters, increased year-round food resources, availability of preferred semi-open habitat, and its tolerance of human presence (Hess 1992).

Second Atlas point count data for the Red-bellied Woodpecker show that the highest densities occur in fragmented and mixed landscapes, including forested residential habitat (appendix D). Moreover, the species is less common at elevations above 250 m (820 ft) and scarce at elevations above 500 m (1,640 ft). The tendency for elevation to constrain the Red-bellied Woodpecker's distribution was reported in the first Atlas (Ickes 1992e), and other states' atlas results indicate a similar relationship (e.g., Van Ness 1996; McGowan 2008q). McGowan (2008q) suggested that lower temperatures at high elevations during winter, rather than summer, may be the species' limiting factor in New York for this permanent resident.

The total Pennsylvania population for the Red-bellied Woodpecker during the second Atlas was estimated to be 355,000 individuals; according to Breeding Bird Survey data, its population increased more than threefold between atlas periods, with a 7.7 percent annual increase from 1966 to 2009 (Sauer et al. 2011). The Red-bellied Woodpecker now vies with the Downy Woodpecker as the most commonly seen woodpecker in much of Pennsylvania.

The extraordinary expansion of the Red-bellied Woodpecker's range in Pennsylvania seems to indicate that its outlook in the state, for the immediate future, is excellent. Because it inhabits a wide variety of forest types and nests in dead portions of live trees, it is more adaptable than most other woodpecker species in the country (Shackelford et al. 2000). However, since woodpecker abundance is primarily influenced by the presence and numbers of live hardwood trees, snags, and logs, forest management programs that actively remove, or do not provide, suitable sizes of these essentials could have a long-term impact on Red-bellied Woodpecker populations (Shackelford and Conner 1997; Giese and Cuthbert 2003).

ROY A. ICKES

Distribution

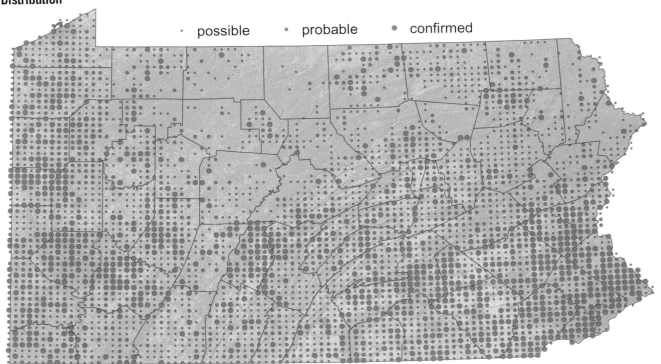

Distribution Change

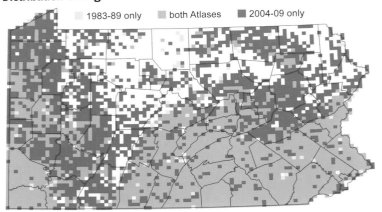

Density

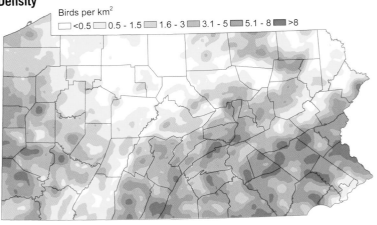

Number of Blocks

	first Atlas 1983–89	second Atlas 2004–09	Change %
Possible	782	1,857	137
Probable	787	991	26
Confirmed	437	888	103
Total	2,006	3,736	86

Population estimate, birds (95% CI):
355,000 (335,000–375,000)

Breeding Bird Survey Trend

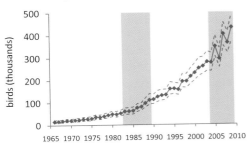

RED-BELLIED WOODPECKER 231

Gerard Dewaghe

Yellow-bellied Sapsucker
Sphyrapicus varius

Although field guides describe the Yellow-bellied Sapsucker as a quiet, inconspicuous bird, its wide repertoire of raucous squawks, squeals, chatters, and arrhythmic drumming reveals it to be a common and noisy breeder in Pennsylvania's northern tier.

The Yellow-bellied Sapsucker is a medium-sized woodpecker notorious for feeding extensively on tree sap obtained from a series of small holes drilled into the cambium layer of various trees, especially aspen, maple, and serviceberry (Eberhardt 2000). Highly vocal on its breeding grounds, the sapsucker also drums like other woodpeckers, albeit with a unique, irregular cadence. Its white-barred back, yellowish underparts, and extensive white in the wing and rump further distinguish the sapsucker from our other black and white woodpeckers. In Pennsylvania, sapsuckers breed commonly in open woodland, edges, and second-growth deciduous and mixed forests, particularly in northern hardwoods. They excavate nest cavities in softer woods such as aspens and Red Maple.

The Yellow-bellied Sapsucker breeds in mixed subboreal and northern forests across much of Canada and south into the northeastern United States as far as northern Pennsylvania. A small disjunct population in the southern Appalachian Mountains, given subspecific status by some taxonomists (*S. v. appalachiensis*), is declining and considered a federal species of concern (Walters et al. 2002). In contrast, northern breeding populations, including those in Pennsylvania, are common, widespread, and increasing, according to Breeding Bird Survey (BBS) data (Sauer et al. 2011). Highly migratory, the Yellow-bellied Sapsucker regularly winters in the southeastern corner of the state and southward as far as Costa Rica.

During the second Atlas period, sapsuckers bred commonly in wooded areas across the northern tier of Pennsylvania above approximately 41.4°N latitude. South of that latitude, atlas volunteers recorded them only sporadically as breeders in a few mountainous areas. They were reported in almost twice as many blocks in the second Atlas period as in the first Atlas, indicating a highly significant 99 percent increase. Similarly, in New York, the number of atlas blocks occupied by sapsuckers increased by over 50 percent, notably so in the western Allegheny Plateau in New York (McGowan 2008r). BBS results also indicate that sapsucker populations in Pennsylvania have increased greatly, at an average rate of 9.2 percent annually, or almost fivefold between atlas periods (Sauer et al. 2011). Historically, the species appears to have been much more numerous in the Allegheny Mountains, but it has long since become scarce there (McWilliams and Brauning 2000), and the atlases show no evidence that the species is regaining that former range, despite the tremendous population increase. Rather, the increase in range since the first Atlas was due to infilling and expansion to the northern, eastern, and western areas of the state.

Although the Yellow-bellied Sapsucker was found in a variety of habitats on point counts, the highest densities were found where more than 70 percent of the land within 150 m (~500 ft) of count locations was forested, possibly suggesting forest area sensitivity. The species was reported only rarely at elevations below 250 m (820 ft). The highest densities of sapsuckers were found in the heavily forested north-central counties of Pennsylvania: about 44 percent of the total estimated state population of 96,000 birds occurred in just Warren, McKean, Potter, and Tioga counties. There, they are often the most abundant woodpecker and can be more common than all other woodpeckers combined (Stoleson, unpublished data).

The Yellow-bellied Sapsucker appears to be thriving in Pennsylvania. Given its broad habitat range, tolerance of edges and younger forests, and relatively short-distance migration, the sapsucker will likely remain a common and conspicuous breeder in our northern counties. The Yellow-bellied Sapsucker migrates at night, and individuals have been killed in collisions with communications towers (Walters et al. 2002). The continuing increase in communications infrastructure and wind turbines may pose a threat in the non-breeding seasons.

SCOTT H. STOLESON

Distribution

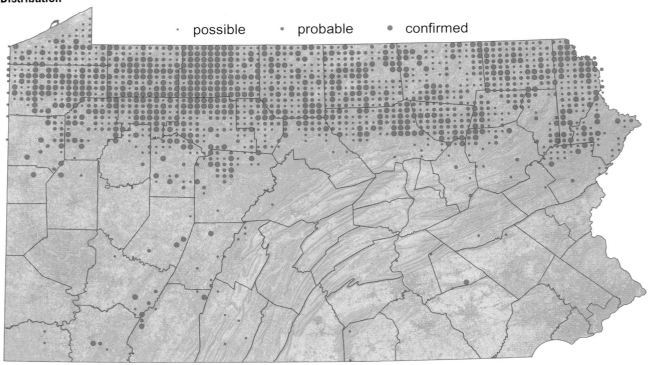

Distribution Change

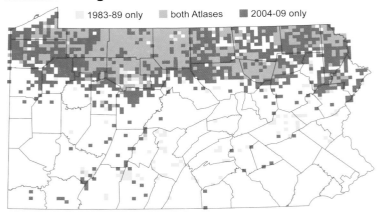

Number of Blocks

	first Atlas 1983–89	second Atlas 2004–09	Change %
Possible	317	513	62
Probable	209	426	104
Confirmed	207	520	151
Total	733	1,459	99

Population estimate, birds (95% CI):
96,000 (91,000–100,000)

Density

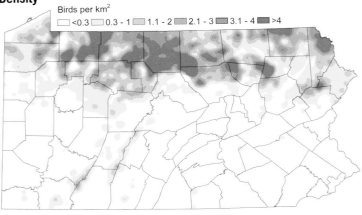

Breeding Bird Survey Trend

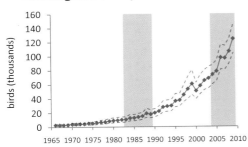

YELLOW-BELLIED SAPSUCKER

*James M. Wedge/*VIREO

Downy Woodpecker
Picoides pubescens

The Downy Woodpecker, the smallest woodpecker in North America, is a common, non-migratory resident found throughout Pennsylvania. The species' range extends from the tree line in northern Canada, south to Florida and the forested sections of Texas, and from Newfoundland in the east to California and Alaska in the west (Jackson and Ouelllet 2002). Given its small size (27 g, ~1 oz), the Downy Woodpecker can excavate a nest cavity in a wide variety of dead woody substrates, as long as the wood is soft with decay. Nest locations are usually in small-diameter dead stubs attached to living trees, and nesting pairs in Pennsylvania have been known to locate their nests less than 100 m (330 ft) from one another (J. Kellam, unpublished data). Given the relative abundance of potential nest sites, even in urban areas, this species is not likely to be limited by nest sites. Indeed, the Downy Woodpecker is a frequent visitor to backyard bird feeders.

Breeding Bird Survey data show that there has been no significant change in the statewide population of Downy Woodpeckers since the 1960s (Sauer et al. 2011). The second Atlas population estimate was 450,000 birds, making this the most numerous of the state's seven woodpeckers. The species was found in 92 percent of atlas blocks, which is not significantly different from the first Atlas. The species was found in every physiographic region of the state. The region with the lowest percentage of blocks (81%) containing the Downy Woodpecker was the Glaciated High Plateau, whereas it was found in 100 percent of blocks in several physiographic sections. The only physiographic region that showed a significant change in occupancy was the Northwestern Glaciated Plateau, where there was an 18 percent increase in the number of blocks occupied from the first Atlas to the second. The explanation for this increase is unknown, as changes in forest cover and developed land in this region since the first Atlas have not been extraordinary (chap. 3). However, the pattern appears to be real, as there was a corresponding but non-significant increase (17%) in the priority blocks within this region.

In the second Atlas, point count detections were highest in developed areas, but this species is a generalist and was found in nearly all habitat types across the state at varying densities (appendix D). Population densities were highest, generally more than six birds per km^2, in the southeastern third of the state. The areas with high Downy Woodpecker densities correspond to low elevations (below 400 m, 1,300 ft) where there is relatively little core forest.

Point count data confirm a steady decrease in Downy Woodpeckers as elevation increases (appendix D). This elevational relationship may be due to the fact that the microclimate of west-facing hillsides and mountain ridges tends to be drier than surrounding valleys and lowlands (Desta et al. 2004). Woodpeckers place nests in wood with fungal infection (Jackson and Jackson 2004), and the rate of wood decay may progress more slowly at higher elevations due to lack of moisture during summer months and greater number of cold winter days (Waltman et al. 1997). In addition, the higher elevations of Pennsylvania may, in general, have greater sources of mortality for the birds; mountainous regions might correlate with greater exposure to cold temperatures.

There is no evidence that threats facing other arboreal bird species, such as forest fragmentation and the deleterious effects of invasive pests (chap. 3), will have any impact on the adaptable Downy Woodpecker. Given the stability of the species' range and population sizes shown by the atlas and Breeding Bird Survey data, respectively, the status of the species gives no cause for concern.

JAMES S. KELLAM

Distribution

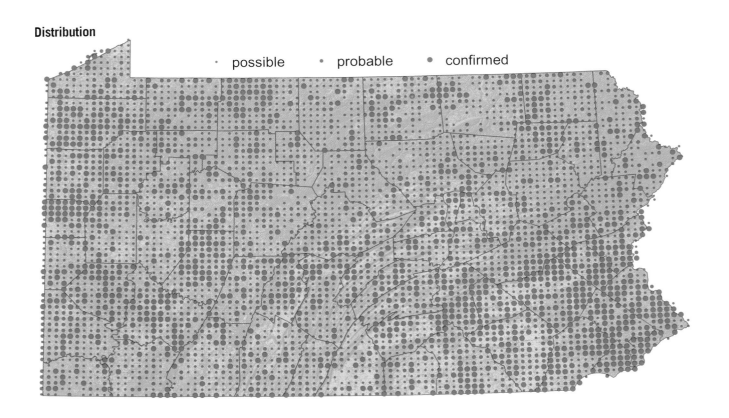

Distribution Change

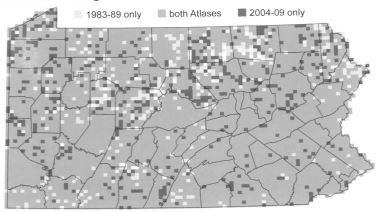

Density

Number of Blocks

	first Atlas 1983–89	second Atlas 2004–09	Change %
Possible	1,671	2,130	27
Probable	1,538	1,224	−20
Confirmed	1,179	1,177	0
Total	4,388	4,531	3

Population estimate, birds (95% CI):
450,000 (430,000–480,000)

Breeding Bird Survey Trend

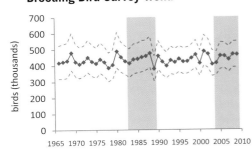

DOWNY WOODPECKER   235

Garth McElroy/VIREO

Hairy Woodpecker
Picoides villosus

The Hairy Woodpecker is distinguished in the field from the similarly clad Downy Woodpecker by its overall larger body size and proportionately longer bill. A year-round resident, it is generally solitary, but it occasionally can be found as a peripheral participant in mixed-species flocks (Jackson et al. 2002). The best available data suggest that the species can have very large home ranges in winter (146 ha, ~360 ac; Kovert-Bratland et al. 2006) and during nesting (44 ha, ~108 ac; Ripper et al. 2005), but no studies have been conducted in Pennsylvania. The species' geographic range is also impressively large. Populations of Hairy Woodpecker are found as far north as the tree line in Canada and south to the Bahamas and Panama (Jackson et al. 2002). The Hairy Woodpecker is considered a keystone species in forested habitats, since it provides nest and roost cavities for numerous birds and mammals (Ripper et al. 2005).

In Pennsylvania, second Atlas data show that the Hairy Woodpecker was found in 73 percent of blocks and was found throughout the state. Accounting for increased observer effort, there was an 8.8 percent increase in occupied blocks between atlases. Statistically significant increases in 3 of 23 physiographic sections were responsible for much of this change: Northwestern Glaciated Plateau (58%), Susquehanna Lowland (41%), and Anthracite Upland (32%). The cause of the increase within these regions is unclear. The Waynesburg Hills was the only physiographic section to show a significant decrease in blocks with Hairy Woodpeckers (−23%), which may have been partly due to reduced observer effort in that area. The highest densities of Hairy Woodpeckers during the second Atlas were found in forested areas in central and northern Pennsylvania. As expected, the lowest densities were in the Piedmont and other areas with limited forest cover.

Atlas breeding data show that probable and confirmed nesting behaviors were documented relatively late in the Hairy Woodpecker nesting cycle. Jackson et al. (2002) report the median fledge date across the geographic range as mid-June. Consistent with this, six nests found in Pennsylvania during the second Atlas years all fledged young by 12 June (J. Kellam, unpublished data). If nesting behavior in May was routinely missed, it is possible that the overall detection rate of this species was relatively low, which might explain the patchy distribution apparent in some areas with suitable forest habitat.

Point count data confirm that the Hairy Woodpecker is an extensive forest obligate. The mean count increased linearly as the percentage of forest within the point count radius increased (appendix D). Point count totals were fairly constant regardless of land elevation. However, since the percentage of forest cover increases with elevation, it appears that Hairy Woodpeckers are proportionately less dense at higher elevations. A similar trend was found for the Downy Woodpecker, which may be due to microclimatic conditions that limit fungal growth (Desta et al. 2004). Woodpeckers are dependent on fungal decay to soften woody substrates used for excavating nests and roost holes (Jackson and Jackson 2004). The population density of Black Bears in Pennsylvania has increased since the late 1980s (Hristienko and McDonald 2007), and bear predation on woodpecker nests can be surprisingly high, depending on the hardness of trunks used as nests (Tozer et al. 2009).

Breeding Bird Survey data show a steady but nonsignificant increase in Hairy Woodpecker population size since the first Atlas (Sauer et al. 2011). The total population size of the Hairy Woodpecker in Pennsylvania, as estimated by second Atlas point counts, was 97,000 birds. Due to its large range and increasing population, the Hairy Woodpecker is not of conservation concern.

JAMES KELLAM

Distribution

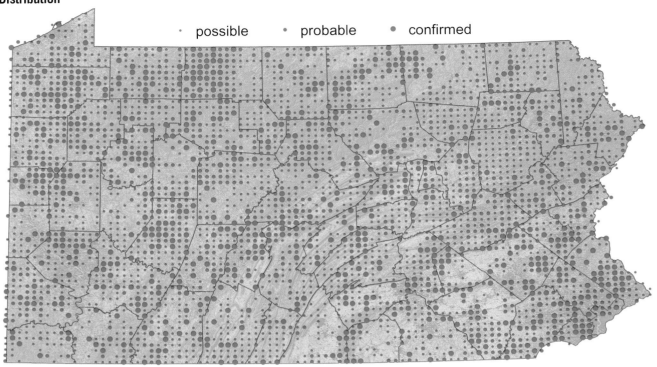

Distribution Change

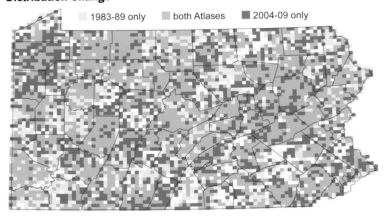

Density

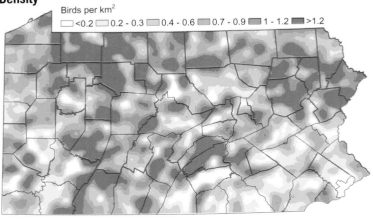

Number of Blocks

	first Atlas 1983–89	second Atlas 2004–09	Change %
Possible	1,519	1,935	27
Probable	1,026	936	−9
Confirmed	570	726	27
Total	3,115	3,597	15

Population estimate, birds (95% CI):
97,000 (90,000–106,000)

Breeding Bird Survey Trend

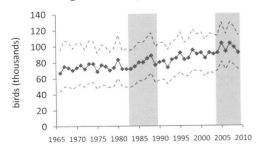

Geoff Malosh

Northern Flicker
Colaptes auratus

The Northern Flicker is a common, partially migratory woodpecker of forest and edge habitats. Populations north of latitude 39°N (including Pennsylvania) are generally migratory, but there is a fair amount of variation in the extent of its movements each year (Wiebe and Moore 2008). Christmas Bird Count (CBC) data show numerous Northern Flickers overwintering in the Commonwealth during the mid-1980s, but almost none for the next 20 years (NAS 2002). Recent CBC surveys have, again, been graced with numerous overwintering flickers. Second Atlas data show most flicker nests are initiated by early June, with fledged young observed a month later. This is consistent with previously published data (Wiebe and Moore 2008).

Pennsylvania has an abundance of breeding Northern Flickers, and relative abundance maps produced by the Breeding Bird Survey (BBS) show that the state contains one of their highest densities east of the Mississippi River (Sauer et al. 2008). In addition to being found in all of Pennsylvania, the species' breeding range is expansive, including forested habitats from the tree line in Canada and south into Latin America and Cuba (Wiebe and Moore 2008).

There is little conservation concern for the Northern Flicker at this time, despite significant population declines observed throughout eastern North America since the mid-twentieth century (Sauer et al. 2011). BBS data from Pennsylvania show that the Northern Flicker population has stabilized following rapid declines between the mid-1960s and mid-1980s. As mentioned in the first Atlas account for the flicker and more recent literature (reviewed by Wiebe and Moore 2008), the European Starling frequently usurps flicker nests, but a continent-wide analysis shows starling impacts on cavity-nesting birds to be minimal (Koenig 2003).

The second Atlas documented the Northern Flicker in 95.8 percent of all blocks across Pennsylvania. This represents an increase in occupancy since the first Atlas, but artifacts of coverage in some areas may have affected the results. Greater effort in the Deep Valleys Section in the second Atlas, compared with the first, resulted in an apparent increase, but when corrected for effort, there is no evidence of an increase. Very few blocks were not occupied in either atlas period, and it could be that the few gaps are merely a result of low atlas effort, especially in areas where flicker densities are low. It seems reasonable to assume that the Northern Flicker is actually present in close to 100 percent of atlas blocks in the state.

Second Atlas point count data show that the flicker frequents a wide variety of habitats but is virtually absent from dense development and coniferous and mixed forest. This is consistent with an analysis by Lawler and Edwards (2006), who found that flickers living in Utah placed nests in locations with the least amount of coniferous forest, selecting aspen trees instead. Foraging sites in open areas near forest edges are chosen because of the abundant ant prey and the short distance to cover when aerial predators threaten (Elchuck and Wiebe 2002). Hence, the flicker is often found in relatively open country, provided there is at least some tree cover for nesting.

Flicker densities, estimated from second Atlas point count surveys, were highest in central parts of the state, where there is more interspersion between forests and open habitat. Densities north of this band were slightly lower and corresponded with an increase in hemlock and other coniferous vegetation that the flicker seems to avoid. The second Atlas estimated total population of Northern Flickers in Pennsylvania at 187,000 birds, but as with all non-songbirds, this estimate may be conservative (chap. 5).

JAMES S. KELLAM AND MICHAEL L. O'REILLY

Distribution

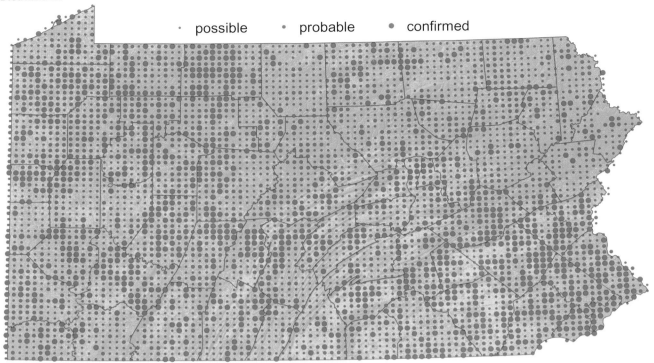

Distribution Change

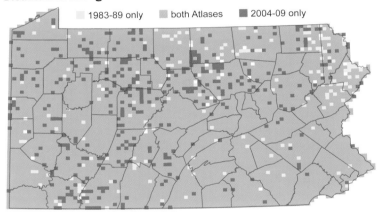

Density

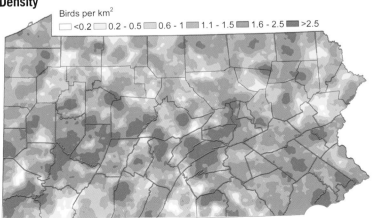

Number of Blocks

	first Atlas 1983–89	second Atlas 2004–09	Change %
Possible	1,553	2,311	49
Probable	1,628	1,384	−15
Confirmed	1,346	1,035	−23
Total	4,527	4,730	4

Population estimate, birds (95% CI):
187,000 (177,000–197,000)

Breeding Bird Survey Trend

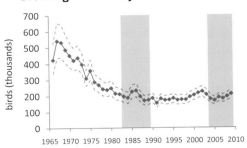

NORTHERN FLICKER 239

Brian Henry/VIREO

Pileated Woodpecker
Dryocopus pileatus

The Pileated Woodpecker is spectacular in appearance, voice, and the extent to which it has recolonized Pennsylvania's woodlands after its historic retreat from forest felling and indiscriminate shooting. Where it was absent for decades at many locations, the Pileated Woodpecker's loud *kuk-kuk kuk-kuk* again resounds. On Pennsylvania Christmas Bird Counts, only three pileateds were reported between 1900 and 1930, and yearly tallies remained in the single digits until 1950. Counts have been steadily increasing since then, to more than 700 statewide in some recent years (NAS 2002), in proportion far beyond the expansion in Christmas Bird Count participation. This population recovery is one of the great success stories among Pennsylvania's bird assemblage, with forest regrowth and protection as contributing factors.

The Pileated Woodpecker's range extends from coast to coast across southern Canada and south to the Gulf coast, east of the plains. Breeding Bird Survey (BBS) data indicate that at the time of the first Atlas, the core of the eastern U.S. population was to the south of Pennsylvania, but the pileated was also concentrated around Lake Superior and in the mountains of New England (Price et al. 1995). BBS data collected in Pennsylvania from 1966 through 2009 indicate a yearly 4 percent increase in Pileated Woodpecker numbers (Sauer et al. 2011), which translates to a population doubling about every 23 years, nearly the elapsed time between atlases. Pennsylvania now supports about 45,000 of these grand woodpeckers, as estimated from the second Atlas's point count survey.

The Pileated Woodpecker requires dead or dying trees for feeding, large-diameter trees for excavating nest cavities, and hollow trees or vacated nest cavities for roosting. Such trees are typically found in older forests, but they can also be found in younger forests in which at least a few older trees remain. Pileated Woodpeckers feed mainly on carpenter ants and beetle larvae, as well as fruits and nuts (Bull and Jackson 1995). Large oval excavations on tree trunks denote where a Pileated Woodpecker sought these wood-boring ants or insects. The Pileated Woodpecker also prefers larger contiguous woodlots, which explains the absence of the species from tracts of the Piedmont, where woodlots are small and often isolated.

In the second Atlas, the Pileated Woodpecker was found in 74 percent of all blocks, concentrated in forested areas, and absent from large agricultural, grassland, or urban areas. This represents a 38 percent increase in block detection over the first Atlas. Many blocks where this woodpecker was undetected during the second Atlas, but present during the first Atlas, are in regions in which there was lower block effort during the second Atlas. This woodpecker was thus probably present in those blocks but not detected. The Pileated Woodpecker has likely recolonized most areas of the state in which its habitat requirements are met.

The highest densities of this species largely coincide with moist, intact forests with large trees. Because the Pileated Woodpecker is never found in high densities, the density map should be considered to be somewhat coarse in scale, due to a relatively low number of detections over large areas.

This woodpecker probably temporarily benefits from the demise of the afflicted ashes, beeches, elms, and hemlocks, though a sudden loss of many trees would not benefit this bird in the long run. Sympathetic forest management practices, such as leaving some large trees during harvesting, should benefit the Pileated Woodpecker, which has proven to be resilient where its key habitat requirements are met and it is free from persecution.

NICHOLAS C. BOLGIANO

Distribution

· possible   · probable   • confirmed

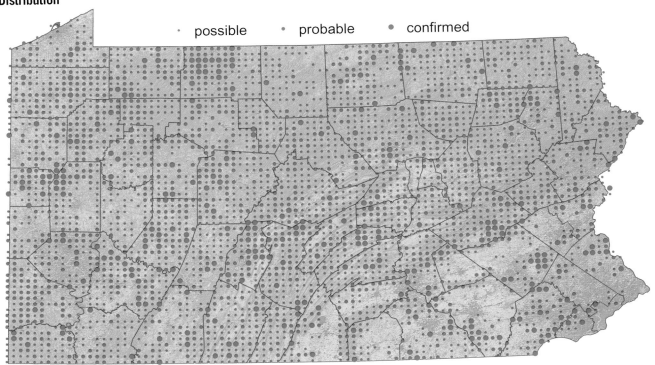

Distribution Change

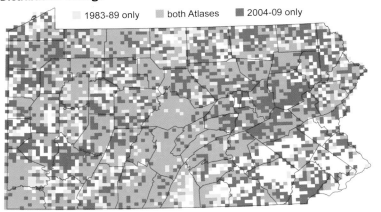

Density

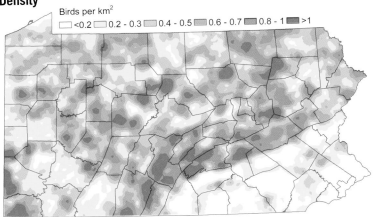

Number of Blocks

	first Atlas 1983–89	second Atlas 2004–09	Change %
Possible	1,394	2,179	56
Probable	1,002	1,163	16
Confirmed	254	311	22
Total	2,650	3,653	38

Population estimate, birds (95% CI):
45,000 (41,000–50,000)

Breeding Bird Survey Trend

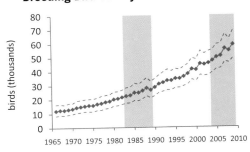

PILEATED WOODPECKER 241

Robert Royce/VIREO

Eastern Wood-Pewee
Contopus virens

The Eastern Wood-Pewee, easily identified by its distinctive, plaintive *pee-a-wee* song, is one of the most widespread and common tyrant flycatchers in Pennsylvania. It has been a regular, statewide nesting species as far back as records exist. Pewees breed throughout the eastern half of the United States to the Great Plains, and they thrive in virtually all of Pennsylvania's woodlands. Because a high proportion of the species' North American population breeds within the state, Pennsylvania has a high stewardship responsibility for maintaining Eastern Wood-Pewee numbers (Rosenberg 2004).

During the first Pennsylvania Atlas, the Eastern Wood-Pewee was found breeding in 88 percent of the state's blocks, encountered least frequently in heavily forested blocks in northern Pennsylvania and the most sparsely wooded blocks of the southeastern counties (Ickes 1992f). As in the first West Virginia atlas (Buckelew and Hall 1994), the pewee's distribution appeared to be negatively correlated with elevation; it was almost absent from some of the higher mountains in north-central Pennsylvania.

Data from the second Atlas indicate that the Eastern Wood-Pewee was still found in 88 percent of Pennsylvania's blocks, with no significant differences between atlas periods in any physiographic province or section (appendix C). Similar to first Atlas results, pewees were absent from some higher elevation blocks in the densely forested northern mountains of the state. Data collected for the second New York atlas indicate a decline of 22 percent in the Adirondacks, and it was suggested that maturation of forests in the Northeast may be affecting pewee populations negatively, as canopies close and create less desirable habitat (McGowan 2008s). The pewee was absent from large and treeless expanses in the Piedmont Province, notably the intensively farmed areas of the Cumberland Valley and central Lancaster County, as well as the most densely urbanized parts of Philadelphia and Pittsburgh. Elsewhere in southern Pennsylvania, there was a general but modest infilling, with increases in block records compared with the first Atlas in all physiographic sections (appendix C). The increase in development in many of those sections (chap. 3) may have created more forest clearings and edges, habitat with which the Eastern Wood-Pewee is frequently associated (McCarty 1996; Sallabanks et al. 2000).

Second Atlas point count data suggest that, in Pennsylvania, the Eastern Wood-Pewee does not exhibit any forest-area sensitivity, although it does tend to avoid coniferous forests (appendix D). Point count results also indicate that, as suggested in the first Atlas, there is a negative relationship between pewee distribution and elevation. The total Pennsylvania population for the Eastern Wood-Pewee during the second Atlas was estimated to be 315,000 singing males, considerably more than the Partners in Flight estimate of 210,000 (Rich et al. 2004).

Counts on Breeding Bird Survey routes have declined by an average of 1.7 percent per year, 1966–2009 (Sauer et al. 2011), but second Atlas results indicate that there has not yet been a concurrent range contraction. This could be because the species is still abundant, but if declines continue, it could be only a matter of time before there is evidence of localized range loss. Because of this, the pewee's outlook in Pennsylvania is secure in the medium term but uncertain in the long term, and further research on the effects of specific ecological factors on its numbers is needed. For example, it was found that high White-tailed Deer densities have a negative impact on the species in northwestern Pennsylvania (deCalesta 1994), while other studies have reported increases in Eastern Wood-Pewee populations after selected forest management practices were conducted (Campbell et al. 2007; Greenberg et al. 2007). Still other research has suggested that forest fragmentation may be influencing species such as the Eastern Wood-Pewee during migration or on the wintering grounds (Somershoe and Chandler 2004; Keller and Yahner 2006).

ROY A. ICKES

Distribution

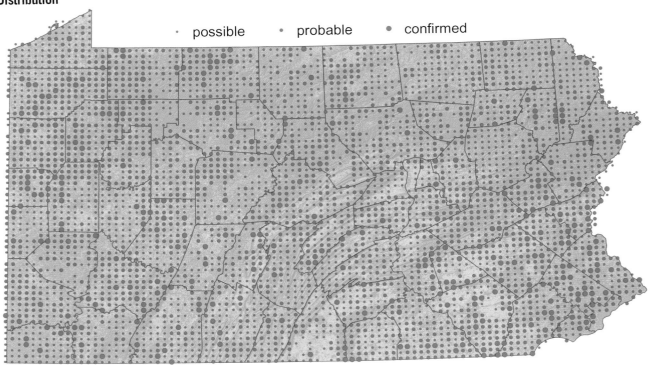

possible · probable · confirmed

Distribution Change

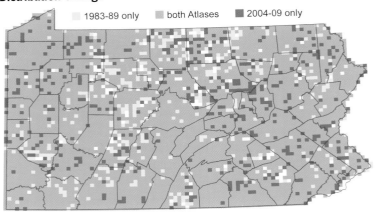

1983–89 only · both Atlases · 2004-09 only

Density

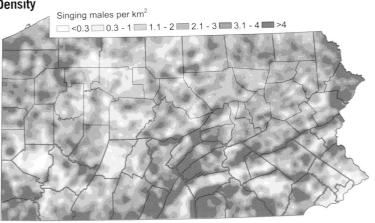

Singing males per km²
<0.3 · 0.3 - 1 · 1.1 - 2 · 2.1 - 3 · 3.1 - 4 · >4

Number of Blocks

	first Atlas 1983–89	second Atlas 2004–09	Change %
Possible	1,635	2,408	47
Probable	2,208	1,678	−24
Confirmed	510	302	−41
Total	4,353	4,388	1

Population estimate, birds (95% CI):
315,000 (300,000–300,000)

Breeding Bird Survey Trend

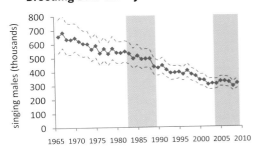

EASTERN WOOD-PEWEE

Douglas A. Gross/PGC

Yellow-bellied Flycatcher
Empidonax flaviventris

A member of the boreal conifer forest community, the Yellow-bellied Flycatcher is the most distinctly plumaged eastern *Empidonax* flycatcher, although it is not the only flycatcher with a yellow belly. The Yellow-bellied Flycatcher is more likely to be heard than seen in its remote, densely vegetated habitat, but its soft voice is often missed or confused with other species (Kauffman and Sibley 2002; Gross and Lowther 2011). Considering the increase in knowledge and awareness of this species, it is surprising that the second Atlas did not provide more reports, but this rare species is easily misidentified or overlooked.

Historically, the Yellow-bellied Flycatcher appears always to have been rare in Pennsylvania (McWilliams and Brauning 2000). The nesting population was rediscovered during the first Atlas, when there were records in only 13 blocks, all in northern counties (Gross 1992b). Primarily a breeding bird of the Canadian boreal forest, it has a range that extends south into the Great Lake states, New England, New York, and northern Pennsylvania, but formerly to North Carolina (Gross and Lowther 2011). The closest nesting population is in the Catskill Mountains. The Yellow-bellied Flycatcher is listed as Endangered in Pennsylvania because of its very small, vulnerable population.

The Yellow-bellied Flycatcher nesting population in Pennsylvania is confined to the Appalachian Plateaus, where it is only found in isolated pockets, mostly in the northeast. In the second Atlas it was again found in only 13 blocks, making it one of the state's rarest regularly nesting birds, especially since each block represents few nesting pairs. It was confirmed nesting only in three blocks, all in the Glaciated High Plateau Section, known as "North Mountain" (including "Dutch Mountain") of Sullivan and Wyoming counties. Coalbed Swamp, in western Wyoming County, is the state's stronghold (Gross 2010b). There were reports of singing males and a possible pair in 10 other blocks scattered across the plateaus, including in McKean and Monroe counties, but these most probably represent solitary males attempting to attract a mate. The Yellow-bellied Flycatcher is among the last species to return to their nesting ground and one of the first to begin their southbound migration, some beginning south in July (Gross and Lowther 2011). Some pairs are on territory as early as 24 May, but others as late as 20 June. This leads to confusion between passage migrants and breeders. A few pairs are double-brooded, with new nests found as late as 2 August (pers. obs.).

The breeding distribution of Yellow-bellied Flycatcher has not changed appreciably since the first Atlas, but it has not included the Poconos since the 1930s (Gross 1991, 2010b). The only continuously occupied locations during the second Atlas were in Sullivan and western Wyoming counties. Other occupied locations were abandoned within the second Atlas period (pers. obs.). Population increases have been found in Ontario and much of New York, but declines have occurred in New Hampshire's White Mountains and New York's Catskills (King et al. 2008; Crins 2007; Peterson 2008a). Pennsylvania comprises a minor part of the species' range, but it is the southernmost regularly occurring population in the Appalachian Mountains.

In Pennsylvania, the Yellow-bellied Flycatcher nests in mossy, cool conifer-forested wetlands, and often in headwater peat lands that are dominated by red spruce or hemlock that also contain many deciduous trees (Gross 2010b). The nesting habitat has a dense shrub and sapling mid-story with many ferns and herbaceous plants. Regularly used sites are in contiguous forests. Nests are among the most difficult to find of any American bird, since they are well concealed on or near the ground in moss, roots, fern clumps, logs, or tip-ups (Gross and Lowther 2011).

The Yellow-bellied Flycatcher persists in Pennsylvania despite habitat reduction from timbering, wetland destruction, and development (Gross 2010b). The formerly extensive conifer forest and its associated birds are still recovering from the timber extraction era, but they face new threats, including forest fragmentation, energy development, climate change, and tree pests (Rotenhouse et al. 2008; Johnson 2010). Some of these threats could be countered with habitat protection and spruce restoration efforts that would benefit a variety of boreal conifer species (Rentch et al. 2007; Gross 2010b).

DOUGLAS A. GROSS

Distribution

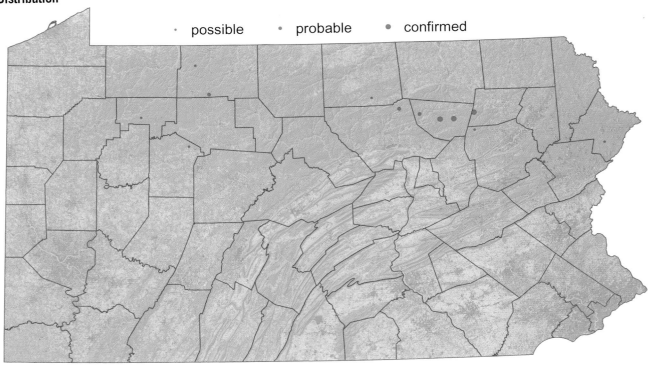

Distribution Change

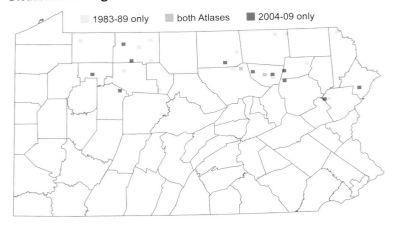

Number of Blocks

	first Atlas 1983–89	second Atlas 2004–09	Change %
Possible	5	7	40
Probable	6	3	−50
Confirmed	2	3	50
Total	13	13	0

Brian E. Small/VIREO

Acadian Flycatcher
Empidonax virescens

The Acadian Flycatcher is the least active of Pennsylvania's four small flycatchers; it often sits quietly on its perch, its wings slightly drooped and still. This quiet demeanor is intermittently interrupted with an explosive *peet-sah* territorial song, heard emanating from stream valleys and ravines across the state. Always difficult to identify unless singing, this species has the longest primaries and largest beak of the five *Empidonax* flycatchers found in Pennsylvania. The Acadian Flycatcher reestablished its range in Pennsylvania and the northeastern states in the early 1900s following a long absence, perhaps due to deforestation in the nineteenth century (Whitehead and Taylor 2002). Pennsylvania lies near the northern limit of the species' current distribution, which extends west to the Mississippi Valley and south to the Gulf Coast and central Florida. It occurs much more frequently in southwestern Pennsylvania on the edge of its core range in West Virginia and the central southeastern United States (Whitehead and Taylor 2002).

During the first Atlas, the Acadian Flycatcher was detected in 35 percent of blocks, the majority within the southwestern portion of the Appalachian Plateaus Province and the Appalachian Mountain Section in south-central Pennsylvania. Localized clusters of detections were found in Erie and Crawford counties, the lower Susquehanna River Valley, Philadelphia County, Lycoming County, and along Kittatinny Ridge (Blue Mountain), through the Delaware Water Gap National Recreation Area and adjacent Glaciated Pocono Plateau. Although the distribution pattern has remained similar, there was a statistically significant 39 percent increase in the number of blocks occupied during the second Atlas, similar to the 48 percent increase in the second New York atlas (Smith 2008b). Most of these blocks are located on the perimeter of previous concentrations and in more recently occupied habitat in the north-central and northwestern portions of the Appalachian Plateaus Province. The species' distribution pattern largely overlaps the distribution of eastern hemlock–dominated streamside habitat, except in southwestern Pennsylvania.

The Acadian Flycatcher's population in Pennsylvania was estimated at 149,000 singing males during the second Atlas. Breeding Bird Survey trend analysis indicates no change in population size in Pennsylvania from 1980 through 2007 in spite of the large range expansion since the first Atlas (Sauer et al. 2011). Counts of singing males were highest in large forest blocks, with 58 percent of detections occurring in mixed or hemlock-dominated forest. Studies across Pennsylvania have shown a pronounced preference for hemlock relative to overall tree species availability in forest stands (Ross et al. 2004; Becker et al. 2008; Allen et al. 2009). However, in some southern portions of the state, the Acadian Flycatcher is found in a wider variety of forest types. Interspecific competition with the closely related Least Flycatcher may help define their respective niches where ranges overlap, as in parts of central and northern Pennsylvania. In the southeastern United States, the Acadian Flycatcher prefers deciduous forest and bald cypress groves (Hamel et al. 1982). Regardless of forest composition, it prefers a closed canopy and open-to-moderate understory (Whitehead and Taylor 2002).

Even though the Acadian Flycatcher's population is currently stable, there is concern that hemlock mortality, caused by Hemlock Woolly Adelgid infestation, will negatively impact its population in the future, especially in the northern and northeastern sections of the state. The infestation has progressed across the eastern two-thirds of Pennsylvania and has reached the Ohio border in Beaver County. Acadian Flycatcher density declined significantly with increasing infestation levels, according to a study comparing nesting success over a range of infestation levels across the state (Allen et al. 2009; Allen and Sheehan 2010). It is hoped that the population expansion detected during the second Atlas will help counteract such declines. Even where hemlocks are preferred, the Acadian Flycatcher occasionally nests in deciduous trees and shrubs, providing evidence of adaptability to changing environmental conditions.

TERRY L. MASTER

Distribution

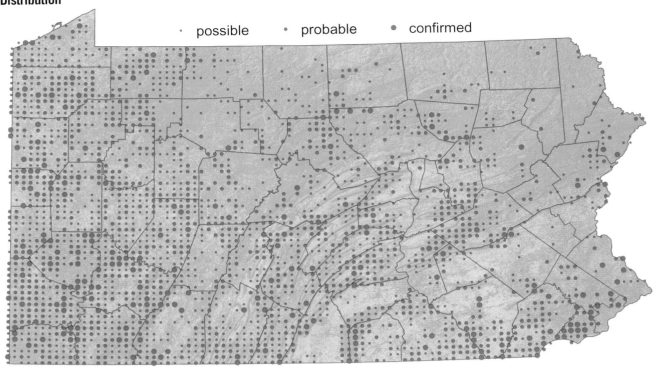

Distribution Change

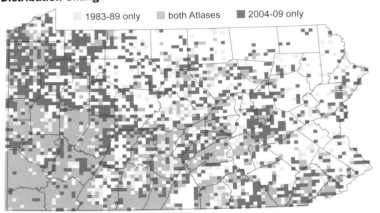

Density

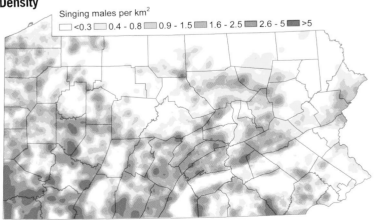

Number of Blocks

	first Atlas 1983–89	second Atlas 2004–09	Change %
Possible	704	1,336	90
Probable	884	909	3
Confirmed	178	209	17
Total	1,766	2,454	39

Population estimate, males (95% CI):
149,000 (140,000–157,000)

Breeding Bird Survey Trend

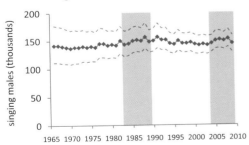

ACADIAN FLYCATCHER

Bob Wood and Peter Wood

Alder Flycatcher
Empidonax alnorum

The Alder Flycatcher is one of the five small, dull-colored flycatchers in the genus *Empidonax* breeding in Pennsylvania that are the bane of most birders. Until 1973, it was considered conspecific with the Willow Flycatcher; unfortunately, its separation as a distinct species did not make it any more separable in the field, except by voice. The alder has a distinctively three-noted song with the accent on the second syllable, *fee-BEE-oh*, while the willow's is a sneezy, two-syllabled *FITZ-bew*. The respective call notes differ as well: the alder's sharp *pip* recalls a Downy Woodpecker and contrasts with the willow's mellifluous *whit* (Lowther 1999). The Alder Flycatcher breeds in a variety of wet shrubby habitats in Pennsylvania, including shrub swamps, alder bogs, and regenerating clear-cuts (Stoleson 2010). In areas of sympatry with the Willow Flycatcher, the Alder Flycatcher prefers denser, wetter, and woodier habitats (Barlow and McGillivray 1983).

The most boreal of the *Empidonax*, the Alder Flycatcher breeds from northern Alaska to Labrador, south to southern Canada, the Great Lakes states, and northern New England. Scattered, isolated populations occur at high elevations in the Appalachian Mountains south to North Carolina and Tennessee (Lowther 1999). Northern Pennsylvania represents the southern edge of the core breeding range and the northern limit of Appalachian populations. The species winters in northern and central South America, further south than our other *Empidonax* (Stotz et al. 1996).

The second Atlas reports Alder Flycatchers from 18 percent of blocks statewide, most of which were located in the glaciated sections of the Appalachian Plateaus. Alders were scarce below 250 m (820 ft) and over 700 m (2,300 ft) elevation (appendix D), as well as in the southern two-thirds of the state. Although considered a species of Conservation Concern in Pennsylvania (PGC-PFBC 2005), the Alder Flycatcher currently appears to be thriving. The 868 occupied blocks represent a 161 percent increase over the first Atlas, with a particularly marked expansion in blocks with Alder Flycatchers in the Ridge and Valley Province (+261%) and the High Plateau (+414%) and Deep Valleys (+344%) sections of the Appalachian Plateaus. Breeding Bird Survey (BBS) data indicate that the Commonwealth's Alder Flycatcher population grew by more than 200 percent since the first Atlas (Sauer et al. 2011), to an estimated 28,000 singing males. Neighboring populations in New York State also grew by 45 percent (Post 2008c), although BBS data reveal no trends for the Northeastern states generally.

Although there has been a considerable range expansion by the Alder Flycatcher, the highest population densities were found within the same areas occupied in the first Atlas. About a third of all point count observations came from Erie and Crawford counties alone, due in part to the extensive wetlands there (Tiner 1990).

The increase in Alder Flycatchers documented in the second Atlas suggests cautious optimism for this species' future in Pennsylvania. Caution is warranted because of this bird's dependence on wooded wetland habitats, which continue to be threatened by habitat degradation and conversion, especially in the glaciated northwest, with its disproportionate share of the state's Alder Flycatchers (Goodrich et al. 2002). Further, the extent of early successional habitats in general has declined across the northeastern states because of changes in forestry and agricultural practices, urban sprawl, wetland loss, and other factors (Goodrich et al. 2002; Brooks 2003). Despite these breeding habitat issues, the alder has increased in the state, suggesting it may be more resilient to anthropogenic effects on habitat than most other early successional species. Concerns on the tropical wintering grounds may be less of an issue for the Alder Flycatcher than for most neotropical migrants, as it prefers non-forest scrub and edge habitats there (Stotz et al. 1996).

SCOTT H. STOLESON

Distribution

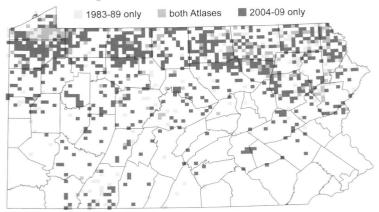

Distribution Change

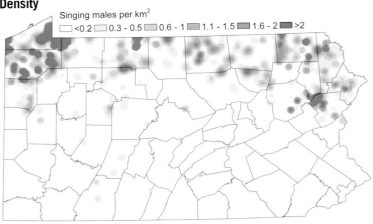

Number of Blocks

	first Atlas 1983–89	second Atlas 2004–09	Change %
Possible	181	521	188
Probable	136	313	130
Confirmed	15	34	127
Total	332	868	161

Population estimate, males (95% CI): 28,000 (22,000–34,000)

Density

Singing males per km^2: <0.2, 0.3 - 0.5, 0.6 - 1, 1.1 - 1.5, 1.6 - 2, >2

ALDER FLYCATCHER 249

Bob Wood and Peter Wood

Willow Flycatcher
Empidonax traillii

Lumped for nearly a century and a half with its virtually identical relative the Alder Flycatcher as "Traill's Flycatcher," the Willow Flycatcher was finally split in 1973 because of its distinctive *fitz-bew* song. It also tolerates drier, more upland habitat than the Alder Flycatcher, although in Pennsylvania it is most often found in brushy wetlands, wet meadows, and streamside thickets below 700 m (2,300 ft) in elevation (Sheehan 2010; appendix D).

The Willow Flycatcher has the widest breeding range of any *Empidonax* flycatcher (Sedgwick 2000), occurring from New Brunswick south into the Appalachians, west to British Columbia, and into the Southwest, where the subspecies *extimus* is listed as federally Endangered. The confusion over its identity makes tracking the Willow Flycatcher's history in Pennsylvania a challenge, but it appears to have been seen only in migration until 1894, when the first "Traill's" nest (most likely that of a Willow Flycatcher) was discovered in Allegheny County (Mulvihill 1992b).

The Willow Flycatcher subsequently overspread much of the Commonwealth in the twentieth century, probably in response to reverting farmland and regenerating woodland (McWilliams and Brauning 2000). The first Atlas documented this expansion, a trend that continued, although not as dramatically as for the Alder Flycatcher, in the two decades thereafter. Breeding Bird Survey (BBS) data, which showed a rapid increase from 1970 through the mid-1980s (Sheehan 2010), have since stabilized (Sauer et al. 2011), while second Atlas data showed a statistically significant increase of 23 percent in Willow Flycatcher block detections across the state. An increase in Willow Flycatcher detections in atlas blocks, with only modest concurrent increases in BBS counts, was noted in neighboring New York state (Post 2008d) and Maryland (Dawson 2010) over a similar time span.

Large increases in detections were apparent across the southeastern part of the state, with a 69 percent increase in block occupancy in the Piedmont and a 45 percent increase in the Ridge and Valley Province. Elsewhere, there was a significant range expansion in the Northwestern Glaciated Plateau Section, increasing from an occupancy rate of 44 percent to 77 percent, the highest region in the state (appendix C). Conversely, in the southwestern corner of the state there were significant declines in the numbers of blocks with Willow Flycatchers in the Allegheny Mountain (−31%) and Waynesburg Hills (−23%) sections.

Greatest densities of singing males were documented in the brushy wetlands of northwestern Pennsylvania, in counties including Mercer, Lawrence, and Erie, where the Willow Flycatcher broadly overlaps with Alder Flycatcher, followed by the agricultural landscape of the Great Valley and southeastern Pennsylvania, Chester, Lebanon, Adams, and Berks counties in particular. The population of singing males was estimated at 135,000 statewide, with Lancaster and York counties together accounting for more than 8 percent of that total. However, Willow Flycatchers remain rare in the highest, most heavily forested regions, with the High Plateau, Deep Valleys, and Glaciated High Plateau sections largely unoccupied.

Given the declines in many young forest and wetland specialists, the Willow Flycatcher's steady increase in the Northeast is unusual and heartening, although any species that depends upon wetlands should be viewed with some caution, and the flycatcher is considered a species of Maintenance Concern under the Pennsylvania Wildlife Action Plan (PGC-PFBC 2005). In appropriate habitat, however, the Willow Flycatcher can be found in high densities, and its preference for disturbed habitat on wintering grounds in Latin America (Sedgewick 2000) may also give it a survival advantage over many neotropical migrants.

SCOTT WEIDENSAUL

Distribution

possible · probable · confirmed

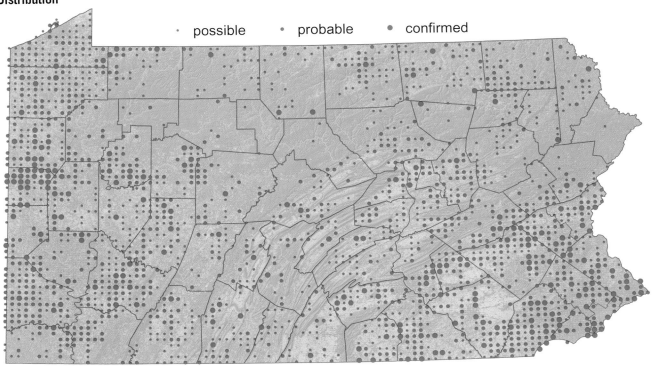

Distribution Change

1983–89 only · both Atlases · 2004-09 only

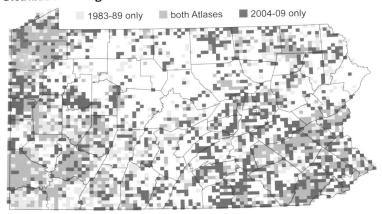

Number of Blocks

	first Atlas 1983–89	second Atlas 2004–09	Change %
Possible	670	995	49
Probable	832	815	–2
Confirmed	150	216	44
Total	1,652	2,026	23

Population estimate, males (95% CI):
135,000 (126,000–144,000)

Density

Singing males per km^2

<0.3 · 0.3 - 1.1 · 1.2 - 1.8 · 1.9 - 2.6 · 2.7 - 3.5 · >3.5

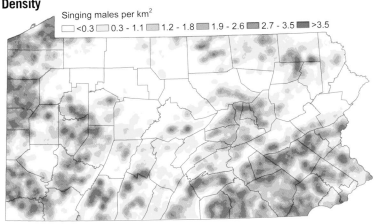

WILLOW FLYCATCHER   251

Rob Curtis/VIREO

Least Flycatcher
Empidonax minimus

The Least Flycatcher is the smallest of Pennsylvania's five, frustratingly similar *Empidonax* flycatchers. While the size distinction may be useless for field identification, the bird's interminably repeated, hiccup-like *chebec* song aptly identifies it. The species breeds in a variety of open deciduous and mixed forests, woodland edges, orchards, and wooded suburbs (Mulvihill 1992c). Where found, they often occur in close proximity to one another, while seemingly identical habitat nearby remains unoccupied; this clustering appears to help deter predators, attract females, or both (Tarof and Briskie 2008; Perry et al. 2008).

The Least Flycatcher breeds from northwestern Canada and southern Newfoundland south to the northern Rocky Mountains, Great Lakes states, and southern New England. Northern Pennsylvania lies at the southern margin of the main breeding range (Tarof and Briskie 2008). In the second Atlas, the species occurred primarily across the northern highlands and in the Allegheny Mountains, with scattered records from the Pittsburgh Low Plateau Section and Ridge and Valley Province, much as in the first Atlas. Overall occupancy of 1,739 blocks was down by 4 percent since the first Atlas, but when corrected for increased survey effort, the decrease was a statistically significant 6.6 percent. As a result, the Acadian Flycatcher overtook the Least Flycatcher to become the state's most widespread *Empidonax*. Block occupancy in the northern half of the state was relatively little changed, but there was a considerable thinning of the species' range in the Pittsburgh Low Plateau, the Ridge and Valley Province, and most of the periphery of its distribution. These changes are strongly suggestive of a northward range contraction, which is supported by decreases in atlas block counts in Maryland (Ellison 2010t), while there was no statistically significant range change in New York (McGowan 2008t).

The highest density of Least Flycatchers was found in extensively forested areas of the Appalachian Plateaus; they were sparse below 400 m (1,300 ft) elevation (appendix D). Breeding Bird Survey data indicate that Least Flycatcher populations in the state have declined by 1.7 percent annually between the first and second atlas periods, for a total drop of 33 percent, with a similar decline across its entire range (Sauer et al. 2011). The statewide population was estimated to be approximately 138,000 singing males, slightly lower than that of the very similar Acadian Flycatcher, which has shown range expansion in many areas in which the Least Flycatcher is in retreat. Although there is no evidence of a causal relationship, the contrast in fortunes of these species, whose respective northern and southern range limits pass through Pennsylvania, is intriguing.

Most forest birds of northern affinities appear to be increasing in abundance in Pennsylvania (Stoleson and Larkin 2010; chap. 6). The Least Flycatcher stands out as an exception to that trend, and the causes for its decline in the state (and across its range) are unclear. Holmes and Sherry (2001) reported that over a 30-year span at the relatively undisturbed Hubbard Brook Experimental Forest in New Hampshire, the Least Flycatcher went from being the most abundant bird to local extirpation. They implicated successional changes in vegetation structure as the forest matured. Pennsylvania's forests have matured since the first Atlas (McWilliams et al. 2007), so concurrent structural changes might explain these population trends. However, Guénette and Villard (2005) found the Least Flycatcher to be relatively insensitive to changes in habitat structure. It is unclear whether these differences in apparent response might be due to local variation, study design, or other factors.

Alternatively, recent work suggests that most North American birds feeding on aerial insects are experiencing widespread declines (Nebel et al. 2010). Such declines probably reflect widespread declines in populations of aerial insects and may indicate large-scale ecosystem changes resulting from climate change, intensified agriculture, and/or acid deposition (Nebel et al. 2010). The broad range of habitats used by Least Flycatchers on their wintering grounds (Stotz et al. 1996; Tarof and Briskie 2008) suggests that habitat issues there are unlikely to limit populations.

SCOTT H. STOLESON

Distribution

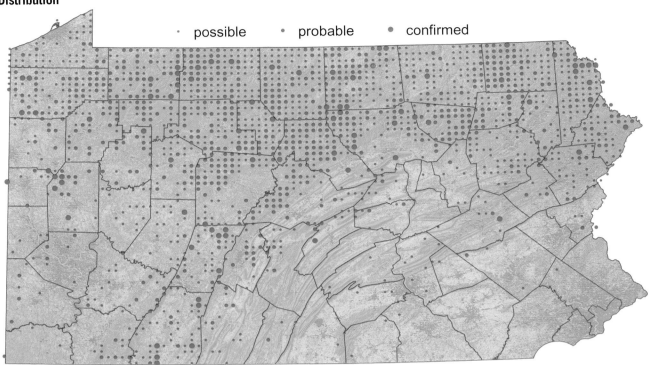

Distribution Change

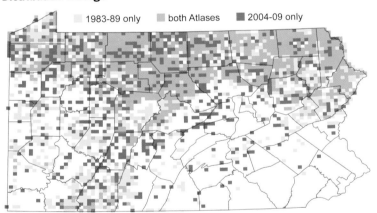

Density

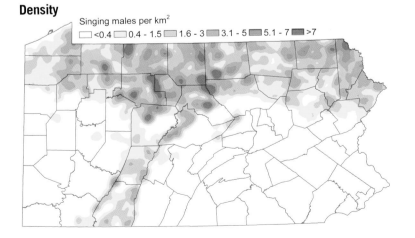

Number of Blocks

	first Atlas	second Atlas	Change
	1983–89	2004–09	%
Possible	920	1,050	14
Probable	770	607	−21
Confirmed	126	82	−35
Total	1,816	1,739	−4

Population estimate, males (95% CI):
138,000 (130,000–148,000)

Breeding Bird Survey Trend

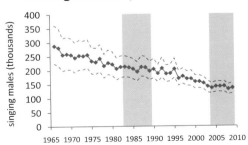

LEAST FLYCATCHER   253

Rob Curtis/VIREO

Eastern Phoebe
Sayornis phoebe

A medium-sized, drab flycatcher, the Eastern Phoebe is the first of the flycatchers to return to Pennsylvania in spring. By the end of March, it is present statewide (Haas and Haas 2005). The species breeds from eastern Canada northwest to British Columbia, the eastern and central United States, south to Georgia, and west to Texas. Its winter range includes the southern part of the breeding range and south into central and eastern Mexico (Weeks 1994).

Having an affinity for building nests on man-made structures, the Eastern Phoebe is found in close proximity to humans and is easily observed. Exhibiting typical flycatcher behavior, it often perches on an exposed branch, sign, or building, calling repeatedly and characteristically bobbing its tail. Picnic pavilions in parks, barns, sheds, porches, and bridges are favored sites that provide horizontal surfaces for nest placement, shelter from weather, and open space that allows easy access to nests. The species also uses natural rock outcroppings on hillsides and along streams, where insects are an abundant food resource. It exhibits strong site fidelity and will occupy the same nest location for successive years (Weeks 1994).

The Eastern Phoebe is usually double-brooded (Weeks 1994), and the range of atlas safe dates (1 May to 31 July) attests to its rather prolonged breeding season. Combined with accessible nest sites, fidelity to these sites, and statewide distribution, opportunities for breeding confirmation were high. Nest occupation was noted as early as 22 March, with a prolonged peak in nesting activity from late May to late July.

The Eastern Phoebe was recorded in 90 percent of blocks during the first Atlas period and in 95 percent during the second Atlas period. It expanded into areas where occupancy was low in the first Atlas, such as in blocks along the Lake Erie shore, urbanized Allegheny County, and intensively farmed central Lancaster County. As a result, the urban cores of Philadelphia and Pittsburgh were the only areas in which phoebes were absent from several contiguous blocks in the second Atlas. Point count data suggest that densities are rather uniform across the state, with lower densities in extensively forested, urbanized, or farmed landscapes, and higher densities in interspersed landscapes typical of the Ridge and Valley Province. Densities were lower at elevations over 600 m (~2,000 ft), likely because high-elevation areas are dominated by contiguous forests; point count data suggest that this species is most abundant where 40 to 70 percent of the immediate landscape is forested (appendix D).

The Eastern Phoebe's statewide population was estimated to be 315,000 singing males, tying with Eastern Wood-Pewee as the most abundant of our nine breeding flycatcher species. Breeding Bird Survey counts from Pennsylvania show that this species' population has fluctuated over time but with no overall trend (Sauer et al. 2011). Sharp declines were noted in 1996, 2003, and 2007, but the population quickly recovered. These periodic declines have previously been attributed to cold winters in the species' southeastern United States winter quarters (Robbins et al. 1986). The Eastern Phoebe remains a common member of the avian community in Pennsylvania. Its acceptance of man-made structures for nesting, as well as its versatility in also using natural sites, including rock outcrops and cliff faces, is undoubtedly an advantage.

Species closely associated with humans, like the Eastern Phoebe, may provide educational opportunities for conservationists. An explanation of the benefits of a nesting pair, such as consuming bothersome insects, may ease concerns of homeowners nervous about the potential mess and disruption caused by a bird's nest on their property. Close observation of the nesting cycle of any bird often results in a feeling of ownership, interest, and pride in homeowners, who then become excited to share their experiences with others.

MIKE FIALKOVICH

Distribution

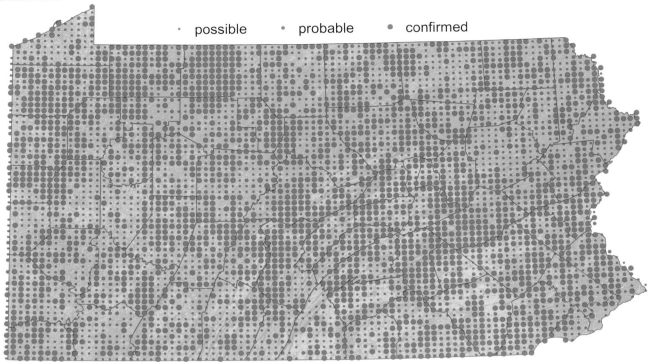

Distribution Change

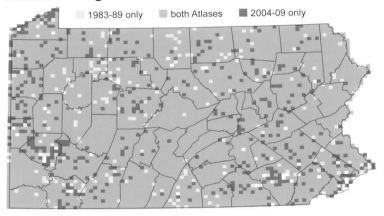

Density

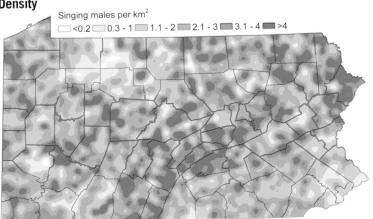

Number of Blocks

	first Atlas 1983–89	second Atlas 2004–09	Change %
Possible	861	1,560	81
Probable	854	858	0
Confirmed	2,731	2,247	–18
Total	4,446	4,665	5

Population estimate, males (95% CI):
315,000 (308,000–323,000)

Breeding Bird Survey Trend

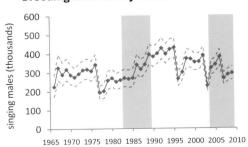

Bob Wood and Peter Wood

Great Crested Flycatcher
Myiarchus crinitus

The only cavity-nesting flycatcher east of the Mississippi, the Great Crested Flycatcher feeds high in the woodland canopy and announces its presence with a loud, distinctive call. The species breeds throughout the eastern half of the United States, and in Pennsylvania it nests in a wide variety of woodland habitats, particularly where dead trees are available. Open areas such as wooded suburbs, old orchards, and small woodlots are favorite sites for the Great Crested Flycatcher.

Historical records show the Great Crested Flycatcher to have been a fairly common breeder in much of Pennsylvania (Poole 1964), except in higher elevations of the northern counties, where it was either absent (Stone 1900; Keim 1905) or considered to be a rare breeder (Dwight 1892). First Atlas results indicated that the species remained widely distributed throughout Pennsylvania, though found in fewer blocks in heavily forested regions (Ickes 1992g). Historically, the Great Crested Flycatcher was reported to be most common along the wooded slopes of the mountain ridges east of the Allegheny Front (Todd 1940), and first Atlas records found that this was still the case.

The second Atlas again demonstrated that the Great Crested Flycatcher remains widely distributed throughout Pennsylvania, showing a relatively unchanged number of blocks between atlas periods, given increased effort in the second Atlas. However, changes in block occupancy varied greatly across the state, with losses in the higher elevations of the north and west and filling in of the range in the southeast, especially in the Piedmont, where block occupancy increased significantly, from 73 percent in the first Atlas to 89 percent in the second. The cause of the apparent increase in the southeast may, in part, be related to the die-off of Eastern Hemlocks due to Hemlock Woolly Adelgid infestation (chap. 3). Great Crested Flycatchers are among those species that may benefit from the increased number of dead trees and canopy gaps associated with adelgid infestation (Becker et al. 2008).

Similar to historical accounts and first Atlas results, the Great Crested Flycatcher was reported less frequently in the heavily forested northern counties of the state during the second Atlas. The localized loss of this species in some of those areas between atlases is somewhat puzzling but may be due to increased forest cover (chap. 3). Point count data show that this species is most abundant in highly interspersed landscapes (appendix D), which have been lost in some higher elevation areas due to forest regrowth on former agricultural lands.

In keeping with this species' preference for mixed landscapes, it is most abundant in the Ridge and Valley Province, the lower Susquehanna Valley, and the northwest corner of the state. High densities at relatively high elevations in Pike County are somewhat anomalous, but they may reflect the expansion of low-density housing developments around the county's many lakes (USBC 2011); second Atlas point count data show that forested residential areas and open water are favored habitat components for this species (appendix D).

The Pennsylvania population of the Great Crested Flycatcher during the second Atlas period was estimated to be 196,000 individuals. Extrapolating from Breeding Bird Survey trends, which show a 1.5 percent annual decrease from 1966 to 2009 (Sauer et al. 2011), there was an estimated 32 percent decline in population size between atlas periods. Even though BBS data suggest that the Great Crested Flycatcher is declining in Pennsylvania, second Atlas results show that there has been no overall range loss. In neighboring Maryland (Ellison 2010u) and New York state (McGowan 2008u), there has been little change in either numbers or distribution, which suggests that the species' prospects in the wider region are secure. Localized population losses in Pennsylvania could be mitigated by supplying nest boxes (White and Seginak 2000; Miller 2010) or ensuring the retention of minimal woody cover (>14%) in the landscape (Perkins et al. 2003), but because this species remains widespread and common, the Great Crested Flycatcher is not a species of conservation concern.

ROY A. ICKES

Distribution

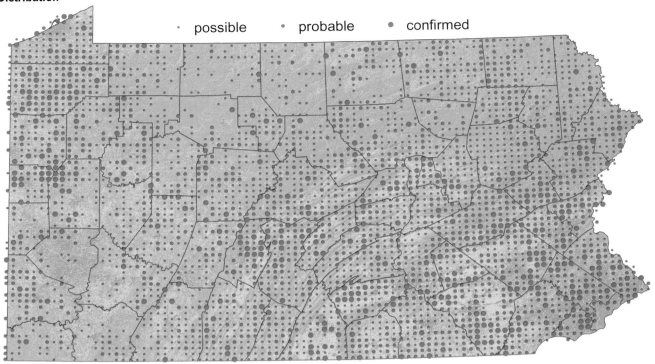

Distribution Change

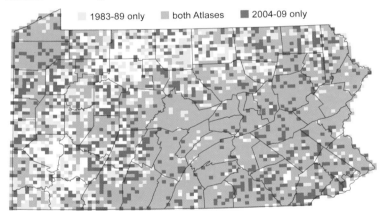

Density

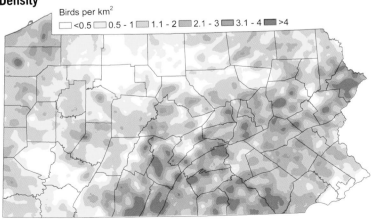

Number of Blocks

	first Atlas 1983–89	second Atlas 2004–09	Change %
Possible	1,561	2,116	36
Probable	1,501	1,109	−26
Confirmed	431	357	−17
Total	3,493	3,582	3

Population estimate, birds (95% CI):
196,000 (184,000–212,000)

Breeding Bird Survey Trend

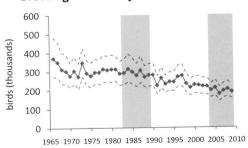

GREAT CRESTED FLYCATCHER

Geoff Malosh

Eastern Kingbird
Tyrannus tyrannus

This large black-and-white flycatcher is often encountered in open country, where it has the distinctive habit of behaving aggressively toward almost any other birds in its territory, hence its English and scientific names. It breeds throughout most of the United States and southern Canada, absent only from the west coast, the desert southwest, and western Rockies, and spends the winter in South America. Because of its wide distribution, Pennsylvania supports a very small proportion of the global population (Rich et al. 2004).

The Eastern Kingbird is the only one of the nine flycatcher species breeding in Pennsylvania that is tied largely to agricultural areas, and as such, it probably benefited from the destruction of primary forests during the nineteenth century. Since that time, its status as a fairly common bird of open areas but largely absent in densely forested country appears to have changed little. It is numerous in the southeastern part of the state, where it frequently is found in suburban and even urban parks. Interestingly, it does not appear to use similar habitats in the west, notably around Pittsburgh (McWilliams and Brauning 2000), where it was absent during the first Atlas.

There has been little statewide change in the number of occupied blocks since the first Atlas; a loss of habitat in areas peripheral to the densly forested plateau and Poconos appears to have compensated for the detection in more blocks in south-central areas. Correcting for increased effort, there was an estimated 4 percent decrease in block occupancy between atlases. The only significant change in block occupancy among the 23 physiographic sections (appendix C) was a 27 percent decrease in the High Plateau, in which there has been some reforestation of former farmland since the first Atlas. The probability of finding Eastern Kingbirds drops off rapidly in atlas blocks that are more than 80 percent forested, as is the case with many blocks from which this species was absent in the Laurel Highlands, Allegheny National Forest, and Sproul and Moshannon state forests.

Point count data confirm that the highest densities are found in agricultural areas, particularly those with pasture, as well as around water and wetlands (appendix D). Breeding Bird Survey data show that the Eastern Kingbird has been in steady decline since the 1960s, losing around one-third of its population in 40 years (Sauer et al. 2011). Similar declines in population size, at a time when range size has been stable, have been noted in New York (McGowan 2008v) and Maryland (Ellison 2010v). Murphy (2001) showed that loss of pasture and grazed land was driving local declines in New York state. While the size of the statewide cattle herd in Pennsylvania has been relatively unchanged in recent years, the local distribution of cattle has changed considerably, with an 80 percent reduction in the number of farms with herds of fewer than 50 cattle between 1987 and 2007 (USDA-NASS 2008). This aggregation of cattle into fewer, larger farms has resulted in a widespread loss of grazed pasture and mixed farming, which has undoubtedly reduced prime kingbird foraging habitats in some areas.

Due to this species' conspicuous nature, it may appear to be more abundant than it actually is, but second Atlas point count data suggest that it is only the sixth most numerous flycatcher in the state, with an estimated population of 110,000 individuals. Estimated densities are highest in the low-lying agricultural areas of the Piedmont, with more than a third of the statewide population south of Kittattiny Ridge. Densities are also high in agriculturally dominated areas in the northwestern and northeastern corners of the state. Despite modest declines in population size, this species remains widespread and moderately common in Pennsylvania.

ANDREW M. WILSON

Distribution

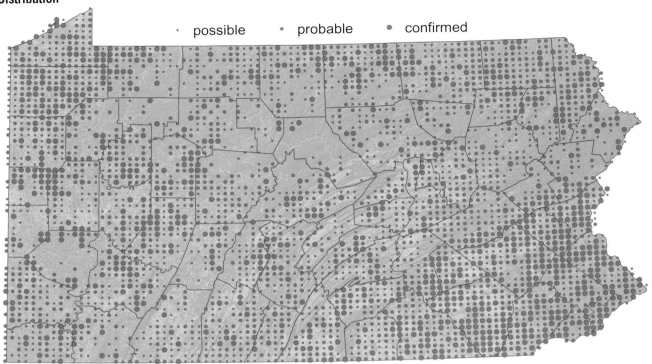

Distribution Change

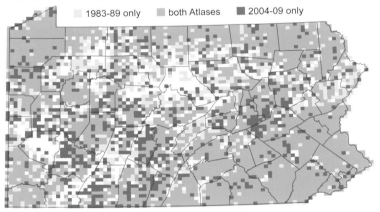

Density

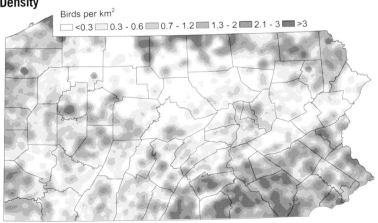

Number of Blocks

	first Atlas 1983–89	second Atlas 2004–09	Change %
Possible	1,147	1,454	27
Probable	1,268	1,133	−11
Confirmed	1,026	889	−13
Total	3,441	3,476	1

Population estimate, birds (95% CI):
110,000 (102,000–120,000)

Breeding Bird Survey Trend

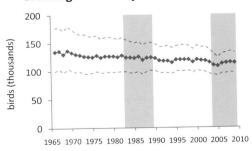

EASTERN KINGBIRD

Denny Bingaman

Loggerhead Shrike
Lanius ludovicianus

When fieldwork for the first Pennsylvania Atlas concluded in 1989, summering Loggerhead Shrikes had been recorded in six blocks in the south-central region of the state but had eluded confirmation as a breeder. At the time, the last known nest had been recorded in Crawford County in 1937 (Brauning 1992g). As it turned out, those sightings during the first Atlas were a precursor of things to come. In 1990, two shrikes were observed in Adams County, one of which was carrying nesting material (Kennell and Kennell 1990). Loggerhead Shrikes were at last confirmed as breeders when a nest with eggs was found in Adams County on 3 May 1992 (Kennell 1992).

Over the next 10 years, the Pennsylvania Game Commission, with help from local birders, conducted surveys in Adams and Franklin counties to search for breeding shrikes (Brauning and Siefken 2006). Surveyors found nests in Adams County every year from 1992 to 1999, and in Franklin in 1993, 1994, and 1996. At their peak in 1994, five nesting pairs were found between both counties. No nests were found after 1999, and the survey was concluded in 2002. Shrikes were reported occasionally in the winter and early spring up until April 2001 (Robinson 2001).

There were only three Loggerhead Shrike sightings during the second Atlas, but two proved to be quite significant. In February 2007, a singing Loggerhead Shrike was found in Adams County very close to one of the Adams nest sites from the 1990s (Robinson 2007). One or two birds were seen periodically until May 2008, adding the species to the second Atlas dataset as a possible breeder. Much more significantly, a Loggerhead Shrike nest, photographed with four young, had been reported to the second Maryland atlas in a block overlapping Pennsylvania. In fact, this observation was within Franklin County, Pennsylvania, in May 2004 (Davidson 2010b). This nest became the first and only confirmation of breeding shrikes during either Pennsylvania atlas effort.

Loggerhead Shrikes prefer to nest low to the ground in dense trees or shrubs, often using species with thorns (Yosef 1996). During the 1990s in Adams and Franklin counties, shrike nests were observed in hawthorns and Eastern Red Cedar (Brauning and Siefken 2000), but they also used other plants, including a Norway Spruce (Henise and Henise 1994). The 2004 nest in Franklin County was found in a magnolia tree 4 m (12 ft) above the ground.

The decline of Loggerhead Shrike in northeastern North America in the first half of the twentieth century is well documented (Yosef 1996; Brauning 1992g). Historically, the Loggerhead Shrike was found primarily in northwestern Pennsylvania (McWilliams and Brauning 2000). In the first Atlas, when shrikes began to appear in south-central counties, it was noted that they had no previous history of breeding in the region (Brauning 1992g). However, Loggerhead Shrikes were documented in Maryland adjacent to south-central Pennsylvania during both Maryland atlases (Davidson 2010). Tellingly, the second Maryland atlas showed a decline in shrike records from 13 blocks to 3, which corresponds closely to the shrike's decline in Adams and Franklin counties since their heyday in the mid-1990s. It seems reasonable to conclude that the recent Pennsylvania population of shrikes is an extension of a small population that may still persist in central Maryland and northern Virginia.

The Loggerhead Shrike is currently listed as Endangered in Pennsylvania. In 1998, the Pennsylvania Game Commission attempted habitat enhancement at three promising sites in Adams County, which consisted of planting up to five Washington Hawthorns and five Eastern Red Cedars at each site (Brauning and Siefken 2006). Unfortunately, the effort was too little, too late, since the species disappeared from the area after the next year. However, the shrike's reappearance in 2004 and 2007 is reason for hope. Indeed, in the summer of 2010, Loggerhead Shrikes were found in a new area in Cumberland County (Gauthier 2010), evidence that the species may yet persist in the region, however tenuously.

GEOFF MALOSH

Distribution

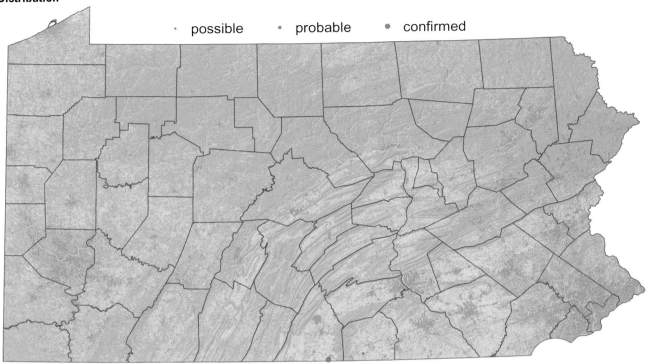

Distribution Change

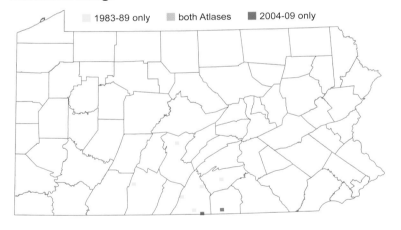

Number of Blocks

	first Atlas 1983–89	second Atlas 2004–09	Change %
Possible	4	1	−75
Probable	2	0	−∞
Confirmed	0	1	∞
Total	6	2	−67

Bob Wood and Peter Wood

White-eyed Vireo
Vireo griseus

While some find it difficult to separate the songs of other North American vireos, the White-eyed Vireo's distinct chattering song is easily distinguished. The complicated and varied song is learned from fathers and, to a lesser degree, neighboring males (Kroodsma and Baylis 1982), and the white-eye often integrates notes from other species into its song repertoire (Adkisson and Conner 1978). True to its name, the White-eyed Vireo has striking white eyes that give it an intense, piercing, almost human gaze. It is an alert and responsive bird that follows and chatters to scold predators.

The White-eyed Vireo breeds in the eastern United States and south into coastal Mexico. Pennsylvania forms the northern limit of its breeding range, where it occurs across the southern half, and in most of the western third, of the state. The White-eyed Vireo was historically rare in Pennsylvania until about the 1950s, when documented breeding records increased (Leberman 1976; Wood 1983) a few decades after the species experienced a northward range expansion (Leberman 1976; Hopp et al. 1995).

Between the two atlas periods, there was little overall change in the White-eyed Vireo's range size in Pennsylvania, but there were some pronounced changes at a local scale. The number of blocks with this species almost tripled in the Northwestern Glaciated Plateau Section, but decreases were apparent in the southeastern and central areas of Pennsylvania. The species was still found predominantly at lower elevations and is rarely encountered at elevations over 400 m (1,300 ft; appendix D). White-eyed Vireos prefer secondary deciduous scrub, abandoned agricultural land, riparian thickets, wooded wetlands, and woodland edges (Bent 1950), especially with thick, shrubby undergrowth and low trees interspersed throughout (James 1971; Hopp et al. 1995). In the southeastern part of the state, expanding urbanization in the Philadelphia area likely has converted White-eyed Vireo habitat into urban and suburban areas dominated by man-made structures with little appropriate shrubby, successional habitat. Farmland abandonment has creating ideal White-eyed Vireo breeding habitat, and this could account for the increase in occupied blocks in Somerset, Bedford, Fulton, Franklin, and Adams counties. The highest densities on point counts were found in wetland habitats, but transitional and shrubby areas in deciduous forests support the bulk of the state's populations overall (appendix D).

The White-eyed Vireo is a relatively short-distance migrant, wintering in the southeastern United States, eastern Mexico, and the Caribbean (Hopp et al. 1995). In the spring, the White-eyed Vireo arrives in Pennsylvania in late April and early May (Leberman 1976) and sets up territories in appropriate deciduous scrub habitat. Females choose mates and nest sites, then both adults build the nest, incubate, and brood (Hopp et al. 1995). The song of the White-eyed Vireo can be heard well into midsummer, when most other migrant songbirds have fallen silent. Atlas volunteers found White-eyed Vireos beginning to nest in late April, and fledglings were noted mostly in late June through early August.

An estimated 34,000 singing male White-eyed Vireos occurred in Pennsylvania during the second Atlas. Breeding Bird Survey data are sparse, but they indicate a shallow increase in White-eyed Vireos in Pennsylvania since 1966 (Sauer et al. 2011), with counts approximately 18 percent higher in the second Atlas period than in the first. Predation, reduced availability of suitable breeding habitat, and Brown-headed Cowbird parasitism may be limiting on the breeding grounds, but, because it is a habitat generalist during the winter, the White-eyed Vireo is unlikely to be limited by habitat changes on the wintering grounds (Hopp et al. 1995). Although it is probably not at risk of experiencing serious declines in population numbers (Reed 1992), it would be prudent to conserve scrubby, transitional, and wooded wetlands habitats, which are important for several declining birds in Pennsylvania.

ANDREA L. CRARY

Distribution

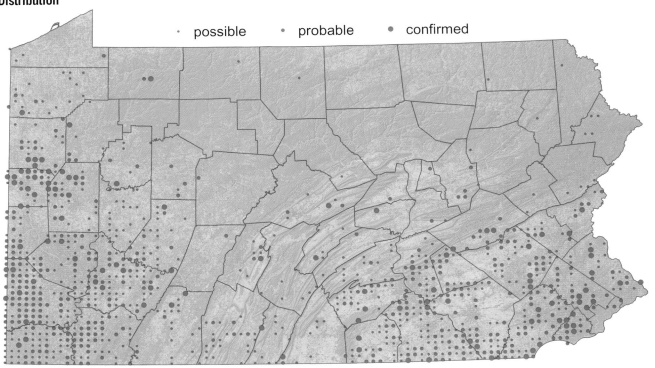

Distribution Change

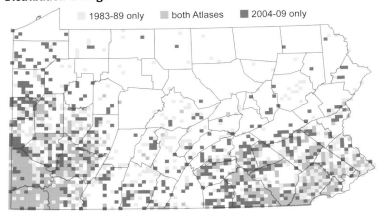

Number of Blocks

	first Atlas 1983–89	second Atlas 2004–09	Change %
Possible	410	531	30
Probable	491	370	−25
Confirmed	121	104	−14
Total	1,022	1,005	−2

Population estimate, males (95% CI):
34,000 (29,000–39,000)

Density

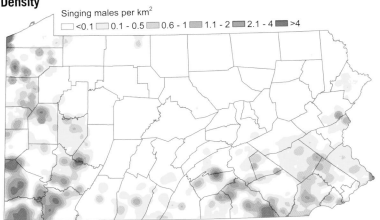

Breeding Bird Survey Trend

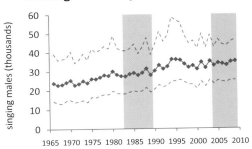

WHITE-EYED VIREO   263

Bob Wood and Peter Wood

Yellow-throated Vireo
Vireo flavifrons

The Yellow-throated Vireo is widely, albeit unevenly, distributed in Pennsylvania, found in tall trees with wide, spreading crowns. While its glowing yellow throat is not easily spotted in the leafy canopy, its loud voice will betray its presence if the listener can distinguish its husky and slower song from the similar song of the seemingly ubiquitous Red-eyed Vireo. The Yellow-throated Vireo occurs throughout most of the eastern United States, from the Great Lakes states and southern New England, south to the Gulf Coast, and west to the eastern edge of the Great Plains (Rodewald and James 1996).

Mature deciduous forests, especially with oaks and maples, are the favored habitat of the Yellow-throated Vireo, where it feeds primarily on caterpillars. It also occurs in mixed deciduous/coniferous forests, but not in pure conifer stands (appendix D). Often found near water, it is most common along rivers lined with tall trees and in wet bottomlands or flood plains. While its congener, the Red-eyed Vireo, occurs throughout forests at high density, the Yellow-throated Vireo seems to prefer edges and is largely absent from the interior of large areas of contiguous forest. This habitat preference, or restriction, may explain why Yellow-throated Vireos are far less common than Red-eyed Vireos (Robbins 1979; Rodewald and James 1996). Furthermore, atlas results show the Yellow-throated Vireo to be most common below about 400 m (~1,300 ft) and nearly absent from areas above 600 m (~1,950 ft).

In Pennsylvania, the distribution of the Yellow-throated Vireo has not changed significantly since the first Atlas. It was most frequently found in the southern Ridge and Valley Province and in the northeastern, northwestern, and southwestern corners of the state. These areas provide lots of favored forest edge habitat at low elevation, often near rivers and streams. Elevation in the Ridge and Valley Province varies as much as 400 m (1,300 ft) over very short distances, but edge habitat is primarily in the valleys; Yellow-throated Vireos (unlike the Red-eyed Vireo) are uncommon to rare on the heavily forested, high slopes of the ridges. The Yellow-throated Vireo is present throughout much of the Piedmont, but at lower density.

During the second Atlas, the number of blocks with Yellow-throated Vireo records was 44 percent higher than during the first Atlas. This increase correlates well with Breeding Bird Survey results, which show sustained increases of 1.2 percent per year since the 1960s (Sauer et al. 2011) and an estimated 20 percent rise in numbers between the atlas periods. The gain in blocks with Yellow-throated Vireos between atlases primarily reflects filling in of the range defined during the first Atlas, rather than significant expansion into new areas. The greatest increases were in the southern Ridge and Valley Province and in the Northwestern Glaciated Plateau, particularly in Crawford, Venango, and Mercer counties. Interestingly, in the early twentieth century, Todd (1940) stated that the greatest concentration of Yellow-throated Vireos in western Pennsylvania was in the glaciated northwest. However, they were reported in relatively few blocks there during the first Atlas, so the current increase may represent recovery of the population there. Despite the overall increase in Yellow-throated Vireo numbers and block occupancy, there was a thinning of the species' range in Fayette and Somerset counties in the southwest, for reasons that are not clear.

Currently, we estimate a population of 48,000 male Yellow-throated Vireos in Pennsylvania. Numbers across the eastern United States appear to be stable. The increasing maturity of forests in Pennsylvania is likely to be favorable for this high canopy-loving species. The most serious threats could be deforestation on Central American wintering grounds, where the Yellow-throated Vireo is thought to be a forest specialist (Lynch 1989; Morton 1992). On the breeding grounds, pesticide spraying was suggested as the reason for the disappearance of Yellow-throated Vireos from tree-lined streets in towns in the first half of the last century (Bent 1950), so the potentially deleterious effects of spraying of forests to control invasive insects should be considered.

GREG GROVE

Distribution

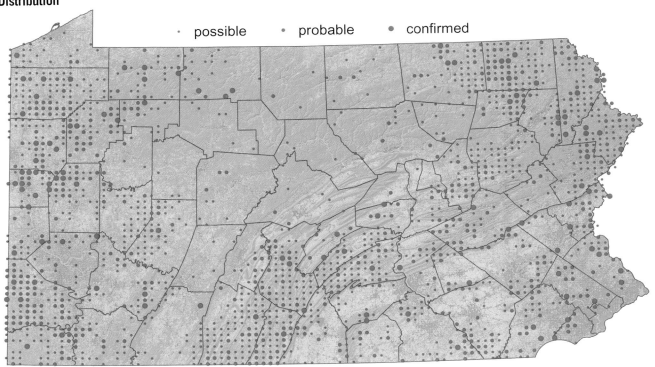

Distribution Change

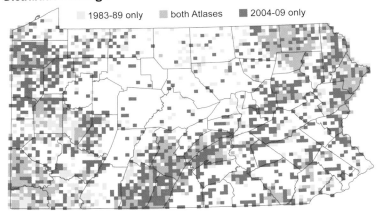

Density

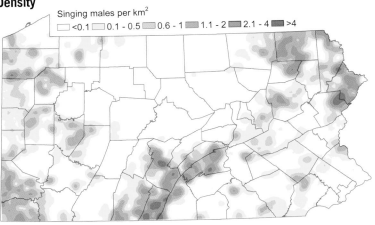

Number of Blocks

	first Atlas 1983–89	second Atlas 2004–09	Change %
Possible	616	1,165	89
Probable	475	483	2
Confirmed	127	112	−12
Total	1,218	1,760	44

Population estimate, males (95% CI):
48,000 (37,000–61,000)

Breeding Bird Survey Trend

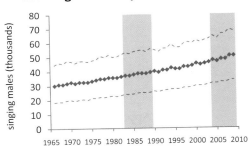

YELLOW-THROATED VIREO

Bob Wood and Peter Wood

Blue-headed Vireo
Vireo solitarius

The Blue-headed Vireo prefers to nest in coniferous and mixed forests, showing an affinity for forests with hemlocks in Pennsylvania (Ross et al. 2004; Allen and Sheehan 2010). Hemlocks are preferred for nest building during late April, when courtship is taking place and deciduous trees are not leafed out. Later nests are built in a variety of understory shrubs and saplings, as well as in low hemlock branches, usually 2–5 m (6–15 ft) above ground. The male's leisurely, mostly pure-toned song is characteristic of such forests from mid-April until nesting ceases in mid-August. Although the Blue-headed Vireo is genetically and socially monogamous (Morton et al. 1998), females abandon fledged broods and renest with new mates quickly (within 5 days); males care for fledglings, but they also renest with new females when fledglings reach independence (Morton et al. 2010). Both males and females may move to new territories, often a kilometer or more from initial territories, between breeding attempts within the same year.

Breeding Blue-headed Vireos are found from eastern British Columbia across the Canadian boreal forest to southwestern Newfoundland, and south into New England states, most of New York and Pennsylvania, then south into the Appalachian Mountains to northern Georgia (James 2007). Historically, the Blue-headed Vireo appears to have been common and widespread in Pennsylvania, but with the logging of forests during the nineteenth century, its range contracted to the northern tier and at higher elevations in the southern counties (McWilliams and Brauning 2000). This distribution was documented during the first Atlas, when it was found in 30 percent of atlas blocks, but rather sparingly in the southern half of the state. Lying within the extension of the Blue-headed Vireo's range into the Appalachian Mountain chain, Pennsylvania is at the southern limit of the species' breeding range. According to Partners in Flight, Pennsylvania hosts less than 3 percent of the species' global breeding population (Rich et al. 2004).

At 5.6 percent per year since 1966 (Sauer et al. 2011), this species is in the top 10 of the most rapidly increasing common birds on state Breeding Bird Survey routes. Given this rapid increase in numbers the highly significant 59 percent increase in the number of blocks containing Blue-headed Vireo is not surprising. Significant increases in block occupancy were noted in nearly all sections of the Appalachian Plateaus and Ridge and Valley provinces (appendix C). The most notable range expansions were in the south, especially in the southern Ridge and Valley, where the species has gone from being localized in the first Atlas to widespread along the higher ridges with suitable habitat in the second. Farther north, there has been an expansion of range, possibly into some lower elevations, although second Atlas point count data show that this species remains scarce below 400 m (1,300 ft) and virtually absent below 200 m (650 ft; appendix D). The increase in Blue-headed Vireos in Pennsylvania is mirrored by similar increases between atlases in Ontario (James 2007), New York (McGowan 2008w), and Maryland (Ellison 2010w). Maturation of forests may be the primary cause of the rebounding fortunes of this and other songbirds typical of northern forests.

Second Atlas data confirm the Blue-headed Vireo's preference for coniferous and mixed forests (appendix D), and its affinity for hemlocks is well documented (Allen and Sheehan 2010). The potential effects of the loss of hemlocks due to the Hemlock Woolly Adelgid on the species' populations are difficult to predict. However, studies in Pennsylvania have shown that breeding densities are more than 10 times higher in hemlocks than in hardwood forests (Allen and Sheehan 2010), while in Connecticut, populations were 2.5 times higher in healthy hemlock stands than in those infested with adelgids (Tingley et al. 2002). Hence, while it is unlikely that this species' resurgence will be completely reversed, it is possible that further expansion into former nesting areas that have yet to be recolonized will be kept in check, as more and more hemlock stands succumb to invasive woolly adelgids.

EUGENE MORTON

Distribution

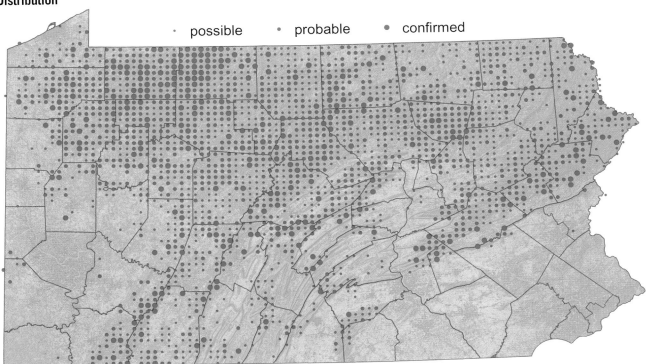

Distribution Change

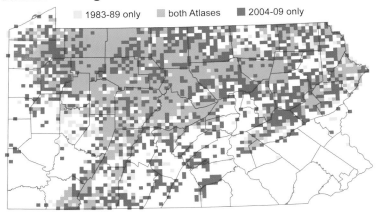

Density

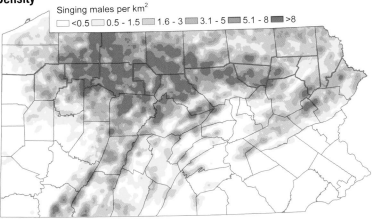

Number of Blocks

	first Atlas 1983–89	second Atlas 2004–09	Change %
Possible	715	1,141	59
Probable	596	925	55
Confirmed	163	274	68
Total	1,474	2,340	59

Population estimate, males (95% CI):
270,000 (255,000–288,000)

Breeding Bird Survey Trend

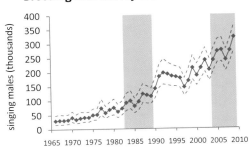

BLUE-HEADED VIREO 267

Dustin Welch

Warbling Vireo
Vireo gilvus

The Warbling Vireo is one of our most persistent singers, its drowsy, warbling song drifting all hours of the day from the canopy and sometimes even from the nest. It is also among our most nondescript birds, being uniformly dull-olive gray with a darker back and slightly lighter breast. The only distinguishing characteristic is a dull white superciliary stripe highlighted by an adjacent darker eye-stripe. It is often found in American Sycamores, cottonwoods, and poplars alongside forested streams and rivers, in deciduous forests with a tall, semi-open canopy, and in disturbed habitats ranging from urban parks to tree lines and abandoned orchards, where it seems to adapt well to human habitat alterations (Garadali and Ballard 2000). This species has the most extensive breeding range of any vireo. It is found in the central and northern half of the United States and ranges northward into western Canada and southward through the mountains of western Mexico (Garadali and Ballard 2000).

The Warbling Vireo displayed an interesting four-corner distribution within Pennsylvania during both atlas projects. The commonality among the state's four corners is high stream density within fragmented forest. In light of this species' preference for riparian habitat, the limited reports in blocks occupied along the Susquehanna River and its branches during the first Atlas remain somewhat of a mystery. Its song is commonly heard on river islands, but such habitats were likely not sampled well, if at all, in either atlas. Linear patterns of block occupancy in the otherwise unoccupied Appalachian Plateaus and Ridge and Valley provinces follow the regions' larger streams and otherwise more open habitats (Master 1992d). During the first Atlas, 22 percent of blocks were occupied throughout the state, but the species was confirmed in only 12 percent of these, probably because it forages and nests in the mid- to high-canopy level.

The distribution pattern of the Warbling Vireo remained similar during the second Atlas, but the number of occupied blocks within traditionally inhabited regions increased considerably. The species was found in 40 percent of blocks, a 76 percent increase from the first Atlas. Most of this increase occurred in the state's four corners, but also in the Great Valley Section of the Ridge and Valley along the Susquehanna River Valley, where block detections increased 131 percent since the first Atlas. Increases were noted in all physiographic provinces and sections (appendix C), but there was considerable turnover in occupied blocks in some regions, notably in the southwestern portion of the state. The Warbling Vireo's distribution within Pennsylvania is similar to that of the Yellow-throated Vireo, another forest canopy inhabitant.

The Warbling Vireo population within the state was estimated at 80,000 singing males, with the highest densities in Crawford, Erie, and Mercer counties, in the northwestern corner of the state. These three counties were estimated to host more than a quarter of the statewide population. The Breeding Bird Survey (BBS) shows a steady, significant population increase on routes in Pennsylvania, with a rise of 3.4 percent per year, 1999–2009 (Sauer et al. 2011). From BBS trends, it is estimated that the population increased by more than 50 percent between atlas periods. The population has also expanded in neighboring Maryland and New York, where block occupancy increased by 14 and 15 percent, respectively, between their atlas periods (Ellison 2010x; McGowan 2008x). Thus, the population of this species is secure and has probably benefited from changes in forest structure, including openings resulting from fragmentation and regeneration.

TERRY L. MASTER

Distribution

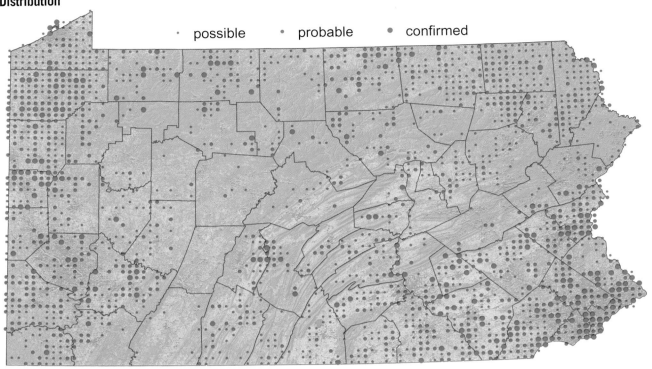

Distribution Change

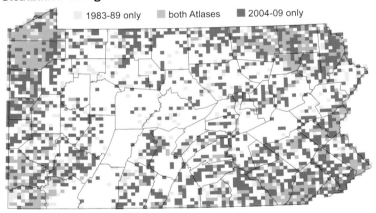

Density

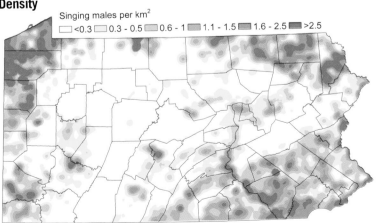

Number of Blocks

	first Atlas 1983–89	second Atlas 2004–09	Change %
Possible	502	1,052	110
Probable	473	697	47
Confirmed	131	202	54
Total	1,106	1,951	76

Population estimate, males (95% CI):
80,000 (75,000–88,000)

Breeding Bird Survey Trend

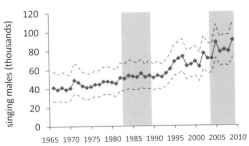

WARBLING VIREO 269

Jacob Dingel

Red-eyed Vireo
Vireo olivaceus

For a beginning birder, it is a thrill to be able to screen the non-stop cacophony of bird sounds and pick out the identity of the singers with any confidence. Many years ago, one of my first "aha" moments came when I could filter out the *see me, up here, aren't I pretty* song of the Red-eyed Vireo. It took a while longer before I could actually see the birds, since they are prone to nesting in the sub-canopy or canopy. Nevertheless, males of this species sing almost incessantly all day and until midsummer, driven, no doubt, by competition among males. Territories are small compared with other forest songbirds, and levels of extra-pair paternity are high (Morton et al. 1998), so there is good reason to sing a lot. An observer watching a nest-building female would not have to wait long to see her mate lurking nearby, followed soon after by a brawl involving an intruder male and the defending male while the female goes about her business.

Because this species does not routinely double-brood, the Red-eyed Vireo has a relatively short breeding season in Pennsylvania, with most breeding completed by the end of July. Atlas records show that incubation spans from late May to early July, with a peak in mid-June, while fledged young were mainly found through the second half of June and throughout July. Records of young being fed were reported as late as 30 August, the late breeding records perhaps relating to replacement clutches (McWilliams and Brauning 2000). There is no evidence of any change in breeding phenology since the first Atlas (Mulvihill 1992d).

Red-eyed Vireos are found wherever there is forest and are common in even small forest fragments, so they are not considered area-sensitive, which contributes to their presence even in blocks dominated by agricultural and residential areas. As far as can be gleaned from historical accounts, the Red-eyed Vireo has always been a common bird across the state (McWilliams and Brauning 2000), although numbers must have tracked vicissitudes in forest cover to a large degree. In the first Atlas, it was found in 93 percent of blocks, absent only from the most urbanized or agricultural blocks, where forest cover was minimal (Mulvihill 1992d). The Red-eyed Vireo was absent from even fewer blocks in the second Atlas, downtown Philadelphia, central Lancaster County, and parts of the Cumberland Valley being the only areas where it was absent in several contiguous blocks.

Given the Red-eyed Vireo's ubiquity in woodland and forest, it is not surprising that the density map of this species closely matches that of forest cover. No less than 36,681 Red-eyed Vireos were tallied on atlas point counts, an average of more than one per stop. The population estimate derived from these data suggest a statewide population of more than 2.8 million singing males, which is an average of 40 singing males per km^2 of wooded land. The Red-eyed Vireo is second only to Song Sparrow in statewide abundance.

The Breeding Bird Survey (BBS) shows an impressive 1.2-percent-per-year increase in population size in Pennsylvania since the 1960s, equating to an increase of 24 percent since the first Atlas (Sauer et al. 2011). Since Red-eyed Vireos are so common to begin with, this BBS increase would not be reflected in an increase in blocks occupied in the second Atlas. Second atlases in neighboring New York (McGowan 2008y) and Maryland (Ellison 2010y) also found that Red-eyed Vireos occupy almost all blocks, more than 96 percent and 93 percent respectively. Increasing forest cover and maturation (McWilliams et al. 2007) likely explain the long-term increase in the species' abundance, but why this long-distance migrant fares so well in contrast to some forest birds, like the Wood Thrush, is not well understood. Red-eyed Vireos migrate to South America during the non-breeding season, rely heavily on easy-to-find fruit, and are not territorial, but no studies have been done on winter habitat quality, deforestation, or over-winter survival of this species.

BRIDGET J. M. STUTCHBURY

Distribution

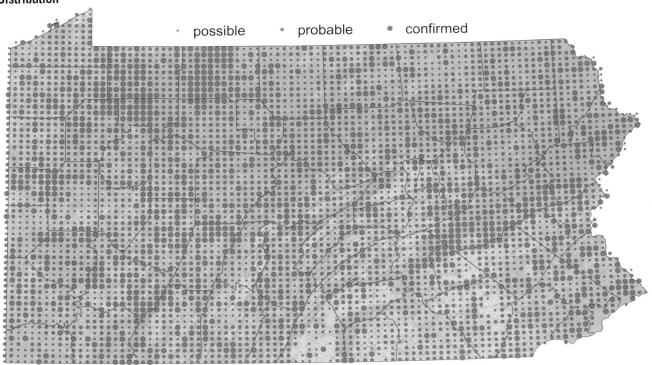

Distribution Change

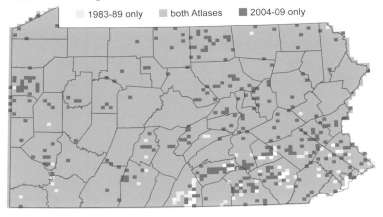

Density

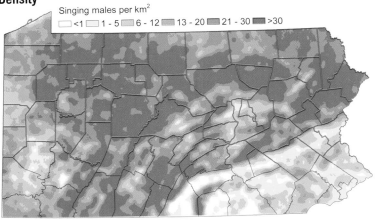

Number of Blocks

	first Atlas 1983–89	second Atlas 2004–09	Change %
Possible	1,078	1,334	24
Probable	2,351	2,357	0
Confirmed	1,157	1,145	−1
Total	4,586	4,836	5

Population estimate, males (95% CI):
2,830,000 (2,690,000–3,000,000)

Breeding Bird Survey Trend

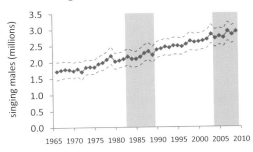

RED-EYED VIREO 271

Bob Wood and Peter Wood

Blue Jay
Cyanocitta cristata

First described scientifically by Pennsylvania resident Alexander Wilson, the brash and beautiful Blue Jay is one of the best-known birds of the Americas (Wilson and Bonaparte 1831; Tarvin and Woolfenden 1999). It is a keystone bird of the eastern deciduous and mixed forest of North America because of its habit of caching tree seeds, inadvertently planting deciduous trees, especially oaks and beeches, in new places (Darley-Hill and Johnson 1981; Johnson and Adkisson 1985; Johnson et al. 1997; Steele et al. 2010b). This is the common jay found east of the Rocky Mountains that has spread westward in recent decades, probably as a response to tree-planting in towns of the prairies (Tarvin and Woolfenden 1999). It is the only American jay that regularly migrates latitudinally, sometimes in remarkable numbers (Broun 1941; Tarvin and Woolfenden 1999).

Unusual for a passerine, Blue Jay pairs do not defend classic multi-purpose nesting territories, and males do not declare territory by singing (Hardy 1961; Gross 1982; Laine 1983). As a result, their behavior is difficult to interpret, and jay populations are a challenge to estimate by standard methods. They have a bewildering variety of vocalizations, belying a complex social system and adaptive behavior of an intelligent bird (Tarvin and Woolfenden 1999).

The Blue Jay is a cosmopolitan species that nests in forests, wood edges, hedgerows, parks, and yards. It flourishes in fragmented habitats because of its resourcefulness, and it takes advantage of human-derived food sources as well as native wild foods. Blue Jays usually nest in trees and can prosper near homes as well as forests and wooded farmland. The massive deforestation of the eighteenth and nineteenth centuries probably reduced them temporarily from large sections of the state, but Blue Jays certainly showed resilience and adaptability, occupying many areas where they apparently had been greatly reduced (Gross 1992c).

The Blue Jay was reported in 4,704 atlas blocks, 98 percent of the total—the twelfth highest block total of all breeding birds in the second Atlas. The absence of Blue Jay records from a few blocks is probably best explained by lack of coverage or detection in those blocks rather than absence of this common bird. Although a brightly colored and easily recognized bird, the Blue Jay is surprisingly quiet and furtive around active nests (Bent 1946; Hardy 1961). For such a conspicuous bird, it was confirmed in only 27 percent of the blocks where found. Volunteers most frequently confirmed nesting by finding the noisy fledged young ($n = 599$), adults carrying food ($n = 359$), or adults feeding young ($n = 356$).

The population of Blue Jays was estimated at 590,000 birds during the second Atlas. The highest densities are found in southeastern counties, particularly Montgomery, Lehigh, and Bucks, where there are scattered woodlots. Blue Jays are less common in the deep and extensive forests of higher elevation areas of the state (appendix D).

Breeding Bird Survey (BBS) counts of Blue Jays in Pennsylvania have been increasing slowly for the last 30 years (Sauer et al. 2011), resulting in an increase of around 11 percent between atlas periods. The Blue Jay is susceptible to West Nile virus, and local declines may be caused by this vulnerability (Komar et al. 2003; LaDeau et al. 2007). However, there is little evidence of a downturn in counts on BBS routes in Pennsylvania since the emergence of West Nile virus; hence, while this species may be vulnerable to the virus, populations seem to be resilient. In studies of the European jay's dispersing of acorns in Stockholm, Sweden, it was estimated that each pair of jays had the value of $22,500, based on the "replacement value" of the services rendered by its seed dispersal in that forest (Hougner et al. 2006). The value of the Blue Jay to eastern forests may be similar. As more tree pests, diseases, and energy development take their toll on our forests, there should be increased appreciation for this flamboyant, but valuable, "airlifter of the oaks" to our forest ecosystems, their wildlife, and economic value (Johnson and Adkisson 1986).

DOUGLAS A. GROSS

Distribution

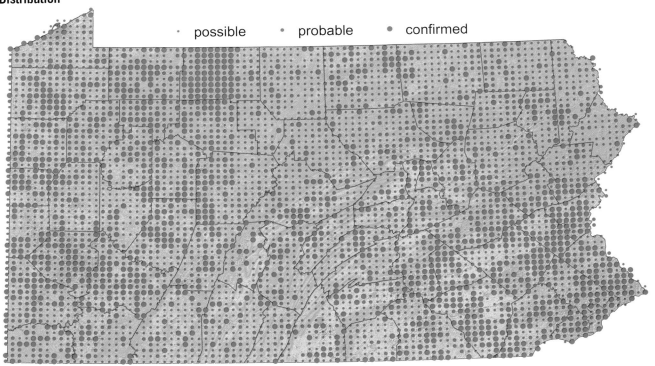

Distribution Change

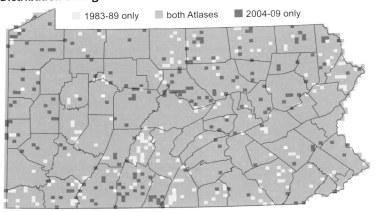

Density

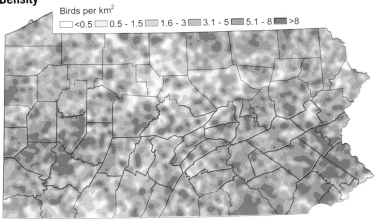

Number of Blocks

	first Atlas 1983–89	second Atlas 2004–09	Change %
Possible	1,521	2,171	43
Probable	1,509	1,255	−17
Confirmed	1,669	1,278	−23
Total	4,699	4,704	0

Population estimate, birds (95% CI):
590,000 (570,000–615,000)

Breeding Bird Survey Trend

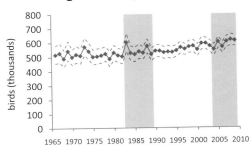

BLUE JAY 273

Rick and Nora Bowers/VIREO

American Crow
Corvus brachyrhynchos

"Intelligent," "crafty," "nuisance," and "pest" are all words used to describe the American Crow. With its large body size and loud, distinctive call, it is readily recognized by birders and non-birders alike. The American Crow is widely distributed across North America, and Pennsylvania is within the core of its range (Verbeek and Caffrey 2002). It was reported from 4,900 blocks (99% of blocks) in the second Atlas, making it the second most widely distributed species in the state.

American Crows have shown no apparent change in abundance or distribution since the first Atlas, when they were found in 98 percent of all blocks and were also the second most widely distributed species (Gross 1992d). The wide distribution results from their use of a broad range of habitat types, including agricultural areas, suburban and urban areas, and forest edge. They are habitat generalists with a preference for open habitats with scattered trees. During the second Atlas, they were reported at their highest abundance in the central region and northeastern corner of the state, both areas with a high interspersion of forest and open habitat. Crows had lower abundances in unfragmented forests and at higher elevations (appendix D), although the latter is also associated with large blocks of forest. The current estimated population size for American Crows in Pennsylvania is approximately 750,000 individuals.

American Crows were confirmed as breeders in 36 percent of blocks and 40 percent of priority blocks, a majority of these as fledged young. Crows will generally nest only once per season, but they will renest if their nesting attempt fails (Verbeek and Caffrey 2002). In Pennsylvania, resident crows begin to breed after the migrants from further north leave in late February and early March. Atlas volunteers reported crows carrying nesting material by the second week in March and reported fledged young by the third week in April. The latest reported date for fledged young was 15 August. Since crows remain together as a family unit through the summer following nesting (Verbeek and Caffrey 2002), groups of more than two crows probably reflected confirmed breeding. However, the possibility remains that the fledged young may have been reported outside the original block containing the nest, so habitat associations of these observations may be obscured.

Although crows are abundant and widely distributed, there have been concerns over the potential effects of West Nile virus on crow populations. By some accounts, mortality rates are close to 100 percent for birds infected with the virus (Wilcox et al. 2007). West Nile was first detected in New York in 1999, and by 2005, crow populations in the Northeast had dropped by 45 percent from 1998 levels (LaDeau et al. 2007). The decline in numbers in Pennsylvania is reflected in Breeding Bird Survey data that show a modest decline between 2003 and 2004, coinciding with the arrival of West Nile virus in the state, followed by a recovery in numbers within the next few years (Sauer et al. 2011). The recovery of the population may be due to surviving individuals acquiring some resistance to the disease, or to a change in the virulence of the disease. The breeding behavior of the crow may also be a factor. In many parts of their range, at least some first-year and older crows forego breeding and instead act as helpers within their parents' territory (McGowan 2001a; Verbeek and Caffrey 2002). If this behavior is flexible and dependent on population density, a greater percentage of young breeders at low population densities could help to buffer population declines and speed up recovery. Crows showed no significant change in block occurrence related to West Nile virus, but this is not surprising, given that the species is numerous almost everywhere. Despite the threat of West Nile virus, crows appear to be doing well in the state. Future predicted changes in habitat, including suburban sprawl and the fragmenting of forests in the north-central region of the state, will tend to favor a generalist like the American Crow.

MARGARET C. BRITTINGHAM

Distribution

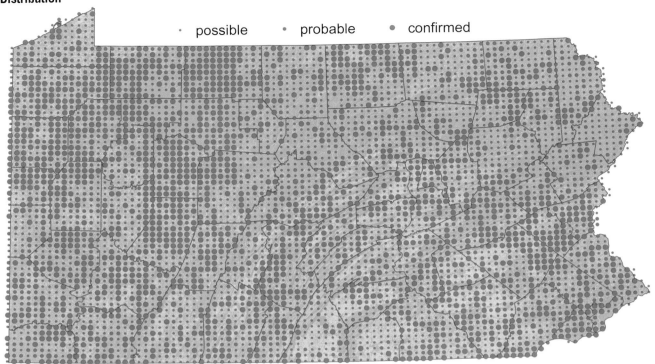

Distribution Change

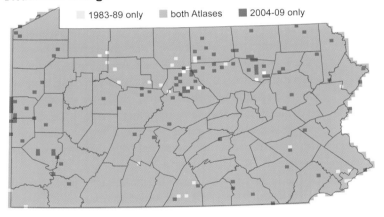

Density

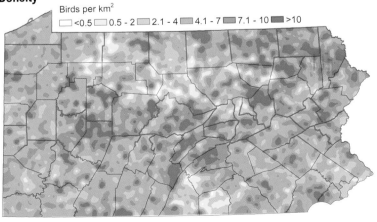

Number of Blocks

	first Atlas 1983–89	second Atlas 2004–09	Change %
Possible	1,732	2,101	21
Probable	1,142	1,023	−10
Confirmed	1,945	1,776	−9
Total	4,819	4,900	2

Population estimate, birds (95% CI):
750,000 (725,000–775,000)

Breeding Bird Survey Trend

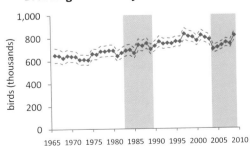

AMERICAN CROW

Marvin R. Hyett/VIREO

Fish Crow
Corvus ossifragus

The nasal *caw* of this crow of rivers and farmland has been heard in more areas in the last few decades as the Fish Crow continued its expansion in Pennsylvania during the second Atlas. It is endemic to the United States, and its breeding range is limited to the Atlantic Coast (McGowan 2001b). Easily confused with the more widespread American Crow, the Fish Crow is best identified by its distinctive call. The Fish Crow is primarily a bird of tidal areas, rivers, and farmland in the southeastern United States and north along the Atlantic Coast to southern Maine (McGowan 2001b). Pennsylvania is on the western edge of its primarily coastal breeding range. The Fish Crow's distribution in the state traces the major rivers and their tributaries that flow to the Atlantic Ocean, primarily the watersheds of the Susquehanna, the Potomac, and the Delaware rivers (Gross 1992e). The Fish Crow not only lives up to its name by its strong association with riverine habitats, but it also is frequently found in agricultural areas.

In Pennsylvania, the Fish Crow is primarily a bird of the southeastern urban, suburban, and agricultural areas stretching up the rivers into central and northern counties along rivers. The highest concentration areas are in the lower Susquehanna River valley, especially Cumberland, Lancaster, and York counties, and the lower Delaware River and its major tributaries such as the Schuylkill. Point count data show that the Fish Crow is strongly associated with pasture in Pennsylvania (appendix D). Because of its primarily southeastern distribution and association with riparian habitats, it is scarce above elevations of 300 m (~1,000 ft), and it is rarely encountered in densely forested areas (appendix D).

The expansion of the Fish Crow into the Ohio River drainage, as noted in the first Atlas (Gross 1992e), continued in the second Atlas period. This expansion has been occurring for over 150 years, perhaps since European settlement (McGowan 2001b). Early in the nineteenth century, it was found only in the lower Susquehanna River valley south of the city of Lancaster and the Delaware River basin (Audubon 1840–1844; Warren 1890). By the second Atlas, the Fish Crow's range extended along the West Branch Susquehanna to Clearfield County and the main branch of the Susquehanna to Wyoming County. Further, there has been an expansion into the Susquehanna and Chemung watersheds, north of the state line into south-central New York, while, along the Upper Delaware River, it has expanded into Wayne County and to New York's Sullivan County (McGowan 2008z). Since the first Atlas, the Fish Crow has become more established in the Juniata River basin and also expanded westward into the Ohio River drainage, with scattered records as far west as Butler and Allegheny counties. It is now established in Indiana County and near Pittsburgh in Allegheny County. It also has become much more established in several southeastern counties, spilling over from riparian corridors into agricultural areas. The net result of these expansions was an 86 percent increase in the number of blocks with Fish Crow records since the first Atlas. Breeding Bird Survey (BBS) counts have increased steadily in Pennsylvania since the 1960s (Sauer et al. 2011). The total population size estimate for the species from point counts in Pennsylvania was 30,000 birds, although this may be conservative (chap. 5). Although it has made advances, it is not nearly as common and widespread as the seemingly ubiquitous American Crow.

The Fish Crow has been found to suffer high rates of mortality due to West Nile virus (Komar et al. 2003), and BBS counts in Pennsylvania plunged by 22 percent in 2000, when the virus was first detected in the state. However, BBS counts subsequently bounced back rapidly, suggesting that Fish Crow populations quickly built resilience to the virus. It is possible that Fish Crows may have indirectly benefited from reduced competition with American Crows (Ellison 2010z), which apparently suffered even higher mortality from West Nile virus (Komar et al. 2003), although BBS data from Pennsylvania do not suggest any significant long-term effects of the virus on populations of either species.

DOUGLAS A. GROSS

Distribution

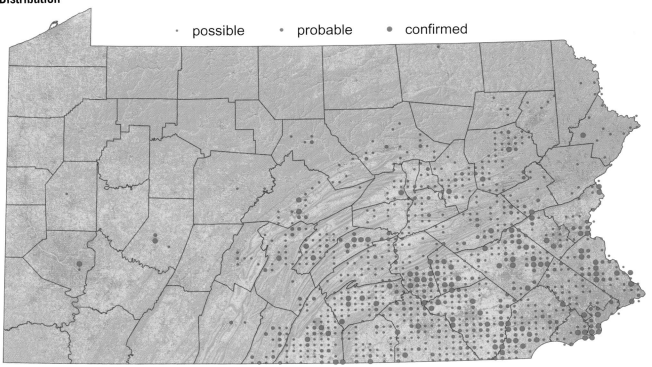

Distribution Change

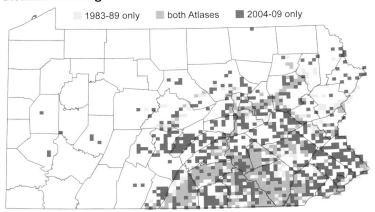

Density

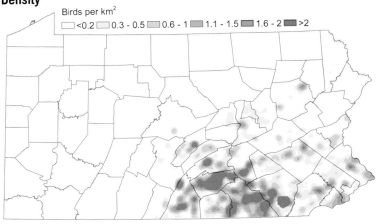

Number of Blocks

	first Atlas	second Atlas	Change
	1983–89	2004–09	%
Possible	337	672	99
Probable	132	178	35
Confirmed	45	106	136
Total	514	956	86

Population estimate, birds (95% CI):
30,000 (27,000–33,000)

Breeding Bird Survey Trend

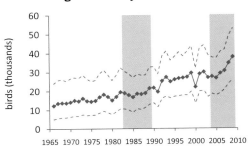

FISH CROW

Tom J. Ulrich/VIREO

Common Raven
Corvus corax

A totem of honor by Vikings and Native Americans, the Common Raven has a certain mystique and cult status in many cultures, and for many good reasons. It is the largest and most robust of our passerines and known for its intelligence and resourcefulness, leading to mischievousness and playfulness (Boarman and Heinrich 1999; Marzluff and Angell 2005). In the eastern United States, the raven is associated with wild areas, although that has become less true lately. The deep croak of this large black corvid is heard much more commonly across the Commonwealth than in decades past. Formerly confined to the state's most remote mountain forests, the Common Raven has continued the increase documented in the first Atlas (Mulvihill 1992e) and has become quite widespread in the state, due in part to its adaptiveness and resiliency.

The range of the Common Raven not only includes the forested central highlands but now extends into agricultural regions of southeastern Pennsylvania, reaching east to the Delaware River and west to the Ohio and West Virginia borders. It recently was confirmed nesting in Jefferson County, Ohio, after a century of absence (ODNR 2008). In pre-settlement forested Ohio, the raven was believed to be the most common large corvid (Peterjohn 1989). The expansion of Common Ravens, observed in the first Atlas, continued unabated in the second Atlas, a significant 114 percent increase in blocks. It could be said that we are witnessing the reclamation of the raven's former domain. This change is echoed by Breeding Bird Survey counts from Pennsylvania, which have increased by an average of 6.5 percent per year since 1966 (Sauer et al. 2011).

Although the raven is now widespread in the state, it is still more abundant in the largely forested and mountainous counties. Highest densities were found in the forested counties of the Appalachian Plateaus, including Potter, McKean, Cameron, Tioga, Clinton, Forest, Elk, and Centre. There also were clusters of higher density in Lycoming, Sullivan, and Somerset counties. Atlas point count data suggest a population of 16,000 ravens in Pennsylvania, about 10 times the estimate previously given by Partners in Flight (Rich et al. 2004).

Several explanations for the recovery of Common Raven populations in Pennsylvania and across the Appalachian Mountains have been proposed. Protection from persecution probably has been a significant factor (Boarman and Heinrich 1999), but some are still killed mistakenly as crows. Its recovery also has been credited to reforestation and an increase of road-killed White-tailed Deer in the state (Bolgiano and Grove 2010). An increased tolerance of humans by some ravens also may be a factor, evidenced by ravens now nesting on human structures, including Beaver Stadium at Penn State and on electric power structures elsewhere, similar to that observed in the American West (Boarman and Heinrich 1999). It has been suggested that ravens commonly associate with wild canids because they can take advantage of carcasses opened up by the large mammal predators (Heinrich 1999; Vucetich et al. 2004). So the expansion of raven populations in Pennsylvania and other northeastern states may follow the expansion of coyote populations (Thurber and Peterson 1991). This bird has taken advantage of wild predators and whatever humans can provide.

In common with other corvids, the Common Raven is susceptible to West Nile virus, with mortality due to the virus recorded from New York south to Tabasco, Mexico (Lanciotti et al. 1999; Estrada-Franco et al. 2003). However, the increase of ravens has continued in Pennsylvania since West Nile virus emerged in the state in 2000 (Sauer et al. 2011), suggesting that its populations may not be as vulnerable as other corvids to this disease. As raven populations continue to increase, so too might conflicts. Ravens are predatory on bird nests of many species, including waterfowl, as well as small game, poultry, and small livestock (Boarman and Heinrich 1999; Marzluff and Angell 2005). Ravens also share cliffs with Peregrine Falcons, which are reestablishing populations in the state. Therefore, its expanding population might have some unforeseen consequences for other species not recently exposed to this clever bird.

DOUGLAS A. GROSS

Distribution

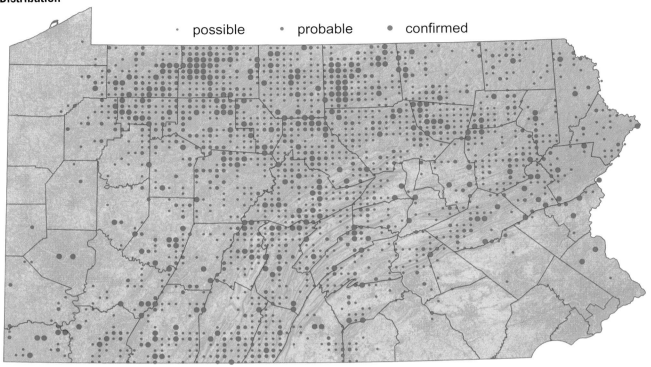

Distribution Change

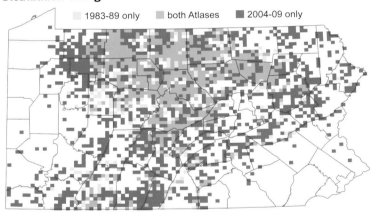

Density

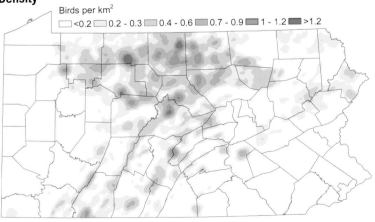

Number of Blocks

	first Atlas	second Atlas	Change
	1983–89	2004–09	%
Possible	551	1,176	113
Probable	177	375	112
Confirmed	122	264	116
Total	850	1,815	114

Population estimate, birds (95% CI):
16,000 (13,000–20,000)

Breeding Bird Survey Trend

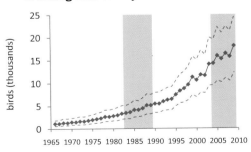

COMMON RAVEN 279

Bob Wood and Peter Wood

Horned Lark
Eremophila alpestris

One of the most widespread birds in North America, the Horned Lark is found in open country; in the eastern United States, it is principally found in overgrazed grassland and agricultural lands. Due to the species' large circumpolar range, Pennsylvania supports an insignificant proportion of the global population (Rich et al. 2004). Until the late nineteenth century, the Horned Lark was known in Pennsylvania only as a wintering bird (the northern subspecies, *E. a. alpestris*), but during the 1890s, the state was rapidly colonized by the expanding population of the Prairie Horned Lark (*E. a. praticola*). Initially a bird of agricultural lands, the species later colonized reclaimed surface mine grasslands and, by the time of the first Atlas, such areas were notable strongholds. Spring-tilled fields are favored within agricultural areas (Castrale 1985). Habitat loss, due to a switch from spring to fall sowing of small grains, may have contributed to the precipitous declines in Breeding Bird Survey (BBS) counts from the 1960s through the 1980s (Sauer et al. 2011).

One of the biggest surprises of the second Atlas was the wholesale shift in range experienced by the Horned Lark since the first Atlas. Former strongholds such as the reclaimed surface mines of Clarion County have been largely abandoned, whereas the range in the southeastern and northwestern counties consolidated. In the first Atlas, the Piedmont and Ridge and Valley provinces held 33 percent of occupied blocks, but by the second Atlas, this had increased to 73 percent. Even so, BBS data suggest that the statewide population is now smaller than at the time of the first Atlas, but the decline seems to have halted around 1999 (Sauer et al. 2011). An almost identical pattern of range shifts and recent population stability was documented by the two Maryland atlases, whose survey periods overlapped those of the Pennsylvania atlases (Ellison 2010za). In contrast, the species showed marked declines in both range and population size in New York (Smith 2008c).

The Horned Lark is area-sensitive (Ribic et al. 2009) and avoids areas close to tree cover (Gremaud 1983), so it is not surprising that it was virtually absent in blocks with less than 120 ha (300 ac) of arable land but was found in nearly all blocks with more than 800 ha (2,000 ac) of arable land. The distributional shift between atlases suggests that the Horned Lark has adapted once again to changing agricultural practices in some of the most intensively farmed areas of the state. A large expansion in soybeans, from 72,000 ha (177,000 ac) in 1987 to 174,000 ha (430,000 ac) in 2007 (USDA-NASS 2008), may have benefited this species; soybeans are typically planted in May (Roth 2010), ensuring a ready supply of tilled and sparsely vegetated ground during the breeding season.

Only one Horned Lark was found on 118 second Atlas point counts within surface mine grasslands, which contrasts sharply with McWilliams and Brauning's (2000) assessment that this species was "most common on reclaimed strip mines." Surface mines probably are suitable to Horned Larks only until grasses and forbs used in reclamation become fully established. The lark's abandonment of minelands likely reflects the widespread maturation of those grasslands.

The Horned Lark's statewide population is estimated to be 40,500 singing males, one-quarter of which are in just three counties: Franklin, Lancaster, and York. Pockets of high densities are found in other areas of open farmland in central and northwestern Pennsylvania. The history of this species in Pennsylvania is strongly tied to farming, and its future status will no doubt be forged by changes in the dynamic agricultural landscape. Because Pennsylvania holds such a small proportion of the global population, this species should not be considered of conservation importance in the state. However, because the Horned Lark is one of very few species adapted to nesting in intensively farmed fields, its loss would be detrimental to otherwise biologically impoverished landscapes, especially during late winter, when the species' song flight is a welcome harbinger of spring.

ANDREW M. WILSON

Distribution

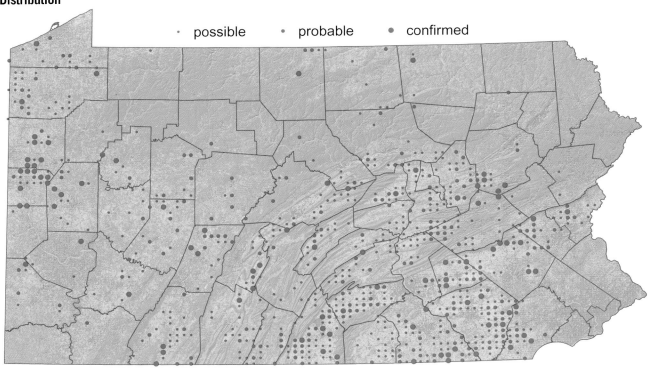

Distribution Change

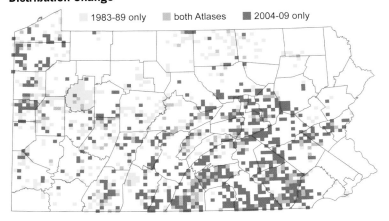

Density

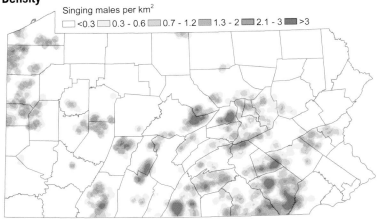

Number of Blocks

	first Atlas 1983–89	second Atlas 2004–09	Change %
Possible	273	479	75
Probable	200	208	4
Confirmed	155	94	–39
Total	628	781	24

Population estimate, males (95% CI):
40,500 (35,000–46,000)

Breeding Bird Survey Trend

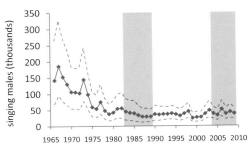

HORNED LARK 281

Rob Curtis/VIREO

Purple Martin
Progne subis

Although it still nests in natural cavities and crevices in parts of western North America, east of the Rocky Mountains the Purple Martin depends entirely on people to provide it with nesting sites in the form of the familiar apartment-type birdhouses or hanging clusters of natural or plastic gourds (Brown 1997). Its nearly complete dependency on human-supplied housing arguably makes this species the most intensively managed migratory songbird in the Western Hemisphere.

The Purple Martin is most abundant in the southeastern United States, but it nests as far north and west as central Alberta. The species' estimated global population, based on Breeding Bird Survey (BBS) data, is about 11 million (Rich et al. 2004), but only an estimated 15,000 (0.13%) nest within Pennsylvania (second Atlas estimate), which is near the northeastern edge of the contiguous range of the species.

When we examine the distribution of Purple Martins observed in both the first and second Atlases, it is reasonable to assume that the observed distributions largely reflect where martins were provided with housing in suitably open landscapes. This conclusion is not new. As early as the first half of the twentieth century, W. E. Clyde Todd (1940) postulated that "the bird-box habit is now so firmly established that it has become a controlling factor in the local distribution of the species."

The distribution of Purple Martins in Pennsylvania contracted dramatically between the first and second Atlases: in that relatively brief 20-year time period, the number of atlas blocks in which the species was recorded decreased by nearly half. The breeding distribution contracted most notably in northeastern and southwestern Pennsylvania and, curiously, in the central counties of Mifflin and Juniata, once a stronghold for martins. A similar contraction (39%) in distribution occurred in New York between 1985 and 2005 (Medler 2008g) and in Ontario (46%) between 1987 and 2005 (Cadman 2007). The significant decline in block occupancy in Pennsylvania is commensurate with a 40 percent decline on BBS routes between the two atlas periods (Sauer et al. 2011).

Adverse weather and competition for nest cavities from the more aggressive European Starling and House Sparrow are commonly presented as reasons for declines in the Purple Martin population (Brown 1997). As a very early spring migrant and obligate aerial insectivore, the species is vulnerable to periods of cold, rainy weather. In 1972, Pennsylvania's Purple Martin population suffered a major setback when Hurricane Agnes brought several days of heavy rain in June (Hall 1972; Tate 1972). Many long-standing colonies were decimated or lost altogether, due to starvation of chicks and adults; many of these colonies never recovered. However, no comparable weather events occurred between the atlases. Pennsylvania's Purple Martins make long-distance migrations to and from their Brazilian wintering grounds (Stutchbury et al. 2009), but migratory and wintering habitat are not known to be limiting; at the landscape level in Pennsylvania, the martin's favored open habitats remain abundant.

Loss of human-supplied housing is the probable cause of the Purple Martin's shrinking distribution and declining numbers in Pennsylvania. Although it tolerates suburban and even urban environments, the species is more common in rural environments, notably where Amish communities occur. Unfortunately, family farms and rural lifestyles in Pennsylvania are in decline, and these human societal changes have resulted in fewer people providing nest sites for Purple Martins. With only 15,000 Purple Martins estimated to breed in Pennsylvania, its population declining, and its distribution contracting, the species should perhaps be formally listed as a Species of Special Concern, and a working group should be formed to address the decline (Tautin et al. 2009). The Purple Martin responds well to management, and with a concerted public and private effort, Pennsylvania should be able to stabilize and grow its population of this once very well-known and popular bird.

JOHN TAUTIN

Distribution

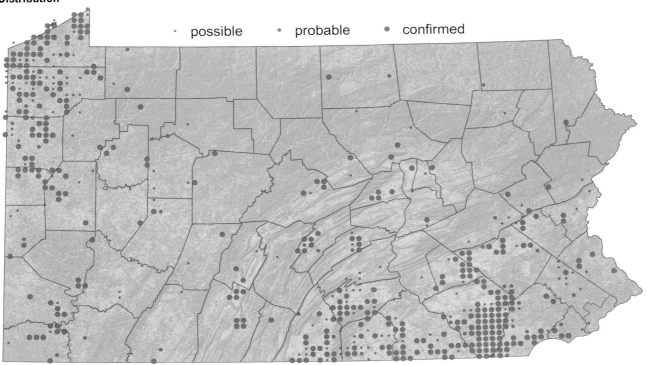

Distribution Change

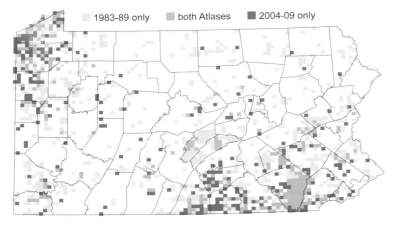

Number of Blocks

	first Atlas 1983–89	second Atlas 2004–09	Change %
Possible	214	138	−36
Probable	91	36	−60
Confirmed	630	346	−45
Total	935	520	−44

Population estimate, birds (95% CI):
15,000 (13,000–18,000)

Breeding Bird Survey Trend

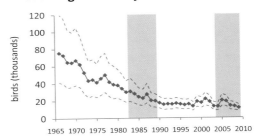

PURPLE MARTIN

Bob Moul

Tree Swallow
Tachycineta bicolor

With a flash of steel-blue and white, the Tree Swallow, darting and soaring over water or fields, has become a familiar sight across Pennsylvania. As a secondary cavity nester, the Tree Swallow needs little more than a hole in a tree and nearby open foraging areas, preferably near water. These conditions are commonly found throughout the state, naturally in northern areas where beaver impoundments lead to standing dead timber. The species readily accepts artificial nest boxes put out for Eastern Bluebirds.

The Tree Swallow's breeding distribution ranges from the tree line in Canada, south to northern Georgia, and nearly from coast to coast (Robertson et al. 1992). Pennsylvania lies well within this range, with breeding records from every county. However, the species was formerly local outside of the northern tier, and it appears to have been much less common during the nineteenth and early twentieth centuries than it is today (McWilliams and Brauning 2000).

During the second Atlas, the Tree Swallow was found in 80 percent of all blocks, representing a highly significant increase in occupancy of 26 percent since the first Atlas period. Increases were noted across most of the state but were most pronounced in southern counties, where there was a marked range consolidation. In the Piedmont, for example, block occupancy almost doubled, from 50 percent in the first Atlas to 92 percent in the second. This southward range expansion was even more marked south of the Mason-Dixon Line in Maryland, where block occupancy increased almost tenfold between atlases (Ellison 2010zb). Despite the range expansion in Pennsylvania, there are still areas with reduced occupancy, including the northeastern corner of the state (Pike and Wayne counties) and parts of the Pittsburgh Plateau Section (Beaver and Allegheny counties) and Deep Valleys Section (Clinton County) of the Appalachian Plateaus Province.

Breeding Bird Survey results indicate a doubling of the population in Pennsylvania between atlas periods, with an annual increase of 3.5 percent (Sauer et al. 2011). Similar increases have occurred in New York, New Jersey, Connecticut, and most of the rest of the United States, except for northern New England (Sauer et al. 2011). The population size, estimated from point count data, is 120,000 individuals, but caution is advised, because the point count protocol may not be suitable for counting swallows (chap. 5). However, it is clear that the Tree Swallow's population size is second only to that of the Barn Swallow among the state's six breeding hirundines.

The Tree Swallow is a relatively early spring migrant. Nest construction was noted by atlas volunteers from mid-April to mid-July. These later nests were most likely second attempts following a failed first nest, since a second brood would be rare in Pennsylvania (Robertson et al. 1992). Evidence of confirmed breeding, such as occupied nests and recently fledged young, was found until the end of July. Atlas point count data indicate that observations of the Tree Swallow were often associated with water and other open habitats, where the species forages. Despite the species' name, point count observations were negatively associated with percentage of tree cover (appendix D). Hence, forest regrowth could exclude this species from atlas blocks where forest cover is close to 100 percent.

Current data indicate that the Tree Swallow is secure in Pennsylvania. This could change, however. Current climate models predict a northward retreat of the Tree Swallow's breeding habitat during the next century (Matthews et al. 2010). In the short term, the impact of invasive insect diseases of trees, such as Hemlock Woolly Adelgid and Emerald Ash Borer, may temporarily provide increased nesting habitat for the Tree Swallow as tree mortality increases.

JERRY SKINNER

Distribution

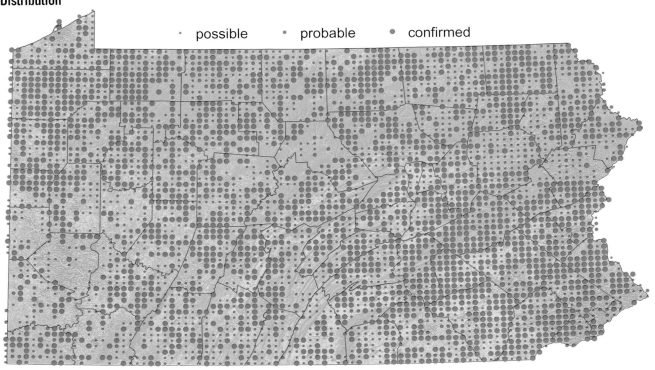

Distribution Change

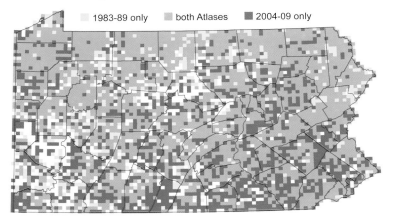

Number of Blocks

	first Atlas 1983–89	second Atlas 2004–09	Change %
Possible	924	1,419	54
Probable	491	434	−12
Confirmed	1,712	2,076	21
Total	3,127	3,929	26

Population estimate, birds (95% CI):
120,000 (115,000–127,000)

Breeding Bird Survey Trend

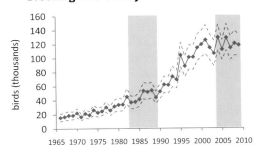

TREE SWALLOW

Bob Wood and Peter Wood

Northern Rough-winged Swallow
Stelgidopteryx serripennis

Although the Northern Rough-winged Swallow lacks the bright plumage of some of its relatives, this "sober-colored little bird" (Dingle 1942) possesses one unique feature: its outermost primary feather on each wing, the leading edge of which is stiff and comb-like. While this provides both the swallow's common name and also its Latin designation (which translates to "scraper-wing saw-feather"), its function is unknown (DeJong 1996). The close resemblance between the Northern Rough-winged Swallow and the slightly smaller, more aerobatic Bank Swallow may have masked the former's presence in the early days of serious bird study, but the rough-wing has clearly expanded its range both in Pennsylvania and continentally since the mid-nineteenth century. Stone (1894) said they were found primarily in the southern half of the state, while Sutton (1928a) called the species "rather rare, somewhat local, and never abundant" in the Commonwealth, "found almost altogether in the more southern and less mountainous counties."

While a birder these days is liable to encounter a Northern Rough-winged Swallow anywhere in Pennsylvania (the species was detected on roughly half of all priority blocks), it remains least common in the mountainous north-central counties and the eastern Poconos, despite evidence of a significant increase in detections there since the first Atlas. More than most swallows, the rough-wing is tied to water, foraging over rivers, streams and impoundments, and nesting near water in crevices, drain pipes, old Bank Swallow or Belted Kingfisher burrows, or any other cavity it can find (DeJong 1996). On second Atlas point counts, it was often found near quarries, mines, or gravel pits, or near open water, although it was found foraging over most habitat types (appendix D).

Artificial nest tubes are readily accepted by the species, and experiments placing such tubes along Pennsylvania bridges in the 1980s were successful (Schwalbe 1992b). There is some disagreement over whether Northern Rough-winged Swallows occasionally excavate their own burrows, but if they do, the behavior is rare (DeJong 1996). Because they are at best weakly colonial, Northern Rough-winged Swallows' nests are more likely to be overlooked than those of other, more gregarious swallows (Schwalbe 1992b).

The second Atlas documented a 43 percent increase in blocks occupied by Northern Rough-winged Swallows since the first Atlas, with the number of occupied blocks more than doubling in 6 of the state's 23 physiographic sections (appendix C). However, the patterns of change were complex, and an expansion in the northern-tier counties of McKean and Warren, where survey effort was also greater during the second Atlas, was the only substantive change in distribution. Elsewhere, there was higher block occupancy in much of the Ridge and Valley Province and in urban Philadelphia; otherwise, a patchwork of gains and losses suggests that this species is an ephemeral, adaptable bird, utilizing suitable nest sites when they become available. Similar patterns of overall increase in rough-wing block occupancy, alongside considerable turnover in occupied blocks, was noted in the New York (McGowan 2008za) and Maryland (Davidson 2010a) atlases.

The Breeding Bird Survey has documented a steady increase in numbers of Northern Rough-winged Swallows in Pennsylvania since the 1960s (Sauer et al. 2011), equating to a 36 percent increase between atlas periods. Point count data suggest a population of 30,000 birds in the state, although this must be treated with caution, because the point count protocol may be deficient for estimating densities of hirundines (chap. 5). This estimate would equate to an average of only seven pairs per occupied block, which suggests that this species is thinly scattered across much of the state; as such, it could be easily missed by atlas volunteers, which might in part explain the high block turnover.

Given its adaptability and positive population trajectory, the Northern Rough-winged Swallow's status in Pennsylvania appears to be secure. Its population is probably limited by the availability of nest sites, but this is one species that may benefit from land disturbances such as mining and construction, which often, inadvertently, provide new nest sites.

SCOTT WEIDENSAUL

Distribution

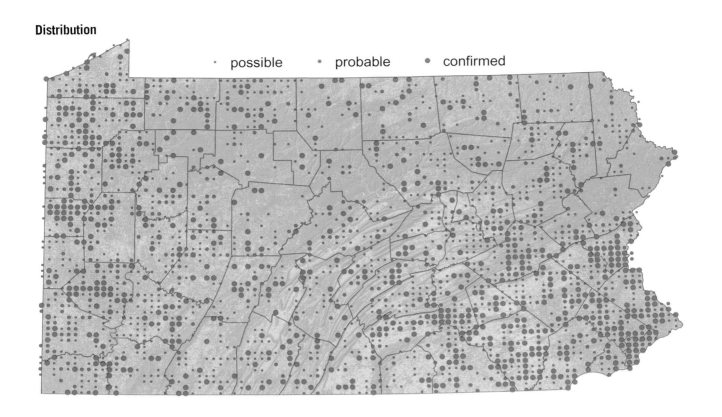

Distribution Change

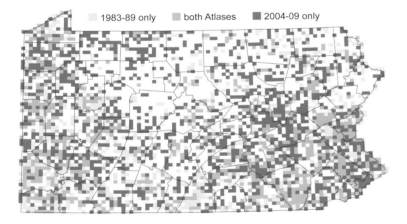

Number of Blocks

	first Atlas 1983–89	second Atlas 2004–09	Change %
Possible	608	1,054	73
Probable	390	393	1
Confirmed	500	691	38
Total	1,498	2,138	43

Population estimate, birds (95% CI):
30,000 (27,000–34,000)

Breeding Bird Survey Trend

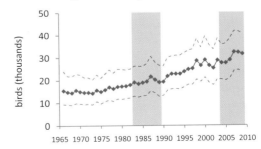

NORTHERN ROUGH-WINGED SWALLOW 287

Bob Wood and Peter Wood

Bank Swallow
Riparia riparia

At an average of about 13.5 grams, the tiny Bank Swallow is Pennsylvania's smallest of this sociable and gregarious family. It is known elsewhere in the world as the Sand Martin, and both common names of this cosmopolitan passerine are appropriate and descriptive. In Pennsylvania and elsewhere throughout its North American range, the Bank Swallow is a summer denizen of the vertical surfaces of river banks, bluffs, and gravel piles, where it excavates its own nesting burrows in sandy, erodible substrates. More so in eastern North America than in the West, the Bank Swallow depends on sand and gravel quarries for nesting sites (Garrison 1999). Due to the ephemeral nature of its required habitat, the species is opportunistic and often does not show fidelity to a particular nesting site year to year, though it will reuse sites if the habitat remains appropriate. Thus, the Bank Swallow has a history of sporadic, unpredictable distribution throughout most regions of Pennsylvania (McWilliams and Brauning 2000).

The Bank Swallow showed a significant statewide decline between the two atlas periods. It was found in only 337 blocks in the second Atlas compared with 521 in the first, a decline of 36 percent. Losses occurred in nearly every physiographic section of the state (appendix C), but as with the Cliff Swallow, the most severe declines were noted in the Glaciated Low Plateau Section in the northeast, where the drop-off was 59 percent, and in the Ridge and Valley Province, which saw a decrease of 45 percent. Meanwhile, and also like the Cliff Swallow, the Bank Swallow showed a minor gain in the northwestern area of the state. The Lake Erie shore itself has a strong history of supporting the Bank Swallow, which continued in the second Atlas period. The species was found in 57 percent of all blocks in the Eastern Lake Section, by far the highest percentage of occupied blocks of any section in the state. Todd (1940) considered the Bank Swallow to be in significant decline throughout western Pennsylvania, but even then he noted an exception at the lakeshore, where the species continues, to the present day, to take advantage of the numerous bluffs and stream outlets for nesting sites. Documented colonies along the lakeshore have reached as many as 500 nests (McWilliams and Brauning 2000). During the second Atlas, colonies in excess of 100 nests were reported in nine different blocks on the lakeshore.

One possible explanation for the statewide declines is the reduction in mining and quarry activity, which make up an important component of Bank Swallow nest sites. Atlas volunteers submitted notes from 64 blocks in which the species was confirmed, and 35 of these (54%) mentioned the swallows using artificial sites, a clear indication of the importance of such sites for the species in Pennsylvania. Changes in existing excavation practices may also be responsible for some declines, since many companies now flatten and grade excavated areas when projects are complete, leaving no suitable habitat for the Bank Swallow. In 2010, at an active gravel excavation in Lawrence County, a thriving colony of at least 50 Bank Swallow pairs (included in second Atlas data) succumbed to this very fate when the sandy slopes they used were leveled at the conclusion of the project (pers. obs.).

Maryland and New York also reported significant declines in Bank Swallow block occupancy: 41 percent and 28 percent, respectively, between first and second atlases (Ellison 2010zc; Freer 2008). The apparent increase in Bank Swallow occupancy in northwestern Pennsylvania and the contraction of the species from eastern parts of Pennsylvania and neighboring states may be early indications of a range shift for the species. Or, again, as with the Cliff Swallow, this apparent shift may well be among the normal machinations of a dynamic and versatile species—in this case, one that enjoys success across the globe.

GEOFF MALOSH

Distribution

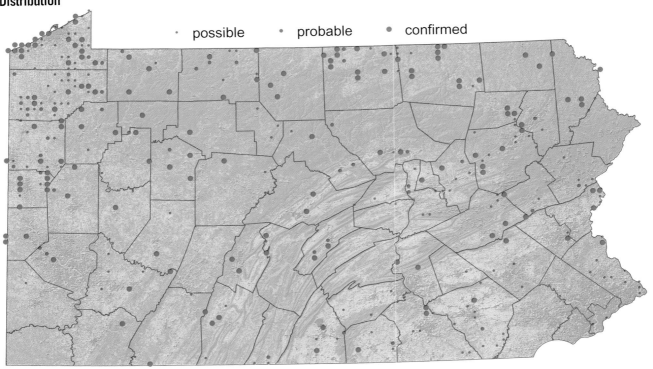

Distribution Change

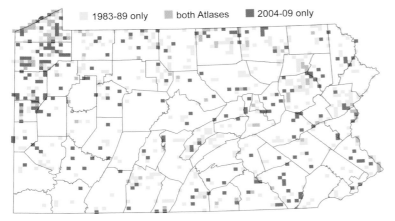

Number of Blocks

	first Atlas 1983–89	second Atlas 2004–09	Change %
Possible	197	150	−24
Probable	81	42	−48
Confirmed	243	145	−40
Total	521	337	−36

Geoff Malosh

Cliff Swallow
Petrochelidon pyrrhonota

If a modern ornithologist were to describe and name the Cliff Swallow based on its current status and distribution in eastern North America, it might aptly be dubbed the "Bridge Swallow." Historically, the Cliff Swallow was a bird of western North America, where it nested under the ledges of cliffs in the many foothills and canyons of the region. It was only in the past 150 years that it dramatically expanded eastward as a breeding species, using bridges, barns, dams, and other artificial structures that became ubiquitous in that time (Brown and Brown 1995). It still uses natural cliff sites in the West, but in Pennsylvania the Cliff Swallow is entirely dependent on man-made structures for nesting sites (McWilliams and Brauning 2000). In this regard, it is one of a few species that has seen a net benefit directly from the activities of modern human civilization.

The Cliff Swallow has a mixed and unpredictable history in Pennsylvania. Not appearing here until about 1830, it quickly, but locally, established itself in most regions of the state (Schwalbe 1992c). Presently, the species appears to be relatively stable in abundance, but it is sporadic in distribution and given to sudden and unexplained abandonment of established nesting areas. The second Atlas found a decline of 14 percent in blocks in which Cliff Swallow was recorded statewide. Much of this decline can be attributed to a steep drop-off in northeastern counties, particularly in the Glaciated Low Plateau Section, with declines primarily centered in Wyoming, Susquehanna, Wayne, and Pike counties. Notable decreases were also evident in the eastern portions of the Ridge and Valley Province and in the Great Valley Section (appendix C). The decrease in the northeastern area, in particular, is perplexing, especially in light of the species' status in nearby New York during its second atlas effort, when confirmations of Cliff Swallows were quite dense throughout an area in New York southwest of the Catskill Peaks (Medler 2008h). This decline is not due to volunteer effort in the area, since coverage of the area was comparable in both Pennsylvania atlases (chap. 6). Additionally, the Cliff Swallow is easily detected when present, so any relative lack of coverage in northeastern Pennsylvania seems an inadequate explanation for the decline there. Meanwhile, the species generally increased in the northwest, though not as significantly as the decline in the northeast. Overall, there was a marked east-to-west shift in occupied blocks statewide, during the second Atlas compared with the first, which must be viewed in the context of the species' erratic history in the state. At first glance, it might illustrate a range contraction to the west, but it could just as likely indicate yet another variation in the range of a dynamic and mobile population.

Atlas volunteers provided details of the structures used by the Cliff Swallow in 181 of the 466 blocks in which the species was confirmed. Of these 181, swallows were found using bridges in 74 blocks (44%), barns in 56 (33%), and churches, houses, dams, and other buildings in the rest. On average, however, bridge colonies held three times as many nests per colony (21) as barns held (7). Though structure data were not reported at every confirmed site, it can still be reasonably concluded that bridges house the most (and the largest) colonies in the state. Very likely, this is due to the reduced competition from House Sparrows at bridge sites compared with sites at barns and other buildings. The House Sparrow has been identified as a threat to the Cliff Swallow in the East (Brown and Brown 1995). In Ohio, for example, House Sparrows were thought to be the primary cause of the near-complete elimination of Cliff Swallows from the state in the first half of the twentieth century, though the species has rebounded in the past 40 years (Peterjohn 2001).

Considering the abundance of suitable man-made structures statewide, particularly bridges, and the Cliff Swallow's willingness to tolerate close association with humans, it can probably be said that the species is not in any particular danger ecologically in Pennsylvania.

GEOFF MALOSH

Distribution

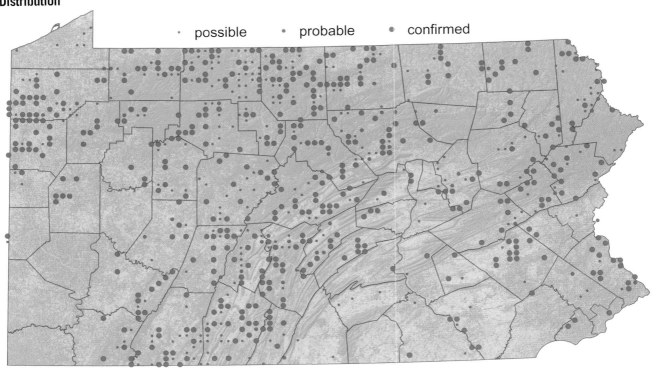

Distribution Change

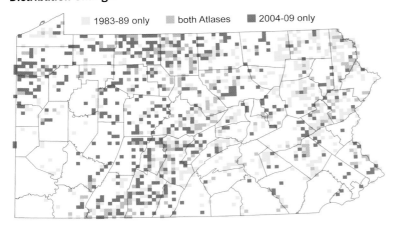

Number of Blocks

	first Atlas 1983–89	second Atlas 2004–09	Change %
Possible	237	256	8
Probable	68	42	−38
Confirmed	578	466	−19
Total	883	764	−14

Breeding Bird Survey Trend

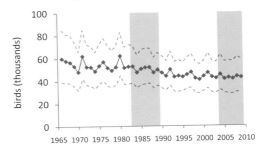

Rob Curtis/VIREO

Barn Swallow
Hirundo rustica

A familiar and well-loved bird to farmers, bird watchers, and many others, the Barn Swallow is a species that is well adapted to humans. It has beautiful, sleek, dark-blue plumage with a maroon face and creamy tan underparts. Its graceful, deeply forked tail has a band of white spots near the tip, and the outer retrices, or tail-streamers, are longer in males than in females (Pyle 1997) and are a good indicator of reproductive success and individual quality (Møller 1994).

The Barn Swallow is the most common swallow, and one of the most widespread songbirds, in the world. In North America, the Barn Swallow breeds across most of the continental United States and southern Canada (Brown and Brown 1999). The European Barn Swallow has associated with humans for over 2,000 years (Møller 1994), and the North American population has undergone a significant range expansion since the start of European settlement (Brown and Brown 1999). The Barn Swallow made an almost complete nest site switch, from mountainous areas with a prevalence of natural caves and crevices to man-made structures, even having nested on Native American dwellings during pre-European settlement (Macoun and Macoun 1909), and it is certainly more prone to nest communally now than it was historically (Brown and Brown 1999). It would seem that the Barn Swallow is one species to which the proliferation of man-made structures is beneficial.

Pennsylvania is located well within the Barn Swallow's breeding range, and the species breeds throughout the state, being absent only from large contiguous forests and highly urbanized areas. There were small range contractions in north-central Pennsylvania and northeast Pennsylvania between atlas periods, resulting in a modest, but significant, reduction in block occupancy, from 92 percent in the first Atlas to 88 percent in the second. This could be due to reforestation reducing the amount of open areas, which the Barn Swallow uses for foraging: it skims low to catch insects, frequently in areas with disturbances like grazing cattle and working farm machinery. Farms and agricultural areas are conducive to the Barn Swallow's nesting, and it is no surprise that it is closely tied to such habitats in Pennsylvania.

An early spring migrant, the Barn Swallow first reaches Pennsylvania in early April, peaking in late April (Leberman 1976; Samuel 1972). It is similarly an early fall migrant, beginning to stage in July and peaking in August (Leberman 1976). Upon arrival to the breeding grounds, the Barn Swallow quickly forms pairs. Atlasers found Barn Swallow nests beginning in early May, eggs in nests peaking throughout mid-May and early June, nestlings in June, and fledglings in late June and early July. The Barn Swallow often raises two broods but completes breeding activities by August.

Pennsylvania's Barn Swallow population, cautiously estimated to be 590,000 birds from atlas point counts, has been in steady decline since the 1960s (Sauer et al. 2011); populations during the second Atlas period were only 70 percent of those in the first Atlas period. Similar declines were noted on Breeding Bird Survey routes in surrounding states (Sauer et al. 2011), part of a more general decline in aerial insectivores in North America (Nebel et al. 2010).

The Barn Swallow has endeared itself to people, and many farmers, who view it as a good luck symbol, protect it and its nest sites. However, the introduction of House Sparrows caused decreases in the Barn Swallow population in the late nineteenth century (Brewster 1906). As nest site competitors, they reduce fledging success in Barn Swallows (Weisheit and Creighton 1989). However, House Sparrow numbers are themselves in steady decline, and recent population decreases may be due to increasing farmland abandonment and forest maturation (McWilliams et al. 2007; Yahner 2003), as well as reductions in the abundance of aerial insects (Nebel et al. 2010). However, the Barn Swallow remains an abundant and widespread species in the Commonwealth.

ANDREA L. CRARY

Distribution

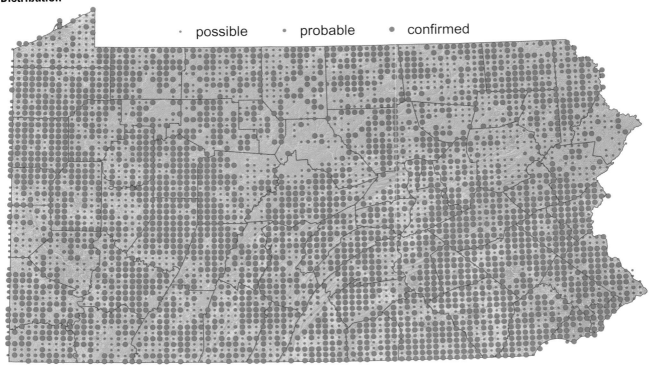

Distribution Change

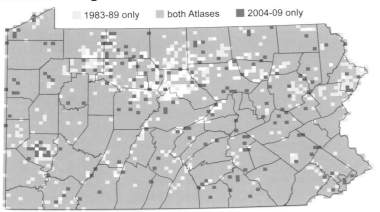

Density

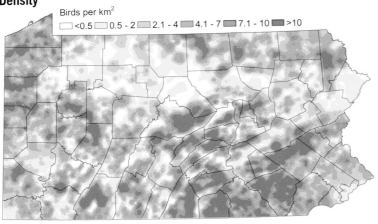

Number of Blocks

	first Atlas 1983–89	second Atlas 2004–09	Change %
Possible	631	1,239	96
Probable	495	449	−9
Confirmed	3,411	2,664	−22
Total	4,537	4,352	−4

Population estimate, birds (95% CI):
590,000 (570,000–610,000)

Breeding Bird Survey Trend

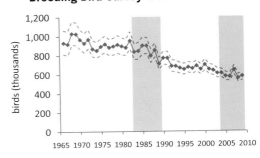

BARN SWALLOW

Doug Wechsler/VIREO

Carolina Chickadee
Poecile carolinensis

One of two chickadee species in Pennsylvania, the Carolina Chickadee differs from its larger close relative, the Black-capped Chickadee, in having a proportionally shorter tail as well as gray rather than white covert feathers on the "shoulder" of the folded wing. Its most common song includes four notes of alternating pitch, *see bee see bay*, unlike the typical song of the Black-capped Chickadee, which is a two-note *fee bee*. Visual similarity of the two chickadees and their ability to learn and produce the song of the other species, especially where the two species overlap and hybridize (Reudink et al. 2007; Curry et al. 2007), complicate analysis of range and abundance. Nevertheless, the Carolina Chickadee is advancing northward in the state and becoming familiar to an increasing percentage of Pennsylvanians.

The Carolina Chickadee is predominantly a bird of forests and mixed landscapes in the South, with a breeding range extending from the Atlantic coast to the Great Plains. The species' northern limit, from New Jersey to Kansas, extends only marginally into Pennsylvania. Accordingly, the state does not contribute significantly to the overall conservation status of this widespread and abundant species.

During the second Atlas period, as in the past, the Carolina Chickadee was concentrated in the southeastern and southwestern corners of Pennsylvania but was absent from higher elevations in the Allegheny Mountain and southern Appalachian Mountain physiographic sections. The distribution overlaps minimally with that of the Black-capped Chickadee: of the blocks inhabited by at least one chickadee species (98% of all blocks), the species occurred together in only 7 percent. The Carolina Chickadee has exhibited one of the largest increases in blocks with records of any Pennsylvania species, a 52 percent increase, indicating a considerable northward range expansion. Atlas results provide little support, however, for lateral shifts that would reflect upslope movement onto ridgetops in Franklin and Fulton counties. The range expansion has come at the expense of the Black-capped Chickadee: of 736 blocks occupied solely by the Carolina Chickadee during the second Atlas period, 18 percent were previously inhabited by the Black-capped Chickadee alone. In addition, the area of range overlap increased significantly between atlas periods. However, notes on some observations submitted to the atlas within the hybrid zone suggest some uncertainty with identification, which may cloud the picture somewhat.

The Carolina Chickadee relies on forested habitats for breeding, but it is remarkably tolerant of human activities and landscape changes. The species can breed in areas with high human density and fragmented forest cover, provided some dead trees or nest boxes are available for nesting. Consistent with its more southerly breeding range in Pennsylvania relative to the Black-capped Chickadee, the Carolina Chickadee begins breeding slightly earlier in the spring, with many nests active by late April. Average dates for breeding confirmation were around 1 week earlier than those for the Black-capped Chickadee (appendix F). Replacement clutches are common, but second broods are rare, so active Carolina Chickadee nests seldom can be found after June.

Abundance of the species during the second Atlas period was highest in scattered parts of the Piedmont Province. Pennsylvania's Carolina Chickadee population was estimated to be 105,000 singing males during the second Atlas period. Breeding Bird Survey (BBS) data suggest little change in abundance, rangewide, 1966–2009 (Sauer et al. 2011). However, the BBS trend for Pennsylvania shows a 6.1 percent annual increase over the same period, and we estimate a population increase of more than 90 percent between atlas periods that corroborates the block expansion.

The Carolina Chickadee's reproductive output can be greatly reduced when House Wrens usurp active nests (Doherty and Grubb 2002a). However, range expansion and mating advantages during hybridization with Black-capped Chickadees (Reudink et al. 2006) both suggest that the Carolina Chickadee will continue to become an increasingly prominent member of Pennsylvania's avifauna.

ROBERT L. CURRY

Distribution

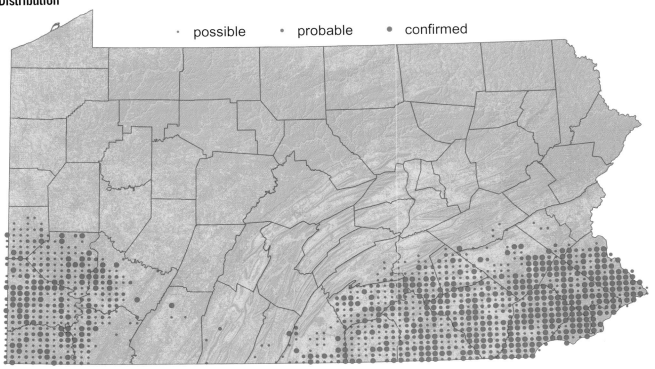

Distribution Change

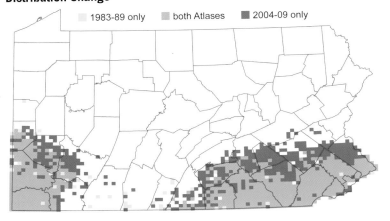

Density

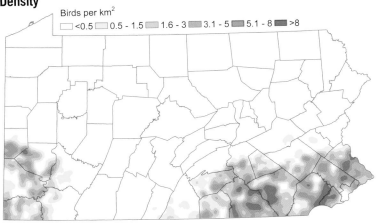

Number of Blocks

	first Atlas 1983–89	second Atlas 2004–09	Change %
Possible	156	380	144
Probable	221	275	24
Confirmed	330	420	27
Total	707	1,075	52

Population estimate, males (95% CI):
105,000 (93,000–119,000)

Breeding Bird Survey Trend

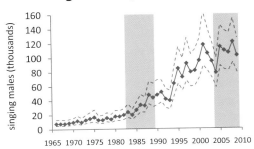

CAROLINA CHICKADEE

Geoff Malosh

Black-capped Chickadee
Poecile atricapillus

Because they are widespread, common, and frequent visitors at bird feeders, chickadees are among the most familiar of forest songbirds throughout Pennsylvania. The Black-capped Chickadee is more widespread and abundant than the Carolina Chickadee. Both are small and active year-round residents that share black, gray, and white plumage patterns; distinguishing features of the Black-capped Chickadee include its relatively long tail and simple two-note song. The two chickadees are not only similar, but they also hybridize where their geographic ranges overlap (Reudink et al. 2006, 2007).

The extensive breeding range of the Black-capped Chickadee spans a broad swath of forested habitats across the northern states and southern Canada, from coast to coast. The species' southern range limit crosses Pennsylvania from east to west, with a dip south of the Mason-Dixon Line following the higher elevations of the Appalachian Mountains. Its range covers most of Pennsylvania, except for the southeastern and southwestern corners, where it is replaced by the Carolina Chickadee. During the second Atlas period, breeding Black-capped Chickadees were recorded in 83 percent of all blocks, and in almost every block within its range. By the first Atlas period, populations had undergone a long-term increase; Breeding Bird Survey data showed gradual increases from the 1960s through the 1980s (Sauer et al. 2011), which continued between the first and second Atlas.

The Black-capped Chickadee's status in Pennsylvania is secure, but the second Atlas documents a contraction at the southern edge of its range where it comes into contact with the Carolina Chickadee. In 59 percent of blocks inhabited by both species, Carolina Chickadees joined Black-capped Chickadees only during the second Atlas period. Contraction of the Black-capped Chickadee's southern range limit was expected, because research has provided evidence of northward movement of the hybrid zone at a rate of roughly 1 km (0.6 miles) per year (Reudink et al. 2007; Curry et al. 2007). The species occupied only 37 fewer blocks—disproportionately along the southern range limit—during the second Atlas period than in the first, a change of less than 1 percent. However, taking into account lower average effort in the first Atlas, we estimate that the actual range change between atlases was closer to 5 percent. Over the same interval, though, the number of blocks with both chickadee species reported increased by 85 percent, mainly due to Carolina Chickadees moving into blocks in which only Black-capped Chickadees were found in the first Atlas period.

Identification based on song is especially problematic in or near the area of overlap with Carolina Chickadees, because both hybrid and genetically pure chickadees of both species are known to be able to learn the song of the other species and thus to produce the "wrong song" (Curry et al. 2007). Extralimital records of Black-capped Chickadees in counties well south of the current position of the hybrid zone (including Chester, Montgomery, and Greene) are suspect and might be attributable to aberrant vocal behavior of hybrid or Carolina Chickadees.

With a statewide population estimated at more than 800,000 pairs, the Black-capped Chickadee ranks fifteenth in overall abundance, just behind its larger close relative, the Tufted Titmouse. Currently, it is most abundant in Pennsylvania's northern and northwestern counties. During the second Atlas period, abundance of the Black-capped Chickadee was positively associated with forest cover (appendix D), and densities were lowest in blocks dominated by non-forest habitat cover. The species is a primary cavity nester, with most nests excavated in dead hardwood trees (especially birch and aspen) beginning in late April; pairs occasionally use existing cavities or nest boxes. These habits allow chickadees to occupy suburban and fragmented landscapes as well as forests. Black-capped Chickadee densities generally increase with elevation, in part because the species is absent from the low-elevation regions occupied by Carolina Chickadees, but also because forest cover tends to increase with elevation.

The Black-capped Chickadee will long be a conspicuous and important component of Pennsylvania's avifauna. However, the species probably will continue to lose ground to the Carolina Chickadee, and the zone of apparent overlap between the two may continue to widen.

ROBERT L. CURRY

Distribution

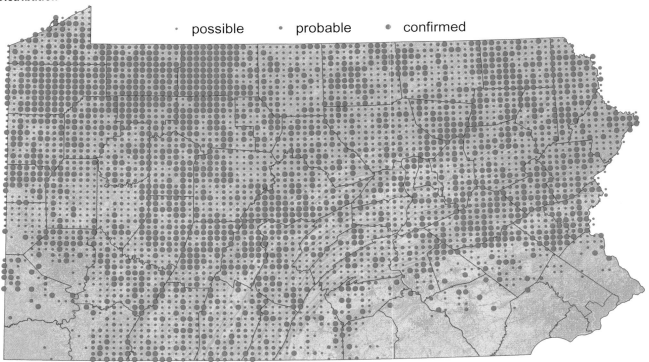

Distribution Change

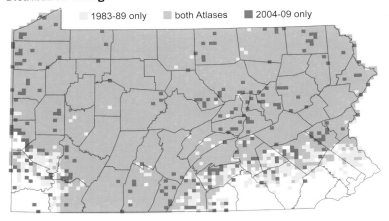

Density

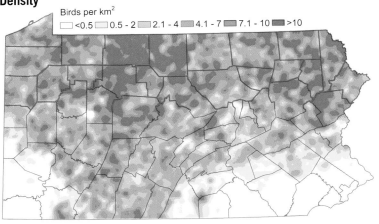

Number of Blocks

	first Atlas 1983–89	second Atlas 2004–09	Change %
Possible	1,085	1,524	40
Probable	1,332	1,066	−20
Confirmed	1,734	1,524	−12
Total	4,151	4,114	−1

Population estimate, males (95% CI):
810,000 (780,000–855,000)

Breeding Bird Survey Trend

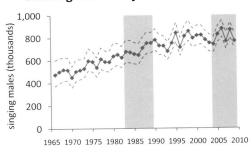

BLACK-CAPPED CHICKADEE 297

Steve Greer/VIREO

Tufted Titmouse
Baeolophus bicolor

One of the most familiar, noisy little birds in all of Pennsylvania, the Tufted Titmouse breeds throughout the state wherever trees are large enough to provide nest holes. Its harsh, scolding calls or *peter, peter* song are unmistakable indicators of its presence. The titmouse is a permanent resident in most of the eastern United States, absent only from extreme northern New York and New England, and was observed in 95 percent of Pennsylvania's blocks in the second Atlas.

The Tufted Titmouse has been a common permanent resident in southeastern and southwestern Pennsylvania for more than 100 years (Baird 1845; Jacobs 1893). It extended its breeding range northward during the 1920s (Burleigh 1931), and by the mid-1960s was present in nearly every county in Pennsylvania (Poole 1964). During the first Atlas, it was recorded throughout the Commonwealth and found in 81 percent of blocks statewide. By that time, it was almost ubiquitous in the southern and central parts of the state, but it occurred in less than 50 percent of the blocks in several northern counties of the Appalachian Plateaus Province, such as McKean, Elk, and Potter (Ickes 1992h).

The second Atlas indicates that the Tufted Titmouse has become still more widespread in the state, expanding across the northern tier, and found in 18 percent more blocks than in the first Atlas. The species did not decrease in block occupancy in any physiographic section (appendix C), and it showed significant increases of 36 to 84 percent in seven sections, including the six northernmost ones. The northward range expansion in Pennsylvania was also noted north of the state border, where blocks with the Tufted Titmouse doubled in number in New York between atlas periods (McGowan 2008zb). Climatic warming and increased bird feeding have been suggested as the primary reasons for the range expansion (Grubb and Pravosudov 1994; Hitch and Leberg 2007).

Point count data for the second Atlas indicate that, in Pennsylvania, the Tufted Titmouse is a habitat generalist, with its highest density recorded in forested residential areas (appendix D). It was most commonly found in fragmented forests/forest edge. Highest average densities, often greater than 10 birds/km^2, were found across the Ridge and Valley Province. Unlike in some neighboring states (Buckelew and Hall 1994; McGowan 2008zb), point count results suggest elevation appears to not be limiting to the titmouse in Pennsylvania. It was slightly less common above 400 m (1,300 ft), but there were still fair numbers of birds recorded on counts above 800 m (2,600 ft).

The Tufted Titmouse population in Pennsylvania was estimated to be 850,000 singing males during the second Atlas. Pennsylvania's Breeding Bird Survey data show a population increase of 37 percent between atlas periods, and the trend for the species since the start of the survey, 1966–2009, was a significant increase of 2.6 percent per year (Sauer et al. 2011). The large decline in 2004 was likely due to increased mortality related to West Nile virus (LaDeau et al. 2007).

The Pennsylvania Tufted Titmouse is on a positive trajectory. Since it is a habitat generalist, differences in density across vegetative zones are difficult to detect (Inman et al. 2002). Two factors that could impact its population numbers are severity of winters and availability of nest cavities. For example, a study using capture–recapture data to model survivorship as a function of winter severity in several bird species reported that Tufted Titmouse survivorship was lower in years with heavy snow cover (Doherty and Grubb 2002b). Additionally, Miller (2010) found that, under certain circumstances, the Tufted Titmouse increased in numbers following the introduction of nest boxes, suggesting that breeding density might be limited by nest cavity availability.

ROY A. ICKES

Distribution

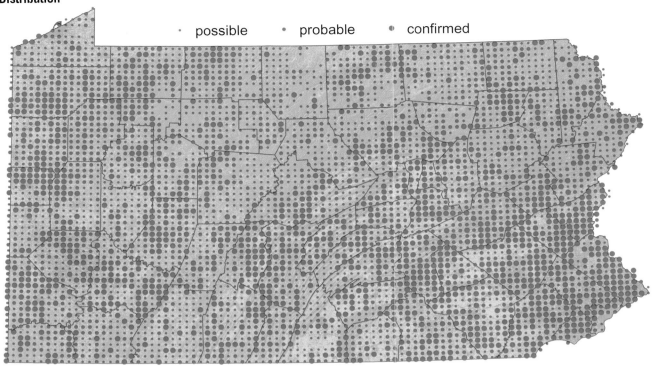

Distribution Change

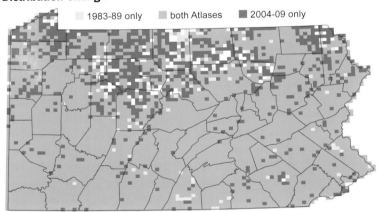

Density

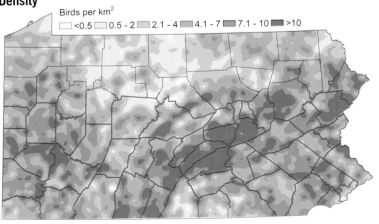

Number of Blocks

	first Atlas 1983–89	second Atlas 2004–09	Change %
Possible	1,237	1,883	52
Probable	1,450	1,432	−1
Confirmed	1,302	1,377	6
Total	3,989	4,692	18

Population estimate, males (95% CI):
850,000 (820,000–890,000)

Breeding Bird Survey Trend

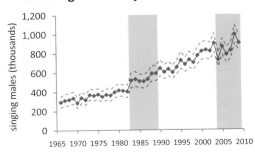

TUFTED TITMOUSE   299

Chuck Musitano

Red-breasted Nuthatch
Sitta canadensis

Formerly known as Canada or Red-bellied Nuthatch, the Red-breasted Nuthatch is a common resident of North America's boreal forests (Ghalambor and Martin 1999). The species' characteristic tinhorn-like *yenk-yenk-yenk* call is heard year-round in coniferous forests from southeastern Alaska, across the boreal forests of Canada and the northern United States, and south into the Rockies and Appalachian Mountains (Ghalambor and Martin 1999). Although resident in most of its breeding range, the Red-breasted Nuthatch withdraws from northernmost areas in most winters, exhibiting synchronized irruptive movements every 2 to 3 years that coincide with a poor crop of conifer seeds on breeding grounds.

The Red-breasted Nuthatch was reported nesting primarily in northern sections of Pennsylvania during the late nineteenth century (Todd 1940), but it probably was a much more widespread breeder before the felling of Pennsylvania's pine and hemlock forests earlier in that century. Poole's (1964) map shows summer records restricted to the state's northern third, with a few records south to northern Berks County. During the 1930s, the Civil Conservation Corp undertook major reforestation projects throughout the state by planting non-native spruces, firs, and pines and native White Pines and Eastern Hemlocks. These conifer plantations became suitable habitat for the Red-breasted Nuthatch and other boreal avian species in the 1960s after these trees matured, resulting in widespread nesting confirmations in the first Atlas project (Santner 1992d).

Although second Atlas data show that the Red-breasted Nuthatch's range is still concentrated in blocks with conifer forests in northern Pennsylvanian counties, this species is also found in plantations farther south. Surface mining reclamation in the 1960s encouraged the planting of spruce, fir, and pine plantations, and the maturation of these plantations enticed not only the Red-breasted Nuthatch to nest in these habitats but also other avian species usually associated with forests in higher altitudes or farther north, such as Golden-crowned Kinglet, Yellow-bellied Sapsucker, Yellow-rumped Warbler, and Winter Wren (Wilhelm 2010). The second Atlas clearly demonstrates a significant expansion of blocks occupied, from 203 in the first Atlas to 554 in the second. Increases in block occupancy were recorded everywhere, but they were especially pronounced in the northern half of the state. The percent of the state's records that came from north of the approximate mid-latitude in Pennsylvania (40.9°N) rose from 74 percent in the first Atlas to 83 percent of records in the second Atlas. A substantial increase in block occupancy was also noted between atlases in New York state's adjacent southern tier (McGowan 2008zc). However, the Red-breasted Nuthatch was found in numerous outposts in the southern half of Pennsylvania; few counties had no records during the second Atlas period.

Breeding Bird Survey data for the Red-breasted Nuthatch are sparse in Pennsylvania, but they suggest an upward trend since the 1960s (Sauer et al. 2011). Despite this increase and the increase in block occupancy between atlases, the species is still thinly distributed, and only 101 were encountered on atlas point counts. From these sparse data, the statewide population was estimated to be in the range of 18,000 to 28,000 individuals. No records of the Red-breasted Nuthatch on point counts occurred below 180 m (600 ft). Point count data confirm that this species is found primarily in coniferous and mixed forests; mean counts in coniferous woodland were more than 50 times higher than those in deciduous woods (appendix D).

Conservation and management of the Red-breasted Nuthatch depend on forestry practices that consider needs of all cavity-nesting birds—retention of some standing dead trees or snags soft enough for nuthatches to excavate—and adequate numbers of cone-producing conifers nearby to provide winter food and resin for smearing nest-cavity entrances. Preliminary studies suggest that resin is an effective barrier against usurping House Wrens and predators (Ghalambor and Martin 1999). With continued forest maturation, it is likely that Pennsylvania's resurgent Red-breasted Nuthatch population will continue to flourish in the near future.

GENE WILHELM

Distribution

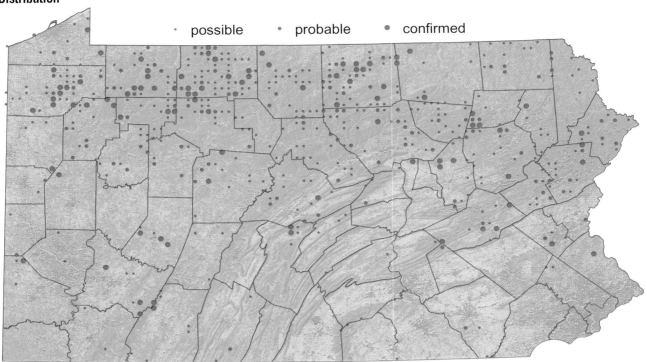

Distribution Change

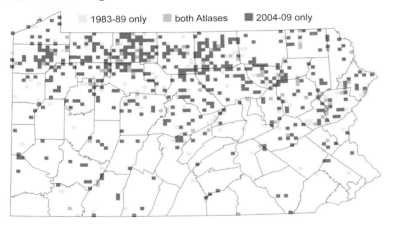

Number of Blocks

	first Atlas 1983–89	second Atlas 2004–09	Change %
Possible	121	351	190
Probable	49	110	124
Confirmed	33	93	182
Total	203	554	173

Population estimate, birds:
18,000–28,000 (approximate)

F. Arthur McMorris/VIREO

White-breasted Nuthatch
Sitta carolinensis

The White-breasted Nuthatch is a year-round resident of Pennsylvania, where it is found in deciduous and mixed forests and is a frequent visitor to wooded suburban yards and bird feeders (McWilliams and Brauning 2000). Pairs stay together throughout the year and maintain year-round territories (Butts 1931), although in some years, populations from northern regions exhibit irruptive movements (McWilliams and Brauning 2000; Grubb and Pravosudov 2008). Nuthatches are bark gleaners, foraging primarily on tree branches and trunks, walking both up and down the trunk in search of food, a characteristic of this family (Grubb and Pravosudov 2008).

Found across most of the continental United States and southern Canada, except treeless areas such as the Great Plains (Grubb and Pravosudov 2008), the White-breasted Nuthatch is common and widespread. Pennsylvania is within the core of its range, being recorded in 4,591 blocks (93% of blocks) in the second Atlas, putting the species in the top 20 most widely distributed breeding species within the state. There was a significant 12 percent increase in block occupancy between the first and second atlases. Increases in block occupancy occurred throughout the state, with infilling within its existing range contributing most of the increase. It is now found in more than 90 percent of blocks in most physiographic sections (appendix C). The most notable range expansion was in the Piedmont, where it was found in 94 percent of blocks, compared with 70 percent in the first Atlas. This coincides with exurban development and a loss of agricultural and grassland habitat (chap. 3) in that region. Although nuthatches are not found in urban areas, they will use suburban areas if larger trees are available. The increase in White-breasted Nuthatch records in southeastern Pennsylvania may be part of wider range expansion; there was a considerable increase in atlas records in the vicinity of the Chesapeake, in Maryland, over a similar time period (Ellison 2010zd).

White-breasted Nuthatches reach their highest densities in small woodlots and forest edges. They are secondary cavity nesters, nesting in natural cavities or those excavated by other species (Grubb and Pravosudov 2008). They often nest in cavities created where a branch broke off or in other natural cavities. Wooded suburban areas are prime habitats, since there are often large trees with cavities in close proximity to bird feeders. The White-breasted Nuthatch population estimate for Pennsylvania was 560,000 individuals. Estimates of population size may be low, due to the early timing of breeding (relative to when most of the bird surveys were conducted) and other artifacts of the sampling method (chap. 5). Breeding Bird Survey (BBS) trends have shown a 2.2 percent annual increase in White-breasted Nuthatch populations in Pennsylvania since the 1960s (Sauer et al. 2011), with BBS counts around 60 percent higher in the second Atlas period than in the first Atlas.

Courtship behavior in White-breasted Nuthatches was observed as early as the end of March. Nest building was recorded from early March to July. Adults were observed carrying food and feeding young from mid-April through 10 August, and fledged young were found as early as mid-May with a mean date of 4 July. Nuthatches have only one brood per season (Grubb and Pravosudov 2008). Although renesting (after nest failure) has not been reported in the literature, the observations of pairs feeding young in late July and August suggest that these were most likely renests. Although the White-breasted Nuthatch was recorded in 93 percent of all atlas blocks, breeding was confirmed in only 22 percent. This low level of confirmation is probably due to the early nesting of nuthatches that occurred before the main atlas season, as well as the difficulty of finding nesting cavities unless the birds are feeding young. The majority of confirmations were of adults carrying food, feeding young, or the presence of fledged young.

Nuthatches appear to be secure in Pennsylvania; due to their preference for forest edges, wooded suburban areas, and small woodlots, this species should continue to do well in the future.

MARGARET C. BRITTINGHAM

Distribution

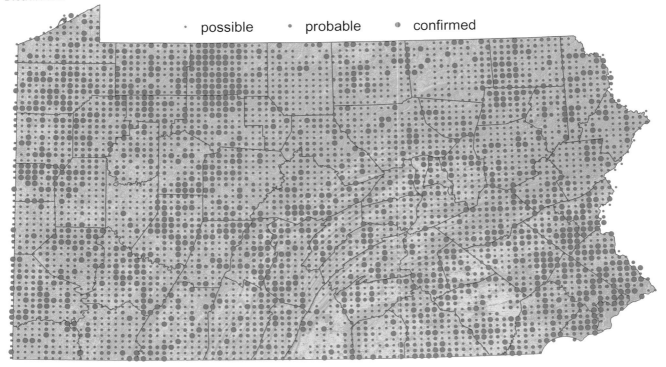

• possible • probable • confirmed

Distribution Change

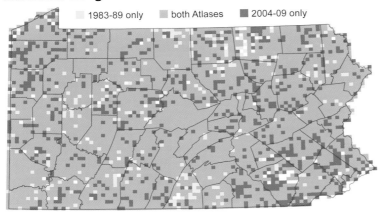

1983–89 only   both Atlases   2004-09 only

Density

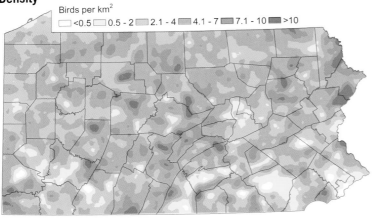

Birds per km²
<0.5 0.5 - 2 2.1 - 4 4.1 - 7 7.1 - 10 >10

Number of Blocks

	first Atlas 1983–89	second Atlas 2004–09	Change %
Possible	1,791	2,298	28
Probable	1,424	1,296	−9
Confirmed	867	997	15
Total	4,082	4,591	12

Population estimate, birds (95% CI):
560,000 (540,000–590,000)

Breeding Bird Survey Trend

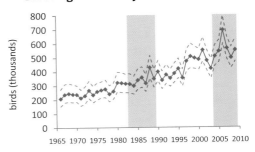

WHITE-BREASTED NUTHATCH

Gerard Dewaghe

Brown Creeper
Certhia americana

Small and brown, clinging tightly to large, rough-barked trees, the Brown Creeper is perhaps the most camouflaged perching bird in Pennsylvania's forests. Spiraling up the trunk in search of spiders and other small organisms that think they are safely hidden in the deep furrows, creepers duck in and out of sight and then fly to the base of the next tree to start upward again. Although its call is high-pitched and not easily heard unless close by, the Brown Creeper has a springtime song that is surprisingly musical and unexpected from such a plainly colored bird.

The Brown Creeper is a widespread breeder in coniferous forests across North America, from Newfoundland to southeastern Alaska, through the Rocky Mountains, and south to Nicaragua (Hejl et al. 2002). Pennsylvania is in the southern part of this range in the eastern United States; the Brown Creeper breeds in the northern half of the state and southward through the Allegheny Mountains. The Brown Creeper breeds in forests with abundant large trees, both living and dead. Usually placing their nests under the bark of dead trees, they forage on the lower, furrowed parts of large trees greater than 30 cm (12 inches) in diameter (Morrison et al. 1987). On the northern Appalachian Plateaus, they reach greatest abundance in old-growth Eastern Hemlock and White Pine forests (Haney 1999). Reflecting the availability of these habitats, the Brown Creeper is generally found at higher elevations (>400 m, ~1,300 ft) in Pennsylvania (appendix D).

Since the first Atlas, the Brown Creeper's range has contracted northward. Although the number of blocks with creeper records decreased by only 4 percent, an increase in observer effort in the northern tier (chap. 6) may have resulted in more detections there. Accounting for this, the adjusted species' distribution has decreased by 16 percent. The number of occupied blocks in the southern half of the state decreased substantially, by 57 percent in the Piedmont and by 34 percent in the Ridge and Valley Province. These losses were offset by a 10 percent increase in blocks detecting creepers in their core range within the Appalachian Plateaus, although this increase was not consistent across sections, with the Allegheny Front and Pittsburgh Low Plateau sections having significantly fewer occupied blocks.

The net result of these changes is that the mean latitude of southern range limit shifted north by about 17 km (10 miles) between atlas periods. This ties in with concurrent range contractions in Maryland (Ellison 2010ze) to the south and range expansions and population increases in New York (McGowan 2008zd) and Ontario (McLaren 2007a) to the north. Hence, while this species may be benefiting from forest maturation in cooler areas such as northern Pennsylvania, other processes, possibly climatic, are simultaneously reducing the suitability of lower elevation forests, especially near the southern range limits. The population size during the second Atlas period was estimated to be 34,000 singing males, with notable strongholds in the Allegheny National Forest and Sproul State Forest.

The future for the Brown Creeper in Pennsylvania is unclear. The distribution of creepers mirrors that of coniferous/hemlock forest cover and core forest distribution in the state. The steady spread of the Hemlock Woolly Adelgid from the southeast will likely lead to a decrease of healthy Eastern Hemlock stands across the state and the species that favor them (Ross et al. 2004). However, greater tree mortality would have mixed implications for the Brown Creeper, temporarily resulting in more nesting sites on dead trees but fewer foraging trees. Because this species is found primarily in interior forests, fragmentation of forest cover is also worrisome. The current boom in drilling for deep shale gas in northern-tier forestland likely will increase fragmentation of contiguous stands and concurrently decrease core forest areas (Brittingham and Goodrich 2010). Such changes in land use, coupled with projected climate change, could result in habitat degradation that could affect the Brown Creeper in coming decades.

JERRY SKINNER

Distribution

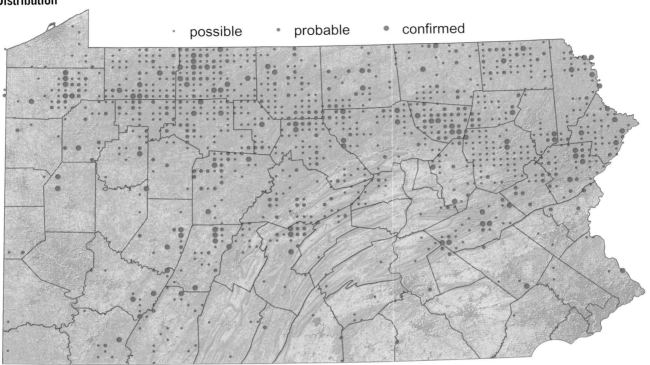

Distribution Change

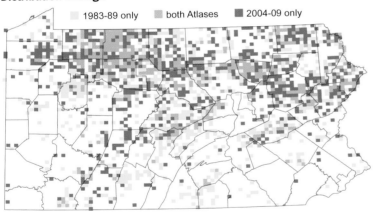

Density

Number of Blocks

	first Atlas 1983–89	second Atlas 2004–09	Change %
Possible	728	770	6
Probable	317	221	−30
Confirmed	94	103	10
Total	1,139	1,094	−4

Population estimate, males (95% CI):
34,000 (28,000–41,000)

Breeding Bird Survey Trend

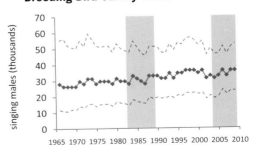

BROWN CREEPER 305

Rob Curtis/VIREO

Carolina Wren
Thryothorus ludovicianus

The Carolina Wren's inquisitive personality, lively behavior, willingness to associate closely with humans, and rollicking *tea-kettle, tea-kettle, tea-kettle* song are endearing to anyone who encounters the species. In its woodland natural habitat, this wren nests in thickets, tangles, hollows in trees, fallen logs, upturned stumps, old woodpecker holes, and other cavities. People in residential areas know the species for its use of any artificial nook and cranny it can find, such as mail boxes, hanging baskets, discarded cans, ledges in outbuildings, and junked cars.

Historically, the Carolina Wren's core range was in the southeastern states, although John J. Audubon (1831) found the species in both southwestern and southeastern Pennsylvania two centuries ago. As recently as the 1950s, publications called southern Pennsylvania its range limit, even though George Miksch Sutton had stated in 1928 that the wren was already expanding here. In 1933, a nest was documented in Warren County, in northern Pennsylvania, although that occurrence was recognized as unusual (Todd 1940).

The first Atlas documented the Carolina Wren's northward range expansion along the state's major river valleys (Ickes 1992i), but there was still much space for the wren to expand further in Pennsylvania. The 3,487 blocks occupied in the second Atlas reflect a 68 percent increase since the first Atlas. Two environmental factors still appear to limit the species' Pennsylvania distribution: dense forests and elevations higher than 400 m (1,300 ft) above sea level. There, the species remains scarce or absent.

The Carolina Wren's northward expansion since the first Atlas was most dramatic in the Appalachian Plateaus, where occupied blocks increased 125 percent, and in the Ridge and Valley Province, where occupancy increased 45 percent. However, even in the Piedmont Province, which was already a Carolina Wren stronghold, it was recorded in 17 percent more blocks. Thus, the species not only advanced northward but also expanded into unoccupied pockets in traditional areas and into some higher elevations in the Allegheny Mountains.

Expansion north of Pennsylvania continues. Occupied blocks increased fourfold between New York's first (1980–1985) and second (2000 to 2005) atlases (Smith 2008d), while in southeastern Ontario, currently the northern breeding frontier, the population grew rapidly during the past two decades (Read 2007).

Bolgiano (2010) demonstrated an exponential increase in Carolina Wren populations in Pennsylvania Christmas Bird Counts since 1940, but the species' numbers are well known to crash during harsh winters and then to rebuild toward high levels. Extreme fluctuations were described for the Pittsburgh area by Hess (1989) and for the state by Bolgiano (2008), who showed that the first Atlas took place when the population was just beginning to recover from a catastrophic collapse during severe winters in the late 1970s. In contrast, the second Atlas came after a decade of fast population growth; Breeding Bird Survey data show a 9 percent annual increase between the two atlas periods (Sauer et al. 2011), equating to a fourfold increase overall. By all accounts, the population reached an all-time high during the second Atlas.

The species' second Atlas population estimate was 270,000 singing males, with the highest densities typically associated with lower elevations in the southern corners of the state. Population densities in areas into which the species expanded its range between atlases tend be lower and will therefore be more vulnerable to local extirpation in cold winters.

The atlas results suggest reasons for the rapid growth and expansion of Carolina Wren populations. First, birds were seen carrying food as early as 27 March and feeding young as late as 10 September (appendix F). The species is a prolific breeder, with up to three broods in a season (Haggerty and Morton 1995). Second, point counts in residential areas averaged more singing males than in any other habitat type. Apparently, the Carolina Wren has been taking advantage of the unintentional provision of nesting sites and shelter from cold weather, as well as food during winter.

PAUL HESS

Distribution

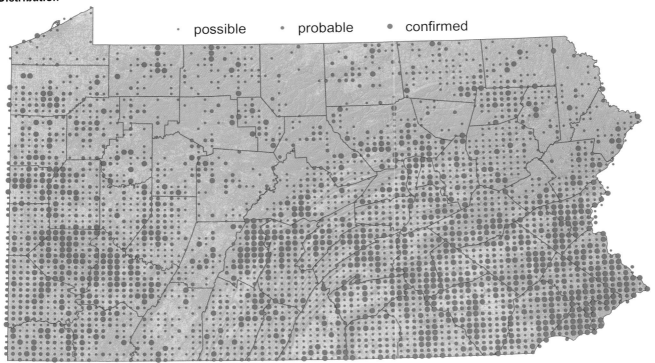

Distribution Change

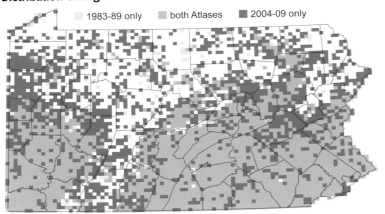

Density

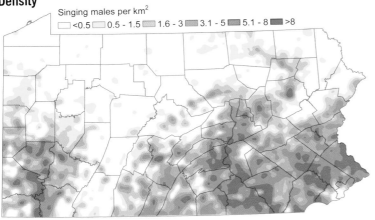

Number of Blocks

	first Atlas 1983–89	second Atlas 2004–09	Change %
Possible	700	1,628	133
Probable	839	1,066	27
Confirmed	531	793	49
Total	2,070	3,487	68

Population estimate, males (95% CI):
270,000 (260,000–282,000)

Breeding Bird Survey Trend

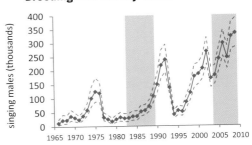

CAROLINA WREN

Bob Wood and Peter Wood

House Wren
Troglodytes aedon

Small, brown, and feisty, the House Wren is a well-known backyard wren in much of North America. Its exuberant and cheerfully bubbly song is easily recognized and often used as the generic suburban songbird song in movies and on television. As a cavity nester, it eagerly nests in birdhouses and boxes provided by people and can be a fierce nest site competitor: some adults will peck and remove eggs or nestlings of other species in cavities or other nests within their territories (Creaser 1925; Sealy 1994).

As its name suggests, the House Wren makes its home close to humans in appropriate open forest or edge habitat, small woodlots, city parks, or residential and urban areas with large trees. It tends to avoid dense core forest, conifers, and higher elevations (appendix D). Prior to European settlement, the House Wren was likely far less abundant in the United States, but it expanded its range rapidly with human movement and clearing of core forest and now seems to benefit from forest fragmentation (Johnson 1998). It builds its nest in old cavities excavated by woodpeckers, but it will nest in a wide array of natural or man-made crevices or cavities (Schutsky 1992d).

The northern House Wren subspecies (*T. a. aedon*) breeds across southern Canada and in most of the continental United States. Pennsylvania is centrally located within the breeding range of the *aedon* group. Those breeding in the northern part of their range migrate to the southern United States and Mexico to overwinter (Johnson 1998). In Pennsylvania, males return first each spring, immediately begin singing to establish territories, and then initiate a variable amount of nest building in cavities. Females arrive shortly after males, choose mates, complete the nest, and begin brooding. In Pennsylvania, atlas volunteers found House Wrens building nests in late May and peaking in early June, nests with eggs in late May through early July, nestlings in June and July, and fledged young peaking in early July (appendix F). House Wrens frequently double-brood, especially those with territories at lower elevations and involving adults with early arrival dates (Johnson 1998).

The House Wren's breeding range has not changed substantially since the first Atlas, but there is evidence of a contraction away from some of the most densely forested blocks on the Appalachian Plateaus of north-central Pennsylvania. However, increases in block records elsewhere, possibly due, in part, to increased survey effort, balanced out those losses. Away from the Appalachian Plateaus, almost all blocks were occupied, indicating how ubiquitous this species is in most of the state. It was one of the most numerous species on point count surveys, reflected in a statewide population estimate of 900,000 singing males, easily making this the most abundant of the five species of wren found in Pennsylvania. It is most abundant in low-elevation areas, especially where there is a mixture of woodlots and low-density housing, such as across parts of the Piedmont and the suburban hinterland of Pittsburgh and Erie. It is rather scarce in more heavily developed areas, such as the centers of the state's largest cities.

In Pennsylvania, the rate of forest maturation is exceeding the rate of tree harvest (McWilliams et al. 2007), and the rate of farmland abandonment and maturation of open habitats into forests had been increasing (Trani et al. 2001). This natural succession could cause a slight decrease in the House Wren population, and data from the Breeding Bird Survey indicate a small, but significant, 0.5 percent decrease per year (Sauer et al. 2011), equating to a 21 percent decline in Pennsylvania between the atlas periods. However, the House Wren remains an abundant species in the state and across its range and has benefited greatly from anthropogenic change; hence, modest declines in its population size are not cause for concern.

ANDREA L. CRARY

Distribution

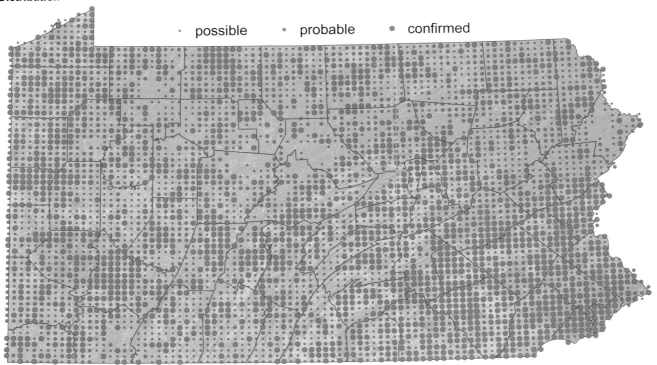

· possible   · probable   · confirmed

Distribution Change

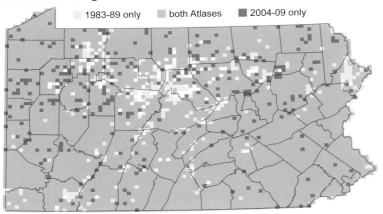

1983-89 only   both Atlases   2004-09 only

Density

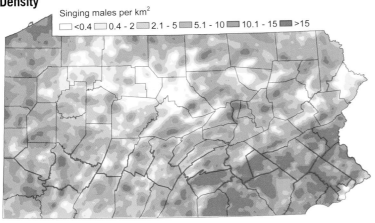

Singing males per km²
<0.4   0.4 - 2   2.1 - 5   5.1 - 10   10.1 - 15   >15

Number of Blocks

	first Atlas 1983–89	second Atlas 2004–09	Change %
Possible	864	1,633	89
Probable	1,232	1,116	−9
Confirmed	2,444	1,782	−27
Total	4,540	4,531	0

Population estimate, birds (95% CI):
900,000 (880,000–925,000)

Breeding Bird Survey Trend

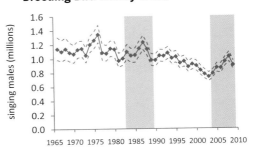

HOUSE WREN   309

Garth McElroy/VIREO

Winter Wren
Troglodytes hiemalis

The Winter Wren's richly varied song, the longest of any North American bird, has been described as "the pinnacle of song complexity" (Kroodsma 1980). It is heard emanating from a streamside tangle or hemlock thicket far more often than the skulky, brown, mouse-like bird is ever seen.

The stub-tailed Winter Wren inhabits the understory of cool, moist forests in Pennsylvania's highlands. Here, the wren may be found where features often associated with old-growth forest, such as large trees, downed logs, tangles, root balls, and snags, provide places for it to feed, nest, and roost. It is most common adjacent to streams or wetlands, often in shady, hemlock-lined ravines (Brauning 1992h).

The Winter Wren, which was split from the Pacific Wren (*T. pacificus*) in 2010 (Chesser et al. 2010), breeds across southern Canada, the Great Lakes, and northeastern United States, extending south throughout the higher Appalachian Mountains; Pennsylvania is at the southern edge of the species' main breeding range. In Pennsylvania, atlas volunteers recorded the Winter Wren from 799 blocks (16.2 percent of all blocks), the majority concentrated in the cool, heavily forested north-central highlands. It is found locally elsewhere in the northern tier and at high elevations in the Allegheny Mountains and in the Ridge and Valley Province. The Winter Wren expanded its range in the state greatly since the first Atlas: block occupancy more than doubled, exceeding increases of other northern species (Blue-headed Vireo, Hermit Thrush, Magnolia Warbler, Black-throated Green Warbler, Blackburnian Warbler) that share its cool, shady habitat. Similarly, recent atlases in New York and Ontario reported significant increases in Winter Wren populations (Brewer 2007a; McGowan 2008ze). Atlas volunteers found the species primarily in coniferous or mixed forest above 400 m (~1,300 ft) elevation (appendix D). Hemlock was a major component in about 70 percent of occupied forests, suggesting that the Winter Wren has one of the strongest hemlock associations of any species in the state.

Pennsylvania supports an estimated 32,000 territorial Winter Wren males, a population that has more than doubled in size since the first Atlas, according to Breeding Bird Survey counts (Sauer et al. 2011). Highest densities of Winter Wrens were found in the High Plateau and Deep Valleys sections, with lesser concentrations in the state's northeastern corner and in the Laurel Hill, in the southwest. The Winter Wren was undoubtedly much more common across Pennsylvania before the extensive deforestation of the late nineteenth and early twentieth centuries, when primeval beech-hemlock forests covered much of the state. Early accounts considered the species locally abundant (e.g., Baily 1896), and it remains so in remnant old-growth stands (Haney 1999). The population increases documented by the atlas project probably represent only a partial recovery to prior levels.

In the State Wildlife Action Plan, the Winter Wren is classified as a species of Maintenance Concern, due to its association with mature conifer forest (PGC-PFBC 2005). Its strong association with hemlock suggests that it may be vulnerable to losses resulting from the spread of the Hemlock Woolly Adelgid (Allen and Sheehan 2010). However, Winter Wren occupancy increased since the first Atlas in the extreme northeastern corner of the state, an area infested by the adelgid over a decade ago (USFS 2009; chap. 3), so it remains to be seen how much of an impact the adelgid might have. Although it is a forest-interior species, the Winter Wren seems to tolerate some degree of logging and has been shown to increase in number after group-selection cutting (Todd 1940; Campbell et al. 2007). As its name suggests, the Winter Wren is a short-distance migrant that winters in a variety of wooded and scrubby habitats across the United States, south to the Gulf Coast; therefore, it seems unlikely that wintering ground issues could limit populations. If an effective Hemlock Woolly Adelgid control program is developed, or if the Winter Wren proves to be resilient to hemlock loss, then this songster should remain a characteristic part of our northern forests.

SCOTT H. STOLESON

Distribution

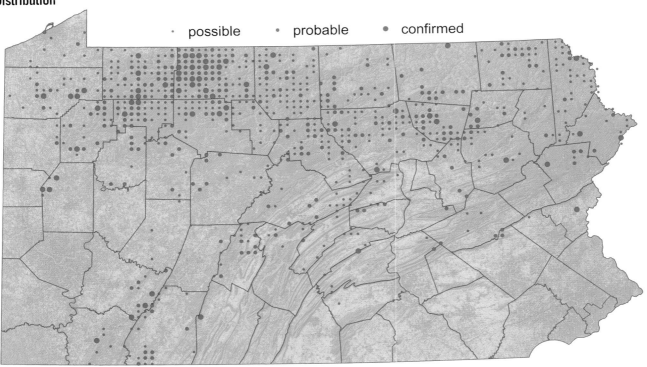

Distribution Change

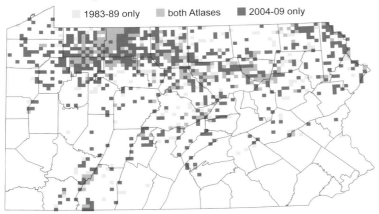

Density

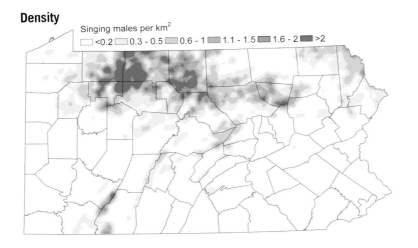

Number of Blocks

	first Atlas 1983–89	second Atlas 2004–09	Change %
Possible	181	496	174
Probable	157	244	55
Confirmed	40	59	48
Total	378	799	111

Population estimate, males (95% CI):
32,000 (27,000–38,000)

Breeding Bird Survey Trend

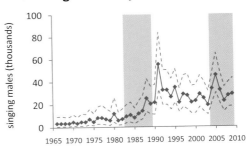

WINTER WREN 311

Gerard Dewaghe

Sedge Wren
Cistothorus platensis

The Sedge Wren, formerly known as the Short-billed Marsh Wren, was renamed to better distinguish it from the closely related Marsh Wren by stressing habitat differences between the two species. Although both occur in wetlands, the Sedge Wren is distinguished by its smaller bill, streaked crown and back, obscure, buff eye line, and distinctive, chattering trill vocalization. It is a widely distributed, polygynous species with many disjunct populations occurring in North, Central, and South America. In North America, the Sedge Wren breeds primarily in the states that formerly constituted the tallgrass prairies. Hence, Pennsylvania is, at best, at the very edge of this species' range, and its presence in the state could be described as extralimital. Historical accounts suggest that it has always been a rare breeding bird in the state (McWilliams and Brauning 2000).

A peculiar, ephemeral short-distance migrant in North America (Herkert et al. 2001), the Sedge Wren prefers moist habitats with rank vegetation like Orchard Grass, sedge meadows, Reed Canary Grass, and hayfields. These vegetative choices differ from the emergent cattail wetlands frequented by the Marsh Wren, although both first Atlas and second Atlas volunteers noted the Sedge Wren using this habitat in Crawford County. Vegetative succession or human disturbances caused by grazing, haying, and planting impart a highly transitory character to most Sedge Wren nesting habitats, making the species one of the most nomadic and least predictable breeding birds (Leberman 1992g).

Despite the Sedge Wren's reputation as an ephemeral breeding bird, some sites occupied in the second Atlas have a long history: one site in Butler County has been used since at least 1970 (Geibel 1975). Breeding evidence increased slightly between the first and second Atlas periods, from four to seven possible, six to eight probable, and three to five confirmed, for a total of 20 blocks. As in the first Atlas, the Sedge Wren was located most frequently in the poorly drained Northwestern Glaciated Plateau Section, where several atlas records involved small colonies, including five to six pairs at two different sites. Elsewhere, the species was rare, with records in four widely scattered blocks in the Lower Susquehanna River Valley representing the only records in the eastern half of the state. Birds were noted carrying food in one of those blocks in both 2004 and 2007, suggesting that not all records away from the northwestern area refer to casual migrants.

Breeding Bird Survey (BBS) data show a significant increase in Sedge Wren numbers in the Eastern Tallgrass Prairie, at the core of its breeding range (Sauer et al. 2011), but BBS data from northeastern states are too sparse to evaluate trends there. There was no significant change in Sedge Wren status between Ontario breeding bird atlases (Brewer 2007b), but there was a 26 percent increase in occupied atlas blocks in New York (McGowan 2008zf). However, in New York, as elsewhere in the Northeast, the Sedge Wren remains rare and is listed as Threatened (McGowan 2008zf). In Pennsylvania, it is currently listed as Endangered (PGC- PFBC 2005), a downgrading since the first Atlas period, when it was listed as Threatened (Leberman 1992g).

As with so many other wetland species, the loss of its habitat is the primary cause of the decline in Sedge Wren population (Gross and Haffner 2010). The specific habitats of the species became scarce due to modern agricultural practices (i.e., frequent mowing, overgrazing, and plowing) and urbanization (Herkert et al. 2001). However, the Conservation Reserve Program in the United States and the North American Waterfowl Management Plan in Canada benefited the species by providing new habitats to colonize (Prescott and Murphy 1999). More sympathetic management practices, such as protecting existing special wetlands, minimizing disturbances (especially spraying pesticides, burning, and mowing), restoring or creating new sedge meadows and moist grasslands, and implementing an educational outreach program about this species for members of the agricultural community are paramount in preventing this endangered species from becoming extirpated in Pennsylvania.

GENE WILHELM

Distribution

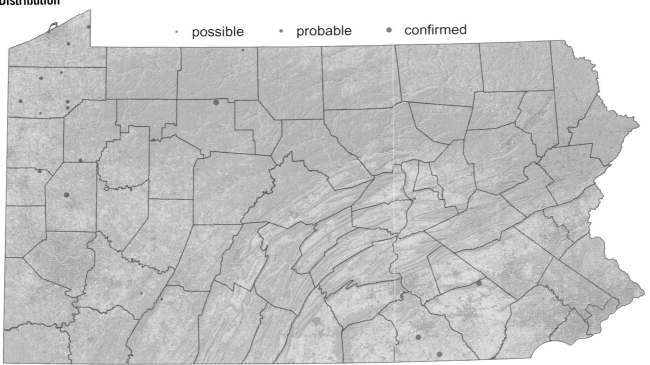

Distribution Change

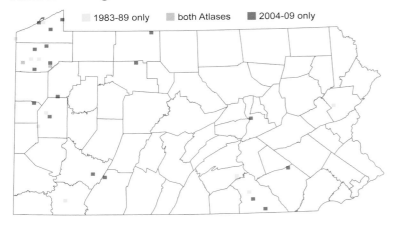

Number of Blocks

	first Atlas 1983–89	second Atlas 2004–09	Change %
Possible	4	7	75
Probable	6	8	33
Confirmed	3	5	67
Total	13	20	54

Gerard Dewaghe

Marsh Wren
Cistothorus palustris

One of the most characteristic sounds that emanates from expansive cattail marshes is the liquid, gurgling, and nearly incessant song of the Marsh Wren. The welfare of this energetic bird is inextricably linked to marshes and tidal rivers, and it has declined dramatically as our wetlands have disappeared. Formerly rather widely distributed across the state, it has been reduced to a fraction of its former range here, and it is now considered a rare bird. Historically, the largest populations occurred in the glaciated northwest and tidal sections of the Delaware River and its tributaries along the state's southeastern corner (Hunt 1904; Harlow 1913; Todd 1940).

The breeding range of the Marsh Wren extends from southern Canada across much of the northern half of the United States, and locally along coastlines, south to southern Mexico. This species is divided into a complex of 14 or more subspecies. Marsh Wrens occurring on the Coastal Plain of southeastern Pennsylvania are assignable to *C. p. palustris*, and those inland to *C. p. dissaeptus* (Kroodsma and Verner 1997). The former is smaller than the latter and has whitish markings as opposed to buff.

The Marsh Wren occurs in a wide array of wetland habitats, including freshwater, brackish, and saltwater marshes, lake and river margins, and tidal creeks (Kroodsma and Verner 1997). Breeding habitat consists primarily of wetlands that include dense stands of emergent vegetation, including cattails, bulrushes, sedges, Common Reed, and other tall, thick-stemmed aquatic plants (Zimmerman et al. 2002c). It is a colonial nester with a preference for larger wetlands of 8 ha (20 ac) or greater (Cashen 1998).

During the first Atlas period, the Marsh Wren was detected in 77 blocks, but it was reported from only 53 blocks during the second Atlas, a 32 percent decline. Notably, the number of blocks for which breeding was confirmed declined from 22 to 12. The Northwestern Glaciated Plateau Section of the Appalachian Plateaus Province accounted for 28 occupied blocks and 7 of the 12 breeding confirmations. More than 50 pairs were reported at John Heinz National Wildlife Refuge (Tinicum) in Delaware and Philadelphia counties, in the same three blocks in which breeding was confirmed during the first Atlas period. Observations declined from 35 to 21 blocks in the remainder of the state between atlases. In contrast, confirmed or probable breeding activity, and reports of appreciable numbers of singing males, in four blocks in the Marsh Creek wetlands ("The Muck") of Tioga County represents an expansion, since Marsh Wrens had not been reported there during the first Atlas. With the exception of the Tinicum records, most records noted that observations occurred in or adjacent to marshes that consisted of or included cattails.

Reid (1992d) provided a summary of the Marsh Wren's decline in the state prior to and at the conclusion of the first Atlas and considered this bird a seriously threatened species. The second Atlas appears to confirm this alarming trend. Crawford County, with observations in 18 blocks, including 5 in which breeding was confirmed, remains the stronghold for this species. The Marsh Wren is still considered common in the extensive Pymatuning and Conneaut Marsh complexes there. Due to their size and protection as state game lands, these wetlands are perhaps the only locations where the Marsh Wren is truly secure, over the long term, in Pennsylvania.

In 1985, the Marsh Wren was listed as Vulnerable (Gill 1985). Human sprawl and the demand for groundwater exacerbate the already epidemic loss and degradation of emergent marshes. If concerted measures are not soon taken to protect the small and medium-sized marshes that constitute much of its occupied range in Pennsylvania, the Marsh Wren seems eligible to become a threatened or endangered species. It currently is listed as a species of High Level Concern in the state's Wildlife Action Plan (PGC-PFBC 2005).

ROBERT W. CRISWELL

Distribution

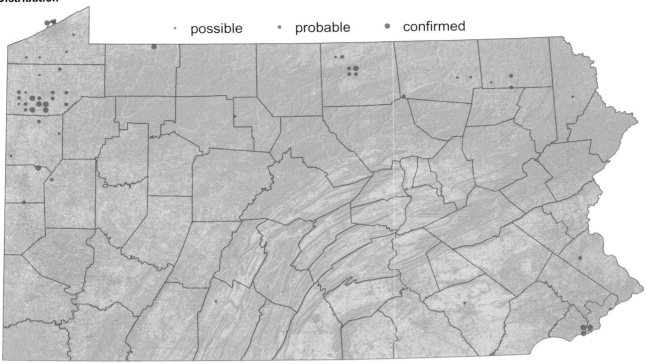

Distribution Change

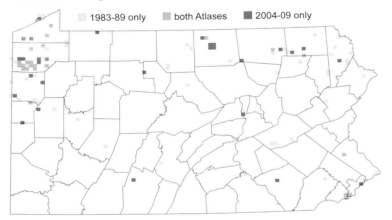

Number of Blocks

	first Atlas 1983–89	second Atlas 2004–09	Change %
Possible	33	19	−42
Probable	22	22	0
Confirmed	22	12	−45
Total	77	53	−32

James R. Woodward/VIREO

Blue-gray Gnatcatcher
Polioptila caerulea

Tiny, dainty, and deceptively frail-looking, the Blue-gray Gnatcatcher is an early spring migrant, returning "before there are many gnats to catch," as Bruce Peterjohn (1989) quipped. In most of Pennsylvania, it comes back in April, fluttering from twig to twig in barely leafed treetops, proclaiming its presence with soft, wheezy contact calls and faint jumbles of *zees*, squeaks, and warbles that make up its complex song.

The gnatcatcher breeds from coast to coast, from the northern United States and south through most of Mexico (AOU 1998). It is among many species whose breeding ranges expanded northward in the eastern United States since the late nineteenth century (Ellison 1992). In Pennsylvania, it was found primarily in the southeastern and southwestern counties in the 1880s (Warren 1890), advanced northward irregularly through the twentieth century (Todd 1940; Poole 1964), and by the first Atlas period was nesting in every county, after what Leberman (1992h) termed a "phenomenal range expansion."

A modest increase in the number of blocks occupied by gnatcatchers between the first and second Atlases masks two sharply different trends: expansion in central and eastern areas, and retraction in western areas. Notable increases were 59 percent in the Piedmont Province and 34 percent in the Ridge and Valley Province. Prominent southwestern decreases were 39 percent in the Allegheny Mountains, 25 percent in the Waynesburg Hills, and 23 percent in the Pittsburgh Low Plateau (appendix C). These declines were corroborated by regional declines in Breeding Bird Survey (BBS) routes in the Ohio Hills region. Ironically, the latter two areas had been traditional strongholds of the gnatcatcher for more than a century. The contrasting trends are not easily explained, because geographic patterns of change in block occupancy between atlases do not consistently match patterns of gains and losses in forestland cover (chap. 3) during the same period. For example, substantial forest-area loss in the Pittsburgh Low Plateau, likely due to urban sprawl, was logically consistent with decreases in occupancy. On the other hand, forest losses in the Piedmont and Ridge and Valley provinces were similarly substantial, but block occupancy actually increased. Only fine-scale ecological studies can determine the factors involved.

Densities of Blue-gray Gnatcatcher were relatively high in the Appalachian Mountain, Great Valley, and easternmost Glaciated Low Plateau sections, low in the Allegheny Mountains and across the northern tier, and patchily high and low in western areas of the state. Although the species nests in a variety of deciduous and mixed woodlands, it is primarily associated with Pennsylvania's oak-hickory forests rather than northern hardwoods (Ross et al. 2001). Point count data in the second Atlas indicate that sites below 300 m (1,000 ft) elevation are strongly favored (appendix D). The gnatcatcher typically shuns the forest interior, preferring to nest near an edge (Leberman 1992h).

BBS data for Pennsylvania show little indication of a trend (Sauer et al. 2011), although such data may be inadequate, because June surveys are late for detecting singing Blue-gray Gnatcatchers. If so, the BBS may be out of synchrony with gnatcatcher detectability: the earliest nest-building date was 2 May in the first Atlas and 17 April in the second Atlas, a 2-week advance, suggesting that this is one of many species nesting earlier in correlation with warming spring climate. The estimated population of 280,000 individuals from atlas point counts could be a considerable underestimate because, as with the BBS, the majority of counts were conducted in June, when gnatcatcher song output is reduced, although the effects of seasonality on detection were taken into account in population models (chap. 5).

Despite no specific conservation concern in Pennsylvania at this time, Stoleson and Larkin (2010) emphasize potential threats to forest ecosystems statewide, including further urban expansion, overabundant deer populations, gas and oil drilling, and climate change. Even such a common forest bird as the Blue-gray Gnatcatcher may require more attention in the future.

PAUL HESS

Distribution

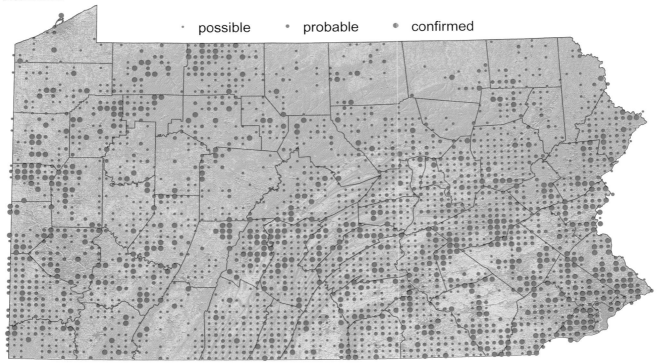

Distribution Change

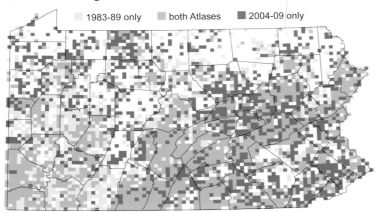

Density

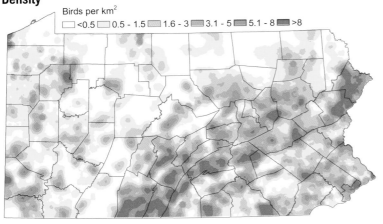

Number of Blocks

	first Atlas 1983–89	second Atlas 2004–09	Change %
Possible	979	1,492	52
Probable	817	631	−23
Confirmed	511	513	0
Total	2,307	2,636	14

Population estimate, birds (95% CI):
280,000 (218,000–370,000)

Breeding Bird Survey Trend

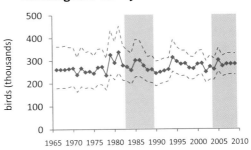

BLUE-GRAY GNATCATCHER 317

Brian E. Small/VIREO

Golden-crowned Kinglet
Regulus satrapa

The Golden-crowned Kinglet, the most diminutive of Pennsylvania's nesting songbirds, expanded its breeding range from boreal conifer forests into a wider variety of locations during the previous century. Termed a "a little gem" by Bent (1949), this very appealing songbird has become much more widespread and better known as a breeding bird in the eastern United States as a result of forest recovery and its adaptiveness (Mulvihill 1992f; Ingold and Galati 1997). It has an ability to colonize remote and disjunct conifer forests throughout its range (Ingold and Galati 1997), a pattern followed in Pennsylvania and documented again by the second Atlas.

The Golden-crowned Kinglet breeds in the North American boreal conifer forest that extends from Alaska, east to Newfoundland, and south in higher elevations of the Appalachians to the Tennessee/North Carolina border (Simpson 1992; Ingold and Galati 1997; Knight 1997). It has nested regularly, and in some places at high densities, in the spruce forests of the central and southern Appalachians, where it has expanded its range in recent decades (Hall 1983; Simpson 1992; Knight 1997). Not too long ago, it was thought that Golden-crowned Kinglet was confined in Pennsylvania to the Poconos and North Mountain as a breeding species (Poole 1964; Mulvihill 1992f). Its range certainly has expanded since the mid-twentieth century.

Rangewide, the Golden-crowned Kinglet is most strongly associated with various spruces and firs, but in Pennsylvania and the Appalachians it is also found in pines and hemlocks (Mulvihill 1992f; Simpson 1992). Although it can be found in other forest types elsewhere in its range, this species qualifies as a conifer forest obligate in Pennsylvania, as shown by atlas point count data (appendix D). Golden-crowned Kinglet has taken advantage of conifer plantings as well as native conifer forests. Norway Spruce, in particular, has been planted as an ornamental, providing new habitat for kinglets. The pattern of finding this species nesting at lower elevations (less than 300 m [~1000 ft]) in Norway Spruce (Mulvihill 1992f) continued and expanded during the second Atlas. With the response of this species to exotic plantings, it is important to note that Golden-crowned Kinglet also is abundant in old-growth hemlock (Haney 1999) and natural spruce and hemlock forests with tall trees (pers. obs.).

Golden-crowned Kinglets were found in 370 blocks in the second Atlas, more than twice as many as in the first Atlas. The increase in blocks occurred across the state, but the most extensive increases were in the northwestern corner of the state (appendix C). There were modest range expansions in the south, notably in the South Mountain Section, and in the west, close to the Ohio border in Beaver and Washington counties. Its population is increasing in Ontario (McLaren 2007b) and New York state, where there are more conifer forests, including exotic species (McGowan 2008zg).

The second Atlas's breeding population estimate of Golden-crowned Kinglet is 16,000 singing males. This is a far cry from the descriptions in historic literature of a very rare and localized breeding bird. Many of the areas that were dominated by conifers in the nineteenth century were, however, timbered before being surveyed by ornithologists, so their avifauna was not documented (Gross 2010b). This kinglet is reclaiming parts of its former breeding range and colonizing conifer woods at lower elevations. Despite the fact that its high-pitched ascending song is easily overlooked or confused with other high-pitched species, as well as the fact that it inhabits the canopy of dense conifer forests and woodlots that are not always easily accessible by atlas volunteers, Pennsylvania's two atlases documented widespread and substantial range expansion of the Golden-crowned Kinglet.

Even with concerns for species with northern distributions in Pennsylvania and the Appalachian Mountains, the Golden-crowned Kinglet's immediate future seems promising because of its adaptiveness and pattern of expansion. Forest management that increases spruce and other conifers could help this and several other conifer-obligate species in Pennsylvania (Gross 2010b).

DOUGLAS A. GROSS

Distribution

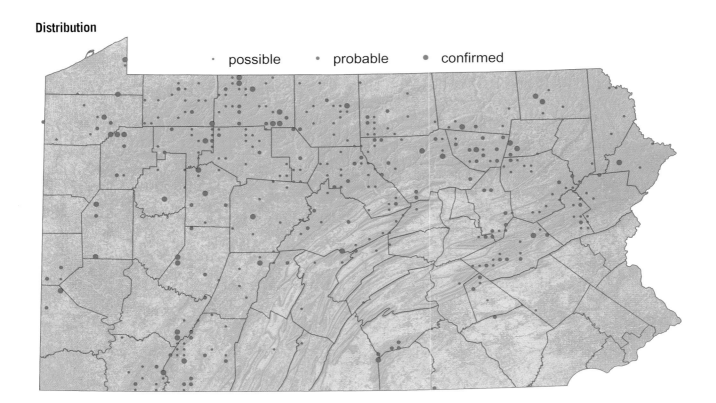

Distribution Change

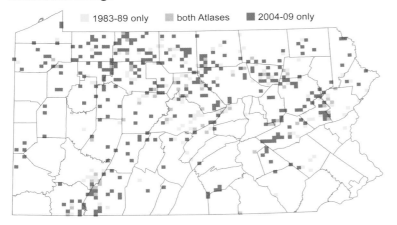

Number of Blocks

	first Atlas 1983–89	second Atlas 2004–09	Change %
Possible	90	231	157
Probable	57	98	72
Confirmed	35	41	17
Total	182	370	102

Population estimate, males (95% CI):
16,000 (12,000–20,000)

Jacob Dingel

Eastern Bluebird
Sialia sialis

The Eastern Bluebird is a familiar and favorite bird found in farmland and open habitats throughout Pennsylvania. It is a secondary cavity nester and is easily attracted to bird houses placed within appropriate habitat. The species breeds throughout eastern North America, and Pennsylvania is within the core of its range. Bluebirds were probably rare in Pennsylvania and local in their breeding distribution prior to European settlement, but they increased in abundance as forests were cleared. By the mid-nineteenth century, they were listed as abundant by many ornithologists (Leberman and Mulvihill 1992; McWilliams and Brauning 2000). Bluebird numbers fluctuated over the next century, with declines resulting from loss of nest sites and competition from European Starlings and House Sparrows, both of which compete with them for nest cavities. In Pennsylvania, the low point in the population cycle occurred in the 1960s (McWilliams and Brauning 2000). Widespread nest box programs have benefited bluebirds, and populations have increased since the 1960s, both within Pennsylvania and also nationally (Sauer et al. 2011).

Bluebirds were reported in 4,219 blocks in the second Atlas period and were found throughout the state, with the exception of highly urbanized areas, such as Pittsburgh and Philadelphia, and heavily forested blocks within the north-central region of the state. Block occupancy was highest in the Ridge and Valley Province, where open valleys adjacent to wooded areas provide optimum habitat, in farmland in the Piedmont Province, and in the Northwestern Glaciated Plateau Section. Bluebirds were found in 9 percent more blocks in the second Atlas than in the first. The largest positive changes in occupancy occurred in the Ridge and Valley Province and particularly in the Piedmont Province, where block occupancy increased from 77 percent to 99 percent. There were modest range contractions from some densely forest areas in the Appalachian Plateaus, largely in areas where forested covered increased between atlas periods (chap. 3).

Bluebirds nest in a wide variety of open habitats including orchards, woodlot/field edges, farmland, and clear-cuts. During the second Atlas, the highest abundance of bluebirds was associated with pasture and farmlands, grassy fields and lawns in suburban areas, and reclaimed surface mines. The Eastern Bluebird population estimate for Pennsylvania is 150,000 breeding pairs. Breeding Bird Survey (BBS) trend data show a 70 percent increase in abundance between the first and second atlases (Sauer et al. 2011). Eastern Bluebird is one of the species that exhibited significant population declines associated with the arrival of West Nile virus (LaDeau et al. 2007). In Pennsylvania, the drop in numbers reported on BBS routes in 2003 could be partly due to West Nile virus, but the Eastern Bluebird is known to be highly susceptible to severe winters (Sauer and Droege 1990), and the winter of 2003/04 was harsh in the Mid-Atlantic states (Coskren 2010a). Whatever the cause of this unusually large decline in numbers, BBS data show that the population recovered by 2007. Clearly, this species has the ability to rebound strongly from population losses and is doing well across the northeastern states; for example, its range increased by 54 percent in New York state between atlas periods (Berner 2008).

Bluebirds have a long breeding season within Pennsylvania, with courtship behavior reported the first week in April and nests with young reported from late April through late August. Bluebirds lay multiple clutches and attempt to raise two, and occasionally three, broods per season (Gowaty and Plissner 1998). Evidence of breeding was reported in 85 percent of all blocks and confirmations in 60 percent. The high rate of confirmations reflects the high use of nest boxes, which were easy for atlas volunteers to find and check for breeding evidence.

Eastern Bluebirds are doing well in Pennsylvania and will presumably continue to do so, benefiting from their ability to use a variety of open habitats as well as the popularity of establishing bluebird houses and trails.

MARGARET C. BRITTINGHAM

Distribution

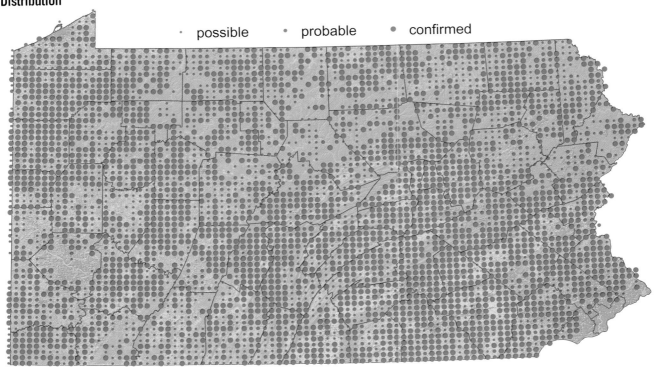

Distribution Change

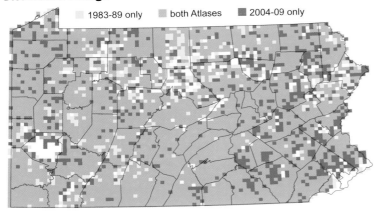

Density

Singing males per km²

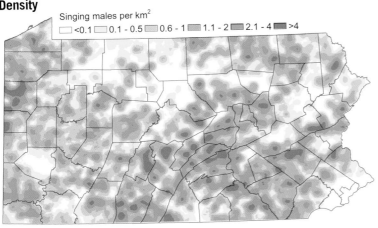

Number of Blocks

	first Atlas 1983–89	second Atlas 2004–09	Change %
Possible	647	1,001	55
Probable	652	692	6
Confirmed	2,567	2,526	−2
Total	3,866	4,219	9

Population estimate, males (95% CI):
150,000 (140,000–161,000)

Breeding Bird Survey Trend

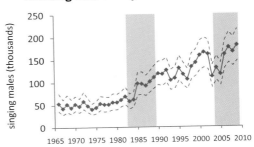

EASTERN BLUEBIRD

Chuck Musitano

Veery
Catharus fuscescens

One of Pennsylvania's four obligate forest thrushes, the Veery has a song that sounds as if it spirals from a deep container, befitting a bird that favors dense understory. The second Atlas confirmed that Veery populations in Pennsylvania appear to be stable and suggests that its habitat preferences may help it to thrive in the presence of competitive thrush species. The breeding range of the Veery includes southern Canada and the northern United States, extending south along the Appalachians to Tennessee and North Carolina (Bevier et al. 2005). Its Pennsylvania distribution spans the breadth of the state's northern tier but narrows in the southern counties. At first glance, this distribution resembles that of the Hermit Thrush, but closer inspection reveals differences.

The Veery prefers damp intermediate-elevation forests with a dense understory. Such forests can be deciduous, conifer, or mixed, with Eastern Hemlock often a component. The Veery is a common thrush of hemlock and laurel ravines, streamsides, and wooded bogs. It may not be as area-sensitive as some forest birds, since it is common in the damp woodlots of the once-glaciated northwestern and northeastern corners of the state.

While the distribution maps of Veery, Hermit Thrush, and Wood Thrush display a considerable degree of overlap, the density maps give more accurate views of their distributions. Where found in the same area, these species tend to partition their habitat selection by elevation, as well as other characteristics (Noon 1981). The Hermit Thrush tends to choose higher and drier elevations, the Wood Thrush the lowest elevations, while the Veery appears not at all limited by elevation, with numbers on point counts roughly proportional to forest extent along the elevation gradient (appendix D). In northern Pennsylvania, Veery density shows areas of high concentration in the tendril-like drainage branches of creeks and rivers. In northeastern Pennsylvania, notably the Pocono Plateau Section, some high Veery concentrations were found on wet mountain tops.

In southwestern Pennsylvania, the Veery is concentrated on the higher ground, including Laurel Hill along the Somerset/Fayette/Westmoreland county boundary and on the Allegheny Front along the Somerset/Bedford county boundary. As was noted in the first Atlas, it is notably absent in lower elevation forests of the state's southwestern corner (Schwalbe 1992d). The Veery is found consistently in the Ridge and Valley Province east of the Susquehanna River, along the Kittatinny Ridge, in the South Mountain Section, and in lower-elevation forest fragments in the Piedmont, although usually in low densities. This disparate distribution is noted south of the state line in Maryland, where it is principally found in lowland forests of the Piedmont, and in high elevation forests of the Appalachian ridges, with a notable gap between (Ellison 2010zf). The statewide population was estimated to be 270,000 singing males, with densities exceeding five males per km^2 in many of the state's extensive forests.

The number of blocks in which Veery was detected increased by 10 percent between atlases; however, this increase can largely be accounted for by the increased effort in the second Atlas. Breeding Bird Survey data from Pennsylvania indicate that the population has been relatively stable since the 1960s (Sauer et al. 2011). Blocks where Veery was detected in one atlas but not the other were generally along the margins of the core distribution, which would be expected when there is a high degree of chance in finding one. It can be concluded that the Veery population in Pennsylvania is stable, in contrast to range-wide declines of 0.9 percent per year (Sauer et al. 2011).

Despite the Veery's current population stability, habitat change on the breeding and wintering grounds gives cause for concern. The loss of Eastern Hemlocks due to Hemlock Woolly Adelgid infestation could seriously affect the moist forest environment that the Veery prefers, while the Veery's wintering grounds, concentrated in south-central and southeastern Brazil, are experiencing rapid deforestation (Bevier et al. 2005).

NICHOLAS C. BOLGIANO

Distribution

possible • probable • confirmed

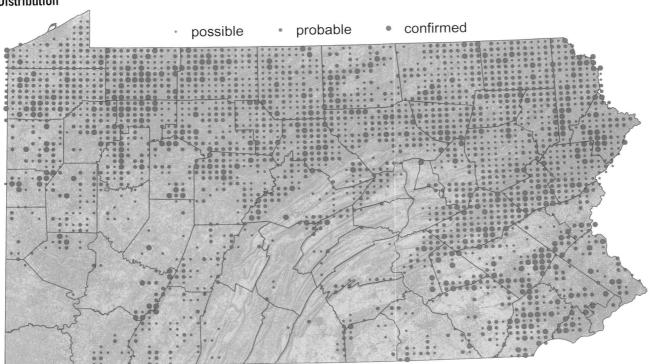

Distribution Change

1983-89 only · both Atlases · 2004-09 only

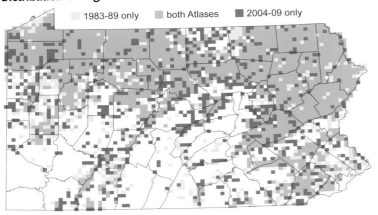

Density

Singing males per km²
<0.5 · 0.5 - 1.5 · 1.6 - 3 · 3.1 - 5 · 5.1 - 8 · >8

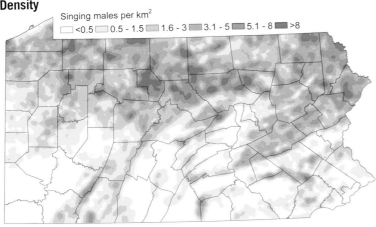

Number of Blocks

	first Atlas 1983–89	second Atlas 2004–09	Change %
Possible	961	1,115	16
Probable	1,021	1,092	7
Confirmed	326	336	3
Total	2,308	2,543	10

Population estimate, males (95% CI):
270,000 (230,000–325,000)

Breeding Bird Survey Trend

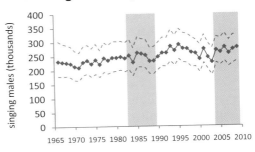

Bob Wood and Peter Wood

Swainson's Thrush
Catharus ustulatus

Besides a disjunct population in the central Appalachian high mountain spruce forests, Pennsylvania is the southern edge of the Swainson's Thrush's breeding range in eastern North America (Mack and Yong 2000). Although it is probably the most common and widespread spotted forest thrush of the continent (Rich et al. 2004), it is the state's rarest nesting thrush, classified as Vulnerable (PGC-PFBC 2005). The second Atlas documented a 128 percent increase in the number of blocks with Swainson's Thrush records since the first Atlas. It may be reclaiming portions of its former range in Pennsylvania, which once had much more extensive conifer forests on the plateaus (Whitney 1990; Gross 2010b). Despite this, the species' range has remained similar, with most records in three clusters: the Allegheny National Forest area, the "Black Forest" area of Potter County, and the North Mountain region of Lycoming, Sullivan, Wyoming, and Luzerne counties. Outside of these areas, block records declined from 18 percent of all records in the first Atlas to 5 percent in the second. However, the concordance of records between atlas periods suggests that the core range is experiencing local nesting success and colonization of similar habitat, rather than expansion into surrounding areas. Some sites in Sullivan County and the Allegheny National Forest are occupied on a year-to-year basis, suggesting site fidelity, but they also seem to vary annually, as also observed in West Virginia (Hall 1983; pers. obs.). The eastern populations, together with those of the mountains of West Virginia and Virginia, are disjunct from northern populations (Bent 1942; Hall 1983; Brauning 1992i; Simpson 1992; Buckelew and Hall 1994). An exception is the population in McKean and Warren, which is contiguous with populations in southern New York (Lowe and Hames 2008a).

The Swainson's Thrush tends to inhabit cool and moist conifer-dominated woods in Pennsylvania, New York, and the rest of the Appalachians (Simpson 1992; Mack and Yong 2000; Lowe and Hames 2008a). Almost all territories are at elevations above 520 m (1,700 ft) and are strongly associated with hemlock, but some have been found in spruce (Brauning 1992i; pers. obs.). Point count records of this species were all within an elevation range of 400 to 700 m (1,300 to 2,300 ft) and almost all in coniferous or mixed forests (appendix D).

The Swainson's Thrush often nests where there is dense mid-story vegetation in mixed or deciduous forest (DeGraaf and Rappole 1995). Historically, it was associated with old-growth forest (Cope 1936), and significantly higher densities were found in old growth than in nearby managed forests (Haney and Schaadt 1996), although they also nest in dense stands of young trees. Territories often include small streams, seeps, and springs where the thrushes forage, sometimes occurring in forest canopy gaps. The population estimate from atlas point count data was 2,600 singing males, making this among Pennsylvania's rarest songbirds. The Pennsylvania population is very small and not a significant percentage of the total population (Rich et al. 2004), but some ornithologists feel that the Appalachian Mountains population is distinct enough to merit subspecies status (Ramos and Warner 1980; Mack and Yong 2000).

The expanding population documented by the second Atlas is part of a long rebound for this species after the timber extraction era. The Swainson's Thrush once was more common in the Appalachians south of Pennsylvania and even locally common in spruce-fir forests of West Virginia mountains (Bent 1949; Hall 1983). Although there have been increases in the pine and hemlock component of Penn's Woods, forests within the Swainson's Thrush range are increasingly threatened by atmospheric acid deposition and conifer tree pests (McWilliams et al. 2007). It will be interesting to see if the resurgence in Swainson's Thrush populations in Pennsylvania can continue with the looming threats of climate change, conifer tree pests and diseases, acid atmospheric deposition, and development by energy development industry in the state's northern forests on the plateaus.

DOUGLAS A. GROSS

Distribution

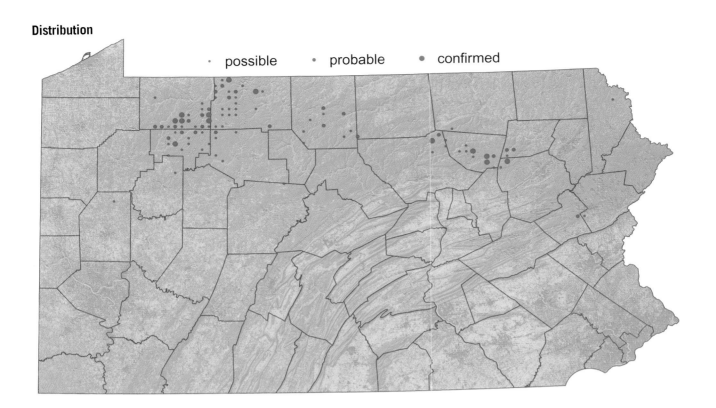

Distribution Change

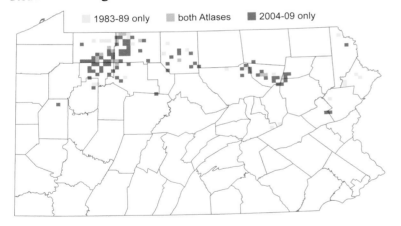

Number of Blocks

	first Atlas 1983–89	second Atlas 2004–09	Change %
Possible	22	60	173
Probable	19	25	32
Confirmed	2	13	550
Total	43	98	128

Population estimate, males (95% CI):
2,600 (1,600–4,100)

Bob Wood and Peter Wood

Hermit Thrush
Catharus guttatus

Those who have heard the Hermit Thrush's song emanating from a Pennsylvania ridgetop or forest cove may justly describe the experience as sublime. The second Atlas found that the source of perhaps the most beautiful song among Pennsylvania's birds is doing quite well. A species decimated by the cutting of Pennsylvania's forests during the nineteenth century, the Hermit Thrush has continued to increase with the regrowth of those forests.

The breeding range of the Hermit Thrush includes the boreal and northern hardwood forest of Canada and the northern United States. This range extends southward along the higher elevations of the Appalachians to West Virginia and Virginia (Jones and Donovan 1996). In Pennsylvania, this species' distribution is typical of northern species whose southern range limit passes through the state—widespread in the north, and restricted to higher elevations south.

The Hermit Thrush's habitat preference for extensive forest at higher elevations defines where it was found during the second Atlas. The highest densities, often exceeding five singing males per km^2, were found in north-central locations where conifers are a larger component of the forest trees, notably the Allegheny National Forest and nearby Elk and Susquehannock state forests. Statewide, the estimated increase in occupied blocks between atlases was 37 percent. Concurrent with the range expansion, Breeding Bird Survey counts in Pennsylvania almost doubled between atlas periods (Sauer et al. 2011).

The largest relative gains were outside the north-central core area. In the Ridge and Valley, occupied blocks increased by 78 percent and were centered in the Bald Eagle and Rothrock state forests of central Pennsylvania and in the southwestern extension of the Pocono Plateau Section into Broad Mountain. In the Allegheny Mountain Section of southern Pennsylvania, there was a 113 percent increase in the number of occupied blocks, with particularly substantial increases on Laurel Hill along the Somerset/Fayette/Westmoreland county boundary. There was an increase of 45 percent in occupied blocks in the Allegheny Front Section, most notably along the Somerset/Bedford county boundary. Hermit Thrush was also found at disjunct ridgetop locations, including the South Mountain Section, where it was discovered breeding for the first time during the first Atlas period (Mulvihill 1992g). The observed pattern of range expansion is consistent with an initial recovery, first filling the core range and then expanding outward; in Pennsylvania, this has resulted in a southward expansion.

Despite the observed range expansion and population increase, the Hermit Thrush's statewide population, estimated to be just over 100,000 singing males, is still considerably less than that of Wood Thrush or Veery. While Hermit Thrush, Veery, and Wood Thrush have overlapping distributions, they tend to partition their habitat selection by elevation and other characteristics, with Hermit Thrush occupying the highest ground and Wood Thrush the lowest (appendix D). Where breeding territories of these species occur in the same general locality, Wood Thrush appears to be able to exclude the other two species from potential nesting territories (Noon 1981). Because acid deposition adversely affects the Wood Thrush (Hames et al. 2002), the Hermit Thrush may have benefited from a decline in Wood Thrush, as has been postulated for New York (Hames and Lowe 2008c). However, the Hermit Thrush remains a species of higher elevations in Pennsylvania; second Atlas point count data suggest that the highest densities are at elevations of 500 to 700 m (~1,600 to 2,300 ft) and that the species is absent below 200 m (650 ft).

Despite favorable trends for the Hermit Thrush, looming habitat changes are of concern. Perhaps the most immediate issue is the effect of Hemlock Woolly Adelgid upon this conifer-loving species, although the Hermit Thrush can also utilize Norway Spruce and Red Pine and hence may prove adaptable enough to avoid large population losses. Also of concern is the threat of ridgetop forest fragmentation, due to development for wind power, and more general forest fragmentation, due to shale gas drilling and other infrastructure development. In the long term, the potential effect of a warming climate on the cool, high-elevation habitat of the Hermit Thrush is also a concern.

NICHOLAS C. BOLGIANO

Distribution

· possible   · probable   · confirmed

Distribution Change

1983–89 only   both Atlases   2004-09 only

Number of Blocks

	first Atlas 1983–89	second Atlas 2004–09	Change %
Possible	637	847	33
Probable	557	821	47
Confirmed	178	209	17
Total	1,372	1,877	37

Population estimate, males (95% CI):
105,000 (95,000–115,000)

Density

Singing males per km^2
<0.1   0.1 - 0.5   0.6 - 1   1.1 - 2   2.1 - 4   >4

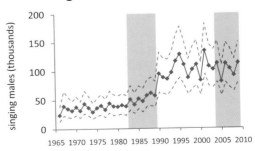

Breeding Bird Survey Trend

HERMIT THRUSH 327

Steve Gosser

Wood Thrush
Hylocichla mustelina

The Wood Thrush produces one of the most beautiful natural sounds of Pennsylvania, with its ringing and unmistakable *eee-o-lay* song that echoes through the forest, especially at dawn and dusk. Many Pennsylvania residents enjoy the serenade of this species, since it is widespread across the state. The less attractive but equally distinctive *put-put-put* call notes are a good sign you may have strayed near a nest, or that some other predator has! Wood Thrushes can be difficult to see, because of their inconspicuous earth tones and shy nature. Their nests are bulky but not necessarily easy to find, since many pairs nest in the sub-canopy or canopy layer (Roth et al. 1996). In northern Pennsylvania, females which have a successful first nest typically attempt second broods, and parents with young occur into early August (appendix F).

As the name implies, Wood Thrush inhabit forests, and the density of singing males was highest in deciduous forest and forested residential areas (appendix D). In Pennsylvania and elsewhere, Wood Thrush are often found in forest fragments, even woodlots of less than 2 ha (5 ac), although these habitats may be suboptimal due to high nest predation rates (Newell 2010).

The Wood Thrush is less common today than a few decades ago. Across its range it has declined dramatically, and according to the Breeding Bird Survey (BBS), populations are down by almost half since 1966, having decreased at a rate of 1.6 percent per year in Pennsylvania (Sauer et al. 2011). The second Atlas does not detect this change, however; there have been no significant declines in the Wood Thrush's range in the state's 23 physiographic sections (appendix C). A similar pattern was observed in neighboring New York (Hames and Lowe 2008d) and Maryland (Ellison 2010zg). This discrepancy between numbers on BBS routes and block records in atlases occurs because, although Wood Thrushes have declined dramatically in population size, they remain widespread, so they are still present in most blocks.

Although there was some turnover of block occupancy between atlases, there was little evidence of expansion or contraction in range, except in the higher elevation forests in the vicinity of Elk and Potter counties, where there is a noticeable cluster of losses since the first Atlas. While this may be partly due to changes in observer effort, this area also has seen some of the highest acid rain deposition rates in North America (Lynch 1998); acid rain has been linked to reduced reproductive success in the Wood Thrush (Hames et al. 2002d). Further, very few Wood Thrushes were found on point count surveys in higher elevation forests, especially in the north-central part of the state. Though absent from the large urban centers and open farmland, the Wood Thrush was otherwise found in high densities, frequently exceeding eight singing males per km^2 in most low-elevation areas, and remains a common bird. Its statewide population was estimated at 660,000 singing males.

Rather than dispersing across the wintering range, Wood Thrushes from a study area in northwestern Pennsylvania migrate to northern Honduras and eastern Nicaragua for the winter (Stutchbury et al. 2009, 2011). This leaves the population vulnerable to deforestation on the wintering grounds. On spring migration, most Wood Thrushes from northwestern Pennsylvania cross the Gulf of Mexico and make landfall in the New Orleans region, suggesting this region is a critical stopover habitat. With vulnerabilities due to habitat on the winter ground and migration routes, the status of the Wood Thrush should be carefully monitored. Due to the fact that Pennsylvania is estimated to support 8.5 percent of the global population (Rich et al. 2004), the state plays an important role in the conservation of this beautiful woodland songbird (PGC-PFBC 2005).

BRIDGET J. M. STUTCHBURY

Distribution

Distribution Change

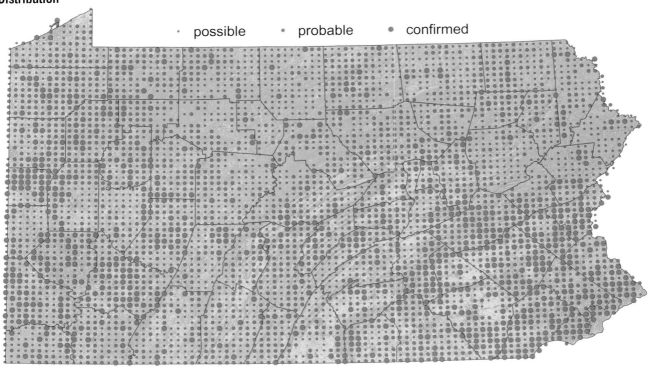

Number of Blocks

	first Atlas 1983–89	second Atlas 2004–09	Change %
Possible	1,464	1,934	32
Probable	2,133	1,917	−10
Confirmed	896	752	−16
Total	4,493	4,603	2

Population estimate, males (95% CI):
660,000 (560,000–780,000)

Density

Singing males per km^2

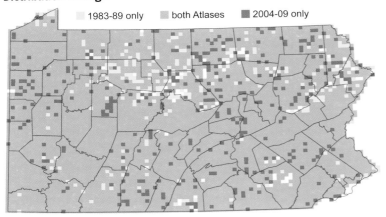

Breeding Bird Survey Trend

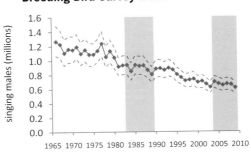

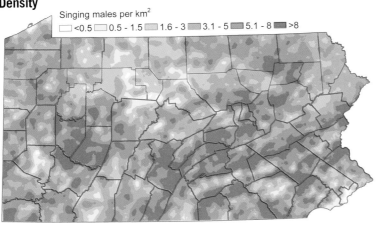

WOOD THRUSH

Bob Wood and Peter Wood

American Robin
Turdus migratorius

"The quiet, homelike beauty of the robin appeals to every American. As the trim bird runs about the dew-drenched lawn, he seems to impart to us his own belief in the goodness of life." So wrote the great George Miksch Sutton in his *Introduction to the Birds of Pennsylvania* (1928a). His colleague W. E. Clyde Todd, writing in *Birds of Western Pennsylvania* (1940), saw it differently: "Unquestionably the Robin is now far too numerous in most places. Finding comparative safety and protection about the abodes of man, it has multiplied to a point where it has become a nuisance and a potential menace to his interests." On one point, though, Sutton and Todd were in solid agreement: there certainly are a lot of American Robins in Pennsylvania.

The American Robin is common to abundant over much of its vast North American breeding range, which extends from Alaska across northern Canada to the Atlantic Ocean, and south well into Mexico (Sallabanks and James 1999). Pennsylvania breeders are of the widespread nominate subspecies (Sallabanks and James 1999). Because they are so numerous and widespread, robins are rarely considered to be of conservation concern. Although some populations may be locally susceptible to pesticides (Sallabanks and James 1999), robins have generally been the beneficiaries of human activities. They have successfully adapted to human-altered habitats in Pennsylvania, and they are abundant in residential districts (McWilliams and Brauning 2000). Partly because of the species' affinity for nesting in yards, there were more American Robin breeding confirmations than for any other species in the second Atlas. In addition to its being abundant and easy to observe, the American Robin has a long nesting season. Nest building was observed as early as 18 March, and a nest with eggs was seen as late as 10 September.

The American Robin was recorded in every priority block and in 99.7 percent of all atlas blocks. Breeding was confirmed in 78 percent of atlas blocks overall. In the first Atlas, too, the American Robin was ubiquitous. Breeding Bird Survey counts in Pennsylvania declined, albeit slowly, from 1966 to 2003, but the population appears to have subsequently recouped some of its previous losses (Sauer et al. 2011). These results are mirrored by those reported by McGowan (2008zh) for the second New York atlas. In New York, the American Robin was reported in more atlas blocks than any other bird species, and its statewide distribution was considered to be unchanged from their first atlas, conducted in the early 1980s.

Practically all writers, from Nuttall (1832) to Bent (1949) to Sallabanks and James (1999) and beyond, have attributed the success of the American Robin to its general adaptability and specific compatibility with humans. And so it is in Pennsylvania at the present time. Atlas volunteers documented American Robins statewide and in every habitat type. According to atlas point count surveys, robins achieve their highest densities in developed areas, while they are least common in forests, but even there they exceed densities of 0.3 singing males per count (appendix D). Simply put, robins occur everywhere in Pennsylvania, from urban districts near sea level in the southeastern corner to mountaintop wildernesses in the state's central and northern districts.

According to atlas point counts, there are an estimated 2.45 million singing male American Robins in Pennsylvania. Add to that total the singing males' mates, their nestlings, and unmated adults, and it seems reasonable to say that, at the peak of the breeding season, there is very roughly one robin in Pennsylvania for each human being in the Commonwealth. Surely, then, there is no cause for conservation concern? That's probably true, but it is important to consider the value of the American Robin as an "early warning" for broader environmental distress. Rachel Carson (1962) said as much, and, more recently, Sallabanks and James (1999) have affirmed the importance of the American Robin as a possible bio-indicator of chemical pollution in the environment.

TED FLOYD

Distribution

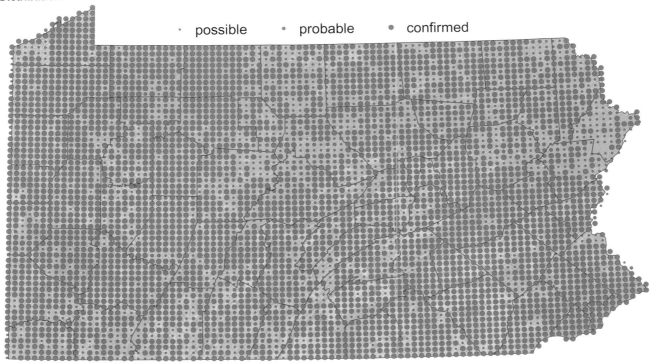

Distribution Change

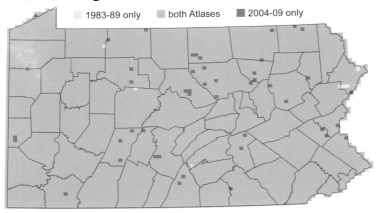

Density

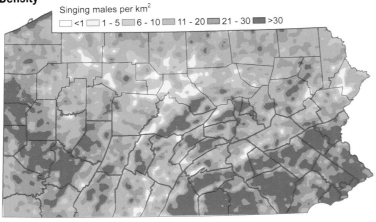

Number of Blocks

	first Atlas 1983–89	second Atlas 2004–09	Change %
Possible	370	640	73
Probable	415	460	11
Confirmed	4,097	3,823	−7
Total	4,882	4,923	1

Population estimate, males (95% CI):
2,450,000 (2,390,000–2,490,000)

Breeding Bird Survey Trend

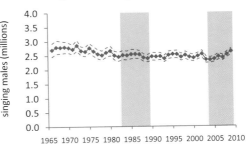

AMERICAN ROBIN

Glenn Bartley/VIREO

Gray Catbird
Dumetella carolinensis

The most common of the three mimic thrushes occurring in Pennsylvania is the Gray Catbird. With its distinctive feline mewing call, extremely variable song, and tolerance of humans, it is a conspicuous species in the state. It breeds across much of the eastern two-thirds of the United States, can be found in a wide variety of Pennsylvania habitats, and during the second Atlas was recorded in more than 98 percent of atlas blocks.

Although not plentiful in several north-central counties historically (Cope 1902), the Gray Catbird has probably not undergone any significant change in its statewide distribution since the beginning of the last century, when it was known as an abundant bird throughout Pennsylvania (Harlow 1913). During the first Atlas, it was the fifth most frequently recorded species in the state; it was found in at least 90 percent of the blocks in all but four of the Commonwealth's counties, three of which have extensive state forests within their boundaries (Ickes 1992j).

Although the Gray Catbird dropped to the thirteenth most frequently recorded species in the state during the second Atlas, it was found in nearly all Pennsylvania blocks; in seven counties, densities were estimated to be greater than 30 singing males per km^2. There were no significant differences in occupancy in any physiographic provinces or sections between atlas periods; however, there appeared to be one slight change in the catbird's state distribution. The tendency, in the first Atlas, to be reported less often in heavily wooded parts of Pennsylvania was not as evident during the second Atlas. For example, the four physiographic sections in which the catbird increased the most in occupied blocks between atlas periods contain extensive forests (chap. 3; appendix D). Point count data indicate that the Gray Catbird is less common in uniformly forested areas and at high elevations, and most common in mixed agricultural areas (appendix D).

During the second Atlas period, the total Pennsylvania population for the Gray Catbird was estimated to be 2.4 million singing males, making it the fifth most numerous songbird in the state. Breeding Bird Survey counts in Pennsylvania increased by 0.6 percent per year from 1966 to 2009, equating to a 12 percent increase between atlas periods (Sauer et al. 2011).

Because the Gray Catbird prefers early successional habitats, throughout the early twentieth century it probably benefited from overgrown abandoned farmlands, suburbanization of wooded areas, increases in ornamental shrubbery in populated areas, and creation of forest edge for roads or utility rights-of-way (Nickell 1965; Cimprich and Moore 1995). Pennsylvania's woodlands continued to mature between the first and second Atlases (McWilliams et al. 2007), contributing to retractions of this and other species of younger forests here as well as in neighboring New York (McGowan 2008zi). Recent research in Pennsylvania and elsewhere further corroborates the positive effect of dense, shrubby vegetation on catbird populations. For example, the species persists, or is more commonly encountered, in relatively fragmented ecosystems (Keller and Yahner 2007; Morgan et al. 2007) and increases in areas with thick herbaceous vegetation and dense, tall shrubs (Rohnke and Yahner 2008). The adaptability of this species to small patches of shrubs in both urbanized and forested areas suggests an excellent future for this lively species.

ROY A. ICKES

Distribution

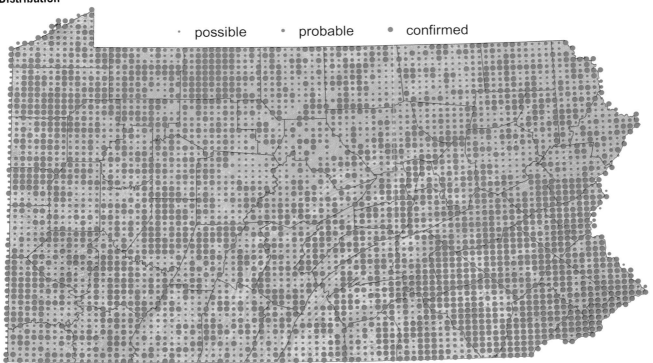

possible • probable • confirmed

Distribution Change

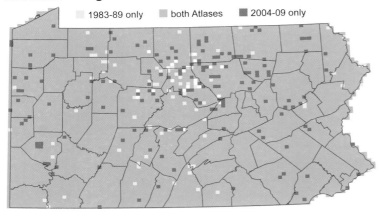

1983–89 only | both Atlases | 2004-09 only

Density

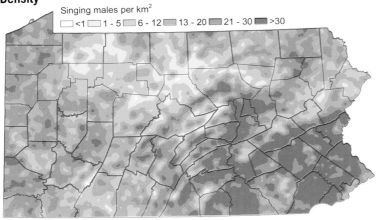

Singing males per km²
<1 | 1 - 5 | 6 - 12 | 13 - 20 | 21 - 30 | >30

Number of Blocks

	first Atlas 1983–89	second Atlas 2004–09	Change %
Possible	829	1,184	43
Probable	1,663	1,495	−10
Confirmed	2,256	2,162	−4
Total	4,748	4,841	2

Population estimate, males (95% CI):
2,380,000 (2,390,000–2,450,000)

Breeding Bird Survey Trend

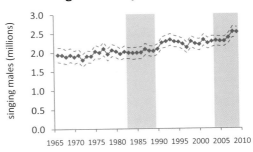

Chuck Musitano

Northern Mockingbird
Mimus polyglottos

The spectacular mimicking ability of the Northern Mockingbird is familiar to most Pennsylvanians, since the species occupies various brushy open habitats in close proximity to people and is conspicuous in its activities, which include imitating other species' songs and calls as well as diverse environmental sounds. Mockingbirds are very common in much of the southern United States, and they have been recorded as far north as southern Canada.

During the last century, the Northern Mockingbird recovered from its drastically reduced population caused by heavy trapping. It was formerly found only irregularly in southeastern Pennsylvania (Reynard 1953) and various western counties (Todd 1940), but since the 1950s it has slowly expanded its range northward (Poole 1964). The first Atlas indicated that mockingbird range expansion continued, since the species was present in almost every Pennsylvania county, was extremely common in the southeast (in over 95 percent of the blocks), and although not present in heavily forested areas of the north was breeding throughout the state (Ickes 1992k).

According to the results of the second Atlas, the Northern Mockingbird is continuing its expansion, with a significant increase of 20 percent in occupied blocks between atlas periods. However, the mockingbird's expansion does not appear to be a simple northward extension of its range. The species' state distribution pattern was somewhat more complex, with northward expansion in the western half and range contraction in the northeastern corner of Pennsylvania. There was substantial range expansion in three physiographic sections on the western part of the state: Pittsburgh Low Plateau (+139%), Allegheny Mountain (+188%), and Northwestern Glaciated Plateau (+700%). An apparent increase in occupied blocks between atlas periods in northeastern Ohio (OBBA 2011) is consistent with the increase in adjacent northwestern Pennsylvania. The only section with a significant decrease in blocks was the Glaciated Pocono Plateau (−56%), although several other northeastern sections saw non-significant decreases. Although Northern Mockingbirds were found in 10 percent more blocks in the New York second atlas than in the first, they remained sparsely distributed in most areas peripheral to Pennsylvania's northern border (McGowan 2008zj). The scarcity of mockingbird reports across the High Plateau Section during this atlas is noteworthy.

During the second Atlas, the species was most numerous in southeastern Pennsylvania—for example, in Adams and York counties, where densities were estimated to average more than six singing males per km^2. Second Atlas results clearly demonstrate that the Northern Mockingbird avoids densely forested areas. Its preference for habitats such as parks and cultivated lands, second growth at low elevations, and suburbs (Derrickson and Breitwisch 1992) probably contributes to this pattern. Additionally, point count analysis suggests a negative relationship with elevation; the species was scarce above 400 m (1,300 ft; appendix D) regardless of habitat, which is suggestive of winter temperatures being a limiting factor for this resident species.

The total Pennsylvania population of Northern Mockingbirds was estimated to be 200,000 singing males during the second Atlas period, and according to Breeding Bird Survey counts, there was little change in population size between atlas periods (Sauer et al. 2011). In the less forested southern half of Pennsylvania, the outlook for the Northern Mockingbird is excellent, whereas in other parts of the state its immediate future appears to be a bit more tenuous. Although severe winters may result in high over-winter mortality in the mountainous and northern portions of the state, as apparently happened in Canada (David et al. 1990), temporarily halting range extension, the results of the second Atlas suggest the mockingbird is still expanding slowly northward in some parts of Pennsylvania.

ROY A. ICKES

Distribution

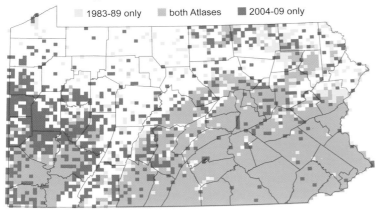

Distribution Change

Density

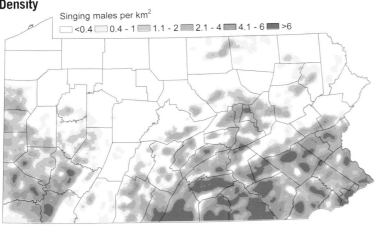

Number of Blocks

	first Atlas 1983–89	second Atlas 2004–09	Change %
Possible	717	1,145	60
Probable	821	846	3
Confirmed	753	757	1
Total	2,291	2,748	20

Population estimate, males (95% CI):
200,000 (192,000–207,000)

Breeding Bird Survey Trend

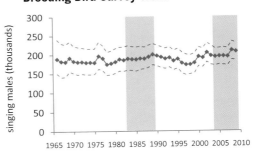

NORTHERN MOCKINGBIRD 335

Bob Wood and Peter Wood

Brown Thrasher
Toxostoma rufum

The least conspicuous of the three mimids breeding in Pennsylvania, the Brown Thrasher is most often seen as a reddish-brown blur flying across a country road. However, unlike its close relatives the Gray Catbird and Northern Mockingbird, which also breed throughout the eastern two-thirds of the United States, the Brown Thrasher is listed as a Maintenance Concern species in Pennsylvania's Wildlife Action Plan (PGC-PFBC 2005). The reason for the state's conservation concern is that Partners in Flight has listed it as a IIA priority species because it is an indicator of early successional habitats and is somewhat area-sensitive (Rich et al. 2004).

Historically, the Brown Thrasher was reportedly common to abundant in much of southern Pennsylvania in the late nineteenth century (Warren 1890), and it extended its breeding range northward by the middle of the last century (Poole 1964). Records from the state's first Atlas indicated that thrashers were found throughout Pennsylvania, although the species was more frequently found in the southern rather than northern areas and tended to be least widespread in the heavily forested areas of the Commonwealth (Ickes 1992l).

Second Atlas data confirm the pattern of avoidance of core forest areas highlighted in the first Atlas (Ickes 1992l). The tendency of thrashers to be less often found in suburban areas (Brittingham 2010) and negatively affected by urban development (Hostetler et al. 2005) can be observed when comparing the change map with a map of development and increases in development (chap. 3).

While there was no overall percentage change in occupied blocks between atlas periods for the Brown Thrasher, there was a significant increase of 23 percent in the Ridge and Valley Province and increases in the Deep Valleys (26%) and Glaciated High Plateau (47%) sections of the Appalachian Plateaus Province. These changes suggest that the Brown Thrasher is continuing to extend its Pennsylvania range slowly northward, at least in the central part of the state. However, these range expansions were balanced by retractions elsewhere, notably in the Glaciated Pocono Plateau (−65%) and Anthracite Valley (−69%) sections in northeastern Pennsylvania. Similarly, New York exhibited a decrease in Brown Thrasher records between its first and second atlas periods in the ecozones adjacent to northeastern Pennsylvania, amid a statewide decrease of 30 percent (McGowan 2008zk). Whether these regional differences in range changes were due to habitat change or differences in detection rates of this often unobtrusive species is difficult to say.

While Breeding Bird Survey (BBS) trends showed a rapid decline in numbers during the 1970s and early 1980s, there appeared to be minimal change in population size between atlas periods (Sauer et al. 2011). The Pennsylvania population of Brown Thrashers was estimated to be 80,000 singing males during the second Atlas period. Population densities are highest in counties along the state's southern border, with the five counties from Bedford east to York accounting for more than 20 percent of the statewide population, as estimated from second Atlas point counts.

Because the Pennsylvania Brown Thrasher population has remained fairly stable between atlas periods, following a rapid decline before then, the outlook for the species in the southern part of the Commonwealth, where densities are highest, would appear to be fine. Interestingly, to the south of Pennsylvania, BBS trends and atlas results in Maryland showed the same stable pattern (Ellison 2010zh), but to the north, in New York, there were continuing declines and localized population losses (McGowan 2008zk). The future of the species in other parts of the state is perhaps a little more tenuous. In the short term, growth of shrubs in reclaimed surface mines is probably contributing to the stability or slight increase in parts of the state. Point count data show that these habitats hold high densities of Brown Thrashers (appendix D). However, populations of Brown Thrasher are likely to rise and fall across the state in response to maturation of shrubs and forests (Cavitt and Haas 2000).

ROY A. ICKES

Distribution

possible · probable · confirmed

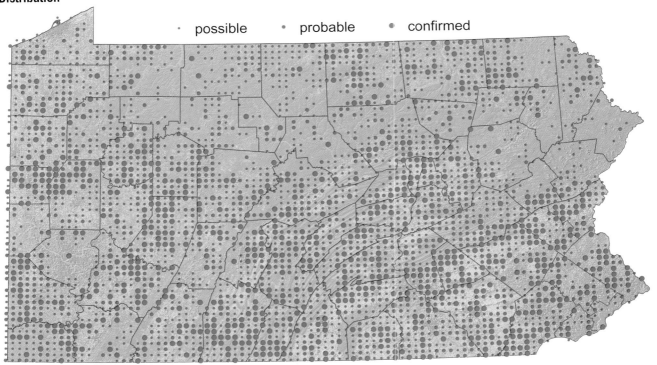

Distribution Change

1983–89 only | both Atlases | 2004-09 only

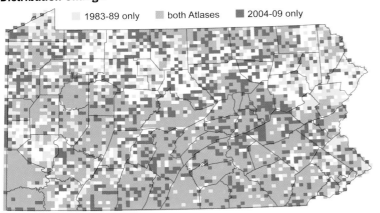

Density

Singing males per km²
<0.3 | 0.3 - 0.8 | 0.9 - 1.5 | 1.6 - 2 | 2.1 - 3 | >3

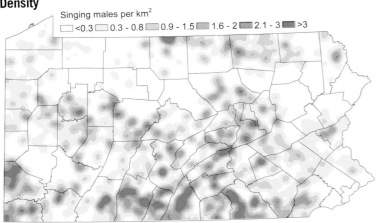

Number of Blocks

	first Atlas 1983–89	second Atlas 2004–09	Change %
Possible	1,302	1,591	22
Probable	1,081	940	−13
Confirmed	690	800	16
Total	3,073	3,331	8

Population estimate, males (95% CI):
80,000 (73,000–87,000)

Breeding Bird Survey Trend

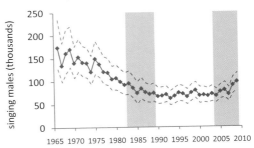

Jacob Dingel

European Starling
Sturnus vulgaris

The European Starling is a non-native species in North America which, following deliberate introductions in the late nineteenth century, adapted so well to expanding urban and farmland habitats that it has become one of the continent's most abundant birds (Cabe 1993). Unpopular among bird lovers due to its competition with native species for nesting cavities, it is generally regarded as an unwelcome addition to the avifauna. Because of this notoriety, coupled with its ubiquity, the starling's intricately marked iridescent plumage and remarkable vocal repertoire are often overlooked.

Following the first documented nesting in 1904, the European Starling's invasion of Pennsylvania was incredibly rapid; within two decades it had reached the northwestern corner of the state, and by the end of the 1920s it was described as one of the most abundant species (Sutton 1928a). As the twentieth century progressed, numbers may have declined in some areas because pasture, a favored foraging habitat, was lost due to farmland abandonment and forest regrowth (Master 1992e). However, during the first Atlas period it was still very widespread, absent only from densely forested areas, particularly on the Appalachian Plateaus.

There is little evidence of much change in distribution between the two atlas periods. Although it was recorded in 3 percent more blocks in the second Atlas than the first Atlas, this increase could be due to increased observer effort. Where this species is abundant, it is not likely to be missed by atlas volunteers, but in blocks that are predominantly forested, a few pairs associated with sparse human habitation could be overlooked, which could explain the scattering of changes in reported block occupancy, particularly in the northern tier. The only localized changes that may be more than random observer effects are in the vicinity of the Delaware State Forest and Poconos, where there are several contiguous blocks of apparent range loss.

Despite the lack of evidence of range changes, Breeding Bird Survey counts of the European Starling in Pennsylvania have declined by an average of 1.6 percent per year since the 1960s (Sauer et al. 2011), with counts around 28 percent lower in the second Atlas period than the first. A decline has been noted throughout North America and also in the starling's native range in Europe, where a decline in invertebrate-rich pasture habitat is thought to be a key driver of the downward trend (Robinson et al. 2005).

Despite the substantial decline, the European Starling remains one of Pennsylvania's most abundant birds, and the second Atlas population estimate of 2 million birds is likely a considerable underestimate (chap. 5). The highest densities are in the Philadelphia metropolitan area and in the farmland of the Piedmont, where the mosaic of development and farmland provides a wealth of nesting and foraging opportunities. Concentrations are found wherever there are towns and cities. Second Atlas point count data confirm that starlings are virtually absent from areas of contiguous forest (appendix D). Hence, it is possible that this species is losing habitat in areas of forest regrowth and consolidation, notably in the southern Allegheny Mountains and on the Appalachain Plateaus (chap. 3). Because this species is so abundant and ubiquitous, the decline in population size has yet to be substantial enough to register changes in atlas block occupancy. The European Starling is among the easiest of birds to confirm breeding: the sight of adults carrying food and noisy families of recently fledged juveniles are very familiar from late May to late July, and hence it is perhaps not surprising that there were more than 5,000 breeding confirmations submitted during the second Atlas. Atlas records also show the starling's breeding season to be protracted: nest building was observed from 17 March, while nests with eggs were noted as late as 11 July.

Because it is a non-native species and often considered to be a pest, the European Starling is offered no protection under United States law. It has long been too abundant for population-scale control programs to be effective (Cabe 1993) or economically viable, and as such, it is likely that the starling will remain a prominent, although often unwelcome, addition to Pennsylvania's avifauna.

ANDREW M. WILSON

Distribution

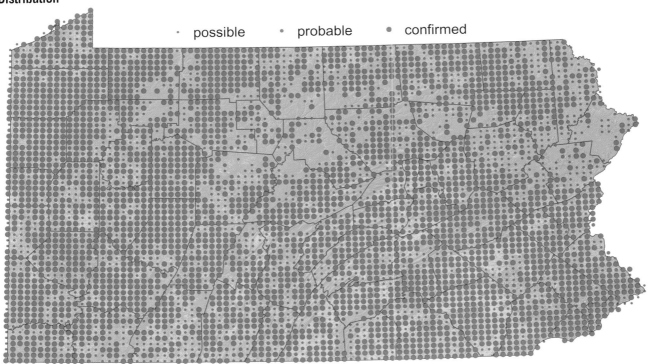

possible   probable   confirmed

Distribution Change

1983-89 only   both Atlases   2004-09 only

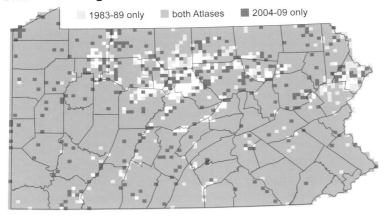

Density

Birds per km²
<0.5   0.5 - 2   2.1 - 5   5.1 - 15   15.1 - 30   >30

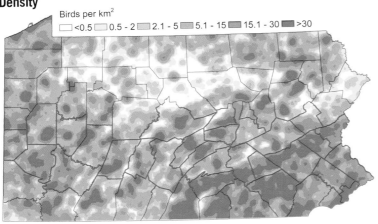

Number of Blocks

	first Atlas 1983–89	second Atlas 2004–09	Change %
Possible	649	1,052	62
Probable	375	218	−42
Confirmed	3,365	3,235	−4
Total	4,389	4,505	3

Population estimate, birds (95% CI):
2,000,000 (1,940,000–2,060,000)

Breeding Bird Survey Trend

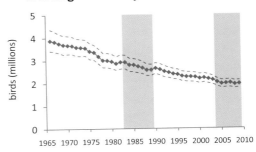

EUROPEAN STARLING

Roger Eriksson/VIREO

Cedar Waxwing
Bombycilla cedrorum

Most backyard birdwatchers know two things about the Cedar Waxwing: it is a stunningly beautiful bird, and it feeds almost exclusively on fruits. Olive-brown on the back and cinnamon-brown on the head, chest, and crest, the waxwing is easily identified by a striking black facial mask, a yellow terminal tail band, and red waxy tips on the secondaries, the feature that lends the species its name (Witmer et al. 1997). The waxwing's frugivory drives nearly every aspect of its life. Although waxwings do consume some insects during summer, their dependence on fruit throughout the year is responsible for many observed behaviors that are atypical of songbirds, including flocking behavior during winter and migration, the late onset of breeding, and low levels of nest site fidelity (Witmer et al. 1997).

The Cedar Waxwing nests throughout Pennsylvania and breeds across the northern United States and southern Canada, from coast to coast (Witmer et al. 1997). It nests in a variety of habitats, including open woods, forest edges, orchards, and suburban yards (Gross 1992f; Witmer et al. 1997). It is a common species across its entire range.

As in the first Atlas, second Atlas volunteers observed the Cedar Waxwing in all of Pennsylvania's 67 counties. The species is now, however, even more widely distributed than during the first Atlas and was detected in 87 percent of all blocks statewide, up from 82 percent during the first Atlas. During the first Atlas, waxwings were much less prevalent in the southeastern portion of the state, including the metropolitan areas of Philadelphia and the Lehigh Valley, and the agricultural landscapes of Lancaster, Lebanon, and York counties (Gross 1992f). Atlas volunteers documented statistically significant increases of 98 and 61 percent, respectively, in block occupancy in Pennsylvania's Atlantic Coastal Plain and Piedmont provinces between the first and second Atlases. A slight southward shift observed in its Pennsylvania breeding range may be the result of the Cedar Waxwing's recolonizing of areas it had previously abandoned (McWilliams and Brauning 2000).

Since fruit availability is highest later in the growing season, the Cedar Waxwing nests substantially later than many songbirds. Some waxwings were observed building nests as early as the second half of April, but the earliest peak in feeding young was much later, 16 June. The species often completes two broods in one nesting season, with nest building for the second brood sometimes starting before the first brood has fledged (Witmer et al. 1997). Atlas records of Cedar Waxwings feeding young showed a distinct double peak, with many records in mid-June, fewer records in late June and early July, and a second peak in mid-July, likely representing second broods.

Atlas point counts may not accurately reflect Cedar Waxwing density and distribution in Pennsylvania, because they were collected in late May to early July. In the early part of this season, many records may be of migrant birds, while later nesting birds would have been missed. While the statewide population estimate of 760,000 individuals should be treated with some caution, this species is clearly a common bird in most of the state.

Despite an increase in Cedar Waxwing block occupancy from the first to second Atlas, the Breeding Bird Survey showed little overall change in numbers during the same time period (Sauer et al. 2011). Interestingly, the species' status was little changed in New York between atlas periods (Witmer 2008), but there was a considerable southward range expansion and population increase between atlas periods in Maryland (Ellison 2010zi). These results suggest that the changes noted in southeast Pennsylvania are part of a region-wide phenomenon, with a notable range expansion into the Piedmont. Results from the second Atlas reflect the Cedar Waxwing's ability to adapt to human-altered landscapes, a trait that will serve the species well in the future.

BRIAN BYRNES

Distribution

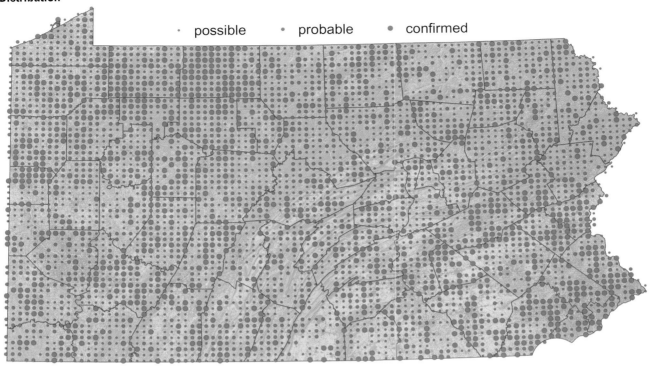

Distribution Change

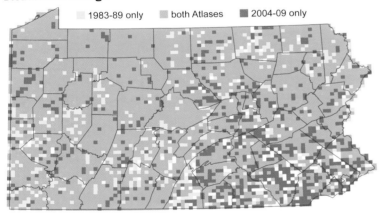

Density

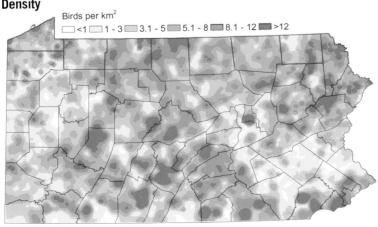

Number of Blocks

	first Atlas 1983–89	second Atlas 2004–09	Change %
Possible	1,521	1,779	17
Probable	1,442	1,443	0
Confirmed	1,099	1,061	–3
Total	4,062	4,283	5

Population estimate, birds (95% CI):
760,000 (730,000–790,000)

Breeding Bird Survey Trend

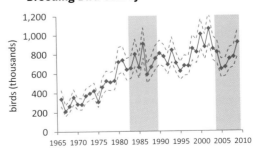

CEDAR WAXWING 341

Bob Wood and Peter Wood

Ovenbird
Seiurus aurocapilla

The *teacher, teacher* song of the Ovenbird resonates across most of Pennsylvania's forests, beginning in mid-May. Ovenbirds nest in large deciduous or mixed forests from southern Canada south to northern Georgia and Alabama. An area-sensitive species, it nests primarily amid the leaf litter and is absent or at low densities in smaller, more isolated forests (Van Horn et al. 1994). Nesting densities are higher in Pennsylvania and to the north than to our south, as well as in older deciduous forests through their range (Van Horn et al. 1994). Research in southeastern Pennsylvania shows lower mating and nesting success in isolated woodlots, with density reduced in woodlots smaller than 100 ha (250 ac; Porneluzi et al. 1993; Brittingham and Goodrich 2010). The scarcity of forest blocks larger than 20 ha (50 ac) in southeastern Pennsylvania (McWilliams et al. 2007; Bishop 2008) undoubtedly affects the distribution of this and other forest-interior species.

Historically one of the most abundant warblers across the state (McWilliams and Brauning 2000), the Ovenbird was found in 4,168 blocks during the second Atlas, 85 percent of the total. Only 16 percent of sightings were of confirmed nesting, however. The species' statewide distribution changed little between atlas periods, with lower occurrence in the southeastern and southwestern regions, where fragmentation undoubtedly limits its distribution. There was, however, a significant 13 percent increase in blocks reporting Ovenbirds between the atlas periods, some of which may be due to increased observer effort. The largest increases were in the Appalachian Plateaus (14%) and Ridge and Valley provinces (15%).

Ovenbird population densities during the second Atlas period were highest at higher elevations and where forest was most contiguous. Pike and Forest counties showed the highest average densities, with more than 25 singing males per km^2. The Pennsylvania Breeding Bird Survey (BBS) trend shows a 34 percent increase between atlas periods and a 1.4 percent annual increase since the 1960s (Sauer et al. 2011). However, BBS counts peaked in the mid-1990s and have stabilized since then. Modest gains in atlas block occupancy, and recent stability on BBS routes after long-term increases, also were noted in neighboring New York (McGowan 2008zl) and Maryland (Coskren 2010b).

Population estimates suggest Pennsylvania is home to 1.6 million males during the nesting season, making this the most abundant warbler and sixth most abundant species overall. Forest maturation across the state is likely one reason for the observed increase (McWilliams et al. 2007). Ovenbird abundance has been found to vary substantially even within the same forest, based on elevation or forest structure (Goodrich et al. 1998). Despite the recent increase, localized declines within contiguous forest habitat have occurred where forest health is compromised by overbrowsing by White-tailed Deer, invasive plants, and other factors (Senner et al. 2009). The Ovenbird was not strongly associated with hemlocks during the second Atlas period.

The Ovenbird's earliest probable nesting evidence occurred on 4 May, with latest evidence recorded on 22 August. Confirmed nesting was recorded as early as 6 May and as late as 27 August; however, nests with young were observed from 25 May through 10 July, which corresponds with the peak of the atlas survey season.

Although Ovenbird populations have now stabilized after a long recovery in Penn's Woods, degradation of forest health and increased fragmentation make this species susceptible to future declines. The Ovenbird is particularly susceptible to predation during nesting (Porneluzi et al. 1993; Van Horn et al. 1994). An increase in utility corridors, road cuts, and gas-line forest cuts across the state could lead to localized declines in Ovenbird density and success, as has been noted elsewhere (Rich et al. 1994). During migration, it is one of the most numerous birds killed by towers, windows, and other man-made obstructions (Taylor and Kershner 1986; Shire et al. 2000; Klem 2010). Declines in wintering areas have been noted where forest loss has been severe (Faaborg and Arendt 1992). Hence, this abundant and widespread forest obligate is an excellent sentinel for monitoring forest conditions across the state and beyond.

LAURIE J. GOODRICH AND
MARGARET C. BRITTINGHAM

Distribution

Distribution Change

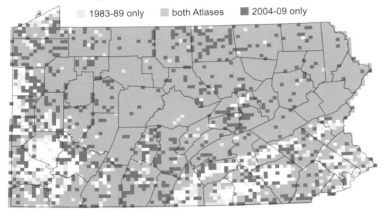

Density

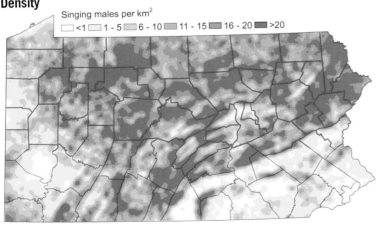

Number of Blocks

	first Atlas 1983–89	second Atlas 2004–09	Change %
Possible	1,192	1,557	31
Probable	1,849	1,924	4
Confirmed	633	687	9
Total	3,674	4,168	13

Population estimate, males (95% CI):
1,600,000 (1,560,000–1,650,000)

Breeding Bird Survey Trend

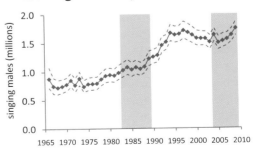

OVENBIRD 343

Bob Wood and Peter Wood

Worm-eating Warbler
Helmitheros vermivorum

The Worm-eating Warbler is known for probing dry leaf clusters for caterpillars, which sometimes have been called worms, hence the bird's name. Despite being an unobtrusive resident of steep hillsides and having a song often difficult to discriminate from a Chipping Sparrow's, the Worm-eating Warbler was found much more often during the second Atlas than during the first.

The Worm-eating Warbler nests in eastern North America where there are large forested tracts with a dense understory, and typically on hillsides. Its breeding range extends from southern New England south through the Appalachians and west to the Ozark and Ouachita mountains of Missouri and Arkansas. Only a handful of atlas detections in New York were north of the Pennsylvania border; most were east of the Delaware River (Smith 2008e). The Worm-eating Warbler also nests in lowland habitat from New Jersey to South Carolina and from Alabama to eastern Texas (Hanners and Patton 1998). In eastern Maryland, this warbler was found along river terraces with a Mountain Laurel understory and on drier islands with an American Holly understory contained within forested wetlands (Stasz 1996; Ellison 2010zj). In southern Delaware, the Worm-eating Warbler breeds in "sites only vaguely resembling a hillside" (Hess et al. 2000). Common to these habitats was the combination of large undisturbed forested tracts and a dense understory.

The Worm-eating Warbler was detected in 1,086 blocks, 22 percent of all Pennsylvania blocks. Nearly two-thirds of the detections were in the Ridge and Valley Province; there, the Worm-eating Warbler was found in about half of all blocks. In the South Mountain Section, it was found in 84 percent of all blocks. On the Reading Prong in eastern Pennsylvania, it was found in 48 percent of all blocks. Most of Pennsylvania's long ridges are located in these regions, and it is on these ridges where the Worm-eating Warbler finds the combination of slope, forest expanse, and thick understory that it prefers.

Elsewhere, the Worm-eating Warbler was primarily found on or near steep hillsides along some of Pennsylvania's larger rivers and streams: the lower Susquehanna, the upper Delaware, the West Branch of the Susquehanna, Pine Creek, and the upper Allegheny. While the Worm-eating Warbler is often associated with dry hillsides, this selection of riparian habitat indicates that its habitat preference spans a range from dry to moist forests, with the characteristic of being well drained.

The habitat requirement of extensive forests with a thick understory explains the historical withdrawal of the Worm-eating Warbler from most of Pennsylvania's southwestern and southeastern counties, as noted by Santner (1992e). However, Worm-eating Warblers remain scarce in western and northern Pennsylvania locations even with the requisite combination of large forested tract, hillside, and dense Mountain Laurel, such as Laurel Ridge. Whether this absence is due to unknown habitat constraints, the result of intra-specific competition with other species, or some other reason, is open to conjecture.

The Worm-eating Warbler was found in 48 percent more blocks during the second Atlas compared with the first Atlas. In contrast, the Breeding Bird Survey in Pennsylvania indicates that the Worm-eating Warbler has experienced no significant change in numbers between the atlases (Sauer et al. 2011). The second Atlas population estimate was 44,000 singing males, making this only the fifteenth most numerous of the 29 breeding warbler species in the state. According to Partners in Flight, Pennsylvania supports more than 6 percent of the global population. The Worm-eating Warbler is on the Watch List of Partners in Flight because its moderately sized population has declined across large parts of its range (Rich et al. 2004). While the population located on Pennsylvania's ridges has fared well compared with other areas (Sauer et al. 2011), the future of the Worm-eating Warbler depends upon the maintenance of large forested tracts. This will become increasingly more difficult if development of wind energy intrudes on the central and southern ridges.

NICHOLAS C. BOLGIANO

Distribution

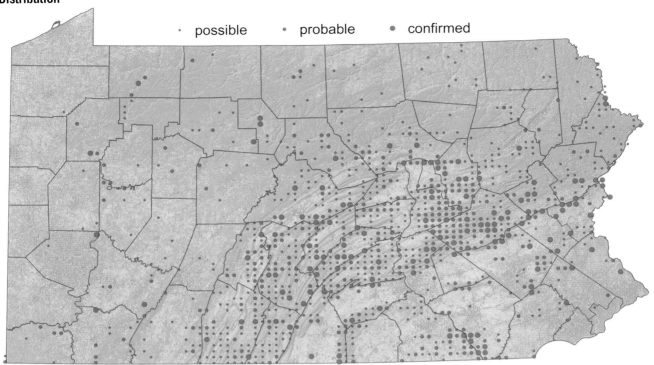

· possible   · probable   • confirmed

Distribution Change

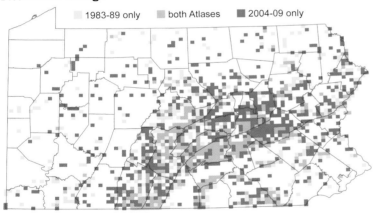

■ 1983-89 only   ■ both Atlases   ■ 2004-09 only

Density

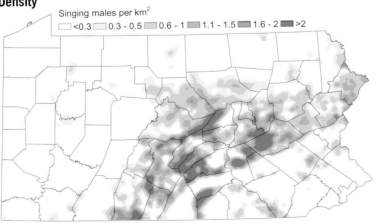

Singing males per km²
<0.3 0.3 - 0.5 0.6 - 1 1.1 - 1.5 1.6 - 2 >2

Number of Blocks

	first Atlas 1983–89	second Atlas 2004–09	Change %
Possible	363	648	79
Probable	271	299	10
Confirmed	101	139	38
Total	735	1,086	48

Population estimate, males (95% CI):
44,000 (38,000–51,000)

Breeding Bird Survey Trend

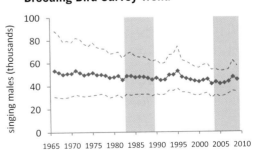

WORM-EATING WARBLER   345

Bob Wood and Peter Wood

Louisiana Waterthrush
Parkesia motacilla

The Louisiana Waterthrush is a signature breeding bird along many of Pennsylvania's picturesque forested mountain streams. Its presence often is revealed, initially, by its clarion song, which can be heard at great distances, even above the noise of cascading water. This atypical wood warbler spends much of its time walking, wading, and rock-hopping in continual tail-bobbing pursuit of aquatic insects, such as mayflies and stoneflies, which it gleans from the water and from partially submerged rocks, logs, and leafy debris, in and alongside fast-flowing streams.

The Louisiana Waterthrush is locally common throughout much of the eastern United States (Mattsson et al. 2009), and Pennsylvania lies near the northern edge of the species' core range in the central Appalachian Mountains. Because Pennsylvania harbors about 8 percent of the species' estimated global population—the fourth highest proportion of any state after West Virginia, Tennessee, and Virginia (Rich et al. 2004)—the Louisiana Waterthrush is designated as a Responsibility Species and a Species of Greatest Conservation Need in the state's Wildlife Action Plan (PGC-PFBC 2005). Recognition of the species as an ecological "umbrella," or bio-indicator, stems, in part, from a substantial body of research in Pennsylvania aimed at gauging its sensitivity to various environmental stressors, particularly stream acidification from industrial air pollution and abandoned mine drainage (e.g., O'Connell et al. 2003; Mulvihill et al. 2008).

During the second Atlas, the Louisiana Waterthrush was observed in 36 percent of all atlas blocks, a significant 29 percent increase compared with the first Atlas. The species was most frequent within the Ridge and Valley Province, where it occurred in 52 percent of blocks, and second in the Piedmont Province (36%). As in the first Atlas, the species was scarce or absent from low-lying areas with sparse forest cover, as well as from large areas of the Appalachian Plateaus Province, which lies toward the northern edge of the species' range. Range expansion in the Northwestern Glaciated Plateau (87%), as well as considerable infilling within the Ridge and Valley Province (46%), accounted for much of the increase in block occupancy.

In general, the Louisiana Waterthrush's Pennsylvania range corresponds to areas characterized by extensive forest cover, high topographic relief, and high stream density within large forest tracts. In addition, waterthrushes have a strong affinity for hemlock; 39 percent of all records submitted by atlas volunteers were in mixed hemlock or hemlock-dominated forests. This tendency was much stronger in the northern half of the state, where 58 percent of records were in such hemlock forests.

Data from second Atlas point count surveys indicate that the Louisiana Waterthrush is most abundant in parts of the Appalachian Mountain and Blue Mountain sections of the Ridge and Valley Province. Based on second Atlas point count data, the population of Louisiana Waterthrush in Pennsylvania is about 35,000 singing males, which is 66 percent higher than earlier estimates based on Breeding Bird Survey (BBS) data (Rich et al. 2004). Currently, the Louisiana Waterthrush appears to be doing well in Pennsylvania. Counts on BBS routes have increased steadily since the 1960s, at a rate of about 1 percent per year (Sauer et al. 2011). However, given its dependence on extensive forest and high-quality streams, the species may be adversely affected by widespread development of the state's shale gas development in the coming years. About 75 percent of its range overlaps the Marcellus shale formation. In addition, Louisiana Waterthrushes have been found to have among the highest level of methylmercury contamination of any songbird in Pennsylvania or New York (Evers et al. 2009), and this has been implicated in an apparent range contraction in New York between its first and second atlases (Rosenberg 2008a). Finally, because the Louisiana Waterthrush also depends on productive stream habitats on its wintering grounds in Central America and the West Indies, environmental impacts on stream water quality in those areas have the potential to adversely affect Pennsylvania's breeding population (Master et al. 2005).

ROBERT S. MULVIHILL

Distribution

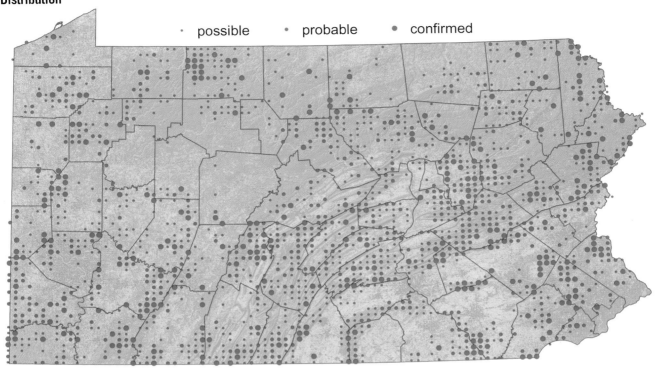

Distribution Change

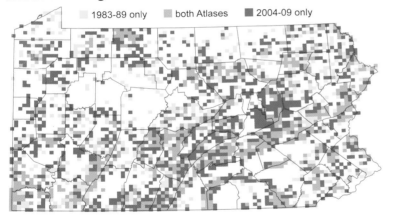

Density

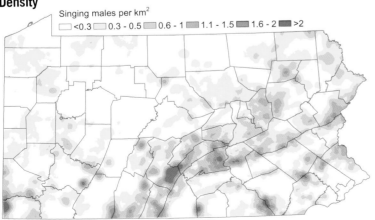

Number of Blocks

	first Atlas 1983–89	second Atlas 2004–09	Change %
Possible	585	943	61
Probable	538	531	−1
Confirmed	263	310	18
Total	1,386	1,784	29

Population estimate, males (95% CI):
35,000 (29,000–46,000)

Breeding Bird Survey Trend

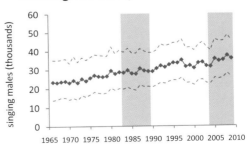

LOUISIANA WATERTHRUSH

Gerard Dewaghe

Northern Waterthrush
Parkesia noveboracensis

A large swamp warbler, the Northern Waterthrush has a cheerful song that can be heard in some of the state's high-elevation forested and shrubby wetlands. Its breeding range extends across the northern part of the continent from Alaska to New England, including much of the boreal forest (Eaton 1995). The Appalachian Mountains of Pennsylvania and West Virginia form this species' southern range limit in the East (Buckelew and Hall 1994; Eaton 1995). Often confused with the similar Louisiana Waterthrush, the Northern Waterthrush has a noticeably smaller bill and subtle differences in plumage, leg color, and vocalizations (Dunn and Garrett 1997).

The Northern Waterthrush nests in forests with slow-moving or still water, including forested wetland and peatlands, thickly vegetated rhododendron swamps, shrub-scrub wetlands, streamside thickets and woods, as well as boreal conifer swamps (Eaton 1995; Gross and Haffner 2010). These habitats are more common across the glaciated portions of the plateaus, especially in the Pocono Mountains region, North Mountain, and the northwest counties, which are reflected in this species' breeding range within Pennsylvania. Whatever the vegetation type, territories tend to be in poorly drained and well-shaded situations. It is found along moving water, but it favors slow-moving streams rather than the fast-moving brooks where the Louisiana Waterthrush is usually found.

There has been a significant 31 percent decline in the number of blocks in which this species was found since the first Atlas. If corrected for the greater field effort in the second Atlas, this decline was an even greater 41 percent. The decrease was larger in the Ridge and Valley Province (−49%) than in the Appalachian Plateaus (−25%), with most of the losses in the northern portions of these provinces. There was considerable turnover of blocks with Northern Waterthrush records from the first Atlas to the second Atlas (chap. 6), possibly due to the difficulty in finding the species, which led to underrepresentation in block coverage in both atlases. It is easily missed because its habitat is often off-road and not easily accessible. The Northern Waterthrush is one of the earliest species to stop singing and leave its breeding grounds (Eaton 1995), giving it a shorter nesting season than many other forest songbirds. Even allowing for detection challenges, there is a strong pattern of decline in the state. Unlike in Pennsylvania, New York's second atlas did not show a significant decline in this species, despite some shifts in range (McGowan 2008zm).

Only 33 Northern Waterthrushes were detected on point count surveys, emphasizing the species' scarcity in Pennsylvania. Most of the point count records were in the Pocono counties of Monroe and Pike and in the northwestern counties of Erie, Crawford, and Warren. These correspond with higher densities of forested wetlands (chap. 3). Breeding was confirmed in only 34 blocks, perhaps due to the species' relative obscurity, but this may also be a reflection of its rarity.

Northern Waterthrush territories are associated with headwater wetlands, an important conservation focal point threatened by conifer loss and climate change (Rotenhouse et al. 2008; Gross and Haffner 2010). Other bird species that thrive in these high-elevation wetlands and forests with high vegetative structure, such as White-throated Sparrow, have declined. Loss of forest vegetation structure, diversity, and regeneration from various factors, including legacy effects of deer browsing or acid rain, are additional stressors. The fate of the Northern Waterthrush lies not only with its wooded wetland breeding habitat but also with its tropical wintering ground, where there are pressures on the mangrove and riparian forests in which it lives (Terborgh 1989; Eaton 1995). The relatively small population of this species, its decline in the state, and its association with forested wetlands and riparian areas are reasons to consider giving the Northern Waterthrush a higher priority for conservation in the state.

DOUGLAS A. GROSS

Distribution

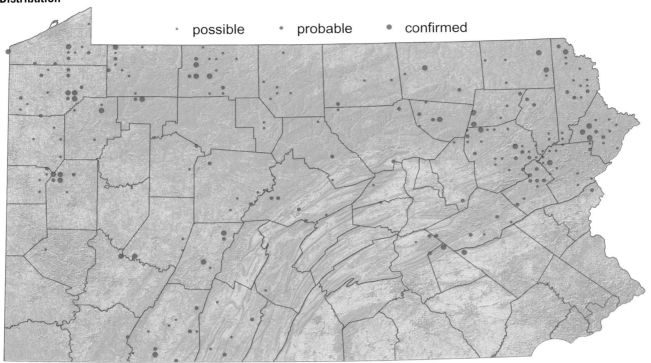

Distribution Change

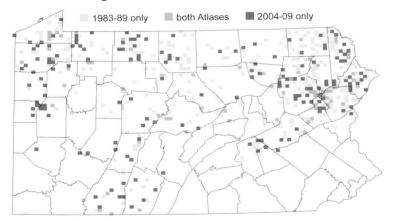

Number of Blocks

	first Atlas 1983–89	second Atlas 2004–09	Change %
Possible	185	124	−33
Probable	105	71	−32
Confirmed	41	34	−17
Total	331	229	−31

Population estimate, males:
5,000 (approximate)

Bob Wood and Peter Wood

Golden-winged Warbler
Vermivora chrysoptera

The Golden-winged Warbler is a neotropical migratory species that winters in parts of Central America and South America and breeds throughout the eastern and upper midwestern United States and into southern Manitoba. In Pennsylvania, the species nests in ephemeral early successional habitat resulting from abandoned farmland, fire disturbance, timber harvest, utility rights-of-way, and managed shrublands. The Golden-winged Warbler is a species experiencing steep population declines across eastern North America and is currently petitioned for federal listing (Buehler et al. 2007). The Breeding Bird Survey has documented precipitous annual declines in Golden-winged Warbler populations throughout the Appalachians, including New York (−5.3%), Pennsylvania (−6.8%), and West Virginia (−8.9%) since the 1960s (Sauer et al. 2011).

As in the first Atlas, the Golden-winged Warbler exhibited a patchy distribution, with most records in a swath extending from the southern Allegheny Mountains, through the Ridge and Valley Province, and sparingly northeast to the Pocono Plateau. The species was all but absent from parts of Pennsylvania south of this core range and was generally scarce farther north and west, where several former strongholds were largely abandoned between atlases. A notable exception was Sproul State Forest, where arson fire in 1990 set back succession on approximately 4,000 ha (10,000 ac) of forest, which was colonized by golden-wings in blocks that were not occupied in the first Atlas. Nonetheless, 20 years post-fire, much of the burned area is becoming unsuitable as breeding habitat for the Golden-winged Warbler.

Due to its long-term decline, the Golden-winged Warbler is now scarce in Pennsylvania, with a statewide population estimate of only 6,300 singing males. Several factors are thought to be driving the decline of this species across most of its historic breeding range, including habitat loss in both the breeding and wintering range, Brown-headed Cowbird parasitism, and hybridization with the Blue-winged Warbler (Buehler et al. 2007). However, causal mechanisms regarding how and to what extent these factors drive Golden-winged Warbler population trends are unclear.

It is widely accepted that hybridization with the Blue-winged Warbler poses a significant challenge to Golden-winged Warbler conservation (Gill et al. 2001; Confer et al. 2003). A recent rangewide genetic study revealed that hybridization with Blue-winged Warbler was not uncommon even in areas thought to support genetically pure Golden-winged Warbler populations, including the aforementioned population in Sproul State Forest (Vallender et al. 2009). Despite the absence of territorial Blue-winged Warblers, cryptic hybridization was confirmed in this population. Therefore, the long-term conservation of the Golden-winged Warbler in Pennsylvania will likely depend on the development and subsequent implementation of habitat management guidelines that create breeding habitat in areas devoid of the Blue-winged Warbler. Such an approach would provide areas where genetic introgression would be limited, thus helping to maintain significant populations of phenotypically pure Golden-winged Warblers. Nonetheless, blue-wings were found in only 30 percent of the atlas blocks from which Golden-winged Warblers were lost between atlases, suggesting that other factors, including habitat availability, are important.

The past abundance of young forest habitat in Pennsylvania was incidental to landscape-scale timber harvest and farmland abandonment. Future breeding habitat for the Golden-winged Warbler in Pennsylvania will likely require a well-designed, long-term management plan that targets significant acreage within the Appalachian and Pocono plateaus and Ridge and Valley regions. Within these areas, habitat management should target landscapes with expansive forest cover at higher elevations. The importance of forest cover and elevation has been identified by the Golden-winged Warbler Working Group, and this hypothesis is currently undergoing rangewide examination. Such an approach would provide areas where genetic introgression with the Blue-winged Warbler would be limited, thus helping to maintain significant populations of phenotypically pure Golden-winged Warblers.

JEFFREY LARKIN AND MARJA BAKERMANS

Distribution

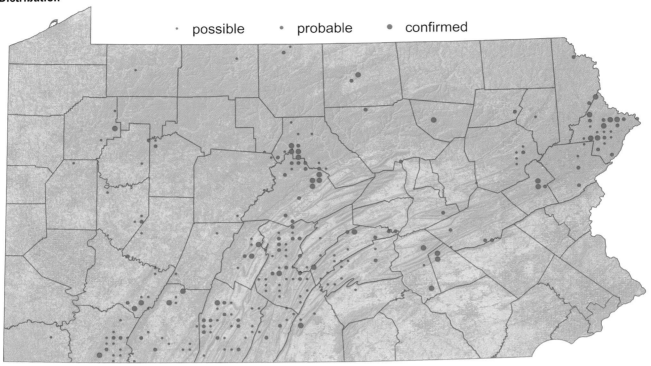

Distribution Change

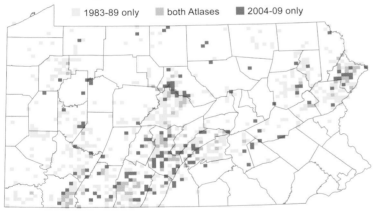

Density

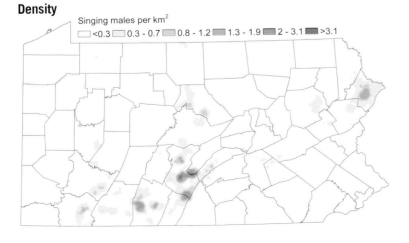

Number of Blocks

	first Atlas 1983–89	second Atlas 2004–09	Change %
Possible	293	131	−55
Probable	236	73	−69
Confirmed	86	36	−58
Total	615	240	−61

Population estimate, males (95% CI):
6,300 (5,000–7,900)

Breeding Bird Survey Trend

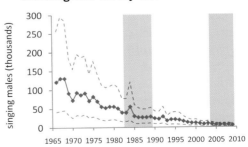

GOLDEN-WINGED WARBLER 351

Bob Wood and Peter Wood

Blue-winged Warbler
Vermivora cyanoptera

The Blue-winged Warbler is a shrubland bird known for its range expansion into the northeastern and upper midwestern portions of North America since European settlement. Now linked to the creation of early successional habitat through agricultural clearing of forests, Blue-winged Warblers have been documented in southeastern Pennsylvania since the mid-nineteenth century (Gill 1980). Blue-winged Warbler expansion began in the western and eastern portions of the state as early as 1915, with movements into the central part of the state in the 1970s (Gill 1980). Its range expansion in Pennsylvania has caused concern over hybridization with the Golden-winged Warbler, a less abundant species. However, in recent history, there is evidence of a contraction of the northeastern range of Blue-winged Warblers, due to loss of habitat through forest maturation and urban sprawl (Gill et al. 2001).

Blue-winged Warblers were recorded in 1,345 (27%) blocks during the second Atlas. Although there was a 4 percent decrease in Blue-winged Warbler occupancy between the first and second Atlas, this change was not statistically significant. While overall block occupancy has remained relatively stable for the species, there have been shifts in their distribution since the first Atlas. The largest species' expansion was in the environs of Allegheny National Forest and Susquehannock State Forest in the northwest of the state, where the number of occupied blocks more than doubled. In contrast, there were notable reductions in block records in the areas surrounding Philadelphia and Pittsburgh, as well as in and around Michaux (Adams County) and Lackawanna (Lackawanna County) state forests. The net result was a 19 percent decline in block records in the eastern half of the state (east of 77.8°W), compared with a 6 percent increase in block records in the western half.

The second Atlas estimates a total population of 52,000 singing males in Pennsylvania, with highest densities in the west. Breeding Bird Survey counts suggest that Blue-winged Warbler populations have remained stable in Pennsylvania since 1980 (Sauer et al. 2011). The distributional shifts within Pennsylvania could be part of continuing range shifts at a regional scale, which is suggested by a 29 percent decline in range in Maryland (Ellison 2010zk) and a northward expansion in New York (Confer 2008) between atlas periods.

Blue-winged Warblers are known to use a wide variety of early successional habitats, and they were most commonly recorded in transitional habitats, woody wetlands, and emergent wetlands during point count surveys (appendix D). Counts were greatest at mid-elevations of 300 to 450 m (~1,000 to 1,500 ft) and decreased at higher elevations; indeed, few Blue-winged Warblers were recorded on point counts above 600 m (~2,000 ft; appendix D). This may be related to the more extensive cover of mature forests at higher elevations in Pennsylvania. Further, an intensive study of Golden-winged and Blue-winged Warbler breeding ecology in north-central Pennsylvania from 2008 through 2010 yielded few observations of breeding Blue-winged Warblers above 500 m (1,640 ft), despite the availability of extensive early successional habitat (Larkin et al. 2011). Average dates for Blue-winged Warbler breeding activities included 28 May for nest building, 21 June for carrying food, and 3 July for fledging young. These are likely biased toward later in the breeding season, when atlasers were most active during the safe date survey period. Because the nesting season is brief (McWilliams and Brauning 2000) and not all suitable habitat is readily accessible, this species' distribution could be underestimated by atlas efforts.

Blue-winged Warbler interaction and hybridization with the Golden-winged Warbler is a major conservation issue for the latter species. The northward expansion, especially in the Golden-winged Warbler stronghold in the central mountains, increases potential contact between these two species. While Blue-winged Warblers currently are restricted to the valleys and Golden-winged Warblers occupy the higher elevation habitats, future expansion of Blue-winged Warblers may place the declining Golden-winged Warbler population in jeopardy.

MARJA BAKERMANS AND JEFFREY LARKIN

Distribution

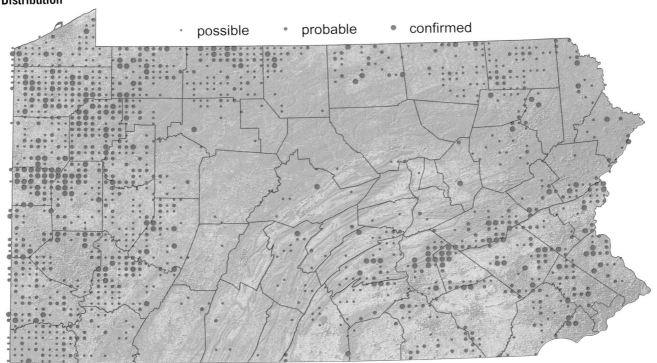

Distribution Change

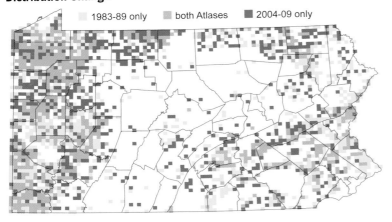

Density

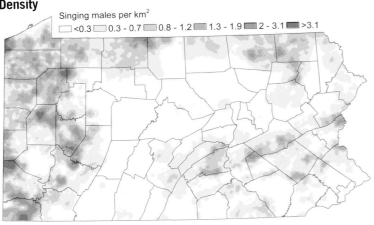

Number of Blocks

	first Atlas 1983–89	second Atlas 2004–09	Change %
Possible	520	753	45
Probable	655	395	−40
Confirmed	228	197	−14
Total	1,403	1,345	−4

Population estimate, males (95% CI):
52,000 (45,000–59,000)

Breeding Bird Survey Trend

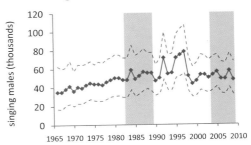

BLUE-WINGED WARBLER 353

Bob Wood and Peter Wood

Brewster's and Lawrence's Warblers

Vermivora cyanoptera x chrysoptera

Perhaps the best-known and most studied hybrid warblers in the eastern United States, the "Brewster's" and "Lawrence's" warblers are both found in Pennsylvania. These hybrids are the cross between Blue-winged and Golden-winged warblers, and the few atlas volunteers who were lucky enough to find breeding records were happy to add these notable sightings to their lists.

The more common hybrid, Brewster's Warbler, results from the mating of a Golden-winged with a Blue-winged Warbler and looks much like a Golden-winged Warbler with a dark, Blue-winged-like eye-stripe, white underparts and wingbars washed with yellow, and gray upperparts (Gill et al. 2001). The much rarer Lawrence's Warbler is the offspring of two adults with introgressed phenotypes, or a backcross of a hybrid with a pure adult, resulting in the recessive phenotype (Golden-winged face pattern) expressed (Curson et al. 1994). The Lawrence's Warbler has the black throat and cheeks of Golden-winged Warblers and the yellow body, green upperparts, and white wingbars of Blue-winged Warblers (Gill et al. 2001). Most plumage characters, with the exception of facial pattern, are controlled by multiple alleles and show a phenotypic variation between Golden-winged and Blue-winged warbler plumages (Parkes 1951). Both hybrids sing typical Golden-winged and Blue-winged warbler songs, and variations thereof.

In Pennsylvania, the hybrid warblers occur in low numbers throughout the state, most often in zones where Golden-winged and Blue-winged warbler ranges overlap, especially in the Appalachian Plateaus and the Ridge and Valley provinces. The second Atlas found Brewster's Warblers in 35 blocks and Lawrence's Warblers in 13—45 percent more than in the first Atlas for both hybrids combined. There was only one block in which hybrids were reported in both atlas periods (in Westmoreland County). The hybrids were distributed fairly evenly throughout the state during the first Atlas but were more concentrated in east-central Pennsylvania during the second Atlas, perhaps a reflection of the contraction in the Golden-winged Warbler's range.

The Golden-winged Warbler population has plunged in recent decades. There was a 61 percent decrease in the number of atlas blocks with Golden-winged Warblers in Pennsylvania between atlas periods, and Breeding Bird Survey counts in the state almost halved in that time (Sauer et al. 2011). This decline can be attributed to loss of habitat, behavior, and hybridization (Confer et al. 2003; Gill 2004; Buehler et al. 2007).

The shrubland habitat upon which Golden-winged Warblers depend is sharply declining due to a decline in young forests and increased urbanization, and, as habitat specialists, they are unable to quickly adapt. As shrublands mature into transitional habitat, the closely related, habitat-generalist Blue-winged Warbler (which has not shown a population decrease in Pennsylvania) is expanding its range northward and upslope into Golden-winged Warbler habitat (Curson 1994; Gill 2004), increasing contact between the two species. Blue-winged and Golden-winged warblers often hold overlapping territories (Gill et al. 2001), which increases the chance of extra-pair fertilizations between species and may drive hybridization. In northeastern Pennsylvania, female Brewster's Warblers prefer male Golden-winged Warblers as mates, and Blue-winged Warbler mitochondrial DNA (DNA that is maternally inherited) is found in Golden-winged Warbler populations, but not the other way around, indicating that female Blue-winged Warblers may drive hybridization (Gill 1997). Once hybrids begin to appear, total replacement of Golden-winged Warblers by Blue-winged Warblers (with both pure and introgressed phenotypes) can occur in as little as 20 to 50 years (Gill 2004; Shapiro et al. 2004).

As contact between Blue-winged and Golden-winged warblers continues to increase, occurrence of Brewster's and Lawrence's Warblers could increase, but a further decline in Golden-winged Warblers may eventually result in fewer opportunities for hybridization.

ANDREA L. CRARY

Distribution

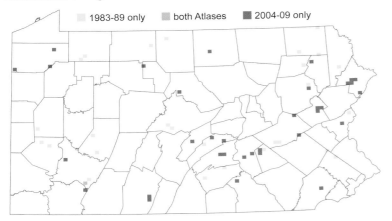

Distribution Change

Number of Blocks

Brewster's Warbler

	first Atlas 1983–89	second Atlas 2004–09	Change %
Possible	18	16	−11
Probable	6	9	50
Confirmed	5	10	100
Total	29	35	21

Lawrence's Warbler

	first Atlas 1983–89	second Atlas 2004–09	Change %
Possible	2	7	250
Probable	2	4	100
Confirmed	0	2	∞
Total	4	13	225

Bob Wood and Peter Wood

Black-and-white Warbler
Mniotilta varia

The Black-and-white Warbler acts like a nuthatch and has the striping of a zebra. It is fairly common in Pennsylvania, ranking eighth among warblers in block detections, although it ranked fifth in the first Atlas and has apparently declined in some parts of its range. The Black-and-white Warbler is a summer resident of mature and second-growth forests in the eastern half of the United States and southern Canada. While the Black-and-white Warbler is found in nearly every Pennsylvania county, the distinct bands of block detections indicate that it has certain habitat preferences. The highest Pennsylvania concentrations were on dry ridgetops, where it commonly forages on the furrowed bark of Chestnut Oaks. It was also found in mixed-forest groves and on riparian slopes. A Pennsylvania study found that Black-and-white Warbler densities tended to be highest where the forest understory was dense (Yahner 1986). During the second Atlas, this species was seldom found away from forested slopes.

The highest concentrations of the Black-and-white Warbler block records were in densely forested areas, which is consistent with the area sensitivity noted for this species (Kricher 1995). The largest contiguous area of high concentration was in north-central Pennsylvania, coincident with the largest contiguous area of intact forest. There were also many detections in Warren and McKean counties, along the Allegheny River, in Bald Eagle State Forest near the center of the state, in the eastern high terrain from the Pocono Plateau Section to the central Susquehanna Valley, on South Mountain, and in the Laurel Highlands.

Block detections of the Black-and-white Warbler increased by a statistically significant 8 percent between atlas periods, but after adjusting for increased volunteer effort, the increase was not significant. New block detections were concentrated along the edges of the core areas—for example, in McKean County in the north and in Greene County in the southwest. Many of the blocks where the Black-and-white Warbler was found during the first Atlas but not during the second were in smaller forest patches across the western third and southern third of Pennsylvania.

Breeding Bird Survey (BBS) data collected in Pennsylvania indicate that a steady decline has occurred at a statistically significant annual rate of 1.5 percent since 1966 (Sauer et al. 2011). From BBS data, it is estimated that the Pennsylvania population declined by 23 percent between atlas periods, with the population in the second Atlas estimated to be 250,000 singing males. The downward trend in Black-and-white Warbler numbers is unusual among the state's common wood warblers. Based on BBS data, it is estimated that it was the fifth most numerous warbler during the first Atlas period, but it has slipped down to ninth most numerous in the second Atlas period. Such declines are not unique to Pennsylvania: the BBS trends for neighboring states show a similar trajectory (Sauer et al. 2011), and there was a 16 percent loss in range in Maryland between atlases (Ellison 2010zl). Both the atlas change map and BBS trend maps (Sauer et al. 2008) indicate that in areas of smaller forests in the western and southern parts of Pennsylvania the Black-and-white Warbler has tended to decline, faring better in the larger forested areas of the north and east.

An area-sensitive species such as the Black-and-white Warbler could be affected by forest fragmentation on either the breeding grounds or on the wintering grounds, which are extensive and include Central America, northern South America, and the West Indies (Kricher 1995). Pennsylvania losses might also be related to changes in forest age or the loss of understory. While the Black-and-white Warbler remains fairly common, the decline in Pennsylvania has been steady, resulting in an approximate halving of the population since the 1960s (Sauer et al. 2011). Pennsylvanians can help this warbler by keeping our forests intact and with a sufficient understory.

NICHOLAS C. BOLGIANO

Distribution

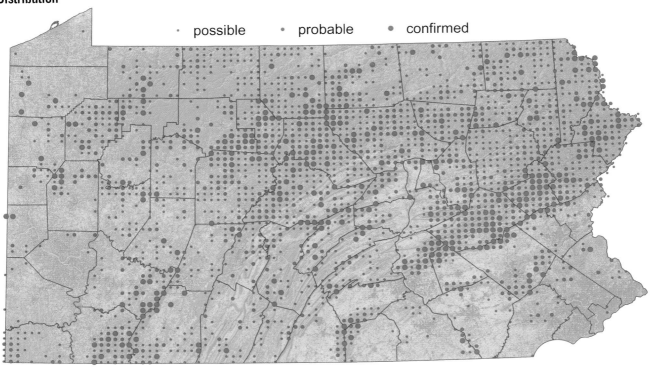

Distribution Change

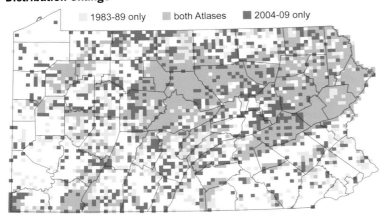

Density

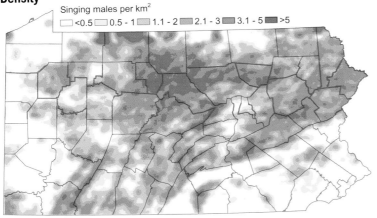

Number of Blocks

	first Atlas 1983–89	second Atlas 2004–09	Change %
Possible	1,018	1,310	29
Probable	693	660	−5
Confirmed	397	311	−22
Total	2,108	2,281	8

Population estimate, males (95% CI):
250,000 (234,000–271,000)

Breeding Bird Survey Trend

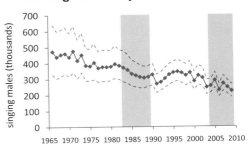

BLACK-AND-WHITE WARBLER 357

Geoff Malosh

Prothonotary Warbler
Prothonotaria citrea

With its brilliant golden-orange plumage, the Prothonotary Warbler is among Pennsylvania's most strikingly beautiful breeding wood warblers, especially when viewed with its reflection mirrored by the dark waters at the edge of its swampy summer home. The only eastern warbler that nests in tree cavities, it is also among the rarer species here, since Pennsylvania is near the northern edge of its breeding range. Historically a bird of southern swamps, flooded bottomlands, borders of lakes, rivers, and ponds (Harrison 1984), this species is a relatively recent addition to Pennsylvania's avifauna. The first evidence of nesting came from Lancaster County in 1924 (Beck 1924), followed by Erie County in 1938 (Todd 1940). While breeding has been observed regularly since the 1950s, the Prothonotary Warbler has remained both rare and local, nesting primarily in the glaciated northwestern counties and along the drainage of the Susquehanna and Delaware rivers in the southeast.

Prothonotary Warbler populations have remained relatively stable in Pennsylvania since the first Atlas, with a non-significant 9 percent drop in the number of blocks in which birds were found by atlas volunteers. In the glacial northwest, numbers were essentially unchanged, with a total of 14 blocks occupied during the first Atlas and 13 during the second. Along the Lake Erie shore, a new site was found at Roderick Wildlife Preserve near the Ohio state line, but at Presque Isle State Park, where several birds were located during the first Atlas, none were seen during the second Atlas, probably because the lower water level of the lake provided fewer flooded woodland ponds on the peninsula. During both atlases, the stronghold of the Prothonotary Warbler was in the Conneaut Marsh and Pymatuning Reservoir area of western Crawford County, where birds were located in eight different blocks; to the south, four blocks were occupied in Mercer County wetlands.

In eastern Pennsylvania, the Prothonotary Warbler's main grouping now lies along the lower Susquehanna River, where it has apparently increased locally due to the placement of nest boxes and nest tubes in suitable habitat. Here, the species is found nesting along the shore in York and Lancaster counties as well as on wooded islands in the river itself. There were also three probable breeding records west of the Susquehanna River, in Cumberland and Franklin counties, and two probable breeding records to the north, near the Juniata River. Elsewhere in eastern Pennsylvania, the species has a rather patchy distribution in the southeastern corner of the state, with several probable breeding records along the Schuylkill River and several confirmed records close to the Delaware River as far north as northern Northampton County. There were also three isolated records north of the Kittatinny Ridge in the Pocono Mountains. Unlike the bottomland nesting areas of the lower Susquehanna River, these northern birds were almost always found in wooded swamps or along slowly moving streams that were dammed by beavers.

Logging within bottomland forests, or the conversion of such wetlands to cultivation or pasturelands, can be a threat to prothonotaries, but much of the warbler's habitat in Pennsylvania already is in the relative safety of public or other conservation lands. While these warblers prefer natural cavities at heights of less than 4 m (12 ft) above standing water (Peterjohn 2001), in areas where there are few such natural sites, the use of nest boxes can be an important tool in enhancing local Prothonotary Warbler populations (Petit 1999). However, the House Wren may outcompete the prothonotary for nest boxes, and predation may also be greater than at natural sites. The destruction of mangrove habitats on their wintering grounds (Petit 1999) may pose an even greater threat to this warbler than does the loss of bottomland forest in their breeding grounds. Despite a northward range expansion, the Prothonotary Warbler has declined across its breeding range since the 1960s (Sauer et al. 2011); therefore, Pennsylvania's small population of this attractive and enigmatic warbler should not be considered secure.

ROBERT C. LEBERMAN AND RICK WILTRAUT

Distribution

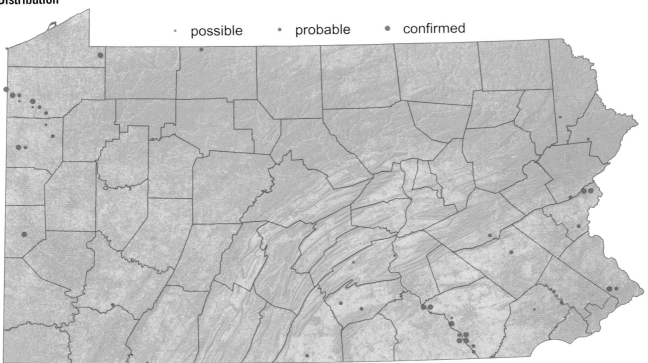

Distribution Change

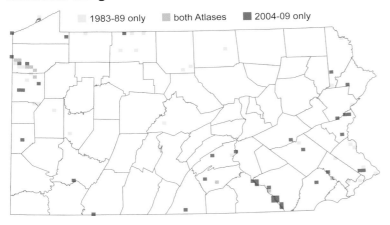

Number of Blocks

	first Atlas 1983–89	second Atlas 2004–09	Change %
Possible	21	13	−38
Probable	14	19	36
Confirmed	8	15	88
Total	43	47	9

Gerard Dewaghe

Swainson's Warbler
Limnothlypis swainsonii

The Swainson's Warbler, a long-billed, relatively large, brown and whitish warbler, was originally thought to be limited to canebrakes, bottomland forests, and swamps on the Atlantic and Gulf coastal plains as far north as Delaware. However, in the 1920s and 1930s a separate population was discovered in the Appalachian Mountains, extending from Georgia and Alabama north to Virginia and West Virginia (Meanley 1971). Diverse Appalachian habitat preferences have been described—dense tangles of rhododendron, mountain laurel, hemlock, and holly along streams in mountain ravines and hollows (Brooks and Legg 1942); thickets of greenbrier, grape, honeysuckle, or Bittersweet (Sims and DeGarmo 1948); cove hardwood forests dominated by Yellow Poplar and other large trees with dense understory (Meanley 1971); and along rivers, including the Ohio in southwestern West Virginia (Hurley 1972).

Sporadic in occurrence in Pennsylvania since the first state record in 1954 (McWilliams and Brauning 2000), Swainson's Warbler maintained its reputation as a tantalizing bird north of its traditionally known breeding range during the second Atlas. Males sang their boisterously loud song at four widely scattered sites, all different from locations in the first Atlas. But again, there was no confirmation of Pennsylvania's first breeding record. Most notable was a persistent singer from 12 May to 17 June 2007 at Settler's Cabin Park in southern Allegheny County. A second bird was reported there for part of that time (Fialkovich 2007), but the existence of a mated pair was not established. The record is classified as probable because of the bird's extended territorial behavior. At the other three locations, males were listed as possible breeders because each was observed singing on one day: 11 June 2006 near Bolivar in Westmoreland County, 19 June 2006 in the Weiser State Forest in Dauphin County, and 1 June 2007 in the Muddy Creek Valley in York County.

The first Atlas listed two territorial males, both in southwestern Pennsylvania in 1989: one at Powdermill Nature Reserve in Westmoreland County, and another at Bear Run Nature Reserve in Fayette County. What may have been the same bird held territory at Bear Run in 1990 and 1991, but there was no evidence of breeding (Krueger and Mulvihill 1992). Southwestern Pennsylvania is closest to extant population in the Appalachians, so it is not surprising that most recent records have been from that area. Interestingly, there were two singing males at separate sites in West Virginia, close to the Pennsylvania state line, in 2009 (Anonymous 2009). As in Pennsylvania, no mates were seen there.

The prospects of finding Swainson's Warbler in eastern Pennsylvania were more remote. The most northerly recent occurrences on the Coastal Plain have been those of a few singing birds in southeastern Maryland (Heckscher and McCann 2006). Swainson's Warblers in Pennsylvania are often described as accidental spring migration "overshoots," but the persistent territorial behavior of individuals in suitable habitat suggests that some of these are not merely vagrants in the usual sense of the word. Krueger (1989) noted that the warbler can be a challenge to find even by experienced observers, because many potential breeding areas are remote and difficult to access. In addition, the species' distribution is localized, and confirmation of breeding is difficult even within the West Virginia counties in which it regularly nests (Buckelew and Hall 1994). Perhaps Swainson's Warbler is more frequent in Pennsylvania than we suppose, although the odds may be low that a male found so far north could attract a mate.

The overall population status of this species is tenuous, and it is seriously threatened by habitat loss in its breeding range and on its wintering grounds in the Caribbean, Mexico, and Central America (Anich et al. 2010). Partners in Flight estimates the global population at 84,000 birds and rates Swainson's Warbler as a Watch List species (Rich et al. 2004). Rangewide population declines probably decrease future chances of breeding confirmation in Pennsylvania, but coordinated surveys of the most suitable habitats could yet reveal this species to be more frequent in the state than is currently thought.

PAUL HESS

Distribution

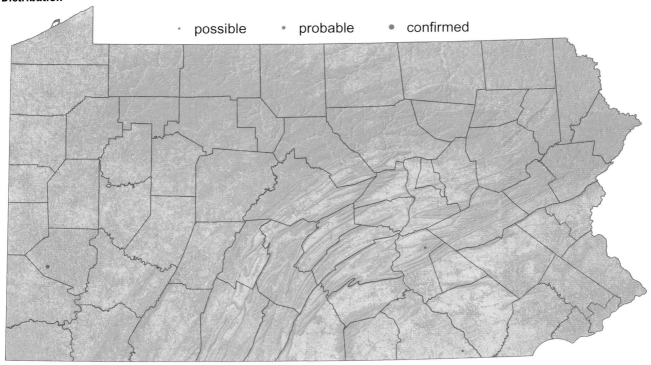

Distribution Change

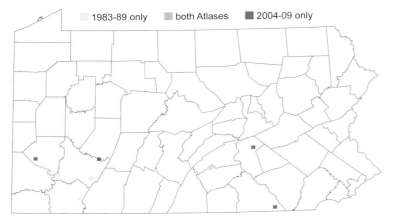

Number of Blocks

	first Atlas 1983–89	second Atlas 2004–09	Change %
Possible	0	3	∞
Probable	2	1	−50
Confirmed	0	0	0
Total	2	4	100

Bob Wood and Peter Wood

Nashville Warbler
Oreothlypis ruficapilla

Alexander Wilson described the Nashville Warbler in 1811, observing it near the city of that name—hence, the otherwise inappropriate name for this warbler with an affinity for northern bogs (Williams 1996). The Nashville Warbler is easily identified by its distinct eye-ring on a gray head, and its bright yellow underparts. Its song, a two-part composition similar to the trills of several other species, is not often heard during nesting season in much of the state, as this is one of Pennsylvania's rarest breeding warblers, ranking 26th of 30 warblers in block detections during the second Atlas.

There are two distinct populations of the Nashville Warbler in North America, one of which breeds in the Pacific Northwest. The range of the eastern population extends from Saskatchewan to the Maritime provinces in Canada, south into the northern Great Lakes states, New England, and down the Appalachians through portions of Pennsylvania into West Virginia (Williams 1996).

With its preference for second-growth forests and bogs, the Nashville Warbler may have been rare prior to the cutting of Pennsylvania's forests; it is perhaps best established in the bogs of the Poconos, still a stronghold today (Master 1992f). Warren (1890) described it as "breeding sparingly" in mountainous regions of Pennsylvania. Todd (1940) mapped a dozen breeding locations in western Pennsylvania, mostly in the north, but also some in the Laurel Highlands in the south, roughly coinciding with the distribution mapped in the first and second Atlases.

Atlas reports were widely scattered and mostly in the northern half of the state. The Nashville Warbler was found in only 4 percent of blocks, a total of 213, compared with 217 during the first Atlas. When corrected for greater effort in the second Atlas, this represents an estimated 20 percent decrease in range. In accord with known habitat preferences, confirmed breeding records that included habitat descriptions mentioned "thicket," "brushy edge," "regrowth," and "hemlock swamp." However, no nests were found; of 24 confirmed records, most were of birds carrying food or attending fledged young.

The most concentrated clusters of blocks with Nashville Warbler records were in the Poconos and the North Mountain area of Sullivan County; many of these blocks were also occupied during the first Atlas. In contrast elsewhere, of blocks occupied in at least one of the two atlases, Nashville Warblers were found in both atlases in only 11 percent of those blocks. This high turnover is not surprising, given the scarcity of this species and the ephemeral nature of its preferred early successional habitat, such as recently logged areas or openings created by fire or wind-throw.

There were fewer records in the south than during the first Atlas, most of which were submitted as possible breeding records, although there were two confirmed records in the Laurel Highlands and one in Perry County. The decline in the south gives the appearance of northward contraction of the breeding range, but this may also be due to fewer misidentifications and better recognition of lingering migrants in southern areas, where nesting is not expected.

This species is more widespread in New York state, where there was a 23 percent decline in blocks with records in the Appalachian Plateaus region (Collins 2008a), which is adjacent to Pennsylvania's northern tier. A significant portion of the Nashville Warbler's Pennsylvania breeding range is on protected public lands. The most immediate possible threat may be loss of winter habitat in the Caribbean (Williams 1996). In the long term, climate warming may adversely affect the favored cool bog environment in the Poconos, and indeed, this species' southern range limit is projected to shift well north of Pennsylvania by the end of the century, even under moderate climate change scenarios (Matthews et al. 2010). On the positive side, the species' preference for brushy second growth means it may benefit from rotational cutting of forests (Williams 1996). In Pennsylvania, Nashville Warbler numbers are too low to allow a confident population estimate from the Breeding Bird Survey or atlas point count data, but there are probably no more than a few thousand breeding pairs in the state.

GREG GROVE

Distribution

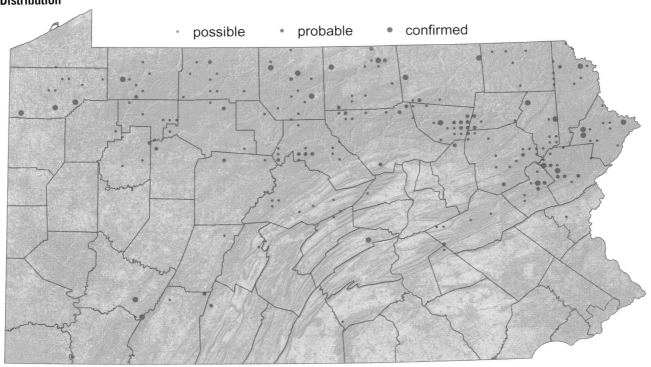

Distribution Change

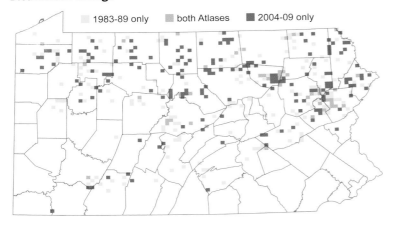

Number of Blocks

	first Atlas 1983–89	second Atlas 2004–09	Change %
Possible	116	147	27
Probable	75	42	−44
Confirmed	26	24	−8
Total	217	213	−2

Bob Wood and Peter Wood

Mourning Warbler
Geothlypis philadelphia

One of the most secretive and sought-after wood warblers is the subtly beautiful Mourning Warbler. The species was first discovered and named by the early ornithologist Alexander Wilson near Philadelphia in 1810, hence the regional reference in the species' scientific name. The black breast reminded Wilson of someone in mourning. The Mourning Warbler is one of the latest migrants of all wood warblers that breed in Pennsylvania, usually arriving on territory in mid- to late May, with a few individuals passing through into early June (Haas and Haas 2005).

The Mourning Warbler is a boreal bird, nesting in northern latitudes and at high elevations. In the first Atlas, it was found primarily in Erie and Crawford counties, and east to Potter County, with only a handful of records elsewhere. During the second Atlas, volunteers found Mourning Warblers in 431 blocks, compared with 235 in the first Atlas (Leberman 1992i), an 83 percent increase.

Although this species was still found primarily in the northern tier of western Pennsylvania, there was a significant eastward range expansion, notably in McKean and Potter counties. Even farther east, there was a substantial range expansion in the vicinity of North Mountain in Lycoming and Sullivan counties as well as an increase in records in the northeastern corner of the state. Records in the northeastern United States are in proximity to well-established populations in the Catskill Mountains in New York (Collins 2008b). Despite the increase in blocks with records, there were decreases in some former strongholds, notably in Erie County. Southward range expansion was limited but evident in both Crawford and Elk counties.

The Mourning Warbler is found in areas of heavy undergrowth, usually regenerating clearings that have been subjected to lumbering activities, fires, windstorms, and power line cuts. At Promised Land (Pike County), several singing Mourning Warblers were found in a regenerating clearing that was the result of an F-2 tornado in May 1998 (G. Dewaghe; R. Wiltraut, pers. obs.), similar to what occurred after tornados in northwestern Pennsylvania in May 1985 (Leberman 1992i). These open areas of second-growth are dominated by tree saplings, ferns, grasses, and brambles. Blackberry seems to be a dominant vegetation type where Mourning Warblers breed in Pennsylvania. Timbered areas with some remaining tall trees are preferred (Dessecker and Yahner 1987), but once the open areas are crowded out by young trees, typically after 10 years or so (Pitocchelli 2011), Mourning Warblers disappear.

The Mourning Warbler usually nests on the ground and builds a bulky nest of dead leaves, with a compact cup of vine stems and grasses (Baicich and Harrison 2005). Atlas volunteers observed adults carrying food as early as 6 June, suggesting that eggs were laid during May, but most confirmed breeding records were from mid-June to late July. Because the Mourning Warbler is a late migrant, some of the possible breeding records entered within safe dates may have been of late migrants passing through.

Detected surprisingly frequently on atlas point counts, the Mourning Warbler's statewide population was estimated at only 10,500 singing males, with as much as half of that total in McKean and Potter counties. Breeding Bird Survey data show an upward trend of 1.7 percent per year in Pennsylvania (Sauer et al. 2011), in keeping with atlas results. In neighboring New York, both the species' range and population appear to be stable (Collins 2008b; Sauer et al. 2001). Although the Mourning Warbler faces the same problems as many other neotropical birds, including habitat loss and collisions with glass and other structures, it appears that the rangewide population is stable (Sauer et al. 2011). Unlike many other neotropical birds, this species appears to have benefited from forest clearings on the breeding grounds as a result of lumbering, fires, power lines, and other factors. The Mourning Warbler has now reclaimed some of its former range in Pennsylvania (McWilliams and Brauning 2000), and while it is restricted by ephemeral habitat availability and shuns low elevations (appendix D), there is ample high-elevation forest where this species could expand yet further.

RICK WILTRAUT

Distribution

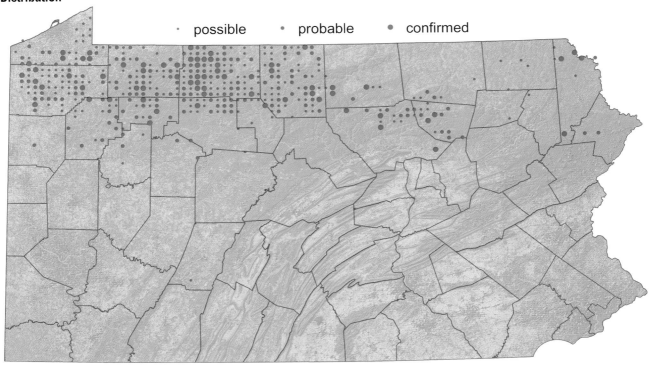

Distribution Change

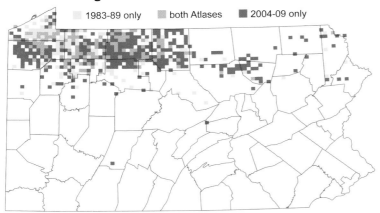

Density

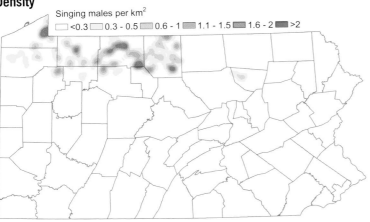

Number of Blocks

	first Atlas 1983–89	second Atlas 2004–09	Change %
Possible	81	207	156
Probable	109	156	43
Confirmed	45	68	51
Total	235	431	83

Population estimate, males (95% CI):
10,500 (8,700–12,200)

Breeding Bird Survey Trend

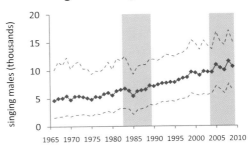

MOURNING WARBLER 365

Bob Wood and Peter Wood

Kentucky Warbler
Geothlypis formosa

Often difficult to observe due to its skulking nature and preference for thick understory vegetation in deciduous forests, the Kentucky Warbler is a fairly common breeder in southwestern Pennsylvania, but it is uncommon to rare elsewhere (McWilliams and Brauning 2000). Its loud, upward-slurred song consisting of three to eight *churree* phrases may be confused with that of the Ovenbird or Carolina Wren. When sighted, the Kentucky Warbler is unmistakable (McDonald 1998).

The Kentucky Warbler is a ground-nesting species of bottomland hardwoods, although it sometimes places its nests just above ground level (Todd 1940). Thick understory vegetation and groundcover are preferred for nesting, especially along streams and at lower elevations (McDonald 1998). The Kentucky Warbler's range stretches from eastern Texas, Oklahoma, and Kansas in the west, south to the Gulf Coast, and north to Wisconsin and New York (McDonald 1998). Pennsylvania is at the northern edge of the range, with few individuals seen north of the state during the breeding season (Rosenberg 2008b).

Prior to 1880, the Kentucky Warbler was considered an uncommon and local breeder in the state (Burns 1901). The overall trend in Pennsylvania and across the mid-Atlantic states during the following century was of northerly expansion, interrupted by periods of range contraction. By the time of the first Atlas, the species was found in 19 percent of atlas blocks, with strongholds in the southwestern and southeastern portions of the state. Despite evidence of northward expansion throughout the Mid-Atlantic region between 1960 and 1980 (Master 1992g), Breeding Bird Survey (BBS) data show a sustained population decline of 2.5 percent per year in Pennsylvania since the 1960s (Sauer et al. 2011). Declines of similar magnitude, if not steeper, were noted in neighboring Delaware, Maryland, New Jersey, and Ohio (Sauer et al. 2011).

The decline in abundance documented by the BBS was mirrored by declines in occupancy between atlases. Losses were documented in almost all physiographic sections from which Kentucky Warblers were reported (appendix C), but they were especially marked in the Piedmont (−46%), where this species is now scarce away from the well-forested lower Susquehanna Valley. Declines in the southwest were almost as large. The Pittsburgh Low Plateau Section, where the decline in occupied blocks was a modest but still statistically significant 18 percent, fared better than in other areas. Despite the thinning of this species' range, there was some evidence of a range expansion in the northwest (e.g., Venango County) and a scatter of records in border counties in the northeast, where this species was not found in the first Atlas.

Despite the overall decline in blocks with Kentucky Warbler in the Appalachian Plateaus, the southwest of this province continues to hold the highest densities of this species in Pennsylvania. Greene, Washington, Westmoreland, Fayette, Bedford, and Somerset counties together host more than half of the state's estimated 17,700 singing males. The vast majority of Kentucky Warblers recorded on point counts were in deciduous forest at elevations of 200 to 600 m (650 to 2,000 ft; appendix D).

Continentwide, as in Pennsylvania, the Kentucky Warbler exhibits a confusing array of regional and localized population increases and decreases, but it is difficult to determine specific causes (Sauer and Droege 1992). The Kentucky Warbler requires large blocks (>500 ha, ~1,200 ac) of deciduous forest with thick understory (Gibbs and Faaborg 1990). Losses of this species between atlases often occurred in areas where there was forest loss, notably due to exurban development in the Philadelphia and Pittsburgh metropolitan areas (chap. 3). In the Piedmont, for example, the number of atlas blocks with 50 percent or more forest cover dropped from 152 in 1994 to 62 in 2005, greatly reducing the number of blocks with the large, unfragmented forests favored by this species. Reproductive output is negatively affected by nest parasitism by Brown-headed Cowbirds (Robinson et al. 1995) and degradation of requisite nesting habitat due to the legacy affects of high deer populations during the last quarter of the twentieth century. Both of these are likely powerful influences on Kentucky Warbler populations in Pennsylvania's more fragmented landscapes.

LEWIS GROVE

Distribution

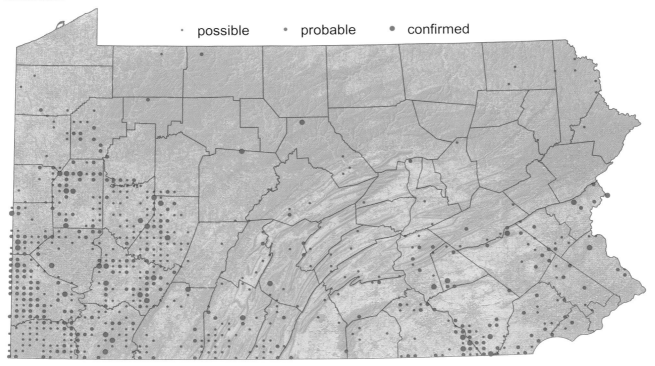

Distribution Change

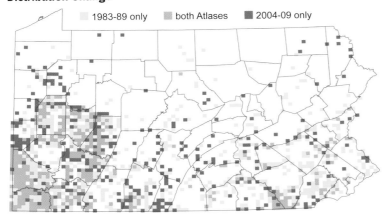

Density

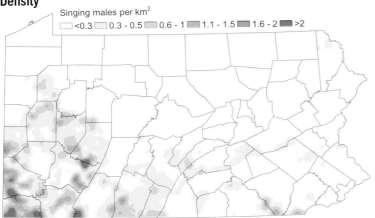

Number of Blocks

	first Atlas 1983–89	second Atlas 2004–09	Change %
Possible	342	393	15
Probable	460	227	−51
Confirmed	127	42	−67
Total	929	662	−29

Population estimate, males (95% CI):
17,700 (13,700–27,000)

Breeding Bird Survey Trend

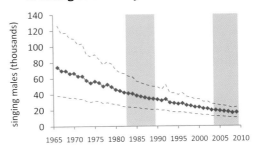

KENTUCKY WARBLER

Chuck Musitano

Common Yellowthroat
Geothlypis trichas

With the male's black mask creating sharp contrast to a brilliant yellow throat and breast, the Common Yellowthroat is a striking bird that robustly betrays its presence from among tangles of dense undergrowth by its loud, rolling *witchity-witchity-witchity* song. It is often found in abundance in wet areas with dense undergrowth, including wet meadows, marshes, and swamps but it is also a characteristic species of untamed upland habitats such as woodland edges, fallow fields, rights-of-way corridors, and overgrown fencerows (Stewart 1953; McWilliams and Brauning 2000). Within these densely vegetated habitats, yellowthroats are found throughout North America, except above tree line in Canada and Alaska and much of the southwestern United States (Guzy and Ritchison 1999). During the non-breeding season, it is found throughout Mexico and Central America. As a neotropical songbird, its breeding season and vocal habits make it easily detected by atlas volunteers.

In Pennsylvania, the Common Yellowthroat is our most widely distributed warbler, with breeding evidence observed in 98 percent of atlas blocks in both atlases. Since it is found in such a wide range of habitats (appendix D), it is not surprising that the Common Yellowthroat is one of our most ubiquitous birds. It generally avoids brushy habitats in urban and suburban areas, unlike some other common habitat associates such as the Eastern Towhee, Carolina Wren, and Song Sparrow, and was therefore least numerous on point count surveys in the most heavily developed urban centers such as the cities and surrounding sprawl of Philadelphia and Pittsburgh. Densities also were low in open farmland, especially in the Piedmont, although this species can be numerous in the rank cover provided by conservation grasslands (Wentworth et al. 2010) and on reclaimed surface mines (appendix D). The highest densities were in forested regions, especially in the Appalachian Plateaus, where densities of more than 12 singing males per km^2 were typical. Point counts were highest where 50 to 80 percent of the surrounding area (within 150 m radius) was forested (appendix D), which infers a preference for forest edges rather than contiguous core forests.

In contrast with many other warbler species, confirmation of breeding by the Common Yellowthroat is often easy to obtain, with more than 2,000 confirmations submitted during the second Atlas, including almost 600 nests in various stages. This wealth of breeding confirmations reflects not only the species' abundance but also its protracted breeding season. The yellowthroat is frequently double-brooded (Guzy and Ritchison 1999); atlas records show that nest building was going strong by the first week of May, and observations of young being fed were evident throughout July.

Though abundant and adaptable, this species may be negatively affected by several factors. The Common Yellowthroat is a frequent host to Brown-headed Cowbird nest parasitism (Lowther 1993), which was considered one of several factors that caused declines in other regions of the country (Dunn and Garrett 1997). Additionally, because the species avoids densely populated areas, continued urbanization will likely continue to cause localized declines, as has already been observed in the expanding sprawl that surrounds our largest cities. Breeding Bird Survey counts showed a strong upward trend between 1966 and the first Atlas period, followed by a long period of stability, with some fluctuations (Sauer et al. 2011). During the second Atlas period, the population was estimated to be more than 1.2 million singing males, making the Common Yellowthroat among the most numerous songbirds in Pennsylvania. A modest decrease in numbers since the first Atlas has likely resulted in the Ovenbird replacing this species as our most numerous species of warbler.

DANIEL P. MUMMERT

Distribution

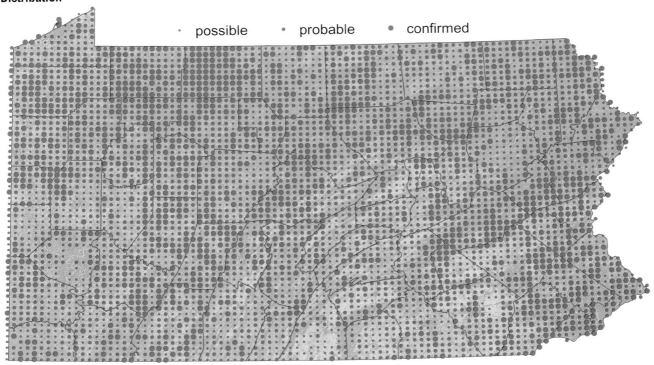

Distribution Change

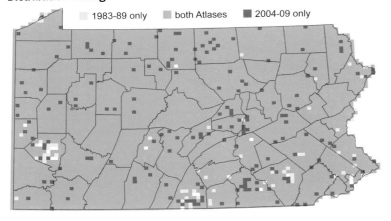

Density

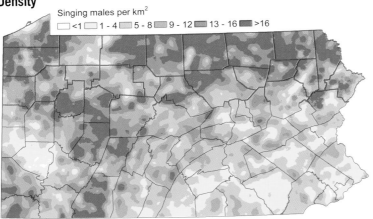

Number of Blocks

	first Atlas 1983–89	second Atlas 2004–09	Change %
Possible	792	1,419	79
Probable	2,328	2,098	−10
Confirmed	1,613	1,329	−18
Total	4,733	4,846	2

Population estimate, males (95% CI):
1,240,000 (1,210,000–1,265,000)

Breeding Bird Survey Trend

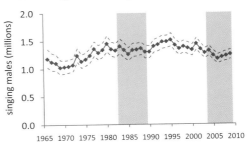

COMMON YELLOWTHROAT

Bob Wood and Peter Wood

Hooded Warbler
Setophaga citrina

The Hooded Warbler's loud song, *weeta weeta weet-ee-o*, and its many variants, was a familiar sound to atlas volunteers through much of the western half of the state and scattered areas eastward. Males are unmistakable in appearance, with a jet-black crown and bib contrasting with bright yellow cheeks. Many females also have a "hood" of sorts, with a partial hood forming an outline trace of the male's head markings. The conspicuous flashing of the outer white feathers while foraging helps to distinguish this species from other similar warblers.

The Hooded Warbler is a forest-dependent gap specialist that is common in mature deciduous forest with a dense understory for nesting sites, readily occupying selectively logged forests (Evans Ogden and Stutchbury 1994). Nests are usually less than a meter off the ground and placed in the dense shrub layer typical of light gaps and forest edges; consequently, Hooded Warblers are frequently parasitized by Brown-headed Cowbirds. Hooded Warblers are often double-brooded, particularly if the first nest of the season fledges young (Evans Ogden and Stutchbury1994), so nesting activity typically continues into late summer. In the second Atlas, singing males were most abundant on point counts in deciduous forest and were also found in mixed forest and woody wetlands, as well as forested residential settings (Appendix D). Though often linked to "forest interior," Hooded Warblers breed at forest edges and are common in small fragments of mature forest (Norris and Stutchbury 2001).

Pennsylvania lies at the northern edge of the Hooded Warbler range, and though abundant in western New York (McGowan 2008zn), this species is uncommon in Ontario (Badzinski 2007). The second Pennsylvania atlas shows a highly significant increase in the number of blocks with the Hooded Warbler; it was found in 29 percent of blocks in the first Atlas and 49 percent of blocks in the second Atlas. This increase was partly due to expansion within the core range in the western half of the state, particularly the northwest, where this species was found in the great majority of blocks. There also were substantial increases in the central counties of the state, where the species was localized in the first Atlas. The Ridge and Valley Province stands out, with a 144 percent increase in blocks with Hooded Warbler between the atlases.

Density estimates derived from second Atlas point counts show that most areas of high abundance (>5 singing males per km^2) are in the state's western historic range, but with pockets of high density in well-forested high-elevation areas of the east. For the time being, the Hooded Warbler is still localized in distribution and scarce in the northeastern and southeastern corners of the state, but given the range expansion between atlases, this could change in coming years. Breeding Bird Survey (BBS) data from Pennsylvania also show a steady upward trend, with an average increase of 3.9 percent per year from 1966 to 2009 (Sauer et al. 2011). From BBS data, we estimate that the statewide population of Hooded Warblers increased by 71 percent between atlas periods.

The Hooded Warbler is evidently in a period of northward range expansion across the region; even greater increases in atlas block occupancy occurred between atlases in New York (+146%; McGowan 2008zn) and Ontario (+400%; Badzinski 2007). The causes of the range expansion and population increases are not fully understood, but maturation of forest, combined with a possible response to climate change, may be important factors (Badzinski 2007; McGowan 2008zn). With an expanding range and increasing population, it is likely that this species' song will become increasingly familiar in most of Pennsylvania in the near future.

BRIDGET J. M. STUTCHBURY

Distribution

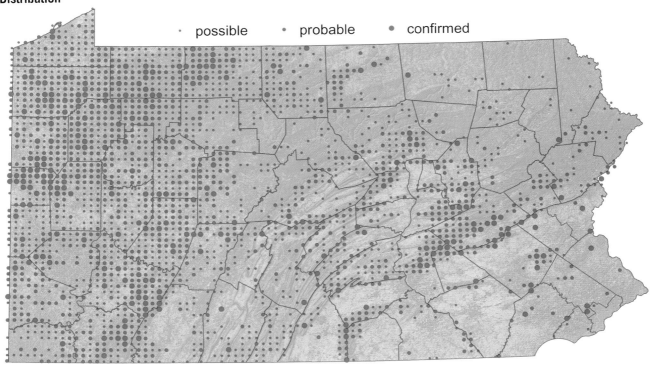

Distribution Change

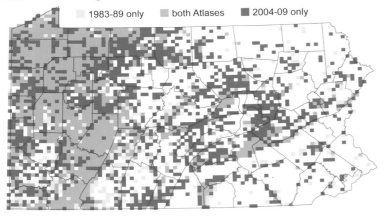

Density

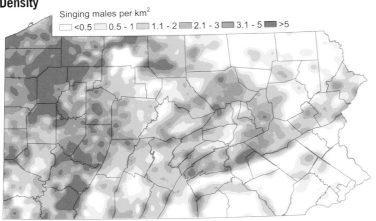

Number of Blocks

	first Atlas 1983–89	second Atlas 2004–09	Change %
Possible	532	1,219	129
Probable	681	948	39
Confirmed	205	256	25
Total	1,418	2,423	71

Population estimate, males (95% CI):
264,000 (251,000–278,000)

Breeding Bird Survey Trend

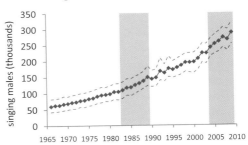

Bob Wood and Peter Wood

American Redstart
Setophaga ruticilla

The male American Redstart is unmistakable, at least older males, which have bold and contrasting orange and black ornaments on the breast, wings, and tail. The weak, high-pitched, and highly variable song is not so distinctive and is sometimes inaudible to older listeners. One-year-old males are distinctive in their own way because they look very much like females, with drab olive-gray and pale yellow wing and tail patches. This fly-catching species fans and flashes its tail while foraging, showing off the orange or yellow patches on the tail, and the telltale rictal bristles at the edges of the mouth reveal its main foraging method.

Nests of the redstart are unusual among Pennsylvania's warblers in that they are built in the crotch of the trunk of a sapling (and sometimes in grapevines) up to 10 m (33 ft) off the ground and blend well with the substrate. Like many migratory songbirds in Pennsylvania, the socially monogamous pairs seen diligently feeding their hungry nestlings do not hint at the high frequency of extra-pair fertilizations that result in 64 percent of broods containing at least one young sired by a neighboring male (Kappes et al. 2009).

As with most of our forest birds, the fortunes of the American Redstart in Pennsylvania have likely tracked changes in forest cover, with its resurgence during the twentieth century as a result of forest regrowth. By the time of the first Atlas, it was found in all 67 counties but was rather local in the southeast and in some areas along the Ohio border (Mulvihill 1992h). During the time between atlases, the redstart's range grew, increasing overall block occupancy from 61 to 75 percent. A 52 percent increase in blocks with redstarts in the Ridge and Valley accounted for much of this change, but there were gains almost everywhere. Despite this, there remain some substantial gaps in the species' distribution, notably in the Piedmont and Pittsburgh areas; indeed, in both of these areas there were signs of localized population losses, possibly associated with the loss of woodland in exurban areas (chap. 3).

Point count data show that the American Redstart is something of a forested habitat generalist, but it is found in highest densities in deciduous forests and forested wetlands (appendix D). It is often considered to be a typical bird of secondary forest and is not considered area-sensitive (Sherry and Holmes 1997), which contributes to its widespread distribution. Hence, the American Redstart is often found where other habitat specialist warblers are absent and, conversely, is often scarce in high-elevation mixed forests where warbler diversity is greatest. With a statewide population estimated to be 730,000 singing males during the second Atlas period, this was the fourth most abundant warbler overall, behind Ovenbird, Common Yellowthroat, and Yellow Warbler.

The Breeding Bird Survey (BBS) shows a 0.7-percent-per-year increase in American Redstarts in Pennsylvania since the 1960s (Sauer et al. 2011), with a modest 16 percent increase between atlas periods, and a rather stable population since the mid-1990s. In contrast, the atlas in neighboring New York (McGowan 2008zo) found that American Redstarts have not increased in block occupancy, and there has been a substantial decline in counts on BBS routes there (Sauer et al. 2011). Declines in some parts of the northeastern United States are thought to be due to forest maturation (Sherry and Holmes 1997). However, to the north in Ontario, redstart populations have remained stable or increased in some regions during their second atlas (McLaren 2007c), while to the south, in Maryland, there was little overall change in its status between atlas periods (Ellison 2010zm). Because it is robust to the effects of forest fragmentation, the American Redstart may be spared the challenges faced by other forest birds in the event of large-scale fragmentation due the development of energy infrastructure. However, forest maturation, coupled with loss of woodlots to development, will likely result in some loss of habitat for this species.

BRIDGET J. M. STUTCHBURY

Distribution

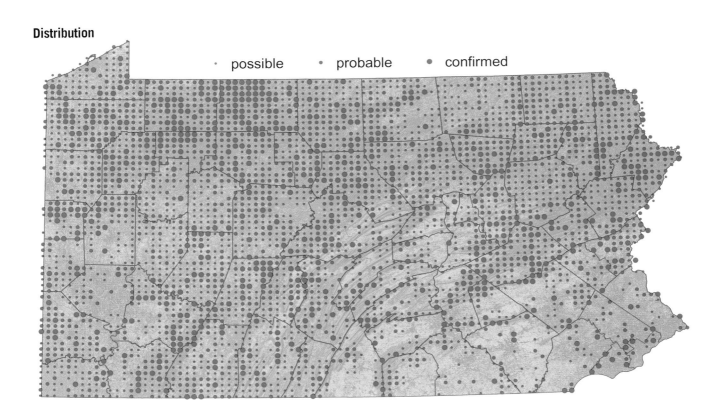

Distribution Change

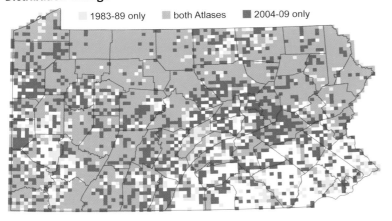

Density

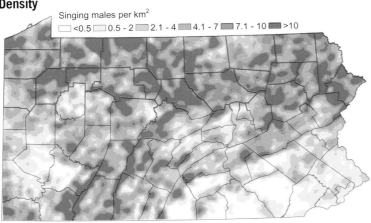

Number of Blocks

	first Atlas 1983–89	second Atlas 2004–09	Change %
Possible	1,084	1,756	62
Probable	1,291	1,335	3
Confirmed	617	625	1
Total	2,992	3,716	24

Population estimate, males (95% CI):
730,000 (700,000–760,000)

Breeding Bird Survey Trend

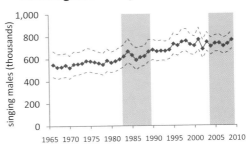

Geoff Malosh

Cerulean Warbler
Setophaga cerulea

A distinctive, buzzy song emanating from high in ridgetop oaks or bottomland forest is often the only indication of the Cerulean Warbler's presence. A good view of the blue, black, and snow-white male, or the pastel greenish-gray and buff female, usually requires considerable effort, patience, or luck.

The Cerulean Warbler breeds in mature deciduous forest throughout eastern North America, as far west as Oklahoma, south to Alabama, north to southern Ontario, and east into New England (Hamel 2000). The core of its breeding range lies in the central Appalachian Mountains; elsewhere it is very local and sparsely distributed. On the breeding grounds, the species occurs in structurally complex, mature forest of two distinct habitat types: oak-hickory on ridgetops and sycamore, elm, or oak along major streams and rivers. It is generally considered to be area-sensitive, avoiding small woodlots or fragmented forest (Hamel 2000). Perhaps ironically, the Cerulean Warbler seems to prefer open canopies or internal gaps within extensive forest. It winters in mid-elevation forests and plantations of the northern Andes from Venezuela to Peru (Stotz et al. 1996).

Historically, within Pennsylvania the Cerulean Warbler was essentially a species of the southwestern counties. Todd (1940) reported it to be generally absent east of the Allegheny Front, and Simpson considered it accidental in Warren County up through the 1920s (Hoover 2003). At the time of the first Atlas, it was still found predominantly in the southwestern corner of the state, but it was found sparsely in most regions and was absent from only a handful of counties.

In the second Atlas, the Cerulean Warbler was documented in only 16 percent of blocks statewide, a small decline from the first Atlas. The Waynesburg Hills and Pittsburgh Low Plateau, once the state's stronghold (McWilliams and Brauning 2000), experienced a drop in block occupancy of more than 50 percent between atlas periods. Despite these significant declines, the borderlands of southwestern Pennsylvania remain a stronghold for the species. In fact, it remains relatively common in western Greene County, which, according to second Atlas point counts, boasts the highest density of any county in the state. Despite declines in Pennsylvania's southwest, there were notable increases to the east and north of the core area, consistent with range expansions north and east of the historic range reported elsewhere for this species (Hamel et al. 2004).

The Cerulean Warbler has experienced long-term declines across its range, 3 percent annually since 1966, according to Breeding Bird Survey data (Sauer et al. 2011). Populations within Pennsylvania have declined significantly since the first Atlas period: 1.8 percent annually, a loss of around 28 percent of the population, which was estimated to be around 24,000 singing males during the second Atlas period. Due to these declines, the Cerulean Warbler is classified as a species of High Level Concern in Pennsylvania's Wildlife Action Plan (PGC-PFBC 2005). It is considered a species of Special Concern in Canada (McCracken 1993) and is on the Partners in Flight Continental Watch List (Rich et al. 2004). Since almost 9 percent of the global population breeds in the state (Rich et al. 2004), Pennsylvania has a high responsibility for this species (Stoleson and Larkin 2010). However, the future of the Cerulean Warbler in the state is not encouraging. Populations continue to decline, despite apparent range expansions and the fact that Pennsylvania's forested areas remain at about 60 percent of total land area (Stoleson and Larkin 2010; chap. 3). Factors increasing fragmentation or structural simplicity of forests may reduce habitat quality. Recent work suggests that some silvicultural practices may benefit the species (Rodewald 2004; Register and Islam 2008). Although its affinity for small forest gaps is well documented, the Cerulean Warbler's response to fragmentation resulting from deep shale gas development remains to be determined. Ongoing research across its breeding and non-breeding ranges should provide insights that guide future conservation efforts.

SCOTT H. STOLESON AND CAMERON RUTT

Distribution

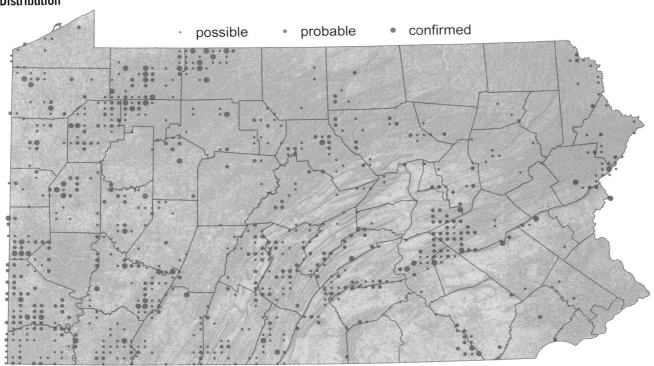

Distribution Change

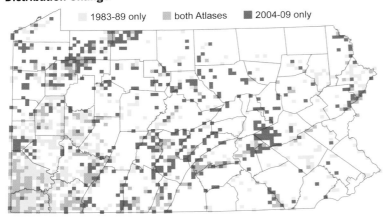

Number of Blocks

	first Atlas 1983–89	second Atlas 2004–09	Change %
Possible	324	437	35
Probable	430	275	−36
Confirmed	82	64	−22
Total	836	776	−7

Population estimate, males (95% CI):
24,000 (20,000–32,000)

Density

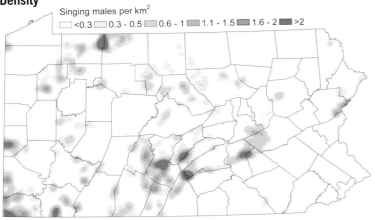

Breeding Bird Survey Trend

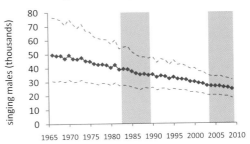

Gerard Dewaghe

Northern Parula
Setophaga americana

The Northern Parula significantly expanded its Pennsylvania population since the first Atlas, spreading north from its former southern base. Though not widespread across the landscape, it is typically found where there are scattered conifers and, often, close to water. The Northern Parula is a summer resident of eastern North America, with geographically separate northern and southern populations. The northern population breeds across the southern part of eastern Canada and the adjacent United States, including northern New England and the Adirondacks, nesting primarily in moist deciduous or coniferous woodlands and constructing nests in beard mosses (tree lichens in the genus *Usnea*). The southern population breeds across the southern United States, nesting primarily in hardwood bottomlands and constructing nests in Spanish Moss. The Pennsylvania population is the northernmost concentration of the southern population, although *Usnea* is the epiphyte that occurs here (Moldenhauer and Rogelski 1996; Brodo et al. 2001).

During the first half of the twentieth century, the Northern Parula largely disappeared from southern New England and New York south of the Adirondacks, the loss generally being attributed to the extirpation of *Usnea* due to air pollution, to which it is sensitive (Moldenhauer and Rogelski 1996). The Northern Parula was also once much more common in northern and eastern Pennsylvania, but it declined during that same time period (Schwalbe 1992e; McWilliams and Brauning 2000). The proximity in time and space of these declines and commonality of *Usnea* species within these regions (Brodo et al. 2001) suggest the same cause of those Northern Parula declines.

In Pennsylvania, the Northern Parula tends to choose either riparian forest, moist forest on steep slopes, or mature spruces in an open setting, often near streams. It is found frequently near the Cerulean Warbler, though the Northern Parula differs in commonly choosing to be near conifers. This predilection for streams and slopes explains why atlas results of Northern Parula often occurred in linear clusters. The Northern Parula also tends to choose edge areas in which *Usnea* grows well and areas in which competitive birds such as the Black-throated Green and Blackburnian warblers tend not to be found (Morse 1967).

During Pennsylvania's first Atlas, the largest clusters of Northern Parula blocks occurred in the state's southernmost counties, with smaller clusters or isolated blocks elsewhere. During the second Atlas, larger clusters were found throughout much of the state, many having mushroomed from isolated clusters of five or fewer contiguous occupied blocks during the first Atlas. Parulas were detected in 22 percent of all blocks during the second Atlas, compared with 10 percent in the first Atlas, a 121 percent increase. Breeding Bird Survey data collected in Pennsylvania from 1966 to 2009 similarly show a 3.4 percent annual increase (Sauer et al. 2011) and a doubling of the population between atlas periods. The population, estimated from second Atlas point counts, was 82,000 singing males.

Clusters of Northern Parula block detections were often associated with specific water or land features: the Monongahela and Youghiogheny River watersheds in the southwestern corner; the Allegheny Front and other ridges where Somerset and Bedford counties meet; the Allegheny River and Clarion River watersheds toward the northwestern corner; the West Branch of the Susquehanna River, the Sinnemahoning Creek and Pine Creek watersheds, and the Allegheny Front in the north-central area; the Juniata River and Susquehanna River watersheds in the south-central and southeast regions; and the Schuylkill River, Lehigh River, and Delaware River watersheds in eastern Pennsylvania. All of these areas possess one or more of the key habitat features typically selected by the Northern Parula.

The Northern Parula can be expected to continue occupying Pennsylvania locations where it currently does not occur but where there is suitable habitat. Second Atlas observations suggest several interesting questions: Is the population expansion caused by cleaner air and healthier *Usnea*, of which there are at least six species in Pennsylvania (Brodo et al. 2001), or are these birds nesting in something other than epiphytes, which is known to occur (Moldenhauer and Rogelski 1996)?

NICHOLAS C. BOLGIANO

Distribution

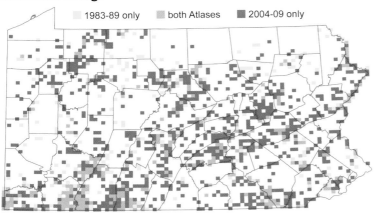

Distribution Change

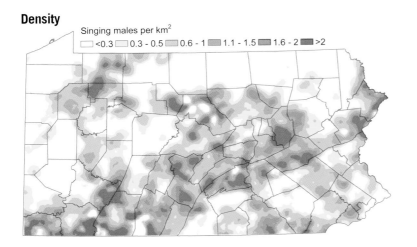

Density

Number of Blocks

	first Atlas 1983–89	second Atlas 2004–09	Change %
Possible	277	719	160
Probable	193	335	74
Confirmed	29	54	86
Total	499	1,108	121

Population estimate, males (95% CI):
82,000 (74,000–92,000)

Breeding Bird Survey Trend

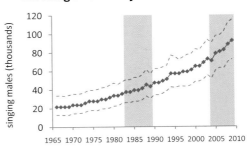

NORTHERN PARULA

Bob Wood and Peter Wood

Magnolia Warbler
Setophaga magnolia

The Magnolia Warbler was called the "Black-and-yellow Warbler" by Alexander Wilson, certainly a more appropriate name for this striking bird of northern conifer forests (Dunn and Hall 2010). Like several other conifer-loving warblers, the Magnolia Warbler has significantly expanded its range in Pennsylvania in the two decades since the first Atlas. Before the logging era, the Magnolia Warbler was probably common where Eastern Hemlock comprised a large portion of the forest; its current resurgence probably represents a recovery of its former range rather than a new expansion.

The Magnolia Warbler is a summer resident of North America's boreal forest, extending south along the Appalachian Mountains into West Virginia and locally into Virginia, North Carolina, and Tennessee. It typically chooses dense stands of young spruces or hemlocks (Dunn and Hall 2010); in some northern Pennsylvania hemlock stands, it is the most common warbler (pers. obs.). However, its preference for young stands is not absolute; in a study of old growth Pennsylvania hemlock, the Magnolia Warbler's highest density was in older trees (Haney 1999).

Like other boreal warblers, the Magnolia Warbler occurs most frequently at elevations above 400 m (1,300 ft) in Pennsylvania. The highest densities on point counts were at elevations of 500–600 m (1,650–1,970 ft; appendix D), where hemlock reaches its highest density along streams and in ravines. At elevations above 600 m, where hemlocks are usually less dense, there were correspondingly fewer Magnolia Warblers. The core of the Pennsylvania distribution for both warbler and hemlock is across the northern counties. The warbler's distribution extends south through the Laurel Highlands and along the Allegheny Front. It also extends south from the Pocono Plateau Section through higher elevations of the Ridge and Valley in eastern Pennsylvania.

The 1,439 block detections of Magnolia Warbler during Pennsylvania's second Atlas represented an increase of 93 percent over the first Atlas. The atlas findings are consistent with the Breeding Bird Survey's estimated 4 percent annual increase in Pennsylvania's breeding Magnolia Warbler population (Sauer et al. 2011), which translates into more than a doubling between atlas periods. The increase in the Magnolia Warbler's distribution in Pennsylvania was consistent with New York's expansion between atlases (McGowan 2008zp).

There was not only an expansion of the Magnolia Warbler's range but also considerable infilling within the range documented in the first Atlas. The Pennsylvania region with the largest relative increase in block detections was the Ridge and Valley, with a 187 percent increase between atlas periods (appendix C). Detections here were mostly in the high elevations of eastern Pennsylvania extending south from the Poconos, as well as patches in Bald Eagle and Rothrock state forests in central Pennsylvania, the latter being a range expansion since the first Atlas. The concentrated area of new block detections in northeastern Pennsylvania was contiguous with an expansion south from the Catskills in New York state (McGowan 2008zp). At the opposite end of the state, there was a southward expansion along the Allegheny Front in Somerset/Bedford counties, the southern range limit expanding by six atlas blocks, about 30 km (18 miles).

With a population estimated at 240,000 singing males, the Magnolia Warbler can be described as a common bird in suitable habitat in Pennsylvania, and its recent population recovery has been impressive. Historically, when vast forests of White Pine and Eastern Hemlock covered parts of northern Pennsylvania's Appalachian Plateaus, where the forest was sometimes called the Black Forest because of its dense conifers, the Magnolia Warbler may have been one of the most common birds. Following forest regeneration, shade-tolerant understory trees, such as hemlock, eventually become overstory trees when small gaps, such as the loss of an individual tree, open the canopy (Whitney 1990; Abrams and Ruffner 1995), once again providing suitable habitat for the Magnolia Warbler. With today's less expansive hemlock groves, the Magnolia Warbler's population may still be only a fraction of its historic size. However, the relentless advance of the Hemlock Woolly Adelgid and its effect upon our state tree, the Eastern Hemlock, may eventually reduce the Magnolia Warbler's ranks once again.

NICHOLAS C. BOLGIANO

Distribution

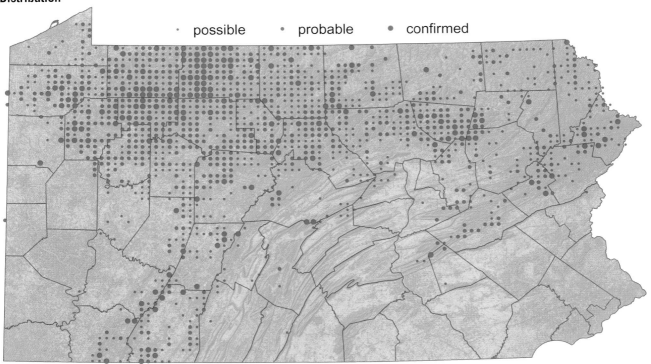

Distribution Change

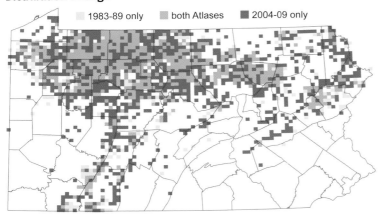

Number of Blocks

	first Atlas 1983–89	second Atlas 2004–09	Change %
Possible	381	701	84
Probable	281	568	102
Confirmed	82	170	107
Total	744	1,439	93

Population estimate, males (95% CI):
240,000 (222,000–258,000)

Density

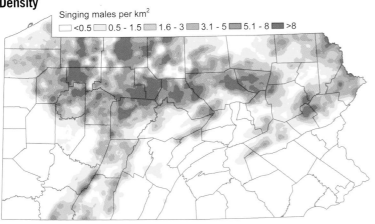

Breeding Bird Survey Trend

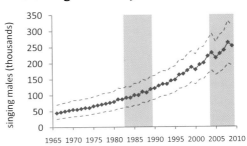

Gerard Dewaghe

Blackburnian Warbler
Setophaga fusca

"In a family in which beauty is the general rule, the Blackburnian Warbler is nevertheless outstanding, and is considered by many to be the loveliest of all the warbler tribe," wrote W. E. C. Todd (1940). Yet the species' predilection for the highest treetops makes it exceedingly difficult for humans to appreciate that beauty.

The Blackburnian Warbler is an active and vocal inhabitant of mature coniferous, mixed, and (locally) deciduous forests across Pennsylvania's highlands. It breeds primarily in the cool forests of the boreal and temperate zones of eastern North America, from Alberta to the Maritimes, and south along the Appalachians to northern Georgia. In the southern half of Pennsylvania, it occurs locally at higher elevations (Morse 2004). Blackburnians winter primarily in forested parts of the Andes of northwestern South America (De La Zerda-Lerner and Stauffer 1998). Pennsylvania's Wildlife Action Plan classifies the Blackburnian Warbler as a priority species of Maintenance Concern (PGC-PFBC 2005).

Atlas volunteers reported the Blackburnian Warbler in 1,580 blocks, or about a third of all blocks in the state, distributed throughout the heavily forested uplands of the Commonwealth. The species was found most frequently in the Appalachian Plateaus (44 percent of the blocks in this province), less commonly in the Ridge and Valley Province (21 percent of blocks), and was virtually absent elsewhere. Point count data show that most blackburnians are found above 400 m (1,312 ft) elevation (appendix D). Similarly, Todd (1940) noted that the "species breeds commonly almost everywhere over 1,500 feet" (~450 m). The Blackburnian Warbler is one of several northern forest birds that have expanded their range since the first Atlas period, resulting in a 72 percent increase in number of blocks occupied. The greatest increases in range occurred in the Ridge and Valley Province (164 percent), reflecting a southward expansion of this species, probably due to the continued maturation of forests. Similar increases occurred in the southern portions of New York (Collins 2008c). Breeding Bird Survey data indicate that Blackburnian Warbler abundance within Pennsylvania increased by about 70 percent since the first Atlas (Sauer et al. 2011). An estimated 360,000 singing males now occur within Pennsylvania, making this the sixth most abundant of all of the state's breeding warblers.

Sometimes referred to as the "hemlock warbler," the Blackburnian Warbler often is tied to the presence of hemlocks or other conifers, including plantations of red pine or exotic spruces (but see Young et al. 2005). In eastern Pennsylvania, the species specializes in mesic hemlock ravine habitats (Ross et al. 2004), and it reaches very high densities in remnant old-growth hemlock stands (Haney 1999). However, the extent of conifer dependence appears to vary across the state. In the eastern third of Pennsylvania, over 70 percent of blocks with Blackburnian Warbler records have significant numbers of hemlocks. In contrast, hemlock was scattered or absent in over half of the reported sites in the north-central region and in 75 percent in the southwestern region, although the latter may simply reflect the paucity of hemlocks there. In those areas with the highest density of Blackburnian Warblers (the High Plateau and Deep Valleys sections, where more than 90 percent of blocks registered the species), they occur commonly in all forest types and are especially abundant in areas with mixed oak (Stoleson, unpublished data).

Numerous factors threaten the Blackburnian Warbler's overall secure status. The species is highly area-sensitive, becoming sparse or absent with even moderate levels of forest fragmentation (Morse 2004). It responds negatively to certain types of logging (Webb et al. 1977) and conventional gas well development (Thomas 2011). The current exponential growth in deep shale gas development seems also likely to affect the species negatively. Unless effective control measures are developed, the continuing spread of Hemlock Woolly Adelgid will likely cause severe Blackburnian Warbler population declines in much of the state (Ross et al. 2004; Allen and Sheehan 2010). Finally, the blackburnian's primary wintering grounds in the forests of the northern Andes continue to be impacted by land conversion to agriculture and urbanization (De La Zerda-Lerner and Stauffer 1998).

SCOTT H. STOLESON

Distribution

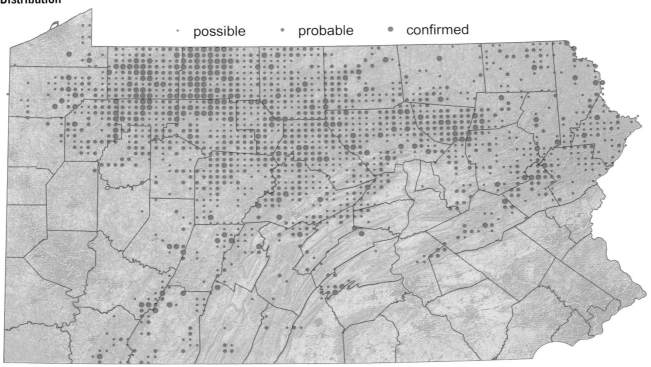

Distribution Change

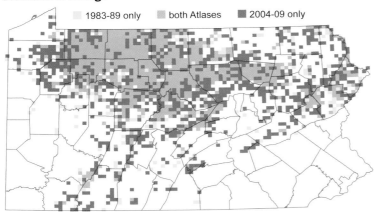

Density

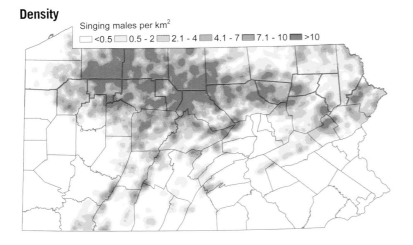

Number of Blocks

	first Atlas 1983–89	second Atlas 2004–09	Change %
Possible	425	793	87
Probable	371	584	57
Confirmed	121	203	68
Total	917	1,580	72

Population estimate, males (95% CI):
360,000 (320,000–405,000)

Breeding Bird Survey Trend

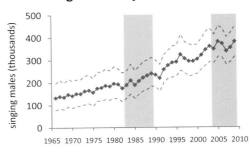

BLACKBURNIAN WARBLER 381

Bob Wood and Peter Wood

Yellow Warbler
Setophaga petechia

With its flashy yellow-gold plumage, enthusiastic *sweet-sweet-sweet-I'm-so-sweet!* song, relatively visible nesting habits, and extensive distribution, the aptly named Yellow Warbler is easily one of the most recognizable of Pennsylvania's warblers.

The Yellow Warbler is among the most abundant breeding birds in North America, its range covering much of the United States and Canada. It is common in Pennsylvania, where it breeds in a wide range of habitats from shrubby wetlands and riparian areas to old fields, early successional forests, and woodland edges, often in association with willows and dogwoods. Although somewhat more of a habitat generalist than other warblers, it does not breed in dense forest and seems to avoid higher elevations in Pennsylvania, which may be an artifact of the prevalence of core forest in such areas (appendix D). Using stable-hydrogen isotope analysis, Boulet et al. (2006) suggest that Pennsylvania Yellow Warbler populations winter in southern Central America and northern South America.

Among the first of neotropical migrants to return to Pennsylvania, the Yellow Warbler arrives from mid-April to early May, quickly establishes territories, and begins nest building. On average, volunteers during the second Atlas found nests with eggs in mid- to late May and early June and nestlings in early to mid-June. The peak of fledging in Pennsylvania is the last week of June and the first week of July. Analysis of atlas point count data show that the Yellow Warbler's song output wanes very quickly during the second half of June (chap. 5), much more so than for other warblers. Breeding is largely completed by mid-July, when fall migration commences. Rarely, some pairs may raise a second brood and conclude breeding by the first week in August. Yellow Warbler breeding efforts are complicated by frequent brood parasitism by Brown-headed Cowbirds, but these cunning hosts have adapted the coping mechanism of burying cowbird eggs under new layers of nest material (Harrison 1984; Lowther et al. 1999).

Atlas volunteers located Yellow Warblers in 4,300 blocks, an 8 percent increase over the first Atlas. There were no declines in blocks with records in any of the Pennsylvanian physiographic provinces, and in the Ridge and Valley Province there was a 16 percent increase in blocks with this species (appendix C). There is little overall pattern to the changes in distribution between the atlases. Some of the turnover may be due to variation in observer effort, but localized gains and losses in the often-ephemeral habitats of this species could also have contributed to changes. During the second Atlas period, the Pennsylvania population of Yellow Warblers was estimated to be 765,000 singing males, third only to Ovenbird and Common Yellowthroat among the state's breeding wood warblers. The highest densities were found along the western border of Pennsylvania, especially in Erie, Crawford, Mercer, Lawrence, Washington, and Greene counties, and in the northeastern corner of the state in Bradford, Susquehanna, and Wayne counties. The high-density areas are typified by highly interspersed mosaics of woodland, farmland, and other habitats, but why densities should be so much higher overall in the west and northeast than in similar landscapes in parts of the southeast is somewhat perplexing.

Breeding Bird Survey data show that populations of the Yellow Warbler in Pennsylvania have fluctuated, but with little overall trend since the 1960s (Sauer et al. 2011). As an abundant breeder in Pennsylvania, with stable populations, the Yellow Warbler is not a species of conservation concern. Although loss of habitat throughout the annual cycle, especially on wintering grounds, can cause declines in populations of migratory songbirds, winter habitat does not seem to be closely related to reproductive success in Yellow Warblers as in other Dendroica species (Lindsay 2008). This would suggest that conservation efforts for this species should be focused equally throughout the Yellow Warbler's annual cycle.

ANDREA L. CRARY

Distribution

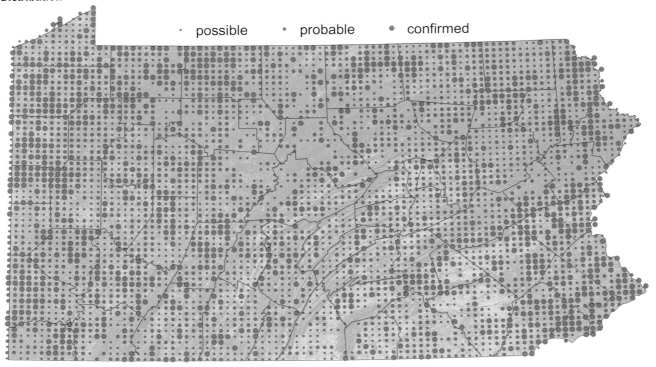

Distribution Change

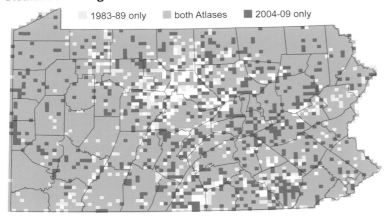

Density

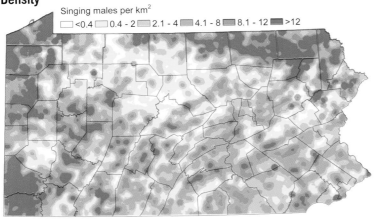

Number of Blocks

	first Atlas 1983–89	second Atlas 2004–09	Change %
Possible	1,132	1,742	54
Probable	1,753	1,554	−11
Confirmed	1,099	1,004	−9
Total	3,984	4,300	8

Population estimate, males (95% CI):
765,000 (730,000–800,000)

Breeding Bird Survey Trend

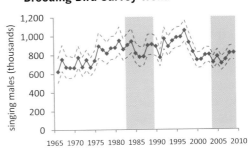

YELLOW WARBLER 383

Bob Wood and Peter Wood

Chestnut-sided Warbler
Setophaga pensylvanica

The Chestnut-sided Warbler is a summer resident of eastern North America, from southern Canada and the northern United States south through the Appalachians to northern Georgia. In the Appalachians south of Pennsylvania, the Chestnut-sided Warbler is generally confined to higher elevations (>450 m, ~1,500 ft; Richardson and Brauning 1995), which is largely true in Pennsylvania (appendix D).

A bird of deciduous forest edges, regenerating timber cuts, and abandoned farms (Richardson and Brauning 1995), the Chestnut-sided Warbler clearly benefits from tree harvesting; regenerating clear-cuts provide the ideal habitat of shrub thickets with high stem densities (Talbott and Yahner 2003; Schill and Yahner 2010). Although the species was probably uncommon in Pennsylvania prior to the felling of primary forests, the resulting regeneration of second- and third-growth forests since the mid-nineteenth century has improved the Chestnut-sided Warbler's fortunes (McWilliams and Brauning 2000). In addition to regenerating stands, open forests with an understory dominated by mountain laurel constitute another important habitat for this species in Pennsylvania.

The 2,718 blocks reporting Chestnut-sided Warbler in the second Atlas represent a 35 percent increase over the first Atlas's results. The Chestnut-sided Warbler was found throughout the Appalachian Plateaus Province, with the exception of less forested, lower-elevation areas of the southwest. An increase in block occupancy around the periphery of the first Atlas range suggests a limited expansion in that province. Block detections in the Ridge and Valley Province more than doubled, especially in Schuylkill and Carbon counties, where this species was very local during the first Atlas. Away from the higher ridgetops, small isolated populations of Chestnut-sided Warblers were found at relatively low elevations in the Piedmont in both atlases, but there was a 37 percent decrease in occupied blocks there, the only physiographic province with a range contraction between atlas periods.

Point count data show that the highest densities of the Chestnut-sided Warbler were at high elevations in northern Pennsylvania, where forest was the predominant land cover (appendix D). Many of the highest-density areas were in state game lands, state forests, or the Allegheny National Forest, where regenerating forest cuts or ridgetops with a mountain laurel understory are common habitats. At some of these locations, the Chestnut-sided Warbler is among the most common of birds (Rodewald and Yahner 2000). With an estimated population of over half a million singing males, the Chestnut-sided Warbler ranked fifth among Pennsylvania's 29 species of breeding wood warbler in terms of both block occupancy and population size.

The increased block detections were consistent with Breeding Bird Survey (BBS) data collected in Pennsylvania, which show an estimated 2.5 percent annual increase between 1966 and 2009, accelerating to 3.1 percent per year since 1999 (Sauer et al. 2011), representing a doubling of counts between atlas periods. Rangewide, BBS data showed that the Chestnut-sided Warbler increased in some regions but decreased in others. While Pennsylvania's Chestnut-sided Warblers have been increasing, negative trends were observed in neighboring regions—much of New England and New York, southern Ontario, Michigan, and the southern Appalachians (Sauer et al. 2011)—although there was a modest increase in block occupancy between atlases in New York state, where forest maturation was suggested to be the cause of the shallow population decline (Post 2008e).

The long-range outlook for the Chestnut-sided Warbler in Pennsylvania appears to be favorable. While large parts of the forest are maturing beyond suitable habitat, timbering operations renew habitat, and ridgetop forests with a mountain laurel understory tend to grow slowly, providing a relatively stable environment. Because many of Pennsylvania's Chestnut-sided Warblers breed on large tracts of state or federal land managed for timber production or wildlife habitat, this species is partially protected against the population declines seen in neighboring regions. This refuge befits the only bird for which Pennsylvania is part of its scientific name.

NICHOLAS C. BOLGIANO

Distribution

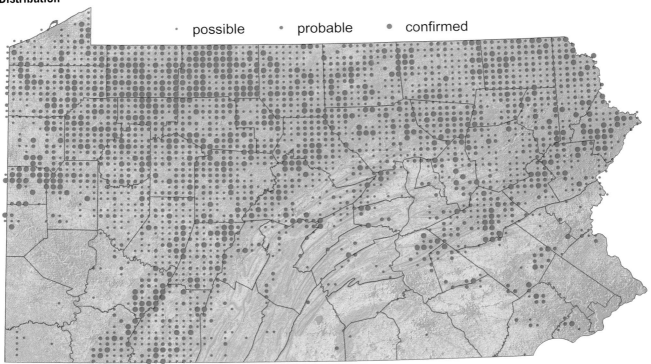

Distribution Change

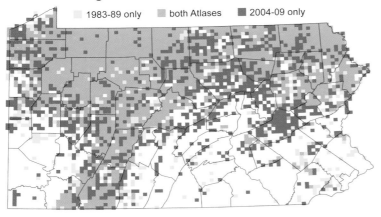

Density

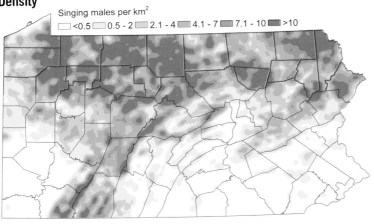

Number of Blocks

	first Atlas 1983–89	second Atlas 2004–09	Change %
Possible	822	1,207	47
Probable	811	969	19
Confirmed	375	542	45
Total	2,008	2,718	35

Population estimate, males (95% CI):
520,000 (495,000–545,000)

Breeding Bird Survey Trend

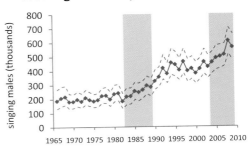

CHESTNUT-SIDED WARBLER 385

Bob Wood and Peter Wood

Blackpoll Warbler
Setophaga striata

A common bird of the North American boreal forest, the Blackpoll Warbler reaches the southern extent of its breeding range in Pennsylvania, where it is considered endangered due to its extreme rarity in the state (Hunt and Eliason 1999; Gross 1994, 2010c). It is one of the species added to the state nesting avifauna since the first Atlas, when there were insufficient data to regard it as a breeding species (Brauning 1992a). The Blackpoll Warbler is notorious for its transatlantic southbound migration from the northeastern United States to South America, one of the longest non-stop migrations of any songbird (Nisbet et al. 1995). Its high-pitched song represents the last wave of northbound spring migrants.

The breeding range of the Blackpoll Warbler is very patchy in the northeastern United States, matching the availability of the boreal conifer forest (Hunt and Eliason 1999; McFarland and Rimmer 2008). The Pennsylvania breeding population is disjunct from the next closest nesting occurrence in New York's Catskill Mountains, but the 160 km (100 mile) distance seems small compared with its long migration (Hunt and Eliason 1999; McFarland and Rimmer 2008).

The breeding population of Blackpoll Warbler was discovered in 1993 and confirmed in 1994 in Coalbed Swamp, Wyoming County (Gross 1994; Davis et al. 1995). From 1994 through 2004, nesting was confirmed each year, primarily by locating dependent young being fed by parents (pers. obs.). As many as 13 territories have been found there in any year, with at least 2 additional territories in nearby Tamarack Swamp (pers. obs.). These swamps are part of the Dutch Mountain wetlands complex in State Game Lands 57, an Important Bird Area (Gross 2004). There may be appropriate habitat elsewhere in northern Pennsylvania (Gross 2010c).

In Pennsylvania, the Blackpoll Warbler's range is confined to boreal conifer swamps and forest in the eastern part of the Allegheny Plateau known as North Mountain of the Glaciated High Plateau Section (Gross 1994, 2010c). Although it inhabits boreal conifer forests of many types across its breeding range, particularly Black Spruce (Bent 1953; Erskine 1977), most Pennsylvania territories are associated with Red Spruce with some larch, Eastern Hemlock, Eastern White Pine, and some northern hardwoods. Many Blackpoll Warbler territories overlapped with Yellow-bellied Flycatcher, but unlike the flycatcher, some blackpolls were found in upland conifer stands. The nest of the Blackpoll Warbler is bulky for a warbler, but well concealed, often next to the tree trunk or where there are criss-crossing limbs (Hunt and Eliason 1999; pers. obs.). Females have high site fidelity, which may explain breeding site tenacity in isolated patches of habitat despite low numbers at any patch in the region.

During the second Atlas, the only nesting confirmations came from the Dutch Mountain wetlands. At least 10 territories were found in the wetland complex in 2004, but searches in 2008 and 2009 failed to locate territories where previously found, suggesting a decline during the atlas period (pers. obs.). There was an additional report in appropriate habitat in Clinton County. Despite interest in this species, there were no reports from the Poconos, where spruce forest exists.

Historically, there was considerably more spruce forest in Pennsylvania than at present, which was not well surveyed by ornithologists before its destruction in the nineteenth century, so pre-logging era Blackpoll Warbler populations may have been overlooked (Gross 2010c). The discovery of nesting Blackpoll Warblers in Coalbed Swamp coincided with an Elm Spanworm outbreak that caused massive defoliation in northern counties, but blackpolls have persisted without this abundant food source. It is not clear whether Blackpoll Warbler populations and subsequent trends are related to spruce budworm outbreaks (Hunt and Eliason 1999; Bolgiano 2004). Spruce is regenerating well in parts of northeastern Pennsylvania, where there is potential for proactive spruce forest management (Rentch et al. 2007; Gross 2010c). Large numbers of Blackpoll Warblers migrate through this region, giving opportunities for colonization and providing potential for continued existence of this boreal conifer species in the state, despite its rarity.

DOUGLAS A. GROSS

Distribution

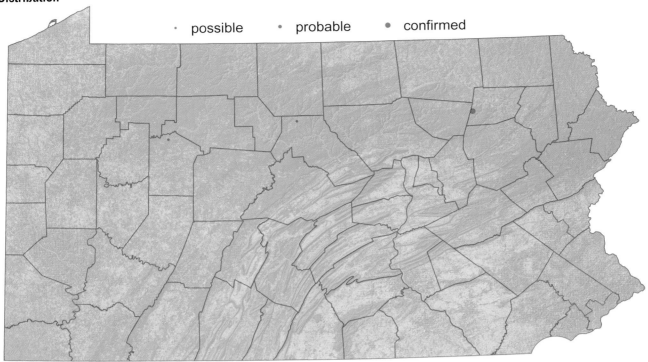

Distribution Change

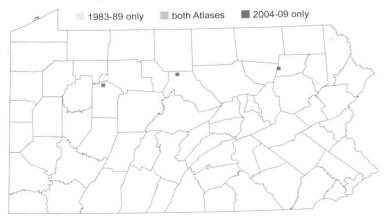

Number of Blocks

	first Atlas 1983–89	second Atlas 2004–09	Change %
Possible	2	2	0
Probable	0	0	0
Confirmed	0	1	∞
Total	2	3	50

BLACKPOLL WARBLER

Bob Wood and Peter Wood

Black-throated Blue Warbler
Setophaga caerulescens

The olive and buff female Black-throated Blue Warbler differs so much in appearance from her blue, black, and white mate that early ornithologists originally described it as a separate species, the "Pine Swamp Warbler." Her drab plumage camouflages her on the nest, hidden in dense shrubbery and often built within a half-meter (~1–2 ft) of the ground. A languorous, buzzy song emanating from a mountain laurel or rhododendron thicket reveals the presence of the Black-throated Blue Warbler. The species breeds in deciduous and mixed forests throughout southeastern Canada and the northeastern United States, west through the Great Lakes states and provinces, and south through the Appalachians to northern Georgia (Holmes et al. 2005). Within this range, it prefers large areas of contiguous forest supporting a dense understory of hobblebush, mountain laurel, conifer seedlings, or other low woody vegetation (Cornell and Donovan 2010).

The second Atlas documented the Black-throated Blue Warbler throughout the more heavily forested areas at higher elevations of the Commonwealth. While the overall distribution of the species did not change between the first and second Atlas periods, the proportion of occupied blocks within that distribution increased greatly, by 72 percent. There was an expansion throughout the range and notable range extensions in southern sections of the Ridge and Valley Province, including into South Mountain and Blue Mountain. The species remains sparse or absent below 500 m (1,650 ft) elevation (appendix D). Breeding Bird Survey data indicate that Black-throated Blue Warbler numbers almost doubled between atlas periods (Sauer et al. 2011). Point count data reveal that the highest densities were found in the north-central part of the state, notably in Sproul and Tiadaghton state forests. An estimated 150,000 pairs were estimated by atlas point counts to breed within the state.

Pennsylvania's forest extent did not change appreciably between the first and second Atlas periods (McWilliams et al. 2007), suggesting that Black-throated Blue Warbler population increases resulted from a change in habitat quality and greater atlasing effort. Forests have matured since the first Atlas, potentially improving habitat for this mature forest obligate (Stoleson and Larkin 2010). Also, increased deer harvests in many areas of the state helped promote widespread recovery of shrubby understory vegetation. Black-throated Blue Warbler populations may be recovering to the levels that occurred before the extensive deforestation of the late nineteenth and early twentieth centuries, since most pre-1900 accounts considered the species abundant (Warren 1890; Baily 1896; Stone 1900).

Pennsylvania's Wildlife Action Plan classifies the Black-throated Blue Warbler as a species of Maintenance Concern and an indicator of high-quality forest, particularly those with high structural complexity. Because it is a forest-interior species that is rarely found in small fragments or woodlots (Holmes et al. 2005), continued fragmentation of Pennsylvania's forests by urban sprawl, energy development, and some types of timber management have the potential to reverse the growth in Black-throated Blue Warbler populations since the first Atlas. Timber harvest practices that retain canopy but promote understory regeneration (e.g., shelterwood and selection cuts) can increase habitat quality for this species (Buford and Capen 1999; Bourque and Villard 2001). In addition, the Black-throated Blue Warbler often occurs at high densities along internal edges created by silviculture, due to understory responses to increased light (Harris and Reed 2002). Its dependence on dense understory and large areas of contiguous forest suggest that this species' future abundance and distribution will likely be affected by timber management practices and factors such as deer herbivory.

While habitat degradation on the wintering grounds can affect populations, little evidence exists to support this as an issue for this species (Sherry and Holmes 1996). Sillett and Holmes (2002) showed that more than 85 percent of apparent mortality in the Black-throated Blue Warbler occurs during migration; therefore, the primary factors limiting populations probably occur outside the Commonwealth.

SCOTT H. STOLESON

Distribution

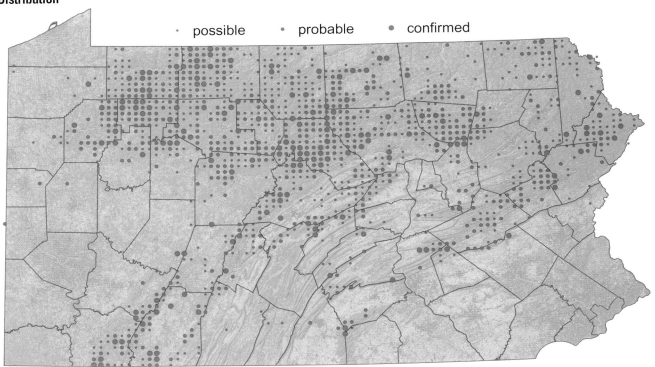

Distribution Change

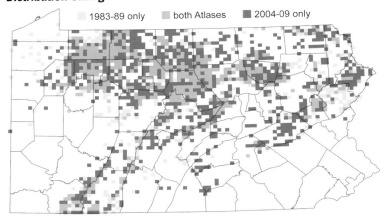

Density

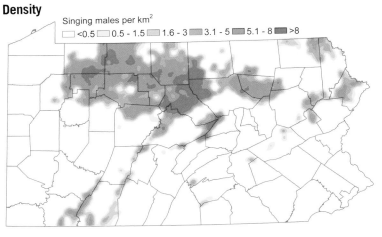

Number of Blocks

	first Atlas 1983–89	second Atlas 2004–09	Change %
Possible	399	621	56
Probable	280	492	76
Confirmed	62	165	166
Total	741	1,278	72

Population estimate, males (95% CI):
150,000 (136,000–171,000)

Breeding Bird Survey Trend

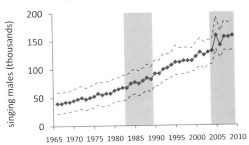

BLACK-THROATED BLUE WARBLER 389

Bob Wood and Peter Wood

Pine Warbler
Setophaga pinus

The Pine Warbler is one of the less well known of Pennsylvania's breeding warblers. Its rather drab yellow plumage does not rival the colors of many other warblers, and its song, although loud, is not distinctive. Furthermore, the Pine Warbler is uncommon or even rare across much of the state. However, unlike some warblers, it is well named, since it is always associated with pine trees.

The core of the Pine Warbler's breeding range extends from Florida and the Gulf Coast north to the mid-Atlantic states; farther north, its range becomes fragmented, but extends into Ontario and northern New England (Rodewald et al. 1999). It is unusual among warblers in that it winters almost completely in the southeastern United States. In Pennsylvania, the Pine Warbler is well distributed in the Ridge and Valley Province and adjoining border of the Appalachian Plateaus. It is absent through most of the Piedmont, except in a handful of scattered pine barrens or pine plantations. It is virtually absent from the southwest and is uncommon across the northwestern and northern tier counties.

The Pine Warbler nests and forages almost exclusively in pine forests, including pine plantations. It also occurs in scattered small clusters of pine trees within otherwise deciduous forests. In Pennsylvania, it favors stiff-needled pines such as Red, Pitch, and Virginia, does not significantly use hemlock or spruce (Rodewald et al. 1999), and prefers relatively mature or at least intermediate-aged forests with somewhat limited and open understory. High densities of Pine Warblers may occur in good habitat—for example, in the extensive pine plantations in Huntingdon County, in pine barrens of Fulton County and the South Mountains, and on ridges in Sproul State Forest in Centre and Clinton counties, where there are numerous, scattered Pitch Pine clusters.

The number of blocks reporting Pine Warblers increased from the first Atlas to the second Atlas by 167 percent. Adjusted for the increased effort in the second Atlas, the increase was just 125 percent, suggesting that it was underreported in some areas during the first Atlas period. The greatest gains were in those portions of the Ridge and Valley and the south-central Appalachian Plateaus, where Pine Warblers were established, though sparse, at the time of the first Atlas. Thus, their increase appears to be primarily the result of filling in of the established range rather than significant extension of that range, with the exception of a modest increase in occupied blocks in the northwest, where there were few records during the first Atlas. The increase between atlas periods correlates well with the Pennsylvania Breeding Birds Survey results; there was a 4.5 percent per year increase from 1996 to 2009 and a population increase of 150 percent between atlas periods (Sauer et al. 2011). The current population of singing males in Pennsylvania is estimated at 42,000.

The Pine Warbler has presumably benefited from the maturation and increase in acreage of Pennsylvania's forests, particularly pine plantations. However, some of the increase found during the second Atlas is probably also due to coverage factors. These include better recognition of the song (a simple trill that is similar to the song of the more common Chipping Sparrow, usually present in the same habitat), increased awareness of its early return date in spring, which may have encouraged volunteers to search for Pine Warblers early in the breeding season, and, finally, better coverage in some remote areas, in particular on the Appalachian Plateaus.

The principal threat to the Pine Warbler may be loss of habitat due to land development and clear-cutting; some studies have shown detrimental effects from logging (Thompson et al. 1992). However, these factors appear not to be a problem in Pennsylvania at the current time. The Pine Warbler is doing well, solidifying its established range and perhaps expanding into new areas.

GREG GROVE

Distribution

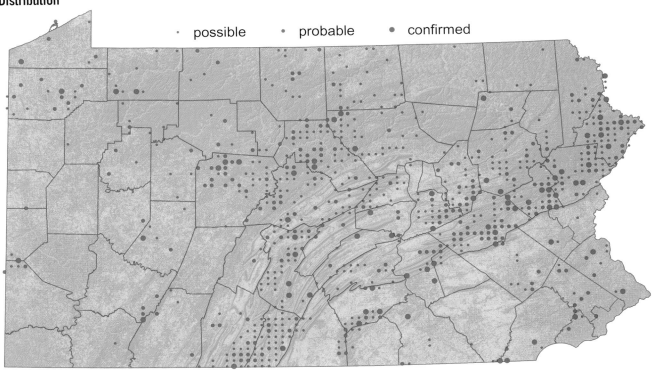

Distribution Change

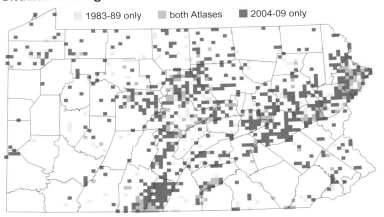

Density

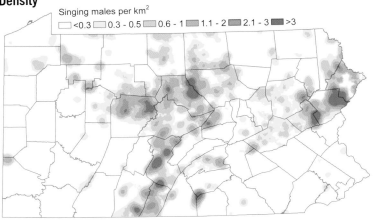

Number of Blocks

	first Atlas 1983–89	second Atlas 2004–09	Change %
Possible	144	489	240
Probable	103	226	119
Confirmed	48	73	52
Total	295	788	167

Population estimate, males (95% CI):
42,000 (46,000–49,000)

Breeding Bird Survey Trend

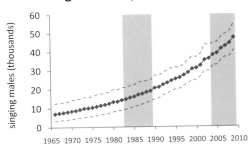

PINE WARBLER 391

Bob Wood and Peter Wood

Yellow-rumped Warbler
Setophaga coronata

The Yellow-rumped Warbler continued to significantly expand its Pennsylvania range and population in the two decades since the first Atlas. In many conifer patches in northern Pennsylvania, it is now possible to hear the junco-like trill and see the contrasting black, yellow, and white breeding colors of the Yellow-rumped Warbler during summer.

The Yellow-rumped Warbler species complex includes two groups once considered to be separate species: Myrtle Warbler in the North and East, and Audubon's Warbler in the West. They were lumped into a single species in 1973, based upon observed hybridization between the groups. Pennsylvania's yellow-rumps are of the Myrtle group, whose population extends from Alaska to Newfoundland, and south through New England and New York into Pennsylvania, with disjunct populations in the mountains of West Virginia and Maryland (Hunt and Flaspohler 1998). The size of the total Yellow-rumped Warbler national population has been estimated at 130 million birds, likely making it the most numerous of the wood warblers (Rich et al. 2004).

In Pennsylvania, the Yellow-rumped Warbler is usually found breeding in conifers growing at higher elevations. It uses native pines and hemlocks, as well as planted or naturalized spruces, pines, and tamaracks. Yellow-rumps may be found in any size patch of conifers, ranging from dense stands to scattered conifers within deciduous forests or semi-open areas (Gross 1992g; Hunt and Flaspohler 1998). Its recent population expansion may have benefited from regrowth of Eastern White Pines, one of Pennsylvania's forest success stories (McWilliams et al. 2007).

The 877 block detections of Yellow-rumped Warbler in Pennsylvania's second Atlas represent a 178 percent increase over the first Atlas, consistent with the Breeding Bird Survey's estimated 4.4-percent annual increase in Pennsylvania's breeding Yellow-rumped Warbler population (Sauer et al. 2011).

Whereas the first Atlas's detections of Yellow-rumped Warbler were mostly from the northeastern corner of the state, the second Atlas documented a major expansion across the higher elevated forests of the north-central plateau region. This is consistent with findings of the New York atlas; during New York's first atlas, the Yellow-rumped Warbler was concentrated in the Adirondacks and Catskills, but it spread farther west by the time of their second atlas (McGowan 2008zq). The highest Yellow-rumped Warbler densities in Pennsylvania, as estimated from point count data, occurred at higher elevations, with breaks between them occurring across valleys where the Yellow-rumped Warbler was uncommon or absent.

In the southern half of Pennsylvania, the Yellow-rumped Warbler was found in numerous small clusters or isolated blocks on ridgetops, where there were only a handful of detections during the first Atlas. Maryland's two atlases, conducted at about the same times as Pennsylvania's, found similar patterns of Yellow-rumped Warbler detection at western ridge locations (Ellison 2010zn).

Before European settlement, White Pines and Eastern Hemlocks comprised a much higher percentage of Pennsylvania's forest trees than currently (Whitney 1990; Nowacki and Abrams 1992; Abrams and Ruffner 1995; Dando 1996). While this habitat would have been well suited to the Yellow-rumped Warbler, other than several reports from Warren (1890) about summering individuals, we know little about the Pennsylvania breeding distribution of the Yellow-rumped Warbler a century or more ago and thus whether the recent range expansion is a reclaiming of former range or a new range expansion.

The Yellow-rumped Warbler in Pennsylvania appears to be poised to consolidate its core distribution and expand along southern ridgetop locations. Because of its catholic use of conifers and the regrowth of White Pine, which grows well at high elevations, the Yellow-rumped Warbler is probably less threatened by the loss of Eastern Hemlock than are some other birds. Because it winters in the southern United States and Mexico and tolerates disturbed habitat at that time, winter habitat loss is not the serious problem that it poses for some birds that winter in the tropics.

NICHOLAS C. BOLGIANO

Distribution

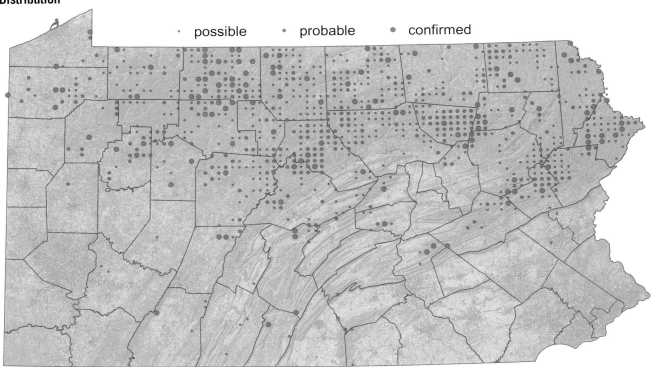

Distribution Change

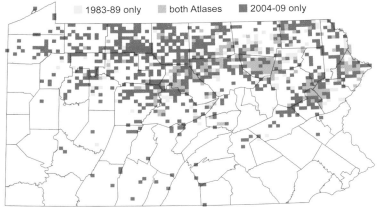

Density

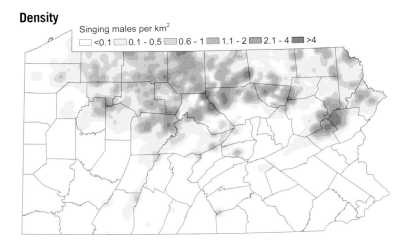

Number of Blocks

	first Atlas	second Atlas	Change
	1983–89	2004–09	%
Possible	153	510	233
Probable	112	247	121
Confirmed	50	120	140
Total	315	877	178

Population estimate, males (95% CI):
45,000 (39,000–51,000)

Breeding Bird Survey Trend

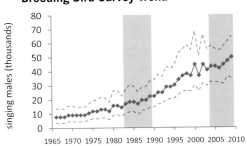

YELLOW-RUMPED WARBLER 393

Greg Lasley/VIREO

Yellow-throated Warbler
Setophaga dominica

The Yellow-throated Warbler, a woodland bird of the southeastern United States, has been expanding its breeding range northward, recovering some of the range from which it retreated (for unknown reasons) in the early twentieth century (Hall 1996). It returns to Pennsylvania in mid-April and prefers to nest in tall American Sycamores along streams and, sometimes, in White Pines.

Prior to the 1970s, the Yellow-throated Warbler was a rare breeder in Pennsylvania, with only scattered records from the southern third of the state. For example, there are four specimens from eastern Pennsylvania in the late eighteenth century (Stone 1894) and several sight records from various areas of the Commonwealth through the 1960s—for example, Lewisburg (Kunkle 1951), Pittsburgh (Parkes 1956), and around Philadelphia (Poole, unpub. ms.). From about the mid-1970s, the Yellow-throated Warbler became considerably more common and widely distributed, with reports from both eastern (Morris et al. 1984) and western (Leberman 1976) areas of Pennsylvania. Data from the first Pennsylvania Atlas indicated that the species had continued to extend its range, mainly westward and northward, and was found in almost half of the state's counties (Ickes 1992m).

According to the results of the second Atlas, the Yellow-throated Warbler is continuing its statewide expansion, with an increase of 29 percent in occupied blocks between atlas periods. The evidence suggests a modest northward expansion; Yellow-throated Warblers were reported in 43 counties, compared with 30 in the first Atlas, and 11 of the newly occupied counties were in the northern half of the state.

The majority of the Yellow-throated Warbler records for the second Atlas continue to be found west of the Allegheny Mountains, mainly in the southwestern corner of the state. However, there was a notable expansion in eastern Pennsylvania—the number of occupied blocks in the southeast quarter of the state increased from 21 to 36. The majority of these were along the lower Susquehanna River. As is typical of scarce species, there was considerable turnover of occupied blocks between atlases, but the northernmost outpost of confirmed breeding in the first Atlas, along Little Pine Creek in Lycoming County, was again occupied in the second Atlas, suggesting that there is an established population there. In the second Atlas, a range expansion upstream along the Allegheny River took the species to Warren County, a short hop to New York state, where it has bred sporadically since the 1980s (McGowan 2008zr).

Point count data for the second Atlas indicate that, in Pennsylvania, the Yellow-throated Warbler is not tied to extensive forest cover, is largely associated with rivers, and is not found above elevations of 400 m (1,300 ft; appendix D). The total Pennsylvania population for the Yellow-throated Warbler during the second Atlas was estimated to be 4,000 singing males, although this is based on a small sample—only 44 singing males were detected on point counts. The Breeding Bird Survey trend for routes in Pennsylvania from 1966 to 2009 was upward, with an average increase of 1.2 percent per year, although sample sizes are small (Sauer et al. 2011).

The continued expansion of the Yellow-throated Warbler in Pennsylvania suggests that the medium-term future of the species in the region is secure. It increased block occupancy in Maryland between atlas periods (Ellison 2010zo), and it is now found occasionally, but with increasing frequency, in New York (McGowan 2008zr). The Yellow-throated Warbler has a history of retraction from the northern limit of its breeding range, followed by a period of expansion of that limit, for reasons that are not apparent (Hall 1996), which makes statements about its long-term future a bit tenuous. Nevertheless, having been reported in almost 65 percent of Pennsylvania's counties during the second Atlas, the Yellow-throated Warbler might be expected, for the immediate future, in suitable habitat anywhere in the state.

ROY A. ICKES

Distribution

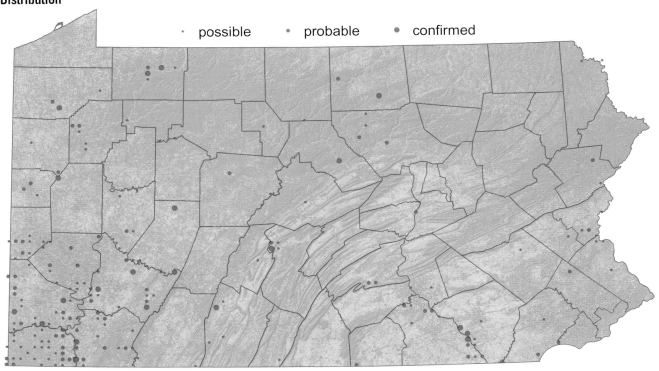

Distribution Change

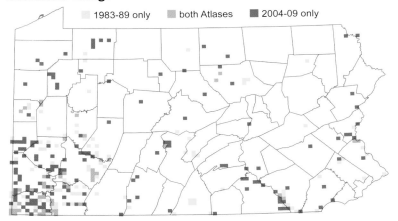

1983–89 only   both Atlases   2004-09 only

Number of Blocks

	first Atlas 1983–89	second Atlas 2004–09	Change %
Possible	55	112	104
Probable	81	63	−22
Confirmed	16	21	31
Total	152	196	29

Population estimate, males (95% CI):
4,000 (2,900–5,600)

YELLOW-THROATED WARBLER

Bob Wood and Peter Wood

Prairie Warbler
Setophaga discolor

The Prairie Warbler is a characteristic species of scrub habitats in many Pennsylvania locales, not the open grasslands the species' name may bring to mind. In Pennsylvania, the species is found in abandoned fields and pastures, regenerating forests, reclaimed surface mines, and pine and scrub oak barrens, including the serpentine barrens of the southeastern corner of the Commonwealth (Leberman 1992j). Most of its breeding range is to our south, but Prairie Warblers nest as far north as southern New England and southern Ontario (Nolan et al. 1999).

Although still relatively common, the Prairie Warbler is listed on Audubon's and the American Bird Conservancy's WatchList as a species of conservation concern, due to declining population trends (NAS 2007). In Pennsylvania, the species is considered of Maintenance Concern, due to evidence of population declines (PGC-PFBC 2005).

The second Atlas shows Prairie Warblers widely distributed in Pennsylvania, with records coming from every county except Philadelphia. However, the species' stronghold was in northeastern Pennsylvania, a dramatic range shift since the first Atlas, when southwestern Pennsylvania was the core range. The former strongholds in the Pittsburgh Low Plateau and Waynesburg Hills sections of the Appalachian Plateaus Province saw statistically significant declines of 34 and 56 percent, respectively, in blocks with Prairie Warblers between first and second Atlases. During the same period, there was a statistically significant increase of 36 percent in blocks with this species in the Ridge and Valley Province.

Habitat changes undoubtedly contributed to the eastward shift. The composition of Pennsylvania's forests has been changing due to forest maturation (McWilliams et al. 2007), a trend that puts Prairie Warbler habitat at a premium. Between 1989 and 2004, there were declines in small trees (<10 cm [4-inch] diameter at breast height) in southwestern Pennsylvania, while little change in the number of small trees occurred in northeastern Pennsylvania (McWilliams et al. 2007). However, it is interesting to note that the Prairie Warbler's range and population declined in Maryland between atlas periods (Coskren 2010c) but increased in New York over a similar period (Smith 2008f), suggesting that larger-scale processes could have contributed to population shifts in Pennsylvania.

Prairie Warblers typically arrive in Pennsylvania between mid-April and early May (Haas and Haas 2005) from their primary wintering grounds in Florida and the Caribbean (Nolan et al. 1999). Due to the species' generally secretive manner, second Atlas volunteers observed nest building only three times, with the latest record coming on 19 June. Most young birds fledge in early July, but fledglings were observed as early as 5 June and as late as 28 July.

Point counts conducted across the Commonwealth suggest a total population of approximately 26,000 singing male Prairie Warblers during the second Atlas period. Breeding Bird Survey results from Pennsylvania show an annual decline of 1.2 percent from 1966 through 2009 (Sauer et al. 2011), a decline that has accelerated somewhat in recent years, resulting in an estimated decline of 30 percent between atlas periods. This suggests that the range expansion in the eastern half of the state has not been sufficient to counterbalance the loss of populations in the western half of the state.

Prairie Warbler populations appear to react quickly to changes in their habitat, as would be expected of a successional-habitat specialist. Since nesting habitat is short-lived without a natural or man-made disturbance, Prairie Warbler populations shift to occupy suitable sites. Increased use of controlled burning as a management tool benefits the species if the land is subsequently not cut or burned for several years. Sustainable forestry practices that produce a matrix of forest stands of varying age also create appropriate habitat, allowing Prairie Warblers to move to younger stands when nesting sites begin to mature. Prairie Warblers should continue to be widespread across the Commonwealth, but changes in land use and management will greatly influence their distribution and abundance.

BRIAN BYRNES

Distribution

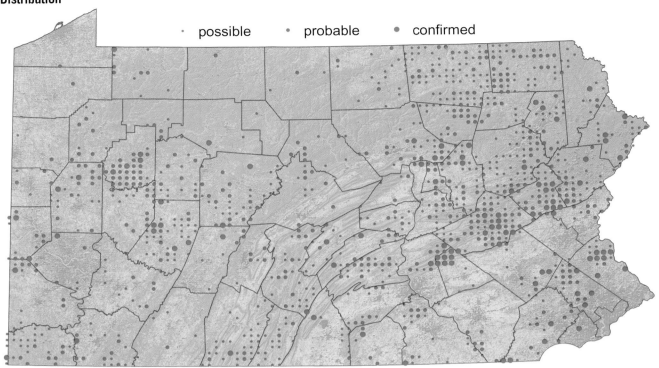

Distribution Change

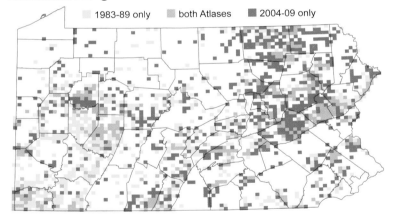

Density

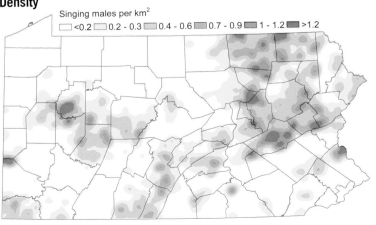

Number of Blocks

	first Atlas 1983–89	second Atlas 2004–09	Change %
Possible	404	609	51
Probable	508	408	−20
Confirmed	123	400	−19
Total	1,035	1,117	8

Population estimate, males (95% CI):
26,000 (23,000–30,000)

Breeding Bird Survey Trend

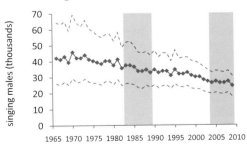

PRAIRIE WARBLER 397

Gerard Dewaghe

Black-throated Green Warbler
Setophaga virens

The persistently buzzy song, *zoo-zee-zoo-zoo-zee*, of the Black-throated Green Warbler is one of the more distinctive, and common, of the wood warbler songs heard in migration and through the summer in Pennsylvanian treetops. Birders, craning their necks, may be treated with the sight of the bright yellow face and black throat pattern. The Black-throated Green Warbler ranges from southern Canada and the northeast United States, west to Minnesota, and south through the Appalachian Mountains to Alabama (Morse and Poole 2005). Pennsylvania lies in the heart of this distribution.

Atlas volunteers and point counters found them most frequently in coniferous and mixed coniferous-deciduous forests (appendix D), although they also appear in deciduous forests. Ross et al. (2004) considered the species to be obligately connected to Eastern Hemlock in the Delaware Water Gap Area. Throughout the state its distribution is nearly identical to that of Eastern Hemlock (chap. 3).

The number of occupied blocks increased dramatically and significantly since the first Atlas, by 48 percent. The Appalachian Plateaus Province continues to be the stronghold for this species, with records in 2,025 blocks. Even in this core area, there was a significant 33 percent increase in block occupancy between atlases, but this was easily surpassed in the Ridge and Valley, where the number of occupied blocks more than doubled, representing a significant southward range expansion throughout the center of the state. Despite the overall expansion, there were localized losses around the western perimeter of the range, although these did little to alter the overall very positive picture for the Black-throated Green Warbler. As in the first Atlas, black-throated greens were effectively absent from the state's southeastern corner and the western border counties, except Crawford County and the Eastern Lake section of Erie. They largely shun elevations below 400 m (1,300 ft), where there are few large contiguous forests (appendix D). Atlas volunteers observed territorial behavior as early as late April. Breeding behavior was confirmed in early May, with observations of birds carrying nest material. Most confirmations were of adults carrying food, with only 26 confirmations (less than a half percent of all evidence) by direct observation of nests. This is not surprising, considering that nests high in evergreens are not easily found.

The Black-throated Green Warbler population in Pennsylvania during the second Atlas period was estimated at 355,000 singing males. Breeding Bird Survey (BBS) data indicate a significant 3.1 percent annual increase from 1966 through 2009 (Sauer et al. 2011), with the population apparently peaking in the first year of the second Atlas, after which there was a rather surprising decline. Nonetheless, on average, BBS counts were more than 80 percent higher in the second Atlas period than in the first, tying in well with the observed range expansion during that time.

Although the Black-throated Green Warbler has increased significantly in Pennsylvania, threats remain. As a forest-interior bird, it is susceptible to the effects of forest fragmentation (Hagen et al. 1996), and much of its Pennsylvania range is within the shale gas field. Another immediate threat to this species is the relentless spread of Hemlock Woolly Adelgid from east to west across the state. This invading insect infests Eastern Hemlock and has already killed vast swaths of them. Affected trees have very high mortality, especially when stressed. The warblers feed on insect prey, mainly on small branches and twigs, which are the first areas of hemlocks to show noticeable damage.

In the long term, current climate models predict a northern retreat and shrinking of the distribution of Eastern Hemlock in Pennsylvania and the Northeast during the next century (Rotenhouse et al. 2008). As hemlocks succumb to disease or shifting climate, the warblers may be forced to shift habitats, perhaps to deciduous trees. Although they show good success in deciduous woods in other parts of their range, such a shift may well lead to a marked population decrease in Pennsylvania in the coming decades.

JERRY SKINNER

Distribution

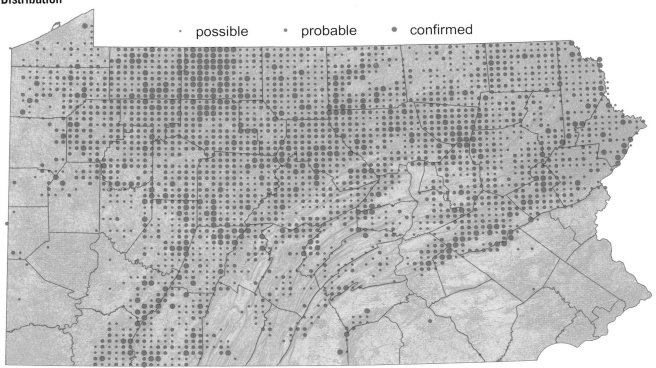

Distribution Change

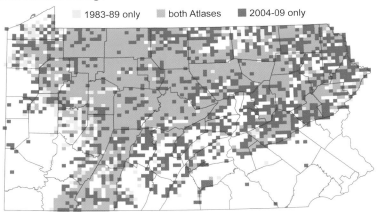

Density

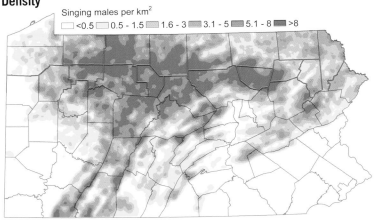

Number of Blocks

	first Atlas 1983–89	second Atlas 2004–09	Change %
Possible	775	1,199	55
Probable	821	1,142	39
Confirmed	207	334	61
Total	1,803	2,675	48

Population estimate, males (95% CI):
355,000 (340,000–370,000)

Breeding Bird Survey Trend

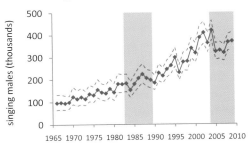

Bob Wood and Peter Wood

Canada Warbler
Cardellina canadensis

With its cheery song and attractive "necklaced" plumage, the Canada Warbler is one of several northern warblers that breed in Pennsylvania. The species is aptly named in that the majority of its breeding range resides within the boreal zone of Canada; however, it also nests in the northeastern United States, the Great Lakes region, and the Appalachian Mountains as far south as northern Georgia. The Canada Warbler winters primarily in mid-elevation forests of the northern Andes, from Venezuela and Colombia, and south to central Peru (Conway 1999).

Like other boreal-zone migrants that nest in Pennsylvania, the Canada Warbler is an uncommon to locally common summer resident of the cooler high-elevation northern counties, with breeding localities in southern Pennsylvania restricted to the mountains. During the second Atlas, it was found most frequently (19 percent of blocks) on the Appalachian Plateaus, and less often (9 percent of blocks) in the Ridge and Valley Province; it was rarely recorded at elevations below 300 m (~1,000 ft). At first glance, its habitat relationships can appear bafflingly complex. In its strongholds in northern Pennsylvania, the Canada Warbler is most abundant in wooded bogs with a partially open mixed canopy of coniferous and deciduous trees, a diverse shrubby understory, an extensive groundcover of sphagnum moss, and an abundance of mossy fern hummocks, favored sites for nest construction. It can also be found in shrubby regenerating mixed forest on blow-downs, clear-cuts, or rights-of-way, particularly those on moist soils adjacent to areas with more mature mixed forest. Finally, the Canada Warbler also breeds in hemlock-dominated ravines with extensive understory thickets of mountain laurel or rhododendron (Mulvihill 1992i). Thus, though its habitat associations may appear to be quite diverse, the common thread for the species in Pennsylvania is dense, shrubby sub-canopy (Hallworth et al. 2007), frequently associated with the presence of hemlocks and other conifers.

Canada Warbler populations rangewide have been in general long-term decline since the 1960s; the species has been placed on the Watch List of Partners in Flight (Rich et al. 2004) and listed as a species of Maintenance Concern in Pennsylvania (PGC-PFBC 2005). Within Pennsylvania, however, its population status appears to be stable. The total of 691 atlas blocks with Canada Warbler records represents a non-significant 9 percent increase over the first Atlas period, and Breeding Bird Survey (BBS) data suggest that Canada Warbler numbers in Pennsylvania have been stable since the 1960s (Sauer et al. 2011). The species is nowhere found in abundance in Pennsylvania; second Atlas point count data suggest that across much of its range it is scarce, with just a few core forest areas in which population densities average more than one singing male per km^2. Because the Canada Warbler is generally scarce, with a statewide population estimated to be 27,000 singing males, sample sizes on BBS routes in the state are small, rendering assessments of population change tenuous.

Although this apparent population stability allows for cautious optimism about the Canada Warbler's prospects in the state, that optimism must be tempered with a note of caution: only 45 percent of the blocks in which the species was recorded during the first Atlas were also occupied during the second Atlas, suggesting that many Pennsylvania populations of Canada Warblers exist only ephemerally. Indeed, given the species' reliance on dense understory or early successional habitats, we might expect Canada Warbler populations to be highly dynamic and subject to pronounced population swings. For example, precipitous between-atlas reductions in atlas block records in New York (−23%; McGowan 2008zs) and Maryland (−22%; Coskren 2010d) have been coupled with similar declines in BBS counts (Sauer et al. 2011). Because loss of understory vegetation could potentially produce similar declines in Pennsylvania, the population dynamics of the Canada Warbler within the state merit continued monitoring.

RONALD L. MUMME, DOUGLAS A. GROSS,
AND SCOTT H. STOLESON

Distribution

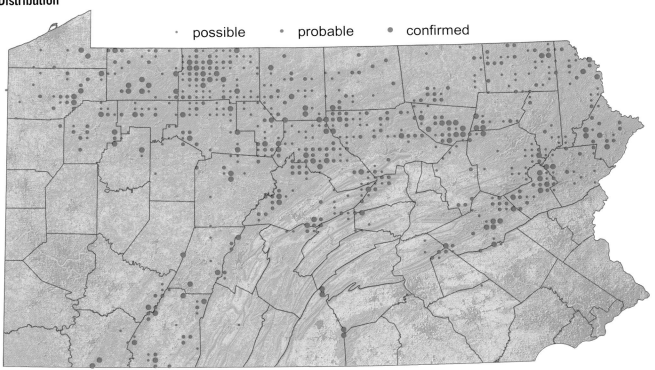

Distribution Change

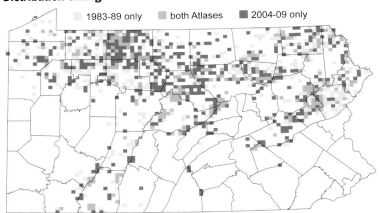

Density

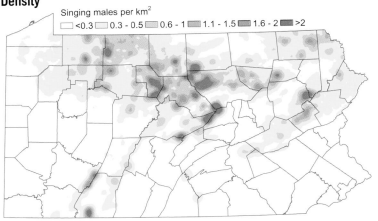

Number of Blocks

	first Atlas 1983–89	second Atlas 2004–09	Change %
Possible	284	346	22
Probable	239	228	−5
Confirmed	111	117	5
Total	634	691	9

Population estimate, males (95% CI):
27,000 (21,500–38,000)

Breeding Bird Survey Trend

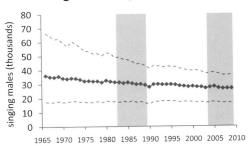

CANADA WARBLER

Bob Wood and Peter Wood

Yellow-breasted Chat
Icteria virens

The Yellow-breasted Chat is unique among the North American wood warblers; indeed, its status as a warbler has been questioned repeatedly, based upon physiological characteristics (Eisenmann 1962), morphological and behavioral observations (Ficken and Ficken 1962), and recent genetic studies (Lovette et al. 2010). Wherever it may ultimately fit into the songbird taxonomy, the chat is a welcome member of Pennsylvania's breeding avifauna. Largely restricted to the southern portion of the state, the chat is spottily distributed and rare in northern Pennsylvania and to the north (Eckerle and Thompson 2001). Because it is an indicator species for extensive thickets and early successional forests, the chat is listed as a species of Maintenance Concern in Pennsylvania's Wildlife Action Plan (PGC-PFBC 2005).

The chat's unique song includes a series of whistles, cackles, mews, chuckles, squawks, gurgles, and rattles, often including some mimicry (Eckerle and Thompson 2001). Floyd (2008) describes the song well as "endless and clownish." Best located by its distinctive song, this skulking songbird breeds in low, dense vegetation with an open canopy of shrubs or scattered trees. In Pennsylvania, it is found primarily in abandoned agricultural fields, scrub barrens, regenerating clear-cuts, power line corridors, revegetated surface mines, and thickets alongside forest edges, wetlands, and streams (Dennis 1958; Leberman 1992k; Dearborn 2010). Chats typically colonize shrublands and thickets 4 to 12 years after disturbance, when woody plants begin to take hold (Shugart and James 1973), and persist until shrub heights exceed 4.5 m (14.7 ft; Crawford et al. 1981). They feed primarily upon small invertebrates, and adults readily incorporate berries into their diet (Eckerle and Thompson 2001). Nest sites often include high densities of blackberry canes (Kroodsma 1982).

Like other shrubland obligates, the Yellow-breasted Chat has experienced a decline in both numbers and range in Pennsylvania since the first Atlas, reflecting the loss of shrubby habitat due to the maturation of forests (McWilliams et al. 2007) and increased development. Statewide occupancy dropped significantly, from 1,242 blocks in the first Atlas to 805 blocks in the second Atlas, a decline of 35 percent. Losses show a pattern of range contraction, with a virtual abandonment of the northern tier (where the species was always scarce) and steep reductions in the number of blocks in former strongholds such as the Pittsburgh Low Plateau Section (−42%; appendix C). However, range contractions were evident even in southern areas, including a virtual loss of this species in the Philadelphia metropolitan area.

From atlas point count data, the state's population of chats is estimated to be 11,200 singing males. Much of the population is now concentrated into a few areas, notably the Waynesburg Hills of the southwestern corner of the state, the southern Ridge and Valley, and the lower Susquehanna Valley. Elsewhere in the state, the Yellow-breasted Chat is now a scarce and localized bird. A contraction in the species' range has also been noted in neighboring states. In New York (where this species was never common) there was a 78 percent drop in the number of blocks with chats between atlases (McGowan 2008zt), while in Maryland there were substantial losses in counties adjacent to Pennsylvania (Ellison 2010zp). These range contractions are also reflected in Breeding Bird Survey data, with a significant annual decline of 5.1 percent in Pennsylvania between 1966 and 2009, and similar declines across the Northeast and Mid-Atlantic (Sauer et al. 2011).

While chats experience a high degree of nest predation (Thompson and Nolan 1973), the greatest threat to populations remains habitat loss. Across much of their western United States range, the greatest threat is increased grazing activity; in Pennsylvania and much of the eastern United States, natural succession is the most direct threat to these areas (Eckerle and Thompson 2001). Researchers have found that chats exhibit a high degree of natural population turnover and will readily and rapidly colonize newly created patches of suitable habitat. Such adaptability will be needed to ensure this species' continued presence in Pennsylvania.

LEWIS GROVE

Distribution

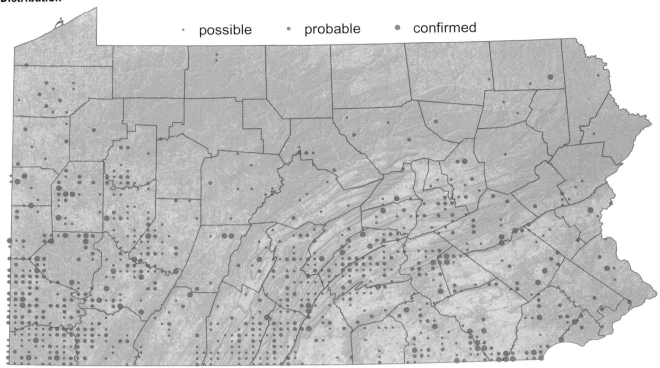

Distribution Change

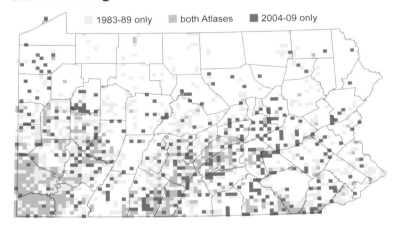

Number of Blocks

	first Atlas 1983–89	second Atlas 2004–09	Change %
Possible	487	434	−11
Probable	604	295	−51
Confirmed	151	76	−50
Total	1,242	805	−35

Population estimate, males (95% CI): 11,200 (8,900–13,500)

Breeding Bird Survey Trend

YELLOW-BREASTED CHAT 403

Chuck Musitano

Eastern Towhee
Pipilo erythrophthalmus

Drink-your-teeeea, a vocalization typically heard in Pennsylvania for the first time in April each year, is a sure sign of spring and the coming neotropical bird migration. Even casual birders can recognize this distinctive song. Equally distinctive is the *chewink* alarm call. Even someone not familiar with the vocalizations can spot and recognize the unique plumage of this member of the sparrow family, even in the dense ground cover where it is often found (Greenlaw 1996).

Eastern Towhees are occupants of mid-to-late stages of secondary succession, most often found in old field thickets, mid-stages of secondary forest growth, and in open forest understory (Lanyon 1981; Morimoto and Wasserman 1991). However, the species is somewhat of a habitat generalist and can be found in almost any area with a dense shrub layer.

Atlas observations indicate that nest building begins in late April to early May, peaking in June. Breeding activity in this potentially double-brooded species continues well into August, with some fledged young observed as late as mid-September. Nests are typically on the ground in relatively dry habitat, usually at the base of small shrubs, small trees, or grass clumps. Late-season nests may be above ground, typically less than 1.5 m (4–5 ft), in dense vegetation such as honeysuckle, briar, or grape vine tangle. Because of its distinct vocalizations and long breeding season, it is one of the easiest species for atlas volunteers to find and is likely to be missed in blocks only if suitable habitat is not surveyed.

Given their general habitat requirements, it is not surprising that breeding Eastern Towhees were found in 95 percent of atlas blocks. Although there was some turnover in occupied blocks between the first and second Atlases, and a slightly higher overall occupancy rate in the second Atlas due to improved coverage, there is little evidence of any significant change in distribution. In both atlases, the only appreciable areas without towhees were central Lancaster County and urban Philadelphia. Point count detections were positively correlated with elevation (appendix D), so it is not surprising that the highest densities were found across the Appalachian Plateaus. Interestingly, densities were generally higher in the western half of the state than in the eastern half, presumably due to the availability of suitable shrubby habitats.

Breeding Bird Survey (BBS) data from Pennsylvania indicate that the current population, estimated by second Atlas point count at 610,000 singing males, is about half that of the 1960s (Sauer et al. 2011). However, most of the decline occurred prior to the first Atlas, and there was no significant change in BBS counts between atlases. Despite previous declines, the Eastern Towhee is not of management concern in Pennsylvania or nationwide. The early years of the BBS likely documented declines from unnaturally high population levels during the middle decades of the twentieth century, when mid-stage forest succession was at its peak. Early avifaunal accounts suggest that the species may have once been rare in northern Pennsylvania, but spread northward with the opening up and subsequent regeneration of forests (McWilliams and Brauning 2000). The Eastern Towhee remains one of the state's most numerous forest birds, and its currently stable population is likely to be maintained by timbering operations, which ensure a constant supply of suitable mid-stage shrubby habitat in forested regions.

MICHAEL CAREY

Distribution

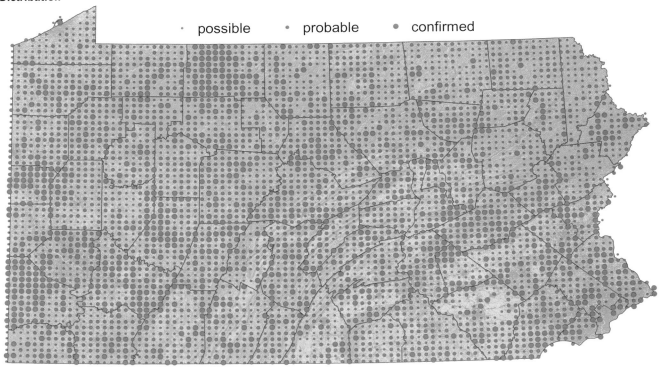

Distribution Change

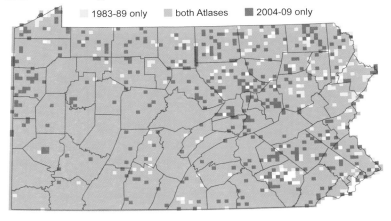

Density

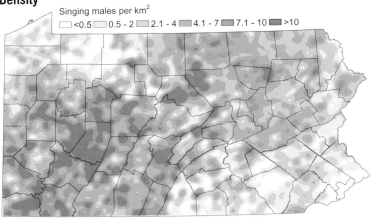

Number of Blocks

	first Atlas 1983–89	second Atlas 2004–09	Change %
Possible	1,174	1,661	41
Probable	2,345	2,039	−13
Confirmed	934	977	5
Total	4,453	4,677	5

Population estimate, males (95% CI):
610,000 (590,000–635,000)

Breeding Bird Survey Trend

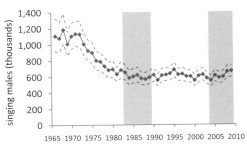

EASTERN TOWHEE 405

Chuck Musitano

Chipping Sparrow
Spizella passerina

When the dry, trilling song of the Chipping Sparrow is heard coming from inside conifer trees in late March, it's a sure sign that spring is right around the corner. This is the sparrow of summer for many people, especially those who live in suburban or urban areas. This small sparrow with a rufous crown and white eye line is the one seen picking at insects in the middle of lawns, perched on fences, or taking a dust bath in the dirt of a driveway.

The male Chipping Sparrow sings from within the cover of trees, often near the top of a conifer, but not on exposed perches. At times, its song is confused with the more musical one of the Pine Warbler, which also sings from conifers, or that of the Dark-eyed Junco, but that species normally sings from a perch out in the open. Migrant Chipping Sparrow numbers peak in late April; by mid-May, they are spread all across Pennsylvania and are already nesting in southern areas.

The Chipping Sparrow breeds across Canada, up into Alaska, in most of the United States (except for parts of the central plains, Southwest, and Florida) and south into Central America. It favors dry habitats with accompanying expanses of open grassy areas, including orchards, woodland edges, parks, golf courses, pastures, and backyards. It often exploits openings in woodlands, including along forest roads, and is sparingly found in riparian habitats and emergent wetlands. The Chipping Sparrow is highly adapted to urban areas because of the abundance of conifer trees, shrubs, and grassy yards found around homes. In 1810, when Alexander Wilson named the Chipping Sparrow, he called it *Fringilla socialis*, the "social sparrow," because of its association with human habitation (Rising and Beadle 1996).

Conifers are preferred nesting sites of the Chipping Sparrow, but it will also use deciduous trees, shrubs, and vine tangles. Its small nest, made of grasses and rootlets, is placed from 1.5 to 10 m (4 to 30 ft) off the ground, but nests as high as 18 m (60 ft) or more have been located, often placed in shrubbery very close to human residences. The Chipping Sparrow likes to line its nest with hair, especially horse hair when it is available, but it will use any animal or human hair it finds (Cassidy 1990).

The Chipping Sparrow has benefited from the early successional, low-growth habitat that evolves as a result of human occupancy (Middleton 1998). Overall, there was no significant increase in distribution between the first Atlas to the second. In the first Atlas, it was found in 97 percent of atlas blocks (Mulvihill 1992j); in the second Atlas, it was found in 99.8 percent of blocks. It is interesting to note that only a handful of blocks were not occupied in either atlas period—in densely urbanized parts of the Philadelphia metropolitan area, alongside the Delaware River. Lowest densities were found in uniformly forested areas at elevations above 200 m (660 ft), in the north-central Deep Valleys area of the Appalachian Plateaus Province and in the aforementioned urban core around Philadelphia.

Breeding Bird Survey counts of the Chipping Sparrow have been in shallow decline, averaging 0.7 percent per year, from 1966 through 2009 (Sauer et al. 2011). Despite this, the chippy remains one of the state's most abundant species, with a population estimated to be almost 3 million singing males in the second Atlas period. The Chipping Sparrow is one of the most familiar birds in Pennsylvania, and it remains well established in our human-altered environment.

ARLENE KOCH

Distribution

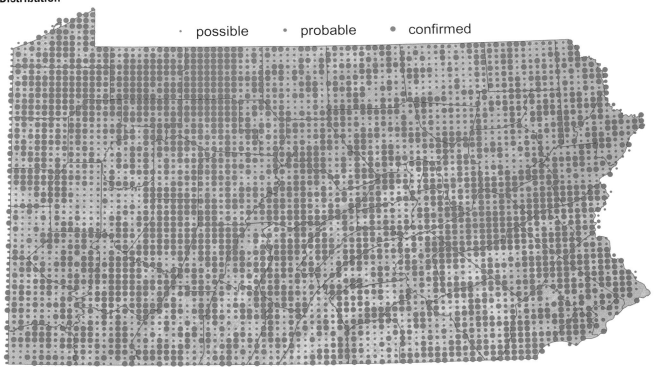

possible • probable • confirmed

Distribution Change

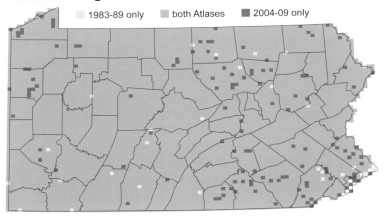

1983–89 only | both Atlases | 2004-09 only

Density

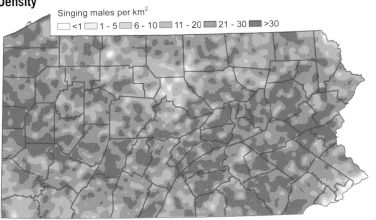

Singing males per km²
<1 | 1 - 5 | 6 - 10 | 11 - 20 | 21 - 30 | >30

Number of Blocks

	first Atlas 1983–89	second Atlas 2004–09	Change %
Possible	701	1,111	58
Probable	1,236	1,254	1
Confirmed	2,849	2,523	−11
Total	4,786	4,888	2

Population estimate, males (95% CI):
2,980,000 (2,900,000–3,050,000)

Breeding Bird Survey Trend

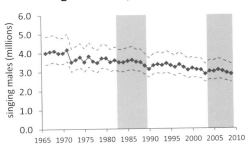

CHIPPING SPARROW 407

Geoff Malosh

Clay-colored Sparrow
Spizella pallida

A close relative of the Chipping Sparrow and often associated with it during migration, the inconspicuous and appropriately named Clay-colored Sparrow would often go unnoticed on the breeding grounds were it not for the distinctive four or five insect-like buzzes of its song. It is a common breeder in the low shrub communities of the Canadian Prairie Provinces and along the north-central United States border. Apparently in response to suitable habitat created by logging and agricultural activities, the Clay-colored Sparrow has extended its breeding range east and north since the nineteenth century (Knapton 1994).

Occasional sightings of the species have been reported throughout Pennsylvania since at least the mid-1940s, but mostly outside of the nesting season (Poole 1964; Wood 1979). A singing male in the Poconos on 25 May 1971 (Street 1976) was the only record of a bird potentially on breeding territory prior to the first Atlas. During the first Atlas, the Clay-colored Sparrow was reported in three blocks in as many northwestern counties, all of which were classified as probable breeding records (Ickes 1992n).

The second Atlas demonstrates that the Clay-colored Sparrow has increased in Pennsylvania. However, it is still rare, and despite its drab appearance and song, finding a Clay-colored Sparrow during atlas fieldwork was a thrill enjoyed by a lucky few atlas volunteers. It was reported in 29 blocks in 15 counties, with three-quarters of records in the northwestern quarter of the state and only four in the eastern half of Pennsylvania. There were 14 probable breeding records from 10 counties and 3 confirmed breeding records in 2 counties. Clarion and Clearfield counties together accounted for 13 occupied blocks. Of the three confirmed breeding records, two were in Clarion and one in nearby Jefferson County.

Ickes (1992n) suggested that, in the first Atlas, the Clay-colored Sparrow arrived from our northwest, with Pennsylvania records reflecting expansion of individuals from a breeding population in New York or directly from Ontario, the likely source of the New York birds. The location of most of the second Atlas records and the expansion of the species in states to the north and west of Pennsylvania continue to support the likelihood of a northwestern origin for the Commonwealth's birds. The Clay-colored Sparrow in New York increased from 23 blocks in the state's first Atlas to 69 in its second (Smith 2008g), while in Ohio there were no records in the first atlas, but 13 blocks by the end of the fifth year of fieldwork in their second atlas (OBBA 2011).

Although the distribution of Clay-colored Sparrow records might point to the origin of birds in Pennsylvania, it also reflects the distribution of a key habitat for this species in the state; brushy, revegetated surface mines (chap. 3). Sixteen of the 47 Clay-colored Sparrow reports in the second Atlas specifically mentioned reclaimed surface mines.

The population of the Clay-colored Sparrow during the second Atlas is likely in low double figures. Assuming that records for all atlas years represented different birds, the number of breeding pairs (based on territorial males) in Pennsylvania would be 37, but many birds were reported in only a single year, so the actual number could be lower. Conversely, this species is almost certainly overlooked, since its preferred habitat—old fields with low shrubs—is found widely in the state, so it is likely that not all suitable sites were visited by atlas volunteers.

The expansion of the Clay-colored Sparrow population between atlas periods suggests that its tenuous foothold in Pennsylvania may be strengthening. The species appears to be susceptible to brood parasitism (Davis 2003; Winter et al. 2004), and management of shrubby grasslands to increase nest cover could increase nest survival (Winter et al. 2005; Grant et al. 2006). However, the Clay-colored Sparrow has not been well studied in Pennsylvania, and we have much to learn about its breeding biology and habitat requirements in an area that is peripheral for this species.

ROY A. ICKES

Distribution

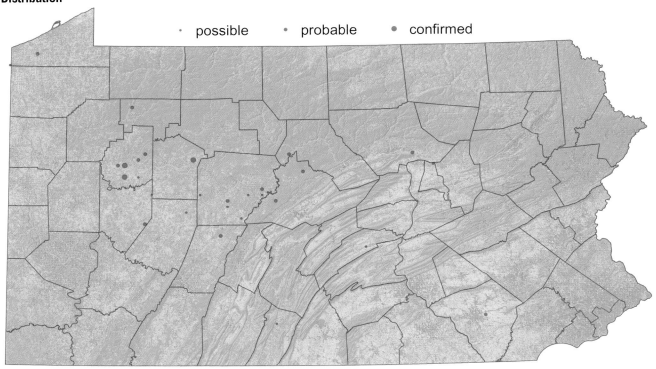

Distribution Change

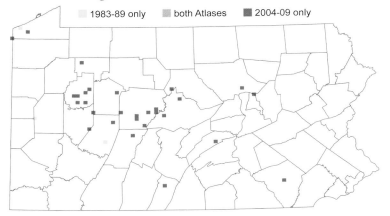

Number of Blocks

	first Atlas 1983–89	second Atlas 2004–09	Change %
Possible	0	12	∞
Probable	3	14	367
Confirmed	0	3	∞
Total	3	29	867

Dustin Welch

Field Sparrow
Spizella pusilla

Marked with a bold white eye-ring and bright pink bill against drab gray and brown plumage, the Field Sparrow often catches the observer's attention as it dives for cover in tangled, low vegetation. But the distinctive song of accelerating clear notes, which is often audible over distances of several hundred meters, announces its presence during the breeding season. This species is widely distributed, breeding throughout most of the United States east of the Rockies and north of the Gulf Coast, with northerly populations moving southward in winter (Carey et al. 2008).

Somewhat poorly named, the Field Sparrow is not typically found in open fields or pastures, although it can be found along the forested or unused edges of such fields. More commonly, it inhabits areas in the earliest stages of ecological succession, fields that are largely open but with some scattered scrub and woody vegetation (Carey et al. 2008). It was found to nest in unused hayfields a year or so after last mowing; populations increased in density for about the next 10 years and declined thereafter. With the exception of an occasional straggler, it was largely absent from areas not mown for 25 years or more (Carey 2010). While lightly settled suburban areas may appear to be usable habitat, the Field Sparrow generally avoids human settlements.

Even with such limited and ephemeral habitat preferences, the Field Sparrow was found breeding across the state in 89 percent of blocks. Some of the highest abundances were found in the Pittsburgh Low Plateau Section, where reclaimed surface mine areas support high densities (appendix D). Clusters of blocks in which no breeding was detected were restricted to uniformly forested regions or heavily urbanized or developed areas of the state. There was no significant change in block occupancy between the first Atlas period and the second Atlas period. Blocks that were occupied in the first Atlas and unoccupied in the second were primarily around expanding urban areas and in Pike and Monroe counties, perhaps due to expanding suburban development there. Blocks that became occupied between the two atlas periods are scattered across the state, with some concentration in the Appalachian Plateaus Province.

According to Breeding Bird Survey (BBS) in Pennsylvania, the Field Sparrow population size has declined by a statistically significant 3 percent per year (Sauer et al. 2011). While block occupancy has remained unchanged between atlas periods, BBS counts suggest that the population size halved (Sauer et al. 2011). During the second Atlas period, the total population size in the state was estimated to be 210,000 singing males. Continued declines at this rate would be expected to be reflected in the species' distribution in the next few decades. But conservation initiatives for American Woodcock and Golden-winged Warbler would likely benefit this species as well. Management recommendations include the management of existing grassland and successional habitats to maintain some woody vegetation, burning to prevent the over-encroachment of woody vegetation, and removing the canopy and thinning shrubs and saplings in forested habitats (Dechant, Sondreal, et al. 1999).

First song of the Field Sparrow is usually detected from mid-March through mid-April. Singing males are detected 1 to 2 weeks earlier than migrant females. Earliest confirmed breeding activity (nest building) was detected on 20 April, with latest activity (fledged young) on 29 August. In one well-studied northeastern Pennsylvania population, earliest first egg dates ranged from 3 May to 16 May, and latest, from 4 July to 27 July. The earliest chick hatch was 19 May, and the latest fledging was August 18 (Carey et al. 2008). Hence, the Field Sparrow, with its loud song and protracted breeding season, was relatively easy to find by atlas volunteers.

Field Sparrows remain commonly observed across the state but are declining in number. As with many species breeding in successional habitats, habitat loss is the primary cause of the decline. Appropriate habitat management is needed to eventually stabilize the breeding population.

MICHAEL CAREY

Distribution

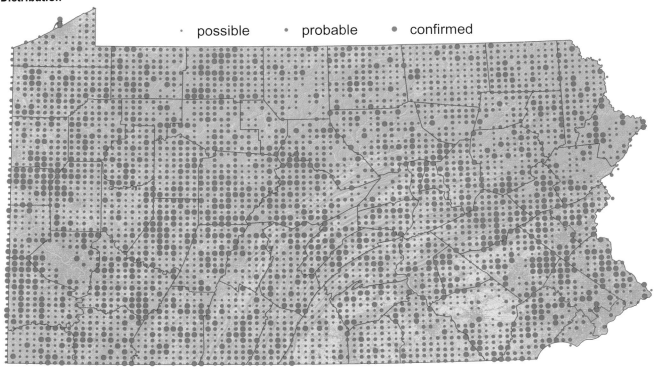

Distribution Change

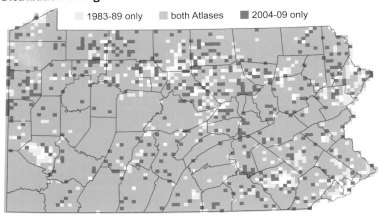

Density

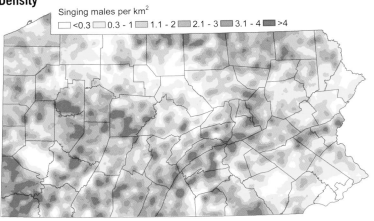

Number of Blocks

	first Atlas 1983–89	second Atlas 2004–09	Change %
Possible	1,237	1,881	52
Probable	1,816	1,613	−11
Confirmed	1,238	878	−29
Total	4,291	4,372	2

Population estimate, males (95% CI):
210,000 (204,000–216,000)

Breeding Bird Survey Trend

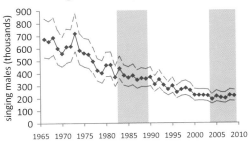

FIELD SPARROW

Bob Wood and Peter Wood

Vesper Sparrow
Pooecetes gramineus

This robust and somewhat nondescript sparrow is named for its habit of singing late into the evening when most other birds have fallen silent. Its musical song can be heard in open country across the northern United States and southern Canada. It winters in the southern United States and Mexico. Often grouped with Savannah, Grasshopper, and Henslow's sparrows in a guild of so-called grassland sparrows, the Vesper Sparrow is, in fact, not a grassland obligate. Favored habitats vary from semi-desert through grassland and cornfields to shrubs and woodland edges, but availability of bare ground for foraging is a consistent requirement (Jones and Cornely 2002). Agricultural lands and reclaimed surface mine grasslands are preferred in Pennsylvania (Santner 1992f), especially where there are elevated song perches, such as trees or utility wires (Dechant et al. 2003).

We really do not know how common this species was in Pennsylvania prior to colonization by European farmers, but avifaunal accounts in the nineteenth century relate that it was a common bird at that time (e.g., Warren 1890). Evidently, this is a species that adapted quickly to agriculture and remained abundant well into the twentieth century. However, numbers on Breeding Bird Survey routes declined rapidly during the late 1960s and 1970s and continued to decline through the second Atlas period. Such declines have been noted across the eastern half of the species' range (Sauer et al. 2011). Increased mechanization and frequency of mowing, resulting in more nest losses, may be key drivers of declines in agricultural areas (Dechant et al. 2003).

By the time of the first Atlas, the Vesper Sparrow was described as locally common (Santer 1992f), but it was found in only 22 percent of atlas blocks, suggesting that its range had already contracted by that time. The species continued to lose ground between the two atlas periods and was found in fewer than 18 percent of blocks in the second Atlas. The 19 percent reduction in blocks with Vesper Sparrows between the two Pennsylvania atlases was, however, a more modest range contraction than that noted in neighboring New York (−49%; Smith 2008h) or Maryland (−39%; Ellison 2010zq) over a similar period.

The Vesper Sparrow is one of the first birds to colonize reclaimed mines (Jones and Cornely 2002), and 20 percent of birds located on atlas point counts in the Appalachian Plateaus Province were on reclaimed surface mines. However, there was 43 percent reduction in the number of occupied blocks in that province in the relatively short time between atlas periods, and this species is now almost gone from former strongholds in the heart of surface mine country. Successional changes and more successful vegetation establishment techniques may have reduced the availability of sparsely vegetated reclaimed sites. It is interesting to note that the Horned Lark, another bird requiring bare ground, showed similar range losses in mining areas.

Once an abundant bird in northeastern Pennsylvania (McWilliams and Brauning 2000), the Vesper Sparrow is now localized in all northern counties. However, range losses in western and northern counties have been somewhat compensated for by range expansions in the Ridge and Valley Province, which is now the core of the Vesper Sparrow's range in Pennsylvania. The majority of birds observed in the Ridge and Valley were in areas dominated by arable farming, but in some areas, such as Adams County, orchards were also used (pers. obs.). The fate of this species in agricultural areas rests with its ability to adapt to changing agricultural practices. Grasslands planted through conservation programs, such as the Conservation Reserve Enhancement Program, are unlikely to benefit this species because they do not provide suitably sparse habitat (Dechant et al. 2003; Wilson and Brittingham 2012).

The estimated statewide population of 17,000 singing males is a far cry from the numbers that must once have been present for the Vesper Sparrow to be described as abundant. Listed as Endangered, Threatened, or of Special Concern in several surrounding states (Whitlock and Carpenter 2007), this species currently does not have any special conservation status in Pennsylvania. Given the sustained downward population trajectory, Vesper Sparrow may now be deserving of more conservation attention.

ANDREW M. WILSON

Distribution

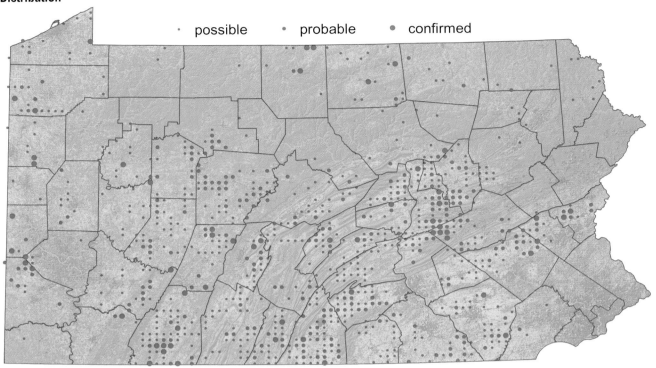

Distribution Change

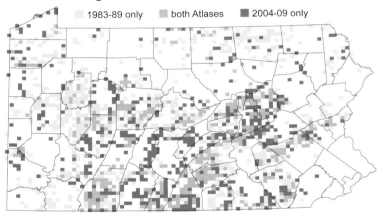

Density

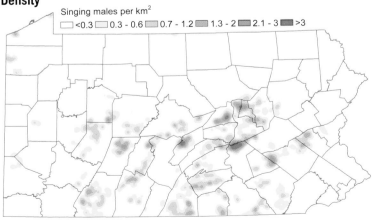

Number of Blocks

	first Atlas 1983–89	second Atlas 2004–09	Change %
Possible	503	582	16
Probable	422	238	−44
Confirmed	162	58	−64
Total	1,087	878	−19

Population estimate, males (95% CI):
17,000 (15,000–19,000)

Breeding Bird Survey Trend

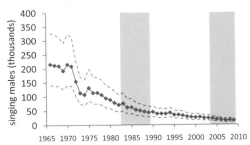

VESPER SPARROW 413

Chuck Musitano

Savannah Sparrow
Passerculus sandwichensis

Named for the town of Savannah, Georgia, where the type specimen was collected, the Savannah Sparrow is an archetypal sparrow, with brown-streaked plumage, unassuming habits, and an understated buzzing song. Hence, it is easily overlooked. It has an extensive distribution, breeding in open habitats throughout the northern half of the United States and most of Canada, and is represented by no fewer than 17 subspecies (Wheelwright and Rising 2008). It winters in the southern United States and Central America.

The Savannah Sparrow appears to have expanded its range in Pennsylvania later than other species—it did not colonize agricultural areas until the early decades of the twentieth century (McWilliams and Brauning 2000). By the time of the first Atlas, it was found in suitable habitat in the western third and northern counties of the state, but it remained scarce in the south. Between the two atlases, the number of blocks occupied by the Savannah Sparrow increased by 26 percent, with notable gains of more than 80 percent in the Ridge and Valley and Piedmont provinces. However, Breeding Bird Survey data show sustained population declines of around 2.8 percent per year since the 1960s (Sauer et al. 2011), amounting to a loss of around 40 percent of the statewide population between atlas periods.

Primarily a bird of hayfields in the northeastern United States (Askins 1999), in Pennsylvania the Savannah Sparrow is also found in arable fields, pasture, and reclaimed surface mine grasslands. Densities on reclaimed surface mines are generally not high; an estimated 1,900 singing males on reclaimed mines in nine counties in western Pennsylvania (Mattice et al. 2005) is a small portion of the statewide population, estimated to be 145,000 singing males. The bulk of the population is found on farmland, where densities are positively correlated with elevation, a relationship that holds true for arable, pastoral, or mixed farmland types. At elevations of greater than 400 m (1,300 ft), densities on farmland sometimes exceed 20 singing males per km^2, estimated from second Atlas point count data. Somerset County, in the Laurel Highlands, supports around 10 percent of the statewide population, while the northwestern corner of the state (Crawford, Erie and Mercer counties) is also a stronghold.

In the Piedmont, the Savannah Sparrow's distribution is patchy; for example, it is widespread in Lancaster County but decidedly scarce in neighboring York County. This could be explained by this species' apparent affinity for Amish farms (pers. obs.), which are managed on a rotational mixed-cropping system that appears to provide prime habitat. Unlike other grassland songbirds, the Savannah Sparrow readily nests in alfalfa (Wheelwright and Rising 2008), although frequent and early cutting of alfalfa fields (chap. 3) must result in considerable losses of nests and eggs.

The Savannah Sparrow is not currently listed in the Pennsylvania Wildlife Action Plan, but its sustained population declines suggest it may get more attention in the future. With a distribution and habitat requirement similar to that of the Bobolink, it is likely that conservation measures for these two species would be highly complementary. For both species, engagement of conservation measures within cropped hayfields, such as delayed cutting or increasing intervals between first and second cuts (Nocera et al. 2005; Perlut et al. 2008), will be needed to maintain populations. For such measures to gain traction, they may require that farmers be compensated for lost hay production. However, Pennsylvania supports considerably less than 1 percent of the global population of Savannah Sparrows (Rich et al. 2004), so advocating for the use of scarce public resources to conserve marginal populations of this species in Penn's Woods is not easily justified. Nocera et al. (2005) suggested that modest delays in hay harvesting could be sufficient to greatly increase fledging success of Savannah Sparrows; such marginal changes offer the greatest promise for combining economically viable agriculture and the provision of habitat for declining grassland wildlife.

ANDREW M. WILSON

Distribution

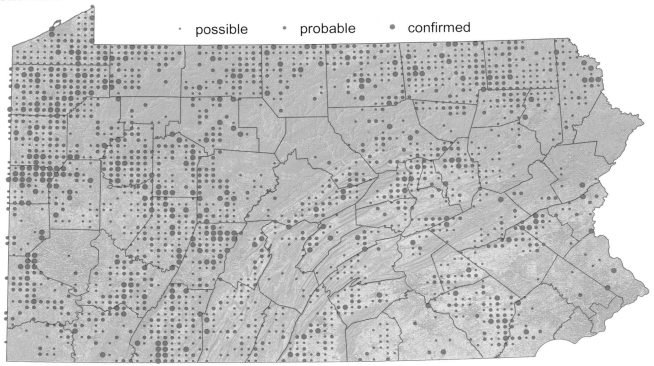

Distribution Change

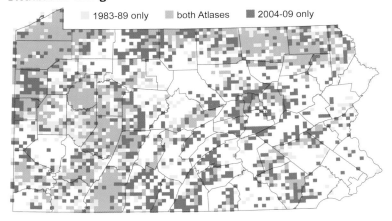

Density

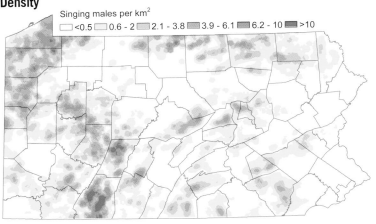

Number of Blocks

	first Atlas 1983–89	second Atlas 2004–09	Change %
Possible	629	1,119	78
Probable	792	690	−13
Confirmed	240	276	15
Total	1,661	2,085	26

Population estimate, males (95% CI):
145,000 (134,000–158,000)

Breeding Bird Survey Trend

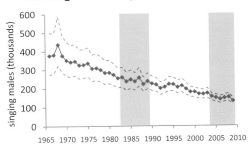

SAVANNAH SPARROW

Bob Wood and Peter Wood

Grasshopper Sparrow
Ammodramus savannarum

Aptly named, the Grasshopper Sparrow is an unobtrusive bird whose presence is typically betrayed by its buzzing insect-like song. It nests in grasslands throughout the United States and southern Canada, absent only from the Rocky Mountains and desert Southwest. It winters in the southern United States and south into Central America.

No doubt scarce and localized in Pennsylvania prior to European settlement, the Grasshopper Sparrow quickly adapted to hayfields and pasture during the nineteenth century and was described as common by the 1880s (Warren 1890). Over the past 100 years, its range has contracted from reforested areas in the northern half of the state and, by the time of the first Atlas, it was scarce in northern tier counties (Santner 1992g). By the second Atlas period, range contractions in the north had progressed further, most notably in the former stronghold of Bradford County, where the species was lost from 65 percent of blocks. However, there was little change overall in block occupancy between the two atlas periods, with modest gains in the Lower Susquehanna Valley compensating for losses farther north.

Despite the relative stability in range, Breeding Bird Survey data show that numbers of this species have been in steep decline since the 1960s (Sauer et al. 2011). Around 70 percent of the population was lost between the two atlas periods, with declines measured at 5.5 percent per year. Such declines have been noted throughout the species' breeding range (Sauer et al. 2011) and are widely attributed to both grassland loss and changes in grassland management (Vickery 1996).

Reclaimed surface mine grasslands support high densities of Grasshopper Sparrows in Pennsylvania, averaging 28 singing males per km^2 in the nine principal surface-mining counties of western Pennsylvania (Mattice et al. 2005). Subsequent analysis suggests that these estimates could be very conservative (Diefenbach et al. 2007), and while such densities are lower than have been reported in remnant native prairies (Ribic et al. 2008), second Atlas point count data suggest that mineland densities are around 10 times the average density on farmland (appendix D). However, because farmland is much more extensive, as much as 90 percent of the state's Grasshopper Sparrow population is found in farmland, and hence the bulk of the population has been subjected to rapid change in agricultural practices in recent decades.

In the first Atlas, most records of fledglings or adults carrying food were after 4 July (Santner 1992g), which contrasts with records in the second Atlas, when half of such breeding confirmations were before the end of June. Whether this signifies a genuine shift in breeding phenology, or merely a change in seasonal emphasis of volunteer effort, is difficult to ascertain. However, given the importance of the timing of nesting in cropped fields, the Grasshopper Sparrow's future in Pennsylvania will depend either on the species' ability to adapt to constantly changing agricultural practices, or through efforts to modify agricultural practices to benefit wildlife.

The State Wildlife Action Plan lists the Grasshopper Sparrow as a species of Maintenance Concern (PGC-PFBC 2005). In western Pennsylvania, where there are source populations on reclaimed surface mines, the status of this species is relatively secure in the short and medium terms. Currently, however, very little reclaimed surface mine grassland is protected (Stauffer et al. 2010); in the long term, unchecked natural or human-aided successional change will result in the disappearance of these grasslands.

In agricultural areas in the southern and eastern parts of the state, the future of the Grasshopper Sparrow will depend on concerted farmland conservation efforts. More sympathetic management of hayfields would likely benefit this species, but this can only be achieved by compensating farmers who are willing to sacrifice some hay yield for wildlife benefits. The creation of more than 40,000 ha (~100,000 ac) of conservation grassland through the Conservation Reserve Enhancement Program does not appear to have halted the decline of this species so far (Wilson and Brittingham 2012), and the future of this species in those areas does not look bright.

ANDREW M. WILSON

Distribution

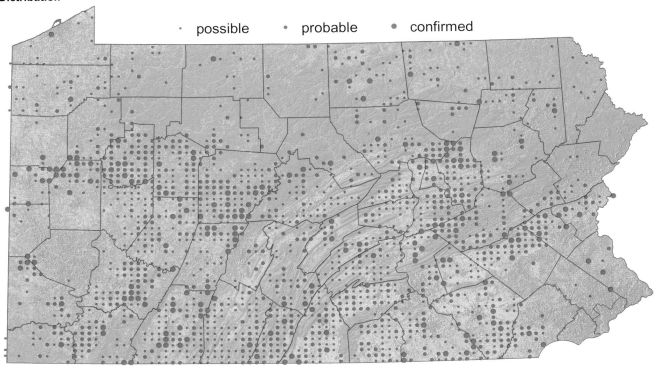

Distribution Change

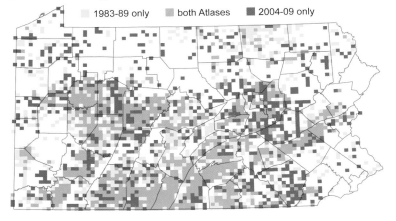

Density

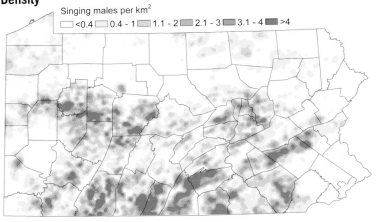

Number of Blocks

	first Atlas	second Atlas	Change
	1983–89	2004–09	%
Possible	655	891	36
Probable	769	599	−22
Confirmed	205	184	−10
Total	1,629	1,674	3

Population estimate, males (95% CI):
92,000 (83,000–102,000)

Breeding Bird Survey Trend

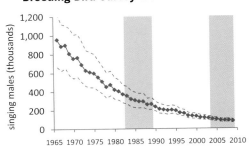

Geoff Malosh

Henslow's Sparrow
Ammodramus henslowii

A native of the tallgrass prairie, the Henslow's Sparrow expanded its range into agricultural grasslands during the nineteenth century, but it has subsequently proved less adaptable to modern agricultural practices than some other grassland obligates. As a result, it has declined throughout its range; it is extirpated from six New England and mid-Atlantic states and endangered in New York, New Jersey, and Maryland (Cooper 2007). It winters in the southeastern United States. Its plumage is drab, typical of the *Ammodramus* sparrows, which, coupled with its pitifully weak *tsi-lick* song, means that it is easily overlooked, even by seasoned birders.

Due to its inconspicuous nature, the Henlow's Sparrow's status in Pennsylvania was something of an enigma for much of the twentieth century. It appears to have colonized the state later than other grassland birds, with the first confirmed nesting in 1913 (McWilliams and Brauning 2000). Between then and the first Atlas, breeding was reported in many counties, but it was considered a scarce bird; "colonies" often appeared to be ephemeral (Reid 1992e). The first Atlas revealed that this species was more widespread than previously thought, with records in 364 blocks, largely in the western third of the state, but with a notable presence in hayfields in the state's northeastern corner.

Most agricultural grasslands do not have a thick layer of dead litter and perennial stalks, which have been identified as a key nesting-habitat requirement for Henslow's Sparrow (Herkert et al. 2002). In Pennsylvania, suitable habitat is found primarily on reclaimed surface mine grasslands, which are largely undisturbed, allowing a buildup of leaf litter. Surveys in a nine-county region of western Pennsylvania in 2001 resulted in an estimate of 4,880 singing males on reclaimed surface mines (Mattice et al. 2005). This corresponds well with the second Atlas statewide population estimate of 7,000 singing males, of which 80 percent was estimated to be within the same nine-county region. This population represents a significant contribution to the global population, which was recently estimated to be as few as 80,000 individuals (Rich et al. 2005).

Despite evidence that populations in Pennsylvania and elsewhere (Cooper 2007) are larger than previously thought, the Breeding Bird Survey shows rangewide declines in abundance since the 1960s (Sauer et al. 2011). Between the first and second Atlases, this species showed a range contraction of 37 percent in Pennsylvania and virtual abandonment of the northeastern area of the state and former strongholds in western counties, such as Crawford. However, there has also been consolidation within the core range and local expansions in Clearfield and Bedford counties. Almost half of occupied blocks and more than 60 percent of the population is now estimated to be in the five contiguous counties of Butler, Clarion, Clearfield, Jefferson, and Venango, the center of surface coal-mining operations in the state (chap. 3).

Around 80 percent of records submitted to the second Atlas were in blocks containing at least 10 ha (22 ac) of surface mine grassland, and many of these records were of multiple birds in "colonies" reported by observers to have been present for several years. Most other records were from abandoned grassland, but it is likely that the persistence of Henslow's Sparrows in Pennsylvania is due almost entirely to populations on surface mine grasslands, which may act as population sources for surrounding areas.

The State Wildlife Action Plan lists the Henslow's Sparrow as a species of High Concern (PGC-PFBC 2005). Conservation of grassland through the Conservation Reserve Program (CRP) is important for this species in the Midwest (Herkert 2007). Unfortunately, most conservation grassland in Pennsylvania is not within the species' range; this spatial mismatch between potential habitat and extant populations appears to have limited the potential of CRP for this species in Pennsylvania (Wilson 2010). Therefore, conservation efforts should focus on maintaining existing habitat on reclaimed surface mine grasslands by managing large areas of grassland in perpetuity, which will require periodic management to prevent encroachment by woody vegetation (Stauffer et al. 2010).

ANDREW M. WILSON

Distribution

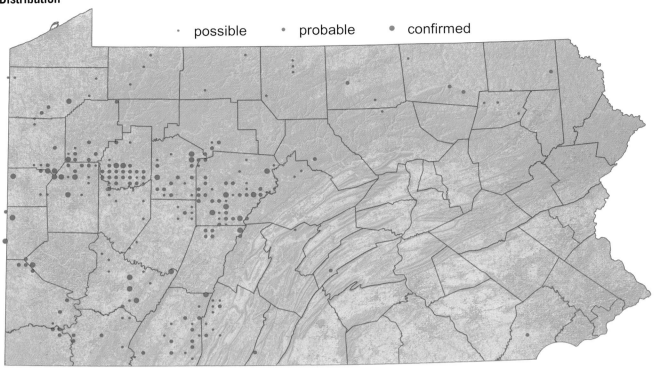

Distribution Change

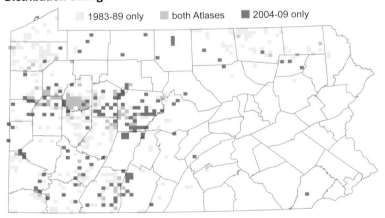

Number of Blocks

	first Atlas 1983–89	second Atlas 2004–09	Change %
Possible	134	95	−29
Probable	178	105	−41
Confirmed	52	29	−44
Total	364	229	−37

Population estimate, males (95% CI):
7,000 (5,400–9,200)

Density

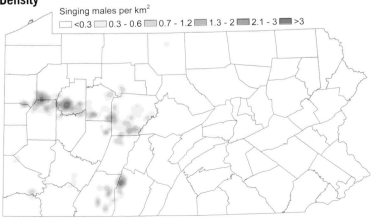

Breeding Bird Survey Trend

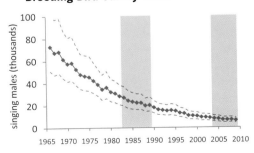

Bob Wood and Peter Wood

Song Sparrow
Melospiza melodia

The familiar Song Sparrow, with its frequently heard song, is one of the most taxonomically diverse and widespread songbirds in North America. There are 24 diagnosable subspecies (Arcese et al. 2002) found across the southern half of Canada and much of the United States, except for the southern Great Plains and Gulf Coast states. Hence, Pennsylvania is in the center of this species' range in eastern North America, and it is one of the state's most widely distributed and common breeding birds. The Song Sparrow is a habitat generalist, nesting in a wide variety of brushy and edge habitats such as overgrown fields, fencerows, second-growth woodlots, wetlands, and suburban gardens.

Historically, the Song Sparrow always has been widely distributed and considered a common-to-abundant breeder in Pennsylvania (McWilliams and Brauning 2000), absent only from dense forests with a closed canopy (Dwight 1892; Burleigh 1932). During the first Atlas, it was the third most widely distributed species, recorded in 97 percent of Pennsylvania's blocks, and found in at least 90 percent of the blocks in every major physiographic region of the state. The few blocks in which the species was missing were located in heavily wooded sections of Pennsylvania (Master 1992h). During the second Atlas, the Song Sparrow had the fourth highest block occupancy of all species, having been recorded in 99 percent of atlas blocks.

The filling in of blocks in which Song Sparrows were missed in the first Atlas probably reflects the increase in survey effort. Analysis of point count data demonstrates that this species' common name is most apt—more than many other songbirds, the Song Sparrow sings consistently throughout the morning and throughout the spring and early summer seasons (chap. 5), making it very easy to find and ideally suited to our methodology. The broad range in breeding confirmations from May through July corresponds to production of two broods per year, making this species even easier to detect.

The population density map indicates that the Song Sparrow is among the most common and uniformly distributed of Pennsylvania's breeding birds, found at more than 20 males per km^2 over much of the state and uniformly found regularly in all habitat types, but numbers are lowest within large contiguous forests (appendix D). The inverse relationship between Song Sparrow abundance and forest cover and elevation reflects the interrelatedness of these two important habitat features (appendix D). No less than 37,797 Song Sparrows were detected on point count surveys, more than one per stop and more than any other species. From these data, the population of Song Sparrows in Pennsylvania was estimated to be 3 million singing males, making it the most numerous breeding species in the state. However, Breeding Bird Survey (BBS) counts in the state decreased by 13 percent between atlas periods (Sauer et al. 2011). Even though BBS data suggest the Song Sparrow is declining in Pennsylvania, second Atlas results indicate that the species is still abundant and ubiquitous, and so it is unlikely that distributional changes would be evident unless much greater declines occurred.

One factor that may have a long-term impact on the species is Brown-headed Cowbird nest parasitism. The negative impact of cowbirds on Song Sparrow populations has been documented since the landmark work of Nice (1937), and recent studies (e.g., Saunders et al. 2003; Wilson and Arcese 2006) have examined the relationships between cowbirds and Song Sparrow reproductive success. Nevertheless, even if cowbirds affect the Song Sparrow negatively, some human activities that create clear-cuts and fragmented forests seem to have a positive effect (e.g., Shulte and Niemi 1998; Keller and Yahner 2007) and could counter losses due to brood parasitism.

ROY A. ICKES

Distribution

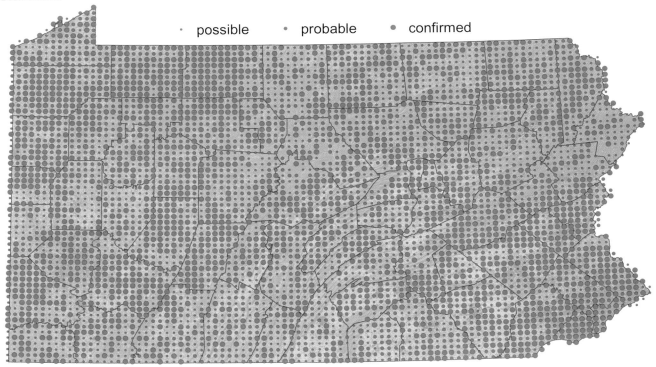

Distribution Change

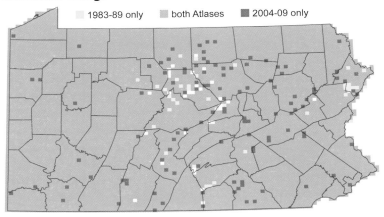

Density

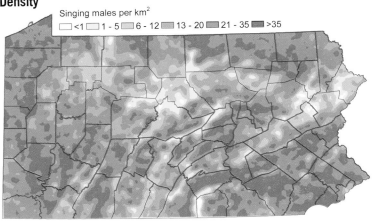

Number of Blocks

	first Atlas 1983–89	second Atlas 2004–09	Change %
Possible	747	1,142	53
Probable	1,637	1,627	−1
Confirmed	2,405	2,098	−13
Total	4,789	4,867	2

Population estimate, males (95% CI):
2,990,000 (2,910,000–3,060,000)

Breeding Bird Survey Trend

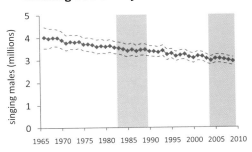

SONG SPARROW

Bob Wood and Peter Wood

Swamp Sparrow
Melospiza georgiana

The Swamp Sparrow is the only one of Pennsylvania's three red-capped sparrow species present seasonally that is characteristic of wetland habitats. These habitats include emergent wetlands, shrub-scrub swamps, beaver ponds, and the borders of wet meadows (Greenberg 1988; Prosser 1998). The Swamp Sparrow is a persistent singer, often vocalizing both day and night during the breeding season, but it is often difficult to see amid the tangle of emergent stems and dense shrubs so typical of its wetland habitat. Periodically, these sparrows emerge from the jumble of vegetation to sing their liquid, trilling song high atop a swaying cattail. The Swamp Sparrow breeds from the northeastern United States and Great Lakes states into the southern half of Canada. Pennsylvania is near the southern edge of its breeding range, which extends west across the mid-Atlantic states to Iowa, where this species is a permanent resident (Mowbray 1997). The nominate subspecies is found in most of Pennsylvania, but the disjunct population of the Delaware River marshes are of the darker "Coastal Plain" subspecies (*M. g. nigrescens*).

During the second Atlas, blocks with Swamp Sparrow were concentrated in areas where emergent wetlands are most prevalent, such as the glaciated sections of northeastern Pennsylvania, and the Northwestern Glaciated Plateau, where the Wisconsin glaciation created numerous ponds, lakes, and depressions now occupied by wetlands. Block occupancy in these physiographic sections ranged from 56 to 73 percent. Other areas with high occupation rates include McKean, Tioga, and Somerset counties, where emergent wetlands are also plentiful. A swath of the Pittsburgh Low Plateau Section, extending southeastward from Clarion to Indiana counties, held many blocks with Swamp Sparrows, despite having a lower density of wetlands (chap. 3). Away from these areas, the species was much more localized, found in only 7 percent of blocks in the Piedmont and 11 percent in the Ridge and Valley Province.

The overall distribution pattern across the state is similar to that observed during the first Atlas (Leberman 1992l), but block occupancy has increased by a highly significant 32 percent. This compares with a modest 6 percent increase in occupancy between atlases in New York (Osborne 2008) and a 10 percent decline in Maryland (Ellison 2010zr). Closer inspection of the change map shows that declines in the southeastern corner of the state, exemplified by a 24 percent drop in block occupancy in the Piedmont, have been more than offset by occupancy increases across the Appalachian Plateaus. Southeastern Pennsylvania has experienced the most pervasive, intense development of all state regions since the first Atlas (chap. 3). Thus, the general development pressure and accompanying loss of small wetlands may be responsible for population declines in this portion of the state. The expansions in the northern tier may be partly due to increased effort or awareness of this species, but long-term statewide Breeding Bird Survey counts for Pennsylvania have shown a steady upward trend (Sauer et al. 2011).

Atlas point count data reveal an estimated Swamp Sparrow population of 43,500 singing males in Pennsylvania, with Crawford, Erie, and Tioga counties supporting important concentrations. High densities were also found in Delaware and Philadelphia counties, where fresh water tidal wetlands along the Delaware River have long been noted as important for this species (Harlow 1918; Miller 1933).

The Swamp Sparrow continues to thrive in the higher elevations of Pennsylvania, perhaps taking advantage of increased beaver activity and the prevalence of roadside ditches, retention ponds, and reservoirs (Brewer et al. 1991) lined with emergent vegetation, the favored habitat of this species. Populations of the Coastal Plain subspecies in the state's southeastern corner may be more vulnerable to future change, especially given that this subspecies is in decline along the Atlantic seaboard (Osborne 2008; Ellison 2010zr; Sauer et al. 2011).

TERRY L. MASTER

Distribution

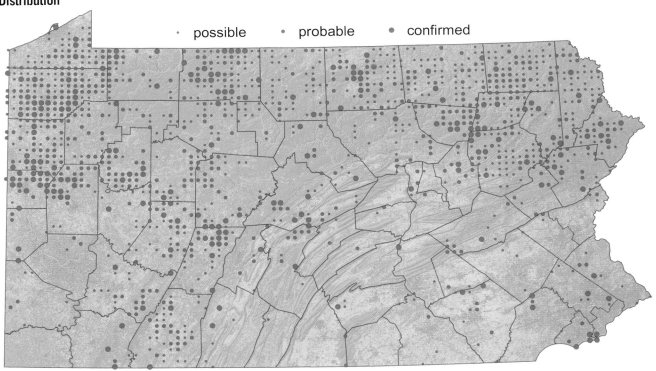

Distribution Change

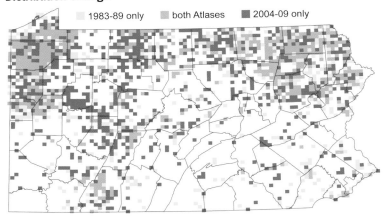

Density

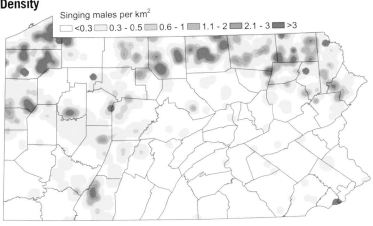

Number of Blocks

	first Atlas 1983–89	second Atlas 2004–09	Change %
Possible	414	632	53
Probable	421	500	19
Confirmed	173	200	16
Total	1,008	1,332	32

Population estimate, males (95% CI):
43,500 (39,000–48,000)

Breeding Bird Survey Trend

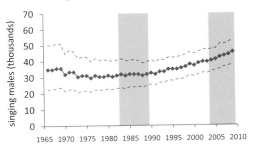

SWAMP SPARROW 423

Bob Wood and Peter Wood

White-throated Sparrow
Zonotrichia albicollis

The pure, distinctive whistles of the White-throated Sparrow's song are familiar to observers during the bird's northbound migration but infrequently encountered on its Pennsylvania breeding grounds. This sparrow was detected in fewer than 3 percent of blocks during the second Atlas, concentrated in forested wetlands and shrublands of Pennsylvania's northern counties.

The White-throated Sparrow's range, here at its southern edge, appears to be contracting. It breeds throughout the boreal and northern hardwood forests of Canada and the northern United States. It is one of the most common birds in many northern forested regions, to such an extent that its song is iconic of those locations (Rich et al. 2004; Falls and Kopachena 2010). Its breeding distribution extends south through New England and New York, where, during the state's second atlas it was found in most blocks in the Adirondacks and many blocks in the Catskills and the Appalachian Plateaus to the south (Peterson 2008b). Though isolated breeding records of the White-throated Sparrow have been reported from West Virginia, and non-breeding birds have been observed summering in the southern Appalachians, the southern extremity of the White-throated Sparrow's breeding range is in northern Pennsylvania (Falls and Kopachena 2010). This is unlike numerous other predominantly boreal birds, whose breeding distributions extend south along the spine of the Appalachians.

Breeding habitat for the White-throated Sparrow is primarily forest openings with dense, shrubby cover, often at the edge of a lake, pond, bog, or beaver meadow (Gross 1992h; Falls and Kopachena 2010). During the second Atlas, this species was found in many of the same blocks as it occurred in the first Atlas. The clusters of occupied blocks during both atlases are notable in northern Susquehanna and Wayne counties; at the juncture of Lackawanna, Luzerne, Carbon, and Monroe counties; on North Mountain in eastern Sullivan and southeastern Wyoming counties; along the border between Wyoming and Bradford counties; and in and near the Allegheny National Forest in McKean County.

The second Atlas found White-throated Sparrow in 33 percent fewer blocks than in the first Atlas. This is consistent with findings from New York state, where a 14 percent overall decline in block records between atlases was especially concentrated near the southern border with Pennsylvania (Peterson 2008b). Clusters of Pennsylvania blocks in which White-throated Sparrows were found during the first Atlas but not during the second were in Pike County, including at the juncture of Pike and Wayne counties and several clusters in Potter County. Second Atlas results suggest that the White-throated Sparrow may be contracting into a few core areas in Pennsylvania. There was evidence of contraction in most areas, with the exception of North Mountain of Sullivan County, where there was an increase in blocks with White-throated Sparrows; the area now stands out as an important stronghold for this species.

The majority of isolated block detections of White-throated Sparrow south of the core areas should be viewed with caution. As Gross (1992h) pointed out, the White-throated Sparrow tends to linger during its spring migration, or may even summer without breeding, and many of the isolated detections south of the main breeding areas may have been of birds exhibiting this behavior. Such points were omitted from the first Atlas's final analysis, but similar second Atlas observations were not screened in the same way. However, one of the southernmost observations, in Mifflin County, was confirmed breeding.

The long-term outlook for the White-throated Sparrow in Pennsylvania is uncertain. While the species is maintaining its numbers in a few core areas, the apparent population contraction along the southern edge of the species' breeding range suggests that this species may face an uncertain future in Pennsylvania. A warming climate (Matthews et al. 2010) and forest maturation are factors suggested as being related to New York state's decline in White-throated Sparrows (Peterson 2008b).

NICHOLAS C. BOLGIANO

Distribution

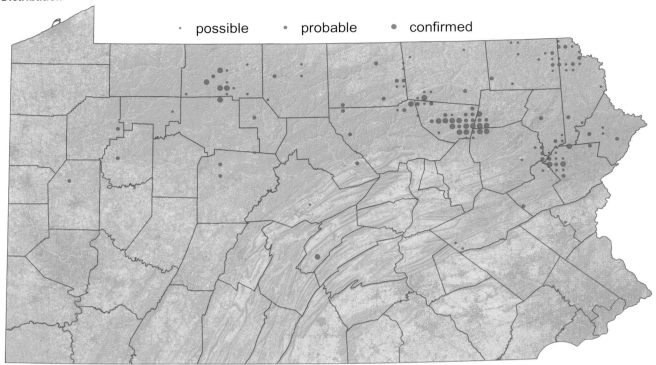

Distribution Change

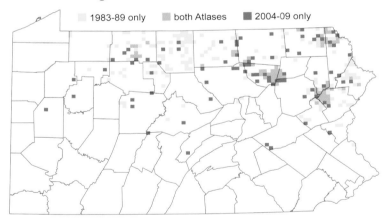

Number of Blocks

	first Atlas 1983–89	second Atlas 2004–09	Change %
Possible	107	65	−39
Probable	74	52	−30
Confirmed	32	25	−22
Total	213	142	−33

Population estimate, males (95% CI):
1,900 (1,400–2,500)

Bob Wood and Peter Wood

Dark-eyed Junco
Junco hyemalis

In many northern Pennsylvania locations, the Dark-eyed Junco is relatively easy to find by its trilling song, its *smack-smack* alarm call, the flash of its white tail feathers, or the activity of a family group. Second Atlas volunteers found it easy to confirm breeding juncos by the conspicuous presence of fledglings, which are often found alongside forest rights-of-way.

The Dark-eyed Junco species complex consists of 15 subspecies in five groups; some groups were recognized as distinct species before 1973. Juncos breed throughout the forested parts of Canada and Alaska, the mountains of the western United States, much of New England and New York, and south from Pennsylvania in the higher elevations of the Appalachians to northern Georgia (Nolan et al. 2002). The Dark-eyed Junco is among the most numerous of North American birds; one estimate of the continental population size is 260 million birds (Rich et al. 2004).

Most of Pennsylvania's breeding juncos are of the widespread slate-colored or *hyemalis* group of northern and eastern North America (Nolan et al. 2002). Many Pennsylvania breeders may be intermediate between the widespread nominate subspecies *J. h. hyemalis* and *J. h. carolinensis* of the southern Appalachians, while southern Pennsylvania's breeding juncos more closely resemble *carolinensis* (Mulvihill 1992k). Compared with *hyemalis*, *carolinensis* tends to be larger with a paler hood, with a bluish rather than pinkish bill, and slightly whiter in the tail (Nolan et al. 2002).

The Dark-eyed Junco is a regular part of Pennsylvania's forests above 400 m (1,300 ft) in elevation (appendix D). It is found in a diverse array of habitats: hemlock ravines, stream sides, the edges of conifer or mountain laurel patches, and in deciduous forest where a few conifers were present. The highest densities are found in coniferous and mixed forests; however, due to their much larger extent, deciduous forests support a significant proportion of the population (appendix D). The statewide population was estimated to be 380,000 singing males.

Like some other northern birds, the Dark-eyed Junco has expanded its Pennsylvania range in recent decades, the 1,831 occupied blocks in the second Atlas representing a statistically significant 37 percent increase over the first Atlas. The second Atlas found that the Dark-eyed Junco's core range now spans most of the northern third of Pennsylvania. Farther south, it was most common on the higher elevations of the Alleghenies, including Laurel Hill and parts of the Allegheny Front, and in Bald Eagle and Rothrock state forests in the center of the state. Despite the overall range expansion, the Dark-eyed Junco was absent from parts of Schuylkill and Carbon counties in the eastern area of the state, as well as Bedford County in the southern area, where it was found during the first Atlas but where Hemlock Woolly Adelgids have killed many of the Eastern Hemlocks (Fergus 2002).

Coupled with the range expansion noted between atlases, the Breeding Bird Survey (BBS) has documented a mean population increase of 1.8 percent per year since 1966 (Sauer et al. 2011), although the rate of increase has slowed since the early 1990s. In New York state, there was a 23 percent range expansion between atlases (Smith 2008i), but counts on BBS routes there have long been in shallow decline. This suggests that within established core areas, Dark-eyed Junco densities may be stable or decreasing, even while range expansions occurred along the edges of the core areas.

The Dark-eyed Junco was commonly found in association with Eastern Hemlock. The range retractions along the southern edge of its Pennsylvania range are probably a preview of what will happen as the adelgids spread north. However, its diverse habitat choices will likely protect the species to some extent, but the loss of the coniferous component in many forests will likely reduce its density, which could be sufficient to cause localized extirpations.

NICHOLAS C. BOLGIANO

Distribution

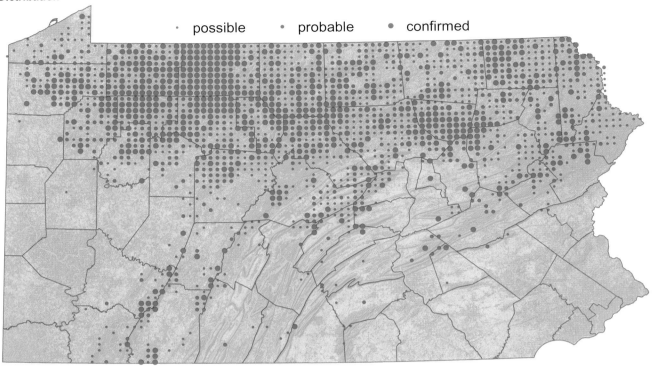

Distribution Change

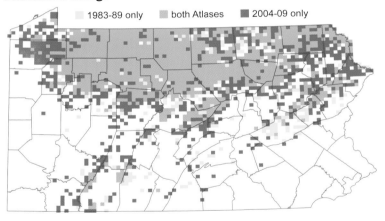

Density

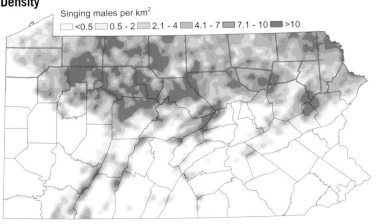

Number of Blocks

	first Atlas 1983–89	second Atlas 2004–09	Change %
Possible	417	700	68
Probable	422	506	20
Confirmed	495	625	26
Total	1,334	1,831	37

Population estimate, males (95% CI):
380,000 (340,000–420,000)

Breeding Bird Survey Trend

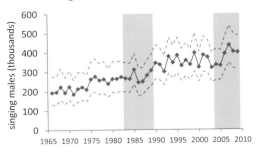

DARK-EYED JUNCO 427

Gerard Dewaghe

Summer Tanager
Piranga rubra

Inhabiting the canopy of dry upland forests, particularly slightly open oak woodlands, the vivid red adult male Summer Tanager is more often heard than seen. It breeds from California across the Southeast, north to Iowa, and east to New Jersey; southern Pennsylvania is the extreme northern edge of its breeding range. Apparently, the Summer Tanager always has been a rare breeder in the state, and it is listed as a High Level Concern species in Pennsylvania's Wildlife Action Plan (PGC-PFBC 2005). The reason for the state's conservation concern is that Partners in Flight lists the Summer Tanager as a IIA priority species in Pennsylvania, due to its very localized breeding distribution in Pennsylvania (Rich et al. 2004).

Historical records for the Summer Tanager's Pennsylvania distribution are minimal. As of the mid-1960s, only nine specimens had been collected and approximately 20 sight records were considered reliable (Poole 1964). All of the specimens were collected in four southeastern counties, and all of the sightings, except three from two western Pennsylvania counties (Todd 1940), were from the same region. During the 1970s and 1980s, Summer Tanagers were reported sporadically in various areas of the state (Morris et al. 1984; Stull et al. 1985).

During the first Atlas, the Summer Tanager was reported in 10 Pennsylvania counties (47 blocks), with the majority of the records from the extreme southwestern corner of the state, including all four confirmed breeding records. Six scattered records (four possible and two probable) came from as many counties east of the Allegheny Mountains (Ickes 1992o).

The second Atlas demonstrated that the Summer Tanager has declined precipitously from its former stronghold in Pennsylvania. It had been a summer resident in Greene County since at least 1970 (Bell 1975) and was found there in 33 blocks during the first Atlas; in the second Atlas there were no sightings from Washington or Greene counties, and only one probable breeding record (a pair) was found in the western half of the state (Allegheny County). The only confirmed breeding record was a female observed collecting nest material in Chester County where a male had been on territory the previous day. The two possible breeding sightings, also from the eastern half of Pennsylvania, were of a juvenile male in Cumberland County and a vocalizing individual in Northampton County.

Apparently, the presence of Summer Tanager in Greene County was short-lived. This decline corresponds with similar patterns in neighboring states. For example, West Virginia nesting records were reported from an area adjacent to part of Greene County, Pennsylvania, before the first Atlas (Hall 1983). However, during the West Virginia atlas (1984–1989), only one Summer Tanager sighting was recorded in the blocks adjacent to Pennsylvania's extreme southwestern corner (Buckelew and Hall 1994). Similarly, although Maryland's second atlas reported a 15 percent increase in blocks with Summer Tanager between atlas periods, very few were in counties adjacent to Pennsylvania (Coskren 2010e). The persistence of the species at the extreme edge of its breeding range may depend in part on the presence of nearby source populations (Goguen 2010a). The role played by the Brown-headed Cowbird on the Summer Tanager's population dynamics is unclear, but high rates of parasitism have been reported (Robinson 1992), and cowbird parasitism may have been particularly significant in the fragmented landscapes of southern Pennsylvania.

Since the Summer Tanager tends to wander northward during spring migration without remaining to nest (Hall 1983), many historical and first Atlas records may have been transient overshoots; the few confirmed breeding records in the southwest during the first Atlas may indicate only a small, temporary northward extension into Pennsylvania. Although the Summer Tanager's general status in several neighboring states has changed minimally between atlases (Coskren 2010e; OBBA 2011), second Atlas data suggest that the Summer Tanager's outlook as a permanent breeder in the state is tenuous and perhaps largely dependent on population changes elsewhere.

ROY A. ICKES

Distribution

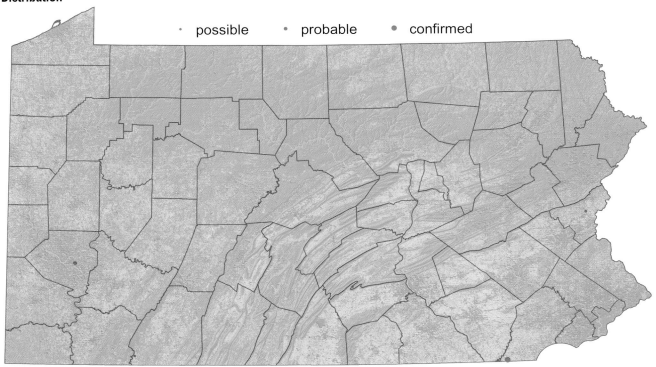

Distribution Change

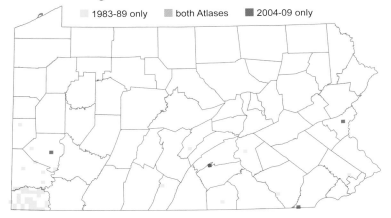

Number of Blocks

	first Atlas 1983–89	second Atlas 2004–09	Change %
Possible	7	2	−71
Probable	36	1	−97
Confirmed	4	1	−75
Total	47	4	−91

Rob Curtis/VIREO

Scarlet Tanager
Piranga olivacea

There are few birds like the Scarlet Tanager, which will make you hold your breath and say "Wow!" every time you see one. The jet-black wings of older males provide a stunning contrast, making the red appear even brighter. When a male is courting a female, he perches beneath his mate and droops his black wings while hunching his back to show off his colors. The irony is that the nesting behavior of Scarlet Tanagers is difficult to observe, because this is a canopy species, although males are very attentive to their mates. While off their nests, females often give a soft, rising *pew, pew, pew* sound overhead, which is a magnet for the male, who, upon hearing it, promptly flies over to perch near his mate and may feed her (Klatt et al. 2008). If you hear a male tanager whisper-singing (a very soft version of his raspy, robin-like song), then he is probably near his nest high in the canopy, with food in his beak, and about to feed his incubating mate.

The Scarlet Tanager's nest is somewhat bulky and usually positioned midway along a branch or, sometimes, in a grapevine. It is difficult to find because of its height above the ground and the secretive behavior of the species near its nest. Of the 762 cases of confirmed breeding evidence in the second Atlas, nests were found in only 6 percent of cases. Nesting activity is common from late May to early July, but it diminishes rapidly thereafter, since this species is single-brooded.

The Scarlet Tanager is widespread and common in Pennsylvania, found in 92 percent of blocks in the second Atlas. This total was 6 percent higher than in the first Atlas, but much of this increase can be attributed to increased survey effort. The Scarlet Tanager is a forest obligate that appears equally at home in deciduous, mixed, and coniferous forests in Pennsylvania (appendix D). Although it is known to be more abundant in large forest patches (Mowbray 1999), atlas point count data (appendix D) suggest that it is not as area-sensitive as some other forest obligates (Robbins et al. 1989a; Roberts and Norment 1999). Male tanagers readily occupy forest fragments, but may have difficulty attracting mates; and fledging success is higher in larger patches (Roberts and Norment 1999). Radio-tracking in northwestern Pennsylvania showed that unmated male Scarlet Tanagers pursue one of two strategies: (1) commute between different forest fragments on a daily basis to increase the chances of finding a female or (2) stay in a single fragment and sing all day (Fraser and Stutchbury 2004). The species is absent only from the least forested parts of the state, such as the agriculturally dominated Great Valley Section and Lancaster County, as well as urban areas such as Philadelphia and Allentown/Bethlehem. Elsewhere, densities of more than five singing males/km^2 are typical wherever there is woodland and forest.

Breeding Bird Survey (BBS) counts in Pennsylvania show that populations of the Scarlet Tanager have been stable since the 1970s (Sauer et al. 2011). In contrast, significant declines in BBS counts have been noted in neighboring Maryland and New York (Sauer et al. 2011), but declines were not of sufficient magnitude to register changes in atlas distributions there (Ellison 2010zs; Lowe and Hames 2008b). The mean count on BBS routes in Pennsylvania is second only to that of West Virginia (Sauer et al. 2011), due to which it has been estimated that Pennsylvania supports as much as 15 percent of the global Scarlet Tanager population (Rich et al. 2004). Thus, Pennsylvania supports a higher proportion of the species' population than that of any other bird species (Stoleson and Larkin 2010); hence, it is listed as a Responsibility Species in the Pennsylvania Wildlife Action Plan (PCG-PFBC 2005). Although it is believed to face an array of threats ranging from forest fragmentation and changes in forest structure to collisions with towers and buildings (Goguen 2010b), the stability of this species' population in Pennsylvania suggests that it is resilient.

BRIDGET J. M. STUTCHBURY

Distribution

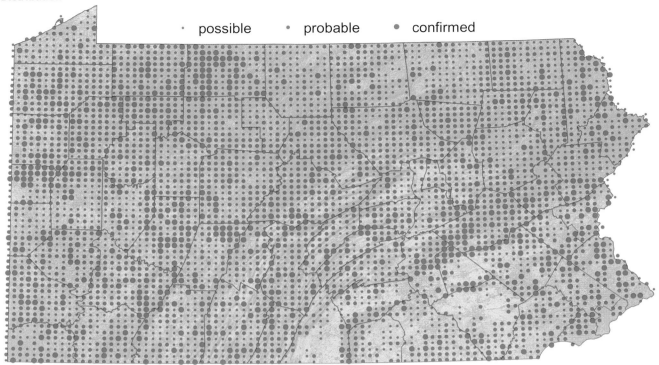

· possible   · probable   · confirmed

Distribution Change

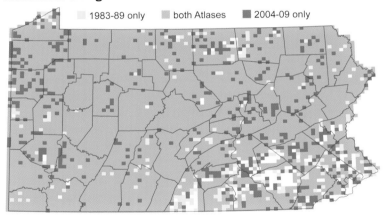

1983–89 only   both Atlases   2004-09 only

Density

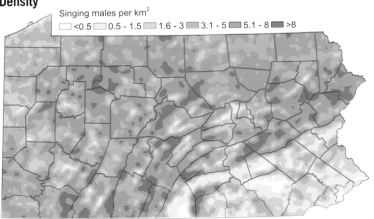

Singing males per km²
<0.5   0.5 - 1.5   1.6 - 3   3.1 - 5   5.1 - 8   >8

Number of Blocks

	first Atlas 1983–89	second Atlas 2004–09	Change %
Possible	1,370	1,812	32
Probable	2,229	2,067	−7
Confirmed	677	664	−2
Total	4,276	4,543	6

Population estimate, males (95% CI):
575,000 (558,000–593,000)

Breeding Bird Survey Trend

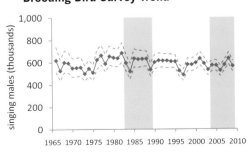

Geoff Malosh

Northern Cardinal
Cardinalis cardinalis

With brilliant red plumage and frequent visits to backyard bird feeders, the Northern Cardinal is one of our most common and beloved birds, and it has doubtless sparked a lifelong interest in birds for many. The Northern Cardinal is so treasured that it is recognized as the most frequently selected state bird, officially recognized in this capacity for seven states, and is used as a mascot for both collegiate and professional sports teams. With its bright plumage, loud song, and long breeding season, the Northern Cardinal was one of the easiest species for atlas volunteers to find and confirm breeding. Nest building was observed as early as 30 March, nests with eggs were found from 4 April to 27 September, and dependent young were noted well into October.

Though frequently associated with backyard bird feeders throughout the year, the Northern Cardinal is more naturally found in shrubby edge habitats, including patches of recently logged forests, old fields containing shrubs and young trees, hedgerows along agricultural fields, and within suburban communities that are well landscaped. Within these habitats, it is found throughout the United States east of the Rocky Mountains as well as much of Arizona and Mexico (Halkin and Linville 1999). Potentially due, in part, to such factors as moderate temperatures and increasing numbers of bird feeders, Northern Cardinals have expanded their range northward since the early nineteenth century and are now found as far north as southeastern Canada (Halkin and Linville 1999). They have also expanded their population in Pennsylvania over the past century (Schutsky 1992e), and were primarily observed in southern Pennsylvania in the late nineteenth and early twentieth centuries (Poole 1964). Throughout the twentieth century, they expanded their range northward, with evidence suggesting that bird feeding played an important role in their expansion (Todd 1940), and by 1960 they could be found in all but the most heavily forested, high-elevation regions of the Appalachian Plateaus (Poole 1964).

By the first Atlas period, the Northern Cardinal was a common resident throughout the state, with breeding evidence recorded in 90 percent of atlas blocks, with gaps predominantly in interior forested sections of the Appalachian Plateaus. A modest expansion into the remaining unoccupied areas by the second Atlas resulted in block occupancy of 95 percent overall, with virtually 100 percent occupancy outside the Appalachian Plateaus (appendix C). The region with the greatest expansion between atlas projects was the Deep Valleys Section of the Appalachian Plateaus, which had a 22 percent increase in the number of occupied second Atlas blocks. Breeding Bird Survey data show a steady 0.8-percent-per-year increase in 1966 to 2009 (Sauer et al. 2011), equating to a 15 percent population increase between atlases. Atlas point count data show that the Northern Cardinal is still less common at higher elevations (appendix D), but this may be due to habitat availability, and there is no longer evidence that this species' range is limited by elevation in Pennsylvania.

The Pennsylvania population of the Northern Cardinal was estimated to be 1.4 million singing males during the second Atlas period, placing it among the ten most numerous songbirds. The highest densities were observed in the heavily suburban counties of the Piedmont, where densities average more than 25 singing males/km^2 in several counties. They were also particularly numerous in the suburbs surrounding Pittsburgh, with Allegheny County having a similarly high density estimated at 23 singing males/km^2.

The Northern Cardinal remains one of our most common birds. With its ability to benefit from suburban development and supplemental feeding at winter bird feeders, it will likely continue to increase throughout the state. If population trends continue, it is likely that the last remaining gaps in the species' range in Pennsylvania will be occupied before too long.

DANIEL P. MUMMERT

Distribution

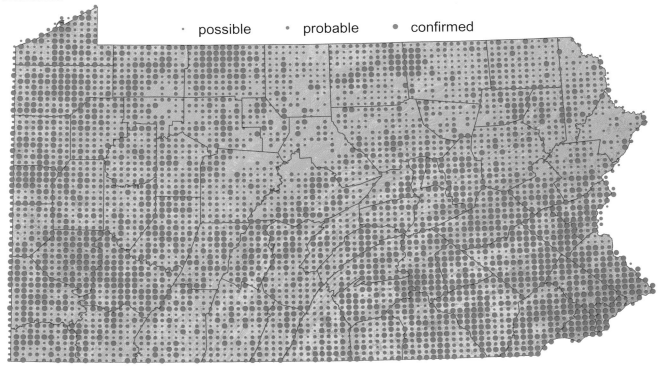

Distribution Change

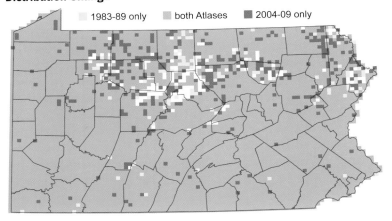

Density

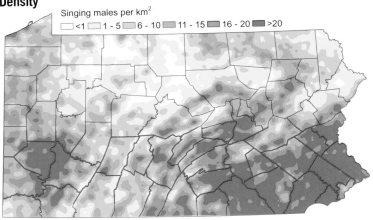

Number of Blocks

	first Atlas 1983–89	second Atlas 2004–09	Change %
Possible	823	1,119	36
Probable	1,881	1,992	6
Confirmed	1,738	1,567	−10
Total	4,442	4,678	5

Population estimate, males (95% CI):
1,400,000 (1,340,000–1,470,000)

Breeding Bird Survey Trend

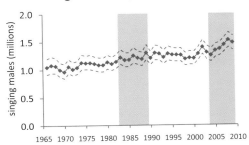

NORTHERN CARDINAL

Bob Wood and Peter Wood

Rose-breasted Grosbeak
Pheucticus ludovicianus

"Melodious" aptly describes the robin-like song of the Rose-breasted Grosbeak, which both sexes may sing. The male's rose-colored breast and otherwise contrasting black-and-white plumage impart striking impressions to incidental observers. Found in 72 percent of all blocks, the Rose-breasted Grosbeak is widely distributed across Pennsylvania, except at lower elevations.

The Rose-breasted Grosbeak is a summer resident across parts of southern Canada, throughout the upper Mississippi Valley and the northeastern United States, and south through the Appalachians to northeastern Georgia. It can be found in a wide variety of habitats, including riparian corridors and second-growth and regenerating woodlands and woodland edges, as well as parks and orchards. Historical accounts from Pennsylvania suggest that populations of this species have long been subject to local fluctuations, likely due to timbering, forest management, and successional change, which determine the prevalence of favored fruit-bearing shrubs (McWilliams and Brauning 2000). On balance, the Rose-breasted Grosbeak has probably benefited from man's management of forested lands, thriving where there is a sufficient amount of its preferred habitat (Leberman 1992m; Wyatt and Francis 2002).

The total of 3,557 blocks occupied by the Rose-breasted Grosbeak represents a statistically significant 13 percent increase from the first Atlas. Many of the second Atlas blocks where it was not found during the first Atlas occurred in western Pennsylvania, especially areas close to the Ohio border, and in the western and northern Ridge and Valley. Many of these blocks were along the edge of the species' range in the first Atlas, suggesting something of a range expansion in those areas. In contrast, there was a reduction in block occupancy in a number of counties in eastern Pennsylvania, especially in Monroe and Carbon, for reasons unknown.

Point count data indicate that, although the Rose-breasted Grosbeak is widely distributed, with an estimated statewide population of 210,000 singing males, it is abundant nowhere. The highest densities were primarily in the Appalachian Plateaus, often along riparian corridors, in forested landscapes with a sufficient amount of young trees, or where the landscape was a mix of forest patches and more open terrain, particularly near water. Few Rose-breasted Grosbeaks were found at elevations below 250 m (820 ft), and the species is largely absent from large urban centers and agricultural areas (appendix D). In southeastern Pennsylvania, it was generally restricted to wooded ridges and was scarce in the Piedmont, and perhaps more surprisingly, was scarce in much of the southern Ridge and Valley, where what superficially appears to be suitable habitat is commonplace.

Breeding Bird Survey (BBS) data collected in Pennsylvania indicate that fluctuations have occurred in the Rose-breasted Grosbeak population since the 1960s but with no overall trend (Sauer et al. 2011). Analysis of BBS data shows that counts were very similar in both atlas periods despite the modest range expansion, suggesting that populations in some parts of the range must have declined. Indeed, there may have been a small population decline since a peak in BBS counts during the 1990s. In New York state, there has been a more shallow downward trend in BBS counts and a decline in block occupancy between atlases (McGowan 2008zu). However, because it is generally found in relatively low densities and populations are known to fluctuate, the Rose-breasted Grosbeak could be an easy bird to miss in atlasing efforts. As a result, some of the apparent turnover of block occupancy between the two Pennsylvania atlases may be due to differences in observer effort.

Locally and statewide, Rose-breasted Grosbeak numbers will likely continue to fluctuate with the availability of the young woodlands that it prefers. While it benefits from forest edges and forest renewal from periodic timbering, the Rose-breasted Grosbeak's reproductive success declines in highly fragmented forests (Wyatt and Francis 2002). The long-term success of this beautiful bird in Pennsylvania will depend upon our stewardship of our forests.

NICHOLAS C. BOLGIANO

Distribution

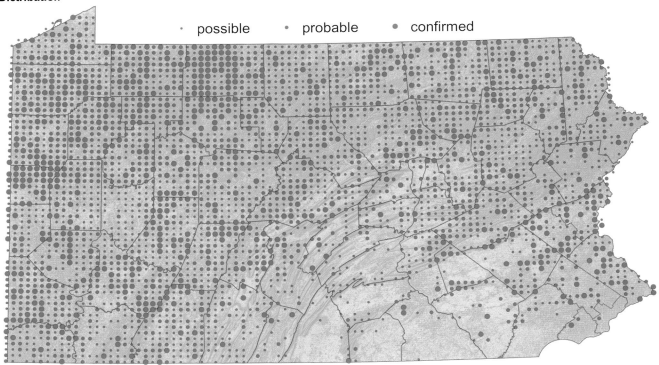

Distribution Change

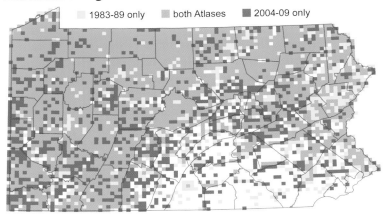

Density

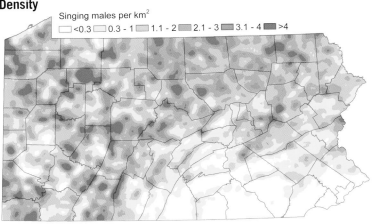

Number of Blocks

	first Atlas 1983–89	second Atlas 2004–09	Change %
Possible	1,197	1,664	39
Probable	1,276	1,219	−4
Confirmed	678	674	−1
Total	3,151	3,557	13

Population estimate, males (95% CI):
210,000 (198,000–220,000)

Breeding Bird Survey Trend

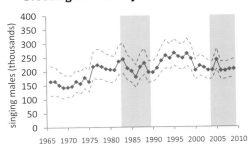

ROSE-BREASTED GROSBEAK   435

Bob Steele/VIREO

Blue Grosbeak
Passerina caerulea

Looking like an Indigo Bunting on steroids, the beautiful male Blue Grosbeak is recognized by its dark blue color, large size, rusty wing bars, and very stout bill. Its loud *pink* call often brings it to the observer's attention, and it has a very beautiful warbling song that is reminiscent of a Purple Finch. The female somewhat resembles a female Brown-headed Cowbird, but its larger bill and rusty wingbars quickly separate it from that species.

A neotropical migratory songbird, the Blue Grosbeak winters in Central America. Its breeding range is mainly in the American South, where it prefers weedy fields and grasslands with scattered shrubs and brushy areas and hedgerows, often bordering row crops (Schutsky 1992f; Lowther and Ingold 2011). Pennsylvania and New Jersey are the northern limit of the species' breeding range in eastern North America. In Pennsylvania, it is mainly found along the southern border in the central and southeastern parts of the state. As in the first Atlas, the majority of second Atlas records came from blocks in southern Fulton and Lancaster counties and adjoining areas.

Of 219 blocks in which the Blue Grosbeak was recorded in the second Atlas, breeding was confirmed in only 27. Since the first Atlas, the Blue Grosbeak has continued to expand its breeding range northward, with confirmed breeding now including Lehigh, Carbon, and Northampton counties. In these areas, grassland restoration sites, such as those on landfills, and weedy areas where housing projects have been abandoned provide favored habitats (pers. obs.). The expansion into these areas began shortly after the first Atlas, when a singing male was found on Blue Mountain in a barren area that was denuded by a zinc smelting operation and subsequently restored with grassland in 1991 (pers. obs.). Block totals in the Great Valley Section (the northern edge of the range during the first Atlas) more than doubled. Elsewhere, the Blue Grosbeak remains scarce and sporadic, but there were several isolated records in western and north-central Pennsylvania, a significant, if tenuous, range expansion. Interestingly, several of these northern outposts were at high elevations, in habitats including an area deforested due to Oak Leaf Rollers and reclaimed surface mine grassland (G. Grove, pers. comm.). The net result of the northward expansion was a near doubling in the number of occupied blocks since the first Atlas.

Despite the expansion of the Blue Grosbeak's range in Pennsylvania, it remains a scarce bird. Only 40 singing males were detected on point count surveys, from which a tenuous population estimate of 1,800 singing males was derived. It is a rather late migrant, typically returning in May but sometimes not until June (McWilliams and Brauning 2000), with breeding extending well into late summer, by which time most atlasing activity was finished. The potential for late arrivals, its general scarcity, and the fact that its song may be less familiar to atlas volunteers outside of its traditional range, suggest that the Blue Grosbeak likely was underrecorded in areas peripheral to its main range. Birders in the northern counties should investigate "Purple Finches" singing in Blue Grosbeak habitat, since the songs of the two species are similar. With a continued, if somewhat slow, range expansion, it is likely that the Blue Grosbeak will become familiar over more of Pennsylvania in years to come.

The Blue Grosbeak has increased across most of its range since the 1960s, especially in states toward the northern edge of its range, including Ohio and Maryland (Sauer et al. 2011). Grassland restoration efforts at landfills and surface mines will undoubtedly benefit this species, and they may facilitate a continued northward range expansion in Pennsylvania. However, the Blue Grosbeak shuns suburbia (Lowther and Ingold 2011), and the loss of favored unkempt grassland and shrubby areas to housing development has been suggested to have caused population losses in some adjacent areas of Maryland (Ellison 2010zt). Hence, optimism about the species' future in Pennsylvania must be tempered with caution, since the species' presence in the state is largely due to the creation of habitat through anthropogenic change.

RICK WILTRAUT

Distribution

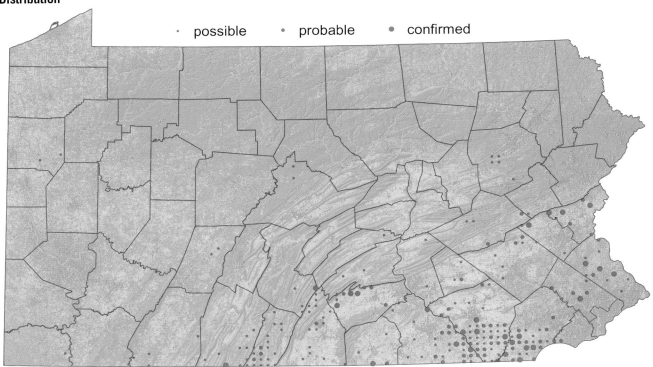

Distribution Change

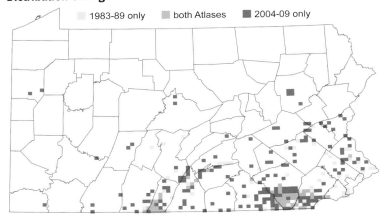

Number of Blocks

	first Atlas 1983–89	second Atlas 2004–09	Change %
Possible	43	132	207
Probable	50	60	20
Confirmed	20	27	35
Total	113	219	94

Population estimate, males (95% CI):
1,800 (1,200–2,600)

Chuck Musitano

Indigo Bunting
Passerina cyanea

In the heat of summer, when most other birds have gone quiet, the male Indigo Bunting sings constantly from the tops of trees, wires, poles, and other perches. His is often the only song heard in the middle of the hottest summer days when other species, even the American Robin, have grown quiet or taken to shady shelter (Pearson 1917). The Indigo Bunting is a small bird that grabs attention because of its intense blue color that can also appear to be turquoise or purplish in certain light. With his distinctive song of five or six paired notes repeated over and over again, the male defends the territory in which the female is nesting somewhere close by. Contrasting with the brilliantly colored male, the female Indigo Bunting has a dull olive-brown color with faint breast streaking and is often mistaken for a sparrow.

The Indigo Bunting normally has two broods per year, nesting in open areas with shrubby thickets, weedy fields, pastures, gravel pits, and forest edges. It breeds in the eastern half of the United States, from southern Canada, south to northern Florida, and sparingly west to central Arizona. It has a long breeding period, and, in the second Atlas, nesting was confirmed as early as 6 May and as late as 11 September. Confirmation of nesting was most often reported as "adults carrying food" or "adults feeding young," no doubt due to the difficulty of actually finding a nest. Of the 1,198 confirmed breeding reports, only 49 were of nests with eggs, 24 were of nests with young, and 24 more were of occupied nests.

In the first Atlas, the Indigo Bunting was described as abundant, which also sums up its status in the second Atlas, when it was among the ten most abundant species on point counts in Pennsylvania, with an estimated population of over 1.5 million singing males. This species has greatly benefited from human modification of the landscape, including the clearing of woodlands and establishment of hedgerows. Such anthropogenic change may have contributed to a northward expansion of its breeding range (Payne 2006). In the second Atlas, with the exception of the Atlantic Coastal Plain and Central Lowlands, the Indigo Bunting was found in over 90 percent of occupied blocks in all physiographic regions (appendix C). Although it was found in 4 percent more blocks than in the first Atlas, most of this was due to infilling within the existing range, and it is possible that the species was missed in some blocks that received minimal coverage in the first Atlas. The only areas from which the Indigo Bunting was absent in the second Atlas were the most urbanized blocks of the Philadelphia metropolitan area, in which there were also some losses since the first Atlas, possibly due to the loss of woodlots and other undeveloped land there (chap. 3).

Second Atlas point count data show that the Indigo Bunting is a habitat generalist that is most abundant in landscapes with interspersed woodland and farmland (appendix D). These preferences are clearly reflected in density estimates, which are highest in the Ridge and Valley Province and other areas with plentiful forest edge or small woodlots. Densities in urban areas, intensively farmed landscapes with few trees (e.g., Lancaster County), and core forest areas are lowest, although this species readily colonizes forest clear-cuts.

Although the Indigo Bunting is still abundant, Breeding Bird Survey shows that it has been declining in numbers in Pennsylvania (Sauer et al. 2011), equating to a 15 percent decrease in population size between atlas periods. Modest population declines have been noted in neighboring states (McGowan 2008zv; Coskren 2010f), the causes of which are not known. Because this species is a long-distance migrant, loss of habitat on its Central American wintering grounds cannot be ruled out. Despite modest declines, the Indigo Bunting currently remains one of Pennsylvania's most common breeding birds.

ARLENE KOCH

Distribution

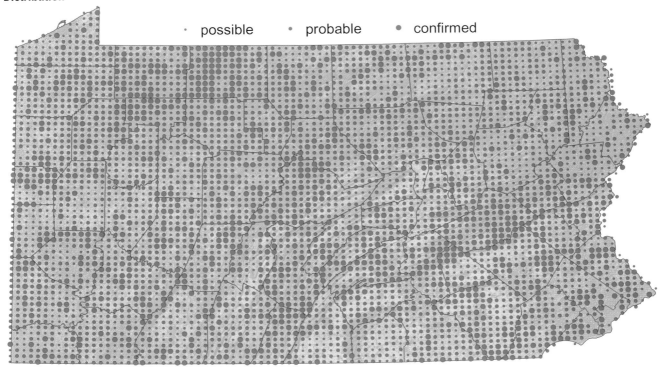

Distribution Change

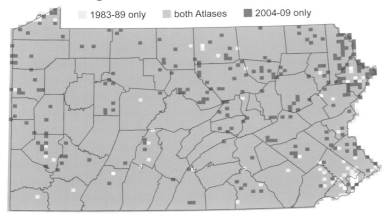

Density

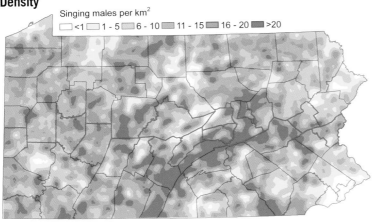

Number of Blocks

	first Atlas 1983–89	second Atlas 2004–09	Change %
Possible	901	1,328	47
Probable	2,576	2,511	−3
Confirmed	1,184	1,012	−15
Total	4,661	4,851	4

Population estimate, males (95% CI):
1,520,000 (1,480,000–1,560,000)

Breeding Bird Survey Trend

INDIGO BUNTING 439

Jeff McDonald

Dickcissel
Spiza americana

The Dickcissel has always been an enigma—a grassland specialist that must have been very rare in Pennsylvania's expansive, pre-colonial forests but which became so common in the farmland of the early nineteenth century that Audubon described it as "abundant in our middle Atlantic districts," while Alexander Wilson said they "abound in the neighborhood of Philadelphia, and seem to prefer level fields covered with rye-grass, timothy, or clover" (Audubon 1838; Wilson and Bonaparte 1831). What early ornithologists called the "black-throated bunting" was common as far north as southern New England, but by the 1880s it had become rare, and by the end of the nineteenth century the Dickcissel had essentially vanished from the East (Rhoads 1903; Gross 1956); the last historic Pennsylvania record was in 1895 (Mulvihill 1992l). Although Audubon (1838) had noted that "their flesh is good, especially that of the young birds," this disappearance seems to have been linked to changing habitat and an inherently nomadic nature rather than persecution, and Rhoads (1903) indicated that the decline began prior to the introduction of mechanized hay mowing.

Through the twentieth century, the Dickcissel was at best a rare breeder anywhere east of the Appalachian Mountains (Gross 1956), although it has been prone to unexpected incursions north and west of its current core breeding range in the central Great Plains and Midwest (Temple 2002). A brief range expansion in the late 1920s and early 1930s temporarily brought the species back to Pennsylvania (Gross 1956; Wentworth 2010), but it did not form a sustained population. The only exception may have been the Maryland border region in Franklin and Fulton counties, where the Dickcissel was reported almost annually beginning in the 1960s (Brauning et al. 1994).

During the first Atlas, however, the Dickcissel was recorded in a surprising total of 45 blocks, most of those records occurring during 1988, when drought in the Midwest seems to have promoted a significant eastward irruption of the species (Mulvihill 1992l). Many of those Dickcissels were found on reclaimed surface mines in the Appalachian Plateaus Province, but in the years since the first Atlas, another smaller irruption occurred in 1996 (McWilliams and Brauning 2000), and the species has bred in many parts of southern Pennsylvania, where a population has persisted in the agricultural grasslands of Adams and Cumberland counties (Wentworth 2010).

During the second Atlas, the Dickcissel was again found in 45 blocks, although its geographic range had shifted markedly; nearly three-quarters of the records in the second Atlas were in the Ridge and Valley and Piedmont provinces, and only 10 in the Appalachian Plateaus. Although the Dickcissel was noted in 23 counties, many of the records were clustered in Cumberland and Franklin counties, as well as in Adams and York counties. Some of these observations involved as many as five singing males at the same site, indicating small colonies.

In tandem with the recent increase in Dickcissel records in Pennsylvania, the second Maryland atlas documented an upswing in reports (Ellison 2010zu), with a now well-established population just south of the Adams County border (A. Wilson, pers. comm.). Following declines in the 1960s and 1970s, the surveywide Breeding Bird Survey (BBS) trend for this species has been stable (Sauer et al. 2011), but sparse BBS data from eastern states suggests a general increase east of the Appalachians.

Given the Dickcissel's historic tendency to dramatically expand and contract its range across wide regions, the species' current breeding status in Pennsylvania should not be taken for granted, and given its tiny population, the species is listed as Endangered by the Commonwealth. Dickcissels are also under threat from targeted persecution on their wintering grounds in Venezuela, and the overall population has declined about one-third since 1966 (Temple 2002). Management of both agricultural grasslands and reclaimed surface mines for this and other grassland obligates should be a priority, since hayfield mowing, among other farming practices, is detrimental to nest survival. But if history has any lesson, it is that the future of the "black-throated bunting" in Pennsylvania cannot be predicted with any certainty.

SCOTT WEIDENSAUL

Distribution

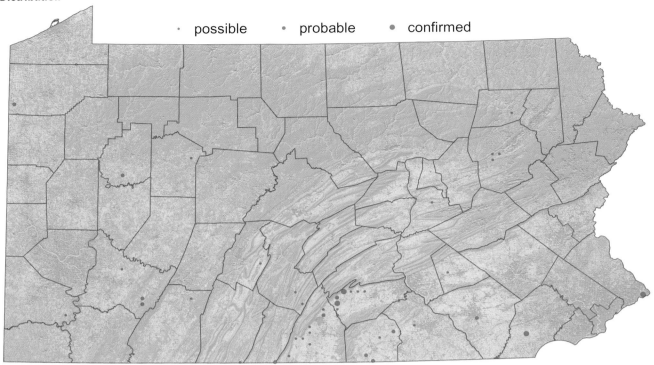

Distribution Change

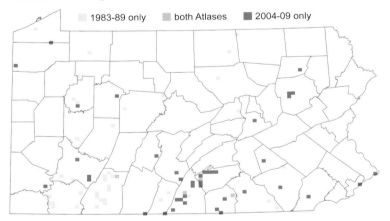

Number of Blocks

	first Atlas 1983–89	second Atlas 2004–09	Change %
Possible	16	30	88
Probable	21	11	−48
Confirmed	8	4	−50
Total	45	45	0

Joe Kosack/PGC

Bobolink
Dolichonyx oryzivorus

Handsome, hyperactive, and vocal, the Bobolink demands attention, in contrast to the skulking and cryptically marked grassland sparrows with which it shares grassland habitats. It breeds across northern United States and southern Canada, wintering in South America. Pennsylvania is near the southern edge of its breeding range.

As with other grassland obligates, Bobolink numbers in Pennsylvania must have increased rapidly when native forests were felled and replaced by farmland. In Pennsylvania, it is a characteristic bird of hayfields; as such, its status is linked to changes in hay-cropping patterns. As long ago as 1940, Todd remarked that this species had declined due to earlier mowing. Early avifaunal accounts suggest that it was most common in northern and western areas of the state (McWilliams and Brauning 2000), a pattern that holds true today. More than 80 percent of occupied blocks in the second Atlas were in the Appalachian Plateaus Province.

The Bobolink's range underwent a modest but statistically significant 14 percent expansion in number of blocks between the first and second Atlases. Notably, there was a 50 percent increase in block occupancy in the low-lying agricultural areas of the Piedmont Province, although the species is still localized there. A 35 percent increase in block occupancy in the Pittsburgh Low Plateau Section may be attributed to the expansion (or discovery) of suitable reclaimed surface mine grasslands, particularly in Clearfield County. In contrast to other grassland obligates, Breeding Bird Survey data do not indicate precipitous declines in populations in recent decades, but the trend is downward (Sauer et al. 2011), with a 22 percent decline in numbers between the first and second Atlas periods.

Second Atlas point count data show the importance of hayfields to the Bobolink, accounting for almost 90 percent of the singing males detected (appendix D). These data also show that densities are strongly correlated with elevation; few were found in hayfields below elevations of 300 m (~1,000 ft), and peak densities occur at the highest elevations, over 600 m (~2,000 ft). A consequence of the association with higher elevations is that the species is often found in landscapes that are mosaics of forest and farmland. In such areas, fields are often small and bordered by forests; this is suboptimal for the Bobolink, which is known to avoid fields close to forested edges (Bollinger and Gavin 2004; Ribic et al. 2009). However, outside of densely forested areas, this species is widespread and, in some places, numerous. The highest densities were found in the Laurel Highlands in Somerset County, which accounts for 13 percent of the statewide population. Other notable concentrations were in the northwestern and northeastern corners of the state. The association with higher elevations is no coincidence; the later growing seasons in those areas preclude early mowing of hayfields and may increase the interval between cuts, providing a sufficient time interval for the Bobolink to nest successfully.

Modifications to hay cropping, such as ensuring a sufficient interval between first and second cuts (Perlut et al. 2008) or leaving unmown patches (Masse et al. 2008), are promising suggestions for conserving populations of Bobolinks and other grassland songbirds in hayfields. To be widely adapted, however, such practices may need to be underpinned by publicly funded compensation to participating farmers, something that has been tested in Vermont (Perlut et al. 2008). The steady loss of traditionally managed hay to more intensively managed forage crops or row crops remains a threat to Bobolink populations. Farm Bill conservation programs, such as the Conservation Reserve Enhancement Program (CREP), could be important in maintaining grasslands by ensuring the economic viability of pastoral farming. Although there is little evidence to suggest that CREP has benefited the Bobolink in southern Pennsylvania (Wilson 2009), tens of thousands of acres of grassland in northern Pennsylvania have been preserved and managed through this program, and the benefits to the Bobolink could be considerable, although as yet unmeasured.

ANDREW M. WILSON

Distribution

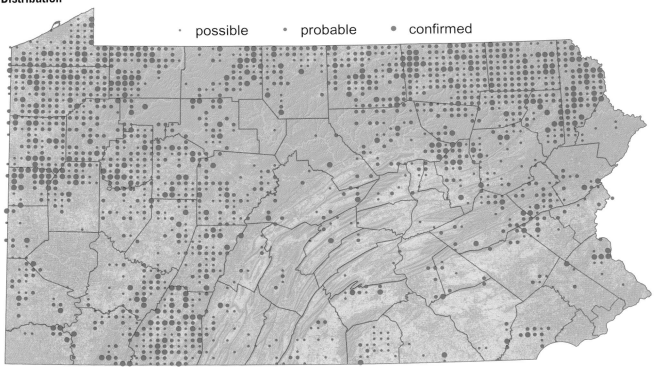

Distribution Change

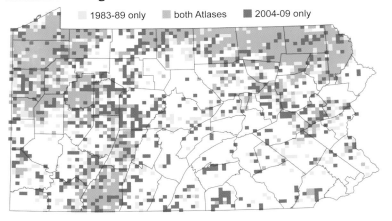

Density

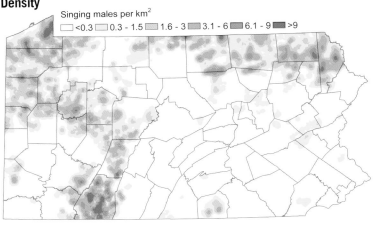

Number of Blocks

	first Atlas	second Atlas	Change
	1983–89	2004–09	%
Possible	507	631	24
Probable	702	790	13
Confirmed	318	318	0
Total	1,527	1,739	14

Population estimate, males (95% CI):
110,000 (104,000–117,000)

Breeding Bird Survey Trend

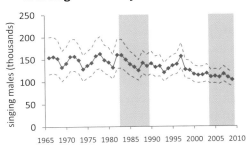

Geoff Malosh

Red-winged Blackbird
Agelaius phoeniceus

With their raucous song, flashing vivid red epaulets, and incessant social interaction, displaying male Red-winged Blackbirds are difficult to miss. The streaked brown females, of which there may be up to 15 in the harem of each male (Yasukawa and Searcy 1995), are much less conspicuous. A seemingly ubiquitous bird of farmland, grassland, and freshwater wetlands across the United States and southern Canada, the Red-winged Blackbird is estimated to be the third most numerous bird in North America, after the American Robin and Dark-eyed Junco (Rich et al. 2004).

A partial migrant in Pennsylvania, red-wings eschew the coldest parts of the state during midwinter but return early, the highly competitive males returning to stake their claim in prime wetland habitats as early as the end of February. The breeding season commences early and is protracted; second Atlas records of nests with eggs spanned the period 26 April to 27 July, with dependent young observed into August. However, the majority of records of nests with eggs were in May and early June, with observations of young from mid-June to mid-July. There was no obvious double peak in nest building, suggestive of double-brooding, as reported in the first Atlas (Gross 1992i), but atlas breeding phenology data are confounded by fieldwork effort, which is not evenly distributed through the breeding season.

Originally a bird of marshes in Pennsylvania, the Red-winged Blackbird showed a remarkable ability to adapt to expanding agricultural habitats in uplands and was described as abundant in early avifaunal accounts (McWilliams and Brauning 2000). Numbers probably peaked in the early twentieth century, when farmland extent was at its zenith, and then declined from the 1960s through 1990s (Sauer et al. 2011). Similar declines in neighboring Ohio were attributed to efficiency and expansion of modern agriculture (Blackwell and Dolbeer 2001). Despite these declines, the Red-winged Blackbird remains one of Pennsylvania's most abundant breeding birds, with a population estimated to be well over 1 million males. We estimate that this was the seventh most numerous bird in the state at the time of the first Atlas, slipping to eleventh most numerous in the second Atlas period.

Found in 93 percent of atlas blocks, the red-wing is absent only from blocks that are nearly 100 percent forested or entirely developed; most blocks have at least some farmland or wetlands, and where these habitats occur, there is a high probability that Red-winged Blackbirds will be found. There was some turnover of block occupancy between atlases, typically in blocks with low amounts of available habitat, such as areas peripheral to Sproul State Forest in north-central Pennsylvania. However, there was little overall pattern of change and no significant net change in block occupancy since the first Atlas. The population density map reflects the areas dominated by open habitats, with the highest densities in lower elevation areas in the southern and northeastern regions of the state. According to second Atlas point count data, the highest densities of Red-winged Blackbird are found in emergent wetlands, followed by hayfields, pasture, and reclaimed surface mine grasslands (appendix D). Due to their greater extent, the combined farmland habitats probably account for around 90 percent of the statewide population. Average densities in farmland are around 30 males/km^2, but densities up to 10 times that have been found in Conservation Reserve Enhancement Program fields (Wentworth et al. 2010), which is probably as high as densities in prime wetland habitats (pers. obs.).

On the wintering grounds, particularly in the southeastern United States, large blackbird flocks are considered agricultural pests; this species is culled in large numbers (Yasukawa and Searcy 1995). However, there is no evidence that culling on wintering grounds has a significant impact on the populations of Red-winged Blackbirds that nest in Pennsylvania. Although Breeding Bird Survey data indicate a population decline of 16 percent between the two atlases, there has been remarkable stability in numbers since the late 1990s. Being such a common and highly adaptable bird, the Red-winged Blackbird is therefore not of conservation concern in Pennsylvania.

ANDREW M. WILSON

Distribution

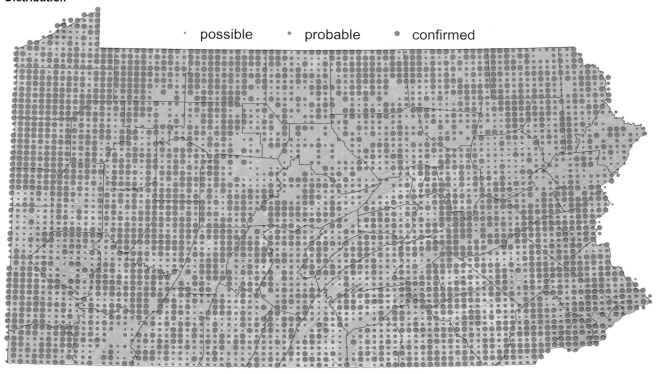

Distribution Change

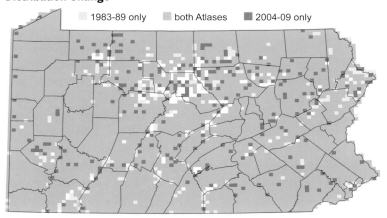

Density

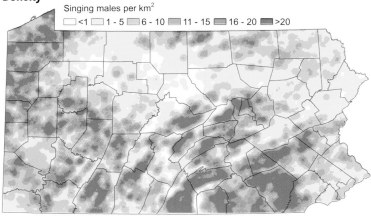

Number of Blocks

	first Atlas 1983–89	second Atlas 2004–09	Change %
Possible	650	859	32
Probable	1,640	1,570	−4
Confirmed	2,247	2,152	−4
Total	4,537	4,581	1

Population estimate, males (95% CI):
1,160,000 (1,130,000–1,190,000)

Breeding Bird Survey Trend

Gerard Dewaghe

Eastern Meadowlark
Sturnella magna

The Eastern Meadowlark, as its name suggests, is a characteristic bird of meadows and other open grasslands. As such, its status in Pennsylvania has always been intrinsically linked with agriculture. Once scarce, the species quickly expanded its range to occupy new grasslands created by the agriculture of early European settlers. Its plaintive song and bright-yellow front are familiar to country dwellers, and it has become something of a cherished and emblematic farmland bird in the eastern United States.

During the first Atlas period, the Eastern Meadowlark was widespread, absent only from the most densely forested regions and largest urban areas. By that time, however, populations had already been in long-term decline; Breeding Bird Survey (BBS) data showed steep declines during the 1970s and early 1980s (Sauer et al. 2011). By the second Atlas period, the population had dropped more than 80 percent. Such declines have been noted across the species' breeding range (Sauer et al. 2011). Concurrent with the population decline, range contractions resulted in an 11 percent reduction in occupied blocks between the first and second atlas Periods; corrected for survey effort, the decline was estimated to be 15 percent. Losses were particularly marked in the Piedmont Province, where 42 percent fewer blocks held meadowlarks. However, the overall reduction in occupied atlas blocks was less than the 25 percent decline in New York state (Smith 2008j) and the 23 percent decline in Maryland (Ellison 2010zv) between atlas periods.

Some of the contraction in the range of the Eastern Meadowlark may be due to the conversion of farmland to developed areas. Between 1969 and 1992, 27 percent of farmland was lost in the Lehigh Valley, 37 percent was lost around Philadelphia (Goodrich et al. 2002), and there have been similar losses around Pittsburgh in recent decades (chap. 3). More generally, population and range losses can be attributed to changes in habitat quality, such as the loss of traditionally managed hayfields and substantial reduction in lightly grazed pasture (PGC-PFBC 2005).

The Pennsylvania population of Eastern Meadowlarks was estimated to be 89,000 singing males during the second Atlas period. Strongholds included the five southern border counties from Adams west to Somerset (which between them support 20 percent of the statewide population), Clarion and Jefferson counties, and Tioga and Bradford counties. Extrapolating from BBS trends, the population was estimated to be approximately 200,000 singing males in the first Atlas period, more than double the current population.

During the second Atlas period, the Eastern Meadowlark was most likely to be found in blocks that were at least 8 percent grassland. Although reclaimed surface mines support higher densities of this species than any other habitat in Pennsylvania, second Atlas point count data suggest that they only support 3 percent of the statewide population; most of the remainder is on agricultural lands. Therefore, maintaining a substantial population of this species in the state is dependent on maintaining suitable nesting habitats within the farmed landscape. Increased fertilizer use and grassland reseeding have resulted in hay meadows that are often too dense, weed-free, and cut too frequently to support this species. Recently fledged young were noted as early as 10 May and as late as 3 September during atlas fieldwork; this protracted breeding season may afford some insurance against nest losses, but it will not protect against a mowing frequency that does not provide a sufficient time window in which to successfully reproduce. Conservation grasslands, such as those in the Conservation Reserve Enhancement Program, appear to have resulted in localized increases in Eastern Meadowlark numbers since 2001, but such increases have not been sufficient to offset continued statewide declines (Wilson and Brittingham 2012).

Unless declines are halted, it is likely that the Eastern Meadowlark's range will continue to contract. It is listed as a species of Conservation Concern by the Pennsylvania Game Commission, due to long-term population declines.

ANDREW M. WILSON

Distribution

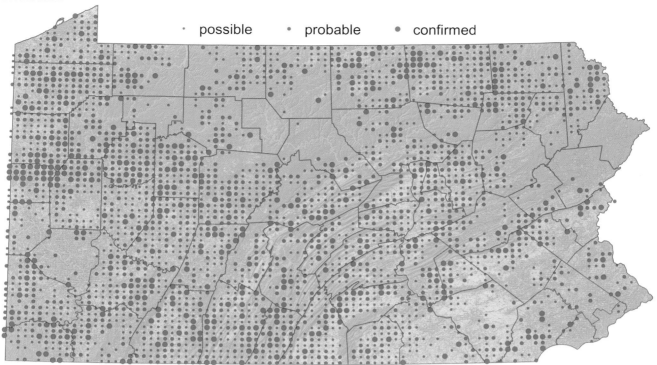

Distribution Change

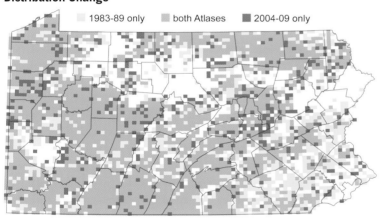

Density

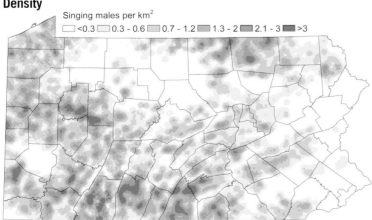

Number of Blocks

	first Atlas 1983–89	second Atlas 2004–09	Change %
Possible	1,174	1,409	20
Probable	1,480	1,050	−29
Confirmed	614	437	−29
Total	3,268	2,896	−11

Population estimate, males (95% CI):
89,000 (85,000–95,000)

Breeding Bird Survey Trend

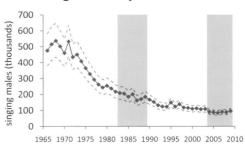

EASTERN MEADOWLARK

Geoff Malosh

Western Meadowlark
Sturnella neglecta

The Western Meadowlark, as its name aptly suggests, is a bird of western North America, occupying grasslands and other open areas in a core range from the Pacific coast in southern Canada, south to northern Baja California and central Mexico, and east to Manitoba, Wisconsin, northern Illinois, Missouri, western Kansas, Oklahoma, and Texas. Scattered breeding populations reach farther east, especially into the Great Lakes region. Breeding has never been confirmed in Pennsylvania; however, a singing male in Butler County, present in 1985 and 1986, was apparently paired with an Eastern Meadowlark female, and both birds were observed carrying food during the first Atlas period (Schwalbe 1992f). Nonetheless, the record was classified only as probable breeding in the first Atlas results. If the birds had truly paired, this would represent a very rare example of hybridization between the two species. During Pennsylvania's second Atlas period, two observations were submitted, both of unpaired singing males and both classified as probable breeding: one in Centre County in July 2004 and one in Dauphin County in 2009. Diagnostic audio recordings were obtained of the Dauphin bird (Pulcinella 2011).

The two atlas records are consistent with all other known records of the Western Meadowlark in Pennsylvania's history: all refer to singing birds present between the months of April and July. It would probably be incorrect to think of territorial Western Meadowlarks in Pennsylvania as "overshoots," the term often applied to, for example, Swainson's Warbler. Rather, Pennsylvania, lying to the south and east of the species' easternmost established breeding populations, is simply one of the farthest outposts of a relatively recent expansion into eastern North America. In the nineteenth century, the species' breeding range was restricted to areas west of the forests of Wisconsin, Illinois, and Missouri, expanding dramatically northeastward into the Great Lakes region in Michigan, northern Indiana and Ohio, western Ontario, and extreme northwestern New York only in response to the clearing of forests and the expansion of agriculture in those areas during the twentieth century (Davis and Lanyon 2008). The first record of the species in Pennsylvania was made in Crawford County in 1935 (Todd 1940), and it has been recorded here approximately 15 times since.

Considering the species' status in the Great Lakes, conventional wisdom might suggest western Pennsylvania to be the most likely area within the state for summering Western Meadowlarks, and some authors have indicated this (e.g., Davis and Lanyon 2008). However, a closer examination of specific records indicates a different pattern: the majority of the acceptable records of the species, particularly in recent years, are from central Pennsylvania. Since 1980, only the aforementioned Butler male in 1985 and 1986 and two records from Westmoreland County, both in 2001, occurred in western Pennsylvania, whereas the species has occurred at least seven times over the same period in central Pennsylvania, in the counties of Centre, Juniata, Mifflin, and Dauphin. During the second New York atlas, the species was notably absent from far western New York, where it had formerly been casual in summer; the only record during the 2000 through 2005 period was of a singing male in the Finger Lakes region (McGowan 2008zw), which is geographically consistent with central Pennsylvania records to the south. The species has not been recorded in southeastern Pennsylvania since 1972, when the last of four known records from that corner of the state occurred in Berks County (McWilliams and Brauning 2000).

Breeding Bird Survey data (Sauer et al. 2011) indicate that the Western Meadowlark is generally declining throughout the continent, with the steepest declines in the northeastern portions of its range, the same areas into which it expanded during the twentieth century. As a result, occurrences of the species in Pennsylvania might well be expected to become even more unusual in summer than they already are, with little hope for the discovery of a breeding pair.

GEOFF MALOSH

Distribution

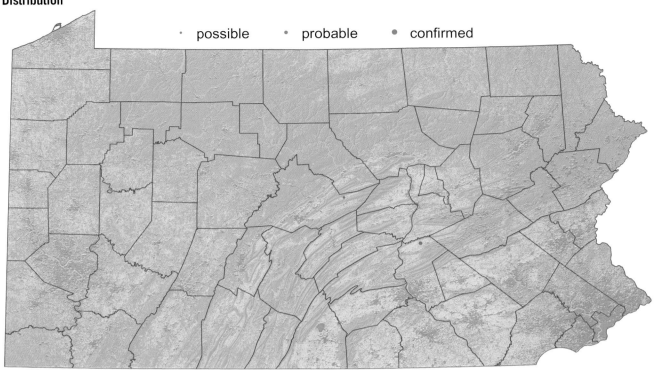

Distribution Change

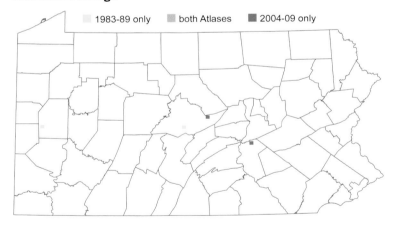

Number of Blocks

	first Atlas 1983–89	second Atlas 2004–09	Change %
Possible	0	1	∞
Probable	2	1	−50
Confirmed	0	0	0
Total	2	2	0

WESTERN MEADOWLARK

Laura C. Williams/VIREO

Common Grackle
Quiscalus quiscula

A large blackbird with a glossy, iridescent head and long, wedge-shaped tail, the Common Grackle is difficult to miss during breeding season. Absent only from the most heavily forested reaches of Pennsylvania, it was recorded in 91 percent of all atlas blocks and is among the most abundant breeding bird species in the Commonwealth.

A highly adaptable species, the Common Grackle's range extends across nearly all of eastern and central North America. The grackle's range has greatly expanded westward in past decades, and it is now found as far west as the Rocky Mountains, with further expansion likely to occur (Peer and Bollinger 1997). Its tendency to form large flocks and cause crop damage has made the Common Grackle a pest in some areas, and efforts to control populations have included lethal methods (Peer and Bollinger 1997).

Short-distance migrants, Common Grackles are among the first species to arrive on their breeding grounds and begin nesting in spring (Peer and Bollinger 1997). Flocks commonly seen flying overhead in February and March, a remnant of the large congregations formed in the winter, signal that the nesting season is coming soon. By April, many individuals are actively nesting. Nest building was observed as early as 16 March, and parents carrying food as early as 2 April. The breeding season peaks in May and June, but some individuals do not complete their nesting cycles until mid- or late July. Being so common, and having a protracted breeding season, the Common Grackle is unlikely to be missed by atlas volunteers. Confirmation is also easy, since breeding adults carrying food and recently fledged young are much in evidence throughout early summer.

Although the Common Grackle is still an abundant species in Pennsylvania—with atlas point counts estimating, conservatively, the population at 1.5 million individuals (chap. 5)—it was once even more common than it is today. From 1966 through 2009, the species declined by an average of 1.9 percent annually (Sauer et al. 2011), but these declines have slowed in recent years. The downward trend of the grackle population in the last few decades in Pennsylvania has also been observed across the species' range. Audubon's analysis of Christmas Bird Count and Breeding Bird Survey data from 1966 through 2005 found that Common Grackle populations declined 61 percent in 40 years, from approximately 190 million to 73 million individuals rangewide (NAS 2007).

Despite the decline in population size, the grackle is still very common and widespread in Pennsylvania. It was found in 91 percent of blocks in the second Atlas, and there was no evidence of any change in overall range size between atlases. There was a thinning in the grackle's distribution in the northeastern area of the state, in which volunteer effort was higher in the second Atlas (chap. 6), suggesting a genuine reduction in block occupancy. In both atlases, there were appreciable gaps in distribution in the densely forested plateau in north-central Pennsylvania and in the Allegheny National Forest; otherwise, grackles were recorded nearly everywhere.

Point count data show the extent to which Common Grackles avoid forests; otherwise, they are habitat generalists, found in abundance in both developed and agricultural habitats (appendix D). This is reflected in the density map, which shows high densities in all suburban and urban areas and across the intensively farmed Piedmont.

Common Grackles are managed as a nuisance species during winter in some agricultural and suburban/urban settings, but unless this practice becomes more widespread, it is unlikely to have a major impact on grackle populations (Peer and Bollinger 1997).

BRIAN BYRNES

Distribution

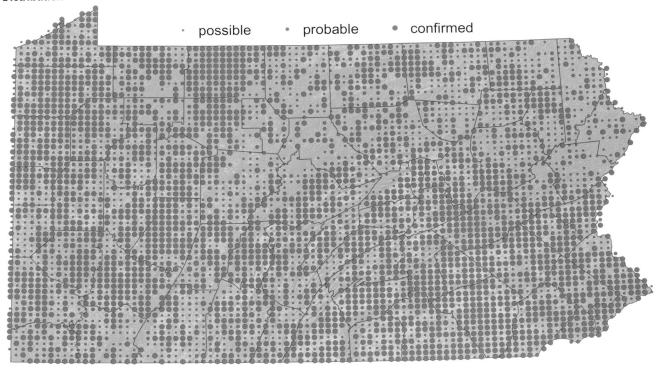

Distribution Change

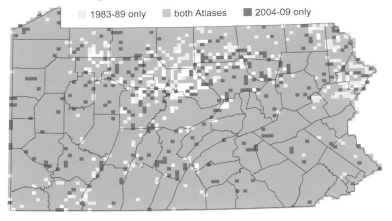

Density

Birds per km²

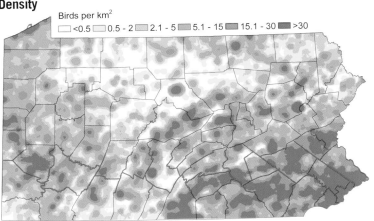

Number of Blocks

	first Atlas 1983–89	second Atlas 2004–09	Change %
Possible	1,093	1,300	19
Probable	797	619	−22
Confirmed	2,571	2,574	0
Total	4,461	4,493	1

Population estimate, birds (95% CI):
1,520,000 (1,490,000–1,550,000)

Breeding Bird Survey Trend

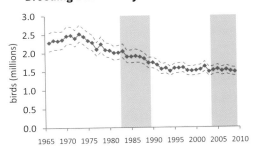

COMMON GRACKLE 451

Michael Patrikeev/VIREO

Brown-headed Cowbird
Molothrus ater

The Brown-headed Cowbird is Pennsylvania's only obligate brood parasite. It never builds a nest of its own and instead lays its eggs in the nests of host species. Historically found in the plains and prairie region west of the Mississippi River, the Brown-headed Cowbird followed the large herds of bison, feeding on the insects and seeds kicked up by their movement. Cowbirds were presumably rare in the forests that covered Pennsylvania before European settlement (Brittingham and Temple 1983). As forests were cleared and land was cultivated, the cowbird moved eastward into Pennsylvania and was reported as a common summer resident by the mid-nineteenth century (Warren 1890). The cowbird now breeds from southeastern Alaska and central British Columbia east to Newfoundland, and south to central Mexico and northern Florida (Lowther 1993). Pennsylvania is within the core of its current range.

As in the first Atlas, cowbirds were ubiquitous across much of the state and reported in 4,510 blocks (91% of the total). There was no substantial change in the number of blocks with atlas records between the first Atlas and the second Atlas. Cowbirds are associated with open habitats with scattered trees, woodlot edges, fields, pastures, orchards, and residential areas (Lowther 1993). They feed on the ground in open habitat, but they will parasitize nests in both open and woodland habitat, with parasitism rates highest at the wood-field ecotone (Lowther 1993). Abundance is negatively correlated with forest cover (appendix D); hence, blocks where this species was not found in the second Atlas were mainly in large expanses of contiguous forest habitat, especially in the Appalachian Plateaus. In most other physiographic provinces and sections, there were atlas records from close to 100 percent of blocks (appendix C).

Cowbird courtship behavior was reported from April to July, with eggs found as early as 10 April and a mean egg date of 5 June. Recently fledged young were found from mid-May to mid-August, with a mean date in late June. Atlasers reported no less than 32 different host species, including 13 species of warblers, with Song Sparrow (27 records), Chipping Sparrow (12), Red-eyed Vireo (8), and Northern Cardinal (8) the most frequently reported.

The Pennsylvania population of Brown-headed Cowbirds was estimated to be 520,000 individuals, with the highest densities in the Piedmont Province in southeastern Pennsylvania. Counts of cowbirds on Breeding Bird Survey (BBS) routes in the state have shown a steady downward trend, averaging 1.8 percent per year since 1966 (Sauer et al. 2011). Even though the downward trend has slowed and even reversed in the last few years, BBS data suggest that the population during the first Atlas period was 35 percent higher than today. Declines have been noted across the eastern part of the species' range, but numbers have been stable farther west (Sauer et al. 2011). Declines on BBS routes in New York have been particularly rapid, but as in Pennsylvania, the species is still common, and the reduced abundance has yet to appear as significant changes in range size (McGowan 2008zx). However, some retreat from blocks in the reforested Adirondacks is an indication that range contractions could occur in Pennsylvania if downward population trends resume.

Declines in cowbird numbers are probably the result of multiple causes, including loss of farmland and open habitat and a concomitant increase in forest cover. Cowbird and blackbird control programs occurring within parts of the cowbird's wintering and breeding range may also affect population size (Lowther 1993), but there have been no studies examining this relationship.

The continual loss of farmland and pasture habitat within Pennsylvania will be detrimental to cowbirds within the southeastern region of the state, but fragmentation of forest habitat associated with natural gas extraction and drilling in the Marcellus Shale formation may enable cowbirds to increase in abundance in the north-central region of the state.

MARGARET C. BRITTINGHAM

Distribution

· possible   · probable   • confirmed

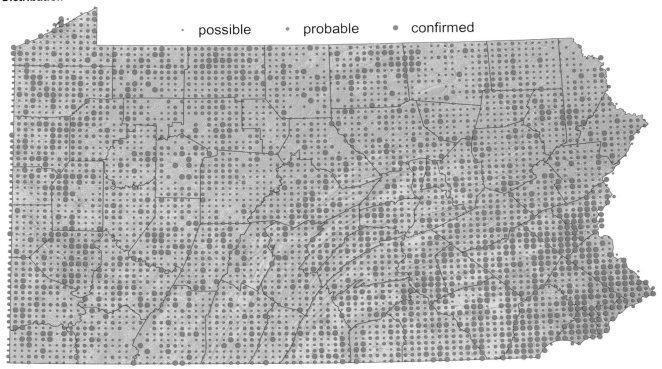

Distribution Change

☐ 1983-89 only   ▨ both Atlases   ■ 2004-09 only

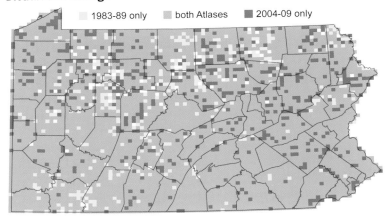

Density

Birds per km²
☐ <0.5 ☐ 0.5 - 2 ▨ 2.1 - 4 ▨ 4.1 - 7 ▨ 7.1 - 10 ■ >10

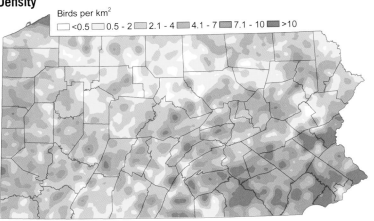

Number of Blocks

	first Atlas 1983–89	second Atlas 2004–09	Change %
Possible	1,517	2,103	39
Probable	1,757	1,599	−9
Confirmed	1,047	808	−23
Total	4,321	4,510	4

Population estimate, birds (95% CI):
520,000 (490,000–550,000)

Breeding Bird Survey Trend

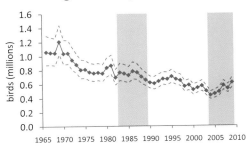

BROWN-HEADED COWBIRD   453

Geoff Malosh

Orchard Oriole
Icterus spurius

Of the two oriole species that breed in Pennsylvania, the Orchard Oriole is less conspicuous and not as familiar as its bright-orange relative, the Baltimore Oriole. It is smaller and darker than most orioles, and it spends only about 3 months out of a year in the state. The Orchard Oriole breeds throughout the eastern two-thirds of the United States, preferring open areas with scattered large trees for nesting.

Prior to the first Atlas, the Orchard Oriole appeared to have a somewhat variable distribution and breeding history in the state. It was reportedly common in the southern tier of counties since at least the late nineteenth century (Jacobs 1893; Stone 1894), declined throughout the state some time during the 1920s and 1930s (Todd 1940), and remained fairly rare in the Commonwealth throughout the 1960s (Poole 1964). Moreover, throughout this time the species was either rare or absent as a breeder in the state's northern and mountainous sections (Harlow 1913; Poole 1964). Results of the first Atlas indicate that the Orchard Oriole had become fairly common again in the southern third of the state and had extended north along certain river valleys, but it was still scarce in the northern and mountainous regions of Pennsylvania (Ickes 1992p).

The second Atlas illustrates the impressive continual expansion of the Orchard Oriole in Pennsylvania, particularly northward. The species increased a highly significant 96 percent in occupied blocks between atlas periods, and is now found nearly statewide. Furthermore, five physiographic sections in which there were significant increases in occupied blocks of 170 to 493 percent (appendix C) are in the central or northwestern parts of the state. Pennsylvania's Allegheny Mountain and Appalachian Mountain sections also saw marked increases in block occupancy (285% and 160%, respectively), suggesting that the Orchard Oriole is extending its breeding range into the state's mountainous areas as well as northward. Additional evidence of the oriole's northward expansion comes from the second New York atlas: an increase of 74 percent in occupied blocks between atlas periods was recorded (McGowan 2008zy).

The Orchard Oriole's northward spread along certain river valleys, reported to occur during the first Atlas (Ickes 1992p), apparently is continuing, since the number of records along several rivers (e.g., the Allegheny and Susquehanna) increased during the second Atlas. The oriole's distinct preference for floodplains, marshes, shorelines of large rivers, and scattered trees along streams (Scharf and Kren 1996) is also reflected in the density map; the great majority of point count observations were at elevations of less than 400 m (1,300 ft; appendix D).

The significant increase in occupied blocks in two other adjoining sections, Gettysburg-Newark Lowland (98%) and Great Valley (89%), is probably related to the Orchard Oriole's status as a member of the "farmland generalists" guild—that is, species associated with woodlots, shrubby areas, and scattered trees in open country (Wilson 2010). Point count data show that this species shuns forested areas more than the Baltimore Oriole, being typically found in predominantly open landscapes with small amounts of tree cover (appendix D). Pennsylvania's Orchard Oriole population during the second Atlas was estimated to be 51,000 singing males. Breeding Bird Survey counts in Pennsylvania increased by 4.6 percent annually from 1966 through 2009, accelerating to an impressive 8.9-percent-per-year increase from 1999 through 2009 (Sauer et al. 2011), equating to more than a doubling of counts between atlas periods.

The extraordinary expansion of the Orchard Oriole population in Pennsylvania seems to indicate that its outlook in the state is excellent. Increases in occupied blocks between atlas periods in neighboring states also suggest a positive future for the species: 74 percent in New York (McGowan 2008zy) and 9 percent in Maryland (Ellison 2010zw). Silvicultural treatments prescribed to enhance wildlife habitat for early successional, gap-dependent species have benefited the Orchard Oriole in the southeastern United States (Twedt and Somershoe 2009), and protection of Pennsylvania's riparian forests should ensure a healthy state population.

ROY A. ICKES

Distribution

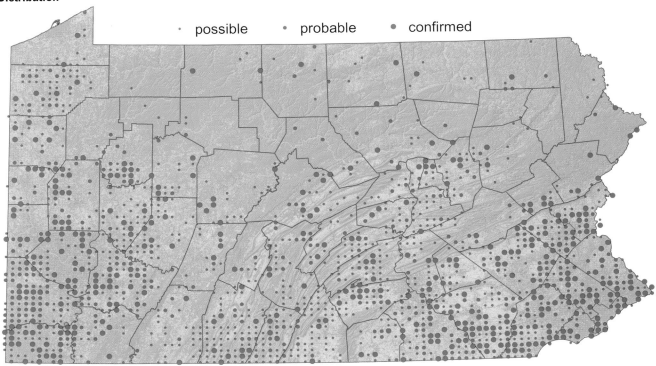

Distribution Change

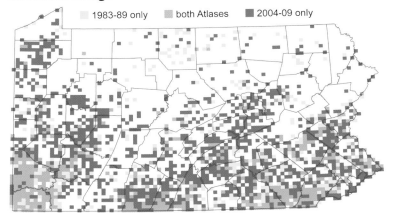

Density

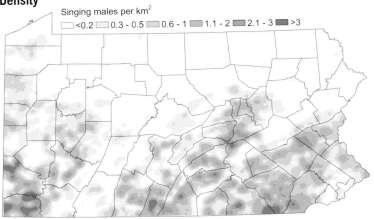

Number of Blocks

	first Atlas 1983–89	second Atlas 2004–09	Change %
Possible	282	798	183
Probable	271	356	31
Confirmed	244	407	67
Total	797	1,561	96

Population estimate, males (95% CI):
51,000 (47,000–57,000)

Breeding Bird Survey Trend

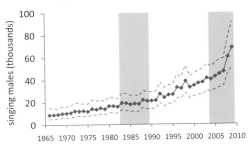

Bob Wood and Peter Wood

Baltimore Oriole
Icterus galbula

No longer a subspecies of the Northern Oriole and once again recognized as a distinct biological species (Monroe et al. 1995), the Baltimore Oriole, with its deep pendant nest and the male's bright orange-and-black plumage, is one of Pennsylvania's most well-known songbirds. Except for the extreme southeastern states, it breeds across the United States east of the Rocky Mountains. In Pennsylvania, it inhabits a variety of shade trees in open deciduous woodland, woodland edges, and residential neighborhoods. A disproportionate fraction of the species' total North American population breeds within Pennsylvania; thus, the state has a high stewardship responsibility for maintaining Baltimore Oriole numbers (Rosenberg 2004).

Although the Baltimore Oriole has reportedly undergone population fluctuations over the years, it appears to have been common through much of the state since at least the early twentieth century (Ickes 1992q; McWilliams and Brauning 2000). During the first Pennsylvania Atlas, it was observed in at least 77 percent of the blocks in every physiographic province. As noted in the first West Virginia atlas (Buckelew and Hall 1994), the species avoids the most densely forested areas in Pennsylvania (Ickes 1992q).

The second Atlas demonstrates that the Baltimore Oriole remains widespread in most regions of Pennsylvania, with records between 89 and 98 percent of blocks in all sections in the Piedmont and Ridge and Valley provinces and between 68 and 92 percent in the more densely forested Appalachian Plateaus Province (appendix C). Second Atlas point count data indicate that the Baltimore Oriole is found at highest densities in interspersed landscapes, which have plentiful forest edge (appendix D). The avoidance of core forests can be seen in the statewide distribution, with an absence of orioles in the most densely forested blocks in the Allegheny Plateau, Laurel Highlands, and Poconos. However, there were non-significant increases in occupied blocks of 10 percent or more in three heavily forested physiographic sections: Anthracite Upland, Deep Valleys, and South Mountain (appendix C). Further, the state's top three counties for estimated oriole density (Bedford, Fulton, and Perry) have many heavily forested blocks, which suggests that, even while avoiding extensive forests, the species may be extending its breeding range into slightly more forested areas of Pennsylvania and exploiting forest openings, such as rivers and roadways.

The Baltimore Oriole often nests around human habitation and over rural roads (Rising and Flood 1998), and this tolerance of humans was evident in the second Atlas. The species was found in almost all atlas blocks in the most developed parts of the state, including the most urbanized blocks in Philadelphia and Pittsburgh. Because of its eye-catchingly-bright plumage and prominent nest, the Baltimore Oriole was one of the easiest species for atlasers to confirm breeding; more than 2,000 confirmed breeding records were submitted, with nest building noted as early as 28 April and occupied nests and dependent young noted through early August, peaking in late June and early July. This species' distinctive, pendulous nest is surprisingly robust, often remaining intact for many months after the breeding season. As a result, atlas volunteers reported 192 unoccupied nests, such records occurring throughout the year; hence, this summer visitor could be confirmed to have nested long after it had left the state for the winter.

Pennsylvania's population of Baltimore Orioles was estimated to be 330,000 singing males during the second Atlas period. Breeding Bird Survey (BBS) data suggest that there has been no overall change in population size between atlas periods (Sauer et al. 2011). The slight increase in the number of blocks with Baltimore Oriole records between atlases, as well as the stability of the population based on BBS data, suggests that the species' outlook in the state is secure. The Baltimore Oriole is commonly encountered in relatively fragmented landscapes (Keller and Yahner 2007), which are found throughout the lower elevations of Pennsylvania. In such areas, fragmentation of existing forests and woodlands has increased, largely due to development (chap. 3), ensuring that this species' favored habitats are plentiful.

ROY A. ICKES

Distribution

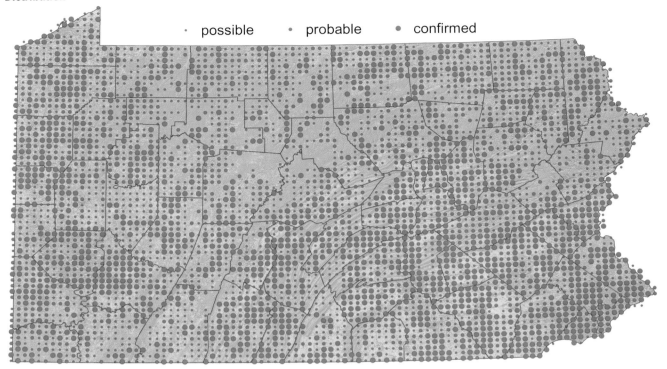

Distribution Change

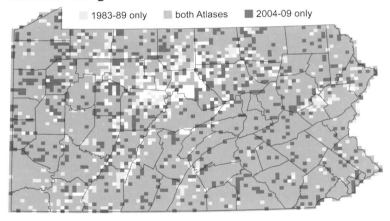

Density

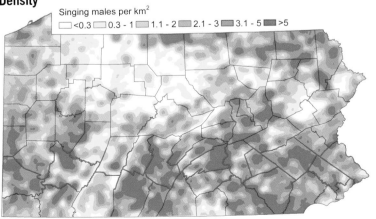

Number of Blocks

	first Atlas 1983–89	second Atlas 2004–09	Change %
Possible	1,014	1,552	53
Probable	967	1,143	18
Confirmed	2,023	1,591	−21
Total	4,004	4,286	7

Population estimate, males (95% CI):
330,000 (306,000–356,000)

Breeding Bird Survey Trend

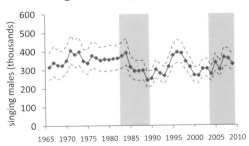

BALTIMORE ORIOLE

Bob Wood and Peter Wood

Purple Finch
Carpodacus purpureus

Roger Tory Peterson captured the essence of the adult male Purple Finch when he described it as "dipped in raspberry juice"; females and second-year males, by contrast, look much like brown sparrows. Like most finches, the Purple Finch is usually associated with conifers, the preferred choice for nest placement. A seed eater in winter, it feeds extensively on buds and blossoms in spring and then insects and fruits as they become available (Wootton 1996).

The Purple Finch nests across the southern half of Canada; in the United States, its range dips south into the Pacific Northwest, the upper Great Lakes states, and the Appalachian Mountains as far south as the highlands of West Virginia and Virginia (Wootton 1996). In Pennsylvania, the breeding range has not changed significantly since the first Atlas. Nesting occurs primarily across the northern tier and at higher elevation further south. In contrast to other species of northern affinity, the Purple Finch is widespread at lower elevations in Erie and Crawford counties, as well as northern parts of the Pittsburgh Low Plateau Section. However, the great majority of atlas records were from elevations above 450 m (~1,500 ft; appendix D). There were a few scattered records in the Piedmont and southern Ridge and Valley provinces, at least some of which could have resulted from misidentification of the House Finch, as occurred also during the first Atlas (Master 1992i).

Purple Finches were reported from a wide variety of habitats, including areas where conifers are not a major component of the habitat. Second Atlas point count data demonstrated their presence to be independent of extent of forest cover; often they are found at edges or even in relatively open habitat—for example, on reclaimed surface mine grasslands (appendix D) where there are scattered pines.

In the Ridge and Valley Province, Purple Finches were found commonly only in the northeast, where they are thought to use a variety of habitat types, including Christmas tree farms, pine plantings, ornamental conifers, and old mineland barrens. Their presence there may be an extension of the population in the Poconos and the North Mountain area. The lack of such a nearby source population may explain why there were very few atlas records in the central Ridge and Valley, where similar habitat exists (D. Gross, pers. comm.). Outside of mountainous areas, the northwest corner of the state has been a stronghold as far back as early in the last century (Todd 1940), possibly due to the northern hardwood forests of maple and beech that provide favored buds for feeding in early spring (Master 1992i).

During the second Atlas, the Purple Finch was recorded in over one-third of all blocks, for a total of 1,748. This represents an increase of 21 percent over the first Atlas, although when corrected for increased survey effort across most of its range (chap. 5), the estimated change is reduced to 8 percent. The greatest increase in block detections was in the Deep Valleys Section in north-central Pennsylvania, partly explained, perhaps, by low coverage there during the first Atlas. Indeed, it is interesting to note that while the range has consolidated considerably in the northern tier, in some peripheral and low lying areas (e.g., Fayette and Schuylkill counties) there is evidence of localized range contraction.

Nonetheless, the range increase correlates with Breeding Bird Survey results, which show a modest, albeit nonsignificant, increase over the last 40 years (Sauer et al. 2011). Second Atlas point count data estimate the current population to be 50,000 singing males. Given the long-term population gain of recent decades and their use of a wide variety of habitats, the Purple Finch seems secure in the state.

GREG GROVE

Distribution

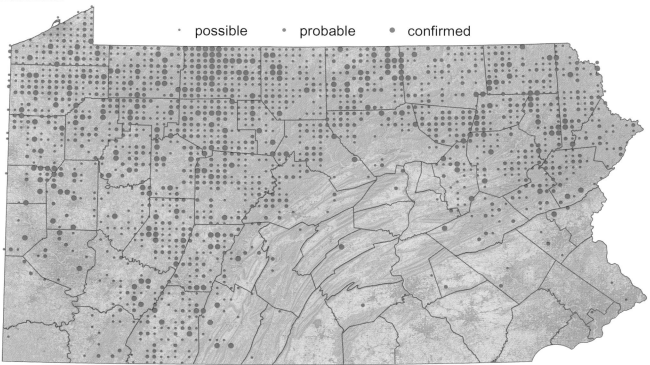

Distribution Change

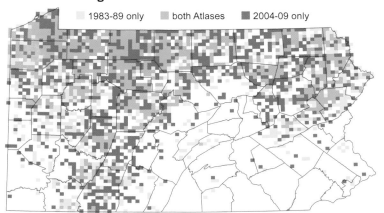

Number of Blocks

	first Atlas 1983–89	second Atlas 2004–09	Change %
Possible	638	931	46
Probable	494	586	19
Confirmed	316	231	−27
Total	1,448	1,748	21

Population estimate, males (95% CI):
50,000 (45,000–56,000)

Density

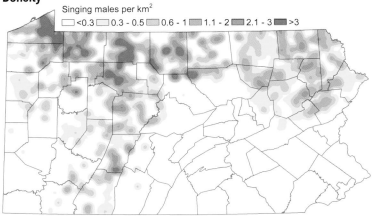

Breeding Bird Survey Trend

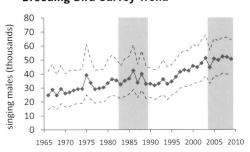

PURPLE FINCH 459

Bob Wood and Peter Wood

House Finch
Carpodacus mexicanus

The story of the House Finch in the eastern United States provides excellent examples of the process of natural selection and of the population dynamics arising from host–parasite associations. Originally a species of the western states, this finch, with its exuberant, warbling song, became a cage favorite. In 1940, birds that had been brought east to Long Island, New York, were released or escaped from captivity (Elliot and Arbib 1953). From there, the House Finch colonized much of the eastern half of the country as numbers increased explosively, until a contagious eye disease entered the population half a century later.

In recent years, eastern birds spreading westward have met the native western population, which has been expanding toward the middle of the country, and the House Finch is now found throughout most of the lower 48 states as well as much of Mexico. The northern edge of its range lies largely within the northern tier of states and extreme southern Canada. In its native West, the House Finch is found in dry grasslands with scattered trees and in open forests, as well as in association with humans and their buildings. In eastern states, however, it associates closely with humans and is rarely found far from human dwellings. The Houses Finch's successful colonization of the East was clearly influenced by the density of the human population and access to numerous winter-feeding stations (Hill 1993).

The House Finch had established populations in Pennsylvania by the early 1970s, and perhaps much earlier (Middleton 1963; Pepper 1972). Pennsylvania Breeding Bird Survey (BBS) data document two decades of explosive growth beginning in the 1970s and continuing through the first Atlas period (Sauer et al. 2004). During the second Atlas, the House Finch was recorded in 81 percent of all blocks, with little change in distribution since the first Atlas. It is found throughout Pennsylvania except in the sparsely populated, heavily forested north-central counties and in the Allegheny National Forest. It is also absent from the state's largest state forests—for example, Rothrock and Bald Eagle in the center of the state, Delaware in the northeast, and Forbes in the southwest. As might be expected, the highest densities coincide with high human density, especially throughout the Piedmont in the southeast, Pittsburgh and surrounding towns, the Scranton/Wilkes-Barre corridor, and in Erie and other northwestern population centers.

Beginning in the Washington, D.C., area in 1993, and then within 2 years in Pennsylvania, House Finches were infected by a highly contagious eye disease, often referred to as conjunctivitis, caused by the bacterium *Mycoplasma gallisepticum* (Ley et al. 1996; Fisher et al. 1997). The disease spread rapidly, facilitated by the high population density; many birds perished, and the overall population declined dramatically within a couple of years, as shown by BBS and Christmas Bird Count data (Hochachka and Dhondt 2000). The disease is not always fatal, and infected survivors develop some immunity, thus acting as a reservoir for continuance of the disease in the population (Faustino et al. 2004; Sydenstricker et al. 2005). By the time of the second Atlas, the counterbalance of disease persistence versus lower numbers (leading to less efficient transmission) appears to have resulted in population equilibrium, at a much lower level than the peak of the previous decade. Based on the point count data from the second Atlas, the current number of singing males in the state is about 420,000, compared with an estimated high of over 700,000 in the mid-1990s.

Despite the population decline, the House Finch remains numerous and will likely remain a common component of Pennsylvania's avifauna. However, given the tenuous nature of the host–parasite relationship, we may see further fluctuations in House Finch numbers—an increase if natural selection produces a strain of birds with greater immunity to *Mycoplasma gallisepticum*, or, conversely, another crash, should a more virulent strain of the bacterium emerge. This host–parasite interaction will continue to be of interest for decades to come.

GREG GROVE

Distribution

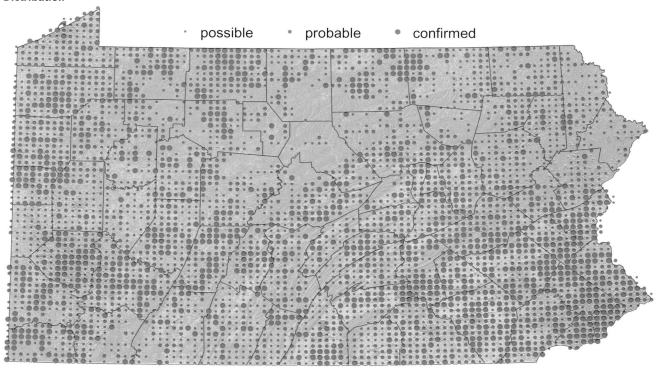

Distribution Change

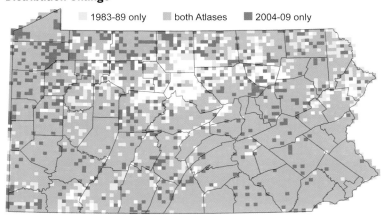

Density

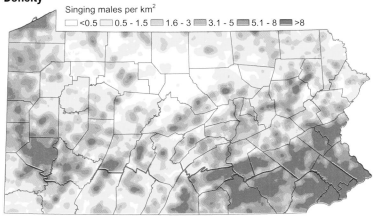

Number of Blocks

	first Atlas 1983–89	second Atlas 2004–09	Change %
Possible	878	1,620	85
Probable	1,207	1,338	11
Confirmed	1,809	1,037	−43
Total	3,894	3,995	3

Population estimate, males (95% CI):
420,000 (395,000–445,000)

Breeding Bird Survey Trend

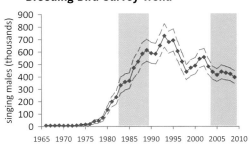

HOUSE FINCH 461

Christopher Wood

Red Crossbill
Loxia curvirostra

A finch that truly is a conifer specialist, the Red Crossbill is named for its fully "crossed" mandibles used to deftly extract seeds from cones. It breeds in conifer forests from Alaska across southern Canada to Newfoundland, south through the Sierras, Rockies, and Appalachians, and can be found in mountain pine forests of Central America (Adkisson 1996). It is a rare breeder in Pennsylvania, but considering the size of Pennsylvania's forests and the remoteness of some conifer stands, Red Crossbills probably are underreported. There is a long history of Red Crossbills nesting occasionally in the higher elevation forests of Pennsylvania and more rarely at lower elevations (Fingerhood 1992a).

Differences in morphology, ecology, and flight call vocalizations separate the species *curvirostra* into 10 "call-Types," or simply Types (Groth 1993a, 1993b; Benkman 1999; Irwin 2010). Each Type has a unique bill depth that is adapted for feeding on key conifers (Benkman 1993a) that are found in a particular core range of each Type (Dickermen 1987; Knox 1992; Kelsey 2008; Young 2011a). Four Types, some likely representing incipient species or possibly full species (Parchman et al. 2006), have occurred in Pennsylvania (Hess et al. 1998; D. Gross, pers. obs.): small-billed Type 3, medium-billed Type 1, slightly smaller medium-billed Type 10 (formerly lumped with Type 4; Irwin 2010), and large-billed Type 2. Medium-billed Type 4 has been documented in adjacent states (Young 2011b), and Type 1, which finds its home in the Appalachians from southern New York southward, is expected to be the most frequent breeder in Pennsylvania (Groth 1988; Gross and Young, pers. obs.).

Historical evidence suggestive of Red Crossbill nesting exists for nearly every county across the northern tier of Pennsylvania (Warren 1890; Pennock 1912; Todd 1940; Fingerhood 1992a; Brauning et al. 1994). Because there were only 6 records in the first Atlas and 11 in the second, it is difficult to draw any conclusions regarding range changes. In Pennsylvania, the Red Crossbill nests when spruce, pine, or hemlock produce adequate cone crops. Breeding occurs mostly in two cycles: January through April, corresponding with the previous year's cone crop, and July through August, correlating with developing spruce cone crops.

Except for a singing bird at Spruce Flats Bog, on Laurel Hill, all records from the second Atlas came from northern counties of the Appalachian Plateaus. Of the 11 records, 6 were of possible breeding, 1 probable, and 4 confirmed. Confirmations consisted of a female feeding dependent young at Algerine Swamp (Lycoming County), tailless fledglings at Pocono Environmental Education Center (Pike County), a female with brood patch at a boreal bog (Monroe County), and dependent young in a Red Pine plantation (McKean County). Three of these confirmed records were associated with spruce in late July through August.

Since Red Crossbills often nest in late winter/early spring or late summer, both the Breeding Bird Survey (BBS) and breeding bird atlases do not adequately sample this species. BBS data for crossbill are unreliable due to small sample sizes and large between-year fluctuations, although they do suggest downward trends (Sauer et al. 2011). Estimates of breeding population size should be viewed with caution; Pennsylvania probably supports a small percentage of Type 1 population (Young 2011b).

Logging of mature coniferous forests that produced the most reliable cone crops (Benkman 1993b) in the late nineteenth century likely caused the decline of Red Crossbills in the northeastern United States (Dickerman 1987). Despite recent forest maturation, logging of conifer forests and pest outbreaks such as the Hemlock Woolly Adelgid will continue to compromise habitat, particularly for Type 1 (Gross 2010d; Young 2011b). Like other conifer associates, crossbills would benefit from management strategies that restore conifers to their historical presence. Type 1 should be surveyed in July through August, when it nests in Red Spruce, and efforts should be taken to record crossbill vocalizations, especially given the future potential for some Types to be awarded full species status, possibly with distinct conservation plans.

MATTHEW A. YOUNG AND DOUGLAS A. GROSS

Distribution

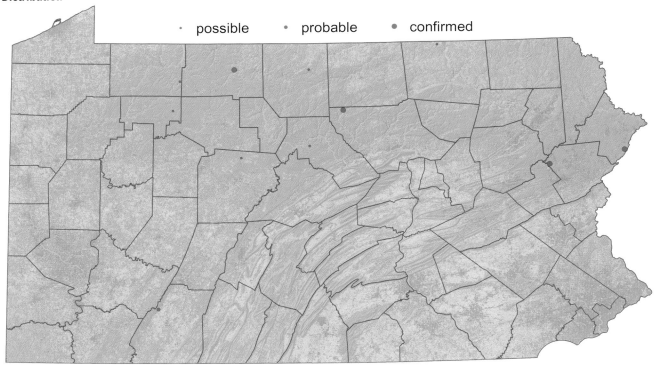

Distribution Change

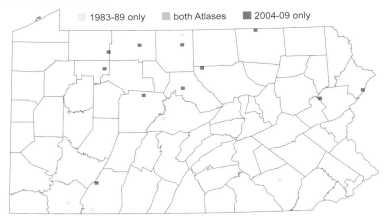

Number of Blocks

	first Atlas 1983–89	second Atlas 2004–09	Change %
Possible	6	6	0
Probable	0	1	∞
Confirmed	0	4	∞
Total	6	11	83

Gerard Bailey/VIREO

Pine Siskin
Carduelis pinus

The nomadic Pine Siskin inhabits coniferous forests in a band from Alaska, across Canada and the upper Great Lakes into the Northeastern states, south through the Appalachians to Georgia, in montane coniferous forests of the western United States, and into Central America (Dawson 1997). Pennsylvania is near the southern extent of its nesting range and supports a very small percentage of its nesting population (Rich et al. 2004). A rare breeder in the state in most years, it can become locally common intermittently, especially near feeder stations (Todd 1940; McWilliams and Brauning 2000). Breeding events are a response to both conifer forest seed crops and artificial feeder stations (Todd 1940; McWilliams and Brauning 2000), especially during larger irruptions that take place every 2 to 5 years (Young 2008b; Bolgiano 2010). Nesting records in northern mountainous counties date back to the late nineteenth century (Warren 1890; Todd 1940), with large numbers in some years (Harlow 1951).

The breeding range of the Pine Siskin changes from year to year, making it difficult to draw definitive conclusions regarding range changes for such an unpredictable breeder. The Pine Siskin was documented in 71 atlas blocks during the first Atlas (Gross 1992j) and 93 blocks in the second Atlas, a 31 percent increase. There were more records in the southern half of the state during the second Atlas, possibly due to variables such as locally fluctuating food sources (i.e., cone crops and feeders) or the distribution of atlas volunteers alert to siskins during years of increased nesting.

The mean date for breeding confirmations in the second Atlas was 20 May (ranging from 3 March to 30 July). Because of the Pine Siskin's early breeding season (Dawson 1997), it is likely that nesting siskins were overlooked outside of the main atlas survey season (mid-May to mid-July); hence, its breeding status may be underreported by bird atlas projects (Gross 1992j). Similarly, although Breeding Bird Survey (BBS) data show a significant decline in Pine Siskin populations since 1966 (Sauer et al. 2011), the BBS likely does not monitor this species adequately, due to the early breeding season and large fluctuations in numbers from year to year (Dawson 1997).

The Pine Siskin nests in areas with well-stocked feeders or in coniferous forests of spruce, Eastern Hemlock, and/or Eastern White Pine. The vast majority of appropriate breeding habitat exists on the Appalachian Plateaus and, to a much lesser extent, the Ridge and Valley sections. Although the siskin was detected during every year of the second Atlas, the irruption years of 2004 and 2009 accounted for the majority of all records: breeding was confirmed in 10 blocks in 2004 and in 12 blocks in 2009, but in only 12 in the four intervening years, with none in 2007. The only two lowland and suburban records from both atlases were likely associated with feeding stations. Perhaps the largest cluster of sightings for both atlases came from the Deep Valleys Section in north-central Pennsylvania, an area where conifers and hamlets with feeders can be found in close proximity.

Like the Red Crossbill (Benkman 1993b), the Pine Siskin relies on the availability of mature conifer forests, which produce the most reliable and largest cone crops. Mature conifer forests and plantations provide a key habitat for these species. Hemlock mortality due to the Hemlock Woolly Adelgid could have a deleterious effect on the siskin. Proactive management for mature conifer forests, including thinning and other techniques to increase growth of conifers, would assist the Pine Siskin and other native conifer breeding birds (Fajvan 2008; Rentch et al. 2007).

MATTHEW A. YOUNG AND DOUGLAS A. GROSS

Distribution

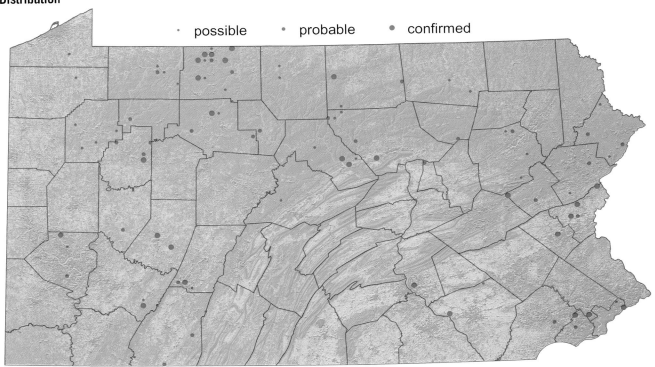

Distribution Change

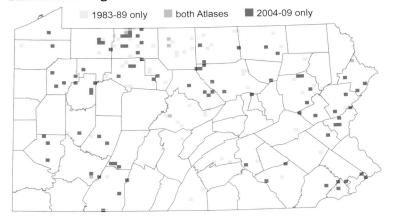

Number of Blocks

	first Atlas 1983–89	second Atlas 2004–09	Change %
Possible	36	38	6
Probable	14	28	100
Confirmed	21	27	29
Total	71	93	31

PINE SISKIN

Bob Wood and Peter Wood

American Goldfinch
Carduelis tristis

One of the most common and easily recognized birds in Pennsylvania is the American Goldfinch. The bright-yellow and black plumage of the adult male, its undulating flight, and the high-pitched flight calls of the "canary bird," as some people still call it, draw attention its way. It is found predominantly in open country with weedy fields, but also along roadsides, on woodland edges, and areas of early successional growth (McGraw and Middleton 2009). It breeds from southern Canada south through the northern three-quarters of the United States, but not in the southwestern states or along the Gulf Coast.

The American Goldfinch is gregarious both on migration, when flocks in the hundreds can be found, and during the breeding season, when group displays by males are performed over breeding territories. This species benefited from the clearing of forests by early settlers and continues to benefit when forests are opened up. Its numbers decline when plants considered to be weeds are removed, since it feeds primarily on seeds of unwanted plants such as burdock and thistles. It often feeds (and, in fact, seems to prefer to feed) while perching atop a stalk or hanging upside down (Coutlee 1964).

The American Goldfinch double- and even triple-broods in some parts of the country (Baicich and Harrison 2005), beginning its first nest in midsummer, when most other species are either done nesting or starting their second brood. Nesting is coordinated with the maturation of thistledown and cattail down, both of which are used to line its small, tightly woven nest of various plant materials (Harrison 1975). The nest is usually placed in the branches of a tree, from only a few feet off the ground to higher than 10 m (30 ft). It has even been documented nesting in the same thistle plant whose down it uses in its nest and whose seeds it eats and feeds to its young (Bent and Austin 1968). It will use whatever trees or shrubs are available and has even been documented nesting in a cornstalk (Zuefle et al. 2009). All nests are close to good food sources, which seem to be more important than the shrub, tree, or plant selected for the nest location.

In the second Atlas, volunteers found the American Goldfinch in 96 percent of blocks statewide. There were only 29 blocks in which this species was found in neither the first or second Atlas; these included some blocks in large contiguous forests and a few in urban areas of Philadelphia and Pittsburgh. Overall, there is little evidence that this species' status has changed between atlas periods, with no significant changes in blocks with goldfinch records at the scale of physiographic provinces or sections (appendix C).

Goldfinch numbers on atlas point counts were relatively uniform across the state, with no evidence of elevational or geographic gradients in abundance. The highest densities were associated with suburban and rural development in all parts of the state, and lowest numbers were in areas that are predominantly forested (appendix D). The statewide population estimate of 315,000 singing males is probably conservative, due to the American Goldfinch's late breeding season, which peaks long after the main point count survey season. The earliest confirmed breeding was a bird carrying nesting material on 2 May, and the latest was of young being fed on 14 October. Most of the confirmed reports of young or occupied nests fell in the period of late June through mid-September, when atlasing activity was minimal.

Breeding Bird Survey counts in Pennsylvania have fluctuated, but they show little overall trend since the 1960s (Sauer et al. 2011). Counts during the second Atlas period were slightly higher than in the first Atlas period; however, due to the apparently cyclical nature of populations, there is no evidence that goldfinch populations have changed in recent decades. At present, the future looks good for the American Goldfinch, one of the most widespread species in Pennsylvania.

ARLENE KOCH

Distribution

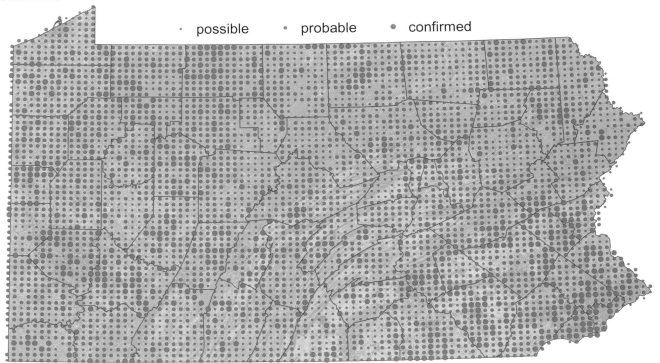

Distribution Change

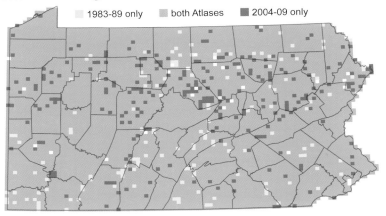

Density

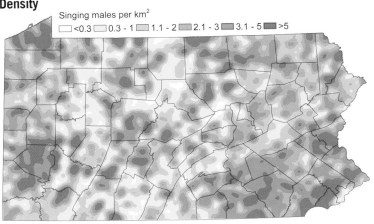

Number of Blocks

	first Atlas 1983–89	second Atlas 2004–09	Change %
Possible	1,183	1,665	41
Probable	2,602	2,488	−4
Confirmed	865	605	−30
Total	4,650	4,758	2

Population estimate, males (95% CI):
315,000 (286,000–338,000)

Breeding Bird Survey Trend

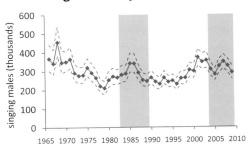

AMERICAN GOLDFINCH

Rob Curtis/VIREO

House Sparrow
Passer domesticus

In the New World, the House Sparrow is a highly invasive species whose populations stem from a deliberate release of 16 birds in Brooklyn in 1851 and other subsequent introductions (Moulton et al. 2010). Maligned as an agricultural pest and threat to native bird species as far back as the 1880s (Moulton et al. 2010), the House Sparrow is now found throughout the United States and temperate Canada and is considered one of the most abundant birds on the North American continent (Rich et al. 2004).

The House Sparrow was well established in Pennsylvania by 1890, and numbers may have peaked shortly afterward, when farmland extent and horse-drawn transport were at their height (McWilliams and Brauning 2000). By the first Atlas, this species had retreated from some reforested areas but was still ubiquitous in urban areas and farmland. During the years between the two atlases, the contraction from densely forested areas continued, notably in the Glaciated Pocono Plateau, where there was a 58 percent reduction in occupied blocks. There were also modest range contractions in the Laurel Highlands and vicinity of the Susquehannock State Forest and Allegheny National Forest.

Second Atlas point count data show that the House Sparrow's highest densities are found in the largest urban areas, notably the Philadelphia and Pittsburgh metropolitan areas. Not surprisingly, the highest overall densities were found in Pennsylvania's most urbanized county— Philadelphia—estimated around 90 birds per km^2. In some rural areas, particularly in the Piedmont, densities are almost as high as in urban areas, a reflection of the large numbers typically found in farmyards. The statewide population is estimated to be 1.5 million birds; however, this could be a considerable underestimate, due to the unsuitability of the point count methods for counting this species (chap. 5).

There can be little doubt that the House Sparrow was once much more abundant than it is now. In the early decades of the twentieth century, when most of Pennsylvania was farmland, it is conceivable that this was the most abundant bird in the state. Breeding Bird Survey data show a sustained population decrease since the mid-1960s (Sauer et al. 2011). However, population declines appear to have leveled off during the mid-1990s.

The substantial declines in House Sparrow numbers in Pennsylvania and elsewhere in the United States (Lowther and Cink 2006; Sauer et al. 2011) are not in isolation. Rapid population declines in the species' native range in Europe have received considerable attention from conservationists and even national newspapers (Summers-Smith 2003). A plethora of hypotheses for the causes of the declines has been proffered, including the resurgence of Eurasian Sparrowhawk populations (Bell et al. 2010) and changes in agricultural practices (Hole et al. 2002). In all likelihood, the causes of the declines in its native range are manifold, and many of them have analogies in the New World. Unique to North America, however, is evidence that the House Sparrows is outcompeted by the House Finch (Cooper et al. 2007), which colonized the eastern United States rapidly during the middle decades of the twentieth century. Interestingly, the slowing of the decline in House Sparrow numbers in Pennsylvania since the mid-1990s coincides with a sharp decline in House Finch numbers due to mycoplasmal conjunctivitis.

The House Sparrow is one of a few bird species offered no protection under U.S. federal statutes; as such, it is actively controlled where it is considered a pest. However, the species breeds almost year-round—during second Atlas fieldwork, the birds were noted carrying food as early as 15 January and feeding young as late as 11 September. It is incredibly fecund, and localized population control will only succeed where maintained indefinitely. The House Sparrow has proven to be among the most adaptable of all bird species, so it seems that this unwanted non-native species will continue to find a home in Pennsylvania for the foreseeable future.

ANDREW M. WILSON

Distribution

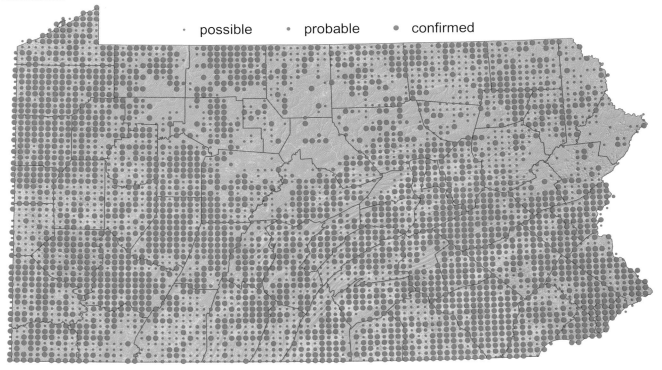

Distribution Change

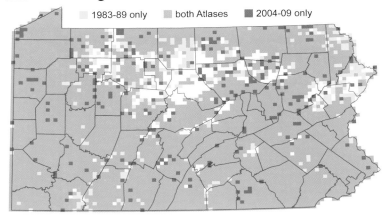

Density

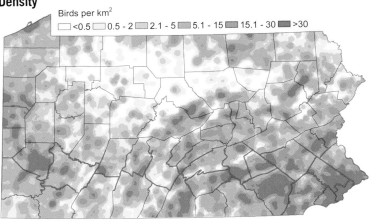

Number of Blocks

	first Atlas 1983–89	second Atlas 2004–09	Change %
Possible	696	1,122	61
Probable	672	649	–3
Confirmed	2,915	2,435	–16
Total	4,283	4,206	–2

Population estimate, birds (95% CI):
1,530,000 (1,460,000–1,600,000)

Breeding Bird Survey Trend

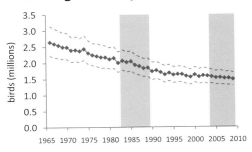

HOUSE SPARROW 469

APPENDIX A | Former Nesting Species

Daniel W. Brauning

All species known to have bred in Pennsylvania are summarized within this volume, either in full species accounts or in this appendix if the species was not reported with at least probable breeding evidence during the second Atlas. Species with no recent breeding activity (since the first Atlas, ending in 1989) are simply listed in Table A.1. Thorough accounts of extirpated birds were provided in the first Atlas (Brauning 1992a) and are not replicated here.

Receiving less attention in this book are the "stragglers," or accidental summer records of species for which errant migration or individuals lingering after normal spring migration resulted in their presence during the summer, possibly even "pairs" throughout the breeding season. A wide array of waterfowl (sometimes incapable of flight), other water birds, or any highly migratory species could fall into this category. This volume's focus is breeding birds, and thus it summarizes the evidence and distribution of species documented breeding within Pennsylvania, but not others.

Some species have nested in Pennsylvania as accidental visitors, well outside their normal breeding range. This may represent early signs of range expansion, opportunistic exploitation of temporarily created habitat, or an anomalous human-assisted event. The expansion into Pennsylvania of nesting Merlins, documented during the second Atlas, is part of a continental expansion in range, whereas historic breeding by Least Terns, on dredge spoil banks in the Delaware River, appears to be an anomaly. Among the stranger examples in Pennsylvania's historic breeding bird history was the inclusion of the Eurasian Jackdaw (*Corvus monedula*) in the first Atlas, a bird which at the time was recognized on the official list of North American Birds (AOU 1985) but which was subsequently determined to be man-assisted and removed from the list of North American birds by the American Ornithologists' Union (AOU 1998).

Newly formed impoundments may at times provide irresistible nesting habitat for migrant waterfowl already paired during the winter or early spring. Such rarities add color and depth to the state's breeding bird list, and certainly excitement for local birders. Various examples of this have been documented in Pennsylvania's history, and a few such examples were observed during the second Atlas. The standard for a species' inclusion in this volume is that it must be a wild species recognized by the American Ornithologists' Union, with confirmed breeding documentation at some point during its history in Pennsylvania. The following brief summaries document the records that failed to meet the confirmed breeding standard during the second Atlas but which warrant a reference to past breeding activity.

Atlas breeding codes attempt to characterize the full array of breeding bird behaviors, but some gray areas remain. For example, the Laughing Gull was reported copulating at landfills in Bucks County (McWilliams and Brauning 2000), but the closest known nesting sites remain along the New Jersey coastline, at least 58 miles to the east. Similarly, the Glossy Ibis was annually seen in southern Chester County during the second Atlas, apparently within a day's round trip from the well-established colony at Pea Patch Island in the state of Delaware (Parsons 1995). These observations suggest that Pennsylvania may play a role in the nesting ecology for species in which nesting is not known, or expected, to occur within the state. Observations of these species by birders may technically have achieved a probable breeding status according to atlas codes, but they do not reflect a local (state) breeding species record. For this reason, a complete species account is not provided for species breeding only in neighboring states, even if Pennsylvania habitats are used during the breeding season. Inclusion in this appendix suffice to document these summer, nesting-season observations within Pennsylvania's borders.

Following is a summary of species observed with probable evidence of nesting during the second Atlas. Also included are species for which breeding has been documented in the past 20 years.

AMERICAN WIGEON
Anas americana

A pair of American Wigeon at Lake Lloyd, Potter County, from 1 to 17 June 2004 provided the species' only probable breeding evidence during the second Atlas. The birds were, according to Chuck Chalfant (pers. comm.), acting "very much like a pair." Lake Lloyd is a small reservoir slightly over 1 mile from the New York border. Twenty-two scattered blocks provided some form of evidence for this species in New York's second atlas between 2000 and 2005 (McGowan 2008zz). The only other second Atlas breeding record in Pennsylvania was of a single male in Crawford County. The American Wigeon regularly nests from Alaska, across Canada, south into the northern prairie pothole region of the midwestern United States, and east into New York. The closest regular breeding locality to Pennsylvania is probably the Erie-Ontario Lake Plain of western New York (McGowan 2008zz). Historically, nesting was documented in Pennsylvania only sporadically (Hartman 1992f), but with the population retracting its range even to the north, this is expected to be only an accidental occurrence in the future.

NORTHERN SHOVELER
Anas clypeata

The Northern Shoveler went unreported during the second Atlas. In the first Atlas, there was a single confirmed breeding record (Hartman 1992g), and this species is believed to have nested successfully in 1990 in the extensive wetlands of Pymatuning and Conneaut Marsh, Crawford County (McWilliams and Brauning 2000). New York's second atlas confirmed scattered nesting events along the Great Lakes plains as well as on Long Island (McGowan 2008zza), far from the predominantly western North American breeding range of this species. Continued searches for nesting activity by this species in large wetlands could be warranted.

RING-NECKED DUCK
Aythya collaris

Two instances of Ring-necked Duck pairs were reported in the second Atlas: one on a tailings pond in Clifford Township, Susquehanna County, and another pair at a small pond near Jonas, in western Monroe County. The pair in Monroe County remained for several days, although the other pair was not re-observed. An additional 32 records, submitted to the second Atlas as observations, did not meeting even the lowest level of breeding activity evidence. These include single males or females, some of which remained for several weeks. None of these observations provide convincing evidence that local nesting likely occurred. The only historic evidence of breeding by the Ring-necked Duck in Pennsylvania occurred in the 1930s, shortly after Pymatuning Reservoir was flooded (Todd 1936). That event lured ring-necks, redheads, and Ruddy Ducks to nest for the first time in the newly created waters. During the twentieth century, the Ring-necked Duck expanded its breeding range southward into New York, where the population continued to expand to the point of being widespread during their second atlas (McGowan 2008zzb). The growing regional population and annual occurrence of summering individuals suggest that this cavity-nesting diving duck should be looked for again as a potential breeding bird in Pennsylvania.

RUDDY DUCK
Oxyura jamaicensis

This distinctive waterfowl, the sole member of its genus regularly occurring in North America, may be found annually summering in Pennsylvania's waters. It appears to be more opportunistic in breeding behavior than other waterfowl that nest incidentally here, far from their native prairie pothole region in central North America. The 23 observations submitted to the second Atlas were scattered across the state, generally involving single birds during June and July. Pairs observed during May were excluded from the second Atlas because of the constraints of safe dates. Persistent observations of pairs, including a female with "five half-grown young," in 2001 (T. Bledsoe, pers. comm.) around the Pymatuning Reservoir near the spillway suggest that nesting may have been a regular occurrence prior to the start of the second Atlas, but this was incompletely documented during the second Atlas. The most recent confirmed breeding of Ruddy Duck in Pennsylvania occurred at Glenn Morgan Lake, Berks County, during 1997 and 1998, and was associated with similar behavior by a number of other rare waterbirds and marsh birds (Keller 1997; McWilliams and Brauning 2000). At this time, however, as in New York (McGowen 2008zzc), it is not clear whether Ruddy Duck has any established population in Pennsylvania.

SNOWY EGRET
Egretta thula

Little has changed in the breeding status of the Snowy Egret since the end of the first Atlas. Pennsylvania's last confirmed breeding event was at Rookery Island, Lancaster County, in 1987, but that colony was abandoned by 1989 (Schutsky 1992g), and no other nesting attempts have been documented in Pennsylvania. This egret nests in the states surrounding Pennsylvania: in New York, around New York City and Long Island (McCrimmon 2008c); in Maryland at various points in the Chesapeake Bay

(Therres 2010c); and in Ohio, on West Sister Island, in the western basin of Lake Erie (OBBA 2011). However, prospects for nesting in Pennsylvania appear dim. Four June observations were submitted to the second Atlas, including an observation of four individuals at Middle Creek Wildlife Management Area in Lancaster County, but none of the records was considered an observation of anything more than non-breeding vagrants. Since Pennsylvania is at the northern edge of the eastern inland breeding range, recolonization here, away from the Wade Island Great Egret colony, appears unlikely.

CATTLE EGRET
Bubulcus ibis

Like the Snowy Egret, the last (and in this case, only) nesting colony of Cattle Egret in Pennsylvania's history was at Rookery Island, Lancaster County, in 1988, after growing to a maximum post-nesting population of 7,500 birds in 1981 (Schutsky 1992h). Since the dramatic abandonment of that colony in 1989, the Cattle Egret has become a rare bird in Pennsylvania. Birds now rarely wander into Pennsylvania from colonies in neighboring states, all of which (except West Virginia) have at least one active colony.

PIPING PLOVER
Charadrius melodus

This species has been an extirpated breeder since the late 1950s, when the Piping Plover abandoned the state's only historic nesting location, the beaches of Presque Isle (Stull et al. 1985). This site, which once supported up to 15 pairs, was part of the Great Lakes subpopulation, which has been federally listed as Endangered by the U.S. Fish and Wildlife Service. The Great Lakes population declined to a low of 17 pairs, restricted to Michigan shorelines in 1986, due to loss of habitat and increasing recreational use of beaches (USFWS 1985; USFWS 2003b). In response to intensive conservation efforts, this meager breeding population began to recover, and in 2009 it reached a total of 71 pairs across three U.S. states and Ontario, Canada (Cuthbert and Roche 2009; USFWS 2009). Recolonization of historic Great Lakes nesting beaches in recent years and continued availability of breeding habitat in the Gull Point Natural Area of Presque Isle State Park suggest that recovery potential exists in Pennsylvania, despite its many challenges (Haffner 2007). In order to increase the likelihood that Piping Plover pairs will return to breed in the Commonwealth, resource managers (under Game Commission guidance) continue to address factors impeding this species' recolonization potential, including vegetation encroachment, habitat disturbance, and high recreational use of Presque Isle's beaches.

BLACK-NECKED STILT
Himantopus mexicanus

This species' first Pennsylvania nest was found by Steve Santner in 1989, during the extended year of the first Atlas, but the last instances were in 1991, all at the Philadelphia sewage treatment plant (Boyle et al. 1992; Santner 1992h). Optimistic expectations following the first Atlas documentation of this unlikely Pennsylvania breeder (Santner 1992h) have proven unrealized. Stilts continue to be reported in Pennsylvania about every other year, but without nesting evidence.

COMMON TERN
Sterna hirundo

No breeding observations of Common Tern were recorded during the first or second Atlases, but a colony was active on Presque Isle from 1926 until 1966. Nesting attempts were documented again in 1995, when two nests with eggs were located, but neither nest was successful (McWilliams 1995). Nesting attempts also occurred on Presque Isle in 2012. Substantial inland colonies persist north of Pennsylvania, in New York's Lake Oneida, and also at Buffalo Harbor, on the eastern shore of Lake Erie, where there is a colony of 1,100 pairs (Richmond 2008b).

LEAST TERN
Sterna antillarum

This species has historically been only an accidental breeder in Pennsylvania and was considered extirpated at the time of the first Atlas (Fingerhood 1992b). A single instance of courtship behavior was observed by experienced birders in the lower Susquehanna River at Washington Boro, Lancaster County. Up to four birds were observed on 1 May 2007, with behavior that appears to have included courtship feeding, although nesting habitat is not expected there. Pennsylvania's two opportunistic breeding events were on dredge spoil islands in the Delaware River, near the Tinicum (John Heinz National Wildlife Refuge) in Philadelphia. The Least Tern nests on beaches, rooftops, and parking lots scattered around the Chesapeake Bay in Maryland and to the north at the Aberdeen Proving Grounds, suggesting that alternate nest sites may be available (Brinker 2010).

OLIVE-SIDED FLYCATCHER
Contopus cooperi

Observations of Olive-sided Flycatcher in the second Atlas include a single probable breeding report, in which a singing male was heard on "at least 12 occasions, always in the same three locations around my home" as late as 11 June 2007 in Potter County (T. McKenrick, pers. comm.). This species is a late migrant, and transients lingering into June were excluded from atlas entry by the 10 June safe date. This probable record remains inconclusive for nesting, since no pairs or other breeding behavior were observed. A late-July observation in the Poconos was intriguing, but it was considered a wandering individual. Olive-sided Flycatcher observation through the summer of 1993 at the Tionesta Natural and Scenic Research area during a year of heightened Elm Spanworm invasion (Gross 2010e) provided one of the most intriguing suggestions of nesting since the previous twentieth-century confirmation of nesting in Warren County, in the 1930s (Poole 1964).

EVENING GROSBEAK
Coccothraustes vespertinus

Within the past decade, the Evening Grosbeak could only be considered a rare Pennsylvania visitor, even in the wintertime. Breeding had been suggested by Pennsylvania ornithologists prior to the first Atlas (Wood 1979), but nesting was not adequately documented during the first Atlas to warrant a reference. A series of observations during the summer of 1994 in Wyoming County led to the discovery, by Skip Conant, of paired grosbeaks, and in mid-July, recently fledged young were seen being fed by adults in Forkston Township, Wyoming County (Conant 1994); no nest was ever located. The state's first breeding confirmation was associated with an Elm Spanworm infestation, which bolstered many birds' populations. The Evening Grosbeak was not observed in that area in the following year, and winter irruptions were waning by the time of this event, so subsequent breeding within Pennsylvania seems highly unlikely.

TABLE A.1 Formerly breeding birds for which no substantive evidence of nesting has been observed since publication of the first Atlas (Brauning 1992a)

Species	Last PA breeding county, and year	Citation
Gadwall (*Anas strepera*)	Conneaut Marsh, Crawford, 1964	Hall 1964
Northern Pintail (*Anas acuta*)	Tinicum Refuge, Philadelphia, 1966	Miller 1966b
Redhead (*Aythya americana*)	Pymatuning Lake, Crawford, 1937	Todd 1940
Common Loon (*Gavia immer*)	Pocono Lake, Monroe, 1946	Street 1956
Greater Prairie-Chicken (*Tympanuchus cupido*)	Poconos, 1870s (extinct subspecies)	Fingerhood 1992c
Black Rail (*Laterallus jamaicensis*)	Centre County, 1986	Brauning 1992
Passenger Pigeon (*Ectopistes migratorius*)	Clinton County, 1892 (extinct)	Townsend 1932
Bewick's Wren (*Thryomanes bewickii*)	Lycoming County, "late 1970s"	Schweinsberg 1988
Bachman's Sparrow (*Aimophila aestivalis*)	Washington County, 1937	Todd 1940
Lark Sparrow (*Chondestes grammacus*)	Alan Seeger, Huntingdon County, 1930	Wood 1932

APPENDIX B Common and Scientific Names of Plants and Animals

PLANTS

alder	*Alnus sp.*	poplar	*Populus sp.*
American Sycamore (sycamore)	*Platanus occidentalis*	Purple Loosestrife	*Lythrum salicaria*
		Red Maple	*Acer rubrum*
Aspen (Quaking Aspen)	*Populus tremuloides*	Red Pine	*Pinus resinosa*
aspen	*Populus sp.*	Red Spruce	*Picea rubens*
Austrian Pine	*Pinus nigra*	Reed Canary Grass	*Phalrus arnudinacea*
Bald Cypress	*Taxodium distichum*	rhododendron	*Rhododendron sp.*
beard moss (old man's beard)	*Usnea sp.*	Scots Pine	*Pinus sylvestris*
		Spatterdock	*Nuphar lutea*
bee balm	*Monarda sp.*	Spotted Jewelweed	*Impatiens capensis*
Bittersweet	*Celastrus scandens*	sedge	*Cyperaceae sp.*
Blackberry	*Rubus sp.*	serviceberry	*Amelanchier sp.*
Black Spruce	*Picea mariana*	Soybean	*Glycine max*
briar	*Smilax sp.*	Spanish Moss	*Tillandsia usneoides*
Bulrush (Lesser Bulrush)	*Typha angustifolia*	sphagnum moss	*Sphagnum sp.*
Buttonbush	*Cephalanthus occidentalis*	spruce	*Picea sp.*
cattail	*Typha sp.*	Tamarack (American Larch)	*Larix laricina*
Chestnut Oak	*Quercus prinus*	viburnum	*Viburnum sp.*
Common Buttonbush	*Cephalanthus occidentalis*	Virginia Pine	*Pinus virginiana*
Common Reed	*Phragmites australis*	Washington Hawthorn	*Crataegus phaenopyrum*
crabapple	*Malus sp.*	Water Willow	*Justicia americana*
dogwood	*Cornus sp.*	White Spruce	*Picea glauca*
Eastern Hemlock	*Tsuga canadensis*	Yellow Poplar	*Liriodendron tulipifera*
Eastern Red Cedar	*Juniperus virginiana*		
Eastern White Pine	*Pinus strobus*		
elm	*Ulmus sp.*		
grape vine	*Vitis sp.*		
greenbrier	*Smilax sp.*		
hawthorn	*Crataegus sp.*		
hickory	*Carya sp.*		
holly	*Ilex sp.*		
honeysuckle	*Lonicera sp.*		
magnolia	*Magnolia grandiflora*		
maple	*Acer sp.*		
Mountain Laurel	*Kalmia latifolia*		
Norway Spruce	*Picea abies*		
oak	*Quercus sp.*		
Orchard Grass	*Dactyus glomerata*		
pine	*Pinus sp.*		
Pitch Pine	*Pinus rigida*		

INVERTEBRATES

cicada	*Cicadidae*
Elm Spanworm	*Ennomos subsignaria*
Emerald Ash Borer	*Agrilus planipennis*
grasshopper	*Caelifera sp.*
Gypsy Moth	*Lymantria dispar*
Hemlock Woolly Adelgid	*Adelges tsugae*
Oak Leaf Roller	*Archips semiferanus*
Rusty Crayfish	*Orconectes rusticus*

FISH

Alewife	*Alosa pseudoharengus*
catfish	*Siluriformes sp.*
trout	*Salmoninae sp.*

REPTILES

Massasauga — *Sistrurus catenatus*

BIRDS

Common Moorhen — *Gallinula chloropus*
Eurasian Sparrowhawk — *Accipiter nisus*
Mexican Whip-poor-will — *Caprimulgus arizonae*
Pacific Wren — *Troglodytes pacificus*

MAMMALS

beaver (North American Beaver) — *Castor canadensis*
bison (American Bison) — *Bison bison*
Black Bear — *Ursus americanus*
Coyote — *Canis latrans*
Fisher — *Martes pennanti*
rabbit — *Sylvilagus sp.*
Raccoon — *Procyon lotor*
rat — *Rattus sp.*
Red Squirrel (American Red Squirrel) — *Tamiasciurus hudsonicus*
Snowshoe Hare — *Lepus americanus*
White-tailed Deer — *Odocoileus virginianus*

APPENDIX C | Summary of Atlas Results by Physiographic Region (Shaded Gray) and Section

How to read this table: 1st = blocks in first Atlas, 2nd = blocks in second Atlas, %change = percentage change in proportion of blocks occupied, sig. = significance of %change using Two-proportion z-test: * = $p < 0.05$, ** = $p < 0.01$, *** = $p < 0.001$, ns = not significant, na = not applicable (gain or loss since first Atlas)

		Appalachian Plateaus	Allegheny Front	Allegheny Mountain	Deep Valleys	Glaciated High Plateau	Glaciated Low Plateau	Glaciated Pocono Plateau	High Plateau	NW Glaciated Plateau	Pittsburgh Low Plateau	Waynesburg Hills	Atlantic Coastal Plain	Central Lowlands	New England	Piedmont	Gettysburg-Newark Lowland	Piedmont Lowland	Piedmont Upland	Ridge and Valley	Anthracite Upland	Anthracite Valley	Appalachian Mountain	Blue Mountain	Great Valley	South Mountain	Susquehanna Lowland
Canada Goose	1st	780	10	60	40	10	139	30	36	208	222	25	19	16	20	365	188	45	132	482	59	14	62	68	166	18	95
	2nd	1946	74	118	203	58	348	37	132	305	531	140	34	29	21	467	217	65	185	1001	122	39	301	70	236	21	212
	%change	149	640	97	408	480	150	23	267	47	139	460	69	67	−4	28	15	44	40	108	107	179	385	3	42	17	123
	sig.	***	***	***	***	***	***	ns	***	***	***	***	*	ns	ns	***	ns	ns	**	***	***	***	***	ns	***	ns	***
Mute Swan	1st	8	0	2	0	0	2	0	0	4	0	0	0	0	0	8	5	0	3	15	1	0	2	1	11	0	0
	2nd	37	0	1	0	0	4	0	3	14	13	2	2	1	1	56	34	7	15	40	2	2	5	8	19	0	4
	%change	363		−50			100		∞	250	∞	∞	∞	∞	∞	600	577	∞	400	167	100	∞	150	700	72		∞
	sig.	***		ns			ns		na	*	na	na	na	na	na	***	***	na	**	***	ns	na	ns	*	ns		na
Trumpeter Swan	1st	0	0	0	0	0	0	0	0	0	0	0	0	0	0	0	0	0	0	0	0	0	0	0	0	0	0
	2nd	0	0	0	0	0	0	0	0	0	0	0	2	3	2	1	1	0	0	1	0	0	0	0	1	0	0
	%change												∞	∞	∞	∞	∞			∞					∞		
	sig.												na	na	na	na	na			na					na		
Wood Duck	1st	916	28	57	105	32	192	32	74	162	180	54	10	11	5	214	103	38	73	442	38	12	161	34	87	17	93
	2nd	1143	35	60	184	49	234	21	85	204	207	64	12	17	12	271	122	42	107	572	52	23	186	48	128	11	124
	%change	25	25	5	75	53	22	−34	15	26	15	19	13	42	119	27	18	11	47	29	37	92	16	41	47	−35	33
	sig.	***	ns	ns	***	ns	*	ns	ns	*	ns	ns	ns	ns	ns	**	ns	ns	*	***	ns	ns	ns	ns	**	ns	*
American Black Duck	1st	167	4	6	12	11	40	22	5	29	19	19	6	2	0	40	13	7	20	82	10	4	15	12	23	2	16
	2nd	63	2	1	8	3	19	7	1	16	6	0	1	1	0	17	10	4	3	29	4	4	4	2	12	0	3
	%change	−62	−50	−83	−33	−73	−53	−68	−80	−45	−68	−∞	−84	−54		−58	−23	−43	−85	−65	−60	0	−73	−83	−48	−∞	−81
	sig.	***	ns	ns	ns	*	**	**	ns	ns	**	na	*	ns		**	ns	ns	***	***	ns	ns	*	**	ns	na	**
Mallard	1st	1460	70	104	155	53	250	43	74	195	402	114	25	22	13	429	205	67	157	949	75	29	331	76	236	24	178
	2nd	1559	56	106	181	47	265	33	88	236	420	127	30	23	20	441	201	68	172	948	116	38	267	73	236	14	204
	%change	7	−20	2	17	−11	6	−23	19	21	4	11	13	−4	40	3	−2	1	10	0	55	31	−19	−4	0	−42	15
	sig.	ns	ns	ns	ns	ns	ns	ns	ns	*	ns	ns	ns	ns	ns	ns	ns	ns	ns	ns	**	ns	**	ns	ns	ns	ns
Blue-winged Teal	1st	71	1	6	3	1	5	1	1	40	13	0	4	3	0	8	4	0	4	14	0	0	3	0	9	0	2
	2nd	25	3	1	0	0	2	0	2	14	3	0	0	1	0	2	2	0	0	4	0	0	1	0	3	0	0
	%change	−65	200	−83	−∞	−∞	−60	−∞	100	−65	−77		−∞	−69		−75	−50		−∞	−71			−67		−67		−∞
	sig.	***	ns	ns	na	na	ns	na	ns	***	*		na	ns	na	ns	ns		na	*			ns		ns		na
Green-winged Teal	1st	18	0	0	3	4	1	0	0	7	2	0	0	1	0	2	2	0	0	2	0	0	1	0	1	0	0
	2nd	9	1	1	0	2	2	0	0	3	0	0	0	0	0	1	0	0	0	4	0	0	0	1	3	0	0
	%change	−50	∞	∞	−∞	−50	100			−∞	−57	−∞		−∞		−∞	−∞			100			−∞	∞	199		
	sig.	ns	na	na	na	ns	ns			na	ns	na		na		na	na			ns			na	na	ns		

(continued)

		Appalachian Plateaus	Allegheny Front	Allegheny Mountain	Deep Valleys	Glaciated High Plateau	Glaciated Low Plateau	Glaciated Pocono Plateau	High Plateau	NW Glaciated Plateau	Pittsburgh Low Plateau	Waynesburg Hills	Atlantic Coastal Plain	Central Lowlands	New England	Piedmont	Gettysburg-Newark Lowland	Piedmont Lowland	Piedmont Upland	Ridge and Valley	Anthracite Upland	Anthracite Valley	Appalachian Mountain	Blue Mountain	Great Valley	South Mountain	Susquehanna Lowland
Hooded Merganser	1st	63	1	0	8	3	8	2	3	36	1	1	1	1	0	1	0	1	0	9	0	0	2	0	2	2	3
	2nd	170	2	3	18	7	32	5	7	71	24	1	0	2	0	10	8	1	1	29	3	0	7	1	11	0	7
	%change	170	100	∞	125	133	300	150	133	97	2300	0	−∞	84		900	∞	0	∞	222	∞		250	∞	448	−∞	133
	sig.	***	ns	na	*	ns	***	ns	ns	***	***	ns	na	ns		**	na	ns	na	**	na		na	*	na	ns
Common Merganser	1st	195	0	0	73	4	56	1	48	5	8	0	0	0	0	4	0	2	2	28	0	2	1	8	6	1	10
	2nd	406	2	5	161	11	94	9	83	16	24	1	2	2	2	22	16	1	5	120	8	17	14	14	23	0	44
	%change	108	∞	∞	121	175	68	800	73	220	200	∞	∞	∞	∞	450	∞	−50	150	329	∞	750	1300	75	282	−∞	340
	sig.	***	na	na	***	ns	**	**	**	*	**	na	na	na	na	***	ns	ns	ns	***	na	***	***	ns	**	na	***
Northern Bobwhite	1st	231	17	19	13	4	26	0	5	15	79	53	4	0	3	212	63	27	122	264	6	4	114	8	84	5	43
	2nd	101	3	13	2	2	12	1	6	12	39	11	0	2	0	41	9	9	23	87	7	0	28	6	28	1	17
	%change	−56	−82	−32	−85	−50	−54	∞	20	−20	−51	−79	−∞	∞	−∞	−81	−86	−67	−81	−67	17	−∞	−75	−25	−67	−80	−60
	sig.	***	**	ns	**	ns	*	na	ns	ns	***	***	na	na	na	***	***	**	***	***	ns	na	***	ns	***	ns	***
Ring-necked Pheasant	1st	886	69	81	33	15	84	2	12	82	355	153	29	1	19	436	198	63	175	825	71	6	264	61	240	16	167
	2nd	491	22	63	16	1	60	2	13	89	163	62	5	0	4	105	55	15	35	452	53	4	142	26	91	5	131
	%change	−45	−68	−22	−52	−93	−29	0	8	9	−54	−59	−84	−∞	−81	−76	−72	−76	−80	−45	−25	−33	−46	−57	−62	−69	−22
	sig.	***	***	ns	*	***	*	ns	ns	ns	***	***	***	ns	***	***	***	***	***	***	ns	ns	***	***	***	**	*
Ruffed Grouse	1st	1942	129	143	363	82	226	44	190	148	497	120	0	5	10	50	35	4	11	775	106	22	371	68	45	28	135
	2nd	1270	88	85	328	71	181	20	159	73	249	16	0	0	0	9	9	0	0	591	107	20	281	35	19	10	119
	%change	−35	−32	−41	−10	−13	−20	−55	−16	−51	−50	−87		−∞	−∞	−82	−74	−∞	−∞	−24	1	−9	−24	−49	−58	−64	−12
	sig.	***	**	***	ns	ns	*	**	ns	***	***	***	na	na	na	***	***	na	na	***	ns	ns	***	***	***	***	ns
Wild Turkey	1st	1671	124	128	331	73	190	35	163	111	423	93	0	7	4	44	30	4	10	676	70	13	376	54	40	28	95
	2nd	2283	103	156	380	93	337	45	184	239	588	158	3	19	18	191	118	10	63	1035	139	40	397	72	141	24	222
	%change	37	−17	22	15	27	77	29	13	115	39	70	∞	149	311	334	292	150	530	53	99	208	6	33	251	−14	134
	sig.	***	ns	ns	ns	ns	***	ns	ns	***	***	***	na	*	***	***	***	ns	***	***	***	***	ns	ns	***	ns	***
Pied-billed Grebe	1st	59	1	2	7	1	7	1	0	24	11	5	1	1	0	4	2	0	2	21	3	0	7	4	5	1	1
	2nd	48	3	2	7	3	7	0	1	15	10	0	4	1	0	4	2	0	2	7	0	0	3	1	2	0	1
	%change	−19	200	0	0	200	0	−∞	∞	−38	−9	−∞	278	−8		0	0		0	−67	−∞		−57	−75	−60	−∞	0
	sig.	ns	ns	ns	ns	ns	ns	na	na	ns	ns	na	ns	ns	na	ns	ns	na	ns	**	na		ns	ns	ns	na	ns
Double-crested Cormorant	1st	3	0	0	0	0	1	0	2	0	0	0	0	0	0	5	3	1	1	1	1	0	0	0	0	0	0
	2nd	36	1	2	4	3	6	2	1	5	11	1	5	3	1	30	13	4	13	36	5	3	2	3	14	0	9
	%change	1100	∞	∞	∞	∞	500	∞	−50	∞	∞	∞	∞	∞	∞	500	331	300	1200	3500	400	∞	∞	∞	∞	∞	∞
	sig.	***	na	na	na	na	ns	na	na	na	na	na	na	na	na	***	*	ns	***	***	ns	na	na	na	na	na	na
American Bittern	1st	38	3	0	10	0	4	0	2	17	2	0	1	2	0	1	1	0	0	11	3	0	1	1	2	1	3
	2nd	24	1	2	4	1	6	0	2	5	1	2	1	0	0	1	1	0	0	5	0	0	2	0	2	0	1
	%change	−37	−67	∞	−60	∞	50		0	−71	−50	∞	−6	−∞		0	0			−55	−∞		100	−∞	0	−∞	−67
	sig.	ns	ns	na	ns	na	ns		ns	**	ns	na	ns	na		ns	ns			ns	na		ns	na	ns	na	ns
Least Bittern	1st	15	0	3	1	0	2	0	0	8	1	0	3	3	0	4	3	1	0	6	1	0	3	0	1	0	1
	2nd	15	1	1	1	0	2	0	0	7	3	0	2	3	0	4	3	0	1	4	0	0	0	1	3	0	0
	%change	0	∞	−67	0		0			−13	200		−37	−8		0	0	−∞	∞	−33	−∞		−∞	∞	199		−∞
	sig.	ns	na	ns	ns		ns			ns	ns		ns	ns		ns	ns	na	na	ns	na		na	na	ns		na
Great Blue Heron	1st	1513	30	27	240	71	337	39	146	253	278	92	11	22	7	207	101	34	72	519	41	16	183	49	126	7	97
	2nd	1658	48	70	255	66	320	33	127	264	359	116	26	25	14	390	184	51	155	795	94	32	252	56	182	11	168
	%change	10	60	159	6	−7	−5	−15	−13	4	29	26	123	4	83	88	81	50	115	53	129	100	38	14	44	57	73
	sig.	*	*	***	ns	ns	ns	ns	ns	ns	**	ns	*	ns	ns	***	***	ns	***	***	***	*	***	ns	**	ns	***
Great Egret	1st	0	0	0	0	0	0	0	0	0	0	0	4	0	0	12	4	6	2	20	3	0	0	0	10	0	7
	2nd	8	1	0	0	0	0	1	0	1	2	3	7	0	0	40	17	7	16	62	6	1	3	4	29	1	18
	%change	∞	∞					∞	∞	∞	∞		65			233	323	17	700	210	100	∞	∞	∞	189	∞	157
	sig.	na	na					na	na	na	na		ns			***	**	ns	***	***	ns	na	na	na	**	na	*

(continued)

		Appalachian Plateaus	Allegheny Front	Allegheny Mountain	Deep Valleys	Glaciated High Plateau	Glaciated Low Plateau	Glaciated Pocono Plateau	High Plateau	NW Glaciated Plateau	Pittsburgh Low Plateau	Waynesburg Hills	Atlantic Coastal Plain	Central Lowlands	New England	Piedmont	Gettysburg-Newark Lowland	Piedmont Lowland	Piedmont Upland	Ridge and Valley	Anthracite Upland	Anthracite Valley	Appalachian Mountain	Blue Mountain	Great Valley	South Mountain	Susquehanna Lowland
Green Heron	1st	1025	38	51	89	38	207	20	53	176	259	94	21	16	12	309	158	44	107	608	49	21	187	64	157	8	122
	2nd	894	24	41	124	32	178	8	49	192	204	42	15	15	12	295	147	38	110	561	54	19	160	51	141	7	129
	%change	−13	−37	−20	39	−16	−14	−60	−8	9	−21	−55	−33	−14	−9	−5	−7	−14	3	−8	10	−10	−14	−20	−11	−13	6
	sig.	**	ns	ns	*	ns	ns	*	ns	ns	*	***	ns	ns	ns	ns	ns	ns	ns	ns	ns	ns	ns	ns	ns	ns	ns
Black-crowned Night-Heron	1st	18	1	1	2	3	3	0	2	2	4	0	5	4	0	82	22	37	23	43	6	6	4	1	23	0	3
	2nd	2	0	0	0	0	2	0	0	0	0	0	6	0	0	34	10	18	6	35	2	4	0	1	25	1	2
	%change	−89	−∞	−∞	−∞	−∞	−33		−∞	−∞	−∞		13	−∞		−59	−55	−51	−74	−19	−67	−33	−∞	0	8	∞	−33
	sig.	***	na	na	na	na	ns		na	na	na		ns	na		*	**	**	ns	ns	ns	ns	ns	ns	na	na
Yellow-crowned Night-Heron	1st	0	0	0	0	0	0	0	0	0	0	0	0	0	0	15	5	9	1	6	0	0	1	0	4	0	1
	2nd	0	0	0	0	0	0	0	0	0	0	0	1	0	0	2	2	0	0	5	0	0	0	0	5	0	0
	%change												∞			−87	−60	−∞	−∞	−17			−∞		25		−∞
	sig.												na			**	ns	na	na	ns			na		ns		na
Black Vulture	1st	13	3	4	1	0	4	1	0	0	0	0	1	0	7	159	74	18	67	172	22	1	51	15	57	14	12
	2nd	25	6	3	4	1	3	1	2	1	3	1	4	0	13	353	154	42	157	416	46	3	113	39	158	17	40
	%change	92	100	−25	300	∞	−25	0	∞	∞	∞	∞	278		70	122	107	133	134	142	109	200	122	160	176	21	233
	sig.	ns	ns	ns	ns	na	ns	ns	na	na	na	na	ns		ns	***	***	***	***	***	**	ns	***	***	***	ns	***
Turkey Vulture	1st	2293	127	169	372	100	334	45	200	251	544	151	16	16	16	433	203	62	168	1147	118	33	449	79	227	31	210
	2nd	2587	123	192	411	101	361	43	218	308	664	166	26	29	23	480	219	66	195	1245	147	41	462	76	243	28	248
	%change	13	−3	14	10	1	8	−4	9	23	22	10	53	67	31	11	7	6	16	9	25	24	3	−4	7	−10	18
	sig.	***	ns	ns	ns	ns	ns	ns	ns	*	***	ns	ns	ns	ns	ns	ns	ns	ns	*	ns	ns	ns	ns	ns	ns	ns
Osprey	1st	86	0	1	23	2	24	20	9	1	6	0	1	0	0	19	7	2	10	36	1	0	4	15	10	1	5
	2nd	169	0	10	23	11	39	18	12	29	26	1	7	1	1	43	21	5	17	48	11	0	10	5	16	2	4
	%change	97		900	0	450	63	−10	33	2800	333	∞	561	∞	∞	126	199	150	70	33	1000		150	−67	59	100	−20
	sig.	***		**	ns	*	ns	ns	ns	***	***	na	*	na	na	**	**	ns	ns	ns	**		ns	*	ns	ns	ns
Bald Eagle	1st	36	0	0	14	0	5	1	2	12	2	0	0	0	0	10	1	2	7	6	1	0	1	1	1	0	2
	2nd	347	2	6	60	3	76	7	57	90	46	0	4	10	2	67	18	16	33	105	8	3	34	10	19	0	31
	%change	864	∞	∞	329	∞	1420	600	2750	650	2200		∞	∞	∞	570	1692	700	371	1650	700	∞	3300	900	1793		1450
	sig.	***	na	na	***	na	***	*	***	***	***		na	na	na	***	***	***	***	***	*	na	***	**	***		***
Northern Harrier	1st	248	7	15	41	20	28	8	8	43	77	1	4	2	1	13	7	1	5	67	5	0	13	4	23	1	21
	2nd	142	7	15	18	17	28	1	5	10	41	0	0	1	2	8	4	1	3	37	2	1	6	4	12	0	12
	%change	−43	0	0	−56	−15	0	−88	−38	−77	−47	−∞	−∞	−54	83	−38	−43	0	−40	−45	−60	∞	−54	0	−48	−∞	−43
	sig.	***	ns	ns	**	ns	ns	*	ns	***	***	na	na	ns	ns	ns	ns	ns	ns	**	ns	na	ns	ns	ns	na	ns
Sharp-shinned Hawk	1st	606	41	47	103	19	68	15	59	54	179	21	2	6	2	45	29	3	13	390	60	12	155	43	48	4	68
	2nd	491	22	42	96	29	51	8	40	58	132	13	1	4	4	61	39	5	17	279	46	8	80	35	58	2	50
	%change	−19	−46	−11	−7	53	−25	−47	−32	7	−26	−38	−53	−39	83	36	34	67	31	−28	−23	−33	−48	−19	20	−50	−26
	sig.	***	*	ns	ns	ns	ns	ns	ns	ns	**	ns	ns	ns	ns	ns	ns	ns	ns	***	ns	ns	***	ns	ns	ns	ns
Cooper's Hawk	1st	767	41	63	103	16	54	19	75	87	237	72	0	4	3	23	10	2	11	251	31	9	80	35	31	3	62
	2nd	690	37	68	79	11	54	7	52	100	206	76	13	6	14	271	125	37	109	518	67	18	113	45	148	11	116
	%change	−10	−10	8	−23	−31	0	−63	−31	15	−13	6	∞	38	326	1078	1145	1750	891	106	116	100	41	29	376	267	87
	sig.	*	ns	ns	ns	ns	ns	*	*	ns	ns	ns	na	ns	**	***	***	***	***	***	***	ns	*	ns	***	*	***
Northern Goshawk	1st	94	3	2	26	10	26	3	17	3	4	0	0	0	0	0	0	0	0	25	5	0	12	4	2	0	2
	2nd	79	1	2	35	8	5	0	24	0	4	0	0	0	0	0	0	0	0	7	3	0	3	0	0	0	1
	%change	−16	−67	0	35	−20	−81	−∞	41	−∞	0									−72	−40		−75	−∞	−∞		−50
	sig.	ns	ns	ns	ns	ns	***	na	ns	na	ns									**	ns		*	na	na		ns
Red-shouldered Hawk	1st	534	13	31	115	27	54	12	98	87	94	3	2	1	1	50	33	4	13	163	26	3	77	14	18	6	19
	2nd	848	28	40	157	36	87	24	141	183	149	3	0	8	4	61	38	3	20	246	29	8	98	21	15	21	54
	%change	59	115	29	37	33	61	100	44	110	59	0	−∞	635	265	22	15	−25	54	51	12	167	27	50	−17	250	184
	sig.	***	*	ns	*	ns	**	*	**	***	***	ns	na	*	ns	ns	ns	ns	ns	***	ns	ns	ns	ns	***	***	***

(continued)

		Appalachian Plateaus	Allegheny Front	Allegheny Mountain	Deep Valleys	Glaciated High Plateau	Glaciated Low Plateau	Glaciated Pocono Plateau	High Plateau	NW Glaciated Plateau	Pittsburgh Low Plateau	Waynesburg Hills	Atlantic Coastal Plain	Central Lowlands	New England	Piedmont	Gettysburg-Newark Lowland	Piedmont Lowland	Piedmont Upland	Ridge and Valley	Anthracite Upland	Anthracite Valley	Appalachian Mountain	Blue Mountain	Great Valley	South Mountain	Susquehanna Lowland
Broad-winged Hawk	1st	1216	71	102	193	48	169	45	127	71	331	59	6	0	9	146	72	14	60	686	109	34	281	68	70	18	106
	2nd	1122	71	94	228	54	152	39	142	74	243	25	2	1	8	49	34	1	14	543	101	28	222	48	27	23	94
	%change	−8	0	−8	18	13	−10	−13	12	4	−27	−58	−69	∞	−19	−66	−53	−93	−77	−21	−7	−18	−21	−29	−62	28	−11
	sig.	ns	ns	ns	ns	ns	ns	ns	ns	ns	***	***	ns	na	ns	***	***	***	***	***	ns	ns	**	ns	***	ns	ns
Red-tailed Hawk	1st	2162	86	155	311	66	322	41	168	276	561	176	13	17	18	395	185	58	152	1028	108	25	366	73	228	22	206
	2nd	2356	106	165	326	82	337	24	184	300	660	172	27	28	23	466	210	67	189	1204	140	39	420	78	247	21	259
	%change	9	23	6	5	24	5	−41	10	9	18	−2	96	51	17	18	13	16	24	17	30	56	15	7	8	−5	26
	sig.	**	ns	ns	ns	ns	ns	*	ns	ns	**	ns	*	ns	ns	*	ns	ns	*	***	*	ns	ns	ns	ns	ns	*
American Kestrel	1st	1518	69	115	123	54	277	12	71	243	423	131	25	12	17	422	197	64	161	945	91	28	301	62	236	14	213
	2nd	1348	73	95	105	51	256	5	52	232	377	102	13	8	16	305	134	50	121	868	70	24	308	48	189	8	221
	%change	−11	6	−17	−15	−6	−8	−58	−27	−5	−11	−22	−51	−39	−14	−28	−32	−22	−25	−8	−23	−14	2	−23	−20	−43	4
	sig.	**	ns	ns	ns	ns	ns	ns	ns	ns	ns	ns	*	ns	ns	***	***	ns	*	ns	ns	ns	ns	ns	*	ns	ns
Merlin	1st	0	0	0	0	0	0	0	0	0	0	0	0	0	0	0	0	0	0	0	0	0	0	0	0	0	0
	2nd	0	0	0	0	0	0	0	0	0	0	0	2	3	2	1	1	0	0	1	0	0	0	0	1	0	0
	%change												∞	∞	∞	∞	∞			∞					∞		
	sig.												na	na	na	na	na			na					na		
Peregrine Falcon	1st	1	0	0	0	0	1	0	0	0	0	0	3	0	0	0	0	0	0	2	0	0	1	0	0	0	1
	2nd	4	0	0	0	0	0	0	0	0	4	0	8	0	0	5	2	0	3	14	0	3	1	1	4	0	5
	%change	300					−∞				∞		152			∞	∞		∞	600		∞	0	∞	∞		400
	sig.	ns					na				na		ns			na	na		na	**		na	na	na	na		ns
King Rail	1st	4	0	0	0	0	1	0	0	2	1	0	1	0	0	0	0	0	0	0	0	0	0	0	0	0	0
	2nd	3	0	0	1	0	0	0	0	2	0	0	0	0	0	2	1	0	1	0	0	0	0	0	0	0	0
	%change	−25			∞		−∞			0	−∞		−∞			∞	∞		∞								
	sig.	ns			na		na			ns	na		na			na	na		na								
Virginia Rail	1st	84	2	3	5	2	26	0	0	36	7	3	3	3	0	13	9	1	3	17	0	2	5	1	3	0	6
	2nd	117	3	3	13	6	31	3	2	45	10	1	1	3	0	7	4	2	1	25	4	1	6	2	3	0	9
	%change	39	50	0	160	200	19	∞	∞	25	43	−67	−69	−8		−46	−56	100	−67	47	∞	−50	20	100	0		50
	sig.	*	ns	ns	ns	ns	ns	na	na	ns	ns	ns	ns	ns	na	ns	ns	ns	ns	ns	na	ns	ns	ns	ns	na	ns
Sora	1st	55	1	5	4	2	11	0	1	25	6	0	2	2	0	15	8	2	5	14	0	1	3	0	4	0	6
	2nd	70	4	2	9	2	15	0	2	27	8	1	1	2	0	8	5	2	1	19	1	1	7	2	4	0	4
	%change	27	300	−60	125	0	36		100	8	33	∞	−53	−8		−47	−38	0	−80	36	∞	0	133	∞	0		−33
	sig.	ns	ns	ns	ns	ns	ns	na	ns	ns	ns	na	ns	ns	na	ns	ns	ns	ns	ns	na	ns	ns	na	ns	na	ns
Common Gallinule	1st	50	0	2	2	1	9	0	0	32	4	0	5	4	0	5	4	0	1	12	1	1	2	0	3	0	5
	2nd	33	2	0	1	0	5	0	0	23	2	0	3	3	0	3	3	0	0	6	1	0	1	0	3	0	1
	%change	−34	∞	−∞	−50	−∞	−44			−28	−50		−43	−31		−40	−25		−∞	−50	0	−∞	−50		0		−80
	sig.	ns	na	na	ns	na	ns			ns	ns		ns	ns		ns	ns		na	ns	ns	na	ns		ns		ns
American Coot	1st	16	0	1	0	0	0	0	0	11	4	0	0	1	0	3	3	0	0	5	1	0	2	0	1	0	1
	2nd	13	1	2	0	0	1	0	0	6	2	1	0	0	0	3	1	1	1	2	0	0	0	0	2	0	0
	%change	−19	∞	100			∞			−45	−50	∞	−∞			0	−67	∞	∞	−60	−∞		−∞		99		−∞
	sig.	ns	na	ns			na			ns	ns	na	na			ns	ns	na	na	ns	na		na		ns		na
Sandhill Crane	1st	0	0	0	0	0	0	0	0	0	0	0	0	0	0	0	0	0	0	0	0	0	0	0	0	0	0
	2nd	0	0	0	0	0	0	0	0	0	0	0	2	3	2	1	1	0	0	1	0	0	0	0	1	0	0
	%change												∞	∞	∞	∞	∞			∞					∞		
	sig.												na	na	na	na	na			na					na		
Killdeer	1st	2166	105	176	199	58	327	28	134	304	660	175	21	27	18	420	199	66	155	1077	87	37	403	69	238	17	226
	2nd	2127	93	169	196	63	331	26	130	314	652	153	23	31	15	400	184	66	150	1125	123	34	399	63	241	12	253
	%change	−2	−11	−4	−2	9	1	−7	−3	3	−1	−13	3	6	−24	−5	−8	0	−3	4	41	−8	−1	−9	1	−29	12
	sig.	ns	ns	ns	ns	ns	ns	ns	ns	ns	ns	ns	ns	ns	ns	ns	ns	ns	ns	ns	*	ns	ns	ns	ns	ns	ns

(continued)

		Appalachian Plateaus	Allegheny Front	Allegheny Mountain	Deep Valleys	Glaciated High Plateau	Glaciated Low Plateau	Glaciated Pocono Plateau	High Plateau	NW Glaciated Plateau	Pittsburgh Low Plateau	Waynesburg Hills	Atlantic Coastal Plain	Central Lowlands	New England	Piedmont	Gettysburg-Newark Lowland	Piedmont Lowland	Piedmont Upland	Ridge and Valley	Anthracite Upland	Anthracite Valley	Appalachian Mountain	Blue Mountain	Great Valley	South Mountain	Susquehanna Lowland
Spotted Sandpiper	1st	383	15	23	43	17	66	7	34	68	72	38	10	9	4	100	50	18	32	216	24	16	50	33	58	2	33
	2nd	235	11	22	23	2	38	6	11	50	59	13	6	13	3	69	46	11	12	152	20	13	38	14	40	1	26
	%change	−39	−27	−4	−47	−88	−42	−14	−68	−26	−18	−66	−43	33	−32	−31	−8	−39	−63	−30	−17	−19	−24	−58	−31	−50	−21
	sig.	***	ns	ns	*	***	**	ns	***	ns	ns	***	ns	ns	ns	*	ns	ns	**	***	ns	ns	ns	**	ns	ns	ns
Upland Sandpiper	1st	33	0	6	0	0	0	0	0	15	10	2	0	1	0	8	6	0	2	12	0	0	4	1	5	0	2
	2nd	17	1	6	0	0	0	0	0	3	7	0	0	1	0	1	1	0	0	4	0	0	2	0	2	0	0
	%change	−48	∞	0						−80	−30	−∞		−8		−88	−83		−∞	−67			−50	−∞	−60		−∞
	sig.	*	na	ns						**	ns	na		ns		*	ns		na	*			ns	na	ns		na
Wilson's Snipe	1st	27	0	2	4	1	5	0	2	11	2	0	0	0	0	4	4	0	0	7	5	0	1	1	0	0	0
	2nd	31	2	4	9	1	4	0	0	10	0	1	0	0	0	0	0	0	0	1	0	0	0	1	0	0	0
	%change	15	∞	100	125	0	−20		−∞	−9	−∞	∞				−∞	−∞			−86	−∞		−∞	0			
	sig.	ns	na	ns	ns	ns	ns		na	ns	na	na				na	na			*	na		na	ns			
American Woodcock	1st	949	66	68	149	39	95	17	76	120	278	41	12	15	5	95	57	5	33	393	48	17	160	42	56	12	58
	2nd	983	39	54	219	50	122	11	124	159	192	13	3	11	1	60	37	2	21	312	38	14	128	27	40	6	59
	%change	4	−41	−21	47	28	28	−35	63	33	−31	−68	−76	−33	−82	−37	−35	−60	−36	−21	−21	−18	−20	−36	−29	−50	2
	sig.	ns	**	ns	***	ns	ns	ns	***	*	***	***	**	ns	ns	**	*	ns	ns	**	ns	ns	ns	ns	ns	ns	ns
Ring-billed Gull	1st	6	0	0	2	0	4	0	0	0	0	0	1	0	0	1	0	1	0	4	1	0	0	1	1	0	1
	2nd	79	1	4	9	4	19	0	6	24	9	3	13	14	0	30	10	6	14	35	5	2	5	4	10	0	9
	%change	1217	∞	∞	350	∞	375		∞	∞	∞	∞	1128	∞		2900	∞	500	∞	775	400	∞	∞	300	896		800
	sig.	***	na	na	*	na	**		na	na	na	na	***	na		***	na	na	na	***	ns	na	na	ns	**		*
Herring Gull	1st	2	0	0	0	0	1	0	0	0	1	0	0	0	0	0	0	0	0	0	0	0	0	0	0	0	0
	2nd	33	0	0	1	0	2	0	1	7	19	3	9	7	0	4	0	2	2	7	3	0	1	1	1	0	1
	%change	1550			∞		100		∞	∞	1800	∞	∞	∞		∞		∞	∞	∞	∞		∞	∞	∞		∞
	sig.	***			na		ns		na	na	***	na	na	na		na		na	na	na	na		na	na	na		na
Great Black-backed Gull	1st	0	0	0	0	0	0	0	0	0	0	0	0	0	0	0	0	0	0	0	0	0	0	0	0	0	0
	2nd	0	0	0	0	0	0	0	0	0	0	0	8	1	0	1	1	0	0	0	0	0	0	0	0	0	0
	%change												∞	∞		∞	∞										
	sig.												na	na		na	na										
Black Tern	1st	10	0	0	0	0	0	0	0	10	0	0	0	1	0	0	0	0	0	0	0	0	0	0	0	0	0
	2nd	2	0	0	0	0	0	0	0	2	0	0	0	1	0	0	0	0	0	0	0	0	0	0	0	0	0
	%change	−80								−80				−8													
	sig.	*								*				ns													
Rock Pigeon	1st	1724	90	151	153	61	292	12	71	273	497	124	29	21	17	463	208	67	188	1120	99	40	403	70	246	22	240
	2nd	1696	71	134	171	53	305	9	84	276	505	88	30	25	20	450	200	66	184	1135	121	38	407	58	246	14	251
	%change	−2	−21	−11	12	−13	4	−25	18	1	2	−29	−2	9	7	−3	−4	−1	−2	1	22	−5	1	−17	0	−36	5
	sig.	ns	ns	ns	ns	ns	ns	ns	ns	ns	ns	*	ns	ns	ns	ns	ns	ns	ns	ns	ns	ns	ns	ns	ns	ns	ns
Eurasian Collared-Dove	1st	0	0	0	0	0	0	0	0	0	0	0	0	0	0	0	0	0	0	0	0	0	0	0	0	0	0
	2nd	0	0	0	0	0	0	0	0	0	0	0	2	3	2	1	1	0	0	1	0	0	0	0	1	0	0
	%change												∞	∞	∞	∞	∞			∞					∞		
	sig.												na	na	na	na	na			na					na		
Mourning Dove	1st	2632	128	200	315	90	429	51	184	311	740	184	33	33	21	483	223	68	192	1316	144	47	484	82	259	31	269
	2nd	2925	135	211	424	113	462	56	223	330	782	189	34	37	23	494	228	69	197	1337	149	49	491	81	260	31	276
	%change	11	5	6	35	26	8	10	21	6	6	3	−3	3	0	2	2	1	3	2	3	4	1	−1	0	0	3
	sig.	***	ns	ns	***	ns	ns	ns	ns	ns	ns	ns	ns	ns	ns	ns	ns	ns	ns	ns	ns	ns	ns	ns	ns	ns	ns
Yellow-billed Cuckoo	1st	1126	78	116	136	21	88	11	109	63	351	153	12	4	13	296	126	47	123	815	72	16	358	54	136	27	152
	2nd	1848	107	138	327	57	249	34	156	162	490	128	9	6	15	307	138	32	137	1053	132	29	416	73	161	28	214
	%change	64	37	19	140	171	183	209	43	157	40	−16	−29	38	5	4	9	−32	11	29	83	81	16	35	18	4	41
	sig.	***	*	ns	***	***	***	**	***	***	***	ns	ns	ns	ns	ns	ns	ns	ns	***	***	*	*	ns	ns	ns	***

(continued)

		Appalachian Plateaus	Allegheny Front	Allegheny Mountain	Deep Valleys	Glaciated High Plateau	Glaciated Low Plateau	Glaciated Pocono Plateau	High Plateau	NW Glaciated Plateau	Pittsburgh Low Plateau	Waynesburg Hills	Atlantic Coastal Plain	Central Lowlands	New England	Piedmont	Gettysburg-Newark Lowland	Piedmont Lowland	Piedmont Upland	Ridge and Valley	Anthracite Upland	Anthracite Valley	Appalachian Mountain	Blue Mountain	Great Valley	South Mountain	Susquehanna Lowland
Black-billed Cuckoo	1st	992	47	111	150	21	104	19	136	84	254	66	6	3	5	125	58	10	57	484	51	8	236	35	72	11	71
	2nd	1143	62	68	234	41	160	24	94	119	288	53	5	4	4	88	43	7	38	459	75	14	193	36	57	4	80
	%change	15	32	−39	56	95	54	26	−31	42	13	−20	−21	23	−27	−30	−26	−30	−33	−5	47	75	−18	3	−21	−64	13
	sig.	**	ns	***	***	**	***	ns	**	*	ns	ns	ns	ns	ns	*	ns	ns	*	ns	*	ns	*	ns	ns	ns	ns
Barn Owl	1st	48	6	2	1	1	10	0	0	5	14	9	6	0	1	71	38	16	17	125	14	0	27	14	50	2	18
	2nd	9	1	1	0	0	2	0	0	1	4	0	0		0	22	16	3	3	86	4	0	17	2	32	1	30
	%change	−81	−83	−50	−∞	−∞	−80			−80	−71	−∞	−∞		−∞	−69	−58	−81	−82	−31	−71		−37	−86	−36	−50	67
	sig.	***	ns	ns	na	na	*			ns	*	na	na		na	***	**	**	**	**	*		ns	**	*	ns	ns
Eastern Screech-Owl	1st	832	77	60	79	20	89	14	30	52	284	127	11	10	15	285	132	40	113	760	64	19	324	54	140	23	136
	2nd	614	38	31	111	7	92	5	52	54	154	70	8	7	10	247	99	23	125	653	77	13	235	64	105	15	144
	%change	−26	−51	−48	41	−65	3	−64	73	4	−46	−45	−31	−36	−39	−13	−25	−43	11	−14	20	−32	−27	19	−25	−35	6
	sig.	***	***	***	**	*	ns	*	*	ns	***	***	ns	ns	ns	ns	*	*	ns	**	ns	ns	***	ns	*	ns	ns
Great Horned Owl	1st	1261	93	76	168	40	150	32	85	117	387	113	15	17	13	299	148	37	114	813	75	21	352	64	127	25	149
	2nd	717	42	48	120	20	93	15	50	87	200	42	17	9	11	280	126	41	113	700	81	23	242	42	148	16	148
	%change	−43	−55	−37	−29	−50	−38	−53	−41	−26	−48	−63	7	−51	−23	−6	−15	11	−1	−14	8	10	−31	−34	16	−36	−1
	sig.	***	***	*	**	**	***	**	**	*	***	***	ns	ns	ns	ns	ns	ns	ns	**	ns	ns	***	*	ns	ns	ns
Barred Owl	1st	753	38	53	196	20	48	16	101	79	169	33	0	3	0	61	41	3	17	259	25	4	140	14	27	25	24
	2nd	841	45	57	247	35	70	9	96	108	141	33	1	1	1	77	46	6	25	341	48	9	167	21	31	22	43
	%change	12	18	8	26	75	46	−44	−5	37	−17	0	∞	−69	∞	26	12	100	47	32	92	125	19	50	14	−12	79
	sig.	*	ns	ns	*	*	*	ns	ns	*	ns	ns	na	ns	na	ns	ns	ns	ns	***	**	ns	ns	ns	ns	ns	*
Long-eared Owl	1st	10	0	1	1	0	4	0	0	0	4	0	0	0	0	2	1	1	0	6	3	0	0	0	2	0	1
	2nd	8	2	3	1	0	0	0	0	1	1	0	0	0	0	1	1	0	0	5	0	0	3	0	1	0	1
	%change	−20	∞	200	0		−∞			∞	−75					−50	0	−∞		−17	−∞		∞		−50		0
	sig.	ns	na	ns	ns		na			na	ns					ns	ns	ns		ns	na		ns		ns		
Short-eared Owl	1st	5	0	0	0	0	0	0	0	0	5	0	1	0	0	0	0	0	0	0	0	0	0	0	0	0	0
	2nd	5	0	0	2	0	0	0	0	1	2	0	0	0	0	0	0	0	0	0	0	0	0	0	0	0	0
	%change	0			∞					∞	−60		−∞														
	sig.	ns			na					na	ns		na														
Northern Saw-whet Owl	1st	66	5	6	21	2	13	2	7	1	9	0	0	1	0	2	2	0	0	27	4	2	7	4	4	2	4
	2nd	230	9	13	118	12	6	4	45	2	21	0	0	0	0	0	0	0	0	53	13	1	21	4	3	4	7
	%change	248	80	117	462	500	−54	100	543	100	133			−∞		−∞	−∞			96	225	−50	200	0	−25	100	75
	sig.	***	ns	ns	***	**	ns	ns	***	ns	*			na		na	na			**	*	ns	**	ns	ns	ns	ns
Common Nighthawk	1st	396	27	25	36	4	15	2	38	50	145	54	16	8	2	56	23	13	20	276	39	26	90	17	64	5	35
	2nd	109	3	4	8	0	0	0	9	9	63	13	5	2	1	13	6	5	2	89	22	18	17	6	18	0	8
	%change	−72	−89	−84	−78	−∞	−∞	−∞	−76	−82	−57	−76	−70	−77	−54	−77	−74	−62	−90	−68	−44	−31	−81	−65	−72	−∞	−77
	sig.	***	***	***	***	na	na	na	***	***	***	***	**	*	ns	***	**	ns	***	***	*	ns	***	*	***	na	***
Chuck-will's-widow	1st	2	0	1	0	0	0	0	0	0	1	0	0	0	0	0	0	0	0	1	0	0	1	0	0	0	0
	2nd	0	0	0	0	0	0	0	0	0	0	0	0	0	0	0	0	0	0	2	0	0	2	0	0	0	0
	%change	−∞		−∞							−∞									100			100				
	sig.	na		na							na									ns			ns				
Eastern Whip-poor-will	1st	438	72	36	49	7	36	15	24	7	162	30	1	0	1	17	9	0	8	405	39	8	236	35	19	25	43
	2nd	196	34	8	66	8	10	7	10	0	50	3	0	2	0	3	1	0	2	295	50	15	167	15	11	15	22
	%change	−55	−53	−78	35	14	−72	−53	−58	−∞	−69	−90	−∞	∞	−∞	−82	−89		−75	−27	28	88	−29	−57	−42	−40	−49
	sig.	***	***	***	ns	ns	***	ns	*	na	***	***	na	na	na	**	*		ns	***	ns	ns	***	**	ns	ns	**
Chimney Swift	1st	1907	94	167	147	43	269	20	121	229	640	177	26	26	17	452	205	67	180	988	90	33	364	59	239	25	178
	2nd	1862	88	176	168	28	173	11	130	264	650	174	27	32	21	449	199	66	184	943	97	35	341	44	232	20	174
	%change	−2	−6	5	14	−35	−36	−45	7	15	2	−2	−2	13	13	−1	−3	−1	2	−5	8	6	−6	−25	−3	−20	−2
	sig.	ns	ns	ns	ns	***	ns	ns	ns	ns	ns	ns	ns	ns	ns	ns	ns	ns	ns	ns	ns	ns	ns	ns	ns	ns	ns

(continued)

		Appalachian Plateaus	Allegheny Front	Allegheny Mountain	Deep Valleys	Glaciated High Plateau	Glaciated Low Plateau	Glaciated Pocono Plateau	High Plateau	NW Glaciated Plateau	Pittsburgh Low Plateau	Waynesburg Hills	Atlantic Coastal Plain	Central Lowlands	New England	Piedmont	Gettysburg-Newark Lowland	Piedmont Lowland	Piedmont Upland	Ridge and Valley	Anthracite Upland	Anthracite Valley	Appalachian Mountain	Blue Mountain	Great Valley	South Mountain	Susquehanna Lowland
Ruby-throated Hummingbird	1st	2235	115	176	313	85	256	39	201	250	630	170	9	22	12	307	151	40	116	933	84	26	399	70	141	27	186
	2nd	2384	108	179	391	81	337	35	205	297	601	150	17	25	18	369	168	45	156	1091	129	30	408	72	193	28	231
	%change	7	−6	2	25	−5	32	−10	2	19	−5	−12	78	4	37	20	11	13	34	17	54	15	2	3	36	4	24
	sig.	*	ns	ns	**	ns	***	ns	ns	*	ns	ns	ns	ns	*	*	ns	ns	*	***	**	ns	ns	ns	**	ns	*
Belted Kingfisher	1st	1756	76	98	280	70	301	34	136	224	387	150	15	25	14	368	159	58	151	885	88	29	316	63	184	17	188
	2nd	1580	73	78	286	65	250	21	141	204	356	106	15	25	17	336	137	52	147	857	80	26	300	60	168	15	208
	%change	−10	−4	−20	2	−7	−17	−38	4	−9	−8	−29	−6	−8	11	−9	−14	−10	−3	−3	−9	−10	−5	−5	−9	−12	11
	sig.	**	ns	ns	ns	ns	*	ns	ns	ns	ns	**	ns	ns	ns	ns	ns	ns	ns	ns	ns	ns	ns	ns	ns	ns	ns
Red-headed Woodpecker	1st	287	17	17	18	3	21	0	11	110	50	40	1	14	5	123	44	35	44	268	4	136	3	77	7	39	
	2nd	126	13	19	5	0	6	0	2	53	12	16	0	12	1	94	39	13	42	141	3	0	47	6	59	5	21
	%change	−56	−24	12	−72	−∞	−71		−82	−52	−76	−60	−∞	−21	−82	−24	−12	−63	−5	−47	−25	−∞	−65	100	−24	−29	−46
	sig.	***	ns	ns	**	na	**		*	***	***	***	na	ns	ns	*	ns	***	ns	***	*	na	***	ns	ns	ns	*
Red-bellied Woodpecker	1st	662	45	44	31	8	43	2	17	100	207	165	14	8	19	455	206	63	186	848	38	4	355	42	229	30	150
	2nd	1908	93	175	142	25	274	25	106	298	586	184	27	32	23	487	225	65	197	1259	136	31	461	79	258	31	263
	%change	188	107	298	358	213	537	1150	524	198	183	12	82	268	11	7	9	3	6	48	258	675	30	88	12	3	75
	sig.	***	***	***	***	**	***	***	***	***	***	ns	*	***	ns	ns	ns	ns	ns	***	***	***	***	***	ns	ns	***
Yellow-bellied Sapsucker	1st	670	15	7	244	75	110	14	107	70	28	0	0	0	0	1	1	0	0	62	14	4	22	7	4	2	9
	2nd	1405	10	15	383	107	341	32	221	201	93	2	0	7	0	3	3	0	0	44	4	5	5	7	0	2	21
	%change	110	−33	114	57	43	210	129	107	187	232	∞		∞		200	199			−29	−71	25	−77	0	−∞	0	133
	sig.	***	ns	ns	***	*	***	**	***	***	***	na		na		ns	ns			ns	*	ns	***	ns	na	ns	*
Downy Woodpecker	1st	2600	127	206	377	99	391	53	206	268	688	185	32	25	19	465	212	65	188	1248	138	45	450	82	256	30	247
	2nd	2662	128	195	395	92	392	50	206	317	709	178	30	33	23	483	222	66	195	1300	151	47	469	83	249	31	270
	%change	2	1	−5	5	−7	0	−6	0	18	3	−4	−11	21	11	4	4	2	4	4	9	4	4	1	−3	3	9
	sig.	ns	ns	ns	ns	ns	ns	ns	ns	*	ns	ns	ns	ns	ns	ns	ns	ns	ns	ns	ns	ns	ns	ns	ns	ns	ns
Hairy Woodpecker	1st	1948	87	145	328	80	276	44	171	168	496	153	14	20	15	280	120	31	129	838	101	30	322	65	140	21	159
	2nd	2189	108	165	350	83	316	47	194	265	544	117	18	19	17	340	150	33	157	1014	133	38	362	70	160	27	224
	%change	12	24	14	7	4	14	7	13	58	10	−24	21	−13	3	21	24	6	22	21	32	27	12	8	14	29	41
	sig.	***	ns	ns	ns	ns	ns	ns	ns	***	ns	*	ns	ns	ns	*	ns	ns	ns	***	*	ns	ns	ns	ns	ns	***
Northern Flicker	1st	2679	131	178	375	108	431	53	202	304	714	183	33	30	21	484	222	68	194	1281	145	47	458	82	258	31	260
	2nd	2842	130	204	434	104	411	50	232	325	767	185	31	36	23	489	227	66	196	1309	146	45	480	79	254	31	274
	%change	6	−1	15	16	−4	−5	−6	15	7	7	1	−11	10	0	1	2	−3	1	2	1	−4	5	−4	−2	0	5
	sig.	*	ns	ns	*	ns	ns	ns	ns	ns	ns	ns	ns	ns	ns	ns	ns	ns	ns	ns	ns	ns	ns	ns	ns	ns	ns
Pileated Woodpecker	1st	1710	102	131	306	69	221	25	147	150	415	144	3	10	7	122	66	8	48	798	62	15	415	54	86	29	137
	2nd	2285	110	159	382	81	337	38	195	254	579	150	1	18	15	237	101	16	120	1097	141	36	450	73	114	31	252
	%change	34	8	21	25	17	52	52	33	69	40	4	−69	65	96	94	52	100	150	37	127	140	8	35	32	7	84
	sig.	***	ns	ns	**	ns	***	ns	**	***	***	ns	ns	ns	ns	***	**	ns	***	***	***	**	ns	ns	ns	*	***
Eastern Wood-Pewee	1st	2657	131	204	379	91	414	54	198	294	715	177	18	33	18	434	195	53	186	1194	134	40	464	81	219	30	226
	2nd	2609	125	184	375	86	418	44	202	319	682	174	20	34	23	460	211	55	194	1242	144	42	460	79	229	31	257
	%change	−2	−5	−10	−1	−5	1	−19	2	9	−5	−2	5	−5	17	6	8	4	4	4	7	5	−1	−2	4	3	14
	sig.	ns	ns	ns	ns	ns	ns	ns	ns	ns	ns	ns	ns	ns	ns	ns	ns	ns	ns	ns	ns	ns	ns	ns	ns	ns	ns
Yellow-bellied Flycatcher	1st	16	0	0	6	2	4	1	2	0	1		0	0	0	0	0	0	0	0	0	0	0	0	0	0	0
	2nd	13	0	0	4	3	2	1	2	0	1		0	0	0	0	0	0	0	0	0	0	0	0	0	0	0
	%change	−19			−33	50	−50	0	0		0																
	sig.	ns			ns	ns	ns	ns	ns																		
Acadian Flycatcher	1st	1065	79	132	101	18	40	2	32	109	377	175	5	9	8	173	78	15	80	506	40	5	288	29	35	26	83
	2nd	1540	91	155	176	20	45	2	125	230	521	175	9	20	2	233	78	19	136	650	76	7	339	33	55	30	110
	%change	45	15	17	74	11	13	0	291	111	38	0	70	104	−77	35	0	27	70	28	90	40	18	14	57	15	33
	sig.	***	ns	ns	***	ns	ns		***	***	***	ns	ns	*	*	**	ns	ns	***	***	***	ns	*	ns	*	ns	ns

(continued)

		Appalachian Plateaus	Allegheny Front	Allegheny Mountain	Deep Valleys	Glaciated High Plateau	Glaciated Low Plateau	Glaciated Pocono Plateau	High Plateau	NW Glaciated Plateau	Pittsburgh Low Plateau	Waynesburg Hills	Atlantic Coastal Plain	Central Lowlands	New England	Piedmont	Gettysburg-Newark Lowland	Piedmont Lowland	Piedmont Upland	Ridge and Valley	Anthracite Upland	Anthracite Valley	Appalachian Mountain	Blue Mountain	Great Valley	South Mountain	Susquehanna Lowland
Alder Flycatcher	1st	297	9	16	36	21	94	17	14	63	26	1	0	7	0	0	0	0	0	28	5	2	5	1	4	1	10
	2nd	757	14	20	160	63	175	22	72	151	79	1	0	7	1	2	2	0	0	101	16	9	22	18	10	0	26
	%change	155	56	25	344	200	86	29	414	140	204	0		−8	∞	∞	∞			261	220	350	340	1700	149	−∞	160
	sig.	***	ns	ns	***	***	***	ns	***	***	***	ns	ns	ns	na	na	na			***	*	*	***	***	ns	na	**
Willow Flycatcher	1st	1042	41	111	50	13	137	8	38	178	315	151	17	18	10	178	75	20	83	387	18	12	110	36	123	5	83
	2nd	1109	34	77	59	23	163	6	32	256	343	116	21	22	12	300	137	30	133	562	47	13	148	50	167	3	134
	%change	6	−17	−31	18	77	19	−25	−16	44	9	−23	17	12	10	69	82	50	60	45	161	8	35	39	35	−40	61
	sig.	ns	ns	*	ns	ns	ns	ns	ns	***	ns	*	ns	ns	ns	***	***	ns	***	***	***	ns	*	ns	**	ns	***
Least Flycatcher	1st	1587	82	97	312	71	381	44	138	156	286	20	0	17	1	2	1	0	1	210	39	27	59	30	14	2	39
	2nd	1537	65	73	395	88	346	28	154	156	221	11	0	13	1	5	5	0	0	183	20	12	75	18	10	3	45
	%change	−3	−21	−25	27	24	−9	−36	12	0	−23	−45		−30	−9	150	398		−∞	−13	−49	−56	27	−40	−29	50	15
	sig.	ns	ns	ns	**	ns	ns	ns	ns	ns	**	ns		ns	ns	ns	ns		na	ns	*	*	ns	ns	ns	ns	ns
Eastern Phoebe	1st	2716	134	186	412	107	455	54	218	272	695	183	12	12	19	419	199	45	175	1268	135	47	483	80	238	31	254
	2nd	2829	135	198	435	103	455	51	212	320	743	177	19	29	23	462	217	58	187	1303	147	48	476	80	248	29	275
	%change	4	1	6	6	−4	0	−6	−3	18	7	−3	50	122	11	10	9	29	7	3	9	2	−1	0	4	−6	8
	sig.	ns	ns	ns	ns	ns	ns	ns	ns	*	ns	ns	ns	**	ns	ns	ns	ns	ns	ns	ns	ns	ns	ns	ns	ns	ns
Great Crested Flycatcher	1st	1902	101	129	269	67	337	50	109	234	477	129	18	29	12	361	180	44	137	1171	125	36	444	77	219	30	240
	2nd	1838	110	133	223	51	363	49	107	277	436	89	21	27	19	441	207	58	176	1236	144	41	447	81	234	30	259
	%change	−3	9	3	−17	−24	8	−2	−2	18	−9	−31	10	−14	45	22	14	32	28	6	15	14	1	5	6	0	8
	sig.	ns	ns	ns	*	ns	ns	ns	ns	ns	ns	**	ns	**	ns	ns	ns	ns	*	ns	ns	ns	ns	ns	ns	ns	ns
Eastern Kingbird	1st	1942	75	91	245	79	405	41	138	281	436	151	30	24	20	457	212	59	186	969	90	34	298	74	239	18	216
	2nd	1901	69	101	211	70	381	33	101	297	497	141	32	29	22	464	211	64	189	1028	110	28	334	70	252	17	217
	%change	−2	−8	11	−14	−11	−6	−20	−27	6	14	−7	1	11	0	2	−1	8	2	6	22	−18	12	−5	5	−6	0
	sig.	ns	ns	ns	ns	ns	ns	ns	*	ns	*	ns	ns	ns	ns	ns	ns	ns	ns	ns	ns	ns	ns	ns	ns	ns	ns
Loggerhead Shrike	1st	1	1	0	0	0	0	0	0	0	0	0	0	0	0	0	0	0	0	5	0	0	2	0	3	0	0
	2nd	0	0	0	0	0	0	0	0	0	0	0	0	1	0	1	1	0	0	1	0	0	0	0	1	0	0
	%change	−∞	−∞											−∞	−∞					−80			−∞		−67		
	sig.	na	na											na	na					ns			na		ns		
White-eyed Vireo	1st	512	15	70	6	3	7	1	11	24	217	158	16	0	10	271	110	21	140	213	9	5	79	18	75	2	25
	2nd	514	11	54	2	0	4	2	5	70	218	148	7	2	8	259	115	20	124	215	17	2	66	20	81	7	22
	%change	0	−27	−23	−67	−∞	−43	100	−55	192	0	−6	−59	∞	−27	−4	4	−5	−11	1	89	−60	−16	11	8	250	−12
	sig.	ns	ns	ns	ns	na	ns	ns	ns	***	ns	ns	*	na	ns	ns	ns	ns	ns	ns	ns	ns	ns	ns	ns	ns	ns
Yellow-throated Vireo	1st	829	30	59	63	11	231	13	44	75	178	125	2	12	5	88	33	4	51	282	28	11	132	29	27	3	52
	2nd	991	29	50	44	9	274	12	78	171	202	122	1	10	10	172	85	11	76	575	59	25	264	39	60	15	113
	%change	20	−3	−15	−30	−18	19	−8	77	128	13	−2	−53	−16	83	95	156	175	49	104	111	127	100	34	121	400	117
	sig.	***	ns	ns	ns	ns	ns	ns	**	***	ns	ns	ns	ns	ns	***	***	***	*	***	***	*	***	ns	***	**	***
Blue-headed Vireo	1st	1243	64	93	297	63	168	32	158	80	288	0	0	3	0	1	1	0	0	228	49	4	113	18	4	4	36
	2nd	1810	93	130	431	100	284	48	226	155	341	2	0	3	0	0	0	0	0	527	109	23	228	33	12	20	102
	%change	46	45	40	45	59	69	50	43	94	18	∞		−8	−∞	−∞				131	122	475	102	83	199	400	183
	sig.	***	*	*	***	**	***	ns	***	***	*	na		ns	na	na	na			***	***	***	***	*	***	*	***
Warbling Vireo	1st	691	18	11	45	16	168	6	37	161	127	102	9	25	1	164	59	20	85	216	5	15	60	14	68	6	48
	2nd	1124	20	20	110	29	314	7	57	260	193	114	29	28	9	311	130	43	138	450	19	18	126	26	158	6	97
	%change	63	11	82	144	81	87	17	54	61	52	12	204	3	722	90	119	115	62	108	280	20	110	86	131	0	102
	sig.	***	ns	ns	***	*	***	ns	*	***	***	ns	***	ns	**	***	***	**	***	***	**	ns	***	ns	***	ns	***
Red-eyed Vireo	1st	2889	131	212	436	106	454	56	234	303	771	186	22	31	17	422	195	49	178	1205	143	47	474	79	190	29	243
	2nd	2994	139	216	455	114	473	56	237	330	788	186	20	36	23	460	214	58	188	1303	150	49	487	82	230	31	274
	%change	4	6	2	4	8	4	0	1	9	2	0	−14	7	24	9	9	18	6	8	5	4	3	4	21	7	13
	sig.	ns	ns	ns	ns	ns	ns	ns	ns	ns	ns	ns	ns	ns	ns	ns	ns	ns	ns	ns	ns	ns	ns	ns	ns	ns	ns

(continued)

		Appalachian Plateaus	Allegheny Front	Allegheny Mountain	Deep Valleys	Glaciated High Plateau	Glaciated Low Plateau	Glaciated Pocono Plateau	High Plateau	NW Glaciated Plateau	Pittsburgh Low Plateau	Waynesburg Hills	Atlantic Coastal Plain	Central Lowlands	New England	Piedmont	Gettysburg-Newark Lowland	Piedmont Lowland	Piedmont Upland	Ridge and Valley	Anthracite Upland	Anthracite Valley	Appalachian Mountain	Blue Mountain	Great Valley	South Mountain	Susquehanna Lowland
Blue Jay	1st	2847	129	199	424	108	448	55	229	310	759	186	33	30	19	481	220	68	193	1289	144	48	461	83	259	31	263
	2nd	2872	128	201	436	110	443	49	232	326	770	177	30	33	23	467	220	60	187	1279	145	48	456	83	248	30	269
	%change	1	−1	1	3	2	−1	−11	1	5	1	−5	−14	1	11	−3	0	−12	−3	−1	1	0	−1	0	−5	−3	2
	sig.	ns	ns	ns	ns	ns	ns	ns	ns	ns	ns	ns	ns	ns	ns	ns	ns	ns	ns	ns	ns	ns	ns	ns	ns	ns	ns
American Crow	1st	2915	139	216	427	102	465	55	236	317	771	187	33	32	21	488	225	69	194	1331	150	48	489	83	256	31	274
	2nd	2981	139	216	443	113	474	56	232	331	789	188	31	36	23	492	228	68	196	1337	151	49	490	83	257	31	276
	%change	2	0	0	4	11	2	2	−2	4	2	1	−11	3	0	1	1	−1	1	0	1	2	0	0	0	0	1
	sig.	ns	ns	ns	ns	ns	ns	ns	ns	ns	ns	ns	ns	ns	ns	ns	ns	ns	ns	ns	ns	ns	ns	ns	ns	ns	ns
Fish Crow	1st	20	2	0	0	0	10	5	0	0	3	0	18	0	2	164	46	32	86	310	22	17	35	26	135	6	69
	2nd	49	3	1	7	0	27	2	0	0	9	0	23	0	13	330	140	51	139	541	57	22	112	31	181	14	124
	%change	145	50	∞	∞		170	−60			200		21		493	101	203	59	62	75	159	29	220	19	34	133	80
	sig.	***	ns	na	na		**	ns			ns	a	ns		***	***	***	*	***	***	***	ns	***	ns	**	ns	***
Common Raven	1st	662	39	39	297	74	31	5	65	0	112	0	0	0	0	0	0	0	0	191	12	4	132	7	3	10	23
	2nd	1229	74	92	386	89	164	29	168	22	179	26	0	0	0	14	12	1	1	572	78	26	304	26	32	17	89
	%change	86	90	136	30	20	429	480	158	∞	60	∞				∞	∞	∞	∞	199	550	550	130	271	963	70	287
	sig.	***	***	***	ns	***	***	***	***	na	***	na				na	na	na	na	***	***	***	***	***	***	ns	***
Horned Lark	1st	415	30	35	25	15	35	1	5	61	192	16	3	4	0	37	6	12	19	169	6	1	59	4	48	4	47
	2nd	214	10	26	6	6	9	1	1	68	81	6	0	4	1	150	31	40	79	412	39	3	115	12	139	1	103
	%change	−48	−67	−26	−76	−60	−74	0	−80	11	−58	−63	−∞	−8	∞	305	414	233	316	144	550	200	95	200	188	−75	119
	sig.	***	**	ns	***	*	***	ns	ns	ns	***	*	na	ns	na	***	***	***	***	***	***	ns	***	*	***	ns	***
Purple Martin	1st	386	8	10	19	7	50	2	32	113	85	60	6	25	4	203	71	31	101	311	21	2	101	36	93	10	48
	2nd	186	1	4	3	0	4	0	7	114	34	19	3	23	1	159	46	29	84	148	2	0	35	9	73	11	18
	%change	−52	−88	−60	−84	−∞	−92	−∞	−78	1	−60	−68	−53	−15	−77	−22	−35	−6	−17	−52	−90	−∞	−65	−75	−22	10	−63
	sig.	***	*	ns	***	na	***	na	***	ns	***	***	ns	ns	*	*	*	ns	ns	***	***	***	***	***	ns	ns	***
Tree Swallow	1st	2025	61	122	311	90	418	53	185	248	463	74	20	19	8	246	114	34	98	809	100	40	262	70	138	9	190
	2nd	2246	92	158	332	82	415	47	179	299	537	105	25	21	21	455	208	62	185	1161	136	45	378	79	245	18	260
	%change	11	51	30	7	−9	−1	−11	−3	21	16	42	18	2	140	85	82	82	89	44	36	13	44	13	77	100	37
	sig.	***	*	*	ns	ns	ns	ns	ns	*	*	*	ns	ns	*	***	***	***	**	***	*	ns	***	ns	***	ns	***
Northern Rough-winged Swallow	1st	803	32	69	59	10	110	11	29	122	245	116	14	18	6	222	94	39	89	435	40	18	130	44	131	7	65
	2nd	1119	44	79	130	26	129	12	78	188	333	100	24	17	15	302	139	47	116	661	84	32	174	56	166	10	139
	%change	39	38	14	120	160	17	9	169	54	36	−14	62	−13	128	36	47	21	30	52	110	78	34	27	26	43	114
	sig.	***	ns	ns	***	**	ns	ns	***	***	***	ns	***	ns	***	***	**	ns	*	***	***	*	*	ns	*	ns	***
Bank Swallow	1st	308	11	9	55	17	82	4	22	70	28	10	6	20	1	36	19	2	15	150	9	16	53	20	25	1	26
	2nd	197	5	5	20	9	34	2	13	82	26	1	5	21	0	31	16	8	7	83	6	10	17	10	17	2	21
	%change	−36	−55	−44	−64	−47	−59	−50	−41	17	−7	−90	−21	−4	−∞	−14	−16	300	−53	−45	−33	−38	−68	−50	−32	100	−19
	sig.	***	ns	ns	***	ns	***	ns	ns	ns	ns	**	ns	ns	na	ns	ns	ns	ns	***	ns	ns	***	ns	ns	ns	ns
Cliff Swallow	1st	515	39	32	123	17	162	21	29	43	47	2	2	3	4	40	27	3	10	319	26	13	143	39	41	1	56
	2nd	512	44	40	146	14	69	9	48	70	72	0	1	1	2	33	21	3	9	215	16	8	112	18	22	0	39
	%change	−1	13	25	19	−18	−57	−57	66	63	53	−∞	−53	−69	−54	−18	−23	0	−10	−33	−38	−38	−22	−54	−47	−∞	−30
	sig.	ns	ns	ns	ns	ns	***	*	*	*	*	na	ns	ns	ns	ns	ns	ns	ns	***	ns	ns	ns	**	*	na	ns
Barn Swallow	1st	2714	126	204	331	101	449	54	200	320	745	184	29	29	19	479	217	67	195	1267	137	45	459	80	256	26	264
	2nd	2548	119	199	312	83	414	44	177	324	706	170	29	29	21	475	217	67	191	1250	139	40	451	74	249	27	270
	%change	−6	−6	−2	−6	−18	−8	−19	−12	1	−5	−8	−6	−8	1	−1	0	0	−2	−1	1	−11	−2	−8	−3	4	2
	sig.	*	ns	ns	ns	ns	ns	ns	ns	ns	ns	ns	ns	ns	ns	ns	ns	ns	ns	ns	ns	ns	ns	ns	ns	ns	ns
Carolina Chickadee	1st	257	1	6	0	0	0	0	0	1	80	169	25	0	1	357	125	42	190	67	0	0	13	1	37	15	1
	2nd	348	1	11	0	0	0	0	0	0	154	182	29	0	14	477	215	66	196	207	6	0	17	12	134	29	9
	%change	35	0	83						−∞	93	8	10		1178	34	71	57	3	209	∞		31	1100	261	93	800
	sig.	***	ns	ns						na	***	ns	ns	na	***	***	***	*	ns	***	na	ns	**	***	*	*	*

(continued)

		Appalachian Plateaus	Allegheny Front	Allegheny Mountain	Deep Valleys	Glaciated High Plateau	Glaciated Low Plateau	Glaciated Pocono Plateau	High Plateau	NW Glaciated Plateau	Pittsburgh Low Plateau	Waynesburg Hills	Atlantic Coastal Plain	Central Lowlands	New England	Piedmont	Gettysburg-Newark Lowland	Piedmont Lowland	Piedmont Upland	Ridge and Valley	Anthracite Upland	Anthracite Valley	Appalachian Mountain	Blue Mountain	Great Valley	South Mountain	Susquehanna Lowland
Black-capped Chickadee	1st	2732	133	214	448	114	453	55	234	303	727	51	0	31	19	159	119	30	10	1210	145	47	469	82	198	29	240
	2nd	2773	138	212	453	113	469	56	237	329	732	34	0	35	17	64	50	11	3	1225	145	49	473	82	183	23	270
	%change	2	4	−1	1	−1	4	2	1	9	1	−33		4	−18	−60	−58	−63	−70	1	0	4	1	0	−8	−21	13
	sig.	ns	ns	ns	ns	ns	ns	ns	ns	ns	ns	ns	ns	ns	ns	***	***	**	*	ns	ns	ns	ns	ns	ns	ns	ns
Tufted Titmouse	1st	2191	125	211	235	47	328	48	142	204	664	187	25	15	18	465	216	60	189	1276	144	38	482	81	248	31	252
	2nd	2800	136	214	391	80	446	55	215	318	755	190	28	30	23	482	220	65	197	1329	151	48	489	82	252	31	276
	%change	28	9	1	66	70	36	15	51	56	14	2	6	84	17	4	1	8	4	4	5	26	1	1	1	0	10
	sig.	***	ns	ns	***	**	***	ns	***	***	*	ns	ns	*	ns	ns	ns	ns	ns	ns	ns	ns	ns	ns	ns	ns	ns
Red-breasted Nuthatch	1st	151	1	7	40	10	23	13	18	14	25	0	0	4	1	4	3	0	1	43	10	0	17	6	2	1	7
	2nd	459	10	11	144	29	68	17	74	54	50	2	0	3	0	7	3	0	4	85	15	4	26	14	5	3	18
	%change	204	900	57	260	190	196	31	311	286	100	∞		−31	−∞	75	0		300	98	50	∞	53	133	149	200	157
	sig.	***	**	ns	***	**	***	ns	***	***	**	na		ns	na	ns	ns		ns	***	ns	na	ns	ns	ns	ns	*
White-breasted Nuthatch	1st	2552	121	194	393	101	383	53	212	246	671	178	8	24	18	345	168	32	145	1137	123	37	438	76	208	29	226
	2nd	2769	131	198	427	94	422	53	232	310	724	178	20	29	22	463	220	53	190	1288	147	44	477	81	240	29	270
	%change	9	8	2	9	−7	10	0	9	26	8	0	136	11	12	34	30	66	31	13	20	19	9	7	15	0	19
	sig.	**	ns	ns	ns	ns	ns	ns	ns	**	ns	ns	*	ns	ns	***	**	*	*	**	ns	ns	ns	ns	ns	ns	*
Brown Creeper	1st	783	40	39	227	40	135	27	103	42	129	1	2	2	2	28	20	1	7	322	55	9	146	23	31	17	41
	2nd	868	21	42	257	52	176	34	130	59	94	3	0	2	0	12	10	0	2	212	47	16	76	18	10	5	40
	%change	11	−48	8	13	30	30	26	26	40	−27	200	−∞	−8	−∞	−57	−50	−∞	−71	−34	−15	78	−48	−22	−68	−71	−2
	sig.	*	*	ns	ns	ns	*	ns	ns	*	ns	na	na	ns	na	*	ns	na	ns	***	ns	ns	***	ns	***	**	ns
Carolina Wren	1st	740	34	55	60	8	69	5	22	25	297	165	19	7	17	421	187	55	179	867	69	15	353	38	199	28	165
	2nd	1666	89	143	157	24	240	12	86	184	543	188	29	15	23	493	228	68	197	1261	137	32	450	80	261	31	270
	%change	125	162	160	162	200	248	140	291	636	83	14	44	97	24	17	21	24	10	45	99	113	27	111	31	11	64
	sig.	***	***	***	***	**	***	ns	***	***	***	ns	*	*	ns	**	***	***	**	***	***	**	***	***	**	ns	***
House Wren	1st	2691	130	201	363	93	451	55	189	302	719	188	33	33	19	482	224	68	190	1282	139	45	468	81	260	29	260
	2nd	2670	117	189	351	98	440	48	187	331	728	181	30	37	23	491	227	69	195	1280	145	46	448	81	259	28	273
	%change	−1	−10	−6	−3	5	−2	−13	−1	10	1	−4	−14	3	11	2	1	1	3	0	4	2	−4	0	−1	−3	5
	sig.	ns	ns	ns	ns	ns	ns	ns	ns	ns	ns	ns	ns	ns	ns	ns	ns	ns	ns	ns	ns	ns	ns	ns	ns	ns	ns
Winter Wren	1st	321	7	23	139	21	41	4	57	12	17	0	0	1	0	1	1	0	0	56	6	6	30	3	1	0	10
	2nd	696	17	39	257	39	103	7	151	47	36	0	0	2	0	0	0	0	0	101	9	6	56	3	3	0	24
	%change	117	143	70	85	86	151	75	165	292	112			84	−∞	−∞				80	50	0	87	0	199		140
	sig.	***	*	*	***	*	***	ns	***	***	**			ns	na	na				***	ns	ns	**	ns	ns		*
Sedge Wren	1st	9	0	0	0	0	0	1	0	5	0	0	0	1	0	1	1	0	2	2	0	0	0	0	2	0	0
	2nd	15	0	2	1	0	0	0	1	9	2	0	0	1	0	3	1	0	2	1	0	0	0	0	0	0	1
	%change	67		∞	∞			−∞	∞	80	−33			−8		200	0		∞	−50					−∞		∞
	sig.	ns		na	na			na	na	ns	ns			ns	na	ns	ns		ns	ns					na		ns
Marsh Wren	1st	58	0	1	5	2	17	1	0	30	2	0	5	4	0	4	2	0	2	6	0	0	1	3	1	0	1
	2nd	44	1	0	4	3	7	0	0	28	1	0	4	2	0	2	1	0	1	1	0	0	0	0	0	0	1
	%change	−24	∞	−∞	−20	50	−59	−∞		−7	−50		−24	−54		−50	−50		−50	−83			−∞	−∞	−∞		0
	sig.	ns	na	na	ns	ns	*	na		ns	ns		ns	ns		ns	ns		ns	ns			na	ns	ns		ns
Blue-gray Gnatcatcher	1st	1293	83	151	154	14	134	20	75	74	416	172	7	5	11	218	110	18	90	774	83	26	343	57	110	26	129
	2nd	1209	70	92	187	17	149	25	107	111	322	129	20	7	13	346	166	31	149	1041	128	25	396	73	168	30	221
	%change	−6	−16	−39	21	21	11	25	43	50	−23	−25	170	29	8	59	50	72	66	34	54	−4	15	28	52	15	71
	sig.	ns	ns	***	ns	ns	ns	ns	*	**	***	*	**	ns	ns	***	***	**	***	***	**	ns	*	*	***	ns	***
Golden-crowned Kinglet	1st	128	4	20	48	10	9	10	10	3	14	0	0	0	1	4	2	0	2	50	10	0	27	8	2	0	3
	2nd	289	7	34	96	18	19	11	46	16	40	2	0	0	1	3	2	0	1	77	30	0	26	9	1	5	6
	%change	126	75	70	100	80	111	10	360	433	186	∞			−9	−25	0		−50	54	200		−4	13	−50	∞	100
	sig.	***	ns	ns	***	ns	ns	ns	***	**	***	na			ns	ns	ns		ns	*	**		ns	ns	ns	na	ns

(continued)

		Appalachian Plateaus	Allegheny Front	Allegheny Mountain	Deep Valleys	Glaciated High Plateau	Glaciated Low Plateau	Glaciated Pocono Plateau	High Plateau	NW Glaciated Plateau	Pittsburgh Low Plateau	Waynesburg Hills	Atlantic Coastal Plain	Central Lowlands	New England	Piedmont	Gettysburg-Newark Lowland	Piedmont Lowland	Piedmont Upland	Ridge and Valley	Anthracite Upland	Anthracite Valley	Appalachian Mountain	Blue Mountain	Great Valley	South Mountain	Susquehanna Lowland
Eastern Bluebird	1st	2344	122	184	308	81	349	37	186	272	630	175	5	17	11	371	170	38	163	1120	105	23	450	71	196	23	252
	2nd	2454	118	172	313	87	402	34	175	321	664	168	1	25	20	469	220	65	184	1250	134	35	454	78	247	29	273
	%change	5	−3	−7	2	7	15	−8	−6	18	5	−4	−81	35	66	26	29	71	13	12	28	52	1	10	26	26	8
	sig.	ns	ns	ns	ns	ns	ns	ns	ns	*	ns	ns	ns	ns	ns	***	*	**	ns	**	ns	ns	ns	*	ns	ns	ns
Veery	1st	1641	39	65	350	77	391	54	196	192	275	2	8	21	17	205	93	19	93	417	103	38	57	67	70	13	69
	2nd	1791	50	79	421	88	400	50	200	246	256	1	4	15	21	204	87	15	102	508	131	40	90	78	59	14	96
	%change	9	28	22	20	14	2	−7	2	28	−7	−50	−53	−34	13	0	−7	−21	10	22	27	5	58	16	−16	8	39
	sig.	*	ns	ns	*	ns	ns	ns	ns	**	ns	ns	ns	ns	ns	ns	ns	ns	ns	**	ns	ns	**	ns	ns	ns	*
Swainson's Thrush	1st	40	0	0	16	3	3	3	14	0	1	0	0	0	0	0	0	0	0	3	2	0	1	0	0	0	0
	2nd	96	0	0	42	10	1	0	40	0	3	0	0	0	0	0	0	0	0	2	0	0	0	2	0	0	0
	%change	140			163	233	−67	−∞	186		200									−33	−∞		−∞	∞			
	sig.	***			***	*	ns	na	***		ns									ns	na		na	na			
Hermit Thrush	1st	1170	53	32	333	65	230	50	162	33	212	0	0	0	0	0	0	0	0	202	53	18	69	26	3	4	29
	2nd	1511	77	68	424	92	310	53	204	54	229	0	0	1	0	6	5	1	0	359	101	28	118	25	8	4	75
	%change	29	45	113	27	42	35	6	26	64	8			∞		∞	∞	∞		78	91	56	71	−4	166	0	159
	sig.	***	*	***	***	*	***	ns	*	*	ns		na	na	na	na	na	na		***	***	ns	***	ns	ns	ns	***
Wood Thrush	1st	2737	128	213	401	97	410	50	222	297	732	187	20	30	16	459	213	54	192	1232	139	42	454	83	234	31	249
	2nd	2759	134	207	385	94	429	38	206	328	753	185	18	29	23	475	223	57	195	1299	141	45	478	83	248	31	273
	%change	1	5	−3	−4	−3	5	−24	−7	10	3	−1	−15	−11	31	3	4	6	2	5	1	7	5	0	6	0	10
	sig.	ns	ns	ns	ns	ns	ns	ns	ns	ns	ns	ns	ns	ns	ns	ns	ns	ns	ns	ns	ns	ns	ns	ns	ns	ns	ns
American Robin	1st	2975	137	215	443	113	470	56	237	330	784	190	34	34	21	492	226	69	197	1327	148	48	486	83	258	31	273
	2nd	2994	139	216	455	114	468	56	236	331	789	190	33	37	23	495	229	69	197	1341	151	49	490	82	262	31	276
	%change	1	1	0	3	1	0	0	0	0	1	0	−8	0	0	1	1	0	0	1	2	2	1	−1	1	0	1
	sig.	ns	ns	ns	ns	ns	ns	ns	ns	ns	ns	ns	ns	ns	ns	ns	ns	ns	ns	ns	ns	ns	ns	ns	ns	ns	ns
Gray Catbird	1st	2871	138	214	396	105	461	56	231	317	765	188	32	34	21	487	226	67	194	1303	143	49	472	83	258	31	267
	2nd	2931	137	214	423	111	467	55	229	329	777	189	33	37	23	494	228	69	197	1323	149	49	475	83	262	30	275
	%change	2	−1	0	7	6	1	−2	−1	4	2	1	−3	0	0	1	0	3	2	2	4	0	1	0	1	−3	3
	sig.	ns	ns	ns	ns	ns	ns	ns	ns	ns	ns	ns	ns	ns	ns	ns	ns	ns	ns	ns	ns	ns	ns	ns	ns	ns	ns
Northern Mockingbird	1st	671	61	32	48	14	185	18	8	12	167	126	30	4	21	486	225	69	192	1079	110	34	356	74	255	26	224
	2nd	1036	62	92	33	11	152	8	8	96	399	175	32	3	22	490	225	69	196	1165	118	31	411	74	257	23	251
	%change	54	2	188	−31	−21	−18	−56	0	700	139	39	1	−31	−4	1	0	0	2	8	7	−9	15	0	0	−12	12
	sig.	***	ns	***	ns	ns	ns	*	ns	***	***	**	ns	ns	ns	ns	ns	ns	ns	ns	ns	ns	*	ns	ns	ns	ns
Brown Thrasher	1st	1733	83	143	156	45	279	31	99	202	529	166	21	17	14	378	166	52	160	910	75	29	346	59	204	22	175
	2nd	1775	94	137	196	66	259	11	100	213	540	159	14	16	16	391	173	52	166	1119	111	9	427	68	228	24	252
	%change	2	13	−4	26	47	−7	−65	1	5	2	−4	−37	−14	4	3	4	0	4	23	48	−69	23	15	11	9	44
	sig.	ns	ns	ns	*	*	ns	***	ns	ns	ns	ns	ns	ns	ns	ns	ns	ns	ns	***	**	***	**	ns	ns	ns	***
European Starling	1st	2572	117	197	283	88	423	44	186	321	727	186	33	27	20	486	225	69	192	1251	134	47	444	79	258	26	263
	2nd	2641	121	202	310	87	414	36	199	330	754	188	36	37	23	489	226	67	196	1279	141	49	458	75	257	24	275
	%change	3	3	3	10	−1	−2	−18	7	3	4	1	3	26	5	1	0	−3	2	2	5	4	3	−5	−1	−8	5
	sig.	ns	ns	ns	ns	ns	ns	ns	ns	ns	ns	ns	ns	ns	ns	ns	ns	ns	ns	ns	ns	ns	ns	ns	ns	ns	ns
Cedar Waxwing	1st	2731	116	207	416	107	421	54	231	293	717	169	11	34	15	238	122	16	100	1033	112	43	419	70	167	24	198
	2nd	2675	116	189	431	98	401	52	227	314	698	149	23	35	21	385	187	48	150	1144	136	42	405	75	216	29	241
	%change	−2	0	−9	4	−8	−5	−4	−2	7	−3	−12	97	−5	28	62	53	200	50	11	21	−2	−3	7	29	21	22
	sig.	ns	ns	ns	ns	ns	ns	ns	ns	ns	ns	ns	*	ns	ns	***	***	***	**	*	ns	ns	ns	ns	*	ns	*
Ovenbird	1st	2316	117	176	408	94	424	56	216	174	565	86	11	12	19	297	144	19	134	1019	142	43	389	82	129	29	205
	2nd	2638	136	191	451	111	468	56	230	246	651	98	11	13	21	312	132	24	156	1173	149	47	459	83	142	30	263
	%change	14	16	9	11	18	10	0	6	41	15	14	−6	0	1	5	−9	26	16	15	5	9	18	1	10	3	28
	sig.	***	ns	ns	ns	ns	ns	ns	ns	***	*	ns	ns	ns	ns	ns	ns	ns	ns	***	ns	ns	*	ns	ns	ns	**

(continued)

		Appalachian Plateaus	Allegheny Front	Allegheny Mountain	Deep Valleys	Glaciated High Plateau	Glaciated Low Plateau	Glaciated Pocono Plateau	High Plateau	NW Glaciated Plateau	Pittsburgh Low Plateau	Waynesburg Hills	Atlantic Coastal Plain	Central Lowlands	New England	Piedmont	Gettysburg-Newark Lowland	Piedmont Lowland	Piedmont Upland	Ridge and Valley	Anthracite Upland	Anthracite Valley	Appalachian Mountain	Blue Mountain	Great Valley	South Mountain	Susquehanna Lowland
Worm-eating Warbler	1st	192	33	23	36	6	49	3	3	4	24	11	1	0	6	83	40	3	40	453	42	13	231	33	39	20	75
	2nd	268	50	26	55	5	61	11	11	2	40	7	0	0	11	101	39	7	55	706	100	17	302	52	53	26	156
	%change	40	52	13	53	−17	24	267	267	−50	67	−36	−∞		67	22	−3	133	38	56	138	31	31	58	35	30	108
	sig.	***	ns	ns	*	ns	ns	*	*	ns	*	ns	na	na	ns	ns	ns	ns	ns	***	***	ns	**	*	ns	ns	***
Louisiana Waterthrush	1st	767	54	99	133	24	150	13	42	38	131	83	4	5	8	128	57	4	67	474	58	12	218	51	50	19	66
	2nd	899	63	96	172	21	147	9	65	71	167	88	1	5	8	177	83	10	84	694	104	13	281	50	71	28	147
	%change	17	17	−3	29	−13	−2	−31	55	87	27	6	−76	−8	−9	38	45	150	25	46	79	8	29	−2	41	47	123
	sig.	**	ns	ns	*	ns	ns	ns	*	**	*	ns	ns	ns	ns	**	*	ns	ns	***	***	ns	**	ns	*	ns	***
Northern Waterthrush	1st	246	9	8	43	13	81	19	18	30	25	0	0	1	0	0	0	0	0	84	25	13	22	12	2	0	10
	2nd	185	7	8	28	7	44	22	10	39	20	0	0	1	0	0	0	0	0	43	15	2	10	5	2	0	9
	%change	−25	−22	0	−35	−46	−46	16	−44	30	−20			−8						−49	−40	−85	−55	−58	0		−10
	sig.	**	ns	ns	ns	ns	***	ns	ns	ns	ns			ns						***	ns	**	*	ns	ns		ns
Golden-winged Warbler	1st	360	47	56	27	1	72	12	25	8	95	17	0	1	0	1	1	0	0	253	23	14	155	14	6	1	40
	2nd	125	17	33	18	1	28	4	6	0	17	1	0	0	0	1	1	0	0	114	8	3	87	7	2	0	7
	%change	−65	−64	−41	−33	0	−61	−67	−76	−∞	−82	−94		−∞		0	0			−55	−65	−79	−44	−50	−67	−∞	−83
	sig.	***	***	*	ns	ns	***	*	***	na	***	***		na		ns	ns			***	**	**	***	ns	ns	na	***
Blue-winged Warbler	1st	940	4	26	44	10	140	8	81	188	292	147	6	13	15	178	110	4	64	251	25	13	35	48	77	13	40
	2nd	982	6	26	88	18	126	4	109	217	276	112	0	12	13	119	71	4	44	219	28	7	40	45	61	5	33
	%change	4	50	0	100	80	−10	−50	35	15	−5	−24	−∞	−15	−21	−33	−36	0	−31	−13	12	−46	14	−6	−21	−62	−18
	sig.	ns	ns	ns	***	ns	ns	ns	*	ns	ns	ns	na	ns	*	***	**	ns	ns	ns	ns	ns	ns	ns	ns	*	ns
Black-and-white Warbler	1st	1359	86	118	298	60	293	54	101	28	286	35	1	1	10	112	63	5	44	625	105	40	229	67	64	25	95
	2nd	1442	80	119	367	76	292	47	101	45	282	33	1	1	9	86	53	6	27	742	140	45	252	78	58	26	143
	%change	6	−7	1	23	27	0	−13	0	61	−1	−6	−6	−8	−18	−23	−16	20	−39	19	33	13	10	16	−10	4	51
	sig.	ns	ns	ns	**	ns	ns	ns	ns	*	ns	ns	ns	ns	ns	ns	ns	ns	*	**	*	ns	ns	ns	ns	ns	**
Prothonotary Warbler	1st	26	1	0	6	0	2	1	0	14	2	0	2	2	0	7	5	1	1	6	0	0	2	1	2	0	1
	2nd	19	0	1	1	0	2	0	0	13	2	0	0	1	0	16	6	3	7	11	0	0	0	2	7	0	2
	%change	−27	−∞	∞	−83		0	−∞		−7	0		−∞	−54		129	19	200	600	83			−∞	100	249		100
	sig.	ns	na	na	ns		ns	na		ns	ns		na	ns		ns	ns	ns	*	ns			ns	na	ns	ns	ns
Swainson's Warbler	1st	2	0	2	0	0	0	0	0	0	0	0	0	0	0	0	0	0	0	0	0	0	0	0	0	0	0
	2nd	2	0	1	0	0	0	0	0	0	0	1	0	0	0	1	0	0	1	1	1	0	0	0	0	0	0
	%change	0		−50								∞				∞			∞	∞	∞						
	sig.	ns		ns								na				na			na	na	na						
Nashville Warbler	1st	159	10	7	23	20	31	23	14	10	21	0	0	0	0	0	0	0	0	60	17	6	21	8	1	1	6
	2nd	180	5	4	49	25	33	22	21	10	11	0	0	0	0	0	0	0	0	33	11	4	9	1	2	0	6
	%change	13	−50	−43	113	25	6	−4	50	0	−48									−45	−35	−33	−57	−88	99	−∞	0
	sig.	ns	ns	ns	**	ns	ns	ns	ns	ns	ns									**	ns	ns	*	*	ns	na	ns
Mourning Warbler	1st	230	0	0	64	0	7	0	73	82	4	0	0	4	0	0	0	0	0	1	0	0	1	0	0	0	0
	2nd	426	0	1	166	22	17	2	108	101	9	0	0	3	0	0	0	0	0	2	0	0	1	0	0	0	1
	%change	85		∞	159	∞	143	∞	48	23	125			−31						100			0				∞
	sig.	***		na	***	na	*	na	**	ns	ns			ns						ns			ns				na
Kentucky Warbler	1st	569	27	103	10	1	2	0	3	13	261	149	7	0	11	165	64	13	88	177	13	5	67	16	46	14	16
	2nd	450	16	69	3	0	7	2	7	24	214	108	0	0	6	89	28	3	58	117	9	0	70	6	20	4	8
	%change	−21	−41	−33	−70	−∞	250	∞	133	85	−18	−28	−∞		−50	−46	−56	−77	−34	−34	−31	−∞	4	−63	−57	−71	−50
	sig.	***	ns	**	ns	na	ns	na	ns	ns	*	**	na		ns	***	***	**	*	***	ns	na	ns	*	**	**	ns
Common Yellowthroat	1st	2914	137	213	437	110	460	56	231	322	764	184	29	33	20	460	212	61	187	1277	145	47	475	81	245	28	256
	2nd	2975	138	216	454	114	470	56	237	331	775	184	29	31	22	476	223	60	193	1313	149	48	487	82	245	27	275
	%change	2	1	1	4	4	2	0	3	3	1	0	−6	−14	0	3	5	−2	3	3	3	2	3	1	0	−4	7
	sig.	ns	ns	ns	ns	ns	ns	ns	ns	ns	ns	ns	ns	ns	ns	ns	ns	ns	ns	ns	ns	ns	ns	ns	ns	ns	ns

(continued)

		Appalachian Plateaus	Allegheny Front	Allegheny Mountain	Deep Valleys	Glaciated High Plateau	Glaciated Low Plateau	Glaciated Pocono Plateau	High Plateau	NW Glaciated Plateau	Pittsburgh Low Plateau	Waynesburg Hills	Atlantic Coastal Plain	Central Lowlands	New England	Piedmont	Gettysburg-Newark Lowland	Piedmont Lowland	Piedmont Upland	Ridge and Valley	Anthracite Upland	Anthracite Valley	Appalachian Mountain	Blue Mountain	Great Valley	South Mountain	Susquehanna Lowland
Hooded Warbler	1st	1124	52	150	86	10	22	4	121	198	410	71	3	21	4	31	15	0	16	235	33	6	105	25	22	18	26
	2nd	1751	64	147	235	22	42	14	218	308	588	113	1	29	10	59	29	4	26	573	95	11	242	47	32	26	120
	%change	56	23	−2	173	120	91	250	80	56	43	59	−69	27	128	90	92	∞	63	144	188	83	130	88	45	44	362
	sig.	***	ns	ns	***	*	*	*	***	***	***	**	ns	ns	ns	**	*	na	ns	***	***	ns	***	**	ns	ns	***
American Redstart	1st	2195	106	177	403	85	372	49	223	195	476	109	11	27	8	136	56	12	68	615	74	40	253	46	73	15	114
	2nd	2575	131	194	438	102	412	48	231	288	595	136	8	31	19	150	81	14	55	933	125	47	353	71	112	23	202
	%change	17	24	10	9	20	11	−2	4	48	25	25	−31	6	117	10	44	17	−19	52	69	18	40	54	53	53	77
	sig.	***	ns	ns	ns	ns	ns	ns	ns	***	***	ns	ns	ns	*	ns	*	ns	ns	***	***	ns	***	*	**	ns	***
Cerulean Warbler	1st	613	28	67	33	7	35	7	37	27	214	158	3	8	4	43	21	2	20	165	18	11	84	14	19	3	16
	2nd	506	31	65	79	1	27	1	74	46	103	79	0	6	1	20	4	2	14	243	37	5	138	21	12	4	26
	%change	−17	11	−3	139	−86	−23	−86	100	70	−52	−50	−∞	−31	−77	−53	−81	0	−30	47	106	−55	64	50	−37	33	63
	sig.	**	ns	ns	***	*	ns	*	***	*	***	***	na	***	ns	**	***	ns	ns	***	**	ns	***	ns	ns	ns	ns
Northern Parula	1st	292	19	80	42	5	32	0	27	8	48	31	1	0	2	63	25	5	33	142	14	2	65	17	15	1	28
	2nd	545	70	103	93	6	53	3	62	23	82	50	5	0	7	142	54	7	81	409	51	5	151	44	38	7	113
	%change	87	268	29	121	20	66	∞	130	188	71	61	372		220	125	115	40	145	188	264	150	132	159	152	600	304
	sig.	***	***	ns	***	ns	*	na	***	**	***	*	ns	na	ns	***	***	ns	***	***	***	ns	***	***	**	*	***
Magnolia Warbler	1st	698	19	57	218	45	55	28	131	58	87	0	0	0	0	0	0	0	0	46	10	4	13	7	1	0	11
	2nd	1304	37	79	387	74	130	39	218	119	220	1	0	2	0	1	1	0	0	132	44	7	37	6	1	0	37
	%change	87	95	39	78	64	136	39	66	105	153	∞		∞		∞	∞			187	340	75	185	−14	0		236
	sig.	***	*	ns	***	**	***	ns	***	***	***	na	na	na	na	na				***	***	*	***	ns	ns		***
Blackburnian Warbler	1st	814	29	28	309	52	75	24	147	30	120	0	0	0	0	0	0	0	0	104	14	5	57	10	1	3	14
	2nd	1305	65	52	424	85	156	42	212	81	188	0	0	0	0	0	0	0	0	275	54	14	139	15	0	1	52
	%change	60	124	86	37	63	108	75	44	170	57									164	286	180	144	50	−∞	−67	271
	sig.	***	***	**	***	**	***	*	***	***	***									***	***	*	***	ns	na	ns	***
Yellow Warbler	1st	2530	106	197	309	83	436	43	199	300	672	185	28	32	16	385	192	43	150	993	90	41	356	71	203	16	216
	2nd	2638	110	182	329	93	455	48	199	327	713	182	27	36	21	430	206	48	176	1148	125	47	393	78	227	11	267
	%change	4	4	−8	6	12	4	12	0	9	6	−2	−9	3	20	12	7	12	17	16	39	15	10	10	11	−31	24
	sig.	ns	ns	ns	ns	ns	ns	ns	ns	ns	ns	ns	ns	ns	ns	ns	ns	ns	ns	***	*	ns	ns	ns	ns	ns	*
Chestnut-sided Warbler	1st	1739	82	132	309	75	312	48	228	176	374	3	1	8	4	51	29	0	22	206	32	31	71	16	18	4	34
	2nd	2225	103	180	427	99	407	46	230	259	465	9	1	17	10	32	16	1	15	433	100	32	128	42	22	4	105
	%change	28	26	36	38	32	30	−4	1	47	24	200	−6	95	128	−37	−45	∞	−32	110	213	3	80	163	22	0	209
	sig.	***	ns	**	***	ns	***	ns	ns	***	**	ns	ns	ns	ns	*	*	na	***	***	***	ns	***	***	ns	ns	***
Blackpoll Warbler	1st	2	0	0	0	0	2	0	0	0	0	0	0	0	0	0	0	0	0	0	0	0	0	0	0	0	0
	2nd	3	0	0	1	1	0	0	1	0	0	0	0	0	0	0	0	0	0	0	0	0	0	0	0	0	0
	%change	50			∞	∞	−∞		∞																		
	sig.	ns			na	na	na		na																		
Black-throated Blue Warbler	1st	627	43	52	201	28	60	26	117	23	77	0	0	0	0	0	0	0	0	115	18	9	51	15	3	0	19
	2nd	1021	63	77	360	61	122	29	177	19	113	0	0	0	0	0	0	0	0	257	67	14	115	17	6	5	33
	%change	63	47	48	79	118	103	12	51	−17	47									123	272	56	125	13	99	∞	74
	sig.	***	*	*	***	***	***	ns	***	ns	**									***	***	ns	***	ns	ns	na	ns
Pine Warbler	1st	140	10	7	35	3	42	1	12	7	23	0	0	0	0	20	6	0	14	135	12	3	64	14	9	10	23
	2nd	359	23	9	93	11	100	20	17	22	64	0	0	2	1	37	21	0	16	389	67	7	167	44	14	18	72
	%change	156	130	29	166	267	138	1900	42	214	178			∞	∞	85	248		14	188	458	133	161	214	55	80	213
	sig.	***	*	ns	***	*	***	***	ns	**	***			na	na	*	**		ns	***	***	ns	***	***	ns	ns	***
Yellow-rumped Warbler	1st	273	3	1	74	44	94	21	13	0	23	0	0	0	0	0	0	0	0	42	6	1	18	7	0	0	10
	2nd	755	14	1	271	65	169	41	79	31	84	0	0	1	0	0	0	0	0	121	40	9	45	4	3	3	17
	%change	177	367	0	266	48	80	95	508	∞	265			∞						188	567	800	150	−43	∞	∞	70
	sig.	***	**	ns	***	*	***	**	***	na	***			na						***	***	**	***	ns	na	na	ns

(continued)

		Appalachian Plateaus	Allegheny Front	Allegheny Mountain	Deep Valleys	Glaciated High Plateau	Glaciated Low Plateau	Glaciated Pocono Plateau	High Plateau	NW Glaciated Plateau	Pittsburgh Low Plateau	Waynesburg Hills	Atlantic Coastal Plain	Central Lowlands	New England	Piedmont	Gettysburg-Newark Lowland	Piedmont Lowland	Piedmont Upland	Ridge and Valley	Anthracite Upland	Anthracite Valley	Appalachian Mountain	Blue Mountain	Great Valley	South Mountain	Susquehanna Lowland
Yellow-throated Warbler	1st	116	1	19	5	0	1	0	2	6	41	41	1	0	0	17	6	1	10	18	1	0	9	0	5	2	1
	2nd	150	4	12	7	1	3	0	10	11	42	60	0	0	0	24	9	3	12	22	0	0	8	1	8	1	4
	%change	29	300	−37	40	∞	200		400	83	2	46	−∞			41	49	200	20	22	−∞		−11	∞	59	−50	300
	sig.	*	ns	ns	ns	na	ns		*	ns	ns	ns	na			ns	ns	ns	ns	ns	na	ns	na	ns	ns	ns	ns
Prairie Warbler	1st	624	16	64	26	7	89	14	12	15	295	86	0	0	1	76	33	1	42	335	45	24	116	48	18	7	77
	2nd	566	14	29	27	25	183	24	15	15	196	38	0	0	3	93	59	1	33	455	98	31	118	58	28	14	108
	%change	−9	−13	−55	4	257	106	71	25	0	−34	−56			174	22	78	0	−21	36	118	29	2	21	55	100	40
	sig.	ns	ns	***	ns	***	***	ns	ns	ns	***	***			ns	**	ns	ns	ns	***	***	ns	ns	ns	ns	ns	*
Black-throated Green Warbler	1st	1521	81	130	351	72	182	38	193	105	369	0	0	2	0	1	1	0	0	281	56	19	124	21	3	4	54
	2nd	2025	109	167	446	105	351	50	228	150	418	1	0	3	0	3	1	1	1	644	121	29	252	59	17	11	155
	%change	33	35	28	27	46	93	32	18	43	13	∞		38		200	0	∞	∞	129	116	53	103	181	465	175	187
	sig.	***	*	*	***	*	***	ns	ns	**	ns	na		ns	na	ns	ns	na	na	***	***	***	***	***	**	**	***
Canada Warbler	1st	546	31	25	131	31	122	36	87	41	42	0	0	1	0	0	0	0	0	88	23	5	32	15	3	0	10
	2nd	575	28	27	226	29	92	27	76	38	32	0	0	0	0	1	1	0	0	115	42	5	48	9	1	2	8
	%change	5	−10	8	73	−6	−25	−25	−13	−7	−24			−∞		∞	∞			31	83	0	50	−40	−67	∞	−20
	sig.	ns	ns	ns	***	ns	*	ns	ns	ns	ns			na		na	na			ns	*	ns	ns	ns	ns	na	ns
Yellow-breasted Chat	1st	652	40	63	25	4	21	2	11	27	290	169	11	0	2	150	57	8	85	427	26	6	246	16	56	12	65
	2nd	379	16	32	11	0	6	1	1	25	168	119	2	1	3	106	37	4	65	314	26	1	166	20	37	5	59
	%change	−42	−60	−49	−56	−∞	−71	−50	−91	−7	−42	−30	−83	∞	37	−29	−35	−50	−24	−26	0	−83	−33	25	−34	−58	−9
	sig.	***	***	**	*	na	**	ns	**	ns	***	**	**	na	ns	**	*	ns	ns	***	ns	*	***	ns	*	ns	ns
Eastern Towhee	1st	2759	132	215	400	101	398	52	222	289	763	187	17	23	19	430	203	44	183	1204	142	44	471	80	217	31	219
	2nd	2867	137	214	428	112	398	47	230	325	788	188	19	23	22	458	216	50	192	1288	148	41	482	80	236	31	270
	%change	4	4	0	7	11	0	−10	4	12	3	1	6	−8	6	7	6	14	5	7	4	−7	2	0	8	0	23
	sig.	ns	ns	ns	ns	ns	ns	ns	ns	ns	ns	ns	ns	ns	ns	ns	ns	ns	ns	ns	ns	ns	ns	ns	ns	ns	*
Chipping Sparrow	1st	2949	137	216	440	111	463	56	234	316	788	188	15	29	20	452	206	62	184	1320	147	48	488	82	259	31	265
	2nd	2988	138	214	453	112	472	56	237	331	787	188	17	36	23	489	225	68	196	1335	150	49	488	82	261	30	275
	%change	1	1	−1	3	1	2	0	1	5	0	0	7	14	5	8	9	10	7	1	2	2	0	0	0	−3	4
	sig.	ns	ns	ns	ns	ns	ns	ns	ns	ns	ns	ns	ns	ns	ns	ns	ns	ns	ns	ns	ns	ns	ns	ns	ns	ns	ns
Clay-colored Sparrow	1st	2	0	1	0	0	0	0	0	0	1	0	0	1	0	0	0	0	0	0	0	0	0	0	0	0	0
	2nd	24	1	0	1	0	0	0	1	2	19	0	0	0	0	1	0	1	0	4	0	0	1	0	0	0	3
	%change	1100	∞	−∞	∞				∞	∞	1800			−∞		∞		∞		∞			∞				∞
	sig.	***	na	na	na				na	na	***			na		na		na		na			na				na
Field Sparrow	1st	2613	128	204	322	88	412	47	189	282	755	186	16	28	19	418	203	43	172	1196	133	45	440	77	238	29	234
	2nd	2679	125	209	352	97	410	43	205	312	744	182	11	24	19	405	202	37	166	1234	144	47	449	76	228	26	264
	%change	3	−2	2	9	10	0	−9	8	11	−1	−2	−35	−21	−9	−3	−1	−14	−3	3	8	4	2	−1	−5	−10	13
	sig.	ns	ns	ns	ns	ns	ns	ns	ns	ns	ns	ns	ns	ns	ns	ns	ns	ns	ns	ns	ns	ns	ns	ns	ns	ns	ns
Vesper Sparrow	1st	604	36	68	29	17	84	0	19	53	280	18	0	1	1	80	24	23	33	401	25	0	121	17	126	4	108
	2nd	347	24	66	13	8	30	0	7	33	159	7	0	8	0	71	17	17	37	452	37	2	154	14	115	6	124
	%change	−43	−33	−3	−55	−53	−64		−63	−38	−43	−61		635	−∞	−11	−29	−26	12	13	48	∞	27	−18	−9	50	15
	sig.	***	ns	ns	*	ns	***		*	*	***	*		*	na	ns	ns	ns	ns	ns	ns	na	*	ns	ns	ns	ns
Savannah Sparrow	1st	1310	36	138	75	41	250	2	70	232	399	67	1	18	3	57	17	17	23	272	6	11	109	9	68	0	69
	2nd	1464	46	126	128	57	271	3	80	296	389	68	0	16	1	106	36	33	37	498	28	10	171	18	119	0	152
	%change	12	28	−9	71	39	8	50	14	28	−3	1	−∞	−18	−70	86	111	94	61	83	367	−9	57	100	74		120
	sig.	**	ns	ns	***	ns	ns	ns	ns	**	ns	ns	na	ns	ns	***	**	*	ns	***	***	ns	***	ns	***		***
Grasshopper Sparrow	1st	837	80	103	27	26	102	2	13	45	384	55	3	2	4	130	63	13	54	653	35	3	278	41	148	9	139
	2nd	820	74	99	43	14	66	3	25	56	397	43	0	4	0	156	59	26	71	694	61	13	241	37	166	6	170
	%change	−2	−8	−4	59	−46	−35	50	92	24	3	−22	−∞	84	−∞	20	−7	100	31	6	74	333	−13	−10	12	−33	22
	sig.	ns	ns	ns	ns	ns	**	ns	*	ns	ns	ns	na	ns	na	ns	ns	*	ns	ns	**	**	ns	ns	ns	ns	ns

(continued)

		Appalachian Plateaus	Allegheny Front	Allegheny Mountain	Deep Valleys	Glaciated High Plateau	Glaciated Low Plateau	Glaciated Pocono Plateau	High Plateau	NW Glaciated Plateau	Pittsburgh Low Plateau	Waynesburg Hills	Atlantic Coastal Plain	Central Lowlands	New England	Piedmont	Gettysburg-Newark Lowland	Piedmont Lowland	Piedmont Upland	Ridge and Valley	Anthracite Upland	Anthracite Valley	Appalachian Mountain	Blue Mountain	Great Valley	South Mountain	Susquehanna Lowland
Henslow's Sparrow	1st	351	12	35	2	8	45	0	6	69	154	20	0	1	0	0	0	0	0	12	0	0	8	0	3	0	1
	2nd	224	20	15	8	2	8	0	5	20	138	8	0	0	0	1	0	0	1	4	0	0	4	0	0	0	0
	%change	−36	67	−57	300	−75	−82		−17	−71	−10	−60	−∞			∞			∞	−67			−50		−∞		−∞
	sig.	***	ns	**	ns	ns	***		ns	***	ns	*		na		na			na	*		na	ns		na		na
Song Sparrow	1st	2928	133	214	417	110	465	52	236	328	785	188	33	34	21	483	221	69	193	1289	142	49	466	80	259	27	266
	2nd	2958	135	216	430	114	463	55	237	331	787	190	35	37	23	495	229	69	197	1319	149	49	475	83	261	26	276
	%change	1	2	1	3	4	0	6	0	1	0	1	0	0	0	2	3	0	2	2	5	0	2	4	0	−4	4
	sig.	ns	ns	ns	ns	ns	ns	ns	ns	ns	ns	ns	ns	ns	ns	ns	ns	ns	ns	ns	ns	ns	ns	ns	ns	ns	ns
Swamp Sparrow	1st	812	15	54	102	48	227	30	36	182	117	1	11	10	2	47	23	2	22	125	11	21	26	14	16	2	35
	2nd	1134	18	66	164	58	268	37	72	243	202	6	7	6	0	36	22	4	10	149	19	14	44	11	12	2	47
	%change	40	20	22	61	21	18	23	100	34	73	500	−40	−45	−∞	−23	−5	100	−55	19	73	−33	69	−21	−25	0	34
	sig.	***	ns	ns	***	ns	ns	ns	***	**	***	ns	ns	ns		ns	ns	ns	*	ns	ns	ns	*	ns	ns		ns
White-throated Sparrow	1st	199	2	0	57	38	58	29	10	2	3	0	0	0	0	1	1	0	0	13	4	3	0	2	2	0	2
	2nd	134	1	0	25	34	40	22	7	0	5	0	0	0	0	0	0	0	0	8	2	1	2	1	1	0	1
	%change	−33	−50		−56	−11	−31	−24	−30	−∞	67			−∞	−∞					−38	−50	−67	∞	−50	−50		−50
	sig.	***	ns	na	***	ns	ns	ns	ns	ns	na	ns		na	na					ns	ns	ns	ns	ns	ns	na	ns
Dark-eyed Junco	1st	1154	40	25	415	91	198	30	197	44	114	0	0	1	0	0	0	0	0	178	46	14	73	12	3	0	30
	2nd	1571	47	61	436	104	338	43	226	130	185	1	0	4	0	1	1	0	0	255	45	25	109	6	4	3	63
	%change	36	18	144	5	14	71	43	15	195	62	∞		268		∞	∞			43	−2	79	49	−50	33	∞	110
	sig.	***	ns	***	ns	ns	***	ns	ns	***	***	na		ns		na	na			***	ns	**	**	ns	ns	na	***
Summer Tanager	1st	41	0	0	0	0	0	0	0	0	2	39	0	0	0	3	1	1	1	3	1	0	2	0	1	0	0
	2nd	1	0	0	0	0	0	0	0	0	1	0	0	0	0	1	0	0	1	2	0	0	1	0	1	0	0
	%change	−98									−50	−∞				−67	−∞	−∞	0	−33	−∞		−50		∞		
	sig.	***									ns	na				ns	na	na	ns	ns	na		ns		na		
Scarlet Tanager	1st	2737	132	211	425	97	422	55	226	258	733	178	11	24	17	334	157	27	150	1153	130	41	471	78	160	31	242
	2nd	2910	137	214	443	106	442	56	237	320	773	182	12	25	22	356	162	27	167	1218	149	46	475	78	173	31	266
	%change	6	4	1	4	9	5	2	5	24	5	2	3	−4	18	7	3	0	11	6	15	12	1	0	8	0	10
	sig.	*	ns	ns	ns	ns	ns	ns	ns	**	ns	ns	ns	ns	ns	ns	ns	ns	ns	ns	ns	ns	ns	ns	ns	ns	ns
Northern Cardinal	1st	2568	125	213	293	79	391	47	183	320	728	189	34	33	20	487	222	69	196	1299	144	43	473	82	259	31	267
	2nd	2764	134	212	356	91	433	49	213	331	756	189	33	37	23	493	229	67	197	1328	150	47	479	83	262	31	276
	%change	8	7	0	22	15	11	4	16	3	4	0	−8	3	5	1	3	−3	1	2	4	9	1	1	1	0	3
	sig.	**	ns	ns	*	ns	ns	ns	ns	ns	ns	ns	ns	ns	ns	ns	ns	ns	ns	ns	ns	ns	ns	ns	ns	ns	ns
Rose-breasted Grosbeak	1st	2341	100	165	400	103	388	50	203	260	557	115	9	25	16	133	86	8	39	626	85	33	209	64	98	7	130
	2nd	2632	125	190	417	100	368	37	225	321	684	165	6	26	18	128	94	7	27	747	121	34	262	59	86	13	172
	%change	12	25	15	4	−3	−5	−26	11	23	23	43	−37	−4	3	−4	9	−13	−31	19	42	3	25	−8	−13	86	32
	sig.	***	ns	ns	ns	ns	ns	ns	ns	*	***	**	ns	ns	ns	ns	ns	ns	ns	**	*	ns	*	ns	ns	ns	*
Blue Grosbeak	1st	0	0	0	0	0	0	0	0	0	0	0	7	0	0	67	16	4	47	39	0	0	24	0	11	0	4
	2nd	11	1	3	1	0	0	0	0	0	2	3	0	0	1	115	25	13	77	92	4	3	40	4	36	1	4
	%change	∞	∞	∞	∞						∞	∞	−∞		∞	72	56	225	64	136	∞	∞	67	∞	226	∞	0
	sig.	na	na	na	na						na	na	na		na	***	ns	*	**	***	na	na	*	na	***	na	ns
Indigo Bunting	1st	2832	135	214	423	103	409	49	232	311	768	188	25	25	18	474	214	68	192	1287	140	45	479	79	249	31	264
	2nd	2977	138	214	451	112	467	54	237	329	787	188	20	29	23	470	219	66	185	1332	151	48	487	83	257	31	275
	%change	5	2	0	7	9	14	10	2	6	2	0	−24	7	17	−1	2	−3	−4	3	8	7	2	5	3	0	4
	sig.	ns	ns	ns	ns	ns	*	ns	ns	ns	ns	ns	ns	ns	ns	ns	ns	ns	ns	ns	ns	ns	ns	ns	ns	ns	ns
Dickcissel	1st	32	0	15	1	1	1	0	2	0	10	2	0	1	0	5	3	0	2	7	0	0	2	0	4	0	1
	2nd	10	0	3	0	0	1	0	0	0	3	1	2	0	0	7	5	1	1	26	0	3	5	0	17	0	1
	%change	−69		−80	−∞	−∞	0		−∞		−70	−50	∞	−∞		40	66	∞	−50	271		∞	150		323		0
	sig.	***		**	na	na	ns		na		ns	ns	na	na		ns	ns	na	ns	***		na	ns		**		ns

(continued)

		Appalachian Plateaus	Allegheny Front	Allegheny Mountain	Deep Valleys	Glaciated High Plateau	Glaciated Low Plateau	Glaciated Pocono Plateau	High Plateau	NW Glaciated Plateau	Pittsburgh Low Plateau	Waynesburg Hills	Atlantic Coastal Plain	Central Lowlands	New England	Piedmont	Gettysburg-Newark Lowland	Piedmont Lowland	Piedmont Upland	Ridge and Valley	Anthracite Upland	Anthracite Valley	Appalachian Mountain	Blue Mountain	Great Valley	South Mountain	Susquehanna Lowland
Bobolink	1st	1261	38	123	105	65	331	14	87	246	226	26	1	10	2	40	25	1	14	212	15	20	39	23	48	0	67
	2nd	1424	45	130	122	65	351	13	103	262	306	27	0	12	0	60	39	1	20	243	34	14	49	30	45	0	71
	%change	13	18	6	16	0	6	−7	18	7	35	4	−∞	10	−∞	50	55	0	43	15	127	−30	26	30	−7		6
	sig.	**	ns	ns	ns	ns	ns	ns	ns	ns	***	ns	na	ns	na	*	ns	ns	ns	ns	**	ns	ns	ns	ns		ns
Red-winged Blackbird	1st	2733	124	201	338	102	450	52	210	327	741	188	30	32	21	475	217	65	193	1245	133	42	442	80	258	24	266
	2nd	2744	121	192	364	101	442	47	213	331	750	183	30	36	23	486	224	68	194	1262	139	47	445	79	258	20	274
	%change	0	−2	−4	8	−1	−2	−10	1	1	1	−3	−6	3	0	2	3	5	1	1	5	12	1	−1	0	−17	3
	sig.	ns	ns	ns	ns	ns	ns	ns	ns	ns	ns	ns	ns	ns	ns	ns	ns	ns	ns	ns	ns	ns	ns	ns	ns	ns	ns
Eastern Meadowlark	1st	1923	89	179	139	67	314	4	94	273	602	162	7	12	11	343	153	45	145	973	69	15	385	56	216	15	217
	2nd	1826	91	164	125	59	300	1	90	284	577	135	2	13	2	198	111	14	73	855	63	16	340	40	171	7	218
	%change	−5	2	−8	−10	−12	−4	−75	−4	4	−4	−17	−73	0	−83	−42	−28	−69	−50	−12	−9	7	−12	−29	−21	−53	0
	sig.	ns	ns	ns	ns	ns	ns	ns	ns	ns	ns	ns	ns	ns	**	***	**	***	***	**	ns	ns	ns	ns	*	ns	ns
Western Meadowlark	1st	1	0	0	0	0	0	0	0	0	1	0	0	0	0	0	0	0	0	1	0	0	1	0	0	0	0
	2nd	0	0	0	0	0	0	0	0	0	0	0	0	0	0	0	0	0	0	2	1	0	1	0	0	0	0
	%change	−∞									−∞									100	∞		0				
	sig.	na									na									ns	na		ns				
Common Grackle	1st	2627	123	199	326	94	421	50	198	318	714	184	32	30	20	485	220	69	196	1266	140	48	450	81	258	29	260
	2nd	2619	123	193	335	92	388	45	196	326	743	178	32	35	23	489	225	68	196	1295	146	46	463	81	258	29	272
	%change	0	0	−3	3	−2	−8	−10	−1	3	4	−3	−6	7	5	1	2	−1	0	2	4	−4	3	0	0	0	5
	sig.	ns	ns	ns	ns	ns	ns	ns	ns	ns	ns	ns	ns	ns	ns	ns	ns	ns	ns	ns	ns	ns	ns	ns	ns	ns	ns
Brown-headed Cowbird	1st	2573	130	211	364	87	390	49	186	275	694	187	27	29	16	458	206	66	186	1220	127	47	449	76	248	28	245
	2nd	2634	125	180	385	94	413	48	189	319	707	174	33	35	23	487	225	65	197	1298	145	48	463	82	259	31	270
	%change	2	−4	−15	6	8	6	−2	2	16	2	−7	15	11	31	6	9	−2	6	6	14	2	3	8	4	11	10
	sig.	ns	ns	ns	ns	ns	ns	ns	ns	ns	ns	ns	ns	ns	ns	ns	ns	ns	ns	ns	ns	ns	ns	ns	ns	ns	ns
Orchard Oriole	1st	356	17	13	30	6	16	0	3	22	105	144	4	3	2	195	64	23	108	237	7	1	82	15	85	6	41
	2nd	657	33	50	18	3	28	0	7	97	283	138	24	2	13	308	127	39	142	557	22	3	213	27	161	2	129
	%change	85	94	285	−40	−50	75		133	341	170	−4	467	−39	493	58	98	70	31	135	214	200	160	80	89	−67	215
	sig.	***	*	***	ns	ns	ns		ns	***	***	ns	***	ns	***	***	***	*	*	***	**	ns	***	ns	***	ns	***
Baltimore Oriole	1st	2325	105	166	282	78	409	43	172	269	618	183	27	32	19	435	202	57	176	1165	123	45	408	79	244	25	241
	2nd	2477	104	172	322	86	430	38	172	309	669	175	33	35	22	471	216	65	190	1248	136	44	439	81	252	28	268
	%change	7	−1	4	14	10	5	−12	0	15	8	−4	15	1	6	8	6	14	8	7	11	−2	8	3	3	12	11
	sig.	*	ns	ns	ns	ns	ns	ns	ns	ns	ns	ns	ns	ns	ns	ns	ns	ns	ns	ns	ns	ns	ns	ns	ns	ns	ns
Purple Finch	1st	1268	36	84	213	58	223	46	134	176	286	12	1	17	1	8	6	0	2	154	38	19	33	24	8	3	29
	2nd	1570	52	104	313	81	261	41	181	223	310	4	1	22	1	4	3	0	1	150	35	19	32	17	7	0	40
	%change	24	44	24	47	40	17	−11	35	27	8	−67	−6	19	−9	−50	−50		−50	−3	−8	0	−3	−29	−13	−∞	38
	sig.	***	ns	ns	***	*	ns	ns	**	*	ns	*	ns	ns	ns	ns	ns		ns	ns	ns	ns	ns	ns	ns	ns	ns
House Finch	1st	2160	111	186	204	55	377	44	132	240	644	167	34	25	21	471	218	67	186	1183	130	46	408	77	251	27	244
	2nd	2197	104	165	209	59	340	33	158	314	645	170	32	35	23	487	225	68	194	1221	135	44	416	75	260	25	266
	%change	2	−6	−11	2	7	−10	−25	20	31	0	2	−11	29	0	3	3	1	4	3	4	−4	2	−3	3	−7	9
	sig.	ns	ns	ns	ns	ns	ns	ns	ns	**	ns	ns	ns	ns	ns	ns	ns	ns	ns	ns	ns	ns	ns	ns	ns	ns	ns
Red Crossbill	1st	4	1	2	0	1	0	0	0	0	0	0	0	0	0	1	0	1	0	1	0	0	1	0	0	0	0
	2nd	11	0	1	4	0	2	1	2	0	1	0	0	0	0	0	0	0	0	0	0	0	0	0	0	0	0
	%change	175	−∞	−50	∞	−∞	∞	∞	∞		∞					−∞		−∞		−∞			−∞				
	sig.	ns	na	ns	na	na	na	na	na		na					na		na		na			na				
Pine Siskin	1st	53	3	3	25	2	7	2	6	1	4	0	0	0	1	1	0	1	0	16	0	0	7	2	3	0	4
	2nd	72	4	8	24	1	8	2	13	3	8	1	1	0	0	7	1	1	5	13	2	0	1	2	6	0	2
	%change	36	33	167	−4	−50	14	0	117	200	100	∞	∞		−∞	600	∞	0	∞	−19	∞		−86	0	99		−50
	sig.	ns	ns	ns	ns	ns	ns	ns	ns	ns	ns	na	na		na	*	na	ns	na	na	na		*	ns	ns		ns

(continued)

492 APPENDIX C

		Appalachian Plateaus	Allegheny Front	Allegheny Mountain	Deep Valleys	Glaciated High Plateau	Glaciated Low Plateau	Glaciated Pocono Plateau	High Plateau	NW Glaciated Plateau	Pittsburgh Low Plateau	Waynesburg Hills	Atlantic Coastal Plain	Central Lowlands	New England	Piedmont	Gettysburg-Newark Lowland	Piedmont Lowland	Piedmont Upland	Ridge and Valley	Anthracite Upland	Anthracite Valley	Appalachian Mountain	Blue Mountain	Great Valley	South Mountain	Susquehanna Lowland
American Goldfinch	1st	2806	127	211	399	106	435	52	223	315	752	186	31	32	21	479	220	67	192	1281	133	48	466	83	257	30	264
	2nd	2893	127	207	430	107	453	55	234	329	776	175	32	36	20	483	224	66	193	1294	146	47	467	83	256	29	266
	%change	3	0	−2	8	1	4	6	5	4	3	−6	−3	3	−13	1	1	−1	1	1	10	−2	0	0	−1	−3	1
	sig.	ns	ns	ns	ns	ns	ns	ns	ns	ns	ns	ns	ns	ns	ns	ns	ns	ns	ns	ns	ns	ns	ns	ns	ns	ns	ns
House Sparrow	1st	2485	113	203	250	76	414	33	167	319	723	187	31	28	21	482	221	67	194	1235	132	48	441	75	255	28	256
	2nd	2374	110	188	217	72	377	14	160	331	718	187	32	36	22	487	222	69	196	1258	135	45	451	72	260	21	274
	%change	−4	−3	−7	−13	−5	−9	−58	−4	4	−1	0	−3	18	−4	1	0	3	1	2	2	−6	2	−4	2	−25	7
	sig.	ns	ns	ns	ns	ns	ns	**	ns	ns	ns	ns	ns	ns	ns	ns	ns	ns	ns	ns	ns	ns	ns	ns	ns	ns	ns
No. blocks	1st	3003	139	216	455	114	475	56	237	331	790	190	34[a]	34[a]	21[a]	494	228[a]	69	197	1342	151	49	491	83	261[a]	31	276
	2nd	3003	139	216	455	114	475	56	237	331	790	190	36[a]	37[a]	23[a]	495	229[a]	69	197	1343	151	49	491	83	262[a]	31	276

[a] due to a change in the number of blocks in some sections (nine partial border blocks were surveyed in the second Atlas but not the first), percentage change values are sometimes not intuitive, for example, 19 occupied blocks in the New England Province (Reading Prong Section) in both atlases would be 19/34 = 56% occupied in first Atlas, 19/37 = 51% occupied in second Atlas.

APPENDIX D   Habitat Associations

This appendix provides additional details of point count data analysis referred to in some species accounts. For 120 species with sufficient data (>30 records of non-flyovers), there are four graphs, presented left to right: (1) relative abundance in each habitat type, (2) relative use of each habitat type, (3) relationship between count and percentage of forest in count circle, and (4) relationship between relative abundance and elevation (intervals of 20 m). We caution that the habitat relationships in particular should be used only as gross measures of habitat associations, for reasons detailed below.

Relative abundance in each habitat type (left)

The point count surveyors assessed the main habitat type (from a list of 18 options) within a 75 m radius of the point count location. Two of these habitat types ("bare rock/sand/clay" and "quarry/active strip mines/gravel pits") are omitted here because samples of both were low. Other habitat types are grouped to reduce the number of categories to 12, for presentation purposes. Note that habitat around most point count locations was a mixture of two or more types, but because only the dominant habitat type was recorded, bird–habitat associations are diluted. For example, there are many records of field birds or forest birds at points described as being in one of the "residential" habitat categories; clearly, most of these bird would not have been in backyards, but rather in other habitats adjacent to the point count location. "Transitional" habitat, described as "e.g., old field, shrub-scrub" in the point count protocol, is likely over-represented in the sample. Also, this habitat category appears to have been used liberally in instances where there was a high diversity of habitats or high interspersion. Hence, we urge that the figures in this appendix be interpreted with caution. However, while acknowledging these limitations, the relationships do provide an indication of the habitats in which each species is found in highest densities and whether each species is a habitat specialist or a habitat generalist.

Relative use of each habitat type (second from left)

The cautions described above are especially important to bear in mind when interpreting these charts. Further, because the point count sample is highly biased toward certain habitats, notably residential development (chap. 5), these charts should not be interpreted as estimates of the percentage of the respective species' population found in each habitat. The percentage of observations in the developed habitat categories, in particular, is likely to be much higher than the percentage of the actual population in those categories. In addition to showing the percentage of birds in each category, we show the percentage of all points in each category (black marker). This provides some indication of whether each species uses a habitat in proportion to its availability (blue bar approximately equal to black marker), whether it uses the habitat in excess of its availability, implying that it is a "preferred" habitat (blue bar > black marker), or whether that habitat is relatively avoided (blue bar < black marker). Again, these relationships should only be considered approximate, and we caution against over-interpretation.

Relationship between count and percent forest in count circle (second from right)

These figures show the relative avoidance (negative relationship) selection for (positive relationship) for forested habitat. We show these relationships for only those birds within 75 m (red marker) and all birds (blue marker).

Relationship between relative abundance and elevation (right)

Elevations were grouped into 20 m intervals to aid interpretation. Polynomial regression lines (2nd, 3rd, or 4th order) are fitted against these data to show the general relationship between bird abundance and elevation. Note that elevation is at the point count location, and hence the actual locations of birds seen or heard could be either lower or higher, depending on topography around each point count location. Sample sizes in elevations intervals above 840 m were very small (<10), so these are excluded from the charts.

There are probably very few species in Pennsylvania that are limited by elevation per se, among them a handful of resident species at the northern edges of the range or conversely, species at the southern edge of their range.

Hence, elevational relationships shown by most species are more likely due to correlations between elevation and land-use/habitat (chap. 3). The proportion of points described as being dominated by forested habitats increases with elevation while the proportion described as one of the developed habitat types decreases. As noted previously, the point count sample is heavily biased toward developed land, which inevitably has some influence on the elevational patterns presented here.

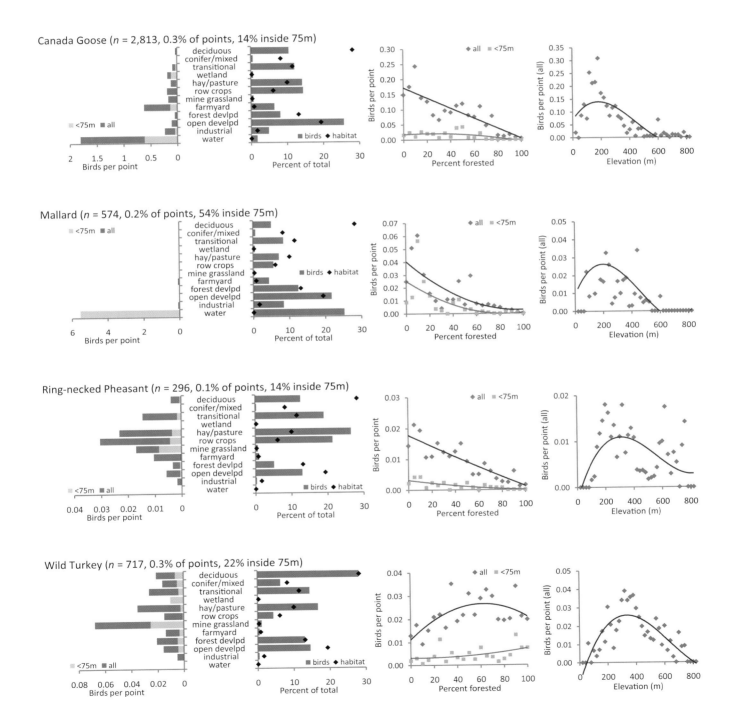

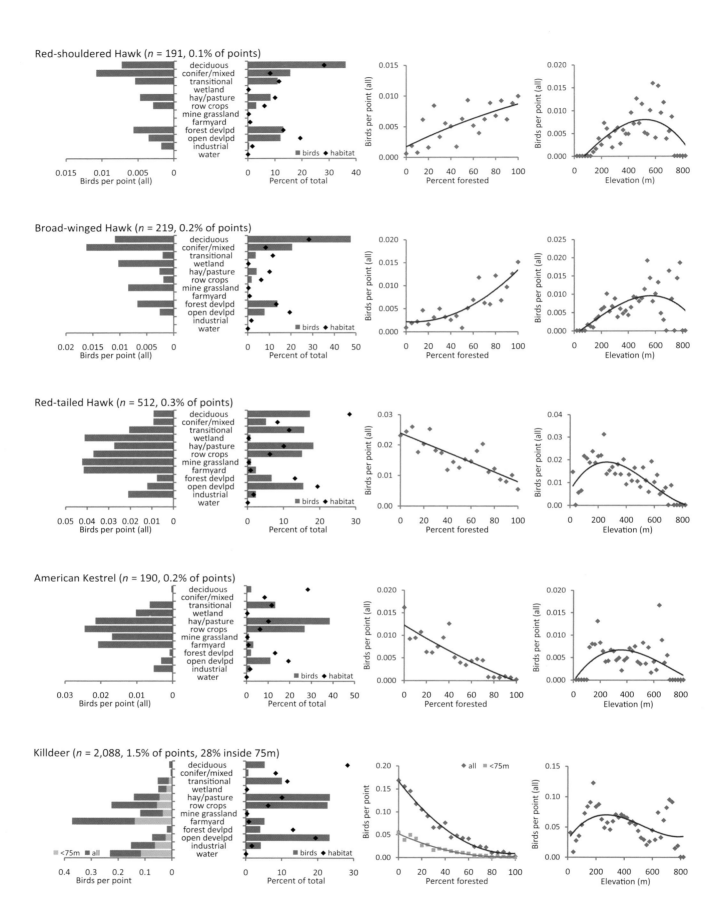

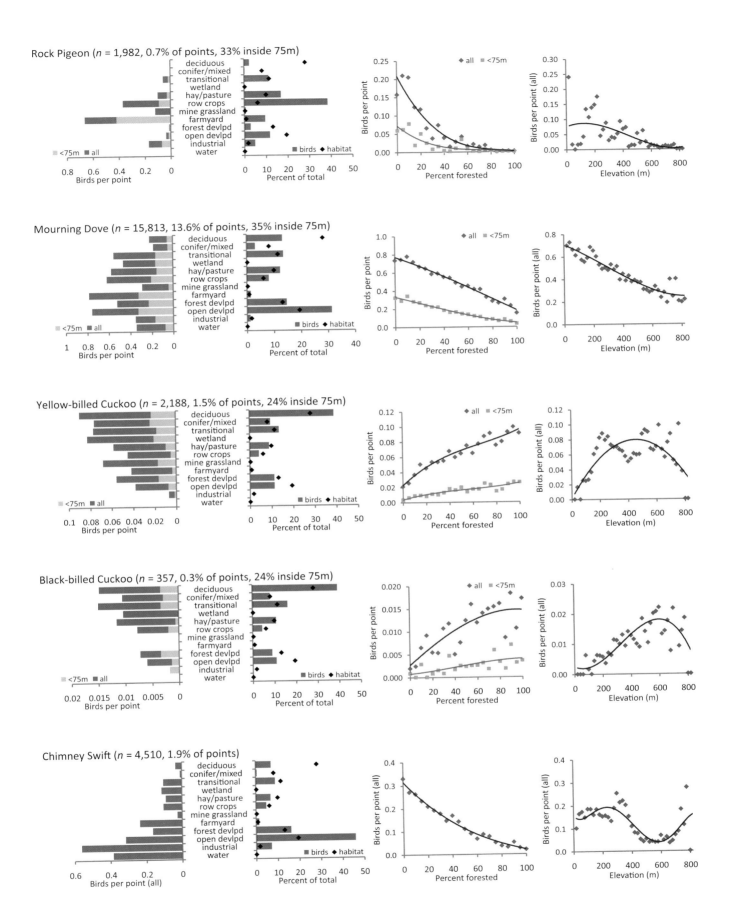

APPENDIX D 497

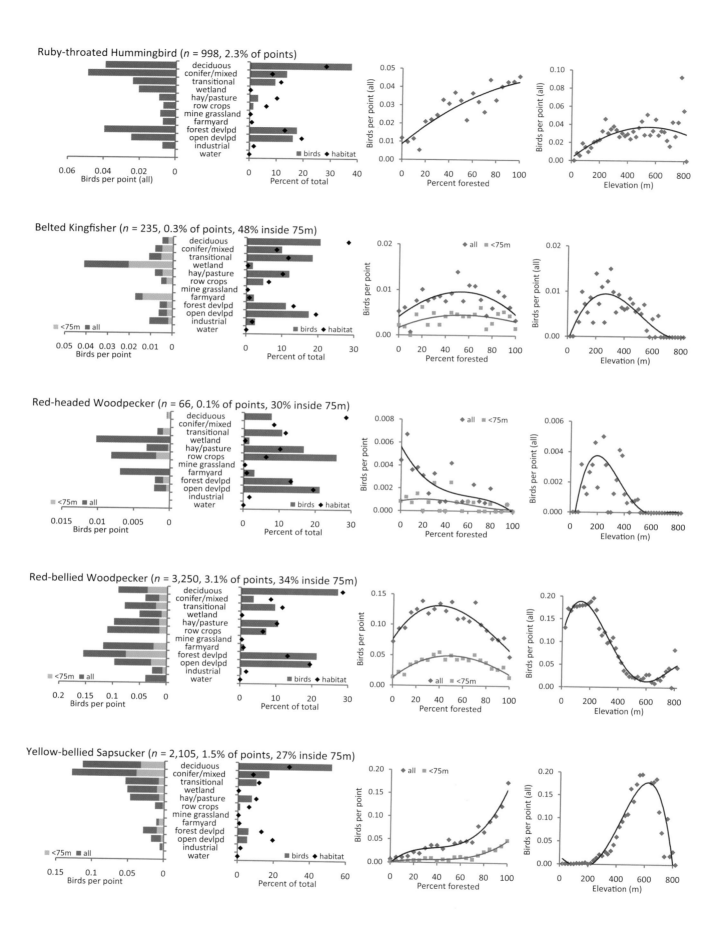

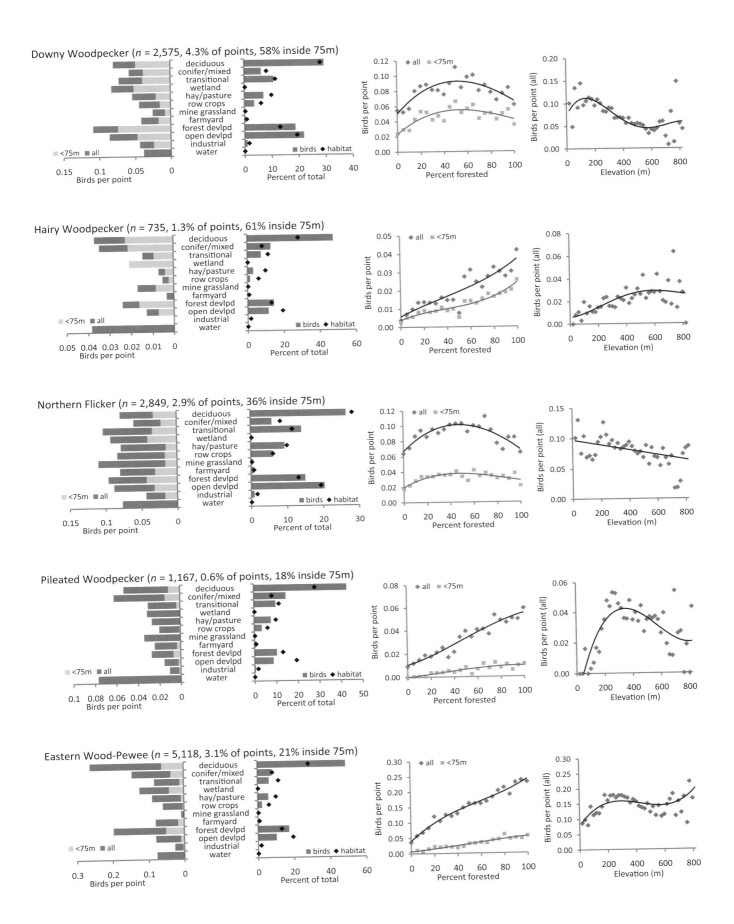

APPENDIX D 499

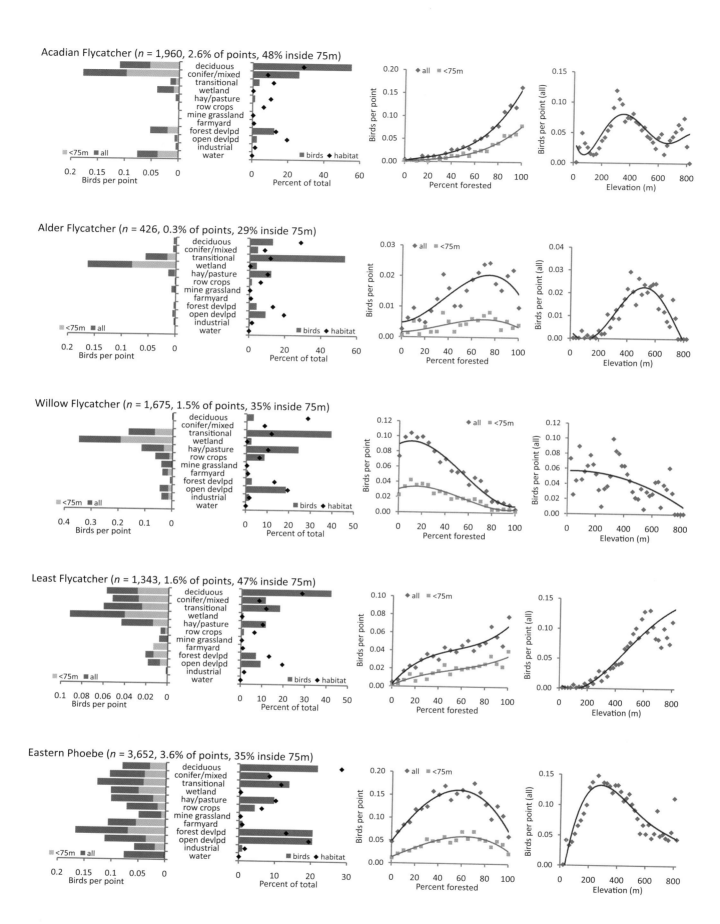

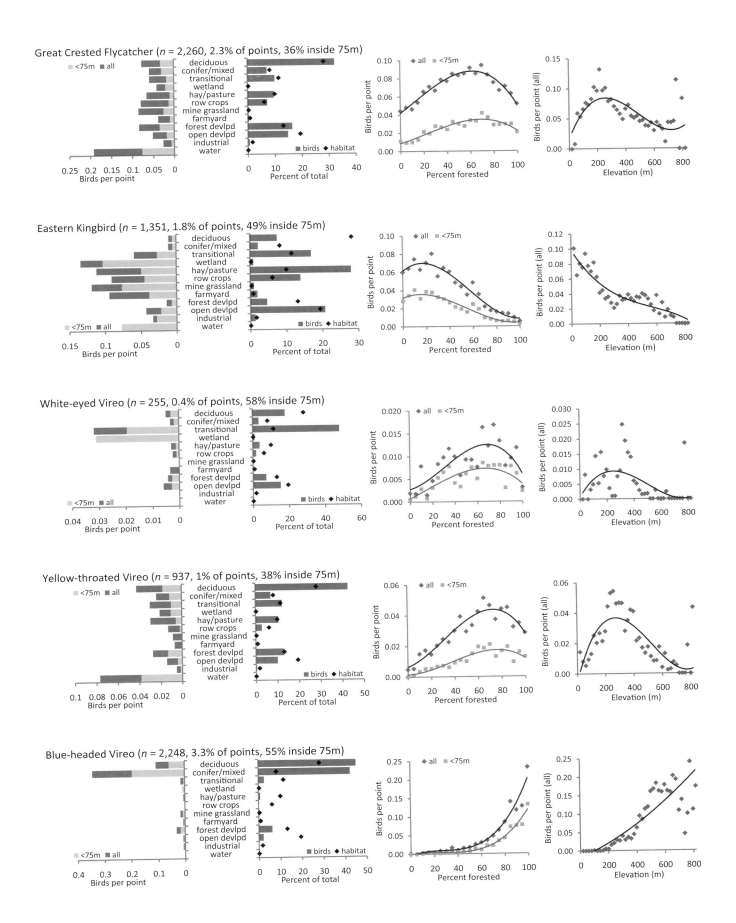

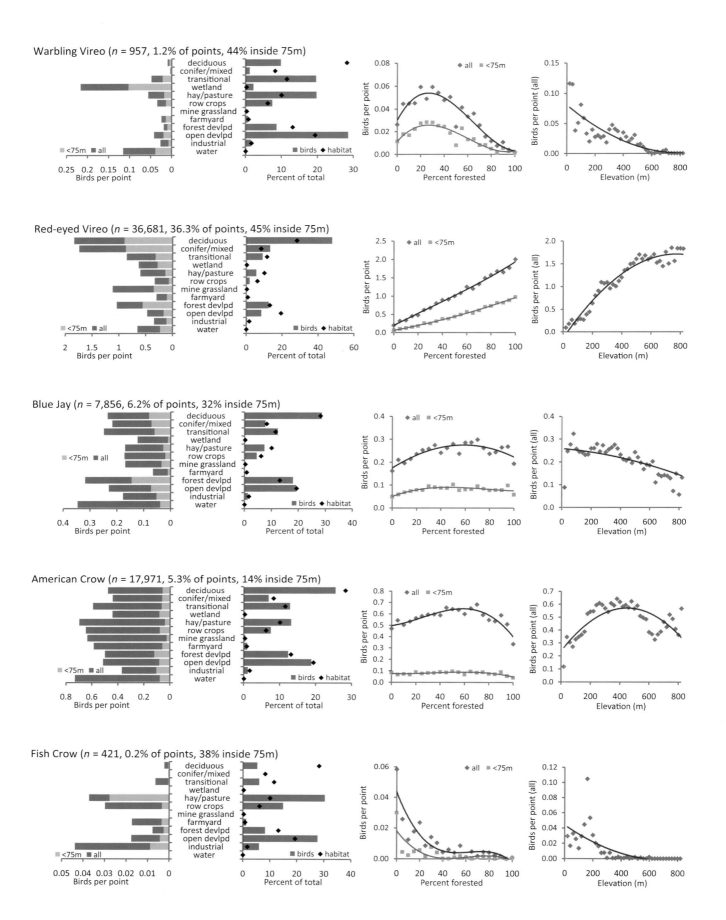

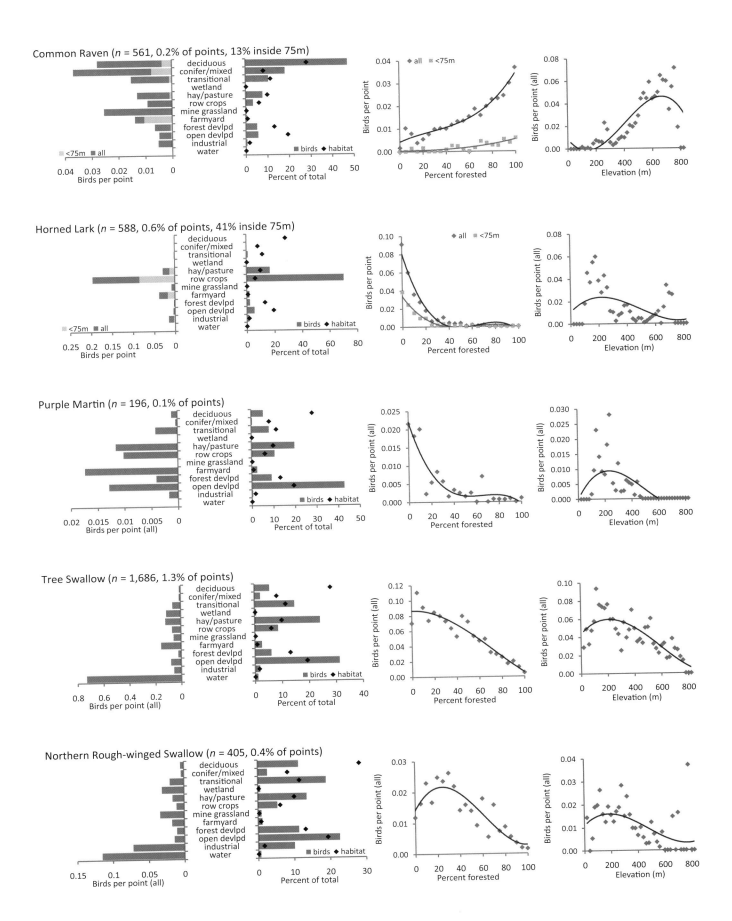

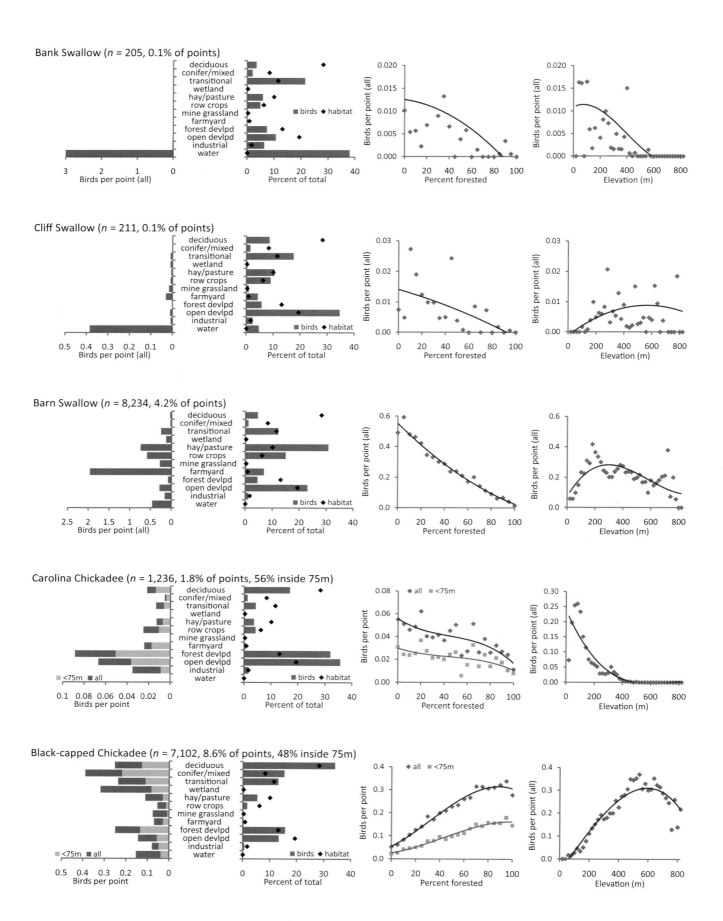

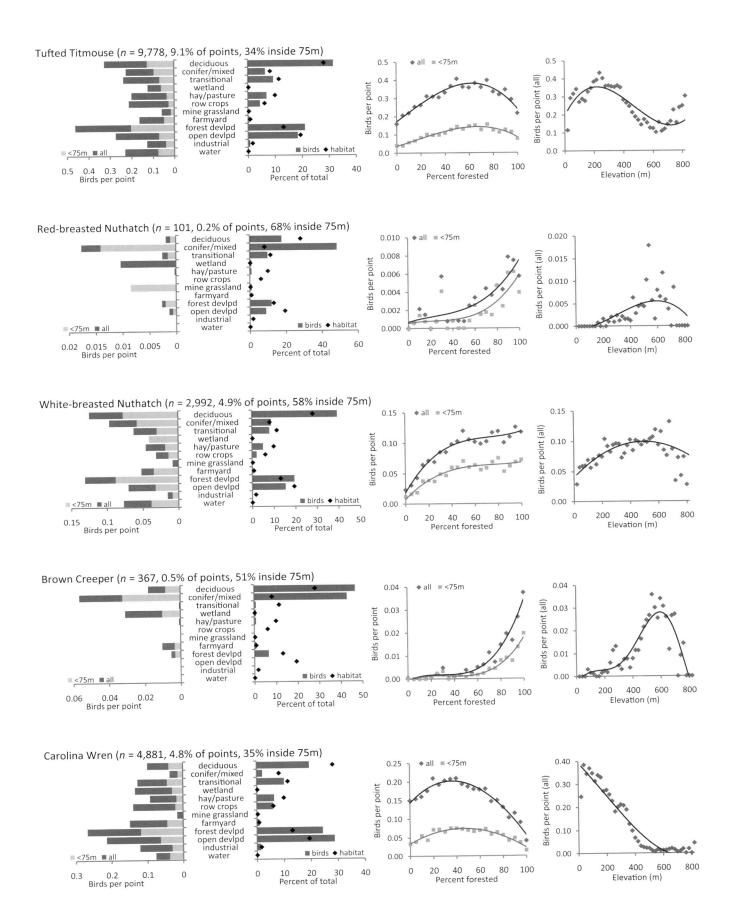

APPENDIX D 505

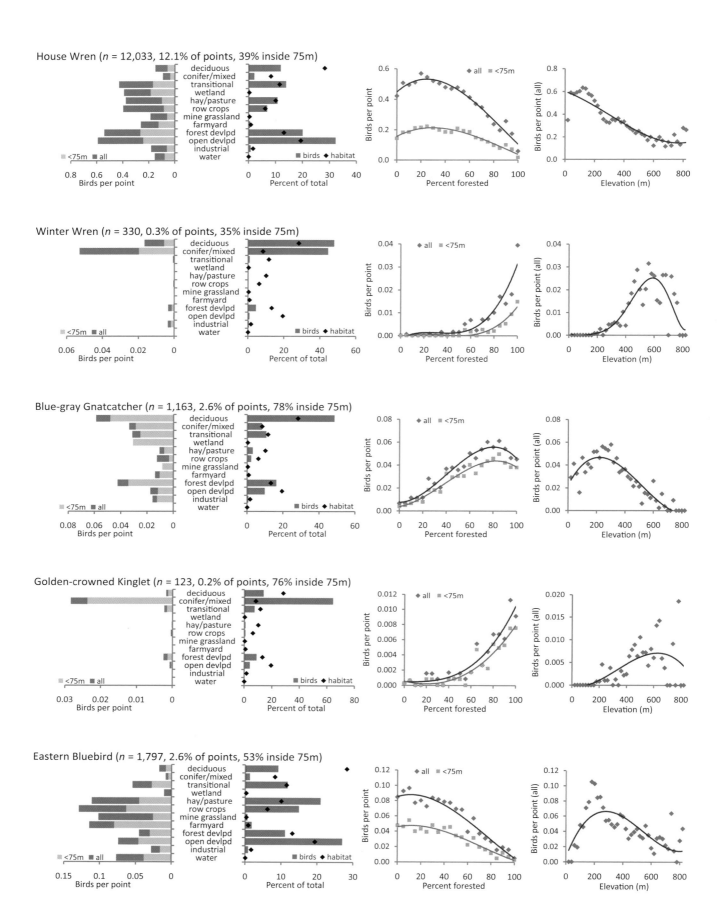

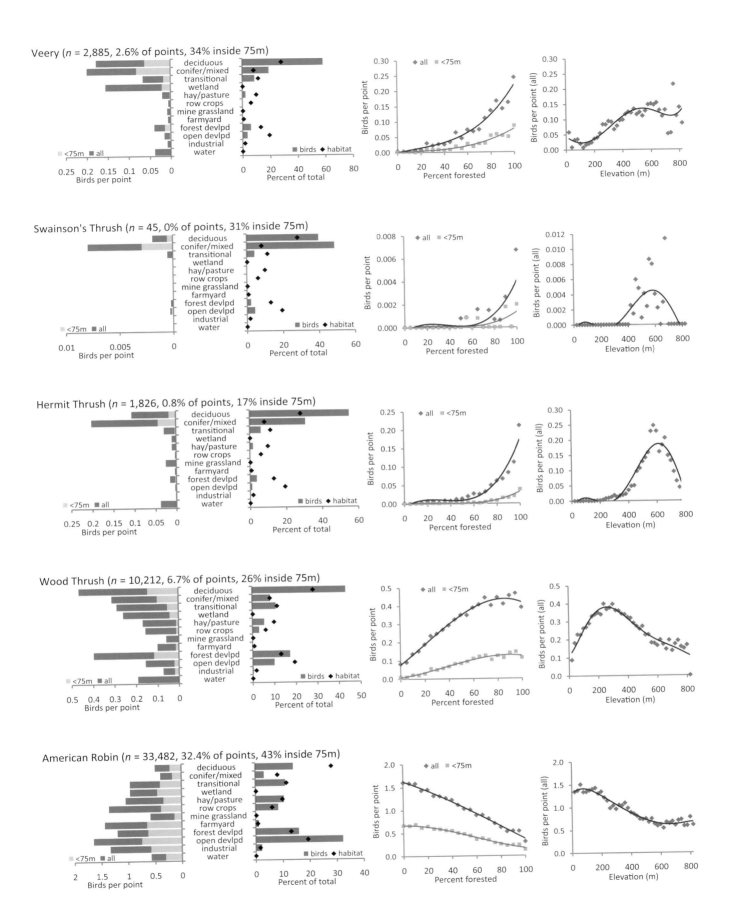

APPENDIX D 507

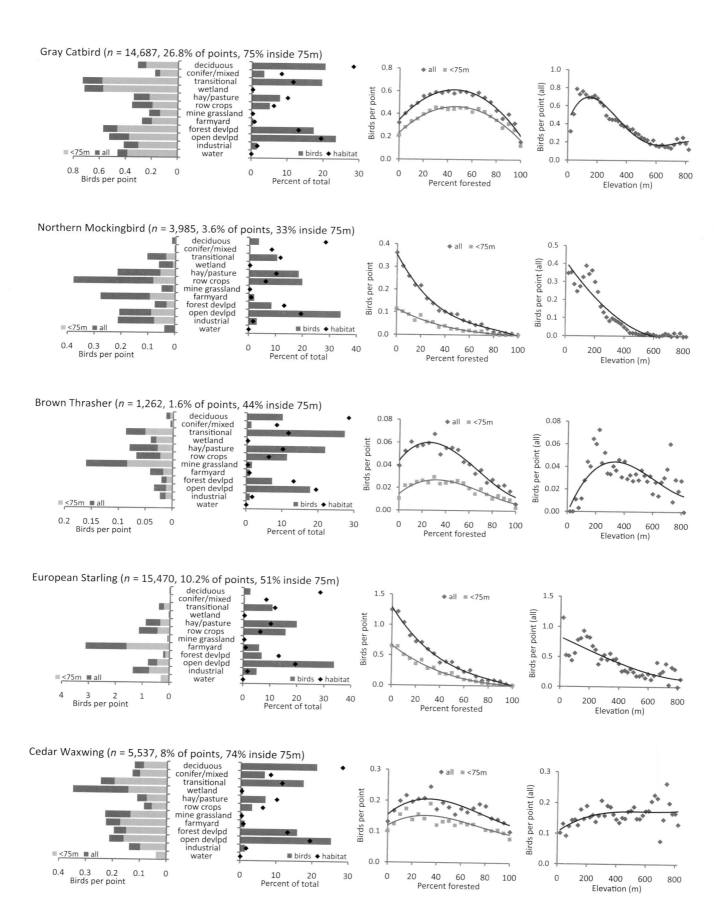

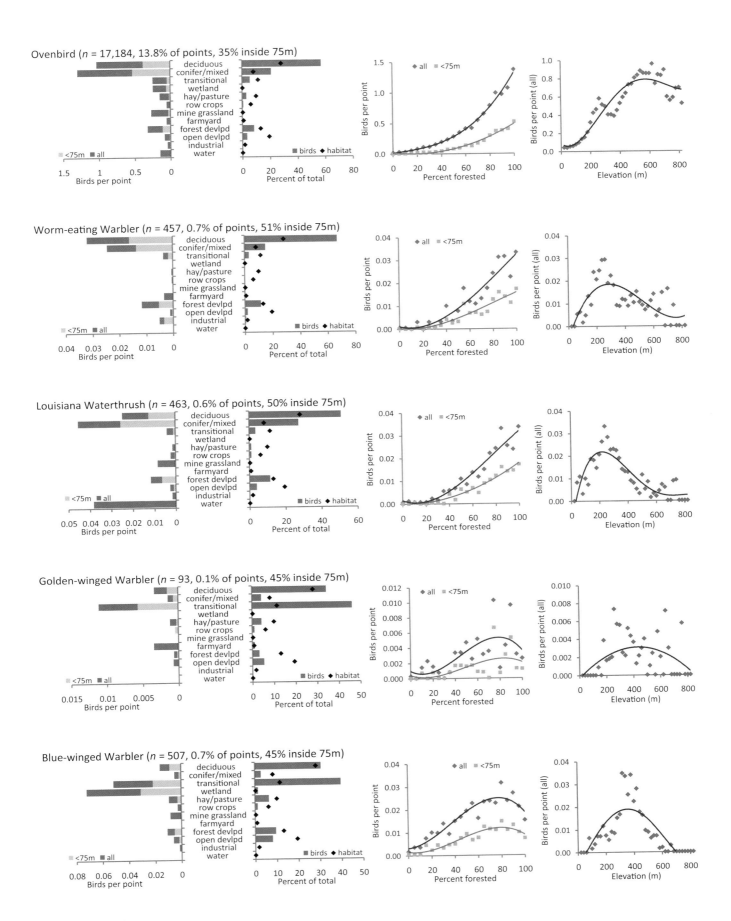

Black-and-white Warbler (*n* = 1,741, 2.8% of points, 58% inside 75m)

Mourning Warbler (*n* = 166, 0.2% of points, 48% inside 75m)

Kentucky Warbler (*n* = 214, 0.2% of points, 37% inside 75m)

Common Yellowthroat (*n* = 19,177, 18.5% of points, 37% inside 75m)

Hooded Warbler (*n* = 2,546, 2.8% of points, 41% inside 75m)

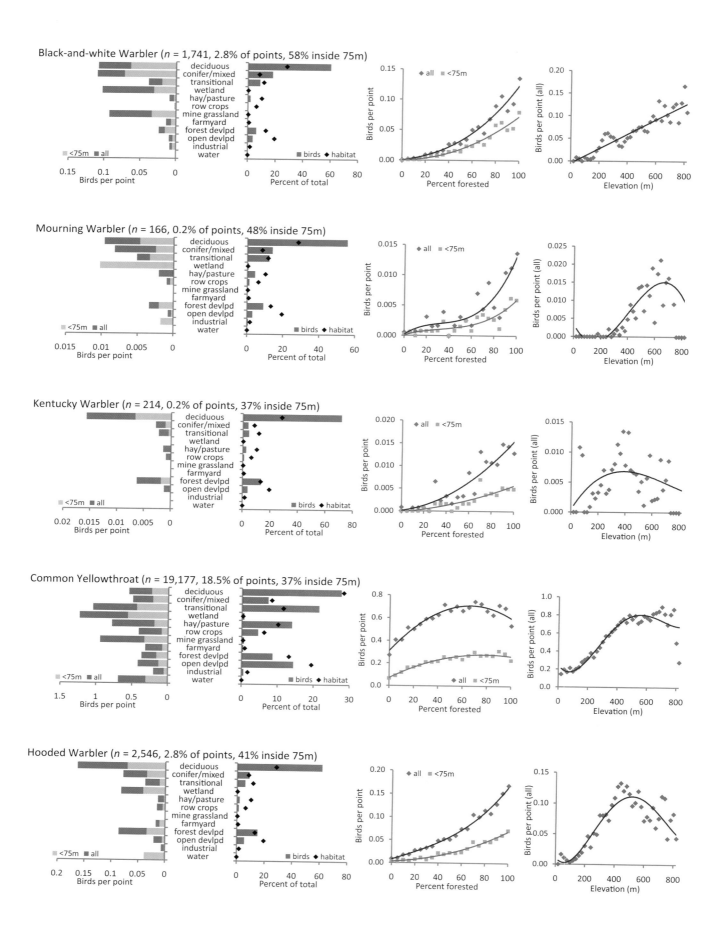

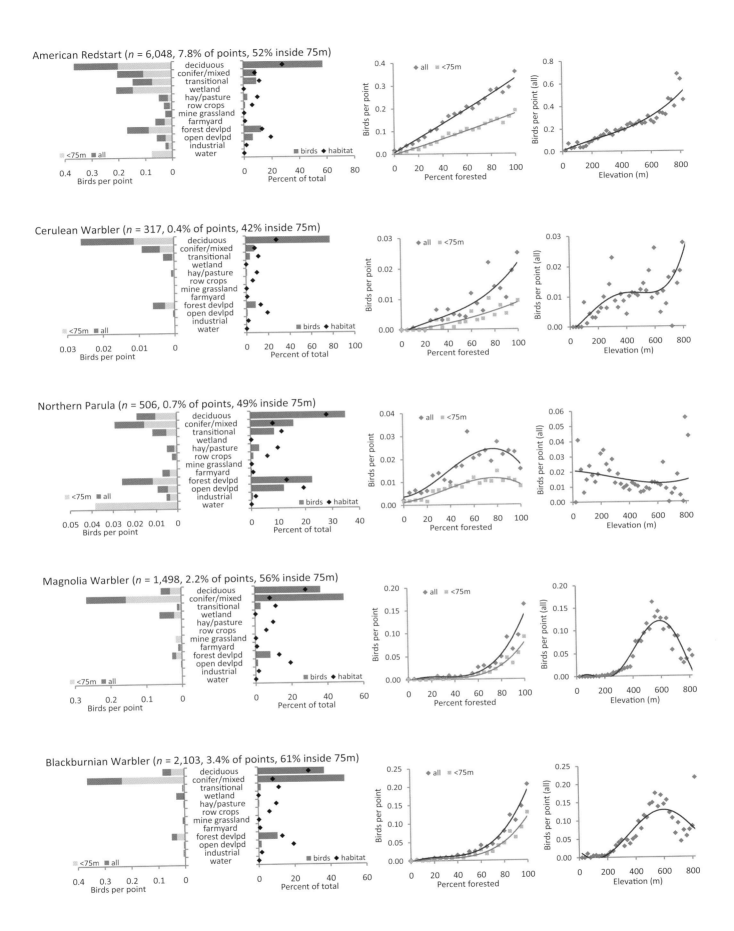

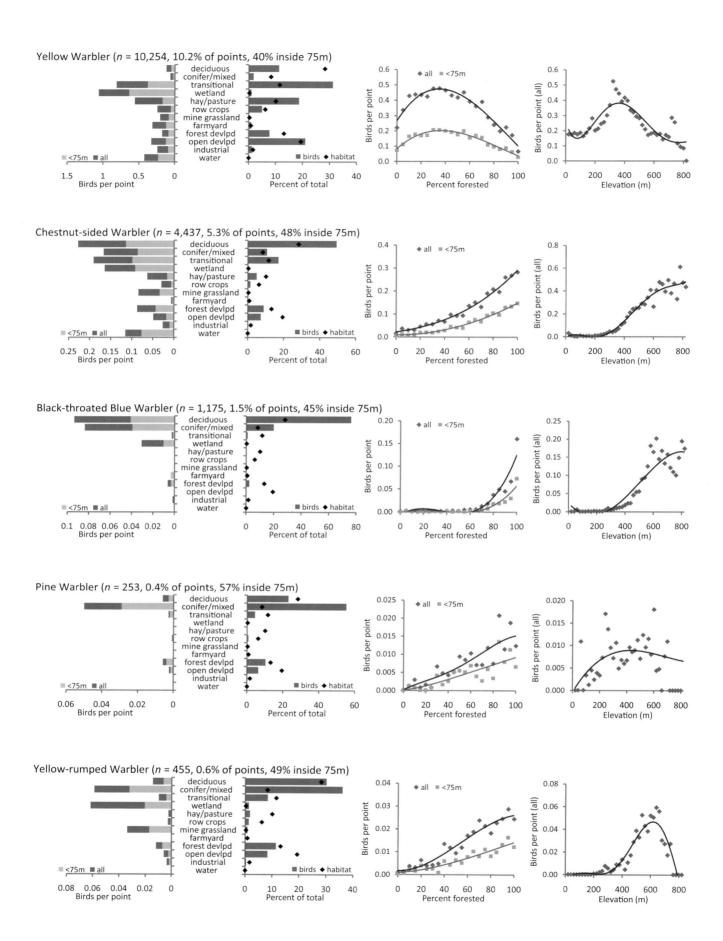

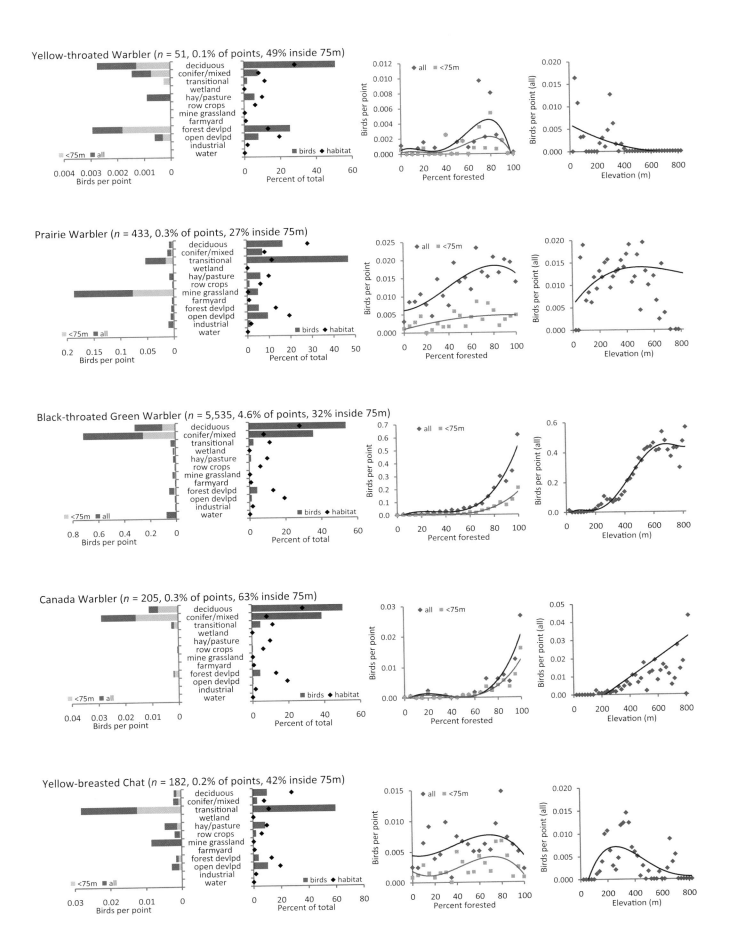

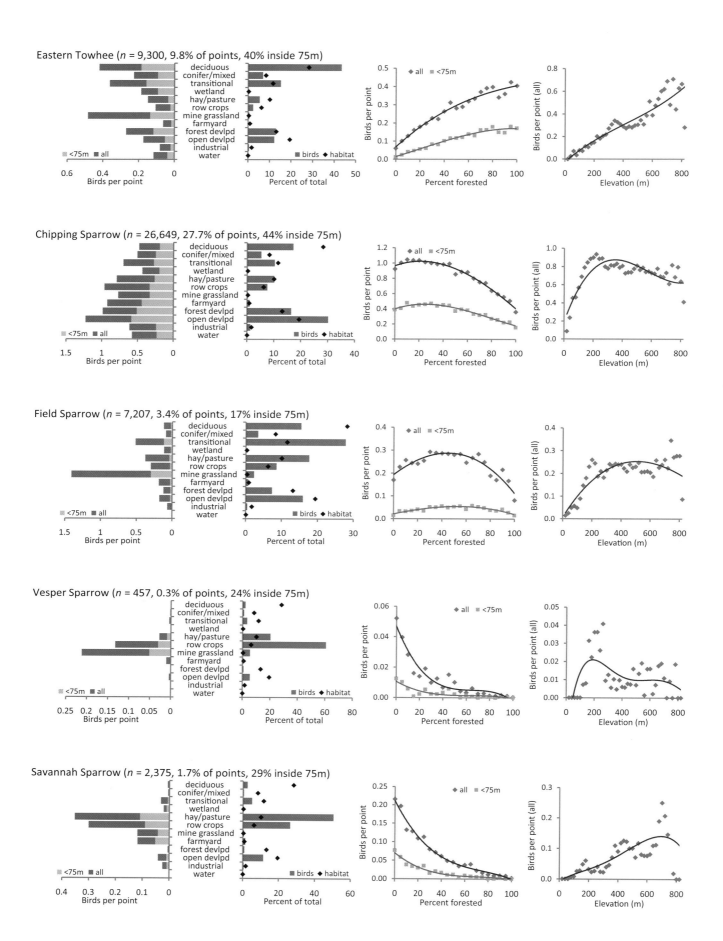

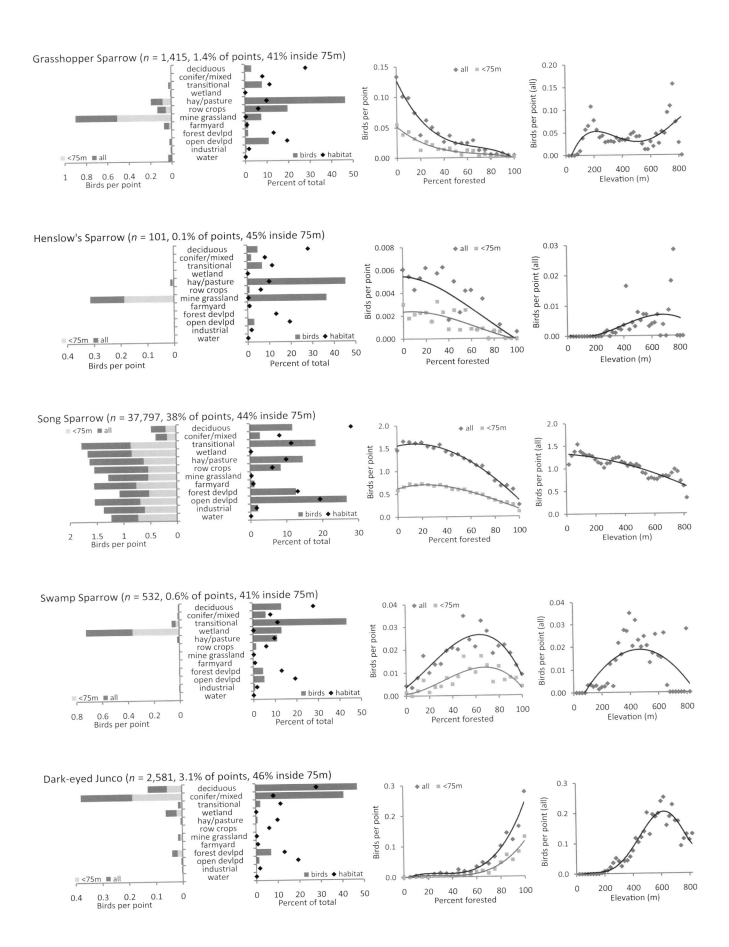

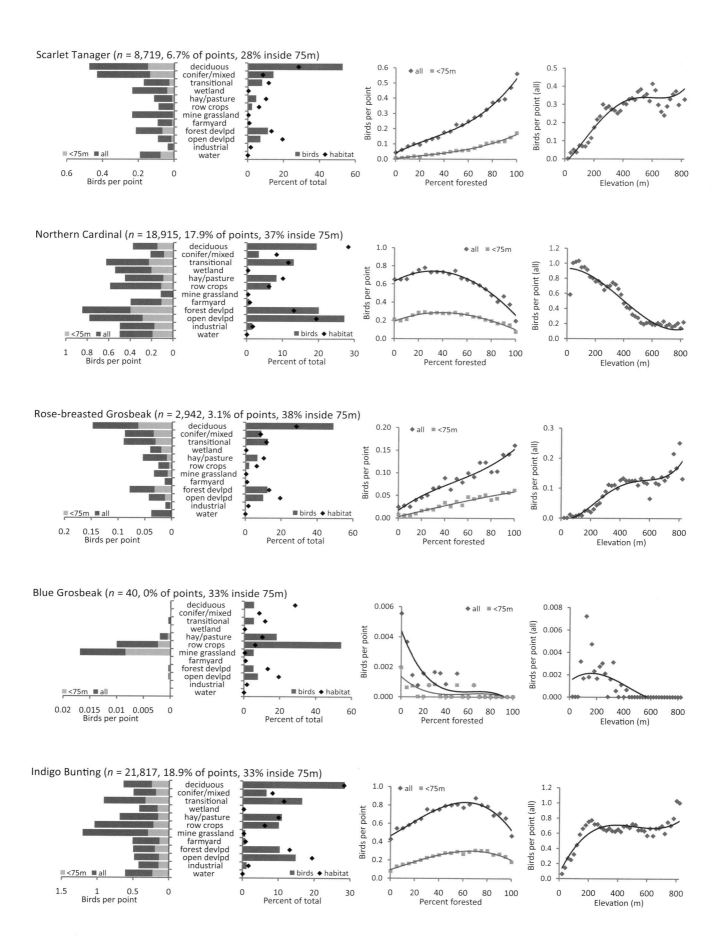

516 APPENDIX D

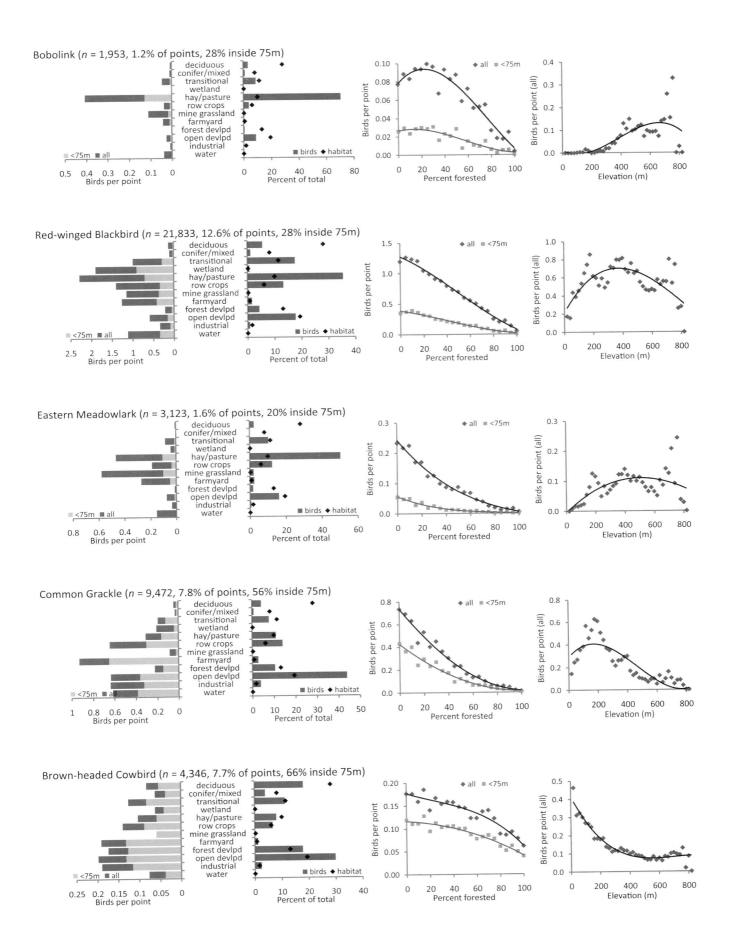

APPENDIX D 517

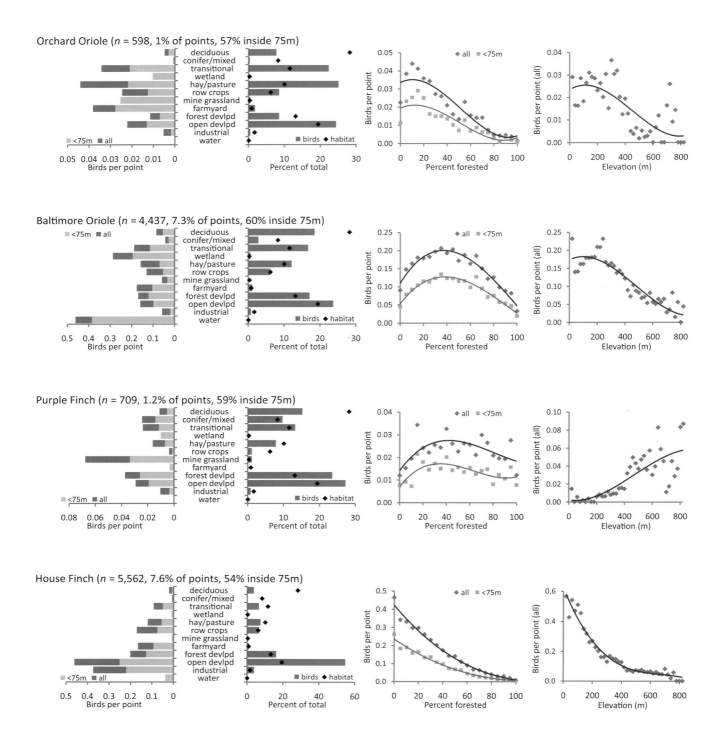

Orchard Oriole (*n* = 598, 1% of points, 57% inside 75m)

Baltimore Oriole (*n* = 4,437, 7.3% of points, 60% inside 75m)

Purple Finch (*n* = 709, 1.2% of points, 59% inside 75m)

House Finch (*n* = 5,562, 7.6% of points, 54% inside 75m)

American Goldfinch (*n* = 5,860, 10.7% of points, 71% inside 75m)

House Sparrow (*n* = 12,706, 9.1% of points, 58% inside 75m)

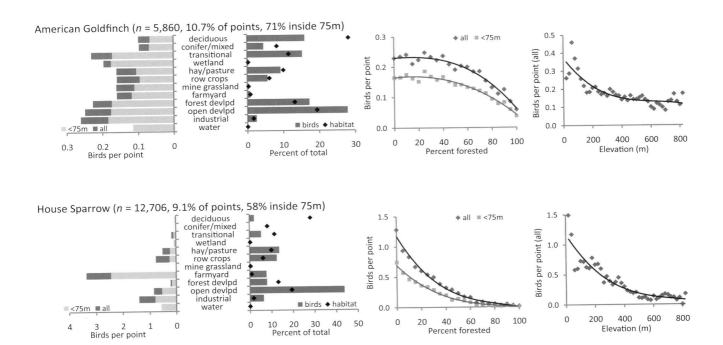

APPENDIX D   519

APPENDIX E — Analytical Methods—Statistical Details

E.1 Changes in block occupancy corrected for survey effort

Predicted probability of occurrence of each species in each block was modeled using Hierarchical Logistic Regression in WinBUGS. The model took the form

$$\text{logit}(p_i) = \alpha + \left(\sum_{k=1}^{s} \beta_k \chi_{ik}\right) + \gamma_i + \delta_i + \varepsilon$$

where p_i is the predicted probability of occurrence in block i, α is the intercept, β_k are the parameter estimate for up to s landscape covariates (typically 3–4), γ_i is the correction due to effort (deviation from a standardized 30 hours), δ_i is the spatial effect, and ε is random error.

Landscape covariates were chosen a priori based on expert knowledge and exploratory data analysis (see appendix D). Because of the large number of models (264), it was not possible to test competing models for each species with time and computing resources available. Landscape metrics included the following:

A. % of block in developed land
B. % of block in deciduous forest
C. % of block in coniferous and mixed forest
D. % of block in coniferous forest
E. % of block in forest (deciduous + coniferous and mixed)
F. % of block in grassland
G. % of block in arable land (row crops)
H. % of block in farmland (grassland + arable)
I. % of block in forested wetland
J. % of block in emergent wetland
K. % of block in wetland (forested + emergent)
L. % of block in open water
M. % of block in non-habitat (habitat strongly avoided, summed across types)
N. Shannon's Evenness Index
O. % of block in reclaimed surface mines
P. length of river (Strahler order 5 or greater) in km
Q. mean elevation of block (m)
R. northing (latitude) in decimal degrees

Land cover data were from Landsat ETM+ derived data, from c. 2005 (Warner 2007). The correction for effort predicted occupancy under a scenario that all blocks received 30 hours coverage in both atlas periods. Effort effects were modeled after Link and Sauer (2007):

$$f(\gamma) = D[(\gamma/30)^C - 1]/C$$

where C and D are parameters that determine the shape of the relationship between hours and probability of detection. The above function provides a multiplier, such that for blocks that received <30 hours of survey effort, the multiplier is >1, increasing the predicted probability of occurrence in blocks with low effort. The function equals 1 for effort of 30 hours.

Spatial effects were incorporated using a Gaussian conditional autoregressive (CAR) model, which incorporates autocorrelation in data. This model assumes that there is spatial correlation between species occurrence in each block and its eight adjacent neighboring blocks (<8 for border blocks). Spatially explicit models are "expected to yield better predictions" (Bahn et al. 2006) than non-spatial models. In ecological terms, such models incorporate important spatial information that may relate to unknown effects of bird population dynamics and underlying environmental variation.

Vague prior distributions (Link et al. 2002) were used to begin the MCMC (Markov chain Monte Carlo) sampling. Parameters were assigned flat normal distributions with mean of 0.0 and variance of 100 (precision = 0.01). Models were fitted with 10,000 iterations following a 10,000 sample burn-in.

E.2 Density estimates

E.2.1 Modeling approach
Bird counts were modeled using Generalized Linear Model (Poisson regression) in WinBUGS. The model took the form

$$\log(\lambda_i) = \left[\alpha + \left(\sum_{k=1}^{s}\beta_k \chi_{ik}\right) + \theta_i + \tau_i + \psi_i + \varepsilon\right] \times \rho_i$$

where λ_i is the count of birds at point i, α is the intercept, β_k are the parameter estimate for s habitat covariates (typically 2–5), θ_i is observer effort, τ_i is the time of season effect, ψ_i is the time of day effect, ε is random error and ρ_i is the probability of occupancy. Further details of each of these model components can be found below.

Vague prior distributions (Link et al. 2002) were used to begin the MCMC sampling. Parameters were assigned flat normal distributions with mean of 0.0 and variance of 100 (precision = 0.01). Models were fitted with 10,000 iterations following a 10,000 sample burn-in.

E.2.2 Habitat covariates

Habitat covariates were chosen a priori based on expert knowledge and exploratory data analysis (see appendix D). Habitat metrics included between two and five of the following:

A. % of count radius in developed land
B. % of count radius in deciduous forest
C. % of count radius in coniferous and mixed forest
D. % of count radius in coniferous forest
E. % of count radius in forest (deciduous + coniferous and mixed)
F. % of count radius in grassland
G. % of count radius in arable land (row crops)
H. % of count radius in farmland (grassland + arable)
I. % of count radius in forested wetland
J. % of count radius in emergent wetland
K. % of count radius in wetland (forested + emergent)
L. % of count radius in open water
M. Shannon Evenness Index
N. % of count radius in reclaimed surface mines
O. elevation at count location (m)
P. Landform at count location:
   a. steep slope
   b. cliff
   c. flat summit/ridgetop
   d. slope crest
   e. hilltop (flat)
   f. hill (gentle slope)
   g. sideslope cooler aspect
   h. sideslope warmer aspect
   i. dry flats
   j. wet flats
   k. valley/toeslope
   l. flat at bottom of steep slope
   m. cove/footslope cooler aspect
   n. cove/footslope warmer aspect
   o. stream/river
   p. water

Land cover data were from Landsat ETM+ derived data, from c. 2005 (Warner 2007). Here, the count radius varied by species, depending on the estimated detection radius (r_m; see below) for that species. Land cover data were calculated for 125 m, 150 m, 175 m, 200 m, and 250 m buffers around each point using ArcGIS. Hence for species with estimated detection radii of <125 m, the land cover data for within the 125 m buffer were used, and so forth, up to species with detection radii >200 m, for which the 250 m buffer was used. Land form data (provided by the Nature Conservancy) describe the topographical type at each sampling location.

E.2.3 Observer effects

Observer effects were incorporated as random observer specific effects, with a mean of zero. Hence, these resulted in a smoothing out of observer effects over space.

E.2.4 Time of season and time of day effects

These were included as power, linear, exponential, or polynomial functions, as illustrated in figure 5.5 and figure 5.6 (chap. 5).

E.2.5 Distribution constraint

Models were constrained by the predicted distribution of each species, based on block level data, corrected for survey effort (see fig. 5.10, chap. 5).

E.2.6 Detectability and detection radius

A new method of analyzing atlas point count data, combining removal sampling with distance information, was developed by George Farnsworth and Duane Diefenbach (Farnsworth et al. 2006). Each model was written in SURVIV by conditioning on the total birds counted.

The model first assumes that birds are distributed with uniform density (D). Each bird has a per interval probability of singing (p). Thus, the probability it fails to sing (and cannot be detected) is

$$q = 1 - p$$

The probability a bird will be detected is a function of the probability it sings and the distance it is from the observer. Another way to look at this is that the probability a bird is not detected is the probability it fails to sing (q) plus an increasing function of distance (r) from the observer such that even if the bird sings, it is likely to be undetected the farther it is from the observer. We define such a function as

$$Q(r) = q + \left(\frac{r}{a}\right)^2$$

where q and a are parameters to be estimated. The maximum detection distance r_m can be estimated from these parameters by setting $Q(r) = 1.0$. It was then possible to solve for the distance where $Q = 1$ (i.e., the distance at which the bird cannot be detected even if it sings; here called r_m for r-maximum).

The number of birds first detected in the jth time interval within a ring defined as between radial distances r_1 and r_2 from the observer can be written as the following integrals:

$$E(X1j) = D\pi \int_0^{rc} Q^{j-1}(1-Q)2r\,dr$$

$$E(X2j) = D\pi \int_{rc}^{rm} Q^{j-1}(1-Q)2r\,dr$$

Here, there are two distance categories. In the first, $r_1 = 0$ and $r_2 = r_c$ (r_c is the boundary between distance categories; in our data, 75 m). In the second, $r_1 = r_c$ and $r_2 = r_m$.

Now let us assume there is heterogeneity between individuals within the population such that some birds sing more frequently than others (and/or perhaps some are louder than others). We can model this heterogeneity by using a two-point mixture model. For this, we divide the population into two via the parameter c such that there are two modeled populations: c and $(1-c)$.

With heterogeneity included, we tested the following models for each species and chose that lowest AIC (Akaike information criterion):

M_f: Full two-point mixture model. This models two groups of birds each with a separate Q—thus, five parameters (q_1, q_2, a_1, a_2, and c).
M_{f1}: No heterogeneity in maximum detectable distance. Constrain $r_{m1} = r_{m2}$.
M_{f2}: No heterogeneity in singing probability. Constrain $q_1 = q_2$.
M_{p1}: Partial mixture. Constrain $q_2 = 0.0$.
M_{p2}: Partial mixture. No heterogeneity in r_m. Constrain $q_2 = 0.0$; $r_{m1} = r_{m2}$.

The conditional probabilities for the statistics were generated in Program SURVIV (White 1992). For most species, model M_{f1} was selected. That is, there was heterogeneity in detectability but no difference between the two groups in detection distance. Exceptions were Yellow-billed Cuckoo, Black-billed Cuckoo, and Chipping Sparrow, for which model M_f was selected.

Hence, these models provided estimates of the detection radius (r_m) and the probability that a bird was not detected during the count period, for each species (see fig. 5.3, chap. 5).

E.2.7 Predictive model of population density
The parameters from the bird count model (see E.2.1) were then used to form a predictive model of counts at a regular-spaced (1.9 km grid) of locations j:

$$log(\lambda_j) = \left[\alpha + \left(\sum_{k=1}^{s} \beta_k X_{jk}\right) + \tau_{max} + \psi_{max} + \xi_j\right] \times \rho_j$$

with k (up to s) habitat parameters (β). Time of season (τ) and time of day (ψ) effects set to the optimal (max) times for each species. Observer effects (θ) and random error drop out because they are standardized around zero. The predicted counts were then used to estimate densities (D = birds per km²) at the same regular grid location (j) using the estimates of proportion detected (p) and detection radius (r_m), as follows:

$$D = [(log(\lambda_j) \times 1{,}000^2) \times (\pi \times r_m^2)] \times p$$

APPENDIX F — Breeding Phenology

	Nest building		Nest with eggs		Nest with young		Fledged Young	
	Median	Range	Median	Range	Median	Range	Median	Range
Canada Goose	1 May	(17 Mar–29 Jul)	1 May	(14 Mar–20 Jul)	4 Jun	(07 Apr–30 Jul)	10 Jun	(25 Mar–04 Aug)
Mute Swan			8 May	(01 Apr–15 May)	4 Jun	(16 May–30 Jun)	18 Jun	(02 May–20 Jul)
Wood Duck	16 May	(30 Mar–29 Jul)	30 May	(07 Mar–13 Jul)	31 May	(01 May–01 Aug)	16 Jun	(21 Apr–05 Aug)
American Black Duck			26 Apr		20 Jun	(15 May 03 Jul)	18–Jun	(21 Apr–11 Aug)
Mallard	8 May	(23 Mar–28 Jul)	19 May	(23 Mar–15 Jul)	6 Jun	(09 Apr–28 Jul)	17 Jun	(02 Apr–25 Aug)
Blue-winged Teal							5 Jul	(26 May–28 Jul)
Green-winged Teal							29 Jun	
Hooded Merganser			31 May	(25 Apr 09 Jul)	7 Jun	(19 Apr 05 Jul)	13 Jun	(19 Apr–09 Aug)
Common Merganser			10 Jun	(22 Mar–16 Jul)	8 Jun	(21 May–15 Jul)	22 Jun	(21 Apr–22 Aug)
Northern Bobwhite	4 May				12 Jul	(04 Jul–20 Jul)	21 Jun	(06 Jun–08 Aug)
Ring-necked Pheasant			6 Jun	(01 May 05 Jul)	11 Jun	(15 May–25 Jul)	28 Jun	(20 Apr 05 Sep)
Ruffed Grouse	26 Jun		15 May	(20 Apr–22 Jun)	10 Jun	(04 Apr–15 Jul)	23 Jun	(02 Apr–23 Aug)
Wild Turkey	19 Jun	(11 May–29 Jul)	27 May	(18 Apr–03 Aug)	16 Jun	(07 May–01 Sep)	28 Jun	(15 Apr–05 Sep)
Pied-billed Grebe	19 Jun		11 Jun		7 Jun	(05 Jun–25 Jun)	22 Jun	(04 May–05 Aug)
Double-crested Cormorant	10 May	(02 May–30 Jul)			23–Jun			
American Bittern							4 Jul	
Least Bittern			25 May	(10 May–25 Jun)			11 Jul	(30 Jun–15 Aug)
Great Blue Heron	7 May	(26 Mar–28 Jun)	3 Jun	(28 Feb–31 Jul)	15 Jun	(15 Apr–30 Jul)	8 Jul	(16 Jun–31 Jul)
Great Egret					23 Jun			
Green Heron	7 Jun	(02 May–07 Jul)	14 Jun	(10 May–13 Jul)	2 Jul	(01 Jun–28 Jul)	7 Jul	(13 May–21 Aug)
Black-crowned Night-Heron	28 May	(22 May–15 Jun)	30 May	(12 May–18 Jun)	22 Jun	(01 Jun–10 Jul)	16 Jun	(01 Jun–13 Jul)
Yellow-crowned Night-Heron			18 May		5 Jul	(21 Jun–23 Jul)		
Black Vulture	29 May	(29 May–29 May)	7 May	(26 Mar–22 Jun)	5 Jun	(27 May–22 Jun)		
Turkey Vulture	3 Jun	(29 May–06 Jun)	10 May	(15 Apr–15 Jul)	25 Jun	(03 May–24 Aug)	17 Jul	(06 Jun–03 Aug)
Osprey	1 Jun	(31 Mar–23 Jul)	4 Jun	(22 Mar–23 Jul)	25 Jun	(03 May–31 Jul)	15 Jul	(14 May–14 Aug)
Bald Eagle	4 Apr	(05 Feb–13 Dec)	6 May	(06 Feb–15 Jul)	4 Jun	(03 Apr–15 Jul)	14 Jul	(01 Jul–30 Jul)
Northern Harrier			16 May	(05 May–08 Jun)	7 Jul	(20 Jun–01 Aug)	18 Jul	(18 Jun–04 Aug)
Sharp-shinned Hawk	25 May	(22 Apr–29 Jul)	4 Jun	(09 May–05 Jul)	10 Jul	(01 Jun–27 Jul)	10 Jul	(28 May–06 Aug)
Cooper's Hawk	26 Apr	(22 Mar–29 Jul)	27 May	(29 Mar–27 Jul)	25 Jun	(16 May–03 Aug)	20 Jul	(10 Jul–30 Jul)
Northern Goshawk	1 May		9 May	(14 Apr–09 Jul)	4 Jun	(16 May–21 Jun)		
Red-shouldered Hawk	15 May	(10 May–28 Jul)	10 May	(14 Mar–16 Jul)	9 Jun	(01 May–13 Aug)	9 Jul	(14 May–13 Aug)
Broad-winged Hawk	10 May	(15 Apr–29 Jul)	12 Jun	(29 Apr–17 Aug)	1 Jul	(18 May–27 Jul)	14 Jul	(07 Jun–10 Aug)
Red-tailed Hawk	24 Apr	(02 Feb–23 Dec)	28 Apr	(24 Feb–15 Jul)	12 Jun	(09 Apr–10 Aug)	3 Jul	(01 May–02 Oct)
American Kestrel	6 Jun	(11 Apr–15 Jul)	9 Jun	(02 Apr–01 Aug)	26 Jun	(14 Apr–08 Aug)	6 Jul	(15 May–11 Sep)
Merlin			28 May	(18 May–28 Jun)	21 Jul	(03 Jun–06 Aug)	30 Jul	(25 Jul–21 Aug)
Peregrine Falcon			26 May	(01 May–01 Jul)	29 May	(27 Apr–23 Jun)	5 Jul	(25 Jun–15 Jul)
Virginia Rail			6 Jun		10 Jun		4 Jul	(19 May–17 Aug)
Sora	23 May	(07 May–08 Jun)	20 May				17 Jul	(23 May–22 Aug)
Common Gallinule			13 Jul		15 Jun		30 Jun	(05 Jun–30 Jul)
American Coot					10 Jul	(05 Jun–15 Aug)	24 Jun	
Sandhill Crane			5 May		15 Jun	(14 Jun–20 Jul)	18 Jun	(06 May–14 Jul)
Killdeer	23 May	(26 Mar–29 Jul)	25 May	(08 Apr–15 Aug)	13 Jun	(30 Apr–20 Jul)	18 Jun	(08 Apr–01 Aug)
Spotted Sandpiper	5 Jun		16 Jun	(23 May–20 Jun)	19 Jun	(11 May–11 Jul)	26 Jun	(04 Jun–30 Jul)
Upland Sandpiper							11 Jun	
Wilson's Snipe					11 Jun			
American Woodcock	29 Jul		16 May	(01 Apr–10 Jul)	8 May	(17 Apr–01 Jul)	11 Jun	(14 Apr–15 Jul)
Ring-billed Gull			10 May					
Herring Gull			10 May	(20 Apr–19 Jun)	8 Jun	(18 May–30 Jun)	15 Jun	(29 May–03 Jul)
Great Black-backed Gull					22 Jun			

(continued)

	Nest building		Nest with eggs		Nest with young		Fledged Young	
	Median	Range	Median	Range	Median	Range	Median	Range
Black Tern			4 Jun					
Rock Pigeon	4 Jun	(24 Mar–06 Nov)	13 Jun	(09 Jan–19 Aug)	3 Jun	(03 Feb–28 Sep)	23 Jun	(03 Feb–16 Aug)
Eurasian Collared-Dove	5 Jun	(12 Apr–30 Jul)						
Mourning Dove	6 Jun	(03 Mar–29 Aug)	30 May	(08 Mar–27 Sep)	9 Jun	(05 Apr–10 Sep)	22 Jun	(15 Mar–13 Aug)
Yellow-billed Cuckoo	5 Jun	(11 May–21 Jul)	18 Jun	(18 May–24 Aug)	25 Jun	(31 May–27 Sep)	3 Jul	(03 Jun–06 Sep)
Black-billed Cuckoo	20 Jun	(30 May–06 Aug)	18 Jun	(21 May–23 Jul)	21 Jun	(24 May–21 Jul)	6 Jul	(04 Jun–25 Aug)
Barn Owl			6 Jun	(10 Apr–20 Oct)	10 Jul	(07 Apr–26 Dec)	30 Jun	(16 May–11 Aug)
Eastern Screech-Owl	19 Jun	(27 May–12 Jul)	10 May	(23 Mar–07 Aug)	15 Jun	(22 Mar–15 Aug)	25 Jun	(20 Apr–15 Aug)
Great Horned Owl			19 Mar	(22 Jan–20 Jun)	14 Apr	(12 Jan–19 Jul)	1 Jun	(29 Jan–21 Sep)
Barred Owl			8 May	(19 Mar–24 Jun)	8 Jun	(13 Feb–15 Aug)	14 Jun	(01 Apr–09 Aug)
Long-eared Owl			27 Feb				21 May	(30 Apr–15 Jul)
Northern Saw-whet Owl			15 Apr		9 Jul		24 Jun	(10 Jun–21 Jul)
Common Nighthawk			21 Jun	(12 Jun–30 Jun)	15 Jul	(03 Jul–27 Jul)	2 Jul	(08 Jun–08 Jul)
Eastern Whip-poor-will			12 Jun	(29 May–27 Jun)	28 Jun	(10 Jun–16 Jul)		
Chimney Swift	8 Jun	(23 May–30 Jun)	17 Jun	(24 May–27 Jul)	4 Jul	(26 May–14 Aug)	3 Jul	(27 May–05 Aug)
Ruby-throated Hummingbird	7 Jun	(01 May–16 Jul)	18 Jun	(09 May–28 Aug)	6 Jul	(21 May–26 Aug)	4 Jul	(10 May–09 Sep)
Belted Kingfisher	20 May	(04 Apr–24 Jul)	6 Jun	(20 Apr–17 Jul)	26 Jun	(19 Apr–31 Jul)	4 Jul	(04 Jun–14 Aug)
Red-headed Woodpecker	30 May	(03 May–14 Jun)	16 Jun	(15 May–15 Jul)	7 Jul	(23 May–23 Jul)	9 Jul	(06 Jun–16 Aug)
Red-bellied Woodpecker	9 May	(24 Mar–09 Jul)	2 Jun	(30 Mar–23 Jul)	18 Jun	(29 Mar–10 Aug)	1 Jul	(06 May–27 Aug)
Yellow-bellied Sapsucker	21 May	(04 Apr–01 Jul)	13 Jun	(10 Apr–12 Jul)	26 Jun	(18 May–31 Jul)	12 Jul	(01 Jun–14 Aug)
Downy Woodpecker	1 May	(18 Mar–28 Jul)	28 May	(18 Apr–12 Jul)	15 Jun	(15 Apr–14 Aug)	27 Jun	(08 May–27 Aug)
Hairy Woodpecker	30 Apr	(27 Mar–12 Jun)	31 May	(03 Apr–10 Jul)	12 Jun	(28 Apr–27 Jul)	30 Jun	(25 Apr–27 Aug)
Northern Flicker	23 May	(10 Apr–29 Jul)	7 Jun	(08 Apr–26 Jul)	24 Jun	(01 May–07 Aug)	10 Jul	(01 May–26 Aug)
Pileated Woodpecker	17 May	(26 Mar–29 Jul)	30 May	(17 Mar–27 Jul)	11 Jun	(01 May–14 Aug)	28 Jun	(03 Apr–20 Aug)
Eastern Wood-Pewee	7 Jun	(25 May–27 Jun)	17 Jun	(28 May–01 Aug)	12 Jul	(01 Jun–20 Sep)	14 Jul	(06 Jun–28 Aug)
Yellow-bellied Flycatcher			2 Aug		1 Jul	(21 Jun–12 Jul)	17 Jul	(13 Jul–21 Jul)
Acadian Flycatcher	5 Jun	(15 May–29 Jul)	14 Jun	(25 May–24 Jul)	8 Jul	(06 Jun–07 Sep)	9 Jul	(01 Jun–31 Jul)
Alder Flycatcher	26 Jun	(16 Jun–28 Jun)	15 Jun	(10 Jun–19 Jun)	7 Jul	(17 Jun–31 Jul)	28 Jul	(23 Jul–04 Aug)
Willow Flycatcher	9 Jun	(10 May–06 Jul)	21 Jun	(02 Jun–07 Jul)	15 Jul	(19 Jun–14 Aug)	8 Jul	(03 Jun–02 Aug)
Least Flycatcher	11 Jun	(12 May–03 Jul)	17 Jun	(01 Jun–04 Jul)	2 Jul	(08 Jun–15 Jul)	6 Jul	(09 Jun–30 Jul)
Eastern Phoebe	12 May	(20 Mar–21 Jul)	7 Jun	(22 Mar–18 Sep)	15 Jun	(22 Apr–01 Aug)	3 Jul	(01 May–09 Aug)
Great Crested Flycatcher	3 Jun	(11 May–27 Jul)	14 Jun	(12 May–20 Jul)	30 Jun	(05 Jun–24 Jul)	4 Jul	(30 May–29 Jul)
Eastern Kingbird	2 Jun	(02 May–29 Jul)	15 Jun	(11 May–15 Jul)	5 Jul	(20 May–26 Aug)	11 Jul	(03 Jun–14 Aug)
Loggerhead Shrike	early May				early May			
White-eyed Vireo	8 Jun	(01 May–28 Jun)	11 Jun	(27 May–25 Jul)	3 Jul	(01 Jun–28 Jul)	8 Jul	(30 May–03 Aug)
Yellow-throated Vireo	5 Jun	(07 May–29 Jun)	2 Jun	(12 May–09 Jul)	26 Jun	(02 Jun–08 Aug)	9 Jul	(15 Jun–08 Aug)
Blue-headed Vireo	1 Jun	(21 Apr–15 Jul)	4 Jun	(19 Apr–28 Jul)	28 Jun	(15 May–26 Jul)	19 Jul	(07 Jun–16 Aug)
Warbling Vireo	2 Jun	(04 Apr–10 Jul)	3 Jun	(08 May–20 Jul)	1 Jul	(23 May–09 Aug)	8 Jul	(13 Jun–09 Aug)
Red-eyed Vireo	5 Jun	(10 May–31 Jul)	15 Jun	(20 May–16 Jul)	4 Jul	(02 Jun–30 Jul)	14 Jul	(01 Jun–08 Aug)
Blue Jay	30 May	(14 Mar–28 Aug)	28 May	(10 Apr–11 Jul)	1 Jul	(05 May–25 Aug)	7 Jul	(17 May–29 Aug)
American Crow	5 May	(05 Mar–31 Jul)	20 May	(27 Mar–18 Jul)	24 Jun	(29 Apr–31 Jul)	28 Jun	(21 Apr–15 Aug)
Fish Crow	1 May	(11 Apr–29 Jun)	14 May	(01 May–15 Jun)	29 Jun	(02 May–02 Aug)	29 Jun	(02 May–28 Jul)
Common Raven	1 Apr	(10 Feb–28 May)	22 Apr	(01 Mar–10 Jun)	4 Jun	(21 Mar–14 Jul)	17 Jun	(08 Apr–28 Aug)
Horned Lark	18 Apr	(15 Mar–11 Jun)	11 May	(03 Apr–16 Jun)	24 Jun	(01 May–17 Jul)	24 Jun	(28 Apr–28 Jul)
Purple Martin	31 May	(23 Apr–24 Jun)	13 Jun	(16 Apr–11 Aug)	29 Jun	(30 May–02 Aug)	23 Jun	(25 May–23 Jul)
Tree Swallow	30 May	(13 Apr–15 Jul)	7 Jun	(08 Apr–30 Jul)	17 Jun	(28 Apr–30 Jul)	28 Jun	(11 May–31 Jul)
Northern Rough-winged Swallow	26 May	(14 Apr–28 Jul)	11 Jun	(12 Apr–16 Jul)	25 Jun	(15 May–17 Jul)	12 Jul	(23 May–30 Jul)
Bank Swallow	3 Jun	(04 May–21 Jul)	12 Jun	(08 May–19 Jul)	5 Jun	(05 Jun–18 Jul)	23 Jun	(07 Jun–31 Jul)
Cliff Swallow	2 Jun	(12 Apr–09 Jul)	15 Jun	(21 Apr–02 Aug)	28 Jun	(02 Jun–26 Aug)	7 Jul	(14 Jun–09 Aug)
Barn Swallow	2 Jun	(18 Apr–29 Jul)	11 Jun	(20 Apr–08 Aug)	26 Jun	(10 May–28 Aug)	30 Jun	(26 Apr–28 Aug)
Carolina Chickadee	1 May	(14 Feb–29 Jul)	25 May	(01 Apr–28 Jul)	14 Jun	(15 Apr–17 Jul)	21 Jun	(06 Apr–10 Aug)
Black-capped Chickadee	6 May	(26 Mar–29 Jul)	30 May	(16 Apr–24 Jul)	22 Jun	(23 Apr–05 Aug)	3 Jul	(15 Jun–14 Aug)
Tufted Titmouse	9 May	(18 Mar–10 Aug)	23 May	(29 Mar–10 Jul)	23 Jun	(07 Apr–15 Aug)	30 Jun	(31 Mar–08 Sep)
Red-breasted Nuthatch	5 May	(16 Apr–15 May)	5 Jun	(05 May–26 Jun)	3 Jul	(13 Jun–08 Aug)	9 Jul	(04 Jun–13 Aug)
White-breasted Nuthatch	3 Jun	(14 Mar–28 Jul)	27 May	(30 Mar–04 Aug)	20 Jun	(05 May–09 Aug)	4 Jul	(15 May–14 Aug)
Brown Creeper	24 May	(10 Apr–23 Jun)	15 Jun	(14 May–11 Jul)	14 Jun	(02 May–27 Aug)	9 Jul	(05 Jun–27 Jul)
Carolina Wren	20 May	(05 Mar–06 Aug)	5 Jun	(01 Apr–13 Aug)	17 Jun	(15 Apr–19 Sep)	24 Jun	(16 Apr–30 Sep)
House Wren	7 Jun	(20 Apr–06 Aug)	16 Jun	(17 Apr–06 Aug)	24 Jun	(12 May–27 Aug)	6 Jul	(01 May–02 Sep)
Winter Wren	19 Jun	(09 May–24 Jul)	10 Jun	(11 May–17 Jul)	17 Jun	(08 Jun–16 Jul)	8 Jul	(13 Jun–07 Aug)
Sedge Wren	16 Jul	(22 Jun–10 Aug)						
Marsh Wren	25 May	(29 Apr–20 Jun)						
Blue-gray Gnatcatcher	12 May	(17 Apr–11 Jul)	27 May	(24 Apr–10 Jul)	23 Jun	(05 May–27 Jul)	1 Jul	(02 Jun–10 Aug)
Golden-crowned Kinglet	22 May		12 Jul	(12 Jul–12 Jul)	1 Jul	(04 Jun–27 Jul)	13 Jul	(22 Jun–02 Aug)

(continued)

	Nest building		Nest with eggs		Nest with young		Fledged Young	
	Median	Range	Median	Range	Median	Range	Median	Range
Eastern Bluebird	22 May	(06 Mar–29 Jul)	9 Jun	(24 Mar–16 Oct)	16 Jun	(30 Apr–24 Aug)	29 Jun	(03 May–26 Aug)
Veery	2 Jun	(12 May–23 Jun)	15 Jun	(23 May–08 Jul)	27 Jun	(08 Jun–30 Jul)	7 Jul	(30 May–05 Aug)
Swainson's Thrush	14 Jun		9 Jun		20 Jul	(16 Jul–24 Jul)	18 Jul	(04 Jul–26 Jul)
Hermit Thrush	14 Jun	(08 May–12 Jul)	6 Jun	(07 May–18 Jul)	28 Jun	(02 Jun–30 Jul)	9 Jul	(31 May–14 Aug)
Wood Thrush	5 Jun	(04 May–16 Jul)	10 Jun	(09 May–03 Aug)	21 Jun	(26 May–02 Aug)	1 Jul	(01 Jun–08 Aug)
American Robin	21 May	(18 Mar–07 Aug)	2 Jun	(10 Apr–20 Aug)	14 Jun	(30 Mar–12 Sep)	25 Jun	(17 Apr–07 Sep)
Gray Catbird	8 Jun	(29 Apr–27 Jul)	15 Jun	(05 May–27 Sep)	26 Jun	(13 May–16 Aug)	7 Jul	(09 May–23 Aug)
Northern Mockingbird	2 Jun	(31 Mar–15 Aug)	15 Jun	(19 Apr–16 Aug)	28 Jun	(06 May–25 Aug)	29 Jun	(29 Apr–26 Aug)
Brown Thrasher	30 May	(13 Apr–05 Jul)	15 Jun	(10 May–10 Jul)	20 Jun	(13 May–28 Jul)	28 Jun	(17 May–27 Aug)
European Starling	8 May	(13 Mar–31 Jul)	2 Jun	(29 Mar–22 Jul)	8 Jun	(28 Mar–01 Aug)	23 Jun	(15 Apr–14 Aug)
Cedar Waxwing	14 Jun	(15 May–10 Aug)	22 Jun	(25 May–17 Aug)	7 Jul	(25 May–27 Sep)	17 Jul	(25 May–20 Sep)
Ovenbird	7 Jun	(07 May–20 Jul)	16 Jun	(18 May–27 Sep)	27 Jun	(25 May–07 Aug)	3 Jul	(09 Jun–07 Aug)
Worm-eating Warbler	2 Jun	(12 May–24 Jun)	26 May	(26 May–26 May)	28 Jun	(05 Jun–31 Jul)	30 Jun	(01 Jun–14 Jul)
Louisiana Waterthrush	5 May	(22 Apr–23 Jun)	26 May	(21 Apr–22 Jun)	16 Jun	(04 May–22 Jul)	21 Jun	(10 May–17 Jul)
Northern Waterthrush	7 Jun	(26 May–24 Jun)	10 Jun		25 Jun	(07 Jun–30 Jul)	7 Jul	(14 Jun–31 Jul)
Golden-winged Warbler	10 Jun	(25 May–20 Jun)	12 Jun	(10 Jun–14 Jun)	12 Jun	(27 May–15 Jul)	19 Jun	(10 Jun–15 Jul)
Blue-winged Warbler	10 Jun	(01 May–20 Jun)	3 Jun	(20 May–18 Jun)	4 Jul	(31 May–24 Jul)	4 Jul	(07 Jun–27 Jul)
Brewster's Warbler (hybrid)					6 Jul	(12 Jun–13 Jul)	14 Jun	(07 Jun–21 Jun)
Lawrence's Warbler (hybrid)	29 May							
Black-and-white Warbler	31 May	(08 May–09 Jul)	12 Jun	(26 May–24 Jun)	2 Jul	(05 Jun–24 Jul)	9 Jul	(04 Jun–07 Aug)
Prothonotary Warbler			13 Jun	(19 May 08 Jul)	19 Jun	(28 May–18 Jul)	2 Jul	(05 Jun–30 Jul)
Nashville Warbler	3 Jun				5 Jul	(17 Jun–24 Jul)	14 Jul	(14 Jun–29 Jul)
Mourning Warbler	11 Jun	(01 Jun–23 Jun)	20 Jun	(06 Jun–25 Jun)	6 Jul	(29 Jun–07 Jul)	17 Jul	(30 Jun–30 Jul)
Kentucky Warbler	15 Jun	(26 May–20 Jun)	16 Jun	(03 Jun–16 Jun)	28 Jun	(15 Jun–12 Jul)	5 Jul	(17 Jun–24 Jul)
Common Yellowthroat	9 Jun	(27 Apr–04 Aug)	14 Jun	(05 May–12 Jul)	4 Jul	(01 Jun–22 Aug)	13 Jul	(01 Jun–13 Sep)
Hooded Warbler	9 Jun	(24 May–27 Jun)	4 Jun	(10 May–14 Jul)	5 Jul	(07 Jun–05 Sep)	8 Jul	(05 Jun–22 Aug)
American Redstart	3 Jun	(08 May–29 Jun)	8 Jun	(10 May–15 Jul)	3 Jul	(30 May–10 Aug)	14 Jul	(03 Jun–11 Aug)
Cerulean Warbler	26 May	(15 May–11 Jul)	11 Jun	(26 May–24 Jun)	24 Jun	(11 Jun–25 Jul)	30 Jun	(16 Jun–31 Jul)
Northern Parula	22 May	(30 Apr–13 Jun)	18 Jun	(31 May–22 Jul)	19 Jun	(01 Jun–21 Jul)	2 Jul	(05 Jun–12 Jul)
Magnolia Warbler	21 May	(17 May–27 Jun)	30 Jun	(28 Jun–02 Jul)	6 Jul	(16 Jun–13 Aug)	19 Jul	(13 Jun–22 Aug)
Blackburnian Warbler	4 Jun	(13 May–11 Jul)	19 Jun	(17 Jun–21 Jun)	5 Jul	(04 Jun–15 Aug)	15 Jul	(26 Jun–31 Jul)
Yellow Warbler	27 May	(28 Apr–03 Jul)	6 Jun	(15 May–06 Jul)	26 Jun	(19 May–31 Jul)	30 Jun	(05 Jun–05 Aug)
Chestnut-sided Warbler	9 Jun	(13 May–13 Jul)	2 Jun	(25 May–02 Jul)	1 Jul	(01 Jun–10 Aug)	19 Jul	(04 Jun–01 Sep)
Blackpoll Warbler					14 Jul	(14 Jul–14 Jul)		
Black-throated Blue Warbler	7 Jun	(18 May–11 Jul)	14 Jun	(31 May–14 Jul)	7 Jul	(07 Jun–19 Aug)	12 Jul	(26 Jun–07 Aug)
Pine Warbler	31 May	(16 Apr–26 Jun)			29 Jun	(26 May–25 Jul)	10 Jul	(27 May–28 Jul)
Yellow-rumped Warbler	27 May	(04 May–02 Jul)	18 Jun	(15 Jun–01 Jul)	1 Jul	(27 May–12 Aug)	15 Jul	(05 Jun–07 Aug)
Yellow-throated Warbler	13 May	(26 Apr–20 Jun)	12 May		22 Jun	(25 May–01 Jul)	2 Jul	
Prairie Warbler	29 May	(18 May–19 Jun)	14 Jun	(05 Jun–20 Jun)	27 Jun	(10 Jun–22 Jul)	6 Jul	(05 Jun–28 Jul)
Black-throated Green Warbler	2 Jun	(02 May–08 Jul)	13 Jun	(21 May–07 Jul)	5 Jul	(09 Jun–31 Jul)	15 Jul	(15 Jun–30 Jul)
Canada Warbler	24 May	(24 May–24 May)	30 May	(23 May–20 Jun)	5 Jul	(13 Jun–30 Jul)	15 Jul	(05 Jun–31 Jul)
Yellow-breasted Chat	11 Jun	(09 May–10 Jul)	1 Jun		22 Jun	(12 Jun–15 Jul)	12 Jul	(20 Jun–20 Aug)
Eastern Towhee	10 Jun	(27 Mar–29 Jul)	16 Jun	(03 May–27 Aug)	29 Jun	(25 May–14 Aug)	10 Jul	(07 May–13 Sep)
Chipping Sparrow	7 Jun	(20 Apr–29 Jul)	10 Jun	(30 Apr–30 Jul)	27 Jun	(28 Apr–17 Sep)	5 Jul	(10 May–27 Aug)
Clay-colored Sparrow	16 May	(12 May–20 May)			6 Jul			
Field Sparrow	1 Jun	(20 Apr–27 Jul)	5 Jun	(26 Apr–28 Jul)	22 Jun	(12 Jun–22 Aug)	7 Jul	(03 May–29 Aug)
Vesper Sparrow	31 May	(20 May–12 Jun)	15 Jun	(06 Jun–15 Jun)	17 Jun	(03 Jun–19 Jul)	9 Jul	(10 Jun–01 Aug)
Savannah Sparrow	12 Jun	(13 May–14 Jul)	12 Jun	(06 Jun–14 Jul)	25 Jun	(30 May–02 Aug)	7 Jul	(01 Jun–04 Aug)
Grasshopper Sparrow	12 Jun	(20 May–27 Jul)	19 Jun	(01 Jun–02 Jul)	3 Jul	(27 May–29 Jul)	6 Jul	(15 May–31 Jul)
Henslow's Sparrow	18 Jun	(05 Jun–20 Jun)	15 Jun	(09 Jun–21 Jun)	2 Jul	(31 May–20 Jul)	12 Jul	(16 Jun–13 Aug)
Song Sparrow	3 Jun	(26 Mar–10 Aug)	5 Jun	(10 Apr–31 Jul)	23 Jun	(30 Apr–16 Sep)	30 Jun	(13 May–07 Oct)
Swamp Sparrow	6 Jun	(27 Apr–23 Jul)	12 Jun	(01 Jun–01 Aug)	8 Jul	(02 Jun–02 Aug)	8 Jul	(06 Jun–04 Aug)
White-throated Sparrow	4 Jun	(27 May–12 Jun)	30 Jun		6 Jul	(13 Jun–28 Jul)	7 Jul	(10 Jun–31 Jul)
Dark-eyed Junco	4 Jun	(26 Apr–15 Jul)	7 Jun	(01 May–17 Jul)	23 Jun	(10 May–08 Aug)	2 Jul	(17 May–15 Aug)
Summer Tanager	30 May							
Scarlet Tanager	4 Jun	(08 May–08 Jul)	13 Jun	(16 May–31 Jul)	4 Jul	(08 Jun–19 Aug)	13 Jul	(01 Jun–14 Aug)
Northern Cardinal	18 May	(14 Jan–23 Jul)	29 May	(01 Apr–27 Sep)	22 Jun	(15 Mar–02 Oct)	30 Jun	(15 Mar–16 Oct)
Rose-breasted Grosbeak	5 Jun	(26 Apr–24 Jul)	10 Jun	(20 May–11 Jul)	2 Jul	(14 Jun–08 Aug)	12 Jul	(27 May–22 Aug)
Blue Grosbeak	1 Jul	(01 Jul–01 Jul)	10 Jul	(24 Jun–26 Jul)	21 Jul	(23 Jun–28 Aug)	5 Jul	(05 Jun–06 Aug)
Indigo Bunting	11 Jun	(06 May–31 Jul)	10 Jun	(28 May–18 Jul)	3 Jul	(02 Jun–25 Aug)	9 Jul	(04 Jun–11 Sep)
Bobolink	9 Jun	(10 May–26 Jun)	9 Jun	(25 May–24 Jun)	23 Jun	(14 May–14 Jul)	30 Jun	(25 May–31 Jul)
Red-winged Blackbird	23 May	(01 Apr–29 Jul)	28 May	(26 Apr–27 Jul)	18 Jun	(20 Apr–01 Aug)	21 Jun	(15 Apr–31 Jul)
Eastern Meadowlark	10 Jun	(28 Apr–29 Jul)	10 Jun	(19 Apr–18 Jul)	26 Jun	(14 May–03 Sep)	5 Jul	(10 May–13 Aug)

(continued)

	Nest building		Nest with eggs		Nest with young		Fledged Young	
	Median	Range	Median	Range	Median	Range	Median	Range
Common Grackle	4 May	(16 Mar–30 Jun)	1 Jun	(14 Apr–01 Jul)	15 Jun	(15 Apr–29 Jul)	19 Jun	(01 Apr–27 Aug)
Brown-headed Cowbird	3 Jun	(10 May–18 Jun)	7 Jun	(10 Apr–17 Jul)	22 Jun	(12 May–22 Aug)	27 Jun	(15 May–12 Aug)
Orchard Oriole	1 Jun	(07 May–04 Jul)	15 Jun	(15 May–09 Jul)	23 Jun	(30 May–26 Jul)	30 Jun	(27 May–28 Jul)
Baltimore Oriole	31 May	(15 Apr–21 Jul)	11 Jun	(04 May–28 Jul)	22 Jun	(29 May–28 Jul)	3 Jul	(01 Jun–11 Aug)
Purple Finch	2 Jun	(30 Apr–20 Jul)	12 Jun	(08 May–18 Jul)	28 Jun	(18 May–13 Aug)	5 Jul	(20 May–09 Aug)
House Finch	29 May	(19 Mar–27 Jul)	3 Jun	(28 Mar–24 Jul)	18 Jun	(15 Apr–17 Sep)	26 Jun	(21 Apr–13 Aug)
Red Crossbill					2 Jul	(21 May–13 Aug)	18 Jul	
Pine Siskin	7 Apr	(03 Mar–06 Jun)	15 May		28 May	(05 May–27 Jun)	20 Jun	(30 Apr–19 Jul)
American Goldfinch	3 Jul	(02 May–27 Aug)	14 Jul	(29 May–18 Aug)	10 Aug	(11 Jun–14 Oct)	29 Jul	(10 Jun–02 Oct)
House Sparrow	22 May	(09 Feb–31 Jul)	11 Jun	(08 Feb–31 Jul)	17 Jun	(26 Mar–11 Sep)	25 Jun	(07 Apr–21 Sep)

There were no such breeding confirmations for Trumpeter Swan, King Rail, Short-eared Owl, Chuck's Whip-poor-will, Swainson's Warbler, Dickcissel, and Western Meadowlark.

APPENDIX G | Atlas Safe Dates

See pages 26–29 and fig 4.2 for explanation of safe dates.

Common name	Begin safe date	End safe date
Canada Goose	01 May	31 Jul
Mute Swan	01 May	31 Aug
Trumpeter Swan	01 May	31 Aug
Wood Duck	01 May	31 Jul
Gadwall	01 Jun	31 Jul
American Wigeon	01 Jun	31 Jul
American Black Duck	01 May	31 Jul
Mallard	01 May	31 Jul
Blue-winged Teal	01 Jun	31 Jul
Northern Shoveler	01 Jun	31 Jul
Northern Pintail	01 Jun	31 Jul
Green-winged Teal	01 Jun	31 Jul
Redhead	01 Jun	15 Aug
Ring-necked Duck	01 Jun	15 Aug
Hooded Merganser	01 Jun	15 Aug
Common Merganser	01 Jun	15 Aug
Red-breasted Merganser	01 Jun	15 Aug
Ruddy Duck	01 Jun	15 Aug
Northern Bobwhite	15 Apr	31 Jul
Ring-necked Pheasant	15 Apr	31 Jul
Ruffed Grouse	01 Apr	31 Jul
Wild Turkey	15 Apr	31 Jul
Common Loon	01 Jun	31 Jul
Pied-billed Grebe	01 Jun	31 Jul
Double-crested Cormorant	01 Jun	31 Jul
American Bittern	01 Jun	31 Jul
Least Bittern	01 Jun	31 Jul
Great Blue Heron	01 Jun	15 Jul
Great Egret	01 Jun	30 Jun
Snowy Egret	01 Jun	30 Jun
Cattle Egret	01 Jun	30 Jun
Green Heron	01 Jun	15 Jul
Black-crowned Night-Heron	01 Jun	30 Jun
Yellow-crowned Night-Heron	01 Jun	30 Jun
Glossy Ibis	01 Jun	30 Jun
Black Vulture	01 May	31 Jul
Turkey Vulture	01 May	31 Jul
Osprey	01 Jun	31 Jul
Bald Eagle	01 May	15 Jul
Northern Harrier	01 Jun	31 Jul
Sharp-shinned Hawk	01 Jun	31 Jul
Cooper's Hawk	01 Jun	31 Jul
Northern Goshawk	01 May	31 Jul
Red-shouldered Hawk	01 May	15 Aug
Broad-winged Hawk	01 Jun	31 Jul
Red-tailed Hawk	01 May	15 Jul
American Kestrel	15 May	31 Jul
Merlin	15 May	31 Jul
Peregrine Falcon	15 May	31 Jul
Yellow Rail	15 May	15 Jul
Black Rail	15 May	15 Aug
King Rail	15 May	15 Aug
Virginia Rail	15 May	15 Aug
Sora	15 May	31 Jul
Common Gallinule	15 May	31 Aug
American Coot	01 Jun	31 Jul
Sandhill Crane	15 May	31 Jul
Piping Plover	15 Jun	15 Jul
Killdeer	01 May	15 Jul
Black-necked Stilt	01 Jun	30 Jun
Spotted Sandpiper	01 Jun	30 Jun
Upland Sandpiper	15 May	30 Jun
Wilson's Snipe	15 May	15 Jul
American Woodcock	01 Apr	15 Jul
Ring-billed Gull	15 May	31 Jul
Herring Gull	15 May	31 Jul
Least Tern	25 May	05 Jul
Black Tern	01 Jun	15 Jul
Common Tern	05 Jun	30 Jun
Rock Pigeon	01 Jan	31 Dec
Eurasian Collared-Dove	01 May	15 Jul
Mourning Dove	01 May	15 Jul

(continued)

Common name	Begin safe date	End safe date
Yellow-billed Cuckoo	05 Jun	31 Jul
Black-billed Cuckoo	05 Jun	31 Jul
Barn Owl	20 Apr	15 Aug
Eastern Screech-Owl	15 Apr	15 Aug
Great Horned Owl	15 Dec	31 Jul
Barred Owl	15 Jan	31 Jul
Long-eared Owl	20 Apr	15 Aug
Short-eared Owl	20 Apr	15 Aug
Northern Saw-whet Owl	20 Apr	15 Aug
Common Nighthawk	05 Jun	31 Jul
Chuck-will's-widow	01 Jun	31 Jul
Eastern Whip-poor-will	01 Jun	31 Jul
Chimney Swift	25 May	31 Jul
Ruby-throated Hummingbird	01 Jun	15 Jul
Belted Kingfisher	15 Apr	15 Jul
Red-headed Woodpecker	25 May	31 Jul
Red-bellied Woodpecker	15 Mar	31 Jul
Yellow-bellied Sapsucker	15 May	31 Jul
Downy Woodpecker	15 Mar	31 Jul
Hairy Woodpecker	15 Mar	31 Jul
Northern Flicker	15 May	31 Jul
Pileated Woodpecker	15 Mar	31 Jul
Olive-sided Flycatcher	10 Jun	31 Jul
Eastern Wood-Pewee	01 Jun	31 Jul
Yellow-bellied Flycatcher	10 Jun	15 Jul
Acadian Flycatcher	25 May	31 Jul
Alder Flycatcher	10 Jun	15 Jul
Willow Flycatcher	10 Jun	15 Jul
Least Flycatcher	05 Jun	15 Jul
Eastern Phoebe	01 May	31 Jul
Great Crested Flycatcher	25 May	31 Jul
Eastern Kingbird	25 May	15 Jul
Loggerhead Shrike	01 May	31 Jul
White-eyed Vireo	25 May	15 Aug
Yellow-throated Vireo	01 Jun	15 Aug
Blue-headed Vireo	25 May	31 Jul
Warbling Vireo	01 Jun	15 Aug
Red-eyed Vireo	01 Jun	31 Jul
Blue Jay	10 Jun	31 Jul
Eurasian Jackdaw	01 May	31 Aug
American Crow	01 May	31 Jul
Fish Crow	01 May	31 Jul
Common Raven	01 Mar	15 Jul
Horned Lark	01 May	31 Jul
Purple Martin	25 May	30 Jun
Tree Swallow	25 May	30 Jun
Northern Rough-winged Swallow	01 Jun	30 Jun
Bank Swallow	01 Jun	30 Jun
Cliff Swallow	05 Jun	05 Jul
Barn Swallow	25 May	30 Jun
Carolina Chickadee	15 Mar	15 Aug
Black-capped Chickadee	15 Apr	15 Aug
Tufted Titmouse	15 Mar	15 Aug
Red-breasted Nuthatch	01 Jun	15 Aug
White-breasted Nuthatch	01 May	15 Aug
Brown Creeper	15 May	31 Jul
Carolina Wren	01 Apr	30 Sep
Bewick's Wren	01 May	15 Aug
House Wren	01 Jun	15 Aug
Winter Wren	15 May	15 Aug
Sedge Wren	01 Jun	15 Aug
Marsh Wren	25 May	31 Jul
Blue-gray Gnatcatcher	15 May	15 Jul
Golden-crowned Kinglet	15 May	15 Aug
Eastern Bluebird	01 May	15 Aug
Veery	01 Jun	31 Jul
Swainson's Thrush	05 Jun	31 Jul
Hermit Thrush	15 May	31 Jul
Wood Thrush	01 Jun	31 Jul
American Robin	01 May	31 Jul
Gray Catbird	01 Jun	31 Jul
Northern Mockingbird	15 May	31 Aug
Brown Thrasher	15 May	31 Jul
European Starling	15 Apr	31 Jul
Cedar Waxwing	15 Jun	31 Jul
Ovenbird	01 Jun	31 Jul
Worm-eating Warbler	25 May	15 Jul
Louisiana Waterthrush	15 Apr	15 Jul
Northern Waterthrush	05 Jun	15 Jul
Golden-winged Warbler	25 May	15 Jul
Brewster's Warbler	25 May	15 Jul
Lawrence's Warbler	25 May	15 Jul
Blue-winged Warbler	25 May	15 Jul
Black-and-white Warbler	01 Jun	31 Jul
Prothonotary Warbler	25 May	15 Jul
Swainson's Warbler	25 May	31 Jul
Nashville Warbler	25 May	31 Jul
Mourning Warbler	15 Jun	31 Jul
Kentucky Warbler	25 May	31 Jul
Common Yellowthroat	25 May	31 Jul
Hooded Warbler	01 Jun	31 Jul
American Redstart	01 Jun	31 Jul
Cerulean Warbler	01 Jun	31 Jul
Northern Parula	25 May	31 Jul

(continued)

Common name	Begin safe date	End safe date
Magnolia Warbler	01 Jun	31 Jul
Blackburnian Warbler	01 Jun	31 Jul
Yellow Warbler	25 May	30 Jun
Chestnut-sided Warbler	01 Jun	31 Jul
Blackpoll Warbler	15 Jun	15 Aug
Black-throated Blue Warbler	01 Jun	31 Jul
Pine Warbler	01 May	15 Aug
Yellow-rumped Warbler	01 Jun	15 Aug
Yellow-throated Warbler	01 May	15 Jul
Prairie Warbler	25 May	31 Jul
Black-throated Green Warbler	01 Jun	31 Jul
Canada Warbler	01 Jun	31 Jul
Yellow-breasted Chat	01 Jun	31 Jul
Eastern Towhee	01 Jun	31 Jul
Bachman's Sparrow	01 Jun	15 Aug
Chipping Sparrow	01 Jun	15 Aug
Clay-colored Sparrow	01 Jun	15 Aug
Field Sparrow	15 May	15 Aug
Vesper Sparrow	15 May	15 Aug
Lark Sparrow	01 Jun	15 Aug
Savannah Sparrow	25 May	15 Aug
Grasshopper Sparrow	01 Jun	15 Aug
Henslow's Sparrow	25 May	15 Aug
Song Sparrow	15 May	15 Aug
Swamp Sparrow	01 Jun	15 Aug
White-throated Sparrow	10 Jun	15 Aug
Dark-eyed Junco	25 May	15 Aug
Summer Tanager	01 Jun	31 Jul
Scarlet Tanager	01 Jun	31 Jul
Northern Cardinal	15 Mar	30 Sep
Rose-breasted Grosbeak	01 Jun	31 Jul
Blue Grosbeak	01 Jun	31 Jul
Indigo Bunting	01 Jun	31 Jul
Dickcissel	01 Jun	31 Jul
Bobolink	15 May	30 Jun
Red-winged Blackbird	15 Apr	30 Jun
Eastern Meadowlark	15 May	31 Jul
Western Meadowlark	15 May	31 Jul
Common Grackle	15 Apr	30 Jun
Brown-headed Cowbird	15 May	15 Jul
Orchard Oriole	01 Jun	31 Jul
Baltimore Oriole	01 Jun	31 Jul
Purple Finch	15 May	31 Jul
House Finch	15 Apr	31 Jul
Red Crossbill	15 May	31 Jul
Pine Siskin	01 Jun	31 Jul
American Goldfinch	10 Jun	31 Aug
Evening Grosbeak	10 Jun	31 Jul
House Sparrow	01 Feb	30 Sep

Literature Cited

ABA (American Birding Association). 2003. American Birding Association Principles of Birding Ethics. American Birding Association Member Handbook. [Online.] Available at http://www.aba.org/about/ethics.html.

Abrams, M. D., and C. M. Ruffner. 1995. Physiographic analysis of witness-tree distribution (1765–1798) and present forest cover through northcentral Pennsylvania. Canadian Journal of Forest Research 25:659–668.

Adkisson, C. S. 1996. Red Crossbill (*Loxia curvirostra*). *In* The Birds of North America Online, no. 256 (A. Poole, Ed.). Cornell Lab of Ornithology, Ithaca, NY.

Adkisson, C. S., and R. N. Conner. 1978. Interspecific vocal imitation in White-eyed Vireos. Auk 95:602–606.

AFC (Atlantic Flyway Council). 2003. Atlantic Flyway Mute Swan Management Plan 2003–2013. Atlantic Flyway Council, Laurel, MD.

———. 2006. Minutes of the 2006 winter meeting of the Atlantic Flyway Council technical section. Atlantic Flyway Council, Louisville, KY.

AFWA (Association of Fish and Wildlife Agencies). 2011. State and Tribal Wildlife Grant Program: 10 years of success. Association of Fish and Wildlife Agencies, Washington, DC.

Alexander, J. D., J. L. Stephens, G. R. Geufel, and T. C. Will. 2009. Decision support tools: Bridging the gap between science and management. Pages 283–291 *in* Proceedings of the Fourth International Partners in Flight Conference: Tundra to Tropics: Connecting Birds, Habitats and People. University of Texas–Pan American Press, McAllen, TX.

Alldredge, M. W., T. R. Simons, and K. H. Pollock. 2007. Factors affecting aural detections of songbirds. Ecological Applications 17:948–955.

Allen, M. C., and J. Sheehan 2010. Eastern hemlock decline and its effect on Pennsylvania's birds. Pages 232–245 *in* Avian Ecology and Conservation: A Pennsylvania Focus with National Implications (S. K. Majumdar, T. L. Master, M. C. Brittingham, R. M. Ross, R. S. Mulvihill, and J. E. Huffman, Eds.). Pennsylvania Academy of Sciences, Easton, PA.

Allen, M. C., J. Sheehan, Jr., T. L. Master, and R. S. Mulvihill. 2009. Responses of Acadian Flycatchers (*Empidonax virescens*) to hemlock woolly adelgid (*Adelges tsugae*) infestation in Appalachian riparian forests. Auk 126:543–553.

Allin, C. C. 1981. Mute Swans in the Atlantic flyway. Pages 149–154 *in* Proceedings of the Fourth International Waterfowl Symposium, New Orleans, LA.

Anders, A. D. 2010. Black-billed Cuckoo (*Coccyzus erythropthalmus*). Pages 262–264 *in* Terrestrial Vertebrates of Concern in Pennsylvania: A Complete Guide to Species of Conservation Concern (M. A. Steele, M. C. Brittingham, T. J. Maret, and J. F. Merritt, Eds.). The Johns Hopkins University Press, Baltimore, MD.

Anders, A. D., and E. Post. 2006. Distribution-wide effects of climate on population densities of a declining migratory landbird. Journal of Animal Ecology 75:221–227.

Anich, N. M., T. J. Benson, J. D. Brown, C. Roa, J. C. Bednarz, R. E. Brown, and J. G. Dickson. 2010. Swainson's Warbler (*Limnothlypis swainsonii*). *In* The Birds of North America Online, no. 126 (A. Poole, Ed.). Cornell Lab of Ornithology, Ithaca, NY.

Anonymous. 2009. The West Virginia Birding List. [Online.] Available at http://list.audubon.org/wa.exe?A1=ind0905&L=wv-bird.

AOU (American Ornithologists' Union). 1985. Thirty-fifth supplement to the American Ornithologists' Union Check-list of North American Birds. Auk 102:680–686.

———. 1998. Check-list of North American Birds, 7th ed. American Ornithologists' Union, Washington, DC.

Arcese, P., M. K. Sogge, A. B. Marr, and M. A. Patten. 2002. Song Sparrow (*Melospiza melodia*). *In* The Birds of North America Online, no. 704 (A. Poole, Ed.). Cornell Lab of Ornithology, Ithaca, NY.

Askins, R. A. 1999. History of grassland birds in eastern North America. Studies in Avian Biology 19:60–71.

Audubon, J. J. 1831. Ornithological Biography, vol. 1. Judah Dobson, Philadelphia, PA.

———. 1838. Ornithological Biography, vol. 4. Adam and Charles Black, Edinburgh, UK.

———. 1840–1844. The Birds of America, vols. 1–7. Chevalier, Philadelphia, PA.

Audubon Pennsylvania. 2011. Audubon Pennsylvania Birds Conservation: The Important Bird Area Program in Pennsylvania. [Online.] Available at http://pa.audubon.org/iba/.

Badzinski, D. S. 2007. Hooded Warbler, Paruline à capuchon (*Wilsonia citrina*). Pages 524–525 *in* Atlas of the Breeding Birds of Ontario, 2001–2005 (M. D. Cadman, D. A. Sutherland, G. G. Beck, D. Lepage, and A. R. Coutrier, Eds.). Bird Studies Canada, Environment Canada, Ontario Field Ornithologists, Ontario Ministry of Natural Resources, and Ontario Nature, Toronto, ON.

Bahn, V., R. J. O'Connor, and W. B. Krohn. 2006. Importance of spatial autocorrelation in modeling bird distributions at a continental scale. Ecography 29:835–844.

Baicich, P. J., and C. J. O. Harrison. 2005. Nests, Eggs, and Nestlings of North American Birds. Princeton University Press, Princeton, NJ.

Baily, W. L. 1896. Summer birds of northern Elk County, Pa. Auk 13:289–297.

Baird, S. F. 1845. Catalogue of birds found in the neighborhood of Carlisle, Cumberland County, Pennsylvania. Literary Record and Journal of the Linnaean Association of Pennsylvania College 1:249–257.

Bakermans, M. H., J. L. Larkin, B. W. Smith, T. M. Fearer, and B. C. Jones. 2011. Golden-winged Warbler Habitat Best Management Practices for Forestlands in Maryland and Pennsylvania. American Bird Conservancy. The Plains, Virginia.

Banko, W. E. 1960. The Trumpeter Swan: Its history, habits, and population in the United States. North American Fauna. U.S. Department of the Interior, U.S. Fish and Wildlife Service, Bureau of Sport Fisheries and Wildlife, Washington, DC.

Bannor, B. K., and E. Kiviat. 2002. Common Moorhen (*Gallinula chloropus*). *In* The Birds of North America, no. 685 (A. Poole and F. Gill, Eds.). The Birds of North America, Inc., Philadelphia, PA.

Barlow, J. C., and W. B. McGillivray. 1983. Foraging and habitat relationships of the sibling species Willow Flycatcher

(*Empidonax traillii*) and Alder Flycatcher (*E. alnorum*) in southern Ontario. Canadian Journal of Zoology 61:1510–1516.

Bart, J., M. Hofschen, and B. G. Peterjohn. 1995. Reliability of the breeding bird survey: Effects of restricting surveys to roads. Auk 112:758–761.

Beck, H. H. 1924. A chapter on the ornithology of Lancaster County, Pennsylvania, with supplementary notes on the mammals. Pages 1–39 *in* Lancaster County, Pennsylvania—A History. Lewis Historical Publishing Company, New York, NY.

Becker, D. A., M. C. Brittingham, and C. B. Goguen. 2008. Effects of hemlock woolly adelgid on breeding birds at Fort Indiantown Gap, Pennsylvania. Northeastern Naturalist 15:227–240.

Bednarz, J. C., D. Klem, Jr., L. J. Goodrich, and S. E. Senner. 1990. Migration counts of raptors at Hawk Mountain, Pennsylvania, as indicators of population trends, 1934–1986. Auk 107:96–109.

Beissinger, S. R., E. C. Steadman, T. Wohlgenant, G. Blate, and S. Zack. 1996. Null models for assessing ecosystem conservation priorities: Threatened birds as titers of threatened ecosystems. Conservation Biology 10:1343–1352.

Bell, C. P., S. W. Baker, N. G. Parkes, M. D. Brooke, and D. E. Chamberlain. 2010. The role of the Eurasian Sparrowhawk (*Accipiter nisus*) in the decline of the House Sparrow (*Passer domesticus*) in Britain. Auk 127:411–420.

Bell, R. K. 1975. Greene County. Pages 25–26 *in* Where to Find Birds in Western Pennsylvania (D. B. Freeland, Ed.). Typecraft Press, Inc., Pittsburgh, PA.

Bellrose, F. C. 1976. Ducks, Geese and Swans of North America, 2nd ed. Stackpole Books, Harrisburg, PA.

Bellrose, F. C., and D. J. Holm. 1994. Ecology and Management of the Wood Duck. Stackpole Books, Mechanicsburg, PA.

Benkman, C. W. 1993a. Adaptation to single resources and the evolution of crossbill (*Loxia*) diversity. Ecological Monographs 63:305–325.

———. 1993b. Logging, conifers, and the conservation of crossbills. Conservation Biology 7:473–479.

———. 1999. The selection mosaic and diversifying co-evolution between crossbills and lodgepole pine. American Naturalist 153:S75–S91.

Bent, A. C. 1929. Life histories of North American shorebirds. U.S. Bulletin of the National Museum 146.

———. 1942. Life histories of North American flycatchers, larks, swallows, and their allies. U.S. Bulletin of the National Museum 179.

———. 1946. Life histories of North American crows, jays, and titmice. U.S. Bulletin of the National Museum 191.

———. 1949. Life histories of North American thrushes, kinglets, and their allies. U.S. Bulletin of the National Museum 196.

———. 1950. Life histories of North American wagtails, shrikes, vireos and their allies. U.S. Bulletin of the National Museum 197.

———. 1953. Life histories of North American wood warblers. U.S. Bulletin of the National Museum 203.

Bent, A. C., and O. L. Austin, Jr. 1968. Life histories of North American cardinals, grosbeaks, buntings, towhees, finches, sparrows, and allies. U.S. Bulletin of the National Museum 237.

Berlanga, H., J. A. Kennedy, T. D. Rich, M. C. Arizmendi, C. J. Beardmore, P. J. Blancher, G. S. Butcher, A. R. Coutrier, A. A. Dayer, D. W. Demarest, W. E. Easton, M. Gustafson, E. Inigo-Elias, E. A. Krebs, A. O. Panjabi, V. Rodriguez Contreras, K. V. Rosenberg, J. M. Ruth, E. S. Castellon, R. M. Vidal, and T. Will. 2010. Saving our shared birds: Partners in Flight tri-national vision for landbird conservation. Cornell Lab of Ornithology, Ithaca, NY.

Berner, K. L. 2008. Eastern Bluebird (*Sialia sialis*). Pages 438–439 *in* The Second Atlas of Breeding Birds in New York State (K. J. McGowan and K. Corwin, Eds.). Cornell University Press, Ithaca, NY.

Bevier, L. R., A. F. Poole, and W. Moskoff. 2005. Veery (*Catharus fuscescens*). *In* The Birds of North America Online, no. 142 (A. Poole, Ed.). Cornell Lab of Ornithology, Ithaca, NY.

Bildstein, K. L. 2006. Migrating Raptors of the World: Their Ecology and Conservation. Cornell University Press, Ithaca, NY.

Bildstein, K. L., M. J. Bechard, P. Porras, E. Campo, and C. J. Farmer. 2007. Seasonal abundances and distribution of Black Vultures (*Coragyps atratus*) and Turkey Vultures (*Cathartes aura*) in Costa Rica and Panama: Evidence for reciprocal migration in the Neotropics. Pages 47–60 *in* Neotropical Raptors (K. L. Bildstein, D. R. Barber, and A. Zimmerman, Eds.). Hawk Mountain Sanctuary, Orwigsburg, PA.

Bildstein, K. L., and K. Meyer. 2000. Sharp-shinned Hawk (*Accipiter striatus*). *In* The Birds of North America, no. 482 (A. Poole and F. Gill, Eds.). The Birds of North America, Inc., Philadelphia, PA.

Bishop, J. A. 2008. Temporal dynamics of forest patch size distribution and fragmentation of habitat types in Pennsylvania. Ph.D. dissertation, The Pennsylvania State University, University Park, PA.

Blackwell, B. F., and R. A. Dolbeer. 2001. Decline of the Red-winged Blackbird population in Ohio correlated to changes in agriculture (1965–1996). Journal of Wildlife Management 65:661–667.

Blancher, P. J., K. V. Rosenberg, A. O. Panjabi, B. Altman, J. Bart, C. J. Beardmore, G. S. Butcher, D. Demarest, R. Dettmers, E. H. Dunn, W. Easton, W. C. Hunter, E. E. Inigo-Elias, D. N. Pashley, C. J. Ralph, T. D. Rich, C. M. Rustay, J. M. Ruth, and T. C. Will. 2007. Guide to the Partners in Flight Population Estimates Database, version: North American Landbird Conservation Plan 2004. Partners in Flight Technical Series No. 5.

Blodgett, K. D., and R. M. Zammuto. 1979. Chimney Swift nest found in hollow tree. Wilson Bulletin 91:154.

Blokpoel, H., and G. D. Tessier. 1986. The Ring-billed Gull in Ontario: A review of a new problem species. Canadian Wildlife Service Occasional Papers 57.

———. 1987. Control of Ring-billed Gull colonies at urban and industrial sites in southern Ontario, Canada. Pages 8–17 *in* Proceedings of Eastern Wildlife Damage Control Conference, Gulf Shores, AL. http://digitalcommons.unl.edu/ewdcc3/2.

Boarman, W. I., and B. Heinrich. 1999. Common Raven (*Corvus corax*). *In* The Birds of North America, no. 476 (A. Poole and F. Gill, Eds.). The Birds of North America, Inc., Philadelphia, PA.

Bolgiano, N. C. 1997. Pennsylvania Christmas Bird Counts of Sharp-shinned and Cooper's Hawks. Pennsylvania Birds 11:134–137.

———. 1999. The story of the Ring-necked Pheasant in Pennsylvania. Pennsylvania Birds 13:2–10.

———. 2000. A history of Northern Bobwhites in Pennsylvania. Pennsylvania Birds 14:58–68.

———. 2004. Cause and effect: Changes in boreal bird irruptions in eastern North America relative to the 1970s' spruce budworm invasion. American Birds 58:26–33.

———. 2005. The 2004–2005 Christmas Bird Count in Pennsylvania. Pennsylvania Birds 19:35–51.

———. 2008. Carolina Wrens in Pennsylvania. Pennsylvania Birds 22:86–87.

———. 2010. The Christmas Bird Count. Pages 148–160 in Avian Ecology and Conservation: A Pennsylvania Focus with National Implications (S. K. Majumdar, T. L. Master, M. C. Brittingham, R. M. Ross, R. S. Mulvihill, and J. E. Huffman, Eds.). Pennsylvania Academy of Science, Easton, PA.

Bolgiano, N. C., and G. Grove. 2010. Birds of Central Pennsylvania. Stone Mountain Publishing, State College, PA.

Bollinger, E. K., and T. A. Gavin. 2004. Responses of nesting Bobolinks (*Dolichonyx oryzivorus*) to habitat edges. Auk 121:767–776.

Both, C., C. A. M. van Turnhout, R. G. Bijlsma, H. Siepel, A. J. Van Strien, and R. P. B. Foppen. 2010. Avian population consequences of climate changes are most severe for long-distance migrants in seasonal habitats. Proceedings of the Royal Society B-Biological Sciences 277:1259–1266.

Boulet, M., H. L. Gibbs, and K. A. Hobson. 2006. Integrated analysis of genetic, stable isotope, and banding data reveal migratory connectivity and flyways in the northern Yellow Warbler (*Dendroica petechia; aestiva* group). Ornithological Monographs 61:29–78.

Bourque, J., and M. A. Villard. 2001. Effects of selection cutting and landscape-scale harvesting on the reproductive success of two Neotropical migrant bird species. Conservation Biology 15:184–195.

Boyd, R. C., and M. J. Cegelski. 2008. Game take and furtaker surveys. Wildlife Management Annual Report. Pennsylvania Game Commission, Harrisburg, PA.

Boyle, W. J., Jr., R. O. Paxton, and D. A. Cutler. 1992. Hudson-Delaware Region. American Birds 46:399.

Bradford, D. F., S. E. Franson, G. R. T. Miller, A. C. Neagle, G. E. Canterbury, and D. T. Heggem. 1998. Bird species assemblages as indicators of biotic integrity in Great Basin rangeland. Environmental Monitoring and Assessment 49:1–22.

Brauning, D. 2004. Birds of Note. Pennsylvania Birds 18:174–176.

Brauning, D., and J. Hassinger. 2010. Management and Recovery of Threatened and Endangered Bird Species. Pages 191–203 in Avian Ecology and Conservation: A Pennsylvania Focus with National Implications (S. K. Mujumdar, T. L. Master, M. C. Brittingham, R. M. Ross, R. S. Mulvhill, and J. E. Huffman, Eds.). The Pennsylvania Academy of Science, Easton, PA.

Brauning, D. W., Ed. 1992a. Atlas of Breeding Birds in Pennsylvania. University of Pittsburgh Press, Pittsburgh, PA.

———. 1992b. Ring-necked Pheasant (*Phasianus colchicus*). Pages 112–113 in Atlas of Breeding Birds in Pennsylvania (D. W. Brauning, Ed.). University of Pittsburgh Press, Pittsburgh, PA.

———. 1992c. Peregrine Falcon (*Falco peregrinus*). Pages 110–111 in Atlas of Breeding Birds in Pennsylvania (D. W. Brauning, Ed.). University of Pittsburgh Press, Pittsburgh, PA.

———. 1992d. King Rail (*Rallus elegans*). Pages 122–123 in Atlas of Breeding Birds in Pennsylvania (D. W. Brauning, Ed.). University of Pittsburgh Press, Pittsburgh, PA.

———. 1992e. Spotted Sandpiper (*Actitis macularius*). Pages 136–137 in Atlas of Breeding Birds in Pennsylvania (D. W. Brauning, Ed.). University of Pittsburgh Press, Pittsburgh, PA.

———. 1992f. Common Nighthawk (*Chordeiles minor*). Pages 168–169 in Atlas of Breeding Birds in Pennsylvania (D. W. Brauning, Ed.). University of Pittsburgh Press, Pittsburgh, PA.

———. 1992g. Loggerhead Shrike (*Lanius ludovicianus*). Pages 285–285 in Atlas of Breeding Birds in Pennsylvania (D. W. Brauning, Ed.). University of Pittsburgh Press, Pittsburgh, PA.

———. 1992h. Winter Wren (*Troglodytes troglodytes*). Pages 254–255 in Atlas of Breeding Birds in Pennsylvania (D. W. Brauning, Ed.). University of Pittsburgh Press, Pittsburgh, PA.

———. 1992i. Swainson's Thrush (*Catharus ustulatus*). Pages 268–269 in Atlas of Breeding Birds in Pennsylvania (D. W. Brauning, Ed.). University of Pittsburgh Press, Pittsburgh, PA.

———. 1998. Grassland breeding bird survey, 1997 annual report. Wildlife Management Annual Report. Pennsylvania Game Commission, Harrisburg, PA.

———. 2010a. Least Bittern (*Ixobrychus exilis*). Pages 187–190 in Terrestrial Vertebrates of Concern in Pennsylvania: A Complete Guide to Species of Conservation Concern (M. A. Steele, M. C. Brittingham, T. J. Maret, and J. F. Merritt, Eds.). The Johns Hopkins University Press, Baltimore, MD.

———. 2010b. Short-eared Owl (*Asio flammeus*). Pages 116–119 in Terrestrial Vertebrates of Concern in Pennsylvania: A Complete Guide to Species of Conservation Concern (M. A. Steele, M. C. Brittingham, T. J. Maret, and J. F. Merritt, Eds.). The Johns Hopkins University Press, Baltimore, MD.

Brauning, D. W., M. C. Brittingham, D. A. Gross, R. C. Leberman, T. L. Master, and R. S. Mulvihill. 1994. Pennsylvania breeding birds of special concern: A listing rationale and status update. Journal of the Pennsylvania Academy of Science 68:3–28.

Brauning, D. W., and D. Siefken. 2000. Loggerhead Shrike nesting survey. Wildlife Management Annual Report. Pennsylvania Game Commission, Harrisburg, PA.

———. 2005. Osprey research/management. Wildlife Management Annual Report. Pennsylvania Game Commission, Harrisburg, PA.

———. 2006. Loggerhead Shrike nesting bird population surveys in Pennsylvania. Wildlife Management Annual Report. Pennsylvania Game Commission, Harrisburg, PA.

Brewer, D. 2007a. Winter Wren, Troglodyte mignon (*Troglodytes troglodytes*). Pages 416–417 in Atlas of the Breeding Birds of Ontario, 2001–2005 (M. D. Cadman, D. A. Sutherland, G. G. Beck, D. Lepage, and A. R. Coutrier, Eds.). Bird Studies Canada, Environment Canada, Ontario Field Ornithologists, Ontario Nature, and Ontario Ministry of Natural Resources, Toronto, ON.

———. 2007b. Sedge Wren, Troglodyte à bec court (*Cistothorus platensis*). Pages 418–419 in Atlas of the Breeding Birds of Ontario, 2001–2005 (M. D. Cadman, D. A. Sutherland, G. G. Beck, D. Lepage, and A. R. Coutrier, Eds.). Bird Studies Canada, Environment Canada, Ontario Field Ornithologists,

Ontario Nature, and Ontario Ministry of Natural Resources, Toronto, ON.

Brewer, G. L. 2010. Mute Swan (*Cygnus olor*). Pages 40–41 *in* Second Atlas of the Breeding Birds of Maryland and the District of Columbia (W. G. Ellison, Ed.). The Johns Hopkins University Press, Baltimore, MD.

Brewer, R., G. A. McPeek, and R. J. Adams. 1991. The Atlas of Breeding Birds of Michigan. Michigan State University Press, East Lansing, MI.

Brewster, J. P. 2007. Spatial and temporal variation in the singing rates of two forest songbirds, the Ovenbird and Black-throated Blue Warbler: Implications for aural counts of songbirds. M.S. thesis, North Carolina State University, Raleigh, NC.

Brewster, W. 1906. The Birds of the Cambridge Region of Massachusetts. Memoirs of the Nuttall Ornithological Club 4.

Brinker, D. F. 2010. Least Tern (*Sternula antillarum*). Pages 158–159 *in* Second Atlas of the Breeding Birds of Maryland and the District of Columbia (W. G. Ellison, Ed.). The Johns Hopkins University Press, Baltimore, MD.

Brinker, D. F., and K. M. Dodge. 1993. Breeding biology of Northern Saw-whet Owls in Maryland. Maryland Birdlife 49:5–15.

Brisbin, I. L., Jr., and T. B. Mowbray. 2002. American Coot (*Fulica americana*). *In* The Birds of North America Online, no. 697a (A. Poole, Ed.). Cornell Lab of Ornithology, Ithaca, NY, Philadelphia, PA.

Brittingham, M. C. 2010. Brown Thrasher (*Toxostoma rufum*). Pages 295–298 *in* Terrestrial Vertebrates of Concern in Pennsylvania: A Complete Guide to Species of Conservation Concern (M. A. Steele, M. C. Brittingham, T. J. Maret, and J. F. Merritt, Eds.). The Johns Hopkins University Press, Baltimore, MD.

Brittingham, M. C., and L. Goodrich. 2010. Habitat fragmentation: A threat to Pennsylvania's forest birds. Pages 204–216 *in* Avian Ecology and Conservation: A Pennsylvania Focus with National Implications (S. K. Majumdar, T. L. Master, M. C. Brittingham, R. M. Ross, R. S. Mulvihill, and J. E. Huffman, Eds.). Pennsylvania Academy of Science, Easton, PA.

Brittingham, M. C., and S. A. Temple. 1983. Have cowbirds caused forest songbirds to decline? BioScience 33:31–35.

Brodo, I. M., S. D. Sharnoff, and S. Sharnoff. 2001. Lichens of North America. Yale University Press, New Haven, CT.

Brooks, M., and W. C. Legg. 1942. Swainson's Warbler in Nicholas County, West Virginia. Auk 59:76–86.

Brooks, R. P., and M. J. Croonquist. 1990. Wetland habitat and trophic response guilds for wildlife species in Pennsylvania. Journal of the Pennsylvania Academy of Science 64:93–102.

Brooks, R. T. 2003. Abundance, distribution, trends, and ownership patterns of early successional forests in the northeastern United States. Forest Ecology and Management 185:65–74.

Broun, M. 1941. Migration of Blue Jays. Auk 58:262–263.

Brown, C. R. 1997. Purple Martin (*Progne subis*). *In* The Birds of North America, no. 287 (A. Poole and F. Gill, Eds.). The Academy of Natural Sciences, Philadelphia, PA, and the American Ornithologists' Union, Washington, DC.

Brown, C. R., and M. B. Brown. 1995. Cliff Swallow (*Petrochelidon pyrrhonota*). *In* The Birds of North America, no. 149 (A. Poole and F. Gill, Eds.). The Academy of Natural Sciences, Philadelphia, PA, and the American Ornithologists' Union, Washington, DC.

———. 1999. Barn Swallow (*Hirundo rustica*). *In* The Birds of North America Online, no. 452 (A. Poole, Ed.). Cornell Lab of Ornithology, Ithaca, NY.

Brown, M., and J. J. Dinsmore. 1986. Implications of marsh size and isolation for marsh bird management. Journal of Wildlife Management 50:392–397.

Brown, S., C. Hickey, B. Harrington, and R. Gill, Eds. 2001. North American Shorebird Conservation Plan, 2nd ed. Manomet Center for Conservation Sciences, Manomet, MA.

Bryant, A. A. 1986. Influence of selective logging on Red-shouldered Hawks, *Buteo lineatus*, in Waterloo Region, Ontario, 1953–1978. Canadian Field Naturalist 100:520–525.

Buckelew, A. R., Jr., and G. A. Hall. 1994. The West Virginia Breeding Bird Atlas. University of Pittsburgh Press, Pittsburgh, PA.

Buckley, N. 1999. Black Vulture (*Coragyps atratus*). *In* The Birds of North America, no. 411 (A. Poole and F. Gill, Eds.). The Academy of Natural Sciences, Philadelphia, PA, and the American Ornithologists' Union, Washington, DC.

Buehler, D. A. 2000. Bald Eagle (*Haliaeetus leucocephalus*). *In* The Birds of North America Online, no. 556 (A. Poole, Ed.). Cornell Lab of Ornithology, Ithaca, NY.

Buehler, D. A., A. M. Roth, R. Vallender, T. C. Will, J. L. Confer, R. A. Canterbury, S. B. Swarthout, K. V. Rosenberg, and L. P. Bulluck. 2007. Status and conservation priorities of Golden-winged Warblers (*Vermivora chrysoptera*) in North America. Auk 124:1439–1445.

Buford, E. W., and D. E. Capen. 1999. Abundance and productivity of forest songbirds in a managed, unfragmented landscape in Vermont. Journal of Wildlife Management 63:180–188.

Bull, E. L., and J. A. Jackson. 1995. Pileated Woodpecker (*Dryocopus pileatus*). *In* The Birds of North America, no. 148 (A. Poole and F. Gill, Eds.). The Academy of Natural Sciences, Philadelphia, PA, and the American Ornithologists' Union, Washington, DC.

Burleigh, T. D. 1931. Notes on the breeding birds of State College, Centre County, Pennsylvania. Wilson Bulletin 43:37–54.

———. 1932. Notes on the breeding birds of Fayette County, Pennsylvania. Cardinal 3:73–83,138–147.

Burns, F. L. 1901. A sectional bird census: Taken at Berwyn, Chester County, Pennsylvania, during the seasons of 1899, 1900, and 1901. Wilson Bulletin 8:84–103.

———. 1907. Alexander Wilson in bird census work. Wilson Bulletin 19:100–102.

Butcher, G. S., D. K. Niven, A. O. Panjabi, D. N. Pashley, and K. V. Rosenberg. 2007. WatchList: The 2007 watch list for the United States birds. American Birds 107:18–25.

Butcher, G. S., B. Peterjohn, and C. J. Ralph. 1993. Overview of national bird population monitoring programs and databases. Pages 192–203 *in* Status and management of Neotropical migratory birds. Rocky Mountain Forest and Range Experiment Station, USDA General Technical Report RM-229, Fort Collins, CO.

Butler, R. W. 1992. Great Blue Heron (*Ardea herodias*). *In* The Birds of North America Online, no. 25 (A. Poole, Ed.). Cornell Lab of Ornithology, Ithaca, NY.

Butts, W. K. 1931. A study of the chickadee and White-breasted Nuthatch by means of marked individuals. Part III: The

White-breasted Nuthatch (*Sitta carolinensis cookei*). Bird-Banding 2:59–76.

Bystrak, D. 1980. Application of mini routes to bird population studies. Maryland Birdlife 36:131–138.

Cabe, P. R. 1993. European Starling (*Sturnus vulgaris*). *In* The Birds of North America Online, no. 48 (A. Poole, Ed.). Cornell Lab of Ornithology, Ithaca, NY.

Cadman, M. D. 2007. Purple Martin, Hirondelle noire (*Progne subis*). Pages 388–389 *in* Atlas of the Breeding Birds of Ontario, 2001–2005 (M. D. Cadman, D. A. Sutherland, G. G. Beck, D. Lepage, and A. R. Coutrier, Eds.). Bird Studies Canada, Environment Canada, Ontario Field Ornithologists, Ontario Nature, and Ontario Ministry of Natural Resources, Toronto, ON.

Cadman, M. D., D. A. Sutherland, G. G. Beck, D. Lepage, and A. R. Courtier, Eds. 2007. Atlas of the Breeding Birds of Ontario: 2001–2005. Bird Studies Canada, Environment Canada, Ontario Field Ornithologists, Ontario Nature, and Ontario Ministry of Natural Resources, Toronto, ON.

Caffrey, C., and C. C. Peterson. 2003. Christmas Bird Count data suggest West Nile virus may not be a conservation issue in northeastern United States. American Birds 57:14–21.

Campbell, S. P., J. W. Witham, and M. L. Hunter, Jr. 2007. Long-term effects of group-selection timber harvesting on abundance of forest birds. Conservation Biology 21:1218–1229.

Canterbury, G. E., T. E. Martin, D. R. Petit, L. J. Petit, and D. F. Bradford. 2000. Bird communities and habitat as ecological indicators of forest condition in regional monitoring. Conservation Biology 14:544–558.

Carey, M. 2010. Early successional and shrubland communities with special reference to Field Sparrow (*Spizella pusilla*) breeding biology. Pages 75–85 *in* Avian Ecology and Conservation: A Pennsylvania Focus with National Implications (S. K. Majumdar, T. L. Master, M. C. Brittingham, R. M. Ross, R. S. Mulvihill, and J. E. Huffman, Eds.). The Pennsylvania Academy of Science, Easton, PA.

Carey, M., D. E. Burhans, and D. A. Nelson. 2008. Field Sparrow (*Spizella pusilla*). *In* The Birds of North America Online, no. 103 (A. Poole, Ed.). Cornell Lab of Ornithology, Ithaca, NY.

Carson, R. 1962. Silent Spring. Houghton Mifflin Co., Boston, MA.

Casalena, M. J. 2006. Management plan for Wild Turkeys in Pennsylvania, 2006–2015. Bureau of Wildlife Management, Pennsylvania Game Commission, Harrisburg, PA.

———. 2010a. Wild Turkey trap and transfer. Wildlife Management Annual Report. Pennsylvania Game Commission, Harrisburg, PA.

———. 2010b. Wild Turkey productivity and harvest trends. Wildlife Management Annual Report. Pennsylvania Game Commission, Harrisburg, PA.

Case, D. J., and D. D. McCool. 2009. Priority Information Needs for Rails and Snipe: A Funding Strategy. Association of Fish and Wildlife Agencies' Migratory Shore and Upland Game Bird Support Task Force, Mishawaka, IN.

Cashen, S. T. 1998. Avian use of restored wetlands in Pennsylvania. M.S. thesis, The Pennsylvania State University, University Park, PA.

Cassidy, J. 1990. Book of North American Birds. The Reader's Digest Association, Inc., Pleasantville, NY.

Castrale, J. S. 1985. Responses of wildlife to various tillage conditions. Transactions of the North American Wildlife and Natural Resources Conference 50:142–149.

Cavitt, J. F., and C. A. Haas. 2000. Brown Thrasher (*Toxostoma rufum*). *In* The Birds of North America Online, no. 557 (A. Poole, Ed.). Cornell Lab of Ornithology, Ithaca, NY.

Chesser, P. T., R. C. Banks, F. K. Barker, C. Cicero, J. L. Dunn, A. W. Kratter, I. J. Lovette, P. C. Rasmussen, J. V. Remsen, Jr., J. D. Rising, D. F. Stotz, and K. Winker. 2010. Fifty-first supplement to the American Ornithologists' Union Checklist of North American Birds. Auk 127:726–744.

———. 2011. Fifty-second supplement to the American Ornithologists' Union Checklist of North American Birds. Auk 128:600–613.

Ciaranca, M. A., C. C. Allin, and G. S. Jones. 1997. Mute Swan (*Cygnus olor*). *In* The Birds of North America Online, no. 273 (A. Poole, Ed.). Cornell Lab of Ornithology, Ithaca, NY.

Cimprich, D. A., and F. R. Moore. 1995. Gray Catbird (*Dumetella carolinensis*). *In* The Birds of North America Online, no. 167 (A. Poole, Ed.). Cornell Lab of Ornithology, Ithaca, NY.

Cink, C. L. 2002. Whip-poor-will (*Caprimulgus vociferus*). *In* The Birds of North America, no. 620 (A. Poole and F. Gill, Eds.). The Birds of North America, Inc., Philadelphia, PA.

Cink, C. L., and C. T. Collins. 2002. Chimney Swift (*Chaetura pelagica*). *In* The Birds of North America, no. 646 (A. Poole and F. Gill, Eds.). The Birds of North America, Inc., Philadelphia, PA.

Clarke, K. E., and L. J. Niles. 2000. Northern Atlantic Regional Shorebird Plan, version 1.0. New Jersey Division of Fish and Wildlife, Woodbine, NJ.

CLO (Cornell Lab of Ornithology). 2011. eBird website. [Online.] Available at http://ebird.org/content/ebird/.

Coleman, J. L., D. Bird, and E. A. Jacobs. 2002. Habitat use and productivity of Sharp-shinned Hawks nesting in an urban area. Wilson Bulletin 114:467–473.

Collins, J. E. 2008a. Nashville Warbler (*Vermivora ruficapilla*). Pages 476–477 *in* The Second Atlas of Breeding Birds in New York State (K. J. McGowan and K. Corwin, Eds.). Cornell University Press, Ithaca, NY.

———. 2008b. Mourning Warbler (*Oporornis philadelphia*). Pages 526–527 *in* The Second Atlas of Breeding Birds in New York State (K. J. McGowan and K. Corwin, Eds.). Cornell University Press, Ithaca, NY.

———. 2008c. Blackburnian Warbler (*Dendroica fusca*). Pages 494–495 *in* The Second Atlas of Breeding Birds in New York State (K. J. McGowan and K. Corwin, Eds.). Cornell University Press, Ithaca, NY.

Colvin, B. A. 1985. Common Barn Owl population decline in Ohio and the relationship to agricultural trends. Journal of Field Ornithology 56:224–235.

Conant, S. 1994. First confirmed Evening Grosbeak nest in Pennsylvania. Pennsylvania Birds 8:133–135.

Confer, J. 2008. Blue-winged Warbler (*Vermivora pinus*). Pages 466–467 *in* The Second Atlas of Breeding Birds in New York State (K. J. McGowan and K. Corwin, Eds.). Cornell University Press, Ithaca, NY.

Confer, J. L., P. E. Allen, and J. L. Larkin. 2003. Effects of vegetation, interspecific competition, and brood parasitism on Golden-winged Warbler nesting success. Auk 121:138–144.

Confer, J. L., P. Hartman, and A. Roth. 2011. Golden-winged Warbler (*Vermivora chrysoptera*). *In* The Birds of North America Online, no. 20 (A. Poole, Ed.). Cornell Lab of Ornithology, Ithaca, NY.

Conklin, W. G. 1938. The Pymatuning State Game Refuge and Museum. Bulletin 19. Pennsylvania Game Commission, Harrisburg, PA.

Conover, M. R., and G. G. Chasko. 1985. Nuisance Canada Goose problems in the eastern United States. Wildlife Society Bulletin 13:228–233.

Conway, C. J. 1995. Virginia Rail (*Rallus limicola*). *In* The Birds of North America Online, no. 173 (A. Poole, Ed.). Cornell Lab of Ornithology, Ithaca, NY.

———. 1999. Canada Warbler (*Wilsonia canadensis*). *In* The Birds of North America, no. 421 (A. Poole and F. Gill, Eds.). The Birds of North America, Inc., Philadelphia, PA.

———. 2011. Standardized North American marsh bird monitoring protocol. Waterbirds 34:319–346.

Conway, C. J., W. R. Eddleman, and S. H. Anderson. 1994. Nesting success and survival of Virginia Rail and Sora. Wilson Bulletin 106:466–473.

Conway, C. J., and J. P. Gibbs. 2005. Effectiveness of call broadcasts surveys for monitoring marsh birds. Auk 122:26–35.

Cooper, C. B., W. M. Hochachka, and A. A. Dhondt. 2007. Contrasting natural experiments confirm competition between House Finches and House Sparrows. Ecology 88:864–870.

Cooper, E. L., and C. C. Wagner. 1973. The effects of acid mine drainage on fish populations. Environmental Protection Agency, Washington, DC.

Cooper, T. R., Ed. 2007. Henslow's Sparrow Conservation Action Plan Workshop Summary, Bloomington, MN.

Cooper, T. R., and K. Parker. 2010. American Woodcock population status. U.S. Fish and Wildlife Service, Laurel, MD.

Cope, F. R., Jr. 1902. Observations on the summer birds of parts of Clinton and Potter counties, Pennsylvania. Cassinia 5:8–21.

Cope, T. M. 1936. Observations of the vertebrate ecology of some Pennsylvania virgin forests. Ph.D. dissertation, Cornell University, Ithaca, NY.

Cornell, K. L., and T. M. Donovan. 2010. Effects of spatial habitat heterogeneity on habitat selection and annual fecundity for a migratory forest songbird. Landscape Ecology 25:109–122.

Corsi, F., E. Dupre, and L. Boitani. 1999. A large-scale model of wolf distribution in Italy for conservation planning. Conservation Biology 13:150–159.

Coskren, T. D. 2010a. Eastern Bluebird (*Sialia sialis*). Pages 296–297 *in* Second Atlas of the Breeding Birds of Maryland and the District of Columbia (W. G. Ellison, Ed.). The Johns Hopkins University Press, Baltimore, MD.

———. 2010b. Ovenbird (*Seiurus aurocapilla*). Pages 356–357 *in* Second Atlas of the Breeding Birds of Maryland and the District of Columbia (W. G. Ellison, Ed.). The Johns Hopkins University Press, Baltimore, MD.

———. 2010c. Prairie Warbler (*Dendroica discolor*). Pages 342–343 *in* Second Atlas of the Breeding Birds of Maryland and the District of Columbia (W. G. Ellison, Ed.). The Johns Hopkins University Press, Baltimore, MD.

———. 2010d. Canada Warbler (*Wilsonia canadensis*). Pages 370–371 *in* Second Atlas of the Breeding Birds of Maryland and the District of Columbia (W. G. Ellison, Ed.). The Johns Hopkins University Press, Baltimore, MD.

———. 2010e. Summer Tanager (*Piranga rubra*). Pages 398–399 *in* Second Atlas of the Breeding Birds of Maryland and the District of Columbia (W. G. Ellison, Ed.). The Johns Hopkins University Press, Baltimore, MD.

———. 2010f. Indigo Bunting (*Passerina cyanea*). Pages 408–409 *in* Second Atlas of the Breeding Birds of Maryland and the District of Columbia (W. G. Ellison, Ed.). The Johns Hopkins University Press, Baltimore, MD.

Coutlee, E. L. 1964. Maintenance behavior of the American Goldfinch. Wilson Bulletin 75:342–357.

Craig, R. J. 2004. Marsh birds of the Connecticut River estuary: A 20-year perspective. Bird Conservation Research Inc., Putnam, CT.

Crawford, H. S., R. G. Hooper, and R. W. Titterington. 1981. Song bird population response to silvicultural practices in central Appalachian hardwoods. Journal of Wildlife Management 45:680–692.

Creaser, C. W. 1925. The egg-destroying activity of the House Wren in relation to territorial control. Bird-Lore 27:163–167.

Crewe, T. L., T. A. Timmermans, and K. E. Jones. 2005. The marsh monitoring program annual report, 1995–2003: Annual indices and trends in bird abundance and amphibian occurrence in the Great Lakes Basin. Bird Studies Canada, Port Rowan, ON.

Crins, W. J. 2007. Yellow-bellied Flycatcher, Moucherolle à ventre jaune (*Empidonax flaviventris*). Pages 342–343 *in* Atlas of the Breeding Birds of Ontario, 2001–2005 (M. D. Cadman, D. A. Sutherland, G. G. Beck, D. Lepage, and A. R. Coutrier, Eds.). Bird Studies Canada, Environment Canada, Ontario Field Ornithologists, Ontario Ministry of Natural Resources, and Ontario Nature, Toronto, ON.

Criswell, R. W. 2010. King Rail (*Rallus elegans*). Pages 207–209 *in* Terrestrial Vertebrates of Concern in Pennsylvania: A Complete Guide to Species of Conservation Concern (M. A. Steele, M. C. Brittingham, T. J. Maret, and J. F. Merritt, Eds.). The Johns Hopkins University Press, Baltimore, MD.

Crocoll, S. 2008a. Northern Goshawk (*Accipiter gentilis*). Pages 196–197 *in* The Second Atlas of Breeding Birds in New York State (K. J. McGowan and K. Corwin, Eds.). Cornell University Press, Ithaca, NY.

———. 2008b. Broad-winged Hawk (*Buteo platypterus*). Pages 200–201 *in* The Second Atlas of Breeding Birds in New York State (K. J. McGowan and K. Corwin, Eds.). Cornell University Press, Ithaca, NY.

Crocoll, S. T. 1994. Red-shouldered Hawk (*Buteo lineatus*). *In* The Birds of North America, no. 107 (A. Poole and F. Gill, Eds.). The Academy of Natural Sciences, Philadelphia, PA, and the American Ornithologists' Union, Washington, DC.

Crossley, G. J., compiler. 1999. A guide to critical bird habitat in Pennsylvania: Pennsylvania Important Bird Areas program. Pennsylvania Audubon Society, Harrisburg, PA.

Cuff, D. J., R. F. Abler, W. Zelinsky, E. K. Muller, and W. J. Young. 1989. Atlas of Pennsylvania. Temple University Press, Philadelphia, PA.

Curry, R. L., L. M. Rossano, and M. W. Reudink. 2007. Behavioral aspects of chickadee hybridization. Pages 95–110 *in* Ecology and Behavior of Chickadees and Titmice: An Integrated Approach (K. Otter, Ed.). Oxford University Press, Oxford, UK.

Curson, J., D. Quinn, and D. Beadle. 1994. Warblers of the Americas. Houghton Mifflin Company, New York, NY.

Curtis, O. E., R. N. Rosenfield, and J. Bielefeldt. 2006. Cooper's Hawk (*Accipiter cooperii*). *In* The Birds of North America Online, no. 75 (A. Poole, Ed.). Cornell Lab of Ornithology, Ithaca, NY.

Cuthbert, F. J., and E. A. Roche. 2009. Piping Plover breeding biology and management in the Great Lakes, 2009. U.S. Fish and Wildlife Service, East Lansing, MI.

Dando, W. 1996. Reconstruction of presettlement forests of northeastern Pennsylvania using original land survey records. M.S. thesis, The Pennsylvania State University, University Park, PA.

Darley-Hill, S., and W. C. Johnson. 1981. Acorn dispersal by the Blue Jay (*Cyanocitta cristata*). Oecologia 50:231–232.

David, N., M. Gosselin, and G. Seutin. 1990. Pattern of colonization by the Northern Mockingbird in Quebec. Journal of Field Ornithology 61:1–8.

Davidson, L. M. 2010a. Northern Rough-winged Swallow (*Stelgidopteryx serripennis*). Pages 260–261 *in* Second Atlas of the Breeding Birds of Maryland and the District of Columbia (W. G. Ellison, Ed.). The Johns Hopkins University Press, Baltimore, MD.

———. 2010b. Loggerhead Shrike (*Lanius ludovicianus*). Pages 234–235 *in* Second Atlas of the Breeding Birds of Maryland and the District of Columbia (W. G. Ellison, Ed.). The Johns Hopkins University Press, Baltimore, MD.

Davis, A. F., J. A. Lundgren, B. Barton, J. R. Belfonti, J. L. Farber, J. R. Kunsman, and A. M. Wilkinson. 1995. A Natural Areas Inventory of Wyoming County, Pennsylvania. Pennsylvania Science Office of the Nature Conservancy, Middletown, PA.

Davis, M. B. 2008. Old growth in the East: A survey. [Online.] Available at http://www.primalnature.org/ogeast/survey.html.

Davis, S. K. 2003. Nesting ecology of mixed-grass prairie songbirds in southern Saskatchewan. Wilson Bulletin 115:119–130.

Davis, S. K., and W. E. Lanyon. 2008. Western Meadowlark (*Sturnella neglecta*). *In* The Birds of North America Online, no. 104 (A. Poole, Ed.). Cornell Lab of Ornithology, Ithaca, NY.

Davis, W. E., Jr., and J. A. Kushlan. 1994. Green Heron (*Butorides virescens*). *In* The Birds of North America, no. 129 (A. Poole and F. Gill, Eds.). The Academy of Natural Sciences, Philadelphia, PA, and the American Ornithologists' Union, Washington, DC.

Dawson, D. K. 2010. Willow Flycatcher (*Empidonax traillii*). Pages 224–225 *in* Second Atlas of the Breeding Birds of Maryland and the District of Columbia (W. G. Ellison, Ed.). The Johns Hopkins University Press, Baltimore, MD.

Dawson, T. P., S. T. Jackson, J. I. House, I. C. Prentice and G. M. Mace. 2011. Beyond predictions: Biodiversity conservation in a changing climate. Science 332:53–58.

Dawson, W. 1997. Pine Siskin (*Carduelis pinus*). *In* The Birds of North America, no. 280 (A. Poole and F. Gill, Eds.). The Academy of Natural Sciences, Philadelphia, PA, and the American Ornithologists' Union, Washington, DC.

De La Zerda-Lerner, S., and D. F. Stauffer. 1998. Habitat selection by Blackburnian Warblers wintering in Colombia. Journal of Field Ornithology 69:457–465.

de Szalay, F., D. Helmers, D. Humburg, S. J. Lewis, B. Pardo, and M. Shieldcastle. 2000. Upper Mississippi Valley/Great Lakes Regional Shorebird Conservation Plan, version 1.0. U.S. Fish and Wildlife Service, Bloomington, MN.

Dearborn, D. C. 2010. Yellow-breasted Chat (*Icteria virens*). Pages 313–316 *in* Terrestrial Vertebrates of Concern in Pennsylvania: A Complete Guide to Species of Conservation Concern (M. A. Steele, M. C. Brittingham, T. J. Maret, and J. F. Merritt, Eds.). The Johns Hopkins University Press, Baltimore, MD.

deCalesta, D. S. 1994. Effect of white-tailed deer on songbirds within managed forests in Pennsylvania. Journal of Wildlife Management 58:711–718.

Dechant, J. E., M. F. Dinkins, D. H. Johnson, L. D. Igl, C. M. Goldade, and B. R. Euliss. 1999. Effects of Management Practices on Grassland Birds: Upland Sandpiper. Northern Prairie Wildlife Research Center, Jamestown, ND.

———. 2003. Effects of Management Practices on Grassland Birds: Vesper Sparrow. Northern Prairie Wildlife Research Center, Jamestown, ND.

Dechant, J. E., M. L. Sondreal, D. H. Johnson, L. D. Igl, C. M. Goldade, B. D. Parkin, and B. R. Euliss. 1999 (rev. 2002). Effects of Management Practices on Grassland Birds: Field Sparrow. Northern Prairie Wildlife Research Center, Jamestown, ND.

DeGraaf, R. M., and J. H. Rappole. 1995. Neotropical Migratory Birds: Natural History, Distribution, and Population Change. Comstock Publishing Associates, Ithaca, NY.

DeJong, M. J. 1996. Northern Rough-winged Swallow (*Stelgidopteryx serripennis*). *In* The Birds of North America, no. 234 (A. Poole and F. Gill, Eds.). The Academy of Natural Sciences, Philadelphia, PA, and the American Ornithologists' Union, Washington, DC.

DeLorme. 2003. The DeLorme Pennsylvania Atlas and Gazetteer, 8th ed. Delorme, Yarmouth, VT.

Dennis, J. V. 1958. Some aspects of the breeding ecology of the Yellow-breasted Chat (*Icteria virens*). Bird-Banding 29:169–183.

Derrickson, K. C., and R. Breitwisch. 1992. Northern Mockingbird (*Mimus polyglottos*). *In* The Birds of North America, no. 7 (A. Poole, P. Stettenheim, and F. Gill, Eds.). The Academy of Natural Sciences, Philadelphia, PA, and the American Ornithologists' Union, Washington, DC.

Dessecker, D. R., G. W. Norman, and S. J. Williamson, Eds. 2006. Ruffed Grouse Conservation Plan. Association of Fish and Wildlife Agencies, Washington, DC.

Dessecker, D. R., and R. H. Yahner. 1987. Breeding-bird communities associated with Pennsylvania northern hardwood clearcut stands. Proceedings of the Pennsylvania Academy of Science 61:170–173.

Desta, F., J. J. Colbert, J. S. Rentch, and K. W. Gottschalk. 2004. Aspect induced differences in vegetation, soil, and microclimatic characteristics of an Appalachian watershed. Castanea 69:92–108.

Detwiler, D. 2008. Habitat use, foraging behavior and competitive interactions of Black-crowned Night-Herons (*Nycticorax nycticorax*) on the Susquehanna River at Wade Island. M.S. thesis, East Stroudsburg University of Pennsylvania, East Stroudsburg, PA.

Dickerman, R. W. 1987. The "old northeastern" subspecies of Red Crossbill. American Birds 41:188–194.

Diefenbach, D. R., M. R. Marshall, J. A. Mattice, and D. W. Brauning. 2007. Incorporating availability for detection in estimates of bird abundance. Auk 124:96–106.

Dill, H. H. 1970. About people and Canada Geese. Pages 3–6 *in* Home Grown Honkers (H. H. Hill and F. B. Lee, Eds.). U.S. Fish and Wildlife Service, Washington, DC.

Dingle, E. V. S. 1942. Rough-winged Swallow. *In* Life Histories of North American Flycatchers, Larks, Swallows, and Their Allies (A. C. Bent, Ed.). Smithsonian Institution, Washington, DC.

Doherty, P. F., Jr., and T. C. Grubb, Jr. 2002a. Nest usurpation is an "edge effect" for Carolina Chickadees (*Poecile carolinensis*). Journal of Avian Biology 33:77–82.

———. 2002b. Survivorship of permanent resident birds in a fragmented forested landscape. Ecology 83:844–857.

Droege, S. 1990. The North American Breeding Bird Survey. Pages 1–4 *in* Survey Designs and Statistical Methods for the Estimation of Avian Population Trends (J. R. Sauer and S. Droege, Eds.). U.S. Fish and Wildlife Service, Washington, DC.

Dunn, E., and G. A. Hall. 2010. Magnolia Warbler (*Setophaga magnolia*). *In* The Birds of North America Online, no. 136 (A. Poole, Ed.). Cornell Lab of Ornithology, Ithaca, NY.

Dunn, J., and K. Garrett. 1997. A Field Guide to the Warblers of North America. Houghton Mifflin Co., New York, NY.

Dunn, J. P. 1992. Pennsylvania's Canada Geese: Giant success or giant dilemma? Pennsylvania Game News 63:16–21.

Dunn, J. P., and K. J. Jacobs. 2000. Special resident Canada Goose hunting seasons in Pennsylvania—management implications for controlling resident Canada Geese *in* Proceedings of the Ninth Wildlife Damage Management Conference (M. C. Brittingham, Ed.), State College, PA.

Dunn, J. P., T. Librandi Mumma, and C. F. Riegner. 2008. Pennsylvania game bird production and release survey. Bureau of Wildlife Management, Pennsylvania Game Commission. Harrisburg, PA.

Dunstan, F. 1985. Definitions of status categories. Page 35 *in* Species of Special Concern in Pennsylvania, Special Publication No. 11 (H. H. Genoways and F. J. Brenner, Eds.). Carnegie Museum of Natural History, Pittsburgh, PA.

Dwight, J., Jr. 1892. Summer birds of the crest of the Pennsylvania Alleghenies. Auk 9:129–141.

Eaton, S. W. 1992. Wild Turkey (*Meleagris gallopavo*). *In* The Birds of North America Online, no. 22 (A. Poole, Ed.). Cornell Lab of Ornithology, Ithaca, NY.

———. 1995. Northern Waterthrush (*Seiurus noveboracensis*). *In* The Birds of North America, no. 182 (A. Poole and F. Gill, Eds.). The Academy of Natural Sciences, Philadelphia, PA, and the American Ornithologists' Union, Washington, DC.

Eberhardt, L. S. 2000. Use and selection of sap trees by Yellow-bellied Sapsuckers. Auk 117:41–51.

Eckerle, K. P., and C. F. Thompson. 2001. Yellow-breasted Chat (*Icteria virens*). *In* The Birds of North America Online, no. 575 (A. Poole, Ed.). Cornell Lab of Ornithology, Ithaca, NY.

Eddleman, W. R., F. L. Knopf, B. Meanly, F. A. Reid, and R. Zembel. 1988. Conservation of North American rallids. Wilson Bulletin 100:458–475.

Eisenmann, E. 1962. On the genus "Chamaethlypis" and its supposed relationship to Icteria. Auk 79:265–267.

Elchuck, C. L., and K. L. Wiebe. 2002. Food and predation risk as factors related to foraging locations of Northern Flickers. Wilson Bulletin 114:349–357.

Elliot, J. J., and R. S. Arbib, Jr. 1953. Origin and status of the House Finch in the eastern United States. Auk 70:31–37.

Ellison, W. G. 1992. Blue-gray Gnatcatcher (*Polioptila caerulea*). *In* The Birds of North America Online, no. 23 (A. Poole, Ed.). Cornell Lab of Ornithology, Ithaca, NY.

———, Ed. 2010a. Second Atlas of the Breeding Birds of Maryland and the District of Columbia. The Johns Hopkins University Press, Baltimore, MD.

———. 2010b. Northern Bobwhite (*Colinus virginianus*). Pages 58–59 *in* Second Atlas of the Breeding Birds of Maryland and the District of Columbia (E. W. G. Ellison, Ed.). The Johns Hopkins University Press, Baltimore, MD.

———. 2010c. Ring-necked Pheasant (*Phasianus colchicus*). Pages 60–61 *in* Second Atlas of the Breeding Birds of Maryland and the District of Columbia (W. G. Ellison, Ed.). The Johns Hopkins University Press, Baltimore, MD.

———. 2010d. Ruffed Grouse (*Bonasa umbellus*). Pages 62–63 *in* Second Atlas of the Breeding Birds of Maryland and the District of Columbia (W. G. Ellison, Ed.). The Johns Hopkins University Press, Baltimore, MD.

———. 2010e. Pied-billed Grebe (*Podilymbus podiceps*). Pages 66–67 *in* Second Atlas of the Breeding Birds of Maryland and the District of Columbia (W. G. Ellison, Ed.). The Johns Hopkins University Press, Baltimore, MD.

———. 2010f. Black Vulture (*Coragyps atratus*). Pages 96–97 *in* Second Atlas of the Breeding Birds of Maryland and the District of Columbia (W. G. Ellison, Ed.). The Johns Hopkins University Press, Baltimore, MD.

———. 2010g. Northern Harrier (*Circus cyaneus*). Pages 104–105 *in* Second Atlas of the Breeding Birds of Maryland and the District of Columbia (W. G. Ellison, Ed.). The Johns Hopkins University Press, Baltimore, MD.

———. 2010h. Broad-winged Hawk (*Buteo platypterus*). Pages 114–115 *in* Second Atlas of the Breeding Birds of Maryland and the District of Columbia (W. G. Ellison, Ed.). The Johns Hopkins University Press, Baltimore, MD.

———. 2010i. American Woodcock (*Scolopax minor*). Pages 150–151 *in* Second Atlas of the Breeding Birds of Maryland and the District of Columbia (W. G. Ellison, Ed.). The Johns Hopkins University Press, Baltimore, MD.

———. 2010j. Eurasian Collared-Dove (*Streptopelia decaocto*). Page 440 *in* Second Atlas of the Breeding Birds of Maryland and the District of Columbia (W. G. Ellison, Ed.). The Johns Hopkins University Press, Baltimore, MD.

———. 2010k. Black-billed Cuckoo (*Coccyzus erythropthalmus*). Pages 176–177 *in* Second Atlas of the Breeding Birds of Maryland and the District of Columbia (W. G. Ellison, Ed.). The Johns Hopkins University Press, Baltimore, MD.

———. 2010l. Barn Owl (*Tyto alba*). Pages 178–179 *in* Second Atlas of the Breeding Birds of Maryland and the District of Columbia (W. G. Ellison, Ed.). The Johns Hopkins University Press, Baltimore, MD.

———. 2010m. Eastern Screech-Owl (*Megascops asio*). Pages 180–181 *in* Second Atlas of the Breeding Birds of Maryland and the District of Columbia (W. G. Ellison, Ed.). The Johns Hopkins University Press, Baltimore, MD.

———. 2010n. Barred Owl (*Strix varia*). Pages 184–185 *in* Second Atlas of the Breeding Birds of Maryland and the District of Columbia (W. G. Ellison, Ed.). The Johns Hopkins University Press, Baltimore, MD.

———. 2010o. Common Nighthawk (*Chordeiles minor*). Pages 192–193 *in* Second Atlas of the Breeding Birds of Maryland and the District of Columbia (W. G. Ellison, Ed.). The Johns Hopkins University Press, Baltimore, MD.

———. 2010p. Chuck-will's-widow (*Caprimulgus carolinensis*). Pages 194–195 *in* Second Atlas of the Breeding Birds of

Maryland and the District of Columbia (W. G. Ellison, Ed.). The Johns Hopkins University Press, Baltimore, MD.

———. 2010q. Eastern Whip-poor-will (*Caprimulgus vociferus*). Pages 196–197 *in* Second Atlas of the Breeding Birds of Maryland and the District of Columbia (W. G. Ellison, Ed.). The Johns Hopkins University Press, Baltimore, MD.

———. 2010r. Ruby-throated Hummingbird (*Archilocus colubris*). Pages 200–201 *in* Second Atlas of the Breeding Birds of Maryland and the District of Columbia (W. G. Ellison, Ed.). The Johns Hopkins University Press, Baltimore, MD.

———. 2010s. Belted Kingfisher (*Megaceryle alcyon*). Pages 202–203 *in* Second Atlas of the Breeding Birds of Maryland and the District of Columbia (W. G. Ellison, Ed.). The Johns Hopkins University Press, Baltimore, MD.

———. 2010t. Least Flycatcher (*Empidonax minimus*). Pages 226–227 *in* Second Atlas of the Breeding Birds of Maryland and the District of Columbia (W. G. Ellison, Ed.). The Johns Hopkins University Press, Baltimore, MD.

———. 2010u. Great Crested Flycatcher (*Myiarchus crinitus*). Pages 230–231 *in* Second Atlas of the Breeding Birds of Maryland and the District of Columbia (W. G. Ellison, Ed.). The Johns Hopkins University Press, Baltimore, MD.

———. 2010v. Eastern Kingbird (*Tyrannus tyrannus*). Pages 232–233 *in* Second Atlas of the Breeding Birds of Maryland and the District of Columbia (W. G. Ellison, Ed.). The Johns Hopkins University Press, Baltimore, MD.

———. 2010w. Blue-headed Vireo (*Vireo solitarius*). Pages 240–241 *in* Second Atlas of the Breeding Birds of Maryland and the District of Columbia (W. G. Ellison, Ed.). The Johns Hopkins University Press, Baltimore, MD.

———. 2010x. Warbling Vireo (*Vireo gilvus*). Pages 242–243 *in* Second Atlas of the Breeding Birds of Maryland and the District of Columbia (W. G. Ellison, Ed.). The Johns Hopkins University Pres, Baltimore, MD.

———. 2010y. Red-eyed Vireo (*Vireo olivaceus*). Pages 244–245 *in* Second Atlas of the Breeding Birds of Maryland and the District of Columbia (W. G. Ellison, Ed.). The Johns Hopkins University Press, Baltimore, MD.

———. 2010z. Fish Crow (*Corvus ossifragus*). Pages 250–152 *in* Second Atlas of the Breeding Birds of Maryland and the District of Columbia (W. G. Ellison, Ed.). The Johns Hopkins University Press, Baltimore, MD.

———. 2010za. Horned Lark (*Eremophila alpestris*). Pages 254–255 *in* Second Atlas of the Breeding Birds of Maryland and the District of Columbia (W. G. Ellison, Ed.). The Johns Hopkins University Press, Baltimore, MD.

———. 2010zb. Tree Swallow (*Tachycineta bicolor*). Pages 258–259 *in* Second Atlas of the Breeding Birds of Maryland and the District of Columbia (W. G. Ellison, Ed.). The Johns Hopkins University Press, Baltimore, MD.

———. 2010zc. Bank Swallow (*Riparia riparia*). Pages 262–263 *in* Second Atlas of the Breeding Birds of Maryland and the District of Columbia (W. G. Ellison, Ed.). The Johns Hopkins University Press, Baltimore, MD.

———. 2010zd. White-breasted Nuthatch (*Sitta carolinensis*). Pages 276–277 *in* Second Atlas of the Breeding Birds of Maryland and the District of Columbia (W. G. Ellison, Ed.). The Johns Hopkins University Press, Baltimore, MD.

———. 2010ze. Brown Creeper (*Certhia americana*). Pages 280–281 *in* Second Atlas of the Breeding Birds of Maryland and the District of Columbia (W. G. Ellison, Ed.). The Johns Hopkins University Press, Baltimore, MD.

———. 2010zf. Veery (*Catharus fuscescens*). Pages 298–299 *in* Second Atlas of the Breeding Birds of Maryland and the District of Columbia (W. G. Ellison, Ed.). The Johns Hopkins University Press, Baltimore, MD.

———. 2010zg. Wood Thrush (*Hylocichla mustelina*). Pages 302–303 *in* Second Atlas of the Breeding Birds of Maryland and the District of Columbia (W. G. Ellison, Ed.). The Johns Hopkins University Press, Baltimore, MD.

———. 2010zh. Brown Thrasher (*Toxostoma rufum*). Pages 310–311 *in* Second Atlas of the Breeding Birds of Maryland and the District of Columbia (W. G. Ellison, Ed.). The Johns Hopkins University Press, Baltimore, MD.

———. 2010zi. Cedar Waxwing (*Bombycilla cedrorum*). Pages 314–315 *in* Second Atlas of the Breeding Birds of Maryland and the District of Columbia (W. G. Ellison, Ed.). The Johns Hopkins University Press, Baltimore, MD.

———. 2010zj. Worm-eating Warbler (*Helmitheros vermivorum*). Pages 352–353 *in* Second Atlas of the Breeding Birds of Maryland and the District of Columbia (W. G. Ellison, Ed.). The Johns Hopkins University Press, Baltimore, MD.

———. 2010zk. Blue-winged Warbler (*Vermivora cyanoptera*). Pages 316–317 *in* Second Atlas of the Breeding Birds of Maryland and the District of Columbia (W. G. Ellison, Ed.). The Johns Hopkins University Press, Baltimore, MD.

———. 2010zl. Black-and-white Warbler (*Mniotilta varia*). Pages 346–347 *in* Second Atlas of the Breeding Birds of Maryland and the District of Columbia (W. G. Ellison, Ed.). The Johns Hopkins University Press, Baltimore, MD.

———. 2010zm. American Redstart (*Setophaga ruticilla*). Pages 348–349 *in* Second Atlas of the Breeding Birds of Maryland and the District of Columbia (W. G. Ellison, Ed.). The Johns Hopkins University Press, Baltimore, MD.

———. 2010zn. Yellow-rumped Warbler (*Dendroica coronata*). Pages 332–333 *in* Second Atlas of the Breeding Birds of Maryland and the District of Columbia (W. G. Ellison, Ed.). The Johns Hopkins University Press, Baltimore, MD.

———. 2010zo. Yellow-throated Warbler (*Dendroica dominica*). Pages 338–339 *in* Second Atlas of the Breeding Birds of Maryland and the District of Columbia (W. G. Ellison, Ed.). The Johns Hopkins University Press, Baltimore, MD.

———. 2010zp. Yellow-breasted Chat (*Icteria virens*). Pages 372–273 *in* Second Atlas of the Breeding Birds of Maryland and the District of Columbia (W. G. Ellison, Ed.). The Johns Hopkins University Press, Baltimore, MD.

———. 2010zq. Vesper Sparrow (*Pooecetes gramineus*). Pages 380–381 *in* Second Atlas of the Breeding Birds of Maryland and the District of Columbia (W. G. Ellison, Ed.). The Johns Hopkins University Press, Baltimore, MD.

———. 2010zr. Swamp Sparrow (*Melospiza georgiana*). Pages 394–395 *in* Second Atlas of the Breeding Birds of Maryland and the District of Columbia (W. G. Ellison, Ed.). The Johns Hopkins University Press, Baltimore, MD.

———. 2010zs. Scarlet Tanager (*Piranga olivacea*). Pages 400–401 *in* Second Atlas of the Breeding Birds of Maryland and the District of Columbia (W. G. Ellison, Ed.). The Johns Hopkins University Press, Baltimore, MD.

———. 2010zt. Blue Grosbeak (*Passerina caerulea*). Pages 406–407 *in* Second Atlas of the Breeding Birds of Maryland and the District of Columbia (W. G. Ellison, Ed.). The Johns Hopkins University Press, Baltimore, MD.

———. 2010zu. Dickcissel (*Spiza americana*). Pages 410–411 *in* Second Atlas of the Breeding Birds of Maryland and the District of Columbia (W. G. Ellison, Ed.). The Johns Hopkins University Press, Baltimore, MD.

———. 2010zv. Eastern Meadowlark (*Sturnella magna*). Pages 416–417 *in* Second Atlas of the Breeding Birds of Maryland and the District of Columbia (W. G. Ellison, Ed.). The Johns Hopkins University Press, Baltimore, MD.

———. 2010zw. Orchard Oriole (*Icterus spurius*). Pages 424–425 *in* Second Atlas of the Breeding Birds of Maryland and the District of Columbia (W. G. Ellison, Ed.). The Johns Hopkins University Press, Baltimore, MD.

Erskine, A. J. 1977. Birds in boreal Canada: Communities, densities, and adaptations. Canadian Wildlife Service Report Series.

———. 1992. Atlas of the Breeding Birds of the Maritime Provinces. Nimbus Publishing and Nova Scotia Museum, Halifax, NS.

Estrada-Franco, H., J. R. Navarro-Lopez, D. W. C. Beasley, L. Coffey, A. Carrara, A. Travassos da Rosa, T. Clements, E. Wang, G. V. Ludwig, A. C. Cortes, P. P. Ramirez, R. B. Tesh, A. D. T. Barrett, and S. C. Weaver. 2003. West Nile virus in Mexico: Evidence of widespread circulation since July 2002. Emerging Infectious Diseases 9:1604–1607.

Evans, D. L. 1997. The influence of broadcast tape-recorded calls on captures of fall migrant Northern Saw-whet Owls (*Aegolius acadicus*) and Long-eared Owls (*Asio otus*). Pages 173–174 *in* Biology and Conservation of Owls of the Northern Hemisphere: 2nd International Symposium (J. R. Duncan, D. H. Johnson, and T. H. Nicholls, Eds.). United States Forest Service, North Central Forest Experiment Station, St. Paul, MN.

Evans Ogden, L. J., and B. J. Stutchbury. 1994. Hooded Warbler (*Wilsonia citrina*). *In* The Birds of North America, no. 110 (A. Poole and F. Gill, Eds.). The Academy of Natural Sciences, Philadelphia, PA, and the American Ornithologists' Union, Washington, DC.

Evers, D. C., M. Duron, D. Yates, and N. Schoch. 2009. An exploratory study of methylmercury availability in terrestrial wildlife of New York and Pennsylvania, 2005–2006. Final Report for New York State Energy and Development Authority, Gorham, ME.

Faaborg, J., and W. J. Arendt. 1992. Puerto Rican dry forest: Which species are in trouble? Pages 57–63 *in* Ecology and Conservation of Neotropical Migrant Landbirds (J. M. Hagan III and D. W. Johnston, Eds.). Smithsonian Institution Press, Washington, DC.

Fajvan, M. A. 2008. The role of silvicultural thinning in eastern forests threatened by hemlock woolly adelgid (*Aldelgies tsugae*). Integrated Restoration of Forested Ecosystems to Achieve Multiresource Benefits: Proceedings of the 2007 National Silviculture Workshop. United States Forest Service, Pacific Northwest Research Station, Portland, OR.

Falls, J. B., and J. G. Kopachena. 2010. White-throated Sparrow (*Zonotrichia albicollis*). *In* The Birds of North America Online, no. 128 (A. Poole, Ed.). Cornell Lab of Ornithology, Ithaca, NY.

Farmer, C. J., R. J. Bell, B. Drolet, L. J. Goodrich, E. Greenstone, D. Grove, D. J. T. Hussell, D. Mizrahi, F. J. Nicoletti, and H. J. Sodergren. 2008a. Trends in autumn counts of migratory raptors in northeastern North America, 1974–2004. Pages 179–215 *in* State of North America's Birds of Prey (K. L. Bildstein, J. P. Smith, E. R. Inzunza, and R. Veit, Eds.). American Ornithologists' Union and Nuttall Ornithological Club, Series in Ornithology, No. 3.

Farmer, C. J., L. J. Goodrich, E. Ruelas Inzunza, and J. P. Smith. 2008b. Conservation status of North American raptors. Pages 303–420 *in* State of North America's Birds of Prey (K. L. Bildstein, J. P. Smith, E. R. Inzunza, and R. Veit, Eds.). American Ornithologists' Union and Nuttall Ornithological Club, Series in Ornithology, No. 3.

Farmer, C. J., and J. P. Smith. 2009. Migration monitoring indicates widespread declines of American Kestrels (*Falco sparverius*) in North America. Journal of Raptor Research 43:263–273.

Farnsworth, G. L., D. R. Diefenbach, R. S. Mulvihill, M. J. Lanzone, M. J., and D. W. Brauning. 2006. Estimating breeding bird abundances with point counts from the "Second Pennsylvania Breeding Bird Atlas." *In* Paper presented at the IV North American Ornithological Conference, Veracruz, Mexico.

Farnsworth, G. L., K. H. Pollock, J. D. Nichols, T. R. Simmons, J. E. Hines, and J. R. Sauer. 2002. A removal model for estimating detection probabilities from point count surveys. Auk 119:414–425.

Faustino, C. R., C. S. Jennelle, V. Connolly, A. K. Davis, E. C. Swarthoiut, A. A. Dhondt, and E. G. Cooch. 2004. *Mycoplasma gallisepticum* infection dynamics in a House Finch population: Seasonal variation in survival, encounter, and transmission rate. Journal of Animal Ecology 73:651–669.

Fergus, C. 2002. Natural Pennsylvania. Stackpole Books, Mechanicsburg, PA.

Fialkovich, M. 2003. Local notes—Allegheny County. Pennsylvania Birds 17:123.

———. 2007. Local notes—Allegheny County. Pennsylvania Birds 21:95.

Ficken, M. S., and R. W. Ficken. 1962. Some aberrant characters of the Yellow-breasted Chat, *Icteria virens*. Auk 79:718–719.

Fike, J. 1999. Terrestrial and Palustrine Plant Communities of Pennsylvania. Pennsylvania Natural Diversity Inventory, Harrisburg, PA.

Fingerhood, E. D. 1992a. Red Crossbill (*Loxia curvirostra*). Pages 437–438 *in* Atlas of Breeding Birds in Pennsylvania (D. W. Brauning, Ed.). University of Pittsburgh Press, Pittsburgh, PA.

———. 1992b. Least Tern (*Sterna antillarum*). Page 433 *in* Atlas of Breeding Birds in Pennsylvania (D. W. Brauning, Ed.). University of Pittsburgh Press, Pittsburgh, PA.

———. 1992c. Greater Prairie-Chicken (*Tympanuchus cupido*). Pages 431–432 *in* Atlas of Breeding Birds in Pennsylvania (D. W. Brauning, Ed.). University of Pittsburgh Press, Pittsburgh, PA.

Fisher, J. R., D. E. Stallknecht, M. P. Luttrell, A. A. Dhondt, and K. A. Converse. 1997. Mycoplasmal conjunctivitis in wild songbirds: The spread of a new contagious disease in a mobile host population. Emerging Infectious Diseases 3:69–72.

Floyd, T. 1994. First breeding colony of Herring Gulls in Pennsylvania. Pennsylvania Birds 8:34.

———. 2008. Smithsonian Field Guide to the Birds of North America. HarperCollins Publishers, New York, NY.

Fowle, C. D. 1946. Notes on the development of the nighthawk. Auk 63:159–162.

Fraser, G., and B. J. M. Stutchbury. 2004. Area-sensitive birds move extensively among forest patches. Biological Conservation 118:377–387.

Freer, V. M. 2008. Bank Swallow (*Riparia riparia*). Pages 398–399 *in* The Second Atlas of Breeding Birds in New York State (K. J. McGowan and K. Corwin, Eds.). Cornell University Press, Ithaca, NY.

Fujisake, I., E. V. Pearlstine, and F. J. Mazzotii. 2010. The rapid spread of invasive Eurasian Collared Doves, *Streptopelia decaocto*, in the continental USA follows human-altered habitats. Ibis 152:622–632.

Gale, G. A., J. A. DeCecco, M. R. Marshall, W. R. McClain, and R. J. Cooper. 2001. Effects of gypsy moth defoliation on forest birds: An assessment using breeding bird census data. Journal of Field Ornithology 72:291–304.

Garadali, T., and G. Ballard. 2000. Warbling Vireo (*Vireo gilvus*). *In* The Birds of North America Online, no. 551 (A. Poole, Ed.). Cornell Lab of Ornithology, Ithaca, NY.

Garrison, B. A. 1999. Bank Swallow (*Riparia riparia*). *In* The Birds of North America Online no. 414 (A. Poole, Ed.). Cornell Lab of Ornithology, Ithaca, NY.

Gauthier, V. 2010. Local notes—Cumberland County. Pennsylvania Birds 24:155.

Geibel, M. 1975. Moraine State Park *in* Where to Find Birds in Western Pennsylvania (D. B. Freeland, Ed.). Typecraft Press, Pittsburgh, PA.

Geissler, P. H., and J. R. Sauer. 1990. Topics in route-regression analysis. Pages 54–57 *in* Survey Designs and Statistical Methods for the Estimation of Avian Population Trends (J. R. Sauer and S. Droege, Eds.). U.S. Fish and Wildlife Service, Washington, DC.

Gelbach, F. R. 1995. Eastern Screech-Owl (*Megascops asio*). *In* The Birds of North America Online, no. 165 (A. Poole, Ed.). Cornell Lab of Ornithology, Ithaca, NY.

Gerlach, T. 2005. Local notes—Bradford County. Pennsylvania Birds 18:183.

Gerstell, R. 1935. The ring-neck pheasant in Pennsylvania. Report to The Board of Commissioners. Pennsylvania Game Commission, Harrisburg, PA.

Ghalambor, C. K., and T. E. Martin. 1999. Red-breasted Nuthatch (*Sitta canadensis*). *In* The Birds of North America, no. 459 (A. Poole and F. Gill, Eds.). The Academy of Natural Sciences, Philadelphia, PA, and the American Ornithologists' Union, Washington, DC.

Gibbons, D. W., J. B. Reid, and R. A. Chapman, Eds. 1993. The New Atlas of Breeding Birds in Britain and Ireland: 1988–1991. Poyser, London, UK.

Gibbs, J. P., and J. Faaborg. 1990. Estimating the viability of Ovenbird and Kentucky Warbler populations in forest fragments. Conservation Biology 4:193–196.

Gibbs, J. P., S. Melvin, and F. A. Reid. 1992. American Bittern (*Botaurus lentiginosus*). *In* The Birds of North America, no. 18 (A. Poole, P. Stettenheim, and F. Gill, Eds.). The Academy of Natural Sciences, Philadelphia, PA, and the American Ornithologists' Union, Washington, DC.

Gibbs, J. P., F. A. Reid, S. M. Melvin, A. F. Poole, and P. Lowther. 2009. Least Bittern (*Ixobrychus exilis*). *In* The Birds of North America Online, no. 17 (A. Poole, Ed.). Cornell Lab of Ornithology, Ithaca, NY.

Giese, C. L. A., and F. J. Cuthbert. 2003. Influence of surrounding vegetation on woodpecker nest tree selection in oak forests of the upper midwest, USA. Forest Ecology and Management 179:523–534.

Gilbert, O. L. 1989. The Ecology of Urban Habitats. Chapman and Hall, New York, NY.

Gill, F. B. 1980. Historical aspects of hybridization between Blue-winged and Golden-winged Warblers. Auk 97:1–18.

———. 1985. Birds. Pages 259–351 *in* Species of Special Concern in Pennsylvania, vol. 11 (H. H. Genoways and F. J. Brenner, Eds.). Carnegie Museum of Natural History, Pittsburgh, PA.

———. 1997. Local cytonuclear extinction of the Golden-winged Warbler. Evolution 51:519–525.

———. 2004. Blue-winged Warblers (*Vermivora pinus*) versus Golden-winged Warblers (*V. chrysoptera*). Auk 121:1014–1018.

Gill, F. B., R. A. Canterbury, and J. L. Confer. 2001. Blue-winged Warbler (*Vermivora pinus*). *In* The Birds of North America, no. 584 (A. Poole and F. Gill, Eds.). The Birds of North America, Inc., Philadelphia, PA.

Glahn, J. 1997. Bird predation and its control at aquaculture facilities in the northeastern Unites States. U.S. Department of Agriculture, Animal Plant Health Inspection Service, Washington, DC.

Goguen, C. B. 2010a. Summer Tanager (*Piranga rubra*). Pages 162–165 *in* Terrestrial Vertebrates of Concern in Pennsylvania: A Complete Guide to Species of Conservation Concern (M. A. Steele, M. C. Brittingham, T. J. Maret, and J. F. Meritt, Eds.). The Johns Hopkins University Press, Baltimore, MD.

———. 2010b. Scarlet Tanager (*Piranga olivacea*). Pages 182–185 *in* Terrestrial Vertebrates of Concern in Pennsylvania: A Complete Guide to Species of Conservation Concern (M. A. Steele, M. C. Brittingham, T. J. Maret, and J. F. Meritt, Eds.). The Johns Hopkins University Press, Baltimore, MD.

Good, T. P. 1998. Great Black-backed Gull (*Larus marinus*). *In* The Birds of North America Online, no. 330 (A. Poole, Ed.). Cornell Lab of Ornithology, Ithaca, NY.

Goodrich, L. 1992a. Northern Harrier (*Circus cyaneus*). Pages 94–95 *in* Atlas of Breeding Birds in Pennsylvania (D. W. Brauning, Ed.). University of Pittsburgh Press, Pittsburgh, PA.

———. 1992b. Sharp-shinned Hawk (*Accipiter striatus*). Pages 96–97 *in* Atlas of Breeding Birds in Pennsylvania (D. W. Brauning, Ed.). University of Pittsburgh Press, Pittsburgh, PA.

———. 1992c. Cooper's Hawk (*Accipiter cooperii*). Pages 98–99 *in* Atlas of Breeding Birds in Pennsylvania (D. W. Brauning, Ed.). University of Pittsburgh Press, Pittsburgh, PA.

Goodrich, L. J., M. C. Brittingham, J. A. Bishop, and P. Barber. 2002. Wildlife habitat in Pennsylvania: Past, present, and future. Report to state agencies. Department of Conservation and Natural Resources. http://www.dcnr.state.pa.us/wlhabitat/toc.aspx.

Goodrich, L. J., S. T. Crocoll, and S. E. Senner. 1996. Broad-winged Hawk (*Buteo platypterus*). *In* The Birds of North America, no. 218 (A. Poole and F. Gill, Eds.). The Academy of Natural Sciences, Philadelphia, PA, and the American Ornithologists' Union, Washington, DC.

Goodrich, L. J., and J. P. Smith. 2008. Raptor migration in North America. Pages 37–149 *in* State of North America's Birds of Prey (K. L. Bildstein, J. P. Smith, E. R. Inzunza, and R. Veit, Eds.). American Ornithologists' Union and Nuttall Ornithological Club, Series in Ornithology, No. 3.

Goodrich, L. J., C. Viverette, S. E. Senner, and K. L. Bildstein. 1998. Long-term use of Breeding Bird Census plots to monitor populations of Neotropical migrants breeding in deciduous forest in eastern Pennsylvania, USA. Pages 149–194 *in* Forest Biodiversity in North Central, and South America and the Caribbean: Research and Monitoring (F. Dallmeier and J. A. Comiskey, Eds.). CRC Press, Boca Raton, FL.

Gowaty, P. A., and J. H. Plissner. 1998. Eastern Bluebird (*Sialia sialis*). *In* The Birds of North America Online, no. 381 (A. Poole, Ed.). Cornell Lab of Ornithology, Ithaca, NY.

Graham, F. R. J. 2011. High Hopes. *In* Audubon Magazine, 113.

Grant, T. A., E. M. Madden, T. L. Shaffer, P. J. Pietz, G. A. Berkey, and N. J. Kadrmas. 2006. Nest survival of Clay-colored and Vesper Sparrows in relation to woodland edge in mixed-grass prairies. Journal of Wildlife Management 70:691–701.

Greenberg, C. H., A. L. Tomcho, J. D. Lanham, T. A. Waldrop, J. Tomcho, R. J. Phillips, and D. Simon. 2007. Short-term effects of fire and other fuel reduction treatments on breeding birds in a southern Appalachian upland hardwood forest. Journal of Wildlife Management 71:1906–1916.

Greenberg, R. 1988. Water as a habitat cue for breeding Swamp and Song Sparrows. Condor 90:420–427.

Greenberg, R., and P. P. Marra, Eds. 2005. Birds of Two Worlds: The Ecology and Evolution of Migration. The Johns Hopkins University Press, Baltimore, MD, and London.

Greenlaw, J. S. 1996. Eastern Towhee (*Pipilo erythrophthalmus*). *In* The Birds of North America, no. 262 (A. Poole and F. Gill, Eds.). The Academy of Natural Sciences, Philadelphia, PA, and the American Ornithologists' Union, Washington, DC.

Gregg, I. D., J. P. Dunn, and K. J. Jacobs. 2000. Waterfowl population monitoring. Wildlife Management Annual Report. Pennsylvania Game Commission. Harrisburg, PA.

Gremaud, G. K. 1983. Factors influencing nongame bird use of rowcrop fields. M.S. thesis, Iowa State University, Ames, IA.

Grimm, J. W., and R. H. Yahner. 1986. Status and management of select species of avifauna in Pennsylvania with emphasis on raptors. Pennsylvania Game Commission, Harrisburg, PA.

Grimm, W. C. 1952. Birds of the Pymatuning Region. Pennsylvania Game Commission, Harrisburg, PA.

Gross, A. O. 1956. The reappearance of the Dickcissel (*Spiza americana*) in eastern North America. Auk 73:66–70.

Gross, D. A. 1982. Residential status and occurrence of helpers at the nest in the Blue Jay (*Cyanocitta cristata*, Linneaus: Corvidae) near Bloomsburg, Pennsylvania. M.S. thesis, Bloomsburg State College, Bloomsburg, PA.

———. 1991. Yellow-bellied Flycatcher nesting in Pennsylvania with a review of its history, distribution, ecology, behavior, and conservation problems. Pennsylvania Birds 5:107–113.

———. 1992a. Hooded Merganser (*Lophodytes cucullatus*). Pages 82–83 *in* Atlas of Breeding Birds in Pennsylvania (D. W. Brauning, Ed.). University of Pittsburgh Press, Pittsburgh, PA.

———. 1992b. Yellow-bellied Flycatcher (*Empidonax flaviventris*). Pages 198–199 *in* Atlas of Breeding Birds in Pennsylvania (D. W. Brauning, Ed.). University of Pittsburgh Press, Pittsburgh, PA.

———. 1992c. Blue Jay (*Cyanocitta cristata*). Pages 228–229 *in* Atlas of Breeding Birds in Pennsylvania (D. W. Brauning, Ed.). University of Pittsburgh Press, Pittsburgh, PA.

———. 1992d. American Crow (*Corvus brachyrhynchos*). Pages 230–231 *in* Atlas of Breeding Birds in Pennsylvania (D. W. Brauning, Ed.). University of Pittsburgh Press, Pittsburgh, PA.

———. 1992e. Fish Crow (*Corvus ossifragus*). Pages 232–233 *in* Atlas of Breeding Birds in Pennsylvania (D. W. Brauning, Ed.). University of Pittsburgh Press, Pittsburgh, PA.

———. 1992f. Cedar Waxwing (*Bombycilla cedrorum*). Pages 282–283 *in* Atlas of Breeding Birds in Pennsylvania (D. W. Brauning, Ed.). University of Pittsburgh Press, Pittsburgh, PA.

———. 1992g. Yellow-rumped Warbler (*Dendroica coronata*). Pages 316–317 *in* Atlas of Breeding Birds in Pennsylvania (D. W. Brauning, Ed.). University of Pittsburgh Press, Pittsburgh, PA.

———. 1992h. White-throated Sparrow (*Zonotrichia albicollis*). Pages 392–393 *in* Atlas of Breeding Birds in Pennsylvania (D. W. Brauning, Ed.). University of Pittsburgh Press, Pittsburgh, PA.

———. 1992i. Red-winged Blackbird (*Agelaius phoeniceus*). Pages 398–399 *in* Atlas of Breeding Birds in Pennsylvania (D. W. Brauning, Ed.). University of Pittsburgh Press, Pittsburgh, PA.

———. 1992j. Pine Siskin (*Carduelis pinus*). Pages 416–417 *in* Atlas of Breeding Birds in Pennsylvania (D. W. Brauning, Ed.). University of Pittsburgh Press, Pittsburgh, PA.

———. 1994. Discovery of a Blackpoll Warbler (*Dendroica striata*) nest: A first for Pennsylvania—Wyoming County. Pennsylvania Birds 8:128–132.

———. 1998a. Birds: Review of status in Pennsylvania. Pages 137–170 *in* Inventory and Monitoring of Biotic Resources in Pennsylvania (J. D. Hassinger, R. J. Hill, G. L. Storm, and R. H. Yahner, Eds.). Pennsylvania Biological Survey, Harrisburg, PA.

———. 1998b. The status and distribution of Yellow-bellied Flycatcher: A threatened member of Pennsylvania's diminished boreal fauna. A Report for the Wild Resource Conservation Fund, Harrisburg, PA.

———. 2000a. A Summary of the Year 2000: Pennsylvania Breeding Survey of Northern Saw-whet Owl (*Aegolius acadicus*), Otherwise Known as Project Toot Route. PSO Newsletter 11 (4): 4–6.

———. 2000b. Pennsylvania breeding survey of Northern Saw-whet Owl (*Aegolius acadicus*): A candidate—undetermined species. Project Toot Route. Pennsylvania Game Commission, Harrisburg, PA.

———. 2004. Avian population and habitat assessment project: Pennsylvania Important Bird Area #48—State Game Lands #57, Wyoming, Luzerne, and Sullivan Counties. Audubon Pennsylvania, Harrisburg, PA.

———. 2009. Bald Eagle breeding and wintering surveys. Wildlife Management Annual Report. Pennsylvania Game Commission, Harrisburg, PA.

———. 2010a. Long-eared Owl (*Asio otus*). Pages 144–148 *in* Terrestrial Vertebrates of Concern in Pennsylvania: A Complete Guide to Species of Conservation Concern (M. A. Steele, M. C. Brittingham, T. J. Maret, and J. F. Merritt, Eds.). The Johns Hopkins University Press, Baltimore, MD.

———. 2010b. Pennsylvania boreal conifer forests and their bird communities: Past, present, and potential. Pages 48–73 *in* Proceedings of the Conference on Ecology and Management of High-Elevation Forests of the Central and Southern Appalachian Mountains (J. S. Rentch, and T. M. Schuler, Eds.). United States Forest Service, Northern Research, Station, Slatyfork, WV.

———. 2010c. Blackpoll Warbler. Pages 218–221 *in* Terrestrial Vertebrates of Pennsylvania: A Complete Guide to Species of Conservation Concern (M. A. Steele, M. C. Brittingham, T. J. Maret, and J. F. Merritt., Eds.). The Johns Hopkins University Press, Baltimore, MD.

———. 2010d. Red Crossbill. Pages 221–224 *in* Terrestrial Vertebrates of Pennsylvania: A Complete Guide to Species of Conservation Concern (M. A. Steele, M. C. Brittingham, T. J. Maret, and J. F. Merritt., Eds.). The Johns Hopkins University Press, Baltimore, MD.

———. 2010e. Olive-sided Flycatcher (*Contopus cooperi*). Pages 119–123 *in* Terrestrial Vertebrates of Concern in Pennsylvania: A Complete Guide to Species of Conservation Concern (M. A. Steele, M. C. Brittingham, T. J. Maret, and J. F. Merritt, Eds.). The Johns Hopkins University Press, Baltimore, MD.

Gross, D. A., and D. W. Brauning. 2010. Bald Eagle Management Plan for Pennsylvania, 2010–2019. Bureau of Wildlife Management, Pennsylvania Game Commission, Harrisburg, PA.

Gross, D. A., and C. D. Haffner. 2010. Wetland bird communities: Boreal bogs to open water. Pages 44–61 *in* Avian Ecology and Conservation: A Pennsylvania Focus with National Implications (S. K. Majumdar, T. L. Master, M. C. Brittingham, R. M. Ross, R. S. Mulvihill, and J. E. Huffman, Eds.). The Pennsylvania Academy of Science, Easton, PA.

———. 2011. Colonial nesting bird study. Wildlife Management Annual Report. Pennsylvania Game Commission, Harrisburg, PA.

Gross, D. A., and P. E. Lowther. 2011. Yellow-bellied Flycatcher (*Empidonax flaviventris*). *In* The Birds of North America Online, no. 566 (A. Poole, Ed.). Cornell Lab of Ornithology, Ithaca, NY.

Groth, J. G. 1988. Resolution of cryptic species in Appalachian Red Crossbills. Condor 90:745–760.

———. 1993a. Call matching and positive assortative mating in Red Crossbills. Auk 110:398–401.

———. 1993b. Evolutionary differentiation in morphology, vocalizations, and allozyme among nomadic sibling species in the North American Red Crossbill (*Loxia curvirostra*) complex. University of California Publication of Zoology 127:1–143.

Grove, G. 2010. Winter raptor survey. Pages 126–136 *in* Avian Ecology and Conservation: A Pennsylvania Focus with National Implications (S. K. Majumdar, T. L. Master, M. C. Brittingham, R. M. Ross, R. S. Mulvihill, and J. E. Huffman, Eds.). Pennsylvania Academy of Science, Easton, PA.

Grubb, T. C., Jr., and V. V. Pravosudov. 1994. Tufted Titmouse (*Baeolophus bicolor*). *In* The Birds of North America Online, no. 86 (A. Poole, Ed.). Cornell Lab of Ornithology, Ithaca, NY.

———. 2008. White-breasted Nuthatch (*Sitta carolinensis*). *In* The Birds of North America Online, no. 54 (A. Poole, Ed.). Cornell Lab of Ornithology, Ithaca, NY.

Guénette, J. S., and M. A. Villard. 2005. Thresholds in forest bird response to habitat alteration as quantitative targets for conservation. Conservation Biology 19:1168–1180.

Guzy, M. J., and G. Ritchison. 1999. Common Yellowthroat (*Geothlypis trichas*). *In* The Birds of North America, no. 448 (A. Poole and F. Gill, Eds.). The Birds of North America, Inc., Philadelphia, PA.

Haas, F. C., and B. M. Haas. 2005. Annotated list of the birds of Pennsylvania. Pennsylvania Biological Survey, Harrisburg, PA.

Haffner, C., and D. Gross. 2008. Colonial nesting bird study. Wildlife Management Annual Report. Pennsylvania Game Commission, Harrisburg, PA.

———. 2009. Colonial nesting bird study. Wildlife Managment Annual Report. Pennsylvania Game Commission, Harrisburg, PA.

———. 2010a. Colonial nesting bird study. Wildlife Managment Annual Report. Pennsylvania Game Commission, Harrisburg, PA.

———. 2010b. Osprey research/management. Wildlife Management Annual Report. Pennsylvania Game Commission, Harrisburg, PA.

Haffner, C. D. 2007. Great Lakes Piping Plover (*Charadrius melodus*) recovery assessment, Presque Isle State Park, Erie Co., Pennsylvania. Pennsylvania Game Commission, Harrisburg, PA.

Haffner, C. D., and D. A. Gross. 2008. Great Blue Heron five-year statewide survey results, 2007–2008. Pennsylvania Birds 22:156–159.

Hagemeijer, E. J. M., and M. J. Blair, Eds. 1997. The EBCC Atlas of European Breeding Birds: Their distribution and abundance. T & A.D. Poyser, London, UK.

Hagen, J. M., W. M. V. Haegen, and P. S. McKinley. 1996. The early development of forest fragmentation effects. Conservation Biology 10:188–202.

Hager, S. 2009. Human-related threats to urban raptors. Journal of Raptor Research 43:210–226.

Haggerty, T. M., and E. S. Morton. 1995. Carolina Wren (*Thryothorus ludovicianus*). *In* The Birds of North America Online, no. 188 (A. Poole, Ed.). Cornell Lab of Ornithology, Ithaca, NY.

Hahn, D. C., N. M. Nemeth, E. Edwards, P. R. Bright, and N. Komar. 2006. Passive West Nile virus antibody transfer from maternal Eastern Screech-Owls (*Megascops asio*) to progeny. Avian Diseases 50:454–455.

Halkin, S., and S. U. Linville. 1999. Northern Cardinal (*Cardinalis cardinalis*). *In* The Birds of North America Online, no. 455 (A. Poole, Ed.). Cornell Lab of Ornithology, Ithaca, NY.

Hall, G. A. 1964. Appalachian region. Audubon Field Notes 18:505–507.

———. 1972. Appalachian region. American Birds 26:857–860.

———. 1983. West Virginia Birds: Distribution and Ecology. Carnegie Museum of Natural History, Pittsburgh, PA.

———. 1996. Yellow-throated Warbler (*Setophaga dominica*). *In* The Birds of North America Online, no. 223 (A. Poole, Ed.). Cornell Lab of Ornithology, Ithaca, NY.

Hallworth, M., P. M. Benham, J. D. Lambert, and L. Reitsma. 2007. Canada Warbler (*Wilsonia canadensis*) breeding ecology in young forest stands compared to a red maple (*Acer rubrum*) swamp. Forest Ecology and Management 255:1353–1358.

Hamel, P. B. 2000. Cerulean Warbler (*Dendroica cerulea*). *In* The Birds of North America, no. 511 (A. Poole and F. Gill, Eds.). The Birds of North America, Inc., Philadelphia, PA.

Hamel, P. B., D. K. Dawson, and P. D. Keyser. 2004. How we can learn more about the Cerulean Warbler (*Dendroica cerulea*). Auk 121:7–14.

Hamel, P. B., H. E. LeGrand, Jr., M. R. Lennartz, and S. A. Gauthreaux, Jr. 1982. Bird-habitat relationships on southeastern forest lands. United States Forest Service General Technical Report SE-22.

Hames, R. S., and J. D. Lowe. 2008a. Sharp-shinned Hawk (*Accipiter striatus*). Pages 192–193 *in* The Second Atlas of Breeding Birds in New York State (K. J. McGowan and K. Corwin, Eds.). Cornell University Press, Ithaca, NY.

———. 2008b. Cooper's Hawk (*Accipiter cooperii*). Pages 194–195 *in* The Second Atlas of Breeding Birds in New York State (K. J. McGowan and K. Corwin, Eds.). Cornell University Press, Ithaca, NY.

———. 2008c. Hermit Thrush (*Catharus guttatus*). Pages 446–447 *in* The Second Atlas of Breeding Birds in New York State (K. J. McGowan and K. Corwin Ed.). Cornell University Press, Ithaca, NY.

———. 2008d. Wood Thrush (*Hylocichla mustelina*). Pages 448–449 *in* The Second Atlas of Breeding Birds in New York State (K. J. McGowan and K. Corwin, Eds.). Cornell University Press, Ithaca, NY.

Hames, R. S., K. V. Rosenberg, J. D. Lowe, S. E. Barker, and A. A. Dhondt. 2002. Adverse effects of acid rain on the distribution of the Wood Thrush (*Hylocichla mustelina*) in North America. Proceedings of the National Academy of Sciences (USA) 99:11235–11240.

Haney, J. C. 1999. Hierarchical comparisons of breeding birds in old-growth conifer-hardwood forest on the Appalachian Plateau. Wilson Bulletin 111:89–99.

Haney, J. C., and C. P. Schaadt. 1996. Functional roles of eastern old growth in promoting forest bird diversity. Pages 76–88 *in* Eastern Old Growth Forests: Prospects for Rediscovery and Recovery (M. B. Davis, Ed.). Island Press, Washington, DC.

Hanners, L. A., and S. R. Patton. 1998. Worm-eating Warbler (*Helmitheros vermivora*). *In* The Birds of North America Online, no. 367 (A. Poole, Ed.). Cornell Lab of Ornithology, Ithaca, NY.

Hardy, J. W. 1961. Studies in behavior and phylogeny of certain New World jays (*Garrulinae*). University of Kansas Science Bulletin 42:13–149.

Harlow, R. C. 1913. The breeding birds of Pennsylvania. M.S. thesis, The Pennsylvania State University, University Park, PA.

———. 1918. Notes on the breeding birds of Pennsylvania and New Jersey. Auk 35:18–29.

———. 1951. Tribal nesting of the pine siskin in Pennsylvania. Cassinia 38:4–9.

Harris, R. J., and J. M. Reed. 2002. Effects of forest-clearcut edges on a forest-breeding songbird. Canadian Journal of Zoology 80:1026–1037.

Harrison, H. H. 1975. A Field Guide to Birds' Nests. Houghton Mifflin Co., Boston, MA.

———. 1984. Wood Warblers' World. Simon and Schuster, New York, NY.

Hartman, F. E. 1989. Pennsylvania waterfowl, shorebirds, and waterbirds. *In* Wetlands Ecology and Conservation: Emphasis in Pennsylvania (S. K. Majumdar, R. P. Brooks, F. J. Brenner, and R. W. Tiner, Eds.). Pennsylvania Academy of Science, Phillipsburg, NJ.

———. 1992a. Canada Goose (*Branta canadensis*). Pages 66–67 *in* Atlas of Breeding Birds in Pennsylvania (D. W. Brauning, Ed.). University of Pittsburgh Press, Pittsburgh, PA.

———. 1992b. American Black Duck (*Anas rubripes*). Pages 72–73 *in* Atlas of Breeding Birds in Pennsylvania (D. W. Brauning, Ed.). University of Pittsburgh Press, Pittsburgh, PA.

———. 1992c. Mallard (*Anas platyrynchos*). Pages 74–75 *in* Atlas of Breeding Birds in Pennsylvania (D. W. Brauning, Ed.). University of Pittsburgh Press, Pittsburgh, PA.

———. 1992d. Blue-winged Teal (*Anas discors*). Pages 76–77 *in* Atlas of Breeding Birds in Pennsylvania (D. W. Brauning, Ed.). University of Pittsburgh Press, Pittsburgh, PA.

———. 1992e. Green-winged Teal (*Anas crecca*). Pages 70–71 *in* Atlas of Breeding Birds in Pennsylvania (D. W. Brauning, Ed.). University of Pittsburgh Press, Pittsburgh, PA.

———. 1992f. American Wigeon (*Anas americana*). Pages 80–81 *in* Atlas of Breeding Birds in Pennsylvania (D. W. Brauning, Ed.). University of Pittsburgh Press, Pittsburgh, PA.

———. 1992g. Northern Shoveler (*Anas clypeata*). Pages 78–79 *in* Atlas of Breeding Birds in Pennsylvania (D. W. Brauning, Ed.). University of Pittsburgh Press, Pittsburgh, PA.

Hatch, J. J., and D. V. Weseloh. 1999. Double-crested Cormorant (*Phalacrocorax auritus*). *In* The Birds of North America Online, no. 411 (A. Poole, Ed.). Cornell Lab of Ornithology, Ithaca, NY.

Heath, S. R., E. H. Dunn, and D. J. Agro. 2009. Black Tern (*Chlidonias niger*). *In* The Birds of North America Online, no. 147 (A. Poole, Ed.). Cornell Lab of Ornithology, Ithaca, NY.

Heckscher, C. M., and J. M. McCann. 2006. Status of Swainson's Warbler on the Delmarva Peninsula. Northeastern Naturalist 13:521–530.

Heinrich, B. 1999. Mind of the Raven: Investigations and Adventures with Wolf-birds. Cliff Street Books, New York, NY.

Hejl, S. J., K. R. Newlon, M. E. Mcfadzen, J. S. Young, and C. K. Ghalambor. 2002. Brown Creeper (*Certhia americana*). *In* The Birds of North America Online, no. 669 (A. Poole, Ed.). Cornell Lab of Ornithology, Ithaca, NY.

Henise, D., and R. Henise. 1994. Local notes—Franklin County. Pennsylvania Birds 8:101.

Herkert, J. R. 2007. Conservation Reserve Program benefits on Henslow's Sparrows within the United States. Journal of Wildlife Management 71:2749–2751.

Herkert, J. R., D. E. Kroodsma, and J. P. Gibbs. 2001. Sedge Wren (*Cistothorus platensis*). *In* The Birds of North America Online, no. 582 (A. Poole, Ed.). Cornell Lab of Ornithology, Ithaca, NY.

Herkert, J. R., P. D. Vickery, and D. E. Kroodsma. 2002. Henslow's Sparrow (*Ammodramus henslowii*). *In* The Birds of North America Online no. 672 (A. Poole and F. Gill, Eds.). Cornell Lab of Ornithology, Ithaca, NY.

Hess, G. K., R. L. West, M. V. Barnhill, III, and L. M. Fleming. 2000. Birds of Delaware. University of Pittsburgh Press, Pittsburgh, PA.

Hess, P. 1989. Carolina Wrens at Pittsburgh, 1970–1988: Persistence in a dangerous environment. Pennsylvania Birds 3:3–7.

———. 1992. The Red-bellied Woodpecker tumbled in 1990 on southeastern Pennsylvania Christmas Bird Counts. Pennsylvania Birds 6:15–17.

Hess, P., M. R. Leahy, and R. M. Ross. 1998. Pennsylvania's crossbill winter of 1997-98. Pennsylvania Birds 12:2–6.

Heusmann, H. W., and J. R. Sauer. 2000. The northeastern states' waterfowl breeding population survey. Wildlife Society Bulletin 28:355–364.

Hickey, J. J., Ed. 1969. Peregrine Falcon Populations: Their Biology and Decline. University of Wisconsin Press, Madison, WI.

Hill, G. E. 1993. House Finch (*Carpodacus mexicanus*). *In* The Birds of North America, no. 46 (A. Poole and F. Gill, Eds.). The Birds of North America, Inc., Philadelphia, PA.

Hill, J. R., III. 1986. First recorded breeding attempt of the Ring-billed Gull in Pennsylvania. Colonial Waterbirds 9:117–118.

Hitch, A. T., and P. L. Leberg. 2007. Breeding distributions of North American bird species moving north as a result of climate change. Conservation Biology 21:534–539.

Hochachka, W. M., and A. A. Dhondt. 2000. Density-dependent decline of host abundance resulting from a new infectious disease. Proceedings of the National Academy of Sciences (USA) 97:5303–5306.

Hochachka, W. M., M. Winter, and R. Charif. 2009. Sources of variation in singing probability of Florida grasshopper sparrows, and implications for design and analysis of auditory surveys. Condor 111:349–360.

Hoekman, S. T., T. S. Gabor, R. Major, H. R. Murkin, and M. S. Lindberg. 2006. Demographics of female Mallards breeding in southern Ontario. Journal of Wildlife Management 70:111–120.

Hole, D. G., M. J. Whittingham, R. B. Bradbury, G. Q. A. Anderson, P. L. M. Lee, J. D. Wilson, and J. R. Krebs. 2002. Widespread local House Sparrow extinctions. Nature 418:931–932.

Holmes, R. T., N. L. Rodenhouse, and T. S. Sillett. 2005. Black-throated Blue Warbler (*Dendroica caerulescens*). *In* The Birds of North America Online, no. 87 (A. Poole, Ed.). Cornell Lab of Ornithology, Ithaca, NY.

Holmes, R. T., and T. W. Sherry. 2001. Thirty-year bird population trends in an unfragmented temperate deciduous forest: Importance of habitat change. Auk 118:589–609.

Hoover, W. N. 2003. Grass Flats: Birds of Warren County, Pennsylvania, Including Notes on Other Species Observed at Presque Isle, Erie County, Pennsylvania. iUniverse, Inc., Lincoln, NE.

Hopp, S. L., A. Kirby, and C. A. Boone. 1995. White-eyed Vireo (*Vireo griseus*). *In* The Birds of North America Online, no. 168 (A. Poole, Ed.). Cornell Lab of Ornithology, Ithaca, NY.

Horn, A. G. 2009. Assessment and Update Status Report on the Least Bittern, *Ixobrychus exilis*, in Canada. Committee on the Status of Endangered Species in Canada, Ottawa, ON.

Hostetler, M., S. Duncan, and J. Paul. 2005. Post-construction effects of an urban development on migrating, resident, and wintering birds. Southeastern Naturalist 4:421–434.

Hothem, R. L., B. E. Brussee, and J. W. E. Davis. 2010. Black-crowned Night-Heron (*Nycticorax nycticorax*). *In* The Birds of North America Online, no. 74 (A. Poole, Ed.). Cornell Lab of Ornithology, Ithaca, NY.

Hottenstein, J., and S. Welch. 1965. Incorporation dates of Pennsylvania municipalities. Bureau of Municipal Affairs, Pennsylvania Department of Internal Affairs, Harrisburg, PA.

Houghton, L. M., and L. M. Rymon. 1997. Nesting distribution and population status of U.S. Ospreys 1994. Journal of Raptor Research 31:44–43.

Hougner, C., J. Colding, and T. Söderqvist. 2006. Economic value of a seed dispersal service in the Stockholm National Urban Park. Ecological Economics 59:364–374.

Houston, C. S., and J. D. E. Bowen. 2001. Upland Sandpiper (*Bartramia longicauda*). *In* The Birds of North America, no. 580 (A. Poole and F. Gill, Eds.). The Birds of North America, Inc., Philadelphia, PA.

Houston, C. S., D. G. Smith, and C. Rohner. 1998. Great Horned Owl (*Bubo virginianus*). *In* The Birds of North America, no. 372 (A. Poole and F. Gill, Eds.). The Academy of Natural Sciences, Philadelphia, PA, and the American Ornithologists' Union, Washington, DC.

Howell, S. N. G. 2002. Hummingbirds of North America. Academic Press, London, UK.

Hristienko, H., and J. J. E. McDonald. 2007. Going into the 21st century: A perspective on trends and controversies in the management of the American black bear. Ursus 18:72–78.

Huffman, J. E., and D. E. Roscoe. 2010. West Nile virus in raptors. Pages 315–322 *in* Avian Ecology and Conservation: A Pennsylvania Focus with National Implications (S. K. Majumdar, T. L. Master, M. C. Brittingham, R. M. Ross, R. S. Mulvihill, and J. E. Huffman, Eds.). Pennsylvania Academy of Science, Easton, PA.

Hughes, J. M. 1997. Taxonomic significance of host-egg mimicry by facultative brood parasites of the avian genus *Coccyzus* (Cuculidae). Canadian Journal of Zoology 75:1380–1386.

———. 1999. Yellow-billed Cuckoo (*Coccyzus americanus*). *In* The Birds of North America, no. 418 (A. Poole and F. Gill, Eds.). The Birds of North America, Inc., Philadelphia, PA.

———. 2001. Black-billed Cuckoo (*Coccyzus erythropthalmus*). *In* The Birds of North America, no. 587 (A. Poole and F. Gill, Eds.). The Birds of North America, Inc., Philadelphia, PA.

Hughes, W. E. 1898. Minutes of the Delaware Valley Ornithological Society, Jan. 20, 1898. Delaware Valley Ornithological Society, Philadelphia, PA.

Hunt, C. J. 1904. That feathered midget of the tidewater—the Long-billed Marsh Wren. Cassinia 8:14–16.

Hunt, P. D., and B. C. Eliason. 1999. Blackpoll Warbler (*Dendroica striata*). *In* The Birds of North America, no. 431 (A. Poole and F. Gill, Eds.). The Birds of North America, Inc., Philadelphia, PA.

Hunt, P. D., and D. J. Flaspohler. 1998. Yellow-rumped Warbler (*Dendroica coronata*). *In* The Birds of North America, no. 376 (A. Poole and F. Gill, Eds.). The Birds of North America, Inc., Philadelphia, PA.

Hunter, M. L., Jr., and A. Hutchinson. 1994. The virtues and shortcomings of parochialism: Conserving species that are locally rare, but globally common. Conservation Biology 8:1163–1165.

Hunter, W. C., D. A. Buehler, R. A. Canterbury, J. L. Confer, P.B. Hamel. 2001. Conservation of disturbance-dependent birds in eastern North America. Wildlife Society Bulletin 29:440–455.

Hunter, W. C., W. Golder, S. Melvin, and J. Wheeler, compilers. 2006. Southeast United States regional waterbird conservation plan. U.S. Fish and Wildlife Service Region 4, Atlanta, GA, and Region 9, Arlington, VA, and North Carolina Audubon Society, Wilmington, NC.

Hurley, G. F. 1972. Swainson's Warbler distribution in West Virginia. Redstart 39:110–112.

Ickes, R. 1992a. Northern Bobwhite (*Colinus virginianus*). Pages 118–119 *in* Atlas of Breeding Birds in Pennsylvania (D. W. Brauning, Ed.). University of Pittsburgh Press, Pittsburgh, PA.

———. 1992b. Pied-billed Grebe (*Podilymbus podiceps*). Pages 44–45 in Atlas of Breeding Birds in Pennsylvania (D. W. Brauning, Ed.). University of Pittsburgh Press, Pittsburgh, PA.

———. 1992c. Yellow-billed Cuckoo (*Coccyzus americanus*). Pages 152–153 in Atlas of Breeding Birds in Pennsylvania (D. W. Brauning, Ed.). University of Pittsburgh Press, Pittsburgh, PA.

———. 1992d. Black-billed Cuckoo (*Coccyzus erythropthalmus*). Pages 150–151 in Atlas of Breeding Birds in Pennsylvania (D. W. Brauning, Ed.). University of Pittsburgh Press, Pittsburgh, PA.

———. 1992e. Red-bellied Woodpecker (*Melanerpes carolinus*). Pages 182–183 in Atlas of Breeding Birds in Pennsylvania (D. W. Brauning, Ed.). University of Pittsburgh Press, Pittsburgh, PA.

———. 1992f. Eastern Wood-Pewee (*Contopus virens*). Pages 196–197 in Atlas of Breeding Birds in Pennsylvania (D. W. Brauning, Ed.). University of Pittsburgh Press, Pittsburgh, PA.

———. 1992g. Great Crested Flycatcher (*Myiarchus crinitus*). Pages 210–211 in Atlas of Breeding Birds in Pennsylvania (D. W. Brauning, Ed.). University of Pittsburgh Press, Pittsburgh, PA.

———. 1992h. Tufted Titmouse (*Parus bicolor*). Pages 242–243 in Atlas of Breeding Birds in Pennsylvania (D. W. Brauning, Ed.). University of Pittsburgh Press, Pittsburgh, PA.

———. 1992i. Carolina Wren (*Thryothorus ludovicianus*). Pages 250–251 in Atlas of Breeding Birds in Pennsylvania (D. W. Brauning, Ed.). University of Pittsburgh Press, Pittsburgh, PA.

———. 1992j. Gray Catbird (*Dumetella carolinensis*). Pages 276–277 in Atlas of Breeding Birds in Pennsylvania (D. W. Brauning, Ed.). University of Pittsburgh Press, Pittsburgh, PA.

———. 1992k. Northern Mockingbird (*Mimus polyglottos*). Pages 278–279 in Atlas of Breeding Birds in Pennsylvania (D. W. Brauning, Ed.). University of Pittsburgh Press, Pittsburgh, PA.

———. 1992l. Brown Thrasher (*Toxostoma rufum*). Pages 280–281 in Atlas of Breeding Birds in Pennsylvania (D. W. Brauning, Ed.). University of Pittsburgh Press, Pittsburgh, PA.

———. 1992m. Yellow-throated Warbler (*Dendroica dominica*). Pages 322–323 in Atlas of Breeding Birds in Pennsylvania (D. W. Brauning, Ed.). University of Pittsburgh Press, Pittsburgh, PA.

———. 1992n. Clay-colored Sparrow (*Spizella pallida*). Pages 376–377 in Atlas of Breeding Birds in Pennsylvania (D. W. Brauning, Ed.). University of Pittsburgh Press, Pittsburgh, PA.

———. 1992o. Summer Tanager (*Piranga rubra*). Pages 358–359 in Atlas of Breeding Birds in Pennsylvania (D. W. Brauning, Ed.). University of Pittsburgh Press, Pittsburgh, PA.

———. 1992p. Orchard Oriole (*Icterus spurius*). Pages 408–409 in Atlas of Breeding Birds in Pennsylvania (D. W. Brauning, Ed.). University of Pittsburgh Press, Pittsburgh, PA.

———. 1992q. Northern Oriole (*Icterus galbula*). Pages 410–411 in Atlas of Breeding Birds in Pennsylvania (D. W. Brauning, Ed.). University of Pittsburgh Press, Pittsburgh, PA.

Iliff, M., L. Salas, E. R. Inzunza, G. Ballard, D. Lepage, and S. Kelling. 2009. The Avian Knowledge Network: A partnership to organize, analyze, and visualize bird observation data for education, conservation, research, and land management. Pages 365–373 in Proceedings of the Fourth International Partners in Flight Conference: Tundra to Tropics: Connecting Birds, Habitats and People. University of Texas–Pan American Press, McAllen, TX.

Ingold, J. L., and R. Galati. 1997. Golden-crowned Kinglet (*Regulus satrapa*). In The Birds of North America, no. 301 (A. Poole and F. Gill, Eds.). The Academy of Natural Sciences, Philadelphia, PA, and the American Ornithologists' Union, Washington, DC.

Inman, R. L., H. H. Prince, and D. B. Hayes. 2002. Avian communities in forested riparian wetlands of southern Michigan, USA. Wetlands 22:647–660.

IPCC (Intergovernmental Panel on Climate Change). 2007. Climate change 2007: The physical science basis, Contribution of Working Group I to the Fourth Assessment Report of the Intergovernmental Panel on Climate Change. (S. Solomon, D. Qin, M. Manning, Z. Chen, M. Marquis, K. B. Averyt, M. Tignor, and H. L. Miller, Eds.). Cambridge University Press, Cambridge, UK.

Irwin, K. 2010. A new and cryptic call type of the Red Crossbill. Western Birds 41:10–25.

IUCN. 2001. IUCN Red List categories and criteria: Version 3.1. IUCN Species Survival Commission. IUCN, Gland, Switzerland and Cambridge, UK. [Online.] Available at www.iucnredlist.org/technical-documents/categories-and-criteria.

Jackson, B. J., and J. A. Jackson. 2000. Killdeer (*Charadrius vociferus*). In The Birds of North America Online, no. 517 (A. Poole, Ed.). Cornell Lab of Ornithology, Ithaca, NY.

Jackson, J. A., and B. J. S. Jackson. 2004. Ecological relationships between fungi and woodpecker cavity sites. Condor 106:37–49.

Jackson, J. A., and H. R. Ouellet. 2002. Downy Woodpecker (*Picoides pubescens*). In The Birds of North America Online, no. 613 (A. Poole, Ed.). Cornell Lab of Ornithology, Ithaca, NY.

Jackson, J. A., H. R. Ouellet, and B. J. Jackson. 2002. Hairy Woodpecker (*Picoides villosus*). In The Birds of North America Online, no. 702 (A. Poole, Ed.). Cornell Lab of Ornithology, Ithaca, NY.

Jacobs, J. W. 1893. Summer Birds of Greene County, Pennsylvania. Republican Book and Job Office, Waynesburg, PA.

Jacobs, K. J., J. P. Dunn, and I. D. Gregg. 2009. Waterfowl population monitoring. Wildlife Management Annual Report. Pennsylvania Game Commission, Harrisburg, PA.

———. 2010. Waterfowl population monitoring. Wildlife Management Annual Report. Pennsylvania Game Commission. Harrisburg, PA.

James, F. C. 1971. Ordinations of habitat relationships among breeding birds. Wilson Bulletin 83:215–236.

James, F. C., D. A. Wiedenfeld, and C. E. McCulloch. 1992. Trends in breeding populations of warblers: Declines in the southern highlands and increases in the lowlands. Pages 43–56 in Ecology and Conservation of Neotropical Migrant Landbirds (J. M. Hagan, III, and D. W. Johnston, Eds.). Smithsonian Institution Press, Washington, DC.

James, R. D. 2007. Blue-headed Vireo, Viréo à tête bleue (*Vireo solitarius*). Pages 368–369 in Atlas of the Breeding Birds of Ontario, 2001–2005 (M. D. Cadman, D. A. Sutherland, G. G. Beck, D. Lepage, and A. R. Coutrier, Eds.). Bird Studies Canada, Environment Canada, Ontario Field Ornithologists, Ontario Ministry of Natural Resources, and Ontario Nature, Toronto, ON.

Jenkins, D. H. 1942. Bobwhite cover in Franklin County, Pennsylvania. M.S. thesis, The Pennsylvania State University, University Park, PA.

Jenkins, D. H., D. A. Devlin, N.C. Johnson, and S. P. Orndorff. 2004. System design and management for restoring Penn's Woods. Journal of Forestry April/May:30–36.

Jennings, O. E., and A. Avinoff. 1953. Wildflowers of Western Pennsylvania and the Upper Ohio Basin. University of Pittsburgh Press, Pittsburgh, PA.

Johnson, L. S. 1998. House Wren (*Troglodytes aedon*). *In* The Birds of North America Online, no. 380 (A. Poole, Ed.). Cornell Lab of Ornithology, Ithaca, NY.

Johnson, N. 2010. Pennsylvania energy impacts assessment. Report 1: Marcellus shale natural gas and wind. The Nature Conservancy, Pennsylvania Chapter and Pennsylvania Audubon, Harrisburg, PA.

Johnson, W. C., and C. S. Adkisson. 1985. Dispersal of beech nuts by Blue Jays in fragmented landscapes. American Midland Naturalist 113:319–324.

———. 1986. Airlifting the oaks. Natural History 95:40–47.

Johnson, W. C., C. S. Adkisson, T. R. Crow, and M. D. Dixon. 1997. Nut caching by Blue Jays (*Cyanocitta cristata* L.): Implications for tree demography. American Midland Naturalist 138:357–370.

Johnston, R. F. 1992. Rock Pigeon (*Columba livia*). *In* The Birds of North America Online, no. 13 (A. Poole, Ed.). Cornell Lab of Ornithology, Ithaca, NY.

Jones, J., P. J. Doran, and R. T. Holmes. 2003. Climate and food synchronize regional forest bird abundances. Ecology 84:3024–3032.

Jones, P. W., and T. M. Donovan. 1996. Hermit Thrush (*Catharus guttatus*). *In* The Birds of North America, no. 261 (A. Poole and F. Gill, Eds.). The Academy of Natural Sciences, Philadelphia, PA, and the American Ornithologists' Union, Washington, DC.

Jones, S. L., and J. E. Cornely. 2002. Vesper Sparrow (*Pooecetes gramineus*). *In* The Birds of North America Online, no. 624 (A. Poole, Ed.). Cornell Lab of Ornithology, Ithaca, NY.

Juniata College. 2009. Raystown Field Station: Osprey Release plan. http://www.juniata.edu/services/station/ospreyreleaseprogram.html.

Kaeser, M. J., P. J. Gould, M. E. McDill, K. C. Steiner, and J. C. Fin. 2008. Classifying patterns of understory vegetation in mixed-oak forests in two ecoregions of Pennsylvania. Northern Journal of Applied Forestry 25:38–44.

Kain, T., Ed. 1987. Virginia's Birdlife: An Annotated Checklist, 2nd ed. Virginia Society of Ornithology, Evington, VA.

Kappes, P. J., B. J. M. Stutchbury, and B. E. Woolfenden. 2009. The relationship between carotenoid-based coloration and pairing, within- and extra-pair mating success in the American Redstart. Condor 111:684–693.

Kaufman, K., and D. Sibley. 2002. The most misidentified birds in North America. Birding 34:136–145.

Keim, T. D. 1905. Summer birds of Port Allegany, McKean County, Pennsylvania. Cassinia 8:36–41.

Keller, C. M. E., and M. R. Fuller. 1995. Comparison of birds detected from roadside and off-road point counts in Shenandoah National Park. Pages 111–115 *in* Monitoring Bird Populations by Point Counts (C. J. Ralph, J. Sauer, and S. Droege, Eds.). USDA Forest Service, General Technical Report PSW-GRT-149, Pacific Southwest Research Station, Albany, CA.

Keller, C. M. E., and J. T. Scallan. 1999. Potential roadside biases due to habitat changes along Breeding Bird Survey routes. Condor 101:50–57.

Keller, G. S., and R. H. Yahner. 2006. Declines of migratory songbirds: Evidence for wintering-ground causes. Northeastern Naturalist 13:83–92.

———. 2007. Seasonal forest-patch use by birds in fragmented landscapes of south-central Pennsylvania. Wilson Journal of Ornithology 119:410–418.

Keller, R. 1997. First nesting of Ruddy Duck (*Oxyura jamaicensis*) in Berks County, Pennsylvania. Pennsylvania Birds 11:142–143.

Kelly, J. F., E. S. Bridge, and M. J. Hamas. 2009. Belted Kingfisher (*Megaceryle alcyon*). *In* The Birds of North America Online, no. 84 (A. Poole, Ed.). Cornell Lab of Ornithology, Ithaca, NY.

Kelsey, T. R. 2008. Biogeography, foraging ecology, and population dynamics of Red Crossbills in North America. Ph.D. dissertation, University of California at Davis, Davis, CA.

Kennamer, J. E., M. Kennamer, and R. Brenneman. 1992. History. Pages 6–17 *in* The Wild Turkey: Biology and Management (J. G. Dickson, Ed.). Stackpole Books, Harrisburg, PA.

Kennell, A. 1992. First confirmed Loggerhead Shrike nesting in Pennsylvania since 1934. Pennsylvania Birds 6:65.

Kennell, A., and E. Kennell. 1990. County reports—Adams County. Pennsylvania Birds 4:57.

Kerlin, N. 2001. County reports—Sullivan County. Pennsylvania Birds 15:116.

Khulman, E. G. 1978. The devastation of American chestnut by blight. *In* American Chestnut Symposium, January 4–5, 1978 (W. L. MacDonald, F. C. Cech, J. Luchok, and C. Smith, Eds.). West Virginia University Books, Morgantown, WV.

Kimmel, J. T., and R. H. Yahner. 1994. The Northern Goshawk in Pennsylvania: Habitat Use, Survey Protocols, and Status. School of Forest Resources, The Pennsylvania State University, University Park, PA.

King, D. I., J. D. Lambert, J. P. Bunnaccord, and L. S. Prout. 2008. Avian population trends in the vulnerable montane forests of the Northern Appalachians, USA. Biodiversity Conservation 17:2691–2700.

Kirby, R. E., G. A. Sargeant, and D. Shutler. 2004. Haldane's rule and American black duck × mallard hybridization. Canadian Journal of Zoology 82:1827–1831.

Kirk, D., and C. Hyslop. 1998. Raptor population status and trends in Canadian raptors: A review. Bird Trends Canada 4:2–9.

Kirk, D. A., and M. J. Mossman. 1998. Turkey Vulture (*Cathartes aura*). *In* The Birds of North America, no. 339 (A. Poole and F. Gill, Eds.). The Academy of Natural Sciences, Philadelphia, PA, and the American Ornithologists' Union, Washington, DC.

Klatt, P. H., B. J. M. Stutchbury, and M. L. Evans. 2008. Incubation feeding in the Scarlet Tanager: A removal experiment. Journal of Field Ornithology 79:1–10.

Klem, D., Jr. 1989. Bird-window collisions. Wilson Bulletin 101:606–620.

———. 2010. Conservation issues: Sheet glass as a principal human-associated avian mortality factor. Pages 276–289 *in* Avian Ecology and Conservation: A Pennsylvania Focus with National Implications (S. K. Majumdar, T. L. Master, M. C. Brittingham, R. M. Ross, R. S. Mulvihill, and J. E. Huffman, Eds.). The Pennsylvania Academy of Science, Easton, PA.

Klimstra, J. D., and P. L. Padding. 2009. Atlantic Flyway harvest and population survey data book. U.S. Fish and Wildlife Service, Laurel, MD.

Klinger, S. 2011. Northern Bobwhite Quail Recovery Plan, 2011–2020. Bureau of Wildlife Management, Pennsylvania Game Commission, Harrisburg, PA.

Klinger, S., J. Dunn, and J. Hassinger. 1998. Game birds in trouble. Keystone Conservationist 4:40–45.

Klinger, S. R., and T. S. Hardisky. 1998. A draft proposal to USDA for Pennsylvania's conservation reserve enhancement program, "Helping Pennsylvania's agriculture improve water quality, reduce soil erosion and enhance wildlife habitat." Pennsylvania Game Commission, Harrisburg, PA.

Klinger, S. R., and C. F. Riegner. 2008. Ring-necked Pheasant Management Plan for Pennsylvania, 2008–2017. Bureau of Wildlife Management, Pennsylvania Game Commission, Harrisburg, PA.

Knapton, R. W. 1994. Clay-colored Sparrow (*Spizella pallida*). *In* The Birds of North America Online, no. 120 (A. Poole, Ed.). Cornell Lab of Ornithology, Ithaca, NY.

Knight, R. L. 1997. Golden-crowned Kinglet (*Regulus satrapa*). Pages 238–240 *in* Atlas of the Breeding Birds of Tennessee (C. P. Nicholson, Ed.). University of Tennessee Press, Knoxville, TN.

Knox, A. G. 1992. Species and pseudospecies: The structure of crossbill populations. Biological Journal of the Linnean Society 47:325–335.

Koenig, W. D. 2003. European Starlings and their effect on native cavity-nesting birds. Conservation Biology 17:1134–1140.

Komar, N., S. Langevin, S. Hinten, N. Nemeth, E. Edwards, D. Hettler, B. Davis, R. Bowen, and M. Bunning. 2003. Experimental infection of North American birds with the New York 1999 strain of West Nile virus. Emerging Infectious Diseases 9:311–322.

Konig, C., F. Weick, and J. H. Becking. 1999. Owls: A Guide to the Owls of the World. Yale University Press, New Haven, CT.

Konze, K. R. 2007. Long-eared Owl, Hibou moyen-duc (*Asio otus*). Pages 300–301 *in* Atlas of the Breeding Birds of Ontario, 2001–2005 (M. D. Cadman, D. A. Sutherland, G. G. Beck, D. Lepage, and A. R. Coutrier, Eds.). Bird Studies Canada, Environment Canada, Ontario Field Ornithologists, Ontario Ministry of Natural Resources, and Ontario Nature, Toronto, ON.

Kovert-Bratland, K. A., W. M. Block, and T. C. Theimer. 2006. Hairy Woodpecker winter ecology in ponderosa pine forests representing different ages since wildfire. Journal of Wildlife Management 70:1379–1392.

Kricher, J. C. 1995. Black-and-white Warbler (*Mniotilta varia*). *In* The Birds of North America, no. 158 (A. Poole and F. Gill, Eds.). The Birds of North America, Inc., Philadelphia, PA.

Kroodsma, D. E. 1980. Winter Wren singing behavior: A pinnacle of song complexity. Condor 82:357–365.

Kroodsma, D. E., and J. R. Baylis. 1982. Acoustic Communication in Birds. Academic Press, New York, NY.

Kroodsma, D. E., and J. Verner. 1997. Marsh Wren (*Cisthorus palustris*). *In* The Birds of North America, no. 308 (A. Poole and F. Gill, Eds.). The Birds of North America, Inc., Philadelphia, PA.

Kroodsma, R. L. 1982. Bird community ecology on power-line corridors in east Tennessee. Biological Conservation 23:79–94.

Krueger, D. 1989. Swainson's Warbler in western Pennsylvania—rare vagrant or breeding species? Pennsylvania Birds 3:86–89.

Krueger, D. E., and R. S. Mulvihill. 1992. Swainson's Warbler (*Limnothlypis swainsonii*). Pages 338–339 *in* Atlas of Breeding Birds in Pennsylvania (D. W. Brauning, Ed.). University of Pittsburgh Press, Pittsburgh, PA.

Kunkle, D. E. 1951. The birds of the Lewisburg region. Bucknell Ornithological Club, Lewisburg, PA.

Kushlan, J. A., M. J. Steinkamp, K. C. Parsons, J. Capp, M. A. Cruz, M. Coulter, and I. Davidson. 2002. Waterbird Conservation for the Americas: The North American Waterbird Conservation Plan, version 1. Waterbird Conservation for the Americas, Washington, DC.

LaDeau, S. L., A. M. Kilpatrick, and P. P. Marra. 2007. West Nile virus emergence and large-scale declines of North American bird populations. Nature 447:710–713.

LaDeau, S. L., P. P. Marra, A. M. Kilpatrick, and C. A. Calder. 2008. West Nile virus revisited: Consequences for North American ecology. BioScience 58:937–946.

Laine, H. 1983. Behavioral aspects of the breeding biology of the Blue Jay (*Cyanocitta cristata*). Ph.D. dissertation, City University of New York, New York, NY.

Lambert, J. D. T. P. H., E. J. Laurent, G. L. Brewer, M. J. Iliff, and R. Dettmers. 2009. The Northeast Bird Monitoring Handbook. American Bird Conservancy, The Plains, VA.

Lanciotti, R. S., J. T. Roehrig, V. Deubel, J. Smith, and M. Parker. 1999. Origin of the West Nile virus responsible for an outbreak of encephalitis in the Northeastern United States. Science 286:2333–2337.

Lanyon, W. E. 1981. Breeding birds and old field succession on fallow Long Island farmland. Bulletin of the American Museum of Natural History 168:1–60.

Larkin, J. L., J. Grata, and M. Frantz. 2011. Golden-winged Warbler breeding ecology and response to habitat manipulation in northcentral Pennsylvania. Final Report, Department of Conservation and Natural Resources, Wild Resource Conservation Program, Harrisburg, PA.

Latta, S. C., and R. S. Mulvihill. 2010. The Louisiana Waterthrush as an indicator of headwater stream quality in Pennsylvania. Pages 246–258 *in* Avian Ecology and Conservation: A Pennsylvania Focus with National Implications (S. K. Mujumdar, T. L. Master, M. C. Brittingham, R. M. Ross, R. S. Mulvihill, and J. E. Huffman, Eds.). The Pennsylvania Academy of Science, Easton, PA.

Laughlin, S. B., J. R. Carroll, and S. M. Sutcliffe. 1990. Standardized breeding criteria codes: Recommendations for North American Breeding Bird Atlas projects *in* Handbook for Atlasing North American Breeding Birds (C. R. Smith, Ed.). Vermont Institute of Natural Science, Quechee, VT.

Lawler, J. J., and T. C. Edwards, Jr. 2006. A variance-decomposition approach to investigating multiscale habitat associations. Condor 108:47–58.

Leberman, R. C. 1976. The Birds of the Ligonier Valley. Carnegie Museum of Natural History Special Publication 3, Pittsburgh, PA.

———. 1992a. American Bittern (*Botaurus lentiginosus*). Pages 46–47 *in* Atlas of Breeding Birds in Pennsylvania (D. W. Brauning, Ed.). University of Pittsburgh Press, Pittsburgh, PA.

———. 1992b. Bald Eagle (*Haliaeetus leucocephalus*). Pages 92–93 *in* Atlas of Breeding Birds in Pennsylvania (D. W. Brauning, Ed.). University of Pittsburgh Press, Pittsburgh, PA.

———. 1992c. American Coot (*Fulica americana*). Pages 130–131 *in* Atlas of Breeding Birds in Pennsylvania (D. W. Brauning, Ed.). University of Pittsburgh Press, Pittsburgh, PA.

———. 1992d. Common Snipe (*Gallinago gallinago*). Pages 140–141 *in* Atlas of Breeding Birds in Pennsylvania (D. W. Brauning, Ed.). University of Pittsburgh Press, Pittsburgh, PA.

———. 1992e. Black Tern (*Chlidonias nigra*). Pages 144–145 *in* Atlas of Breeding Birds in Pennsylvania (D. W. Brauning, Ed.). University of Pittsburgh Press, Pittsburgh, PA.

———. 1992f. Barred Owl (*Strix varia*). Pages 160–161 *in* Atlas of Breeding Birds in Pennsylvania (D. W. Brauning, Ed.). University of Pittsburgh Press, Pittsburgh, PA.

———. 1992g. Sedge Wren (*Cistothorus platensis*). Pages 256–257 *in* Atlas of Breeding Birds in Pennsylvania (D. W. Brauning, Ed.). University of Pittsburgh Press, Pittsburgh, PA.

———. 1992h. Blue-gray Gnatcatcher (*Polioptila caerulea*). Pages 262–263 *in* Atlas of Breeding Birds in Pennsylvania (D. W. Brauning, Ed.). University of Pittsburgh Press, Pittsburgh, PA.

———. 1992i. Mourning Warbler (*Oporornis philadelphia*). Pages 348–349 *in* Atlas of Breeding Birds in Pennsylvania (D. W. Brauning, Ed.). University of Pittsburgh Press, Pittsburgh, PA.

———. 1992j. Prairie Warbler (*Dendroica discolor*). Pages 326–327 *in* Atlas of Breeding Birds in Pennsylvania (D. W. Brauning, Ed.). University of Pittsburgh Press, Pittsburgh, PA.

———. 1992k. Yellow-breasted Chat (*Icteria virens*). Pages 356–357 *in* Atlas of Breeding Birds in Pennsylvania (D. Brauning, Ed.). University of Pittsburgh Press, Pittsburgh, PA.

———. 1992l. Swamp Sparrow (*Melospiza georgiana*). Pages 390–391 *in* Atlas of Breeding Birds in Pennsylvania (D. W. Brauning, Ed.). University of Pittsburgh Press, Pittsburgh, PA.

———. 1992m. Rose-breasted Grosbeak (*Pheucticus ludovicianus*). Pages 364–365 *in* Atlas of Breeding Birds in Pennsylvania (D. Brauning, Ed.). University of Pittsburgh Press, Pittsburgh, PA.

Leberman, R. C., and R. S. Mulvihill. 1992. Eastern Bluebird (*Sialia sialis*). Pages 264–265 *in* Atlas of Breeding Birds in Pennsylvania (D. Brauning, Ed.). University of Pittsburgh Press, Pittsburgh, PA.

Leopold, A. 1931. Game Survey of the North Central States. Sporting Arms and Ammunition Manufacturer's Institute, Madison, WI.

Ley, D. H., J. E. Berkhoff, and J. M. McLaren. 1996. *Mycoplasma gallisepticum* isolated from House Finches (*Carpodacus mexicanus*) with conjunctivitis. Avian Diseases 40:480–483.

Lindsay, A. M. 2008. Seasonal events and associated carry-over effects in a Neotropical migratory songbird, the Yellow Warbler (*Dendroica petechia*). M.S. thesis, Ohio State University, Columbus, Ohio.

Link, W. A., E. Cam, J. D. Nichols, and E. G. Cooch. 2002. Of bugs and birds: Markov chain Monte Carlo for hierarchical modeling in wildlife research. Journal of Wildlife Management 6:277–291.

Link, W. A., and J. R. Sauer. 2007. Seasonal components of avian population change: Joint analysis of two large-scale monitoring programs. Ecology 88:49–55.

Liscinsky, S. A. 1965. The American Woodcock in Pennsylvania. Pittman Robertson Federal Aid Project W-50-R. Pennsylvania Game Commission, Harrisburg, PA.

Lokemoen, J. T., H. F. Duebbert, and D. E. Sharp. 1990. Homing and reproductive habits of Mallards, Gadwalls, and Blue-winged Teal. Wildlife Monographs 106:1–28.

Longcore, J. R., D. G. Mcauley, G. R. Hepp, and J. M. Rhymer. 2000. American Black Duck (*Anas rubripes*). *In* The Birds of North America, no. 481 (A. Poole and F. B. Gill, Eds.). The Birds of North America, Inc., Philadelphia, PA.

Lor, S., and R. A. Malecki. 2006. Breeding ecology and nesting habitat associations of five marsh bird species in western New York. Waterbirds 29:427–436.

Lovallo, M. J. 2008. Status and management of fisher (*Martes pennanti*) in Pennsylvania, 2008–2017. Bureau of Wildlife Management, Pennsylvania Game Commission, Harrisburg, PA.

Lovallo, M. J., and T. S. Hardisky. 2009. Furbearer population and harvest monitoring. Wildlife Management Annual Report. Pennsylvania Game Commission, Harrisburg, PA.

Lowe, J. D., and R. S. Hames. 2008a. Swainson's Thrush (*Catharus ustulatus*). Pages 444–445 *in* The Second Atlas of Breeding Birds in New York State (K. J. McGowan and K. Corwin, Eds.). Cornell University Press, Ithaca, NY.

———. 2008b. Scarlet Tanager (*Piranga olivacea*). Pages 540–541 *in* The Second Atlas of Breeding Birds in New York State (K. J. McGowan and K. Corwin, Eds.). Cornell University Press, Ithaca, NY.

Lowther, P. E. 1993. Brown-headed Cowbird (*Molothrus ater*). *In* The Birds of North America Online, no. 47 (A. Poole, Ed.). Cornell Lab of Ornithology, Ithaca, NY.

———. 1999. Alder Flycatcher (*Empidonax alnorum*). *In* The Birds of North America Online, no. 446 (A. Poole, Ed.). Cornell Lab of Ornithology, Ithaca, NY.

Lowther, P. E., C. Celada, N. Klein, C. C. Rimmer, and D. A. Spector. 1999. Yellow Warbler (*Dendroica petechia*). *In* The Birds of North America, no. 454 (A. Poole and F. Gill, Eds.). The Birds of North America, Inc., Philadelphia, PA.

Lowther, P. E., and C. L. Cink. 2006. House Sparrow (*Passer domesticus*). *In* The Birds of North America Online, no. 12 (A. Poole, Ed.). Cornell Lab of Ornithology, Ithaca, NY.

Lowther, P. E., and J. L. Ingold. 2011. Blue Grosbeak (*Passerina caerulea*). *In* The Birds of North America Online, no. 79 (A. Poole, Ed.). Cornell Lab of Ornithology, Ithaca, NY.

Lumsden, H. G. 2007. Trumpeter Swan, Cygne trompette (*Cygnus buccinator*). Pages 66–67 *in* Atlas of the Breeding Birds of Ontario, 2001–2005 (M. D. Cadman, D. A. Sutherland, G. G. Beck, D. Lepage, and A. R. Coutrier, Eds.). Bird Studies Canada, Environment Canada, Ontario Field Ornithologists, Ontario Ministry of Natural Resources, Ontario Nature, Toronto, ON.

Lumsden, H. G., and M. C. Drever. 2002. Overview of the Trumpeter Swan Reintroduction Program in Ontario, 1982–2000. Waterbirds 25:301–312.

Lunn, D. J., A. Thomas, N. Best, and D. Spiegelhalter. 2000. WinBUGS—a Bayesian modelling framework: Concepts, structure, and extensibility. Statistics and Computing 10:325–337.

Lynch, J. A. 1998. Atmospheric deposition in Pennsylvania. Pages 245–257 *in* The Effects of Acidic Deposition on Pennsylvania's Forests, vol. 1 (W. S. Sharpe and J. R. Drohan, Eds.). Environmental Resources Research Institute, The Pennsylvania State University, University Park, PA.

Lynch, J. F. 1989. Distribution of overwintering Nearctic migrants in the Yucatán Peninsula, I: General patterns of occurrence. Condor 91:515–544.

Mack, D. E., and W. Yong. 2000. Swainson's Thrush (*Catharus ustulatus*). *In* The Birds of North America, no. 540 (A. Poole and F. Gill, Eds.). The Birds of North America, Inc., Philadelphia, PA.

Macoun, J., and J. H. Macoun. 1909. Catalogue of Canadian Birds. Canada Department of Mines Geological Survey, Ottawa, Canada.

MacWhirter, R. B., and K. L. Bildstein. 1996. Northern Harrier (*Circus cyaneus*). *In* The Birds of North America, no. 210 (A. Poole and F. Gill, Eds.). The Birds of North America, Inc., Philadelphia, PA.

Mallory, M., and K. Metz. 1999. Common Merganser (*Mergus merganser*). *In* The Birds of North America Online, no. 442 (A. Poole, Ed.). Cornell Lab of Ornithology, Ithaca, NY.

Malosh, G. 2009. County Reports—Lawrence County. Pennsylvania Birds 23:69.

Mandel, J. T., K. L. Bildstein, G. Bohrer, and D. W. Winkler. 2008. Movement ecology of migration in Turkey Vultures. Proceedings of the National Academy of Sciences (USA) 105:19102–19107.

Mank, J. E., J. E. Carlson, and M. C. Brittingham. 2004. A century of hybridization: Decreasing genetic distance between American Black Ducks and Mallards. Journal of Conservation Genetics 5:395–403.

Marks, J. S., D. L. Evans, and D. W. Holt. 1994. Long-eared Owl (*Asio otus*). *In* The Birds of North America, no. 133 (A. Poole and F. Gill, Eds.). The Birds of North America, Inc., Philadelphia, PA.

Marti, C. D. 1992. Barn Owl (*Tyto alba*). *In* The Birds of North America, no. 1 (A. Poole, P. Stettenheim, and F. Gill., Eds.). The Academy of Natural Sciences, Philadelphia, PA, and the American Ornithologists' Union, Washington, DC.

Marti, C. D., P. W. Wagner, and K. W. Denne. 1979. Nest boxes for the management of Barn Owls. Wildlife Society Bulletin 7:145–148.

Martin, F. W., and J. R. Sauer. 1993. Population characteristics and trends in the Eastern Management Unit. Pages 281–304 *in* Ecology and Management of the Mourning Dove (T. S. Baskett, M. W. Sayre, R. E. Tomlinson, and R. E. Mirarchi, Eds.). Stackpole Books, Harrisburg, PA.

Marzilli, V. 1989. Up on the roof. Maine Fish and Wildlife 31:25–29.

Marzluff, J. M., and T. Angell. 2005. In the Company of Crows and Ravens. Yale University Press, New Haven, CT.

Masse, R. J., A. M. Strong, and N. G. Perlut. 2008. The potential of uncut patches to increase the nesting success of grassland songbirds in intensively managed hayfields: A preliminary study from the Champlain Valley of Vermont. Northeastern Naturalist 15:445–452.

Master, T. L. 1992a. Green-backed Heron (*Butorides striatus*). Pages 58–59 *in* Atlas of Breeding Birds in Pennsylvania (D. W. Brauning, Ed.). University of Pittsburgh Press, Pittsburgh, PA.

———. 1992b. Short-eared Owl (*Asio flammeus*). Pages 164–165 *in* Atlas of Breeding Birds in Pennsylvania (D. W. Brauning, Ed.). University of Pittsburgh Press, Pittsburgh, PA.

———. 1992c. Belted Kingfisher (*Ceryle alcyon*). Pages 178–179 *in* Atlas of Breeding Birds in Pennsylvania (D. W. Brauning, Ed.). University of Pittsburgh Press, Pittsburgh, PA.

———. 1992d. Warbling Vireo (*Vireo gilvus*). Pages 294–295 *in* Atlas of Breeding Birds in Pennsylvania (D. W. Brauning, Ed.). University of Pittsburgh Press, Pittsburgh, PA.

———. 1992e. European Starling (*Sturnus vulgaris*). Pages 286–287 *in* Atlas of Breeding Birds in Pennsylvania (D. W. Brauning, Ed.). University of Pittsburgh Press, Pittsburgh, PA.

———. 1992f. Nashville Warbler (*Vermivora ruficapilla*). Pages 304–305 *in* Atlas of Breeding Birds in Pennsylvania (D. W. Brauning, Ed.). University of Pittsburgh Press, Pittsburgh, PA.

———. 1992g. Kentucky Warbler (*Oporornis formosus*). Pages 346–347 *in* Atlas of Breeding Birds in Pennsylvania (D. W. Brauning, Ed.). University of Pittsburgh Press, Pittsburgh, PA.

———. 1992h. Song Sparrow (*Melospiza melodia*). Pages 388–389 *in* Atlas of Breeding Birds in Pennsylvania (D. W. Brauning, Ed.). University of Pittsburgh Press, Pittsburgh, PA.

———. 1992i. Purple Finch (*Carpodacus purpureus*). Pages 412–413 *in* Atlas of Breeding Birds in Pennsylvania (D. W. Brauning, Ed.). University of Pittsburgh Press, Pittsburgh, PA.

———. 2010a. Great Egret (*Ardea alba*). Pages 190–192 *in* Terrestrial Vertebrates of Concern in Pennsylvania: A Complete Guide to Species of Conservation Concern (M. A. Steele, M. C. Brittingham, T. J. Maret, and J. F. Merritt, Eds.). The Johns Hopkins University Press, Baltimore, MD.

———. 2010b. Yellow-crowned Night-Heron (*Nyctanassa violacea*). Pages 195–197 *in* Terrestrial Vertebrates of Concern in Pennsylvania: A Complete Guide to Species of Conservation Concern (M. A. Steele, M. C. Brittingham, T. J. Maret, and J. F. Merritt, Eds.). The Johns Hopkins University Press, Baltimore, MD.

Master, T. L., R. S. Mulvihill, R. C. Leberman, J. Sánchez, and E. Carman. 2005. A preliminary survey of riparian songbirds in Costa Rica, with emphasis on wintering Louisiana Waterthrush. Pages 528–532 *in* Bird Conservation Implementation and Integration in the Americas: Proceedings of the Third International Partners in Flight Conference, 20–24 March 2002, Asilomar, CA, vol. 1 (C. J. Ralph and T. D. Rich, Eds.). U.S. Department of Agriculture, Forest Service, Pacific Southwest Research Station, Albany, CA.

Matray, P. 1974. Broad-winged Hawk nesting and ecology. Auk 91:307–324.

Matthews, S. N., L. R. Iverson, A. M. Prasad, and M. P. Peters. 2010. A climate change atlas for 147 bird species of the eastern United States [database] 2007– ongoing. United States Forest Service, Northern Research Station.

———. 2011. Changes in potential habitat of 147 North American breeding bird species in response to redistribution of trees and climate following predicted climate change. [Online.]. Ecography 34.

Mattice, J. A., D. W. Brauning, and D. R. Diefenbach. 2005. Abundance of grassland songbirds on reclaimed surface mines in western Pennsylvania. Pages 504–510 *in* Bird Conservation Implementation and Integration in the Americas: Proceedings of the Third International Partners in Flight Conference, 20–24 March 2002, Asilomar, CA, vol. 1 (C. J. Ralph and T. D. Rich, Eds.). U.S. Department of Agriculture, Forest Service, Pacific Southwest Research Station, Albany, CA.

Mattsson, B. J., T. L. Master, R. S. Mulvihill, and W. D. Robinson. 2009. Louisiana Waterthrush (*Seiurus motacilla*). *In* The Birds of North America Online, no. 151 (A. Poole, Ed.). Cornell Lab of Ornithology, Ithaca, NY.

Mazur, K. M., and P. C. James. 2000. Barred Owl (*Strix varia*). *In* The Birds of North America Online, no. 508 (A. Poole, Ed.). Cornell Lab of Ornithology, Ithaca, NY.

Mazzocchi, I., and S. Muller. 2008. Black Tern (*Chlidonias nigra*). Pages 266–267 *in* The Second Atlas of Breeding Birds in New York State (K. J. McGowan and K. Corwin, Eds.). Cornell University Press, Ithaca, NY.

McCarty, J. P. 1996. Eastern Wood-Pewee (*Contopus virens*). *In* The Birds of North America, no. 245 (A. Poole and F. Gill, Eds.). The Birds of North America, Inc., Philadelphia, PA.

McConnell, S. 2003. Nest site vegetation characteristics of Cooper's Hawks in Pennsylvania. Journal of the Pennsylvania Academy of Science 76:72–76.

McCracken, J. 2008. Are aerial insectivores being bugged out? BirdWatch Canada 42:4–7.

McCracken, J. D. 1993. Status report on the Cerulean Warbler (*Dendroica cerulea*) in Canada. Committee on the Status of Endangered Wildlife in Canada, Ottawa, ON.

McCrimmon, D. A., Jr. 2008a. Great Egret (*Ardea alba*). Pages 162–163 *in* The Second Atlas of Breeding Birds in New York State (K. J. McGowan and K. Corwin, Eds.). Cornell University Press, Ithaca, NY.

———. 2008b. Black-crowned Night-Heron (*Nycticorax nycticorax*). Pages 174–175 *in* The Second Atlas of Breeding Birds in New York State (K. J. McGowan and K. Corwin, Eds.). Cornell University Press, Ithaca, NY.

———. 2008c. Snowy Egret (*Egretta thula*). Pages 164–165 *in* The Second Atlas of Breeding Birds in New York State (K. J. McGowan and K. Corwin, Eds.). Cornell University Press, Ithaca, NY.

McCrimmon D. A., Jr., J. C. Ogden, and G. T. Bancroft. 2001. Great Egret (*Ardea alba*). *In* The Birds of North America Online, no. 570 (A. Poole, Ed.). Cornell Lab of Ornithology, Ithaca, NY.

McDonald, M. V. 1998. Kentucky Warbler (*Oporornis formosus*). *In* The Birds of North America Online, no. 324 (A. Poole, Ed.). Cornell Lab of Ornithology, Ithaca, NY.

McFarland, K., and C. C. Rimmer. 2008. Blackpoll Warbler (*Dendroica striata*). Pages 506–507 *in* The Second Atlas of Breeding Birds in New York State (K. J. McGowan and K. Corwin, Eds.). Cornell University Press, Ithaca, NY.

McGovern, D. 2006. Local notes—Delaware County. Pennsylvania Birds 20:96.

McGowan, K. J. 2001a. Demographic and behavioral comparisons of suburban and rural crows. Pages 365–381 *in* Avian Ecology and Conservation in an Urbanizing World (J. M. Marzluff, R. Bowman, and R. Donelly, Eds.). Kluwer Academic Press, Norwell, MA.

———. 2001b. Fish Crow (*Corvus ossifragus*). *In* The Birds of North America, no. 589 (A. Poole and F. Gill, Eds.). The Birds of North America, Inc., Philadelphia, PA.

———. 2008a. Mute Swan (*Cygnus olor*). Pages 90–91 *in* The Second Atlas of Breeding Birds in New York State (K. J. McGowan and K. Corwin, Eds.). Cornell University Press, Ithaca, NY.

———. 2008b. Trumpeter Swan (*Cygnus buccinator*). Pages 92–93 *in* The Second Atlas of Breeding Birds in New York State (K. J. McGowan and K. Corwin, Eds.). Cornell University Press, Ithaca, NY.

———. 2008c. Pied-billed Grebe (*Podilymbus podiceps*). Pages 150–151 *in* The Second Atlas of Breeding Birds in New York State (K. J. McGowan and K. Corwin, Eds.). Cornell University Press, Ithaca, NY.

———. 2008d. American Bittern (*Botaurus lentiginosus*). Pages 156–157 *in* The Second Atlas of Breeding Birds in New York State (K. J. McGowan and K. Corwin, Eds.). Cornell University Press, Ithaca, NY.

———. 2008e. Black Vulture (*Coragyps atratus*). Pages 182–183 *in* The Second Atlas of Breeding Birds in New York State (K. J. McGowan and K. Corwin, Eds.). Cornell University Press, Ithaca, NY.

———. 2008f. Turkey Vulture (*Cathartes aura*). Pages 184–185 *in* The Second Atlas of Breeding Birds in New York State (K. J. McGowan and K. Corwin, Eds.). Cornell University Press, Ithaca, NY.

———. 2008g. Sandhill Crane (*Grus canadensis*). Pages 228–229 *in* The Second Atlas of Breeding Birds in New York State (K. J. McGowan and K. Corwin, Eds.). Cornell University Press, Ithaca, NY.

———. 2008h. Spotted Sandpiper (*Actitis macularius*). Pages 238–239 *in* The Second Atlas of Breeding Birds in New York State (K. J. McGowan and K. Corwin, Eds.). Cornell University Press, Ithaca, NY.

———. 2008i. Yellow-billed Cuckoo (*Coccyzus americanus*). Pages 284–285 *in* The Second Atlas of Breeding Birds in New York State (K. J. McGowan and K. Corwin, Eds.). Cornell University Press, Ithaca, NY.

———. 2008j. Black-billed Cuckoo (*Coccyzus erythropthalmus*). Pages 286–287 *in* The Second Atlas of Breeding Birds in New York State (K. J. McGowan and K. Corwin, Eds.). Cornell University Press, Ithaca, NY.

———. 2008k. Barn Owl (*Tyto alba*). Pages 290–291 *in* The Second Atlas of Breeding Birds in New York State (K. J. McGowan and K. Corwin, Eds.). Cornell University Press, Ithaca, NY.

———. 2008l. Great Horned Owl (*Bubo virginianus*). Pages 294–295 *in* The Second Atlas of Breeding Birds in New York State (K. J. McGowan and K. Corwin, Eds.). Cornell University Press, Ithaca, NY.

———. 2008m. Barred Owl (*Strix varia*). Pages 296–297 *in* The Second Atlas of Breeding Birds in New York State (K. J. McGowan and K. Corwin, Eds.). Cornell University Press, Ithaca, NY.

———. 2008n. Ruby-throated Hummingbird (*Archilocus colubris*). Pages 314–315 *in* The Second Atlas of Breeding Birds in New York State (K. J. McGowan and K. Corwin, Eds.). Cornell University Press, Ithaca, NY.

———. 2008o. Belted Kingfisher (*Megaceryle alcyon*). Pages 316–317 *in* The Second Atlas of Breeding Birds in New York State (K. J. McGowan and K. Corwin, Eds.). Cornell University Press, Ithaca, NY.

———. 2008p. Red-headed Woodpecker (*Melanerpes erythrocephalus*). Pages 320–321 *in* The Second Atlas of Breeding Birds in New York State (K. J. McGowan and K. Corwin, Eds.). Cornell University Press, Ithaca, NY.

———. 2008q. Red-bellied Woodpecker (*Melanerpes carolinus*). Pages 322–323 *in* The Second Atlas of Breeding Birds in New York State (K. J. McGowan and K. Corwin, Eds.). Cornell University Press, Ithaca, NY.

———. 2008r. Yellow-bellied Sapsucker (*Sphyrapicus varius*). Pages 324–325 *in* The Second Atlas of Breeding Birds in New York State (K. J. McGowan and K. Corwin, Eds.). Cornell University Press, Ithaca, NY.

———. 2008s. Eastern Wood-Pewee (*Contopus virens*). Pages 342–343 *in* The Second Atlas of Breeding Birds in New York State (K. J. McGowan and K. Corwin, Eds.). Cornell University Press, Ithaca, NY.

———. 2008t. Least Flycatcher (*Empidonax minimus*). Pages 352–353 *in* The Second Atlas of Breeding Birds in New York State (K. J. McGowan and K. Corwin, Eds.). Cornell University Press, Ithaca, NY.

———. 2008u. Great Crested Flycatcher (*Myiarchus crinitus*). Pages 356–357 *in* The Second Atlas of Breeding Birds in New York State (K. J. McGowan and K. Corwin, Eds.). Cornell University Press, Ithaca, NY.

———. 2008v. Eastern Kingbird (*Tyrannus tyrannus*). Pages 358–359 *in* The Second Atlas of Breeding Birds in New York State (K. J. McGowan and K. Corwin, Eds.). Cornell University Press, Ithaca, NY.

———. 2008w. Blue-headed Vireo (*Vireo solitarius*). Pages 368–369 *in* The Second Atlas of Breeding Birds in New York State (K. J. McGowan and K. Corwin, Eds.). Cornell University Press, Ithaca, NY.

———. 2008x. Warbling Vireo (*Vireo gilvus*). Pages 370–371 *in* The Second Atlas of Breeding Birds in New York State (K. J. McGowan and K. Corwin, Eds.). Cornell University Press, Ithaca, NY.

———. 2008y. Red-eyed Vireo (*Vireo olivaceus*). Pages 374–375 *in* The Second Atlas of Breeding Birds in New York State (K. J. McGowan and K. Corwin, Eds.). Cornell University Press, Ithaca, NY.

———. 2008z. Fish Crow (*Corvus ossifragus*). Pages 384–385 *in* The Second Atlas of Breeding Birds in New York State (K. J. McGowan and K. Corwin, Eds.). Cornell University Press, Ithaca, NY.

———. 2008za. Northern Rough-winged Swallow (*Stelgidopteryx serripennis*). Pages 396–397 *in* The Second Atlas of Breeding Birds in New York State (K. J. McGowan and K. Corwin, Eds.). Cornell University Press, Ithaca, NY.

———. 2008zb. Tufted Titmouse (*Baeolophus bicolor*). Pages 410–411 *in* The Second Atlas of Breeding Birds in New York State (K. J. McGowan and K. Corwin, Eds.). Cornell University Press, Ithaca, NY.

———. 2008zc. Red-breasted Nuthatch (*Sitta canadensis*). Pages 412–413 *in* The Second Atlas of Breeding Birds in New York State (K. J. McGowan and K. Corwin, Eds.). Cornell University Press, Ithaca, NY.

———. 2008zd. Brown Creeper (*Certhia americana*). Pages 416–417 *in* The Second Atlas of Breeding Birds in New York State (K. J. McGowan and K. Corwin, Eds.). Cornell University Press, Ithaca, NY.

———. 2008ze. Winter Wren (*Troglodytes troglodytes*). Pages 424–425 *in* The Second Atlas of Breeding Birds in New York State (K. J. McGowan and K. Corwin, Eds.). Cornell University Press, Ithaca, NY.

———. 2008zf. Sedge Wren (*Cistothorus platensis*). Pages 426–427 *in* The Second Atlas of Breeding Birds in New York State (K. J. McGowan and K. Corwin, Eds.). Cornell University Press, Ithaca, NY.

———. 2008zg. Golden-crowned Kinglet (*Regulus satrapa*). Pages 430–431 *in* The Second Atlas of Breeding Birds in New York State (K. J. McGowan and K. Corwin, Eds.). Cornell University Press, Ithaca, NY.

———. 2008zh. American Robin (*Turdus migratorius*). Pages 450–451 *in* The Second Atlas of Breeding Birds in New York State (K. J. McGowan and K. Corwin, Eds.). Cornell University Press, Ithaca, NY.

———. 2008zi. Gray Catbird (*Dumetella carolinensis*). Pages 454–455 *in* The Second Atlas of Breeding Birds in New York State (K. J. McGowan and K. Corwin, Eds.). Cornell University Press, Ithaca, NY.

———. 2008zj. Northern Mockingbird (*Mimus polyglottos*). Pages 456–457 *in* The Second Atlas of Breeding Birds in New York State (K. J. McGowan and K. Corwin, Eds.). Cornell University Press, Ithaca, NY.

———. 2008zk. Brown Thrasher (*Toxostoma rufum*). Pages 458–459 *in* The Second Atlas of Breeding Birds in New York State (K. J. McGowan and K. Corwin, Eds.). Cornell University Press, Ithaca, NY.

———. 2008zl. Ovenbird (*Seiurus aurocapilla*). Pages 518–519 *in* The Second Atlas of Breeding Birds in New York State (K. J. McGowan and K. Corwin, Eds.). Cornell University Press, Ithaca, NY.

———. 2008zm. Northern Waterthrush (*Seiurus noveboracensis*). Pages 520–521 *in* The Second Atlas of Breeding Birds in New York State (K. J. McGowan and K. Corwin, Eds.). Cornell University Press, Ithaca, NY.

———. 2008zn. Hooded Warbler (*Wilsonia citrina*). Pages 530–531 *in* The Second Atlas of Breeding Birds in New York State (K. J. McGowan and K. Corwin, Eds.). Cornell University Press, Ithaca, NY.

———. 2008zo. American Redstart (*Setophaga ruticilla*). Pages 512–513 *in* The Second Atlas of Breeding Birds in New York State (K. J. McGowan and K. Corwin, Eds.). Cornell University Press, Ithaca, NY.

———. 2008zp. Magnolia Warbler (*Dendroica magnolia*). Pages 484–485 *in* The Second Atlas of Breeding Birds in New York State (K. J. McGowan and K. Corwin, Eds.). Cornell University Press, Ithaca, NY.

———. 2008zq. Yellow-rumped Warbler (*Dendroica coronata*). Pages 490–491 *in* The Second Atlas of Breeding Birds in New York State (K. J. McGowan and K. Corwin, Eds.). Cornell University Press, Ithaca, NY.

———. 2008zr. Yellow-throated Warbler (*Dendroica dominica*). Pages 496–497 *in* The Second Atlas of Breeding Birds in New York State (K. J. McGowan and K. Corwin, Eds.). Cornell University Press, Ithaca, NY.

———. 2008zs. Canada Warbler (*Wilsonia canadensis*). Pages 534–535 *in* The Second Atlas of Breeding Birds in New York State (K. J. McGowan and K. Corwin, Eds.). Cornell University Press, Ithaca, NY.

———. 2008zt. Yellow-breasted Chat (*Icteria virens*). Pages 536–537 *in* The Second Atlas of Breeding Birds in New York State (K. J. McGowan and K. Corwin, Eds.). Cornell University Press, Ithaca, NY.

———. 2008zu. Rose-breasted Grosbeak (*Pheucticus ludovicianus*). Pages 578–579 *in* The Second Atlas of Breeding Birds in New York State (K. J. McGowan and K. Corwin, Eds.). Cornell University Press, Ithaca, NY.

———. 2008zv. Indigo Bunting (*Passerina cyanea*). Pages 582–583 *in* The Second Atlas of Breeding Birds in New York State (K. J. McGowan and K. Corwin, Eds.). Cornell University Press, Ithaca, NY.

———. 2008zw. Western Meadowlark (*Sturnella neglecta*). Pages 632–633 *in* The Second Atlas of Breeding Birds in New York State (K. J. McGowan and K. Corwin, Eds.). Cornell University Press, Ithaca, NY.

———. 2008zx. Brown-headed Cowbird (*Molothrus ater*). Pages 600–601 *in* The Second Atlas of Breeding Birds in New York State (K. J. McGowan and K. Corwin, Eds.). Cornell University Press, Ithaca, NY.

———. 2008zy. Orchard Oriole (*Icterus spurius*). Pages 602–603 *in* The Second Atlas of Breeding Birds in New York State (K. J. McGowan and K. Corwin, Eds.). Cornell University Press, Ithaca, NY.

———. 2008zz. American Wigeon (*Anas americana*). Pages 98–99 *in* The Second Atlas of Breeding Birds in New York State (K. J. McGowan and K. Corwin, Eds.). Cornell University Press, Ithaca, NY.

———. 2008zza. Northern Shoveler (*Anas clypeata*). Pages 108–109 *in* The Second Atlas of Breeding Birds in New York State (K. J. McGowan and K. Corwin, Eds.). Cornell University Press, Ithaca, NY.

———. 2008zzb. Ring-necked Duck (*Aythya collaris*). Pages 118–119 *in* The Second Atlas of Breeding Birds in New York State (K. J. McGowan and K. Corwin, Eds.). Cornell University Press, Ithaca, NY.

———. 2008zzc. Ruddy Duck (*Oxyura jamaicensis*). Pages 130–131 *in* The Second Atlas of Breeding Birds in New York State (K. J. McGowan and K. Corwin, Eds.). Cornell University Press, Ithaca, NY.

McGowan, K. J., and K. Corwin, Eds. 2008. The Second Atlas of Breeding Birds in New York State. Cornell University Press, Ithaca, NY.

McGraw, K. J., and A. L. Middleton. 2009. American Goldfinch (*Spinus tristis*). *In* The Birds of North America Online, no. 80 (A. Poole, Ed.). Cornell Lab of Ornithology, Ithaca, NY.

McLaren, M. A. 2007a. Brown Creeper, Grimpereau brun (*Regulus satrapa*). Pages 410–411 *in* Atlas of the Breeding Birds of Ontario, 2001–2005 (M. D. Cadman, D. A. Sutherland, G. G. Beck, D. Lepage, and A. R. Coutrier, Eds.). Bird Studies Canada, Environment Canada, Ontario Field Ornithologists, Ontario Ministry of Natural Resources, Ontario Nature, Toronto, ON.

———. 2007b. Golden-crowned Kinglet, Roitelet a couronne doree (*Regulus satrapa*). Pages 422–423 *in* Atlas of the Breeding Birds of Ontario, 2001–2005 (M. D. Cadman, D. A. Sutherland, G. G. Beck, D. Lepage, and A. R. Coutrier, Eds.). Bird Studies Canada, Environment Canada, Ontario Field Ornithologists, Ontario Ministry of Natural Resources, Ontario Nature, Toronto, ON.

———. 2007c. American Redstart, Paruline flamboyante (*Setophaga ruticilla*). Pages 504–505 *in* Atlas of the Breeding Birds of Ontario, 2001–2005 (M. D. Cadman, D. A. Sutherland, G. G. Beck, D. Lepage, and A. R. Coutrier, Eds.). Bird Studies Canada, Environment Canada, Ontario Field Ornithologists, Ontario Ministry of Natural Resources, and Ontario Nature, Toronto, ON.

McMorris, F. A., and D. W. Brauning. 2010. Peregrine Falcon research and management. Wildlife Management Annual Report. Pennsylvania Game Commission, Harrisburg, PA.

McShea, W. J., and J. H. Rappole. 2000. Managing the abundance and diversity of breeding bird populations through manipulation of deer populations. Conservation Biology 14:1161–1170.

McWilliams, G. M., and D. W. Brauning. 2000. The Birds of Pennsylvania. Cornell University Press, Ithaca, NY.

McWilliams, J. 1995. Attempted nesting of three species of Laridae at Presque Isle State Park, 1995. Pennsylvania Birds 9:79–80.

———. 1999. The first successful Ring-billed Gull nesting in Pennsylvania. Pennsylvania Birds 13:62.

McWilliams, W. H., S. P. Cassell, C. L. Alerich, B. J. Butler, M. L. Hoppus, S. B. Horsley, A. J. Lister, T. W. Lister, R. S. Morin, S. Randall, C. H. Perry, J. A. Westfall, E. H. Wharton, and C. W. Woodall. 2007. Pennsylvania's Forest, 2004. Northeastern Research Station Resource Bulletin NRS-20. United States Forest Service, Newtown Square, PA.

Meanley, B. 1971. Natural History of the Swainson's Warbler. North American Fauna. U.S. Department of the Interior, Bureau of Sport Fisheries and Wildlife, Washington, DC.

Medica, D. L., R. Clauser, and K. L. Bildstein. 2007. Prevalence of West Nile virus in a breeding population of American Kestrels in Pennsylvania. Journal of Wildlife Diseases 43:538–541.

Medler, M. D. 2008a. King Rail (*Rallus elegans*). Pages 218–219 *in* The Second Atlas of Breeding Birds in New York State (K. J. McGowan and K. Corwin, Eds.). Cornell University Press, Ithaca, NY.

———. 2008b. Common Moorhen (*Gallinula chloropus*). Pages 224–225 *in* The Second Atlas of Breeding Birds in New York State (K. J. McGowan and K. Corwin, Eds.). Cornell University Press, Ithaca, NY.

———. 2008c. American Coot (*Fulica americana*). Pages 226–227 *in* The Second Atlas of Breeding Birds in New York State (K. J. McGowan and K. Corwin, Eds.). Cornell University Press, Ithaca, NY.

———. 2008d. Long-eared Owl (*Asio otus*). Pages 298–299 *in* The Second Atlas of Breeding Birds in New York State (K. J. McGowan and K. Corwin, Eds.). Cornell University Press, Ithaca, NY.

———. 2008e. Common Nighthawk (*Chordeiles minor*). Pages 306–307 *in* The Second Atlas of Breeding Birds in New York State (K. J. McGowan and K. Corwin, Eds.). Cornell University Press, Ithaca, NY.

———. 2008f. Whip-poor-will (*Caprimulgus vociferus*). Pages 310–311 *in* The Second Atlas of Breeding Birds in New York State (K. J. McGowan and K. Corwin, Eds.). Cornell University Press, Ithaca, NY.

———. 2008g. Purple Martin (*Progne subis*). Pages 392–393 *in* The Second Atlas of Breeding Birds in New York State (K. J. McGowan and K. Corwin, Eds.). Cornell University Press, Ithaca, NY.

———. 2008h. Cliff Swallow (*Petrochelidon pyrrhonota*). Pages 400–401 *in* The Second Atlas of Breeding Birds in New York State (K. J. McGowan and K. Corwin, Eds.). Cornell University Press, Ithaca, NY.

Melvin, S. M., and J. P. Gibbs. 1996. Sora (*Porzana carolina*). *In* The Birds of North America, no. 250 (A. Poole and F. Gill, Eds.). The Birds of North America, Inc., Philadelphia, PA.

Merendino, T. M., C. D. Ankney, and D. G. Dennis. 1993. Increasing Mallards, decreasing American Black Ducks: More evidence for cause and effect. Journal of Wildlife Management 57:199–208.

Meyer, R. 2006. *Porzana carolina*. United States Forest Service, Rocky Mountain Research Station, Fire Sciences Laboratory. http://www.fs.fed.us/database/feis/.

Michener, E. 1863. Insectivorous birds of Chester County, Pennsylvania. U.S. Agriculture Report. United States Department of Agriculture, Washington, DC.

Middleton, A. L. 1998. Chipping Sparrow (*Spizella passerina*). *In* The Birds of North America Online, no. 334 (A. Poole, Ed.). Cornell Lab of Ornithology, Ithaca, NY.

Middleton, R. J. 1963. Probable breeding of the House Finch in Pennsylvania. Cassinia 47:40.

Miller, E. W. 1989. A concise historical atlas of Pennsylvania. Temple University Press, Philadelphia, PA.

Miller, J. C. 1966a. The first Short-eared Owl nesting record for the Tinicum area. Cassinia 49:30.

———. 1966b. A nesting record of the pintail at Tinicum Wildlife Preserve. Cassinia 49:30.

———. 1999. Update of the status of birds in the John Heinz National Wildlife Refuge at Tinicum and adjacent areas in Philadelphia and Delaware counties, Pennsylvania, 1988–1994. Cassinia 67:3–8.

Miller, K. E. 2010. Nest-site limitation of secondary cavity-nesting birds in even-age southern pine forests. Wilson Journal of Ornithology 122:126–134.

Miller, R. F. 1933. The breeding birds of Philadelphia. Pennsylvania Oologist 50:86–95.

Miller, T. A. 2007. Habitat modeling, validation and creation using Second Pennsylvania Breeding Bird Atlas data. M.S. thesis, The Pennsylvania State University, University Park, PA.

Mirabella, A. 2005. Local notes—Bucks County. Pennsylvania Birds 19:63.

Mitchell, C. D. 1994. Trumpeter Swan (*Cygnus buccinator*). *In* The Birds of North America, no. 105 (A. Poole and F. Gill, Eds.). The Birds of North America, Inc., Philadelphia, PA.

Mitra, S. S. 2008. Chuck-will's-widow (*Caprimulgus carolinensis*). Pages 308–309 *in* The Second Atlas of Breeding Birds in New York State (K. J. McGowan and K. Corwin, Eds.). Cornell University Press, Ithaca, NY.

Moldenhauer, R. R., and D. J. Regelski. 1996. Northern Parula (*Parula americana*). *In* The Birds of North America, no. 215 (A. Poole and F. Gill, Eds.). The Birds of North America, Inc., Philadelphia, PA.

Møller, A. P. 1994. Sexual Selection and the Barn Swallow. Oxford University Press, Oxford, UK.

Monroe, B. L., Jr., R. C. Banks, J. W. Fitzpatrick, T. R. Howell, N. K. Johnson, H. Ouellet, J. V. Remsen, and R. W. Storer. 1995. Fortieth supplement to the American Ornithologists' Union Check-list of North American birds. Auk 112:819–830.

Moore, S., J. R. Narwot, and J. P. Severson. 2009. Wetland scale habitat determinants influencing Least Bittern use of created wetlands. Waterbirds 321:16–24.

Morgan, T. C., C. A. Bishop, and T. D. Williams. 2007. Yellow-breasted Chat and Gray Catbird productivity in a fragmented western riparian system. Wilson Journal of Ornithology 119:494–498.

Morimoto, D. C., and F. E. Wasserman. 1991. Dispersion patterns and habitat associations of Rufous-sided Towhees, Common Yellowthroats, and Prairie Warblers in the southeastern Massachusetts pine barrens. Auk 108:264–276.

Morrell, T. E., and R. H. Yahner. 1994. Habitat characteristics of Great Horned Owls in southcentral Pennsylvania. Journal of Raptor Research 28:164–170.

Morris, B., R. Wiltraut, and F. Brock. 1984. Birds of the Lehigh Valley area. Lehigh Valley Audubon Society, Emmaus, PA.

Morrison, M. L., K. A. With, I. C. Timossi, W. M. Block, and K. A. Milne. 1987. Foraging behavior of bark-foraging birds in the Sierra Nevada. Condor 89:201–204.

Morrison, R. I. G., B. J. McCaffery, R. E. Gill, S. K. Skagen, S. L. Jones, G. W. Page, C. L. Gratto-Trevor, and B. A. Andres. 2006. Population estimates of North American shorebirds, 2006. Wader Study Group Bulletin 111:67–85.

Morse, D. H. 1967. Competitive relationships between Parula Warblers and other species during the breeding season. Auk 84:490–502.

———. 2004. Blackburnian Warbler (*Setophaga fusca*). *In* The Birds of North America Online, no. 102 (A. Poole, Ed.). Cornell Lab of Ornithology, Ithaca, NY.

Morse, D. H., and A. F. Poole. 2005. Black-throated Green Warbler (*Setophaga virens*). *In* The Birds of North America Online, no. 55 (A. Poole, Ed.). Cornell Lab of Ornithology, Ithaca, NY.

Morton, E. S. 1992. What do we know about the future of neotropical migrant landbirds? Pages 579–589 *in* Ecology and Conservation of Neotropical Migrant Landbirds (J. M. Hagan, III, and D. W. Johnston, Eds.). Smithsonian Institution Press, Washington, DC.

Morton, E. S., B. J. M. Stutchbury, and I. Chiver. 2010. Parental conflict and brood desertion by females in Blue-headed Vireos. Behavioral Ecology and Sociobiology 64:947–954.

Morton, E. S., B. J. M. Stutchbury, J. S. Howlett, and W. H. Piper. 1998. Genetic monogamy in Blue-headed Vireos, and a comparison with a sympatric vireo with extrapair paternity. Behavioral Ecology 9:515–524.

Moulton, M. P., W. P. Cropper, Jr., M. L. Avery, and L. E. Moulton. 2010. The earliest House Sparrow introductions to North America. Biological Invasions 12:2955–2958.

Mowbray, T. B. 1997. Swamp Sparrow (*Melospiza georgiana*). *In* The Birds of North America Online, no. 279 (A. Poole, Ed.). Cornell Lab of Ornithology, Ithaca, NY.

———. 1999. Scarlet Tanager (*Piranga olivacea*). *In* The Birds of North America, no. 479 (A. Poole and F. Gill, Eds.). The Birds of North America, Inc., Philadelphia, PA.

Mowbray, T. B., C. R. Ely, J. S. Sedinger, and R. E. Trost. 2002. Canada Goose (*Branta canadensis*). *In* The Birds of North America, no. 682 (A. Poole and F. Gill, Eds.). The Birds of North America, Inc., Philadelphia, PA.

Mueller, H. 1999. Wilson's Snipe (*Gallinago delicata*). *In* The Birds of North America Online, no. 417 (A. Poole, Ed.). Cornell Lab of Ornithology, Ithaca, NY.

Muller, M. J., and R. W. Storer. 1999. Pied-billed Grebe (*Podilymbus podiceps*). *In* The Birds of North America Online, no. 410 (A. Poole, Ed.). Cornell Lab of Ornithology, Ithaca, NY.

Mulvihill, R. S. 1992a. Ruby-throated Hummingbird (*Archilocus colubris*). Pages 176–177 *in* Atlas of Breeding Birds in Pennsylvania (D. W. Brauning, Ed.). University of Pittsburgh Press, Pittsburgh, PA.

———. 1992b. Willow Flycatcher (*Empidonax traillii*). Pages 204–205 *in* Atlas of Breeding Birds in Pennsylvania (D. W. Brauning, Ed.). University of Pittsburgh Press, Pittsburgh, PA.

———. 1992c. Least Flycatcher (*Empidonax minimus*). Pages 206–207 *in* Atlas of Breeding Birds in Pennsylvania (D. W. Brauning, Ed.). University of Pittsburgh Press, Pittsburgh, PA.

———. 1992d. Red-eyed Vireo (*Vireo olivaceus*). Pages 296–297 *in* Atlas of Breeding Birds in Pennsylvania (D. W. Brauning, Ed.). University of Pittsburgh Press, Pittsburgh, PA.

———. 1992e. Common Raven (*Corvus corax*). Pages 236–237 *in* Atlas of Breeding Birds in Pennsylvania (D. W. Brauning, Ed.). University of Pittsburgh Press, Pittsburgh, PA.

———. 1992f. Golden-crowned Kinglet (*Regulus satrapa*). Pages 260–261 *in* Atlas of Breeding Birds in Pennsylvania (D. W. Brauning, Ed.). University of Pittsburgh Press, Pittsburgh, PA.

———. 1992g. Hermit Thrush (*Catharus guttatus*). Pages 270–271 *in* Atlas of Breeding Birds in Pennsylvania (D. W. Brauning, Ed.). University of Pittsburgh Press, Pittsburgh, PA.

———. 1992h. American Redstart (*Setophaga ruticilla*). Pages 332–333 *in* Atlas of Breeding Birds in Pennsylvania (D. W. Brauning, Ed.). University of Pittsburgh Press, Pittsburgh, PA.

———. 1992i. Canada Warbler (*Wilsonia canadensis*). Pages 354–355 *in* Atlas of Breeding Birds in Pennsylvania (D. W. Brauning, Ed.). University of Pittsburgh Press, Pittsburgh, PA.

———. 1992j. Chipping sparrow (*Spizella passerina*). Pages 374–375 *in* Atlas of Breeding Birds in Pennsylvania (D. W. Brauning, Ed.). University of Pittsburgh Press, Pittsburgh, PA.

———. 1992k. Dark-eyed Junco (*Junco hyemalis*). Pages 394–395 *in* Atlas of Breeding Birds in Pennsylvania (D. W. Brauning, Ed.). University of Pittsburgh Press, Pittsburgh, PA.

———. 1992l. Dickcissel (*Spiza americana*). Pages 370–371 *in* Atlas of Breeding Birds in Pennsylvania (D. W. Brauning, Ed.). University of Pittsburgh Press, Pittsburgh, PA.

Mulvihill, R. S., F. L. Newell, and S. C. Latta. 2008. Effects of acidification on the breeding ecology of a stream-dependent songbird, the Louisiana waterthrush (*Seiurus motacilla*). Freshwater Biology 53:2158–2169.

Murphy, M. T. 2001. Source-sink dynamics of a declining Eastern Kingbird population and the value of sink habitats. Conservation Biology 15:737–748.

Myers, W., J. Bishop, R. Brooks, T. O'Connell, D. Argent, G. Storm, J. Stauffer, and R. Carline. 2000. Pennsylvania Gap Analysis Project: Leading Landscapes for Collaborative Conservation. School of Forest Resources, Cooperative Fish and Wildlife Research Unit, and Environmental Resources Research Institute, The Pennsylvania State University, University Park, PA.

NABCI (North American Bird Conservation Initiative). 2007. Opportunities for improving bird monitoring. U.S. North American Bird Conservation Initiative Report. Available from the Division of Migratory Bird Management. U.S. Department of Interior, Fish and Wildlife Service, Arlington, VA. [Online.] Available at http://www.nabci-us.org/.

NABCI-US (North American Bird Conservation Initiative–United States). 2011. [Online.] Available at http://www.nabci-us.org/.

NAS (National Audubon Society). 2002. The Christmas Bird Count historical results. National Audubon Society. [Online.] Available at http://www.audubon.org/bird/cbc.

———. 2007. Common birds in decline. National Audubon Society. New York, NY. [Online.] Available at http://stateofthebirds.audubon.org/cbid/.

———. 2010. The Christmas Bird Count historical results. National Audubon Society. [Online.] Available at http://www.audubon.org/bird/cbc.

NatureServe. 2006. Strategic plan, 2007–2011: Guiding conservation to action. NatureServe, Arlington, VA.

NatureServeExplorer. 2011. [Online.] Available at http://www.natureserve.org/explorer/.

NAWMP (North American Waterfowl Management Plan). 2004. North American Waterfowl Management Plan, 2004. Strategic guidance: Strengthening the biological foundation. Canadian Wildlife Service, Gatineau, QC, U.S. Fish and Wildlife Service, Arlington, VA, and Secretaria de Medio Ambiente y Recursos Naturales, Colonia Tlacaopac, Mexico.

Nebel, S., A. Mills, J. D. McCracken, and P. D. Taylor. 2010. Declines of aerial insectivores in North America follow a geographic gradient. Avian Conservation and Ecology 5:1–14. [Online.] Available at http://www.ace-eco.org/vol15/iss12/art11/.

Nemeth, N. M., D. C. Hahn, D. H. Gould, and R. A. Bowen. 2006. Experimental West Nile virus infection in Eastern Screech-Owls (*Megascops asio*). Avian Diseases 50:252–258.

Newell, F. 2010. Wood Thrush (*Hylocichla mustelina*). Pages 171–174 *in* Terrestrial Vertebrates of Concern in Pennsylvania: A Complete Guide to Species of Conservation Concern (M. A. Steele, M. C. Brittingham, T. J. Maret, and J. F. Merritt, Eds.). The Johns Hopkins University Press, Baltimore, MD.

Nice, M. M. 1937. Studies in the life history of the Song Sparrow, Pt. 1. Transactions of the Linnaean Society of New York 4:1–247.

Nickell, W. P. 1965. Habitats, territory, and nesting of the catbird. American Midland Naturalist 73:433–478.

Nisbet, I. C., T. D. B. McNair, W. Post, and T. C. Williams. 1995. Transoceanic migration of the Blackpoll Warbler: Summary of scientific evidence and response to criticisms by Murray. Journal of Field Ornithology 66:612–622.

Nocera, J. J., G. J. Parsons, G. R. Milton, and A. H. Fredeen. 2005. Compatibility of delayed cutting regime with bird breeding and hay nutritional quality. Agriculture Ecosystems & Environment 107:245–253.

Nolan V. Jr., E. D. Ketterson, and C. A. Buerkle. 1999. Prairie Warbler (*Setophaga discolor*). *In* The Birds of North America Online, no. 455 (A. Poole, Ed.). Cornell Lab of Ornithology, Ithaca, NY.

Nolan V. Jr., E. D. Ketterson, D. A. Cristol, C. M. Rogers, E. D. Clotfelter, R. C. Titus, S. J. Schoech, and E. Snajdr. 2002. Dark-eyed Junco (*Junco hyemalis*). *In* The Birds of North America Online, no. 716 (A. Poole, Ed.). Cornell Lab of Ornithology, Ithaca, NY.

Nolan V. Jr., and C. F. Thompson. 1975. The occurrence and significance of anomalous reproductive activities in two North American nonparasitic cuckoos *Coccyzus* spp. Ibis 117:496–503.

Noon, B. R. 1981. The distribution of an avian guild along a temperate elevational gradient: The importance and expression of competition. Ecological Monographs 51:105–124.

Norris, D. R., and B. J. M. Stutchbury. 2001. Extraterritorial movements of a forest songbird in a fragmented landscape. Conservation Biology 15:729–739.

Noss, R. F., E. T. LaRoe, and J. M. Scott. 1995. Endangered Ecosystems of the United States: A Preliminary Assessment of Loss and Degradation. National Biological Service, Washington, DC.

Nowacki, G. J., and M. D. Abrams. 1992. Community, edaphic, and historical analysis of mixed oak forests of the Ridge and Valley Province in central Pennsylvania. Canadian Journal of Forest Research 22:790–800.

Nuttall, T. 1832. A Manual of the Ornithology of the United States and of Canada: Volume 1—Land Birds. Hilliard, Gray, and Company, Boston, MA.

Nye, P. E. 2008. Osprey (*Pandion haliaetus*). Pages 186–187 *in* The Second Atlas of Breeding Birds in New York State (K. J. McGowan and K. Corwin, Eds.). Cornell University Press, Ithaca, NY.

NYNHP. 2009. Online conservation guide for *Ixobrychus exilis*. New York Natural Heritage Program (NYNHP). http://www.acris.nynhp.org/guide.php?id=6751.

OBBA. 2011. Ohio Breeding Bird Atlas II—Current Atlas Results. The Ohio State University, School of Environment and Natural Resources and the Ohio Department of Natural Resources-Division of Wildlife. http://www.ohiobirds.org/obba2.

O'Connell, T., J. Bishop, R. Mulvihill, M. Lanzone, T. Miller, and R. Brooks. 2004. Sampling design for Pennsylvania's Second Breeding Bird Atlas, 2004–2009. Unpublished report submitted to the Pennsylvania Game Commission, Harrisburg, PA.

O'Connell, T. J., L. E. Jackson, and R. P. Brooks. 2000. Bird guilds as indicators of ecological condition in the central Appalachians. Ecological Applications 10:1706–1721.

O'Connell, T. J., R. P. Brooks, S. E. Laubscher, R. S. Mulvihill, and T. E. Master. 2003. Using Bioindicators to Develop a Calibrated Index of Regional Ecological Integrity for Forested Headwater Ecosystems. Penn State Cooperative Wetlands Center, The Pennylvania State University, University Park, PA.

O'Connor, R. J., E. Dunn, D. H. Johnson, S. I. Jones, D. Petit, K. Pollock, C. R. Smith, J. L. Trapp, and E. Welling. 2000. A programmatic review of the North American Breeding Bird Survey: Report of a peer review panel. [Online.] Available at www.pwrc.usgs.gov/bbs/bbsreview/bbsfinal.pdf.

ODNR (Ohio Department of Natural Resources). 2008. Common Ravens nest in Ohio for the first time in more than 100 years. Ohio Department of Natural Resources. News Release 28 May 2008.

Oksanen, J., F. G. Blanchet, R. Kindt, P. Legendre, P. R. Minchin, R. B. O'Hara, G. L. Simpson, P. Solymos, M. H. H. Stevens, and H. Wagner. 2011. Vegan: Community Ecology Package. R package version 2.0-2.

Oring, L. W., E. M. Gray, and J. M. Reed. 1997. Spotted Sandpiper (*Actitis macularius*). *In* The Birds of North America Online, no. 289 (A. Poole, Ed.). Cornell Lab of Ornithology, Ithaca, NY.

Ortego, B. 2005. Summary of highest counts of individuals for the United States. American Birds 105:123–128.

Osborne, C. 2008. Swamp Sparrow (*Melospiza georgiana*). Pages 568–569 *in* The Second Atlas of Breeding Birds in New York State (K. J. McGowan and K. Corwin, Eds.). Cornell University Press, Ithaca, NY.

Osnas, E. E. 2003. The role of competition and local habitat conditions for determining occupancy patterns in grebes. Waterbirds 26:209–216.

Otis, D. L., J. H. Schulz, and D. P. Scott. 2008. Mourning Dove (*Zenaida macroura*) harvest and population parameters derived from a national banding study. Biological Technical Publication. U.S. Department of the Interior, Fish and Wildlife Service, Washington, DC.

Otto, M. C., and J. R. Sauer. 2007. Contiguous 48 states Bald Eagle breeding pair survey design. Draft post-delisting monitoring plan for the Bald Eagle (*Haliaeetus leucocephalus*). U.S. Fish and Wildlife Service, Washington, DC.

Pabian, S. E., and M. C. Brittingham. 2007. Terrestrial liming benefits birds in an acidified forest in the Northeast. Ecological Applications 17:2194–2194.

———. 2011. Soil calcium availability limits forest songbird productivity and density. Auk 128:441–447.

PADEP (Pennsylvania Department of Environmental Protection). 2010. Wetlands Net Gain Strategy. Pennsylvania Department of Environmental Protection. [Online.] Available at http://www.portal.state.pa.us/portal/server.pt/community/wetlands/10635/net_gain_strategy/554350.

Palmer, R. S. 1988. Handbook of North American Birds, vol. 4. Yale University Press, New Haven, CT.

Panjabi, A. O., E. H. Dunn, P. J. Blancher, W. C. Hunter, B. Altman, J. Bart, C. J. Beardmore, H. Berlanga, G. S. Butcher, S. K. Davis, D. W. Demarest, R. Dettmers, W. Easton, H. Gomez de Silva Garza, E. E. Iñigo-Elias, D. N. Pashley, C. J. Ralph, T. D. Rich, K. V. Rosenberg, C. M. Rustay, J. M. Ruth, J. S. Wendt, and T. C. Will. 2005. The Partners in Flight handbook on species assessment, version 2005. Partners in Flight Technical Series No. 3. Rocky Mountain Bird Observatory. [Online.] Available at http://www.rmbo.org/pubs/downloads/Handbook2005.pdf.

Parchman, T. L., C. W. Benkman, and S. C. Britch. 2006. Patterns of genetic variation in the adaptive radiation of New World crossbills. Molecular Ecology 15:1873–1887.

Parkes, K. C. 1951. The genetics of the Golden-winged and Blue-winged Warbler complex. Wilson Bulletin 63:5–15.

———. 1956. A field list of birds of the Pittsburgh region. Carnegie Museum of Natural History, Pittsburgh, PA.

Parkhurst, J. A., R. P. Brooks, and D. E. Arnold. 1992. Assessment of predation at trout hatcheries in central Pennsylvania. Wildlife Society Bulletin 29:411–419.

Parsons, K. C. 1995. Heron nesting at Pea Patch Island, upper Delaware Bay, USA: Abundance and reproductive success. Colonial Waterbirds 18:69–78.

Payne, R. B. 2006. Indigo Bunting (*Passerina cyanea*). *In* The Birds of North America Online, no. 4 (A. Poole, Ed.). Cornell Lab of Ornithology, Ithaca, NY.

Pearson, T. G. 1917. Birds of America. Garden City Books, Garden City, NY.

Peer, B. D., and E. K. Bollinger. 1997. Common Grackle (*Quiscalus quiscula*). *In* The Birds of North America Online, no. 271 (A. Poole, Ed.). Cornell Lab of Ornithology, Ithaca, NY.

Pelham, P. H., and J. G. Dickson. 1992. Physical characteristics. Pages 32–45 *in* The Wild Turkey: Biology and Management (J. G. Dickson, Ed.). Stackpole Books, Harrisburg, PA.

Pennock, C. J. 1912. Crossbills in Chester County, Pennsylvania in summer. Auk 29: 245–246.

Pepper, W. E. 1972. Nesting House Finch in Pennsylvania. Cassinia 53:46.

Perkins, A. W., B. J. Johnson, and E. E. Blankenship. 2003. Response of riparian avifauna to percentage and pattern of woody cover in an agricultural landscape. Wildlife Society Bulletin 31:642–660.

Perlut, N. G., A. M. Strong, T. M. Donovan, and N. J. Buckley. 2008. Regional population viability of grassland songbirds: Effects of agricultural management. Biological Conservation 141:3159–3151.

Perry, E. F., J. C. Manolis, and D. E. Andersen. 2008. Reduced predation at interior nests in clustered all-purpose territories of Least Flycatchers (*Empidonax minimus*). Auk 125:643–650.

Peterjohn, B. G. 1989. The Birds of Ohio. Indiana University Press, Bloomington, IN.

———. 2001. The Birds of Ohio: With Ohio Breeding Bird Atlas Maps. The Wooster Book Company, Wooster, OH.

Peterjohn, B. G., and D. L. Rice. 1991. The Ohio Breeding Bird Atlas. Ohio Department of Natural Resources, Columbus, OH.

Peterson, J. M. C. 2008a. Yellow-bellied Flycatcher (*Empidonax flaviventris*). Pages 344–345 *in* The Second Atlas of Breeding Birds in New York State (K. J. McGowan and K. Corwin, Eds.). Cornell University Press, Ithaca, NY.

———. 2008b. White-throated Sparrow (*Zonotrichia albicollis*). Pages 570–571 *in* The Second Atlas of Breeding Birds in New York State (K. J. McGowan and K. Corwin, Eds.). Cornell University Press, Ithaca, NY.

Petit, L. J. 1999. Prothonotary Warbler (*Prothonotaria citrica*). *In* The Birds of North America, no. 408 (A. Poole and F. Gill, Eds.). The Birds of North America, Inc., Philadelphia, PA.

PGC-PFBC (Pennsylvania Game Commission–Pennsylvania Fish and Boat Commission). 2005. Pennsylvania Wildlife Action Plan (formerly Comprehensive Wildlife Conservation Strategy) (L. Williams, Ed.). Pennsylvania Game Commission and Pennsylvania Fish and Boat Commission, revised 2008, Harrisburg, PA.

Pierotti, R. J., and T. P. Good. 1994. Herring Gull (*Larus argentatus*). *In* The Birds of North America Online, no. 124 (A. Poole, Ed.). Cornell Lab of Ornithology, Ithaca, NY.

Pitocchelli, J. 2011. Mourning Warbler (*Oporornis philadelphia*). *In* The Birds of North America Online, no. 72 (A. Poole, Ed.). Cornell Lab of Ornithology, Ithaca, NY.

PNHP. 2011. Pennsylvania Natural Heritage Program (PNHP). [Online.] Available at http://www.naturalheritage.state.pa.us/.

Poling, T. D., and S. E. Hayslette. 2006. Dietary overlap and foraging competition between Mourning Doves and Eurasian Collared-Doves. Journal of Wildlife Management 70:998–1004.

Poole, A. F., L. R. Bevier, C. A. Marantz, and B. Meanley. 2005. King Rail (*Rallus elegans*). *In* The Birds of North America Online, no. 3 (A. Poole, Ed.). Cornell Lab of Ornithology, Ithaca, NY.

Poole, E. L. 1964. Pennsylvania Birds: An Annotated List. Livingston Publishing Co., Narberth, PA.

Porej, D. 2003. Vegetation cover and wetland complex size as predictors of bird use of created wetlands in Ohio. Pages 151–160 *in* The Olentangy River Wetland Research Park at the Ohio State University (W. J. Mitsch, L. Zhang, and C. Anderson, Eds.). Department of Evolution, Ecology, and Organismal Biology.

Porneluzi, P., J. Bednarz, L. Goodrich, J. Hoover, and N. Zawada. 1993. Reproductive performance of territorial ovenbirds occupying forest fragments and a contiguous forest in Pennsylvania. Conservation Biology 7:618–622.

Post, T. J. 2008a. Ruffed Grouse (*Bonasa umbellus*). Pages 138–139 *in* The Second Atlas of Breeding Birds in New York State (K. J. McGowan and K. Corwin, Eds.). Cornell University Press, Ithaca, NY.

———. 2008b. Northern Harrier (*Circus cyaneus*). Pages 190–191 *in* The Second Atlas of Breeding Birds in New York State (K. J. McGowan and K. Corwin, Eds.). Cornell University Press, Ithaca, NY.

———. 2008c. Alder Flycatcher (*Empidonax alnorum*). Pages 348–349 *in* The Second Atlas of Breeding Birds in New York State (K. J. McGowan and K. Corwin, Eds.). Cornell University Press, Ithaca, NY.

———. 2008d. Willow Flycatcher (*Empidonax traillii*). Pages 350–351 *in* The Second Atlas of Breeding Birds in New York State (K. J. McGowan and K. Corwin, Eds.). Cornell University Press, Ithaca, NY.

———. 2008e. Chestnut-sided Warbler (*Dendroica pensylvanica*). Pages 482–483 *in* The Second Atlas of Breeding Birds in New York State (K. J. McGowan and K. Corwin, Eds.). Cornell University Press, Ithaca, NY.

Pottie, J. J., and H. W. Heusmann. 1979. Taxonomy of resident Canada Geese in Massachusetts. Transactions Northeast Fish and Wildlife Conference 36:132–137.

Pough, R. H. 1951. Audubon Water Bird Guide. Doubleday & Company, Garden City, NY.

Poulin, R. G., S. D. Grindal, and R. M. Brigham. 1996. Common Nighthawk (*Chordeiles minor*). *In* The Birds of North America, no. 213 (A. Poole and F. Gill, Eds.). The Academy of Natural Sciences, Philadelphia, PA, and the American Ornithologists' Union, Washington, DC.

Prescott, D. R. C., and A. J. Murphy. 1999. Bird populations of seeded grasslands in the aspen parkland of Alberta. Studies in Avian Biology 19:203–210.

Preston, C. R., and R. D. Beane. 2009. Red-tailed Hawk (*Buteo jamaicensis*). *In* The Birds of North America Online, no. 52 (A. Poole, Ed.). Cornell Lab of Ornithology, Ithaca, NY.

Price, J., S. Droege, and A. Price. 1995. The Summer Atlas of North American Birds. Academic Press, New York, NY.

Prosser, D. J. 1998. Avian use of different successional stage beaver ponds in Pennsylvania. M.S. thesis, The Pennsylvania State University, University Park, PA.

Pulcinella, N. 1997. Eighth report of the Pennsylvania Ornithological Records Committee, June 1997. Pennsylvania Birds 11:129.

———. 2011. Eighteenth report of the Pennsylvania Ornithological Records Committee. Pennsylvania Birds 25:2–8.

Pyle, P. 1997. Identification guide to North American birds, Part 1: Columbidae to Ploceidae. Slate Creek Press, Bolinas, CA.

Raftovich, R. V., K. A. Wilkins, K. D. Richkus, S. S. Williams, and H. L. Spriggs. 2009. Migratory bird hunting activity and harvest during the 2007 and 2008 hunting seasons. U.S. Fish and Wildlife Service, Laurel, MD.

Ralph, C. J., S. Droege, and J. R. Sauer. 1995. Managing and Monitoring Birds Using Point Counts: Standards and Applications. United States Forest Service, Albany, CA.

Ramos, M. A., and D. W. Warner. 1980. Ecological aspects of migrant bird behavior in Veracruz, Mexico. Pages 353–394 *in* Migrant Birds in the Neotropics: Ecology, Behavior, Distribution, and Conservation (A. Keast and E. S. Morton, Eds.). Smithsonian Institute Press, Washington, DC.

Rasmussen, J. L., S. G. Sealy, and R. J. Cannings. 2008. Northern Saw-whet Owl (*Aegolius acadicus*). *In* The Birds of North

America Online, no. 42 (A. Poole, Ed.). Cornell Lab of Ornithology, Ithaca, NY.

Read, P. 2007. Carolina Wren, Troglodyte de Caroline (*Thryothorus ludovicianus*). Pages 412–413 *in* Atlas of the Breeding Birds of Ontario, 2001–2005 (M. D. Cadman, D. A. Sutherland, G. G. Beck, D. Lepage, and A. R. Coutrier, Eds.). Bird Studies Canada, Environment Canada, Ontario Field Ornithologists, Ontario Nature, and Ontario Ministry of Natural Resources, Toronto, ON.

Reed, J. M. 1992. A system for ranking conservation priorities for Neotropical migrant birds based on relative susceptibility to extinction. Pages 524–536 *in* Ecology and Conservation of Neotropical Migrant Landbirds (J. M. Hagan III and D. W. Johnston, Eds.). Smithsonian Institution Press, Washington, DC.

Reese, J. G. 1996. Chuck-will's-widow (*Caprimulgus carolinensis*). Pages 192–193 *in* Atlas of the Breeding Birds of Maryland and the District of Columbia (C. S. Robbins and E. A. T. Blom, Eds.). University of Pittsburgh Press, Pittsburgh, PA.

Register, S. M., and K. Islam. 2008. Effects of silvicultural treatments on Cerulean Warbler (*Dendroica cerulea*) abundance in southern Indiana. Forest Ecology and Management 255:3502–3505.

Reid, W. 1992a. Common Merganser (*Mergus merganser*). Pages 84–85 *in* Atlas of Breeding Birds in Pennsylvania (D. W. Brauning, Ed.). University of Pittsburgh Press, Pittsburgh, PA.

———. 1992b. Sora (*Porzana carolina*). Pages 126–127 *in* Atlas of Breeding Birds in Pennsylvania (D. W. Brauning, Ed.). University of Pittsburgh Press, Pittsburgh, PA.

———. 1992c. Eastern Screech-Owl (*Otus asio*). Pages 156–157 *in* Atlas of Breeding Birds in Pennsylvania (D. W. Brauning, Ed.). University of Pittsburgh Press, Pittsburgh, PA.

———. 1992d. Marsh Wren (*Cistothorous palustris*). Pages 258–259 *in* Atlas of Breeding Birds in Pennsylvania (D. W. Brauning, Ed.). University of Pittsburgh Press, Pittsburgh, PA.

———. 1992e. Henslow's Sparrow (*Ammodramus henslowii*). Pages 386–387 *in* Atlas of Breeding Birds in Pennsylvania (D. W. Brauning, Ed.). University of Pittsburgh Press, Pittsburgh, PA.

Reitz, S. L., and B. Nelson. 2010. Red-shouldered Hawk (*Buteo lineatus*). Pages 241–243 *in* Terrestrial Vertebrates of Concern in Pennsylvania: A Complete Guide to Species of Conservation Concern (M. A. Steele, M. C. Brittingham, T. J. Maret, and J. F. Merritt, Eds.). The Johns Hopkins University Press, Baltimore, MD.

Rentch, J. S., T. M. Schuler, W. M. Ford, and G. J. Nowacki. 2007. Red spruce stand dynamics, simulations, and restoration opportunities in the central Appalachians. Restoration Ecology 15:440–452.

Reudink, M. W., S. G. Mech, and R. L. Curry. 2006. Extrapair paternity and mate choice in a chickadee hybrid zone. Behavioral Ecology and Sociobiology 17:56–62.

Reudink, M. W., S. G. Mech, S. P. Mullen, and R. L. Curry. 2007. Structure and dynamics of the hybrid zone between Black-capped Chickadees (*Poecile atricapillus*) and Carolina Chickadees (*Poecile carolinensis*) in southeastern Pennsylvania. Auk 124:463–478.

Reynard, G. B. 1953. Nesting mockingbirds in the lower Delaware Valley. Cassinia 40:27–32.

Rhoads, S. N. 1903. Exit the Dickcissel—a remarkable case of local extinction. Cassinia 7:17–28.

Ribic, C. A., M. J. Gusy, and D. W. Sample. 2008. Grassland bird use of remnant prairie and Conservation Reserve Program fields in an agricultural landscape in Wisconsin. American Midland Naturalist 161:110–122.

Ribic, C. A., R. R. Koford, J. R. Herkert, D. H. Johnson, N. D. Niemuth, D. E. Naugle, K. K. Bakker, D. W. Sample, and R. B. Renfrew. 2009. Area sensitivity in North American grassland birds: Patterns and processes. Auk 126:233–244.

Ribic, C. A., S. J. Lewis, S. Melvin, J. Bart, and B. Peterjohn. 1999. Proceedings of the Marshbird Monitoring Workshop. Region 3, U.S. Department of the Interior, Fish and Wildlife Service, Fort Snelling, MN.

Rich, A. C., D. S. Dobkin, and L. J. Niles. 1994. Defining forest fragmentation by corridor width: The influence of narrow forest-dividing corridors on forest-nesting birds in southern New Jersey. Conservation Biology 8:1109–1121.

Rich, T. D., C. J. Beardmore, H. Berlanga, P. J. Blancher, M. S. W. Bradstreet, G. S. Butcher, D. W. Demarest, E. H. Dunn, W. C. Hunter, E. E. Inigo-Elias, J. A. Kennedy, A. M. Martell, A. O. Panjabi, D. N. Pashley, K. V. Rosenberg, C. M. Rustay, J. S. Wendt, and T. C. Will. 2004. Partners in Flight North American Landbird Conservation Plan. Cornell Lab of Ornithology. [Online.] Available at http://www.partnersinflight.org/cont_plan/.

Richards, K. C. 1976. Some declining bird species of southeastern Pennsylvania. Cassinia 55:33–36.

Richardson, M., and D. W. Brauning. 1995. Chestnut-sided Warbler (*Dendroica pensylvanica*). *In* The Birds of North America, no. 190 (A. Poole and F. Gill, Eds.). The Birds of North America, Inc., Philadelphia, PA.

Richmond, M. E. 2008a. Ring-billed Gull (*Laurus delawarensis*). Pages 254–255 *in* The Second Atlas of Breeding Birds in New York State (K. J. McGowan and K. Corwin, Eds.). Cornell University Press, Ithaca, NY.

———. 2008b. Common Tern (*Sterna hirundo*). Pages 270–271 *in* The Second Atlas of Breeding Birds in New York State (K. J. McGowan and K. Corwin, Eds.). Cornell University Press, Ithaca, NY.

Ripper, D., J. C. Bednarz, and D. E. Varland. 2005. Landscape use by Hairy Woodpeckers in managed forests of northwestern Washington. Journal of Wildlife Management 71:2612–2623.

Rising, J. D., and D. D. Beadle. 1996. A Guide to the Identification and Natural History of the Sparrows of the United States and Canada. Academic Press, San Diego, CA.

Rising, J. D., and N. J. Flood. 1998. Baltimore Oriole (*Icterus galbula*). *In* The Birds of North America Online, no. 384 (A. Poole, Ed.). Cornell Lab of Ornithology, Ithaca, NY.

Robbins, C. S. 1979. Effects of forest fragmentation on breeding bird populations. Pages 198–213 *in* Management of North Central and Northeastern Forests for Nongame Birds (R. M. DeGraff and K. E. Evans, Eds.). United States Forest Service, General Technical Report NC-51, St. Paul, MN.

———. 1990. Use of Breeding Bird Atlases to monitor population change. Pages 18–22 *in* Survey Designs and Statistical Methods for the Estimation of Avian Population Trends (J. R. Sauer and S. Droege, Eds.). U.S. Fish and Wildlife Service, Washington, DC.

Robbins, C. S., D. Bystrak, and P. H. Geissler. 1986. The Breeding Bird Survey: Its first fifteen years, 1965–1979. U.S. Department of the Interior Fish and Wildlife Service, Washington, DC.

Robbins, C. S., D. K. Dawson, and B. A. Dowell. 1989a. Habitat area requirements of breeding forest birds of the Middle Atlantic States. Wildlife Monographs 103:1–34.

Robbins, C. S., S. Droege, and J. R. Sauer. 1989b. Monitoring bird populations with Breeding Bird Survey and atlas data. Annnales Zoolologici Fennici 26:297–304.

Robbins, C. S., and P. H. Geissler. 1990. Survey Methods and Mapping Grids. In Handbook for Atlasing North American Breeding Birds (C. R. Smith, Ed.). Vermont Institute of Natural Science, Quechee, VT.

Roberts, C., and C. J. Norment. 1999. Effects of plot size and habitat requirements on breeding success of Scarlet Tanagers. Auk 116:73–82.

Robertson, R. J., B. J. Stutchbury, and R. R. Cohen. 1992. Tree Swallow (*Tachycineta bicolor*). In The Birds of North America Online, no. 11 (A. Poole, Ed.). Cornell Lab of Ornithology, Ithaca, NY.

Robinson, P. 2001. Local notes—Adams County. Pennsylvania Birds 15:100.

———. 2007. Local notes—Adams County. Pennsylvania Birds 21:48.

Robinson, R. A., G. M. Siriwardena, and H. Q. P. Crick. 2005. Status and population trends of Starling (*Sturnus vulgaris*) in Great Britain. Bird Study 52:252–260.

Robinson, S. K. 1992. Population dynamics of breeding Neotropical migrants in a fragmented Illinois landscape. Pages 408–418 in Ecology and Conservation of Neotropical Migrant Landbirds (J. Hagan and D. W. Johnston, Eds.). Smithsonian Institute Press, Washington, DC.

Robinson, S. K., F. R. Thompson, III, T. M. Donovan, D. R. Whitehead, and J. Faaborg. 1995. Regional forest fragmentation and the nesting success of migratory birds. Science 267:1987–1990.

Rodewald, A. D. 2004. Landscape and local level influences of forest management on Cerulean Warblers in Pennsylvania. Pages 472–477 in Proceedings of the 14th Central Hardwoods Forest Conference (D. A. Yaussy, D. M. Hix, R. P. Long, and P. C. Goebel, Eds.). United States Forest Service, Northeastern Research Station, Wooster, OH.

Rodewald, A. D., and R. H. Yahner. 2000. Bird communities associated with harvested hardwood stands containing residual trees. Journal of Wildlife Management 69:924–932.

Rodewald, P. G., and R. D. James. 1996. Yellow-throated Vireo (*Vireo flavifrons*). In The Birds of North America, no. 247 (A. Poole and F. Gill, Eds.). The Birds of North America, Inc., Philadelphia, PA.

Rodewald, P. G., M. J. Santiago, and A. D. Rodewald. 2005. Habitat use of breeding Red-headed Woodpeckers on golf courses in Ohio. Wildlife Society Bulletin 33:448–453.

Rodewald, P. G., J. H. Withgott, and K. G. Smith. 1999. Pine Warbler (*Setophaga pinus*). In The Birds of North America Online, no. 438 (A. Poole, Ed.). Cornell Lab of Ornithology, Ithaca, NY.

Rodgers, N. W. 1994. More notes on the Sandhill Cranes in Lawrence County. Pennsylvania Birds 8:137–138.

Rohnke, A. T., and R. H. Yahner. 2008. Long-term effects of wastewater irrigation on habitat and a bird community in central Pennsylvania. Wilson Journal of Ornithology 120:146–152.

Romagosa, C. M., and R. F. Labisky. 2000. Establishment and dispersal of the Eurasian Collared-Dove in Florida. Journal of Field Ornithology 71:159–166.

Romano, W. B. 2008. Habitat selection and foraging behavior of Great Egrets (*Ardea alba*) in a riparian environment on the Susquehanna River in Harrisburg, Pennsylvania. M.S. thesis, East Stroudsburg University of Pennsylvania, East Stroudsburg, PA.

Root, T., and E. S. Goldsmith. 2010. Global climate change: Conservation challenges and impacts on birds. Pages 292–299 in Avian Ecology and Conservation: A Pennsylvania Focus with National Implications (S. K. Majumdar, T. L. Master, M. C. Brittingham, R. M. Ross, R. S. Mulvihill, and J. E. Huffman, Eds.). Pennsylvania Academy of Sciences, Easton, PA.

Root, T. L., J. T. Price, K. R. Hall, S. H. Schneider, C. Rosenweig, and J. A. Pounds. 2003. Fingerprints of global warming on wild animals and plants. Nature 421:57–60.

Rosenberg, K. V. 2004. Partners in Flight continental priorities and objectives defined at the state and Bird Conservation Region levels: Pennsylvania. Cornell Lab of Ornithology, Ithaca, NY.

Rosenberg, K. V. 2008a. Louisiana Waterthrush (*Seiurus motacilla*). Pages 522–523 in The Second Atlas of Breeding Birds in New York State (K. J. McGowan and K. Corwin, Eds.). Cornell University Press, Ithaca, NY.

———. 2008b. Kentucky Warbler (*Oporornis formosus*). Pages 524–525 in The Second Atlas of Breeding Birds in New York State (K. J. McGowan and K. Corwin, Eds.). Cornell University Press, Ithaca, NY.

Rosenberg, K. V., J. D. Lowe, and A. A. Dhondt. 1999. Effects of forest fragmentation on breeding tanagers: A continental perspective. Conservation Biology 13:568–583.

Rosenberg, K. V., and J. V. Wells. 1995. Importance of geographic areas to Neotropical migrant birds in the Northeast. Final report to the U.S. Fish and Wildlife Service, Region 5. Hadley, MA.

Rosenburg, C., G. Hammerson, M. Koenen, and D. W. Hehlman. 1992. Barn Owl. Species Management Abstract. The Nature Conservancy, Arlington, VA.

Rosenfield, R. N., J. Bielefeldt, J. L. Affeldt, and D. J. Beckmann. 1996. Urban nesting biology of Cooper's Hawks in Wisconsin. Pages 41–44 in Raptors in Urban Landscapes (D. Bird, D. Varland, and J. Negro, Eds.). Academic Press, London, UK.

Ross, B. D., M. L. Morrison, W. Hoffman, T. S. Fredericksen, R. J. Sawicki, E. Ross, M. B. Lester, J. Beyea, and B. N. Johnson. 2001. Bird relationships to habitat characteristics created by timber harvesting in Pennsylvania. Journal of the Pennsylvania Academy of Science 74:71–84.

Ross, K. 2007. American Black Duck, Canard noir (*Anas rubripes*). Pages 76–77 in Atlas of the Breeding Birds of Ontario, 2001–2005 (M. D. Cadman, D. A. Sutherland, G. G. Beck, D. Lepage, and A. R. Coutrier, Eds.). Bird Studies Canada, Environment Canada, Ontario Field Ornithologists, Ontario Nature, and Ontario Ministry of Natural Resources, Toronto, ON.

Ross, R. M. 2010a. Colonially nesting waders. Pages 259–273 in Avian Ecology and Conservation: A Pennsylvania Focus with National Implications (S. Majumdar, T. L. Master, M. C. Brittingham, R. M. Ross, R. S. Mulvihill, and J. E. Huffman, Eds.). Pennsylvania Academy of Science, Easton, PA.

———. 2010b. Great Blue Heron (*Ardea herodias*). Pages 234–237 *in* Terrestrial Vertebrates of Concern in Pennsylvania: A Complete Guide to Species of Conservation Concern (M. A. Steele, M. C. Brittingham, T. J. Maret, and J. F. Merritt, Eds.). The Johns Hopkins University Press, Baltimore, MD.

Ross, R. M., and J. H. Johnson. 1999. Fish losses to Double-crested Cormorant predation in eastern Lake Ontario, 1992–1997. Pages 61–70 *in* Symposium on Double-crested Cormorants: Population Status and Management Issues in the Midwest, U.S. Department of Agriculture Technical Bulletin No. 1879 (M. E. Tobin, Ed.). U.S. Department of Agriculture, Milwaukee, WI.

Ross, R. M., L. A. Redell, R. M. Bennett, and J. A. Young. 2004. Mesohabitat use of threatened hemlock forests by breeding birds of the Delaware River Basin in northeastern United States. Natural Areas Journal 24:307–315.

Rotenhouse, N. L., S. N. Matthews, K. P. McFarland, J. D. Lambert, L. R. Iverson, A. Prasad, T. S. Sillett, and R. T. Holmes. 2008. Potential effects of climate change on birds of the Northeast. Mitigation and Adaptation Strategies for Global Change 13:517–540.

Roth, G. W. 2010. Section 6, Soybeans. Crop Management Extension Group. College of Agricultural Sciences, The Pennsylvania State University. [Online.] Available at http://extension.psu.edu/agronomy-guide/cm/sec6/sec61.

Roth, R. R., M. S. Johnson, and T. J. Underwood. 1996. Wood Thrush (*Hylocichla mustelina*). *In* The Birds of North America, no. 246 (A. Poole and F. Gill, Eds.). The Birds of North America, Inc., Philadelphia, PA.

Ruefenacht, B., M. V. Finco, M. D. Nelson, R. L. Czaplewski, E. H. Helmer, J. A. Blackard, G. R. Holden, A. J. Lister, D. Salajanu, D. Weyermann, and K. Winterberger. 2008. Conterminous U.S. and Alaska forest type mapping using Forest Inventory and Analysis data. Photogrammetric Engineering & Remote Sensing 74:1379–1388.

Ruelas Inzuna, E., L. J. Goodrich, S. W. Hoffman, E. Martínez, J. P. Smith, E. Peresbarbosa, R. Rodríguez, K. L. Scheuermann, S. L. Mesa, Y. Cabrera, C. N. Ferriz, R. Straub, M. M. Peñaloza, and J. G. Barrios. 2009. Long-term conservation of migratory birds in México: The Veracruz River of raptors project. Pages 577–589 *in* Proceedings of the Fourth International Partners in Flight Conference: Tundra to Tropics: Connecting Birds, Habitats and People (T. D. Rich, C. Arizmendi, D. Demarest, and C. Thompson, Eds.). University of Texas–Pan American Press, McAllen, TX.

RUPRI (Rural Policy Research Institiute). 2006. Demographic and Economic Profile: Pennsylvania, updated 2006. Rural Policy Research Institute, University of Missouri-Columbia. [Online.] Available at http://www.rupri.org/Forms/Pennsylvania.pdf.

Ryder, J. P. 1993. Ring-billed Gull (*Larus delawarensis*). *In* The Birds of North America Online, no. 33 (A. Poole, Ed.). Cornell Lab of Ornithology, Ithaca, NY.

Ryman, L. 2006. Bald Eagles nest successfully on Osprey platform. Journal of Raptor Research 40:306–307.

Rymon, L. M. 1989. The restoration of Ospreys (*Pandion haliaetus*) to breeding status in Pennsylvania by hacking (1980–1986). Pages 359–362 *in* Raptors in the Modern World, Proceedings of the III World Conference on Birds of Prey and Owls (B. U. Meyburg and R. D. Chancellor, Eds.). World Working Group on Birds of Prey and Owls, Berlin, Germany.

———. 1992. Osprey (*Pandion haliaetus*). Pages 90–91 *in* Atlas of Breeding Birds in Pennsylvania (D. W. Brauning, Ed.). University of Pittsburgh Press, Pittsburgh, PA.

Sallabanks, R., and F. C. James. 1999. American Robin (*Turdus migratorius*). *In* The Birds of North America, no. 462 (A. Poole and F. Gill, Eds.). The Birds of North America, Inc., Philadelphia, PA.

Sallabanks, R., J. R. Walters, and J. A. Collazo. 2000. Breeding bird abundance in bottomland hardwood forests: Habitat, edge, and patch size effects. Condor 102:748–758.

Samuel, D. E. 1972. Notes on Barn and Cliff Swallows. Eastern Bird Banding Association News 35:76–84.

Sanders, T. A., and K. Parker. 2010. Mourning Dove population status, 2010. U.S. Fish and Wildlife Service, Division of Migratory Bird Management, Washington, DC.

Sandilands, A. 2007. Northern Harrier, Busard Saint-Martin (*Circus cyaneus*). Pages 172–173 *in* Atlas of the Breeding Birds of Ontario, 2001–2005 (M. D. Cadman, D. A. Sutherland, G. G. Beck, D. Lepage, and A. R. Coutrier, Eds.). Bird Studies Canada, Environment Canada, Ontario Field Ornithologists, Ontario Ministry of Natural Resources, and Ontario Nature, Toronto, ON.

Santner, S. 1992a. Killdeer (*Charadrius vociferus*). Pages 132–133 *in* Atlas of Breeding Birds in Pennsylvania (D. W. Brauning, Ed.). University of Pittsburgh Press, Pittsburgh, PA.

———. 1992b. Long-eared Owl (*Asio otus*). Pages 162–163 *in* Atlas of Breeding Birds in Pennsylvania (D. W. Brauning, Ed.). University of Pittsburgh Press, Pittsburgh, PA.

———. 1992c. Chuck-will's-widow (*Caprimulgus carolinensis*). Pages 170–171 *in* Atlas of Breeding Birds in Pennsylvania (D. W. Brauning, Ed.). University of Pittsburgh Press, Pittsburgh, PA.

———. 1992d. Red-breasted Nuthatch (*Sitta canadensis*). Pages 244–245 *in* Atlas of Breeding Birds in Pennsylvania (D. W. Brauning, Ed.). University of Pittsburgh Press, Pittsburgh, PA.

———. 1992e. Worm-eating Warbler (*Helmitheros vermivorus*). Pages 336–337 *in* Atlas of Breeding Birds in Pennsylvania (D. W. Brauning, Ed.). University of Pittsburgh Press, Pittsburgh, PA.

———. 1992f. Vesper Sparrow (*Pooecetes gramineus*). Pages 380–381 *in* Atlas of Breeding Birds in Pennsylvania (D. W. Brauning, Ed.). University of Pittsburgh Press, Pittsburgh, PA.

———. 1992g. Grasshopper Sparrow (*Ammodramus savannarum*). Pages 384–385 *in* Atlas of Breeding Birds in Pennsylvania (D. W. Brauning, Ed.). University of Pittsburgh Press, Pittsburgh, PA.

———. 1992h. Black-necked Stilt (*Himantopus mexicanus*). Pages 134–135 *in* Atlas of Breeding Birds in Pennsylvania (D. W. Brauning, Ed.). University of Pittsburgh Press, Pittsburgh, PA.

Sargent, R. R. 1999. Ruby-throated Hummingbird. Stackpole Books, Mechanicsburg, PA.

Sauer, J. R., and S. Droege. 1990. Recent Population trends in of the Eastern Bluebird. Wilson Bulletin 102:239–252.

———. 1992. Geographic Patterns in Population Trends of Neotropical Migrants in North America. Pages 26–42 *in* Ecology and conservation of neotropical migrant landbirds (J. M. Hagan III and D. W. Johnston, Eds.). Smithsonian Institute Press, Washington, DC.

Sauer, J. R., J. E. Hines, and J. Fallon. 2004. The North American Breeding Bird Survey results and analysis, 1966–2003. USGS Patuxent Wildlife Research Center, Laurel, MD.

———. 2008. The North American Breeding Bird Survey results and analysis, 1966–2007. USGS Patuxent Wildlife Research Center, Laurel, MD.

Sauer, J. R., J. E. Hines, J. E. Fallon, K. L. Pardieck, D. J. Ziolkowsk, Jr., and W. A. Link. 2011. The North American Breeding Bird Survey results and analysis, 1966–2009. USGS Patuxent Wildlife Research Center. http://www.mbr-pwrc.usgs.gov/bbs/.

Sauer, J. R., and W. A. Link. 2011. Analysis of the North American Breeding Bird Survey using hierarchical models. Auk 128:87–98.

Sauer, J. R., B. G. Peterjohn, and W. A. Link. 1994. Observer differences in the North American Breeding Bird Survey. Auk 11:50–62.

Saunders, C. A., P. Arcese, and K. D. O'Conner. 2003. Nest site characteristics in the Song Sparrow and parasitism by Brown-headed Cowbirds. Wilson Bulletin 115:24–28.

Schaadt, C. P., and L. M. Rymon. 1983. The restoration of Ospreys by hacking. Pages 299–305 in Biology and Management of Bald Eagles and Ospreys (D. M. Bird, Ed.). Harpell Press, Ste. Anne de Bellevue, QC.

Scharf, J. R., and J. Kren. 1996. Orchard Oriole (*Icterus spurius*). In The Birds of North America, no. 255 (A. Poole and F. Gill, Eds.). The Birds of North America, Inc., Philadelphia, PA.

Schill, K. L., and R. H. Yahner. 2010. Nest-site selection and nest survival of early succesional birds in central Pennsylvania. Wilson Journal of Ornithology 121:476–484.

Schorger, 1944. The quail in early Wisconsin. Transactions of the Wisconsin Academy of Sciences, Arts and Letters 36:77–103.

———. A. 1952. Introduction of the domestic pigeon. Auk 69:462–463.

Schutsky, R. M. 1992a. Great Egret (*Casmerodius albus*). Pages 52–53 in Atlas of Breeding Birds in Pennsylvania (D. W. Brauning, Ed.). University of Pittsburgh Press, Pittsburgh, PA.

———. 1992b. Black-crowned Night-Heron (*Nycticorax nycticorax*). Pages 60–61 in Atlas of Breeding Birds in Pennsylvania (D. W. Brauning, Ed.). University of Pittsburgh Press, Pittsburgh, PA.

———. 1992c. Yellow-crowned Night-Heron (*Nycticorax violaceus*). Pages 62–63 in Atlas of Breeding Birds in Pennsylvania (D. W. Brauning, Ed.). University of Pittsburgh Press, Pittsburgh, PA.

———. 1992d. House Wren (*Troglodytes aedon*). Pages 252–253 in Atlas of Breeding Birds in Pennsylvania (D. W. Brauning, Ed.). University of Pittsburgh Press, Pittsburgh, PA.

———. 1992e. Northern Cardinal (*Cardinalis cardinalis*). Pages 362–363 in Atlas of Breeding Birds in Pennsylvania (D. W. Brauning, Ed.). University of Pittsburgh Press, Pittsburgh, PA.

———. 1992f. Blue Grosbeak (*Guiraca caerulea*). Pages 366–367 in Atlas of Breeding Birds in Pennsylvania (D. W. Brauning, Ed.). University of Pittsburgh Press, Pittsburgh, PA.

———. 1992g. Snowy Egret (*Egretta thula*). Pages 54–55 in Atlas of Breeding Birds in Pennsylvania (D. W. Brauning, Ed.). University of Pittsburgh Press, Pittsburgh, PA.

———. 1992h. Cattle Egret (*Bubulcus ibis*). Pages 56–57 in Atlas of Breeding Birds in Pennsylvania (D. W. Brauning, Ed.). University of Pittsburgh Press, Pittsburgh, PA.

Schwalbe, P. W. 1992a. Great Horned Owl (*Bubo virginianus*). Pages 158–159 in Atlas of Breeding Birds in Pennsylvania (D. W. Brauning, Ed.). University of Pittsburgh Press, Pittsburgh, PA.

———. 1992b. Northern Rough-winged Swallow (*Stelgidopteryx serripennis*). Pages 220–221 in Atlas of Breeding Birds in Pennsylvania (D. W. Brauning, Ed.). University of Pittsburgh Press, Pittsburgh, PA.

———. 1992c. Cliff Swallow (*Hirundo pyrrhonota*). Pages 224–225 in Atlas of Breeding Birds in Pennsylvania (D. W. Brauning, Ed.). University of Pittsburgh Press, Pittsburgh, PA.

———. 1992d. Veery (*Catharus fuscescens*). Pages 266–267 in Atlas of Breeding Birds in Pennsylvania (D. W. Brauning, Ed.). University of Pittsburgh Press, Pittsburgh, PA.

———. 1992e. Northern Parula (*Parula americana*). Pages 306–307 in Atlas of Breeding Birds in Pennsylvania (D. W. Brauning, Ed.). University of Pittsburgh Press, Pittsburgh, PA.

———. 1992f. Western Meadowlark (*Sturnella neglecta*). Pages 402–403 in Atlas of Breeding Birds in Pennsylvania (D. W. Brauning, Ed.). University of Pittsburgh Press, Pittsburgh, PA.

Schwalbe, P. W., and R. M. Ross. 1992. Great Blue Heron (*Ardea herodias*). Pages 50–51 in Atlas of Breeding Birds in Pennsylvania (D. W. Brauning, Ed.). University of Pittsburgh Press, Pittsburgh, PA.

Schweinsberg, A. R. 1988. Birds of the Central Susquehanna Valley, Lewisburg, PA.

Sealy, S. G. 1994. Observed acts of egg destruction, egg removal and predation on nests of passerine birds at Delta Marsh, Manitoba. Canadian Field Naturalist 108:41–51.

Sebastiani, J. 2008. First Delaware record of nesting Sharp-shinned Hawk (*Accipiter striatus*) at Ashland Nature Center. Delmarva Ornithologist 37:27–54.

Sechler, F. C., Jr. 2010. Northern Harrier (*Circus cyaneus*). Pages 133–136 in Terrestrial Vertebrates of Concern in Pennsylvania: A Complete Guide to Species of Conservation Concern (M. A. Steele, M. C. Brittingham, T. J. Maret, and J. F. Merritt, Eds.). The Johns Hopkins University Press, Baltimore, MD.

Sedgwick, J. A. 2000. Willow Flycatcher (*Empidonax traillii*). In The Birds of North America, no. 533 (A. Poole and F. Gill, Eds.). The Birds of North America, Inc., Philadelphia, PA.

Senner, N. R., L. J. Goodrich, D. R. Barber, and M. W. Miller. 2009. Ovenbird nest site selection within a large, contiguous forest in eastern Pennsylvania: Variations in microhabitat at used and unused sites. Journal of the Pennsylvania Academy of Science 83:3–8.

Senner, S. E., and L. Goodrich. 1992. Broad-winged Hawk (*Buteo platypterus*). Pages 104–105 in Atlas of Breeding Birds in Pennsylvania (D. W. Brauning, Ed.). University of Pittsburgh Press, Pittsburgh, PA.

Shackelford, C. E., R. E. Brown, and R. N. Conner. 2000. Red-bellied Woodpecker (*Melanerpes carolinus*). In The Birds of North America, no. 500 (A. Poole and F. Gill, Eds.). The Birds of North America, Inc., Philadelphia, PA.

Shackelford, C. E., and R. N. Conner. 1997. Woodpecker abundance and habitat use in three forest types in eastern Texas. Wilson Bulletin 109:614–629.

Shapiro, L. H., R. A. Canterbury, D. M. Stover, and R. C. Fleischer. 2004. Reciprocal introgression between Golden-winged Warblers (*Vermivora chrysoptera*) and Blue-winged

Warblers (*V. pinus*) in eastern North America. Auk 121:1019–1030.

Sharrock, J. T. R., Ed. 1976. The Atlas of the Breeding Birds in Britain and Ireland. T. & A. D. Poyser, Hertfordshire, UK.

Shea, R. E., H. K. Nelson, L. N. Gillette, J. G. King, and D. K. Weaver. 2002. Restoration of Trumpeter Swans in North America: A century of progress and challenges. Waterbirds 25:296–300.

Sheaffer, S. E., and R. A. Malecki. 1996. Atlantic Flyway Progress Report: Adaptive harvest management for eastern Mallards. U.S. Fish and Wildlife Service, Laurel, MD.

———. 1998. Status of Atlantic Flyway resident nesting Canada Geese. Pages 67–70 *in* Biology and Management of Canada Geese (D. H. Rusch, M. D. Samuel, D. D. Humburg, and B. D. Sullivan, Eds.). Proceedings International Canada Goose Symposium, Milwaukee, WI.

Sheehan, J. 2010. Willow Flycatcher (*Empidonax trailii*). Pages 283–285 *in* Terrestrial Vertebrates of Concern in Pennsylvania: A Complete Guide to Species of Conservation Concern (M. A. Steele, M. C. Brittingham, T. J. Maret, and J. F. Merritt, Eds.). The Johns Hopkins University Press, Baltimore, MD.

Shepherd, D. D. 1992. Monitoring Ontario's owl populations: A recommendation. Ontario Ministry of Natural Resources Report. Ontario Ministry of Natural Resources, Port Rowan, ON.

Sherry, T. W., and R. T. Holmes. 1996. Winter habitat quality, population limitation, and conservation of Neotropical-Nearctic migrant birds. Ecology 77:36–48.

———. 1997. American Redstart (*Dendroica ruticilla*). *In* The Birds of North America, no. 277 (A. Poole and F. Gill, Eds.). The Academy of Natural Sciences, Philadelphia PA, and the American Ornithologists' Union, Washington, DC.

Shire, G., K. Brown, and G. Winegrad. 2000. Communications towers: A deadly hazard for birds. American Bird Conservancy, Washington, DC.

Shortle, J., D. Abler, S. Blumsack, R. Crane, Z. Kaufman, M. McDill, R. Najjar, R. Ready, T. Wagener, and D. Wardrop. 2009. Pennsylvania climate impact assessment: Report to the Department of Environmental Protection. Environment and Natural Resources Institute, The Pennsylvania State University, University Park, PA.

Showalter, C. R., and R. C. Whitmore. 2002. The effect of gypsy moth defoliation on cavity-nesting bird communities. Forest Science 48:273–281.

Shugart, H. H., Jr., and D. James. 1973. Ecological succession of breeding bird populations in northwestern Arkansas. Auk 90:62–77.

Shulte, L. A., and G. J. Niemi. 1998. Bird communities of early successional burned and logged forest. Journal of Wildlife Management 62:1418–1429.

Sillett, T. S., and R. T. Holmes. 2002. Variation in survivorship of a migratory songbird throughout its annual cycle. Journal of Animal Ecology 71:296–308.

Sillett, T. S., R. T. Holmes, and T. W. Sherry. 2000. Impacts of a global climate cycle on population dynamics of a migratory songbird. Science 288:2040–2042.

Simpson, M. R., Jr. 1992. Birds of the Blue Ridge Mountains. University of North Carolina, Chapel Hill, NC.

Sims, E., and W. R. DeGarmo. 1948. A study of Swainson's Warbler in West Virginia. Redstart 16:1–8.

Smallwood, J. A., and D. M. Bird. 2002. American Kestrel (*Falco sparverius*). *In* The Birds of North America, no. 602 (A. Poole and F. Gill, Eds.). The Birds of North America, Inc., Philadelphia, PA.

Smallwood, J. A., M. F. Causey, D. H. Mossop, J. R. Klucsarits, B. Robertson, S. Roberston, J. Mason, M. J. Maurer, R. J. Melvin, R. D. Dawson, G. R. Bortolotti, J. W. Parrish, T. F. Breen, and K. Boyd. 2009a. Why are American Kestrel (*Falco sparverius*) populations declining in North America? Evidence from nest-box programs. Journal of Raptor Research 43:274–282.

Smallwood, J. A., P. Winkler, G. I. Fowles, and M. Craddock. 2009b. American Kestrel breeding habitat: The importance of patch size. Journal of Wildlife Management 43:308–314.

Smith, C. R. 2008a. Eastern Screech-Owl (*Megascops asio*). Pages 292–293 *in* The Second Atlas of Breeding Birds in New York State (K. J. McGowan and K. Corwin, Eds.). Cornell University Press, Ithaca, NY.

———. 2008b. Acadian Flycatcher (*Empidonax virescens*). Pages 346–347 *in* The Second Atlas of Breeding Birds in New York State (K. J. McGowan and K. Corwin, Eds.). Cornell University Press, Ithaca, NY.

———. 2008c. Horned Lark (*Eremophila alpestris*). Pages 390–391 *in* The Second Atlas of Breeding Birds in New York State (K. J. McGowan and K. Corwin, Eds.). Cornell University Press, Ithaca, NY.

———. 2008d. Carolina Wren (*Thryothorus ludovicianus*). Pages 420–421 *in* The Second Atlas of Breeding Birds in New York State (K. J. McGowan and K. Corwin, Eds.). Cornell University Press, Ithaca, NY.

———. 2008e. Worm-eating Warbler (*Helmitheros vermivorum*). Pages 516–517 *in* The Second Atlas of Breeding Birds in New York State (K. J. McGowan and K. Corwin, Eds.). Cornell University Press, Ithaca, NY.

———. 2008f. Prairie Warbler (*Dendroica discolor*). Pages 500–501 *in* The Second Atlas of Breeding Birds in New York State (K. J. McGowan and K. Corwin, Eds.). Cornell University Press, Ithaca, NY.

———. 2008g. Clay-colored Sparrow (*Spizella pallida*). Pages 548–549 *in* The Second Atlas of Breeding Birds in New York State (K. J. McGowan and K. Corwin, Eds.). Cornell University Press, Ithaca, NY.

———. 2008h. Vesper Sparrow (*Pooecetes gramineus*). Pages 552–553 *in* The Second Atlas of Breeding Birds in New York State (K. J. McGowan and K. Corwin, Eds.). Cornell University Press, Ithaca, NY.

———. 2008i. Dark-eyed Junco (*Junco hyemalis*). Pages 572–573 *in* The Second Atlas of Breeding Birds in New York State (K. J. McGowan and K. Corwin, Eds.). Cornell University Press, Ithaca, NY.

———. 2008j. Eastern Meadowlark (*Sturnella magna*). Pages 592–593 *in* The Second Atlas of Breeding Birds in New York State (K. J. McGowan and K. Corwin, Eds.). Cornell University Press, Ithaca, NY.

Smith, J. P., C. J. Farmer, S. W. Hoffman, G. S. Kaltenecker, K. Z. Woodruff, and P. Sherrington. 2008. Trends in autumn counts of migratory raptors in Western North America, 1974–2004. Pages 217–251 *in* State of North America's Birds of Prey (K. L. Bildstein, J. P. Smith, E. R. Inzunza, and R. Veit, Eds.). American Ornithologists' Union and Nuttall Ornithological Club, Series in Ornithology, No. 3.

Smith, K. G., J. H. Withgott, and P. G. Rodewald. 2000. Red-headed Woodpecker (*Melanerpes erythrocephalus*). *In* The Birds of North America Online, no. 518 (A. Poole, Ed.). Cornell Lab of Ornithology, Ithaca, NY.

Solomon, S., D. Qin, M. Manning, Z. Chen, M. Marquis, K. B. Averyt, M. Tignor, and H. L. Miller, Eds. 2007. Climate Change 2007: The Physical Science Basis. Contribution of Working Group I to the Fourth Assessment Report of the Intergovernmental Panel on Climate Change (IPCC). Cambridge University Press, Cambridge, UK.

Somershoe, S. G., and C. R. Chandler. 2004. Use of oak hammocks by neotropical migrant songbirds: The role of area and habitat. Wilson Bulletin 116:56–63.

Squires, J. R., and R. T. Reynolds. 1997. Northern Goshawk (*Accipiter gentilis*). *In* The Birds of North America, no. 298 (A. Poole and F. Gill, Eds.). The Birds of North America, Inc., Philadelphia, PA.

Stasz, J. L. 1996. Worm-eating Warbler (*Helmitheros vermivorus*). Pages 352–353 *in* Atlas of the Breeding Birds of Maryland and the District of Columbia (C. S. Robbins and E. A. T. Blom, Eds.). University of Pittsburgh Press, Pittsburgh, PA.

Stauffer, G. E., M. R. Marshall, D. R. Diefenbach, and D. W. Brauning. 2010. Reclaimed surface mine habitat and grassland bird populations. Pages 86–96 *in* Avian Ecology and Conservation: A Pennsylvania Focus with National Implications (S. K. Majumdar, T. L. Master, M. C. Brittingham, R. M. Ross, R. S. Mulvihill, and J. E. Huffman, Eds.). Pennsylvania Academy of Science, Easton, PA.

Steele, M. A., M. C. Brittingham, T. J. Maret, and J. F. Merritt, Ed. 2010a. Terrestrial Vertebrates of Pennsylvania: A Complete Guide to Species of Conservation Concern. The Johns Hopkins University Press, Baltimore, MD.

Steele, M. A., N. Lichti, and R. K. Swihart. 2010b. Avian-mediated seed dispersal: An overview and synthesis with an emphasis on temperate forests of central and eastern U.S. Pages 28–43 *in* Avian Ecology and Conservation: A Pennsylvania Focus with National Implications (S. K. Majumdar, T. L. Master, M. C. Brittingham, R. M. Ross, R. S. Mulvihill, and J. E. Huffman, Eds.). Pennsylvania Academy of Science, Easton, PA.

Steiner, K. C., and B. J. Joyce. 1999. Survival and growth of a *Quercus rubra* regeneration cohort during the five years following masting. Pages 255–257 *in* 12th Central Hardwood Forest Conference (W. Stringer and D. L. Loftis, Eds.). United States Forest Service, General Technical Report SRS-24, Southern Research Station, Lexington, KY.

Stempka, J. J. 2009. Factors influencing utilization of artificial nesting cylinders by Mallards and Wood Ducks in Northwest Pennsylvania and Southern Ontario. M.S. thesis, University of Western Ontario, London, ON.

Stewart, R. E. 1949. Ecology of a nesting Red-shouldered Hawk population. Wilson Bulletin 61:26–35.

———. 1953. A life history study of the Yellow-throat. Wilson Bulletin 65:99–115.

Stewart, S. N., E. O. Wilson, J. A. McNeely, R. A. Mittermeier, and J. P. Rodriguez. 2010. The barometer of life. Science 328:177.

Stoleson, S. H. 2010. Alder Flycatcher (*Empidonax alnorum*). Pages 280–283 *in* Terrestrial Vertebrates of Concern in Pennsylvania: A Complete Guide to Species of Conservation Concern (M. A. Steele, M. C. Brittingham, T. J. Maret, and J. F. Merritt, Eds.). The Johns Hopkins University Press, Baltimore, MD.

Stoleson, S. H., and J. L. Larkin. 2010. Breeding birds of Pennsylvania: Forest communities. Pages 14–27 *in* Avian Ecology and Conservation: A Pennsylvania Focus with National Implications (S. K. Majumdar, T. L. Master, M. C. Brittingham, R. M. Ross, R. S. Mulvihill, and J. E. Huffman, Eds.). Pennsylvania Academy of Science, Easton, PA.

Stone, W. 1894. Birds of eastern Pennsylvania and New Jersey. Delaware Valley Ornithological Club, Philadelphia, PA.

———. 1900. The summer birds of the higher parts of Sullivan and Wyoming County, Pennsylvania. Cassinia 3:20–23.

Stotz, D. F., J. W. Fitzpatrick, T. A. Parker, and D. K. Moskovits. 1996. Neotropical Birds: Ecology and Conservation. University of Chicago Press, Chicago, IL.

Stout, W. E., S. A. Temple, and J. R. Cary. 2006. Landscape features of Red-tailed Hawk nesting habitat in an urban/suburban environment. Journal of Raptor Research 40:181–192.

Strahler, A. N. 1952. Hypsometric (area-altitude) analysis of erosional topology. Geological Society of America Bulletin 63:1117–1142.

Straight, C. A., and R. J. Cooper. 2000. Chuck-will's-widow (*Caprimulgus carolinensis*). *In* The Birds of North America, no. 499 (A. Poole and F. Gill, Eds.). The Birds of North America, Inc., Philadelphia, PA.

Stratford, J. A. 2010. The effects of envirornmental contaminants on avian populations. Pages 340–358 *in* Avian Ecology and Conservation: A Pennsylvania Focus with National Implications (S. K. Majumdar, T. L. Master, M. C. Brittingham, R. M. Ross, R. S. Mulvihill, and J. E. Huffman, Eds.). The Pennsylvania Academy of Science, Easton, PA.

Street, J. F. 1956. Birds of the Pocono Mountains, Pennsylvania. Delaware Valley Ornithology Club, Philadelphia, PA.

Street, P. B. 1976. Birds of the Pocono Mountains, 1955–1975. Cassinia 55:3–16.

Strode, P. K. 2003. Implications of climate change for North American wood warblers (Parulidae). Global Change Biology 9:1137–1144.

Studds, C. S., and P. P. Marra. 2011. Rainfall-induced changes in food availability modify the spring departure programme of a migratory bird. Proceedings of the Royal Society B-Biological Sciences 278:3437–3443.

Stull, J., J. A. Stull, and G. M. McWilliams. 1985. Birds of Erie County, Pennsylvania, including Presque Isle. Allegheny Press, Elgin, PA.

Stutchbury, B. J. M., E. A. Gow, T. Done, M. MacPherson, J. W. Fox, and V. Afanasyev. 2011. Effects of post-breeding moult and energetic condition on timing of songbird migration into the tropics. Proceedings of the Royal Society of London B 278:131–137.

Stutchbury, B. J. M., S. A. Tarof, T. Done, E. Gow, P. M. Kramer, J. Tautin, J. W. Fox, and V. Afanasyev. 2009. Tracking long-distance songbird migration using geolocators. Science 323:896.

Sullivan, B., L. S. T. Kelling, C. L. Wood, M. J. Iliff, D. Fink, M. Herog, D. Moody, and G. Ballard. 2009. Data exploration through Visualization Tools. Pages 415–418 *in* Proceedings of the Fourth International Partners in Flight Conference: Tundra to Tropics: Connecting Birds, Habitats and People. University of Texas–Pan American Press, McAllen, TX.

Summers-Smith, J. D. 2003. The decline of the House Sparrow: A review. British Birds 96:439–446.

Sutherland, D. A., and W. J. Crins. 2007. Sandhill Crane, Grue du Canada (*Grus canadensis*). Pages 208–209 *in* Atlas of the Breeding Birds of Ontario, 2001–2005 (M. D. Cadman, D. A. Sutherland, G. G. Beck, D. Lepage, and A. R. Coutrier, Eds.). Bird Studies Canada, Environment Canada, Ontario Field Ornithologists, Ontario Ministry of Natural Resources, and Ontario Nature, Toronto, ON.

Sutton, C., and P. Sutton. 1994. How to Spot An Owl. Chapters Publishers Ltd., Shelburne, VT.

Sutton, G. M. 1928a. An Introduction to the Birds of Pennsylvania. J. Horace McFarland Company, Harrisburg, PA.

———. 1928b. The birds of Pymatuning Swamp and Conneaut Lake, Crawford County, Pennsylvania. Annals of the Carnegie Museum 18:19–239.

Suzuki, R., and H. Shimodaira. 2005. Pvclust: Hierarchical clustering with P-values. R package version 1.0-3.

Swift, B. 2008. American Black Duck (*Anas rubripes*). Pages 100–101 *in* The Second Atlas of Breeding Birds in New York State (K. J. McGowan and K. Corwin, Eds.). Cornell University Press, Ithaca, NY.

Sydenstricker, K. V., A. A. Dhondt, D. H. Ley, and G. V. Kollias. 2005. Re-exposure of captive House Finches that recovered from *Mycoplasma gallisepticum* infection. Journal of Wildlife Diseases 41:326–333.

Tacha, T. C., S. A. Nesbitt, and P. A. Vohs. 1992. Sandhill Crane (*Grus canadensis*). *In* The Birds of North America, no. 31 (A. Poole, P. Stettenheim, and F. Gill, Eds.). The Birds of North America, Inc., Philadelphia, PA.

Takats, D. L., C. M. Francis, G. L. Holroyd, J. R. Duncan, K. M. Mazur, R. J. Cannings, R. J. Harris, and D. Holt. 2001. Guidelines for nocturnal owl monitoring in North America. Beaverhill Bird Observatory and Bird Studies, Edmonton, AB.

Talbott, S. C., and R. H. Yahner. 2003. Temporal and spatial use of even-aged reproduction stands by bird communities in central Pennsylvania. Northern Journal of Applied Forestry 20:117–123.

Tapley, J. l., R. K. Abernethy, and J. E. Kennamer. 2007. Status and distribution of the Wild Turkey in 2004. Proceedings of the National Wild Turkey Symposium 9:21–31.

Tarof, S., and J. V. Briskie. 2008. Least Flycatcher (*Empidonax minimus*). *In* The Birds of North America Online, no. 99 (A. Poole, Ed.). Cornell Lab of Ornithology, Ithaca, NY.

Tarvin, K. A., and G. E. Woolfenden. 1999. Blue Jay (*Cyanocitta cristata*). *In* The Birds of North America, no. 469 (A. Poole and F. Gill, Eds.). The Birds of North America, Inc., Philadelphia, PA.

Tate, J. L., Jr. 1972. The changing seasons. American Birds 26:828–831.

Tatu, K. S., J. T. Anderson, L. J. Hindman, and G. Seidel. 2007. Mute Swans' impact on submerged aquatic vegetation in Chesapeake Bay. Journal of Wildlife Management 71:1431–1439.

Tautin, J. 2010. American Black Duck (*Anas rubripes*). Pages 226–230 *in* Terrestrial Vertebrates of Concern in Pennsylvania: A Complete Guide to Species of Conservation Concern (M. A. Steele, M. C. Brittingham, T. J. Maret, and J. F. Merritt, Eds.). The Johns Hopkins University Press, Baltimore, MD.

Tautin, J., B. Cousens, K. Kostka, S. Kotska, and D. A. Airola. 2009. Addressing regional declines in Purple Martin populations. Pages 82–87 *in* Proceedings of the Fourth International Partners in Flight Conference: Tundra to Tropics: Connecting Birds, Habitats and People. University of Texas–Pan American Press, McAllen, TX.

Taylor, I. 1994. Barn Owls: Predator–Prey Relationships and Conservation. Cambridge University Press, Cambridge, UK.

Taylor, W. K., and M. A. Kershner. 1986. Migrant birds killed at the vehicle assembly building (VAB), John F. Kennedy Space Center. Journal of Field Ornithology 57:142–154.

Temple, S. A. 2002. Dickcissel (*Spiza americana*). *In* The Birds of North America, no. 703 (A. Poole and F. Gill, Eds.). The Birds of North America, Inc., Philadelphia, PA.

Terborgh, J. 1989. Where Have All the Birds Gone? Princeton University Press, Princeton, NJ.

Terres, J. K. 1980. The Audubon Society Encyclopedia of North American Birds. Alfred A. Knopf, New York, NY.

Therres, G. D. 1999. Wildlife species of conservation concern in the northeastern United States. Northeast Wildlife 54:93–100.

———. 2010a. Great Egret (*Ardea alba*). Pages 78–79 *in* Second Atlas of the Breeding Birds of Maryland and the District of Columbia (W. G. Ellison, Ed.). The Johns Hopkins University Press, Baltimore, MD.

———. 2010b. King Rail (*Rallus elegans*). Pages 126–127 *in* Second Atlas of the Breeding Birds of Maryland and the District of Columbia (W. G. Ellison, Ed.). The Johns Hopkins University Press, Baltimore, MD.

———. 2010c. Snowy Egret (*Egretta thula*). Pages 80–81 *in* Second Atlas of the Breeding Birds of Maryland and the District of Columbia (W. G. Ellison, Ed.). The Johns Hopkins University Press, Baltimore, MD.

Therres, G., and D. Brinker. 2004. Mute Swan interaction with other birds in Chesapeake Bay. Pages 43–46 *in* Mute Swans and their Chesapeake Bay Habitats: Proceedings of a Symposium, USGS/BRD/ITR-2004-0005 (M. C. Perry, Ed.), Reston, VA.

Thogmartin, W. E., J. A. Fitzgerald, and M. T. Jones. 2009. Conservation design: Where do we go from here? Pages 426–436 *in* Proceedings of the Fourth International Partners in Flight Conference: Tundra to Tropics: Connecting Birds, Habitats and People. University of Texas–Pan American Press, McAllen, TX.

Thomas, C. D., and J. J. Lennon. 1999. Birds extend their ranges northward. Nature 399:213.

Thomas, E. H. 2011. Effect of oil and gas well development on songbird abundance in the Allegheny National Forest. M.S. thesis, The Pennsylvania State University, University Park, PA.

Thompson, C. F., and V. Nolan, Jr. 1973. Population biology of the Yellow-breasted Chat (*Icteria virens* L.) in southern Indiana. Ecological Monographs 43:145–171.

Thompson, F. R., III, W. D. Dijak, T. G. Kulowiec, and D. A. Hamilton. 1992. Breeding bird populations in Missouri Ozark forests with and without clearcutting. Journal of Wildlife Management 56:23–30.

Thurber, J. M., and R. O. Peterson. 1991. Changes in body size associated with range expansion in the coyote (*Canis latrans*). Journal of Mammalogy 72:750–755.

Timmermans, S. T. A. 2007a. American Bittern, Butor d'Amérique (*Botaurus lentiginosus*). Pages 154–155 *in* Atlas of the Breeding Birds of Ontario, 2001–2005 (M. D. Cadman, D. A. Sutherland, G. G. Beck, D. Lepage, and A. R. Coutrier, Eds.). Bird Studies Canada, Environment Canada, Ontario Field Ornithologists, Ontario Ministry of Natural Resources, and Ontario Nature, Toronto, ON.

———. 2007b. Common Moorhen, Gallinule poule-d'eau (*Gallinula chloropus*). Pages 204–205 *in* Atlas of the Breeding Birds of Ontario, 2001–2005 (M. D. Cadman, D. A. Sutherland, G. G. Beck, D. Lepage, and A. R. Coutrier, Eds.). Bird Studies Canada, Environment Canada, Ontario Field Ornithologists, Ontario Ministry of Natural Resources, and Ontario Nature, Toronto, ON.

Timmermans, S. T. A., and G. E. Craigie. 2002. The marsh monitoring program 2002 report: Monitoring Great Lakes wetlands and their amphibian and bird inhabitants. Bird Studies Canada program report to Environment Canada and the U.S. Environmental Protection Agency, Toronto, ON.

Tiner, R. W. 1990. Pennsylvania's wetlands: Current status and recent trends. Bureau of Water Resources Management, Harrisburg, PA.

Tingley, M. W., D. A. Orwig, R. Field, and G. Motzkin. 2002. Avian response to removal of a forest dominant: Consequences of hemlock woolly adelgid infestations. Journal of Biogeography 29:1505–1516.

Titus, K., M. R. Fuller, D. F. Stauffer, and J. R. Sauer. 1989. The status and distribution of *Buteo* hawks in the northeastern U.S. Pages 53–64 *in* Proceedings of the Northeast Raptor Management Symposium and Workshop. National Wildlife Federation, Washington, DC.

Titus, K., and J. Mosher. 1981. Nest-site habitat selected by woodland hawks in the central Appalachians. Auk 98:270–281.

Todd, W. E. C. 1936. The Redhead and Ring-necked Duck breeding at Pymatuning Lake. Auk 53:440.

———. 1940. Birds of Western Pennsylvania. University of Pittsburgh, Pittsburgh, PA.

Townsend, C. W. 1932. Passenger Pigeon. *In* Life Histories of North American Gallinaceous Birds (A. C. Bent, Ed.). U.S. National Museum Bulletin, No. 162, Washington, DC.

Tozer, D. C. 2007. American Coot, Foulque d'Amérique (*Fulica americana*). Pages 206–207 *in* Atlas of the Breeding Birds of Ontario, 2001–2005 (M. D. Cadman, D. A. Sutherland, G. G. Beck, D. Lepage, and A. R. Coutrier, Eds.). Bird Studies Canada, Environment Canada, Ontario Field Ornithologists, Ontario Ministry of Natural Resources, and Ontario Nature, Toronto, ON.

Tozer, D. C., E. Nol, D. M. Burke, K. A. Elliot, and K. L. Falk. 2009. Predation by bears on woodpecker nests: Are nestling begging and habitat choice risky business? Auk 126:300–309.

Trani, M. K., R. T. Brooks, T. L. Schmidt, V. A. Rudis, and C. M. Gabbard. 2001. Patterns and trends of early successional forests in the eastern United States. Wildlife Society Bulletin 29:413–424.

Trevor, C. L., and B. A. Andres. 2006. Population estimates of North American shorebirds, 2006. Wader Study Group Bulletin 111:67–85.

Tumer, C., J. Hill, and D. W. Johnston. 1984. Chimney Swift nesting in a hollow tree. Raven 55:12.

Twedt, D. J., and S. G. Somershoe. 2009. Bird response to prescribed silvicultural treatments in bottomland hardwood forests. Journal of Wildlife Management 73:1140–1150.

UCS (Union of Concerned Scientists). 2008. Climate change in Pennsylvania: Impacts and solutions for the Keystone State. Union of Concerned Scientists. [Online.] Available at http://www.climatechoices.org/pa.

USBC (U.S. Bureau of the Census). 1989. Census of Agriculture, 1987. Volume 1, Pennsylvania state and county data. U.S. Government Printing Office, Washington, DC.

———. 1994. Census of Agriculture, 1992. Volume 1, Pennsylvania state and county data. U.S. Government Printing Office, Washington, DC.

———. 2011. State and County QuickFacts. U.S. Bureau of the Census. [Online.] Available at http://quickfacts.census.gov/qfd/states/42000.html.

USDA-FSA (U.S. Department of Agriculture—Farm Service Agency). 2009. Conservation Reserve Program annual summary and enrollment statistics, FY 2009. U.S. Department of Agriculture, Farm Service Agency, Washington, DC.

USDA-NASS (U.S. Department of Agriculture–National Agriculture Statistics Service). 1999. Census of Agriculture 1997. Pennsylvania State and County Data. Volume 1, State level data. U.S. Department of Agriculture, National Agriculture Statistics Service, Washington, DC.

———. 2004. Census of Agriculture 2002. Pennsylvania State and County Data. Volume 1, State level data. U.S. Department of Agriculture, National Agriculture Statistics Service, Washington, DC.

———. 2008. Crop progress reports for 2008. 2008 Annual Reports. U.S. Department of Agriculture, National Agriculture Statistics Service, Harrisburg, PA.

———. 2009. Census of Agriculture 2007. Pennsylvania State and County Data. Volume 1, State level data. U.S. Department of Agriculture, National Agriculture Statistics Service, Washington, DC.

USFS (U.S. Forest Service). 2009. Hemlock wooly adelgid infestation by state and county. U.S. Forest Service, Northeastern Area State and Private Forestry. Washington, DC. [Online.] Available at http://na.fs.fed.us/fhp/hwa/infestations/hwa_infestations09.

———. 2010. Hemlock woolly adelgid: Risk, detection, and spread. United States Forest Service, Northeastern Area State and Private Forestry. Washington, DC. [Online.] Available at http://www.nrs.fs.fed.us/disturbance/invasive_species/hwa/risk_detection_spread/.

USFWS (U.S. Fish and Wildlife Service). 1985. Determination of endangered and threatened status for the Piping Plover. Federal Register, no. 50. U.S. Fish and Wildlife Service, Washington, DC.

———. 1999. Endangered and threatened wildlife and plants; final rule to remove the Peregrine Falcon in North America from the list of endangered and threatened wildlife, and to remove the similarity of appearance provision for free-flying peregrines in the coterminous United States. Federal Register. U.S. Fish and Wildlife Service, Washington, DC.

———. 2003a. Final environmental impact statement: Double-crested Cormorant management. U.S. Fish and Wildlife Service, Washington, DC.

———. 2003b. Recovery plan for the Great Lakes Piping Plover (*Charadrius melodus*). U.S. Fish and Wildlife Service, Ft. Snelling, MN.

———. 2005. Mourning Dove national strategic harvest management plan. Unpublished report. U.S. Fish and Wildlife Service, Washington, DC.

———. 2007. A blueprint for the future of migratory birds: Migratory Bird Program strategic plan, 2004–2014. U.S. Fish and Wildlife Service, Migratory Bird Program, Washington, DC.

———. 2008. Birds of conservation concern 2008. U.S. Fish and Wildlife Service, Division of Migratory Bird Management, Arlington, VA.

———. 2009. Piping Plover (*Charadrius melodus*) 5-year review: Summary and evaluation. U.S. Fish and Wildlife Service, Ft. Snelling, MN, and Hadley, MA.

———. 2010. Draft interim American Woodcock harvest strategy. U.S. Fish and Wildlife Service. [Online.] Available at http://www.fws.gov/migratorybirds/NewReportsPublications/SpecialTopics/SpecialTopics.html.

USGS (U.S. Geological Survey). 2010. National Land Cover, Version 1. U.S. Geological Survey, National Biological Information Infrastructure, Gap Analysis Program (GAP).

Vallender, R., S. L. V. Wilgenburg, L. P. Bulluck, A. Roth, R. Canterbury, J. Larkin, R. Fowlds, and I. J. Lovette. 2009. Extensive rangewide mitochondrial introgression indicates substantial cryptic hybridization in the Golden-winged Warbler (*Vermivora chrysoptera*). [Online.] Available at http://www.ace-eco.org/vol4/iss2/art4/.

Van Fleet, K. 2008. Marsh bird monitoring conservation project: January 2006 through December 2007. Unpublished report. Pennsylvania Game Commission, Harrisburg, PA.

Van Horn, M. A., and T. M. Donovan. 1994. Ovenbird (*Seiurus aurocapillus*). *In* The Birds of North America Online, no. 88 (A. Poole, Ed.). Cornell Lab of Ornithology, Ithaca, NY.

Van Ness, K. D., Jr. 1996. Red-bellied Woodpecker. Pages 204–205 *in* Atlas of the Breeding Birds of Maryland and the District of Columbia (C. S. Robbins and E. A. T. Blom, Eds.). University of Pittsburgh Press, Pittsburgh, PA.

Verbeek, N. A. M., and C. Caffrey. 2002. American Crow (*Corvus brachyrhynchos*). *In* The Birds of North America, no. 647 (A. Poole and F. Gill, Eds.). The Birds of North America, Inc., Philadelphia, PA.

Vickery, P. D. 1996. Grasshopper Sparrow (*Ammodramus savannarum*). *In* The Birds of North America Online, no. 239 (A. Poole, Ed.). Cornell Lab of Ornithology, Ithaca, NY.

Viverette, C. B., S. Struve, L. J. Goodrich, and K. L. Bildstein. 1996. Decreases in migrating Sharp-shinned Hawks (*Accipiter striatus*) at traditional raptor migration watchsites in eastern North America. Auk 113:32–40.

Vucetich, J. A., R. O. Peterson, and T. A. Waite. 2004. Raven scavenging favours group foraging in wolves. Animal Behaviour 67:1117–1126.

Walsh, J., V. Elia, R. Kane, and T. Halliwell. 1999. Birds of New Jersey. New Jersey Audubon Society, Bernardsville, NJ.

Walters, E. L., E. H. Miller, and P. E. Lowther. 2002. Yellow-bellied Sapsucker (*Sphyrapicus varius*). *In* The Birds of North America Online, no. 662 (A. Poole, Ed.). Cornell Lab of Ornithology, Ithaca, NY.

Waltman, W. J., E. J. Ciolkosz, M. J. Mausbach, M. D. Svoboda, D. A. Miller, and P. J. Kolb. 1997. Soil Climate Regimes of Pennsylvania. Pennsylvania State University Agricultural Experiment Station Bulletin. The Pennsylvania State University Agricultural Experiment Station, University Park, PA.

Warkentin, I. G., N. S. Sodhi, R. H. M. Espie, A. F. Poole, L. W. Oliphant, and P. C. James. 2005. Merlin (*Falco columbarius*). *In* The Birds of North America Online, no. 44 (A. Poole, Ed.). Cornell Lab of Ornithology, Ithaca, NY.

Warner, E. D. 2007. PAMAP Program Land Cover for Pennsylvania, 2005. The Pennsylvania State University. [Online.] Available at ftp://www.pasda.psu.edu/pub/pasda/orser.

Warren, B. H. 1890. Report on the Birds of Pennsylvania, 2nd ed. State Board of Agriculture, Harrisburg, PA.

Warren, C. R. 1950. Survey of Pennsylvania Migratory Waterfowl. Final report Pittman-Robertson Project 30-R. Pennsylvania Game Commission, Harrisburg, PA.

Warren, R. J., II. 2008. Mechanisms driving understory evergreen herb distributions across slope aspects: As derived from landscape position. Plant Ecology 198:297–308.

Waters, N. M. 2010. Impacts of environmental contaminants on bird population ecology. Pages 332–339 *in* Avian Ecology and Conservation: A Pennsylvania Focus with National Implications (S. K. Majumdar, T. L. Master, M. C. Brittingham, R. M. Ross, R. S. Mulvihill, and J. E. Huffman, Eds.). Pennsylvania Academy of Science, Easton, PA.

Watts, B. D. 1995. Yellow-crowned Night-Heron (*Nyctanassa violacea*). *In* The Birds of North America, no. 161 (A. Poole and F. Gill, Eds.). The Birds of North America, Inc., Philadelphia, PA.

Watts, B. D., and A. E. Duerr. 2010. Nest turnover rates and list frame decay in Bald Eagles: Implications for the national monitoring plan. Journal of Wildlife Management 74:940–944.

Watts, B. D., and B. J. Paxton. 2007. Ospreys of the Chesapeake Bay: Population recovery, ecological requirements, and current threats. Waterbirds 30:39–49.

Weaver, M., and R. C. Boyd. 2008. Game take and furtaker surveys. Wildlife Management Annual Report. Pennsylvania Game Commission, Harrisburg, PA.

Webb, W. L., D. F. Behrend, and B. Saisorn. 1977. Effect of logging on songbird populations in a northern hardwood forest. Wildlife Monographs 55:1–35.

Weeks, H. P., Jr. 1994. Eastern Phoebe (*Sayornis phoebe*). *In* The Birds of North America Online, no. 94 (A. Poole, Ed.). Cornell Lab of Ornithology, Ithaca, NY.

Weidensaul, S. 2010. Migration and wintering ecology of Northern Saw-whet Owls. Pages 137–147 *in* Avian Ecology and Conservation: A Pennsylvania Focus with National Implications (S. Majumdar, T. L. Master, M. C. Brittingham, R. M. Ross, R. S. Mulvihill, and J. E. Huffman, Eds.). Pennsylvania Academy of Science, Easton, PA.

Weisheit, A. S., and P. D. Creighton. 1989. Interference by House Sparrows in nesting activities of Barn Swallows. Journal of Field Ornithology 60:323–328.

Weller, M. W. 1961. Breeding biology of the Least Bittern. Wilson Bulletin 73:11–35.

Wells, J. V. 1995. Investigations into the distribution and abundance of species. Ph.D. dissertation, Cornell University, Ithaca, NY.

Wells, J. V., B. Robertson, K. V. Rosenberg, and D. W. Mehlman. 2010. Global versus local conservation focus of U.S. state agency Endangered bird species lists. PloS ONE 5:1–5.

Wentworth, K. 2010. Dickcissel (*Spiza americana*). Pages 165–167 *in* Terrestrial Vertebrates of Concern in Pennsylvania: A Complete Guide to Species of Conservation Concern (M. A. Steele, M. C. Brittingham, T. J. Maret, and J. F. Merritt, Eds.). The Johns Hopkins University Press, Baltimore, MD.

Wentworth, K. L., M. C. Brittingham, and A. M. Wilson. 2010. Conservation reserve enhancement program fields: Benefits for grassland and shrub-scrub species. Journal of Soil and Water Conservation 65:50–60.

Weseloh, C. 2007. Black Tern, Guifette noire (*Chlidonias nigra*). Pages 268–269 *in* Atlas of the Breeding Birds of Ontario, 2001–2005 (M. D. Cadman, D. A. Sutherland, G. G. Beck, D. Lepage, and A. R. Coutrier, Eds.). Bird Studies Canada, Environment Canada, Ontario Field Ornithologists, Ontario Ministry of Natural Resources, and Ontario Nature, Toronto, ON.

Wheelwright, N. T., and J. D. Rising. 2008. Savannah Sparrow (*Passerculus sandwichensis*). *In* The Birds of North America Online, no. 45 (A. Poole, Ed.). Cornell Lab of Ornithology, Ithaca, NY.

Whitcomb, R. F., C. S. Robbins, J. F. Lynch, B. L. Whitcomb, M. K. Klimkiewicz, and D. Bystrak. 1981. Effects of forest fragmentation on avifauna of the eastern deciduous forest. Pages 125–205 *in* Forest Island Dynamics in Man-dominated Landscapes (R. L. Burgess and B. M. Sharpe, Eds.). Springer-Verlag, New York, NY.

White, C. M., N. J. Clum, T. J. Cade, and W. G. Hunt. 2002. Peregrine Falcon (*Falco peregrinus*). *In* The Birds of North America Online, no. 660 (A. Poole, Ed.). Cornell Lab of Ornithology, Ithaca, NY.

White, D. H., and J. T. Seginak. 2000. Nest box use and productivity of Great Crested Flycatchers in prescribed-burned longleaf pine forests. Journal of Field Ornithology 71:147–152.

White, G. C. 1992. PC SURVIV User's Manual. Department of Fishery and Wildlife Biology, Colorado State University, Fort Collins, CO.

Whitehead, D. R., and T. Taylor. 2002. Acadian Flycatcher (*Empidonax virescens*). *In* The Birds of North America Online, no. 614 (A. Poole, Ed.). Cornell Lab of Ornithology, Ithaca, NY.

Whitlock, A. L., and L. L. Carpenter. 2007. A comprehensive list of Species of Greatest Conservation Need from the Northeast States' Wildlife Action Plans. Unpublished report for the Northeast Wildlife Diversity Technical Committee of the Northeastern Association of Fish and Wildlife Agencies, Hadley, MA.

Whitney, G. G. 1990. The history and status of the hemlock hardwood forests of the Allegheny Plateau. Journal of Ecology 78:443–458.

Wiebe, K. L., and W. S. Moore. 2008. Northern Flicker (*Colaptes auratus*). *In* The Birds of North America Online, no. 166a (A. Poole, Ed.). Cornell Lab of Ornithology, Ithaca, NY.

Wiggins, D. A., D. W. Holt, and S. M. Leasure. 2006. Short-eared Owl (*Asio flammeus*). *In* The Birds of North America Online, no. 62 (A. Poole, Ed.). Cornell Lab of Ornithology, Ithaca, NY.

Wilcox, B. R., M. J. Yabsley, A. E. Ellis, D. E. Stallknecht, and S. E. J. Gibbs. 2007. West Nile virus antibody prevalence in American Crows (*Corvus brachyrhynchos*) and Fish Crow (*Corvus ossifragus*) in Georgia, USA. Avian Diseases 51:125–128.

Wilhelm, G. 1992. Sandhill Cranes in Mercer County. Pennsylvania Birds 6:105–106.

———. 1993a. King Rail breeding in western Pennsylvania. Pennsylvania Birds 7:89–90.

———. 1993b. First breeding record of Sandhill Crane for Pennsylvania. Pennsylvania Birds 7:91–92.

———. 1993c. Sandhill Cranes, Lawrence County. Pennsylvania Birds 7:8.

———. 1994. Second breeding record of Sandhill Crane for Pennsylvania. Pennsylvania Birds 8:136–137.

———. 1995. Scenario of the Upland Sandpiper in western Pennsylvania. Pennsylvania Birds 8:204–206.

———. 2010. Local notes—Butler County. Pennsylvania Birds 23:218.

Williams, J. M. 1996. Nashville Warbler (*Vermivora ruficapilla*). *In* The Birds of North America Online, no. 205 (A. Poole, Ed.). Cornell Lab of Ornithology, Ithaca, NY.

Williams, L. 2010. Pennsylvania's Wildlife Action Plan. Pages 20–25 *in* Terrestrial Vertebrates of Concern in Pennsylvania: A Complete Guide to Species of Conservation Concern (M. A. Steele, M. C. Brittingham, T. J. Maret, and J. F. Merritt, Eds.). The Johns Hopkins University Press, Baltimore, MD.

Williamson, S. L. 2001. A Field Guide to the Hummingbirds of North America. Houghton Mifflin Co., New York, NY.

Wilson, A., and C. L. Bonaparte. 1831. American Ornithology. Porter and Coates, Philadelphia, PA.

Wilson, A. M. 2009. Bird population responses to conservation grasslands in Pennsylvania. Ph.D. dissertation, The Pennsylvania State University, University Park, PA.

———. 2010. The status and conservation status of farmland birds in Pennsylvania. Pages 217–231 *in* Avian Ecology and Conservation: A Pennsylvania Focus with National Implications (S. K. Majumdar, T. L. Master, M. C. Brittingham, R. M. Ross, R. S. Mulvihill, and J. E. Huffman, Eds.). The Pennsylvania Academy of Science, Easton, PA.

Wilson, A. M., and M. C. Brittingham. 2012. Initial response of bird populations to conservation grasslands in southern Pennsylvania. Journal of Soil and Water Conservation 67:59–67.

Wilson, A. M., M. C. Brittingham, and G. Grove. 2010. Association of wintering raptors with Conservation Reserve Enhancement Program grasslands in Pennsylvania. Journal of Field Ornithology 81:361–372.

Wilson, D. M., and J. Bart. 1985. Reliability of singing bird surveys: Effects of song phenology during the breeding season. Condor 87:69–73.

Wilson, S., and P. Arcese. 2006. Nest depredation, brood parasitism, and reproductive variation in island populations of Song Sparrows (*Melospiza melodia*). Auk 123:784–794.

Wilson, S., S. L. LaDeau, A. P. Tøttrup, and P. P. Marra. 2011. Range-wide effects of breeding- and nonbreeding-season climate on the abundance of a Neotropical migrant songbird. Ecology 92:1789–1798.

Wink, J., S. E. Senner, and L. J. Goodrich. 1987. Food habits of Great Horned Owls in Pennsylvania. Journal of the Pennsylvania Academy of Science 61:133–137.

Winter, M., D. H. Johnson, and J. A. Shaffer. 2005. Variability in vegetation effects on density and nesting success of grassland birds. Journal of Wildlife Management 69:185–197.

Winter, M., D. H. Johnson, J. A. Shaffer, and W. D. Svedarsky. 2004. Nesting biology of three grassland passerines in the northern tallgrass prairie. Wilson Bulletin 116:211–223.

Wires, L. R., S. J. Lewis, G. J. Soulliere, S. W. Matteson, D. V. "Chip" Weseloh, R. P. Russell, and F. J. Cuthbert. 2010. Upper Mississippi Valley / Great Lakes Waterbird Conservation Plan. A plan associated with the Waterbird Conservation for the Americas Initiative. Final Report submitted to the U.S. Fish and Wildlife Service, Fort Snelling, MN.

Wisniewski, T. 2011. The Pennsylvania State Climatologist. The Pennsylvania Climate Office. http://climate.met.psu.edu/www_prod/data.

Witmer, M. 2008. Cedar Waxwing (*Bombycilla cedrorum*). Pages 462–463 *in* The Second Atlas of Breeding Birds in New York State (K. J. McGowan and K. Corwin, Eds.). Cornell University Press, Ithaca, NY.

Witmer, M. C., D. J. Mountjoy, and L. Elliot. 1997. Cedar Waxwing (*Bombycilla cedrorum*). *In* The Birds of North America Online, no. 309 (A. Poole, Ed.). Cornell Lab of Ornithology, Ithaca, NY.

WMI (Wildlife Management Institute). 2010. The deer management program of the Pennsylvania Game Commission: A comprehensive review and evaluation. A Report to the Executive Director of the Pennsylvania Legislative Budget and Finance Committee. Wildlife Management Institute, Washington, DC.

Wood, M. 1932. Eastern Lark Sparrow breeding in central Pennsylvania. Auk 49:98.

———. 1979. Birds of Pennsylvania: When and Where to Find Them. The Pennsylvania State University, College of Agriculture, University Park, PA.

———. 1983. Birds of Central Pennsylvania, 3rd ed. The Pennsylvania State University, University Park, PA.

Wood, P. B., C. B. Viverette, L. J. Goodrich, M. Pokras, and C. Tibbott. 1996. Environmental contaminant levels in Sharp-shinned Hawks from the eastern United States. Journal of Raptor Research 30:136–144.

Wootton, J. T. 1996. Purple Finch (*Carpodacus purpureus*). *In* The Birds of North America Online, no. 208 (A. Poole, Ed.). Cornell Lab of Ornithology, Ithaca, NY.

Wünschmann, A., J. Shivers, J. Bender, L. Carroll, S. Fuller, M. Saggese, A. v. Wettere, and P. Redig. 2004. Pathologic findings in Red-tailed Hawks (*Buteo jamaicensis*) and Cooper's Hawks (*Accipiter cooperii*) naturally infected with West Nile virus. Avian Diseases 48:570–580.

———. 2005. Pathologic and immunohistochemical findings in Goshawks (*Accipiter gentilis*) and Great Horned Owls (*Bubo virginiana*) naturally infected with West Nile virus. Avian Diseases 49:252–259.

Wunz, G. A. 1978. The Wild Turkey, our all-American bird. Pennsylvania Game News September:7–18.

Wunz, G. A., and A. Hayden. 1981. No more game-farm turkey stocking. Pennsylvania Game News March:24–27.

WVDNR (West Virginia Department of Natural Resources). 2010. Rare, threatened and endangered species: Osprey. West Virginia Department of Natural Resources. [Online.] Available at http://www.wvdnr.gov/wildlife/RETSpecies.asp.

Wyatt, V. E., and C. M. Francis. 2002. Rose-breasted Grosbeak (*Pheucticus ludovicianus*). *In* The Birds of North America, no. 692 (A. Poole and F. Gill, Eds.). The Birds of North America, Inc., Philadelphia, PA.

Yahner, R. H. 1986. Structure, seasonal dynamics, and habitat relationships of avian communities in small even-aged forest stands. Wilson Bulletin 98:61–82.

———. 2003. Terrestrial vertebrates in Pennsylvania: Status and conservation in a changing landscape. Northeastern Naturalist 10:343–360.

Yasukawa, K., and W. A. Searcy. 1995. Red-winged Blackbird (*Agelaius phoeniceus*). *In* The Birds of North America, no. 184 (A. Poole and F. Gill, Eds.). The Birds of North America, Inc., Philadelphia, PA.

Yosef, R. 1996. Loggerhead Shrike (*Lanius ludovicianus*). *In* The Birds of North America Online, no. 231 (A. Poole, Ed.). Cornell Lab of Ornithology, Ithaca, NY.

Young, L., M. G. Betts, and A. W. Diamond. 2005. Do Blackburnian Warblers select mixed forest?: The importance of spatial resolution in defining habitat. Forest Ecology and Management 214:358–372.

Young, M. A. 2008a. Red Crossbill (*Loxia curvirostra*). Pages 612–613 *in* The Second Atlas of Breeding Birds in New York State (K. J. McGowan and K. Corwin, Eds.). Cornell University Press, Ithaca, NY.

———. 2008b. Pine Siskin (*Carduelis pinus*). Pages 616–617 *in* The Second Atlas of Breeding Birds in New York State (K. J. McGowan and K. Corwin, Eds.). Cornell University Press, Ithaca, NY.

———. 2011a. Red Crossbill (*Loxia curvirostra*) call-types of New York: Their taxonomy, flight call vocalizations and ecology. Kingbird 61:106–123.

———. 2011b. Status and distribution of Type 1 Red Crossbill (*Loxia curvirostra*): An Appalachian call type? North American Birds 65:554–561.

Zimmerman, A. L., J. A. Dechant, B. E. Jamison, D. H. Johnson, C. M. Goldade, J. O. Church, and B. R. Euliss. 2002a. Effects of management practices on wetland birds: Virginia Rail. Northern Prairie Wildlife Research Center, Jamestown, ND.

Zimmerman, A. L., J. A. Dechant, D. H. Johnson, C. M. Goldade, J. O. Church, and B. R. Euliss. 2002c. Effects of management practices on wetland birds: Marsh Wren. Northern Prairie Wildlife Research Center, Jamestown, ND.

Zimmerman, A. L., E. Jamison, J. A. Dechant, D. H. Johnson, C. M. Goldade, J. O. Church, and B. R. Euliss. 2002b. Effects of management practices on wetland birds: Sora. Northern Prairie Wildlife Research Center, Jamestown, ND.

Zuckerberg, B. 2008. Long-term changes in the distributions of breeding birds in response to regional reforestation and climate change in New York State. Ph.D. dissertation, State University of New York College of Environmental Science and Forestry, Syracuse, NY.

Zuefle, M. E., W. P. Brown, and T. E. Dutt. 2009. An unusual American Goldfinch nest in a corn plant. Pennsylvania Birds 24:33–35.

Index

Bold indicates species accounts. All references to the species are provided for entries under bird scientific names (italics).

Abbreviations, list of, 84
Abundance estimation, 1, 39–45, 520–522
Acadian Flycatcher. *See* Flycatcher, Acadian
Accipiter cooperii, 55, 60, 61, **146–147**, 156, 228, 479, 523, 527
    A. gentilis, 27, 55, 59, 68, 73, **148–149**, 479, 523, 527
    A. striatus, 27, 55, 59, 73, **144–145**, 479, 523, 527
Acid deposition, 20, 79
    and Great Blue Heron, 124
    and Hermit Thrush, 326
    and Least Flycatcher, 252
    and Louisiana Waterthrush, 346
    and Northern Waterthrush, 348
    and Sharp-shinned Hawk, 144
    and Swainson's Thrush, 324
    and Wood Thrush, 326, 328
Acid mine drainage. *See* Pollution, mine drainage
Acid precipitation. *See* Acid deposition
Acid rain. *See* Acid deposition
Acidification, 20
Acknowledgments, xi–xxiv
Acorn, 228, 272
Actitis macularius, 55, 100, **176–177**, 481, 523, 527
Adelgid, Hemlock Woolly, x, 2, 14, 29, 81
    and Acadian Flycatcher, 246
    and Barred Owl, 208
    and Blackburnian Warbler, x, 380
    and Black-throated Green Warbler, 398
    and Blue-headed Vireo, 266
    and Brown Creeper, 304
    and Dark-eyed Junco, 426
    and Great Crested Flycatcher, 256
    and Hermit Thrush, 326
    and Magnolia Warbler, 378
    and Northern Goshawk, 148
    and Northern Saw-whet Owl, 214
    and Pine Siskin, 464
    and Red Crossbill, 462
    and Tree Swallow, 284
    and Veery, 322
    and Winter Wren, 310
Aegolius acadicus, 27, 30, 32, 56, 68, **214–215**, 482, 524, 528
Aeral insectivores, 67, 77
Aerial insects
    and Barn Swallow, 292
    and Chimney Swift, 67, 222
    and Common Nighthawk, 67
    and Least Flycather, 252
    and Purple Martin, 67, 282
    and Red-headed Woodpecker, 228
Aerial photography, 29, 35
Agelaius phoeniceus, 58, 63, **444–445**, 492, 517, 525, 529

Agricultural abandonment, 10, 15, 17, 21
    and American Woodcock, 182
    and Barn Swallow, 292
    and Brown-headed Cowbird, 452
    and European Starling, 338
    and Golden-winged Warbler, 350
    and Gray Catbird, 332
    and Great Crested Flycatcher, 256
    and House Wren, 308
    and Northern Bobwhite, 108
    and Ruffed Grouse, 112
    and Upland Sandpiper, 178
    and White-breasted Nuthatch, 302
    and White-eyed Vireo, 262
    and Wild Turkey, 114
    and Yellow Warbler, 382
    and Yellow-breasted Chat, 402
Agricultural expansion, 9, 10, 20
    and American Goldfinch, 466
    and Blue-winged Warbler, 352
    and Bobolink, 442
    and Brown-headed Cowbird, 452
    and Dickcissel, 440
    and Eastern Meadowlark, 446
    and Grasshopper Sparrow, 416
    and Henslow's Sparrow, 418
    and Northern Bobwhite, 108
    and Red-winged Blackbird, 444
    and Savannah Sparrow, 414
    and Upland Sandpiper, 178
    and Vesper Sparrow, 412
    and Western Meadowlark, 448
    and Wild Turkey, 114
Agricultural intensification, 17–18
    and Alder Flycatcher, 248
    and Barn Owl, 202
    and Eastern Kingbird, 258
    and Eastern Meadowlark, 446
    and Green Heron, 128
    and Henslow's Sparrow, 418
    and House Sparrow, 468
    and Indigo Bunting, 438
    and Least Bittern, 122
    and Northern Bobwhite, 108
    and Northern Harrier, 142
    and Red-winged Blackbird, 444
    and Ring-necked Pheasant, 110
    and Savannah Sparrow, 416
    and Sedge Wren, 312
    and Short-eared Owl, 212
    and Upland Sandpiper, 178
    and Vesper Sparrow, 412
Agriculture, 9, 15–18, 21, 62, 66–68, 73
    and Alder Flycatcher, 248
    and American Crow, 274
    and Barn Owl, 202
    and Barn Swallow, 292
    and Barred Owl, 208

    and Blue Grosbeak, 436
    and Bobolink, 442
    and Canada Goose, 88
    and Cedar Waxwing, 340
    and Clay-colored Sparrow, 408
    and Common Grackle, 450
    and Common Raven, 278
    and Common Yellowthroat, 368
    and Dickcissel, 440
    and Eastern Bluebird, 320
    and Eastern Kingbird, 258
    and Eastern Meadowlark, 446
    and European Starling, 338
    and Fish Crow, 276
    and Grasshopper Sparrow, 416
    and Gray Catbird, 332
    and Great Horned Owl, 206
    and Henslow's Sparrow, 418
    and Horned Lark, 280
    and House Sparrow, 468
    and Killdeer, 174
    and Mallard, 98
    and Mourning Dove, 196
    and Mute Swan, 90
    and Northern Bobwhite, 108
    and Northern Harrier, 142
    and Orchard Oriole, 454
    and Pileated Woodpecker, 240
    and Red-tailed Hawk, 154
    and Red-winged Blackbird, 444
    and Ring-necked Pheasant, 110
    and Rock Pigeon, 192
    and Ruby-throated Hummingbird, 224
    and Sandhill Crane, 172
    and Savannah Sparrow, 414
    and Short-eared Owl, 212
    and Upland Sandpiper, 178
    and Vesper Sparrow, 412
    and Western Meadowlark, 448
    and Wild Turkey, 114
    and Yellow Warbler, 382
Aimophila aestivalis, 28, 474
Aix sponsa, 55, 64, **94–95**, 104, 106, 477, 523, 527
Alder, 182, 248
Alder Flycatcher. *See* Flycatcher, Alder
Alewife, 118
Alfalfa, 17, 18, 414
Allegheny Front, 7
Allegheny Front Physiographic section, 5, 12, 52, 477–493
Allegheny Mountain Physiographic section, 5, 12, 52, 477–493
Allegheny National Forest, xxii, 78, 79
Allegheny River, 4
American Bittern. *See* Bittern, American
American Black Duck. *See* Duck, American Black
American Coot. *See* Coot, American

American Crow. *See* Crow, American
American Goldfinch. *See* Goldfinch, American
American Kestrel. *See* Kestrel, American
American Ornithologists' Union (AOU), 81, 471
American Redstart. *See* Redstart, American
American Robin. *See* Robin, American
American Wigeon. *See* Wigeon, American
American Woodcock. *See* Woodcock, American
Amish farms, 178, 282, 414
Ammodramus henslowii, 28, 29, 57, 64, 73, 76, 78, 412, **418–419**, 491, 515, 525, 529
    A. savannarum, x, 57, 62, 64, 73, 412, **416–417**, 490, 515, 525, 529
Amphibians, 114, 152, 204
Anas acuta, 474
    A. americana, 27, 55, **471–472**, 527
    A. clypeata, 27, 55, **472**, 527
    A. crecca, 27, 55, 74, **102–103**, 477, 523, 527
    A. discors, 55, 58, **100–101**, 477, 523, 527
    A. platyrhynchos, 55, 64, 94, 96, **98–99**, 106, 477, 495, 523, 527
    A. rubripes, 55, 60, 74, **96–97**, 106, 477, 523, 527
    A. strepera, 474, 527
Ant, Carpenter, 240
Anthracite Upland Physiographic section, 5, 12, 52, 477–493
Anthracite Valley Physiographic section, 5, 12, 52, 477–493
Appalachian Mountain Physiographic section, 5, 12, 52, 477–493
Appalachian Mountains, 3, 4, 7, 9, 71
Appalachian Plateaus Physiographic province, 5, 12, 52, 477–493
Archeological, evidence for Trumpeter Swan, 92
Archilochus colubris, ix, 56, 59, 63, **224–225**, 483, 498, 524, 528
Ardea alba, 27, 55, 58, 59, 60, 73, 118, **126–127**, 130, 473, 478, 523, 527
    A. herodias, 27, 49, 55, 60, 61, 73, 82, **124–125**, 128, 172, 478, 523, 527
Area-sensitivity
    and American Bittern, 120
    and American Redstart, 372
    and Black-and-white Warbler, 356
    and Bobolink, 442
    and Cerluean Warbler, 374
    and Eastern Wood-Pewee, 242
    and Horned Lark, 280
    and Kentucky Warbler, 366
    and Marsh Wren, 314
    and Ovenbird, 342
    and Red-eyed Vireo, 270
    and Scarlet Tanager, 430
    and Veery, 322
    and Yellow-bellied Sapsucker, 232
Ash, x, 10, 13, 240
Asio flammeus, 27, 29, 30, 32, 56, 73, **212–213**, 482, 524, 528
    A. otus, 27, 29, 30, 32, 56, 59, 60, 73, **210–211**, 482, 524, 528

Aspect, 10, 521
Aspen, 13
    and American Woodcock, 182
    and Black-capped Chickadee, 296
    and Northern Flicker, 238
    and Ruffed Grouse, 112
    and Yellow-bellied Sapsucker, 232
Atlantic Coastal Plain Physiographic province, 4, 5, 12, 52, 477–493
Atlantic Flyway Breeding Waterfowl Survey, 88, 94, 95, 98, 99, 100, 102, 104, 106
Audubon Society of Pennsylvania, xxii, 23, 71, 74. *See also* National Audubon Society
Authors, list of, xxiii–xxiv
Autocorrelation, spatial, 38
Avian Knowledge Network, 75
Aythya americana, 474, 527
    A. collaris, **472**, 527

Baeolophus bicolor, 56, 60, 63, 296, **298–299**, 486, 505, 524, 528
Bald Eagle. *See* Eagle, Bald
Balm, Bee, 224
Baltimore Oriole. *See* Oriole, Baltimore
Banding, 16, 92, 94, 96
    and American Black Duck, 96
    and Bald Eagle, 140
    and Great Black-backed Gull, 188
    and Mallard, 98
    and Peregrine Falcon, 160
    and Ruby-throated Hummingbird, 224
    and Trumpeter Swan, 92
    and Wood Duck, 94
Bank Swallow. *See* Swallow, Bank
Bare ground, 216, 412
Barley, 17
Barn, 35, 202, 254, 290
Barn Owl. *See* Owl, Barn
Barn Swallow. *See* Swallow, Barn
Barred Owl. *See* Owl, Barred
Barrens, 75
    and Blue Grosbeak, 436
    and Chuck-will's-widow, 218
    and Eastern Whip-poor-will, 220
    and Pine Warbler, 390
    and Prairie Warbler, 396
    and Yellow-breasted Chat, 402
Bartramia longicauda, 27, 50, 55, 73, 76, **178–179**, 481, 523, 527
Bat, food item for Merlin, 158
BBS. *See* Breeding Bird Survey
Bear, Black, 236
Beaver, North American, 20
    and American Woodcock, 182
    and Hooded Merganser, 104
    and Prothonotary Warbler, 358
    and Swamp Sparrow, 422
    and Tree Swallow, 284
    and Wood Duck, 94
Beech, American, 10, 13, 80
    and Blue Jay, 272
    and Pileated Woodpecker, 240
    and Purple Finch, 458

    and Red-headed Woodpecker, 228
    and Winter Wren, 310
Belted Kingfisher. *See* Kingfisher, Belted
Biases, 38–45
Biofuel, and Ruffed Grouse, 112
Birch, 10, 13, 80, 296
Birch, Black, 10
Bird Conservation Regions (BCR), 71, 72
Bird house. *See* Nest boxes
Bison, 178, 452
Bittern, American, 27, 30, 31, 55, 74, 76, 102, **120–121**, 168, 170, 190, 478, 523, 527
Bittern, Least, 27, 30, 31, 55, 59, 74, 76, 102, **122–123**, 168, 170, 190, 478, 523, 527
Bittersweet, 360
Black Rail. *See* Rail, Black
Black Tern. *See* Tern, Black
Black Vulture. *See* Vulture, Black
Black-and-white Warbler. *See* Warbler, Black-and-white
Blackberry, 364, 402
Black-billed Cuckoo. *See* Cuckoo, Black-billed
Blackbird, Red-winged, 58, 63, **444–445**, 492, 517, 525, 529
Blackburnian Warbler. *See* Warbler, Blackburnian
Black-capped Chickadee. *See* Chickadee, Black-capped
Black-crowned Night-Heron. *See* Night-Heron, Black-crowned
Black-necked Stilt. *See* Stilt, Black-necked
Blackpoll Warbler. *See* Warbler, Blackpoll
Black-throated Blue Warbler. *See* Warbler, Black-throated Blue
Black-throated Green Warbler. *See* Warbler, Black-throated Green
Block, border, 24, 25
Block, definition of, 24, 25
Block, normal, 24, 25
Block, priority, 24, 25, 36
Block completion, 35–36
Block owner, 25
Block turnover, calculation of, 45–46
Blue Grosbeak. *See* Grosbeak, Blue
Blue Jay. *See* Jay, Blue
Blue Mountain Physiographic section, 5, 12, 52, 477–493
Bluebird, Eastern, 7, 41, 42, 57, 63, 284, **320–321**, 487, 506, 525, 528
Blue-gray Gnatcatcher. *See* Gnatcatcher, Blue-gray
Blue-headed Vireo. *See* Vireo, Blue-headed
Blue-winged Teal. *See* Teal, Blue-winged
Blue-winged Warbler. *See* Warbler, Blue-winged
Bobolink, 58, 64, 73, 414, **442–443**, 492, 517, 525, 529
Bobwhite, Northern, x, 27, 55, 58, 73, **108–109**, 478, 523, 527
Bog, 20, 212, 214, 248, 322, 362, 424
Bombycilla cedrorum, 57, 59, 63, 158, **340–341**, 487, 508, 525, 528
Bonasa umbellus, 28, 55, 58, 60, 68, **112–113**, 148, 478, 523, 527

Borer, Emerald Ash, x, 284
Botaurus lentiginosus, 27, 30, 31, 55, 74, 76, **120–121**, 168, 170, 190, 478, 523, 527
Brambles, 364
Branta canadensis, ix, 54, 55, 59, 64, **88–89**, 477, 495, 523, 527
Breeding Bird Survey, 7, 36, 47, 48–50, 62, 68, 75, 77, 81, 83
Breeding codes, 25–26, 28, 35, 36, 49, 54, 82
Breeding phenology, 523–526
Brewster's Warbler. *See* Warbler, Brewster's
Briar, 404
Bridge, nesting substrate
    for Cliff Swallow, 290
    for Eastern Phoebe, 254
    for Herring Gull, 186
    for Northern Rough-winged Swallow, 286
    for Peregrine Falcon, 160
    for Rock Pigeon, 192
Broad-winged Hawk. *See* Hawk, Broad-winged
Brood parasitism. *See* Cowbird, Brown-headed
Brown Creeper. *See* Creeper, Brown
Brown Thrasher. *See* Thrasher, Brown
Brown-headed Cowbird. *See* Cowbird, Brown-headed
Bubo virginianus, 30, 32, 56, **206–207**, 208, 482, 524, 528
Bubulcus ibis, 55, 58, 60, 473, 527
Budworm, Spruce, 144, 386
Building, nesting substrate
    for American Kestrel, 156
    for Barn Owl, 202
    for Barn Swallow, 292
    for Black Vulture, 134
    for Carolina Wren, 306
    for Chimney Swift, 222
    for Cliff Swallow, 290
    for Common Nighthawk, 216
    for Common Raven, 278
    for Eastern Phoebe, 254
    for Great Black-backed Gull, 188
    for Herring Gull, 186
    for House Finch, 460
    for Peregrine Falcon, 160
    for Ring-billed Gull, 184
    for Rock Pigeon, 192
    for Turkey Vulture, 136
Buildings, 21–22
    collisions with. *See* Windows, collisions with
Bulrush, 20, 120, 314
Bunting, Indigo, 58, 63, 436, **438–439**, 491, 516, 525, 529
Burdock, 466
Burrow, nesting, and Belted Kingfisher, 226
Buteo jamaicensis, 55, 64, 150, **154–155**, 206, 480, 496, 523, 527
    B. lineatus, 27, 55, 68, 73, **150–151**, 479, 496, 523, 527
    B. platypterus, 55, 60, 61, 64, 73, **152–153**, 480, 496, 523, 527

    B. virescens, 55, **128–129**, 226, 479, 523, 527
Buttonbush, Common, 190

Canada Goose. *See* Goose, Canada
Canada Warbler. *See* Warbler, Canada
Cankerworm, 12
Canopy cover, 12, 13, 14
    and Acadian Flycatcher, 246
    and Black-throated Blue Warbler, 388
    and Broad-winged Hawk, 152
    and Canada Warbler, 400
    and Cerulean Warbler, 374
    and Eastern Wood-Pewee, 242
    and Field Sparrow, 410
    and Golden-crowned Kinglet, 318
    and Great Crested Flycatcher, 256
    and Magnolia Warbler, 378
    and Mourning Warbler, 364
    and Red-eyed Vireo, 270
    and Scarlet Tanager, 430
    and Song Sparrow, 420
    and Summer Tanager, 428
    and Swainson's Thrush, 324
    and Warbling Vireo, 268
    and Wood Thrush, 328
    and Yellow-breasted Chat, 402
    and Yellow-throated Vireo, 264
Caprimulgus carolinensis, 27, 56, 58, **218–219**, 482, 528
    C. vociferus, 27, 32, 56, 73, 77, **220–221**, 482, 524, 528
Captive bred. *See* Introductions/stocking
Cardellina canadensis, 57, 59, 61, 64, 68, **400–401**, 490, 512, 525, 529
Cardinal, Northern, 58, 63, **432–433**, 491, 516, 525, 529
Cardinalis cardinalis, 58, 63, **432–433**, 491, 516, 525, 529
Carnegie Museum of Natural History, xi, 24, 37
Carolina Chickadee. *See* Chickadee, Carolina
Carolina Wren. *See* Wren, Carolina
Carpodacus mexicanus, 58, 63, 83, 458, **460–461**, 468, 492, 518, 526, 529
    C. purpureus, 58, 64, 436, **458–459**, 492, 518, 526, 529
Catbird, Gray, ix, 41, 57, 58, 62, 63, 78, 332–333, 336, 487, 508, 525, 528
Cathartes aura, 55, 64, 134, **136–137**, 479, 523, 527
Catharus fuscescens, 57, 63, **322–323**, 326, 487, 507, 525, 528
    C. guttatus, 40, 41, 42, 43, 57, 59, 60, 61, 64, 68, 310, 322, **326–327**, 487, 507, 525, 528
    C. ustulatus, 27, 57, 68, 73, **324–325**, 487, 507, 525, 528
Cattail, 20, 30, 31
    and American Bittern, 120
    and American Goldfinch, 466
    and King Rail, 162
    and Least Bittern, 122
    and Marsh Wren, 314

    and Sedge Wren, 312
    and Swamp Sparrow, 422
    and Virginia Rail, 164
Catterpillar. *See* Larvae
Cattle Egret. *See* Egret, Cattle
Cavity nesting, 11, 12, 28, 35
    and American Kestrel, 156
    and Barn Swallow, 292
    and Black Vulture, 134
    and Black-capped Chickadee, 296
    and Carolina Wren, 306
    and Chimney Swift, 222
    and Common Merganser, 106
    and Downy Woodpecker, 234
    and Eastern Bluebird, 320
    and Eastern Phoebe, 254
    and European Starling, 338
    and Great Crested Flycatcher, 256
    and Hairy Woodpecker, 236
    and Hooded Merganser, 104
    and House Wren, 308
    and Northern Flicker, 238
    and Northern Rough-winged Swallow, 286
    and Pileated Woodpecker, 240
    and Prothonotary Warbler, 358
    and Purple Martin, 282
    and Red-bellied Woodpecker, 230
    and Red-breasted Nuthatch, 300
    and Red-headed Woodpecker, 228
    and Ring-necked Duck, 472
    and Tree Swallow, 284
    and Tufted Titmouse, 298
    and Turkey Vulture, 136
    and White-breasted Nuthatch, 302
    and Wood Duck, 94
    and Yellow-bellied Sapsucker, 232
Cedar, Eastern Red, 260
Cedar Waxwing. *See* Waxwing, Cedar
Central Lowlands Physiographic province, 12
Certhia americana, 56, 59, 60, 64, 68, **304–305**, 486, 505, 524, 528
Cerulean Warbler. *See* Warbler, Cerulean
Chaetura pelagica, 27, 56, 63, 73, **222–223**, 482, 497, 524, 528
Charadrius melodus, 21, 74, 186, **473**, 527
    C. vociferus, 26, 27, 55, 64, **174–175**, 480, 496, 523, 527
Chat, Yellow-breasted, x, 57, 60, 61, 64, 74, **402–403**, 490, 512, 525, 529
Cherry, Black, 10
Chesapeake Bay, 4, 90
Chestnut, American, 12
Chestnut-sided Warbler. *See* Warbler, Chestnut-sided
Chickadee, Black-capped, 56, 63, 294, **296–297**, 486, 504, 524, 528
    Carolina, 56, 59, 60, 64, **294–295**, 296, 485, 504, 524, 528
Chimney Swift. *See* Swift, Chimney
Chipping Sparrow. *See* Sparrow, Chipping
Chlidonias niger, 27, 55, 58, 74, 168, **190–191**, 481, 524, 527
Chondestes grammacus, 28, 474, 529

INDEX   571

Chordeiles minor, 27, 32, 56, 58, 59, 67, 73, 77, **216–217**, 482, 524, 528
Christmas Bird Count, 75, 186, 188, 192, 206, 208, 238, 240, 450, 460
Chuck-will's-widow, 27, 56, 58, **218–219**, 482, 528
Cicada, food item for Red-headed Woodpecker, 228
Circus cyaneus, 27, 55, 60, 73, 100, **142–143**, 479, 523, 527
Cistothorus palustris, 27, 57, 74, 77, 102, 168, 190, 312, **314–315**, 486, 524, 529
  C. platensis, 27, 57, 73, **312–313**, 486, 524, 528
Clay-colored Sparrow. *See* Sparrow, Clay-colored
Cliff Swallow. *See* Swallow, Cliff
Cliffs, nesting site
  for Cliff Swallow, 290
  for Common Raven, 278
  for Eastern Phoebe, 254
  for Herring Gull, 186
  for Peregrine Falcon, 160
  for Rock Pigeon, 192
Climate, 4–8
  and Downy Woodpecker, 234
  and Red-bellied Woodpecker, 230
Climate change, x, 7, 59–62, 70, 76, 79–80
  and Black-billed Cuckoo, 60, 200
  and Black-throated Green Warbler, 398
  and Blue-gray Gnatcatcher, 316
  and Brown Creeper, 60, 304
  and Hermit Thrush, 60, 326
  and Hooded Warbler, 60, 370
  and Least Flycatcher, 252
  and Nashville Warbler, 60, 362
  and Northern Saw-whet Owl, 214
  and Northern Waterthrush, 60, 348
  and Swainson's Thrush, 324
  and Tree Swallow, 284
  and Tufted Titmouse, 60, 298
  and White-throated Sparrow, 424
  and Yellow-bellied Flycatcher, 244
Clover, 17, 18, 440
Coccothraustes vespertinus, 28, 474, 529
Coccyzus americanus, 41, 56, 60, 64, **198–199**, 200, 481, 497, 522, 524, 528
  C. erythropthalmus, 40, 41, 56, 59, 60, 64, 73, 198, **200–201**, 482, 497, 522, 524, 528
Colaptes auratus, 42, 56, 64, 156, **238–239**, 483, 499, 524, 528
Colinus virginianus, x, 27, 55, 58, 73, **108–109**, 478, 523, 527
Collared-Dove, Eurasian, x, 54, 56, **194–195**, 481, 524, 527
Collisions. *See* Vehicles, collisions with; Windows, collisions with; Wires, collisions with
Colony nesting, 27–28, 37, 59, 75
  and Bank Swallow, 288
  and Black-crowned Night-Heron, 118, 126, 130

  and Cliff Swallow, 290
  and Dickcissel, 440
  and Double-crested Cormorant, 59, 118
  and Great Black-backed Gull, 188
  and Great Blue Heron, 124
  and Great Egret, 59, 126
  and Henslow's Sparrow, 418
  and Herring Gull, 186
  and Marsh Wren, 314
  and Northern Rough-winged Swallow, 286
  and Purple Martin, 282
  and Ring-billed Gull, 59, 184, 186
  and Sedge Wren, 312
  and Upland Sandpiper, 178
  and Yellow-crowned Night-Heron, 132
Columba livia, 21, 55, 63, **192–193**, 481, 497, 524, 527
Common Gallinule. *See* Gallinule, Common
Common Grackle. *See* Grackle, Common
Common Merganser. *See* Merganser, Common
Common Nighthawk. *See* Nighthawk, Common
Common Raven. *See* Raven, Common
Common Tern. *See* Tern, Common
Common Yellowthroat. *See* Yellowthroat, Common
Communication towers, collisions with. *See* Wires, collisions with
Competition
  between Acadian Flycatcher and Least Flycatcher, 246, 252
  between Common Raven and Peregrine Falcon, 278
  between Eastern Bluebird and European Starling, 320
  between Eastern Bluebird and House Sparrow, 320
  between Eurasian Collared-Dove and Mourning Dove, 194
  between European Starling and native species, 338
  between Northern Flicker and European Starling, 238
  between Prothonotary Warbler and House Wren, 358
  between Red-breasted Nuthatch and House Wren, 300
  between Red-headed Woodpecker and European Starling, 228
  between Red-shouldered Hawk and Red-tailed Hawk, 150
  between Sora and Virginia Rail, 166
  between Wood Thrush and Hermit Thrush, 328
  between Wood Thrush and Veery, 328
  between Yellow-billed and Black-billed Cuckoo, 198, 200
Confirmed breeding, definition of, 28
Conflicts with humans
  and Canada Goose, 88
  and Common Grackle, 450

  and Eurasian Collared-Dove, 194
  and Herring Gull, 186
  and House Sparrow, 468
  and Red-winged Blackbird, 444
  and Ring-billed Gull, 184
  and Rock Pigeon, 192
Coniferous trees, 11, 43, 44, 46, 58, 73
Conjunctivitis, mycoplasmal, 83, 460, 468
Conneaut Marsh, 18, 20
Conservation, 70–80
Conservation concern, species of, 15, 29, 71, 73–74, 75, 446
Conservation easement, 166, 172
Conservation grassland. *See* Conservation Reserve Enhancement Program (CREP)
Conservation Reserve Enhancement Program (CREP), 17, 20, 48–50, 36, 79
  and Blue-winged Teal, 100
  and Bobolink, 442
  and Common Yellowthroat, 368
  and Eastern Meadowlark, 446
  and Grasshopper Sparrow, 416
  and Henslow's Sparrow, 418
  and Mallard, 98
  and Northern Harrier, 162
  and Red-winged Blackbird, 444
  and Ring-necked Pheasant, 110
  and Sedge Wren, 312
  and Vesper Sparrow, 412
Conservation Reserve Program. *See* Conservation Reserve Enhancement Program (CREP)
Conservation status, 27–28, 72, 76–77
Contaminants
  and Broad-winged Hawk, 152
  and Cooper's Hawk, 146
  and Great Blue Heron, 124
  and Louisiana Waterthrush, 346
  and Sharp-shinned Hawk, 144
  and Spotted Sandpiper, 176
Contopus cooperi, 27, 56, 73, **474**, 528
  C. virens, 56, 59, 63, **242–243**, 254, 483, 499, 524, 528
Contributers, list of, xi–xxiv
Cooper's Hawk. *See* Hawk, Cooper's
Coot, American, 27, 30, 31, 55, 74, 168, **170–171**, 190, 480, 523, 527
Coragyps atratus, 55, 59, 60, **134–135**, 136, 479, 523, 527
Cormorant, Double-crested, 54, 55, 58, 59, **118–119**, 126, 130, 478, 523, 527
Corn, 17, 192, 412
Cornell Lab of Ornithology, xxii, 24, 29, 35, 36, 37, 75
Corvus brachyrhynchos, 56, 58, 63, **274–275**, 485, 502, 524, 528
  C. corax, ix, x, 56, 59, 64, 158, **278–279**, 485, 503, 524, 528
  C. ossifragus, 56, 64, 158, **276–277**, 485, 502, 524, 528
Cottonwood, 10, 268
County boundaries, 8, 24, 82
Coverage, 48–51

572 INDEX

Cowbird, Brown-headed, 58, 59, 63, **452–453**, 492, 517, 526, 529. *See also* Parasitism
Coyote, 278
Crabapple, 128
Crane, Sandhill, 27, 54, 55, 124, **172–173**, 480, 523, 527
Crayfish, Rusty, 132
Creeper, Brown, 56, 59, 60, 64, 68, **304–305**, 486, 505, 524, 528
Crop damage, 88
Crops, 17
    and Savannah Sparrow, 414
Crops, changes in
    and Horned Lark, 280
    and Northern Harrier, 142
    and Ring-necked Pheasant, 110
Crossbill, Red, 28, 50, 54, 58, 68, **462–463**, 464, 492, 526, 529
    White-winged, 69
Crow, American, 56, 58, 63, **274–275**, 485, 502, 524, 528
    Fish, 56, 64, 158, **276–277**, 485, 502, 524, 528
Crustaceans, and Yellow-crowned Night-Heron, 132
Cuckoo, Black-billed, 40, 41, 56, 59, 60, 64, 73, 198, **200–201**, 482, 497, 522, 524, 528
    Yellow-billed, 41, 56, 60, 64, **198–199**, 200, 481, 497, 522, 524, 528
Cyanocitta cristata, ix, 56, 58, 63, **272–273**, 485, 502, 524, 528
Cygnus buccinator, x, 54, 55, **92–93**, 477, 527
    C. olor, 37, 55, **90–91**, 124, 477, 523, 527
Cypress, Bald, and Acadian Flycatcher, 246

Dark-eyed Junco. *See* Junco, Dark-eyed
Data, sources of, 48
Data entry, 35, 49
Data management, 36
DDT. *See* Insecticides
Deep Valleys Physiographic section, 5, 12, 52, 477–493
Deer, 15, 134, 316, 366, 388
Deer, White-tailed, x, 15, 81, 134, 136, 154, 242, 278, 342
Deer browsing, 15, 35, 342, 348, 388
Defoliation, 12, 13, 14
    and Black-billed Cuckoo, 200
    and Blackpoll Warbler, 386
    and Yellow-billed Cuckoo, 198
Delaware River, 3, 4, 20, 24, 79
Delaware Water Gap National Recreation Area, 79
Department of Conservation and Natural Resources (DCNR), xxii, xxiii, 12, 13, 78
Detectability, effect of time of day on bird, 39, 42, 44, 45, 521
    effect of time of season on bird, 39, 41, 44, 45, 521
Detection issues
    and American Bittern, 102
    and Barred Owl, 208
    and Belted Kingfisher, 226
    and Brown Thrasher, 336
    and Eastern Screech-Owl, 204
    and Eastern Whip-poor-will, 220
    and Eurasian Collared-Dove, 194
    and European Starling, 338
    and Great Horned Owl, 206
    and Green-winged Teal, 102
    and Hooded Merganser, 104
    and Least Bittern, 122
    and Long-eared Owl, 210
    and Northern Harrier, 142
    and Northern Saw-whet Owl, 214
    and Pileated Woodpecker, 240
    and Rose-breasted Grosbeak, 434
    and Sharp-shinned Hawk, 144
    and Sora, 166
    and Spotted Sandpiper, 176
    and Turkey Vulture, 136
    and Virginia Rail, 164
    and Wilson's Snipe, 180
    and Yellow-bellied Flycatcher, 244
Developed land, ix, 21–22
Development
    and American Crow, 274
    and American Kestrel, 156
    and American Redstart, 372
    and American Robin, 330
    and Baltimore Oriole, 456
    and Barn Owl, 202
    and Barn Swallow, 292
    and Black-throated Blue Warbler, 388
    and Blue Grosbeak, 436
    and Blue Jay, 272
    and Blue-gray Gnatcatcher, 316
    and Blue-winged Warbler, 352
    and Broad-winged Hawk, 152
    and Brown Thrasher, 336
    and Brown-headed Cowbird, 452
    and Carolina Chickadee, 294
    and Cedar Waxwing, 340
    and Chimney Swift, 222
    and Chipping Sparrow, 406
    and Common Gallinule, 168
    and Common Grackle, 450
    and Common Yellowthroat, 368
    and Cooper's Hawk, 146
    and Downy Woodpecker, 234
    and Eastern Bluebird, 320
    and Eastern Kingbird, 258
    and Eastern Meadowlark, 446
    and Eastern Phoebe, 254
    and European Starling, 338
    and Field Sparrow, 410
    and Fish Crow, 276
    and Golden-winged Warbler, 350
    and Great Crested Flycatcher, 256
    and Green Heron, 128
    and House Finch, 460
    and House Sparrow, 468
    and House Wren, 308
    and Indigo Bunting, 438
    and Kentucky Warbler, 366
    and Killdeer, 174
    and King Rail, 162
    and Least Bittern, 122
    and Mallard, 98
    and Marsh Wren, 314
    and Northern Bobwhite, 108
    and Northern Cardinal, 432
    and Northern Flicker, 238
    and Northern Harrier, 142
    and Northern Mockingbird, 334
    and Northern Rough-winged Swallow, 286
    and Northern Waterthrush, 348
    and Pine Warbler, 390
    and Purple Martin, 282
    and Red-bellied Woodpecker, 230
    and Red-eyed Vireo, 270
    and Red-tailed Hawk, 154
    and Red-winged Blackbird, 444
    and Ruby-throated Hummingbird, 224
    and Short-eared Owl, 212
    and Song Sparrow, 420
    and Swamp Sparrow, 422
    and Tufted Titmouse, 298
    and Virginia Rail, 164
    and Warbling Vireo, 268
    and White-breasted Nuthatch, 302
    and White-eyed Vireo, 262
    and Wood Thrush, 328
    and Yellow-breasted Chat, 402
Dickcissel, 28, 58, 59, 61, 73, **440–441**, 491, 525, 529
Disease, ix, 12, 77, 79, 204, 284, 324, 460
Distance sampling, 40, 521–522
Disturbance, human
    and Black Tern, 190
    and Northern Harrier, 142
    and Pied-billed Grebe, 116
    and Piping Plover, 473
    and Sandhill Crane, 172
    and Sedge Wren, 312
    and Short-eared Owl, 212
Disturbance, noise, 33, 35, 40
Dogwood, 182, 190, 382
Dolichonyx oryzivorus, 58, 64, 73, 414, **442–443**, 492, 517, 525, 529
Double-crested Cormorant. *See* Cormorant, Double-crested
Dove, Eurasian Collared. *See* Collared-Dove, Eurasian
    Mourning, 56, 63, **196–197**, 481, 497, 524, 527
    Rock. *See* Pigeon, Rock
Downy Woodpecker. *See* Woodpecker, Downy
Drought, 164, 180, 440
Dryocopus pileatus, x, 56, 58, 64, 82, **240–241**, 483, 499, 524, 528
Duck, 168, 170
    American Black, 55, 60, 74, **96–97**, 106, 477, 523, 527
    Ring-necked, **472**, 527
    Ruddy, **472**, 527
    Wood, 55, 64, **94–95**, 104, 106, 477, 523, 527

Dumetella carolinensis, ix, 41, 57, 58, 62, 63, 78, **332–333**, 336, 487, 508, 525, 528
Dune, 20–21

Eagle, Bald, x, 27, 37, 47, 49, 50, 54, 55, 59, 73, 76, 79, 124, 138, **140–141**, 479, 523, 527
Earthworm, 132, 182
Eastern Bluebird. *See* Bluebird, Eastern
Eastern Kingbird. *See* Kingbird, Eastern
Eastern Lake Physiographic section, 4, 12
Eastern Meadowlark. *See* Meadowlark, Eastern
Eastern Phoebe. *See* Phoebe, Eastern
Eastern Screech-Owl. *See* Screech-Owl, Eastern
Eastern Towhee. *See* Towhee, Eastern
Eastern Whip-poor-will. *See* Whip-poor-will, Eastern
Eastern Wood-Pewee. *See* Wood-Pewee, Eastern
Ebird, 36, 75
Ectopistes migratorius, 474
Effort, 49–53
    analysis, 38–39, 49–58
    documenting of, 35, 36
Egg dates, 18
Egret, Cattle, 55, 58, 60, **473**, 527
    Great, 27, 55, 58, 59, 60, 73, 118, **126–127**, 130, 473, 478, 523, 527
    Snowy, 27, 55, 58, 60, 126, **472–473**, 527
Egretta thula, 27, 55, 58, 60, 126, **472–473**, 527
Electrocution, and Red-tailed Hawk, 154
Elevation, 3, 4, 5, 6, 7, 9, 10, 38, 42, 43, 44, 45, 46, 59, 61, 65, 66, 77, 494–519, 521
    and Alder Flycatcher, 248
    and American Crow, 274
    and American Redstart, 372
    and Belted Kingfisher, 226
    and Black-and-white Warbler, 356
    and Blackburnian Warbler, 380
    and Black-capped Chickadee, 296
    and Black-throated Blue Warbler, 388
    and Black-throated Green Warbler, 398
    and Blue Grosbeak, 436
    and Blue-gray Gnatcatcher, 316
    and Blue-headed Vireo, 266
    and Blue-winged Warbler, 352
    and Bobolink, 442
    and Brown Creeper, 304
    and Canada Goose, 88
    and Canada Warbler, 400
    and Carolina Chickadee, 294
    and Carolina Wren, 306
    and Chestnut-sided Warbler, 384
    and Chipping Sparrow, 406
    and Chuck-will's-widow, 218
    and Dark-eyed Junco, 426
    and Downy Woodpecker, 234
    and Eastern Phoebe, 254
    and Eastern Screech-Owl, 204
    and Eastern Towhee, 404
    and Eastern Wood-Pewee, 242
    and Fish Crow, 276
    and Golden-crowned Kinglet, 318
    and Golden-winged Warbler, 350
    and Gray Catbird, 332
    and Great Crested Flycatcher, 256
    and Hairy Woodpecker, 236
    and Hermit Thrush, 326
    and Hooded Warbler, 370
    and House Wren, 308
    and Kentucky Warbler, 366
    and Least Flycatcher, 252
    and Magnolia Warbler, 378
    and Mourning Warbler, 364
    and Northern Cardinal, 432
    and Northern Goshawk, 148
    and Northern Mockingbird, 334
    and Northern Saw-whet Owl, 214
    and Northern Waterthrush, 61, 348
    and Orchard Oriole, 454
    and Ovenbird, 342
    and Purple Finch, 458
    and Red Crossbill, 462
    and Red-bellied Woodpecker, 230
    and Red-headed Woodpecker, 228
    and Rose-breasted Grosbeak, 434
    and Savannah Sparrow, 414
    and Song Sparrow, 420
    and Swainson's Thrush, 324
    and Swamp Sparrow, 422
    and Tufted Titmouse, 298
    and Veery, 322
    and White-eyed Vireo, 262
    and White-throated Sparrow, 61
    and Willow Flycatcher, 250
    and Wilson's Snipe, 180
    and Winter Wren, 310
    and Wood Thrush, 328
    and Yellow Warbler, 382
    and Yellow-bellied Sapsucker, 232
    and Yellow-rumped Warbler, 392
    and Yellow-throated Vireo, 264
    and Yellow-throated Warbler, 394
Elm, American, 10, 12, 13, 240, 374
Emerald Ash Borer. *See* Borer, Emerald Ash
Empidonax alnorum, x, 27, 41, 56, 61, 64, 73, **248–249**, 250, 484, 500, 524, 528
    E. flaviventris, 27, 50, 56, 59, 68, 73, **244–245**, 386, 483, 524, 528
    E. minimus, 56, 63, 246, **252–253**, 484, 500, 524, 528
    E. traillii, 56, 63, 73, 248, **250–251**, 484, 500, 524, 528
    E. virescens, 41, 45, 56, 58, 60, 63, 68, 73, **246–247**, 252, 483, 500, 524, 528
Endangered species, 26–28, 29, 70, 73–74, 76, 160, 162, 178, 190, 212, 244, 312, 440, 473
Energy development, x, 79
    and American Redstart, 372
    and American Woodcock, 182
    and Barred Owl, 208
    and Belted Kingfisher, 226
    and Blackburnian Warbler, 380
    and Black-throated Blue Warbler, 388
    and Blue Jay, 272
    and Blue-gray Gnatcatcher, 316
    and Brown Creeper, 304
    and Brown-headed Cowbird, 452
    and Cerulean Warbler, 374
    and Great Blue Heron, 124
    and Hermit Thrush, 326
    and Louisiana Waterthrush, 346
    and Northern Saw-whet Owl, 214
    and Ovenbird, 342
    and Swainson's Thrush, 324
    and Worm-eating Warbler, 344
    and Yellow-bellied Sapsucker, 232
Eremophila alpestris, 50, 56, 59, 60, 64, **280–281**, 412, 485, 503, 524, 528
Erie, 21
Erie National Wildlife Refuge, 79
Erosion, riverbanks and Great Egret, 126
Eurasian Collared-Dove. *See* Collared-Dove, Eurasian
European Starling. *See* Starling, European
Evening Grosbeak. *See* Grosbeak, Evening
Extirpated species, 27–28

Falco columbarius, x, 54, 55, **158–159**, 471, 480, 523, 527
    F. peregrinus, 27, 37, 47, 54, 55, 59, 73, 138, **160–161**, 192, 480, 523, 527
    F. sparverius, ix, 39, 55, 60, 64, **156–157**, 480, 496, 523, 527
Falcon, Peregrine, 27, 37, 47, 54, 55, 59, 73, 138, **160–161**, 192, 480, 523, 527
Fallow field, 172, 368
Farmland. *See* Agriculture
Feeder
    and Black-capped Chickadee, 296
    and Carolina Wren, 306
    and Downy Woodpecker, 234
    and Eurasian Collared-Dove, 194
    and House Finch, 460
    and Northern Cardinal, 432
    and Pine Siskin, 464
    and Red-bellied Woodpecker, 230
    and Ruby-throated Hummingbird, 224
    and Sharp-shinned Hawk, 144
    and Tufted Titmouse, 298
    and White-breasted Nuthatch, 302
Fern, 364
    mossy, 400
Fertilizer, 446
Field Sparrow. *See* Sparrow, Field
Finch, House, 58, 63, 83, 458, **460–461**, 468, 492, 518, 526, 529
    Purple, 58, 64, 436, **458–459**, 492, 518, 526, 529
Fir, 300, 318
Fire, 15, 350, 364
Fish, food item
    for Bald Eagle, 140
    for Belted Kingfisher, 226
    for Common Merganser, 106
    for Double-crested Cormorant, 118
    for Great Blue Heron, 124
    for Green Heron, 128

Fish Crow. *See* Crow, Fish
Fisher, 148
Fishery conflicts
    and Black-crowned Night-Heron, 130
    and Double-crested Cormorant, 118
    and Green Heron, 128
Fledgling dates, 523–526
Flicker, Northern, 42, 56, 64, 156, **238–239**, 483, 499, 524, 528
Flicker, Yellow-shafted. *See* Flicker, Northern
Flood control, 20
Flycatcher, Acadian, 41, 45, 56, 58, 60, 63, 68, 73, **246–247**, 252, 483, 500, 524, 528
    Alder, x, 27, 41, 56, 61, 64, 73, **248–249**, 250, 484, 500, 524, 528
    Great Crested, 56, 59, 64, **256–257**, 484, 501, 524, 528
    Least, 56, 63, 246, **252–253**, 484, 500, 524, 528
    Olive-sided, 27, 56, 73, **474**, 528
    Willow, 56, 63, 73, 248, **250–251**, 484, 500, 524, 528
    Yellow-bellied, 27, 50, 56, 59, 68, 73, **244–245**, 386, 483, 524, 528
Forest, 9–15
Forest, bottomland
    and Cerulean Warbler, 374
    and Kentucky Warbler, 366
    and Northern Parula, 376
    and Prothonotary Warbler, 358
    and Swainson's Warbler, 360
Forest, coniferous
    and Barred Owl, 208
    and Blackburnian Warbler, 380
    and Blackpoll Warbler, 386
    and Black-throated Green Warbler, 398
    and Blue-headed Vireo, 266
    and Brown Creeper, 304
    and Canada Warbler, 400
    and Dark-eyed Junco, 426
    and Eastern Wood-Pewee, 242
    and Golden-crowned Kinglet, 318
    and Hermit Thrush, 326
    and House Wren, 308
    and Long-eared Owl, 210
    and Merlin, 158
    and Northern Flicker, 238
    and Northern Goshawk, 148
    and Northern Saw-whet Owl, 214
    and Pine Siskin, 464
    and Purple Finch, 458
    and Red Crossbill, 462
    and Red-breasted Nuthatch, 300
    and Scarlet Tanager, 430
    and Sharp-shinned Hawk, 144
    and Swainson's Thrush, 324
    and Veery, 322
    and Winter Wren, 310
    and Yellow-bellied Flycatcher, 244
    and Yellow-rumped Warbler, 392
Forest, contiguous, 9, 10, 20
    and American Crow, 274
    and Barn Swallow, 292
    and Black-and-white Warbler, 356
    and Black-throated Blue Warbler, 388
    and Blue-gray Gnatcatcher, 316
    and Broad-winged Hawk, 152
    and Brown Creeper, 304
    and Brown Thrasher, 336
    and Brown-headed Cowbird, 452
    and Carolina Wren, 306
    and Chimney Swift, 222
    and Comnon Grackle, 450
    and Comnon Yellowthroat, 368
    and Cooper's Hawk, 146
    and Downy Woodpecker, 234
    and Eastern Kingbird, 258
    and Eastern Phoebe, 254
    and Eastern Screech-Owl, 204
    and Eastern Whip-poor-will, 220
    and Golden-winged Warbler, 350
    and Gray Catbird, 332
    and Great Horned Owl, 206
    and Hooded Warbler, 370
    and Kentucky Warbler, 366
    and Killdeer, 174
    and Louisiana Waterthrush, 346
    and Northern Goshawk, 148
    and Northern Mockingbird, 334
    and Northern Saw-whet Owl, 214
    and Ovenbird, 342
    and Pileated Woodpecker, 240
    and Red-shouldered Hawk, 144
    and Red-tailed Hawk, 154
    and Red-winged Blackbird, 444
    and Ruby-throated Hummingbird, 224
    and Sharp-shinned Hawk, 144
    and Song Sparrow, 420
    and Tree Swallow, 284
    and Worm-eating Warbler, 344
    and Yellow Warbler, 382
    and Yellow-bellied Flycatcher, 244
    and Yellow-bellied Sapsucker, 232
Forest, deciduous
    and American Redstart, 372
    and Baltimore Oriole, 456
    and Blackburnian Warbler, 380
    and Black-throated Green Warbler, 398
    and Canada Warbler, 400
    and Cerulean Warbler, 374
    and Chestnut-sided Warbler, 384
    and Dark-eyed Junco, 426
    and Hooded Warbler, 370
    and Kentucky Warbler, 366
    and Ovenbird, 342
    and Scarlet Tanager, 430
    and Veery, 322
    and White-breasted Nuthatch, 302
    and Wood Thrush, 328
Forest, mixed
    and American Redstart, 372
    and Barred Owl, 208
    and Black-and-white Warbler, 356
    and Blackburnian Warbler, 380
    and Black-throated Green Warbler, 398
    and Blue-headed Vireo, 266
    and Dark-eyed Junco, 426
    and Hooded Warbler, 370
    and Northern Flicker, 238
    and Ovenbird, 342
    and Red-breasted Nuthatch, 300
    and Swainson's Thrush, 324
    and Veery, 322
    and White-breasted Nuthatch, 302
    and Winter Wren, 310
    and Yellow-throated Vireo, 264
Forest, old growth, 11, 78
    and Blackburnian Warbler, 380
    and Golden-crowned Kinglet, 318
    and Magnolia Warbler, 378
    and Swainson's Thrush, 324
    and Winter Wren, 310
Forest, primary. *See* Forest, old growth
Forest, secondary/second growth, 11, 112
    and American Redstart, 372
    and Black-and-white Warbler, 356
    and Chestnut-sided Warbler, 384
    and Eastern Towhee, 404
    and Nashville Warbler, 362
    and Rose-breasted Grosbeak, 434
    and Song Sparrow, 420
    and Yellow-bellied Sapsucker, 232
Forest, young, 11, 15, 78, 79
    and American Woodcock, 182
    and Gray Catbird, 332
    and Magnolia Warbler, 378
    and Rose-breasted Grosbeak, 434
    and Ruffed Grouse, 112
Forest clearance, 17, 61
    and American Woodcock, 182
    and Barred Owl, 208
    and Blackpoll Warbler, 386
    and Blue Jay, 272
    and Blue-gray Gnatcatcher, 316
    and Blue-headed Vireo, 266
    and Blue-winged Warbler, 352
    and Common Raven, 278
    and Eastern Bluebird, 320
    and Eastern Kingbird, 258
    and Eastern Towhee, 404
    and Golden-crowned Kinglet, 318
    and Indigo Bunting, 438
    and Red Crossbill, 462
    and Red-breasted Nuthatch, 300
    and Wood Duck, 94
Forest clearcut
    and Black-throated Blue Warbler, 388
    and Canada Warbler, 400
    and Chestnut-sided Warbler, 384
    and Eastern Bluebird, 320
    and Indigo Bunting, 438
    and Pine Warbler, 390
    and Song Sparrow, 420
    and Yellow-breasted Chat, 402
Forest clearings. *See* Forest gaps

Forest edge, 15
- and American Crow, 274
- and American Goldfinch, 466
- and Baltimore Oriole, 456
- and Black-throated Blue Warbler, 388
- and Blue Jay, 272
- and Cedar Waxwing, 340
- and Chestnut-sided Warbler, 384
- and Chipping Sparrow, 406
- and Common Yellowthroat, 368
- and Eastern Bluebird, 320
- and Eastern Wood-Pewee, 242
- and Gray Catbird, 332
- and Great Horned Owl, 206
- and Hooded Warbler, 370
- and House Wren, 308
- and Indigo Bunting, 438
- and Least Flycatcher, 252
- and Nashville Warbler, 362
- and Northern Cardinal, 432
- and Northern Flicker, 238
- and Rose-breasted Grosbeak, 434
- and Tufted Titmouse, 298
- and White-breasted Nuthatch, 302
- and White-eyed Vireo, 262
- and Yellow Warbler, 382
- and Yellow-breasted Chat, 402
- and Yellow-throated Vireo, 264

Forest fragmentation, x, 10, 79
- and American Crow, 274
- and American Redstart, 372
- and Baltimore Oriole, 456
- and Barred Owl, 208
- and Black-and-white Warbler, 356
- and Black-billed Cuckoo, 200
- and Blackburnian Warbler, 380
- and Black-throated Green Warbler, 398
- and Blue Jay, 272
- and Broad-winged Hawk, 152
- and Carolina Chickadee, 294
- and Cerulean Warbler, 374
- and Eastern Wood-Pewee, 242
- and Gray Catbird, 332
- and Great Horned Owl, 206
- and House Wren, 308
- and Kentucky Warbler, 366
- and Northern Saw-whet Owl, 214
- and Ovenbird, 342
- and Scarlet Tanager, 430
- and Sharp-shinned Hawk, 144
- and Song Sparrow, 420
- and Summer Tanager, 428
- and Tufted Titmouse, 298
- and Warbling Vireo, 268
- and Wood Thrush, 328
- and Yellow-billed Cuckoo, 198

Forest gaps
- and Baltimore Oriole, 456
- and Cerulean Warbler, 374
- and Hooded Warbler, 370
- and Magnolia Warbler, 378
- and Orchard Oriole, 454
- and White-throated Sparrow, 424

Forest interior. *See* Forest, contiguous
Forest maturation, 11
- and American Redstart, 372
- and Barn Swallow, 292
- and Blackburnian Warbler, 380
- and Black-throated Blue Warbler, 388
- and Blue-headed Vireo, 266
- and Blue-winged Warbler, 352
- and Brown Creeper, 304
- and Brown Thrasher, 336
- and Chestnut-sided Warbler, 384
- and Common Merganser, 106
- and Eastern Whip-poor-will, 220
- and Eastern Wood-Pewee, 242
- and Hooded Warbler, 370
- and House Wren, 308
- and Least Flycatcher, 252
- and Ovenbird, 342
- and Pileated Woodpecker, 240
- and Pine Warbler, 390
- and Prairie Warbler, 396
- and Red Crossbill, 462
- and Red-breasted Nuthatch, 300
- and Red-eyed Vireo, 270
- and Ruffed Grouse, 112
- and White-throated Sparrow, 424
- and Yellow-breasted Chat, 402

Forest regeneration, 10, 11, 15, 61
- and Alder Flycatcher, 248
- and American Woodcock, 182
- and Chestnut-sided Warbler, 384
- and Eastern Whip-poor-will, 220
- and Mourning Warbler, 364
- and Nashville Warbler, 362
- and Rose-breasted Grosbeak, 434
- and Warbling Vireo, 270

Forest understory, x, 12, 14, 15, 35
- and Acadian Flycatcher, 246
- and Black-and-white Warbler, 356
- and Black-throated Blue Warbler, 388
- and Blue-headed Vireo, 266
- and Canada Warbler, 400
- and Chestnut-sided Warbler, 384
- and Eastern Towhee, 404
- and Eastern Whip-poor-will, 220
- and Hooded Warbler, 370
- and Kentucky Warbler, 366
- and Magnolia Warbler, 378
- and Northern Saw-whet Owl, 214
- and Pine Warbler, 390
- and Swainson's Warbler, 360
- and Veery, 322
- and Wild Turkey, 114
- and Winter Wren, 310
- and Worm-eating Warbler, 344

Forms, recording, 24, 26, 31, 32, 34
Fruit, food item
- for Cedar Waxwing, 340
- for Purple Finch, 458
- for Rose-breasted Grosbeak, 434

Fulica americana, 27, 30, 31, 55, 74, 168, **170–171**, 190, 480, 523, 527

Gadwall, 474, 527
Gallinago delicata, 27, 55, 59, 74, **180–181**, 481, 523, 527
Gallinula galeata, 27, 30, 31, 55, 74, **168–169**, 190, 480, 523, 527
Gallinule, Common, 27, 30, 31, 55, 74, **168–169**, 190, 480, 523, 527
GAP analysis, 29, 36
GAP model, ix
Gas development. *See* Energy development
Gavia immer, 474, 527
Gazetteer, 84–86
Geographic Information System (GIS), ix, xxii, xxiii, 24, 34, 45
Georeferencing of bird records, 2, 35, 36, 48, 59, 72
Geothlypis formosa, 27, 57, 64, 73, **366–367**, 488, 510, 525, 528
- *G. philadelphia*, 57, 64, **364–365**, 488, 510, 525, 528
- *G. trichas*, 57, 58, 63, **368–369**, 372, 382, 488, 510, 525, 528

Gettysburg National Military Park, 79
Gettysburg-Newark Lowland Physiographic section, 5, 12, 52, 477–493
Glaciated High Plateau Physiographic section, 5, 12, 52, 477–493
Glaciated Low Plateau Physiographic section, 5, 12, 17, 52, 477–493
Glaciated Pocono Plateau Physiographic section, 5, 12, 52, 477–493
Glaciation, 3
Gnatcatcher, Blue-gray, 57, 63, **316–317**, 486, 506, 524, 528
Golden-crowned Kinglet. *See* Kinglet, Golden-crowned
Golden-winged Warbler. *See* Warbler, Golden-winged
Goldfinch, American, 58, 63, **466–467**, 493, 519, 526, 529
Goose, Canada, ix, 54, 55, 59, 64, **88–89**, 477, 495, 523, 527
Goshawk, Northern, 27, 55, 59, 68, 73, **148–149**, 479, 523, 527
Grackle, Common, 58, 63, **450–451**, 492, 517, 526, 529
Grape/grapevine, 360, 372, 404
Grass, Orchard, 312
- Reed Canary, 312
Grasshopper, food item for Red-headed Woodpecker, 228
Grasshopper Sparrow. *See* Sparrow, Grasshopper
Grassland, x, 10, 15–18, 46, 61, 62, 66–68, 79, 82
- and American Kestrel, 156
- and Barn Owl, 202
- and Blue Grosbeak, 436
- and Blue-winged Teal, 100
- and Bobolink, 442
- and Clay-colored Sparrow, 408
- and Dickcissel, 440
- and Eastern Meadowlark, 446
- and Field Sparrow, 410
- and Grasshopper Sparrow, 416

and Henslow's Sparrow, 418
and Horned Lark, 280
and Mallard, 90
and Northern Harrier, 142
and Prairie Warbler, 396
and Red-winged Blackbird, 444
and Ring-necked Pheasant, 110
and Sandhill Crane, 172
and Savannah Sparrow, 414
and Sedge Wren, 312
and Short-eared Owl, 212
and Upland Sandpiper, 178
and Vesper Sparrow, 412
Gravel pit, 82, 286, 288, 438
Gray Catbird. *See* Catbird, Gray
Great Black-backed Gull. *See* Gull, Great Black-backed
Great Blue Heron. *See* Heron, Great Blue
Great Crested Flycatcher. *See* Flycatcher, Great Crested
Great Egret. *See* Egret, Great
Great Horned Owl. *See* Owl, Great Horned
Great Valley Physiographic section, 5, 12, 52, 477–493
Grebe, Pied-billed, 27, 30, 31, 55, 59, 74, **116–117**, 190, 478, 523, 527
Green Heron. *See* Heron, Green
Greenbrier, 360
Green-winged Teal. *See* Teal, Green-winged
Grosbeak, Blue, 58, 60, **436–437**, 491, 516, 525, 529
Evening, 28, **474**, 529
Rose-breasted, 58, 63, **434–435**, 491, 516, 525, 529
Ground-nesting
and Kentucky Warbler, 366
and Mallard, 98
and Mourning Warbler, 364
and Ring-billed Gull, 184
and Sandhill Crane, 172
Grouse, Ruffed, 28, 55, 58, 60, 68, **112–113**, 148, 478, 523, 527
Grus canadensis, 27, 54, 55, 124, **172–173**, 480, 523, 527
Guild (species guild), 46–47, 55–58, 62–68, 79
Gull, Great Black-backed, x, 54, 55, **188–189**, 481, 524
Herring, 54, 55, 58, 184, **186–187**, 481, 523, 527
Laughing, 471
Ring-billed, 54, 55, 59, **184–185**, 481, 523, 527

Habitat, 3, 9–22
Hairy Woodpecker. *See* Woodpecker, Hairy
Haliaeetus leucocephalus, x, 27, 37, 47, 49, 50, 54, 55, 59, 73, 76, 79, 124, 138, **140–141**, 479, 523, 527
Handbook, atlas project, 24
Hare, Snowshoe, 148
Harrier, Northern, 27, 55, 60, 73, 100, **142–143**, 479, 523, 527
Harvesting
and American Woodcock, 182
and Mallard, 98

and Mourning Dove, 196
and Sora, 166
and Wild Turkey, 114
and Wilson's Snipe, 180
and Wood Duck, 94
Hawk
Broad-winged, 55, 60, 61, 64, 73, **152–153**, 480, 496, 523, 527
Cooper's, 55, 60, 61, **146–147**, 156, 228, 479, 523, 527
Red-shouldered, 27, 55, 68, 73, **150–151**, 479, 496, 523, 527
Red-tailed, 55, 64, 150, **154–155**, 206, 480, 496, 523, 527
Sharp-shinned, 27, 55, 59, 73, **144–145**, 479, 523, 527
Hawk Mountain Sanctuary, xxi
Hawthorn, 182, 260
Washington, 260
Hay, 17, 18
cutting. *See* Mowing
Hayfield
and Barn Owl, 202
and Bobolink, 442
and Dickcissel, 440
and Eastern Meadowlark, 446
and Field Sparrow, 410
and Grasshopper Sparrow, 416
and Henslow's Sparrow, 418
and Mallard, 98
and Northern Harrier, 142
and Red-winged Blackbird, 444
and Savannah Sparrow, 414
and Sedge Wren, 312
Haying. *See* Mowing
Hedgerow, 210, 272, 432, 436, 438
Helmitheros vermivorum, 57, 64, 73, 78, **344–345**, 488, 509, 525, 528
Hemlock, Eastern, ix, x, 2, 10, 11, 29, 35, 45, 68, 80, 81
and Acadian Flycatcher, 68, 208, 246
and Barred Owl, 68, 208
and Blackburnian Warbler, 68, 380
and Blackpoll Warbler, 68, 386
and Black-throated Blue Warbler, 68
and Black-throated Green Warbler, 68, 398
and Blue-headed Vireo, 68, 266
and Brown Creeper, 68, 304
and Canada Warbler, 68, 400
and Cooper's Hawk, 146
and Dark-eyed Junco, 68, 426
and Golden-crowned Kinglet, 68, 318
and Great Crested Flycatcher, 256
and Hermit Thrush, 68, 326
and Louisiana Waterthrush, 68, 346
and Magnolia Warbler, 68, 378
and Nashville Warbler, 68, 362
and Northern Flicker, 238
and Northern Goshawk, 68, 148
and Northern Saw-whet Owl, 68, 214
and Northern Waterthrush, 68
and Ovenbird, 342
and Pileated Woodpecker, 240

and Pine Siskin, 464
and Pine Warbler, 390
and Red Crossbill, 68, 462
and Red-breasted Nuthatch, 68, 300
and Red-shouldered hawk, 68
and Ruffed Grouse, 68
and Sharp-shinned Hawk, 144
and Swainson's Thrush, 68, 324
and Swainson's Warbler, 360
and Tree Swallow, 284
and Veery, 322
and White-throated Sparrow, 68, 114
and Wild Turkey, 114
and Winter Wren, 68, 310
and Yellow-bellied Flycatcher, 68, 114
and Yellow-rumped Warbler, 68, 392
Henslow's Sparrow. *See* Sparrow, Henslow's
Herbicides, 108, 212
Hermit Thrush. *See* Thrush, Hermit
Heron, Great Blue, 27, 49, 55, 60, 61, 73, 82, **124–125**, 128, 172, 478, 523, 527
Green, 55, **128–129**, 226, 479, 523, 527
Herring Gull. *See* Gull, Herring
Hickory, 80
High Plateau Physiographic section, 5, 12, 52, 477–493
Himantopus mexicanus, 55, 58, 473, 527
Hirundo rustica, ix, 56, 63, 158, 284, **292–293**, 485, 504, 524, 528
Hobblebush, 388
Hole nesting. *See* Cavity nesting
Holly, 344, 360
Honeysuckle, 360, 404
Hooded Merganser. *See* Merganser, Hooded
Hooded Warbler. *See* Warbler, Hooded
Horned Lark. *See* Lark, Horned
House Finch. *See* Finch, House
House Sparrow. *See* Sparrow, House
House Wren. *See* Wren, House
Hummingbird, Ruby-throated, ix, 56, 59, 63, **224–225**, 483, 498, 524, 528
Hunting. *See* Harvesting; Persecution
Hurricane, 282
Hybridization
American Black Duck × Mallard, 96
Audubon's Warbler × Myrtle Warblers, 392
Black-capped Chickadee × Carolina Chickadee, 294, 296
Blue-winged Warbler × Golden-winged Warblers, x, 350, 352, 354–355
Eastern Meadowlark × Western Meadowlark, 448
Hylocichla mustelina, 57, 59, 63, 73, 77, 78, 79, 82, 326, **328–329**, 487, 507, 525, 528

Ibis, Glossy, 471, 527
Icteria virens, x, 57, 60, 61, 64, 74, **402–403**, 490, 512, 525, 529
Icterus galbula, 58, 63, 454, **456–457**, 492, 518, 526, 529
I. spurius, ix, 58, 60, 64, **454–455**, 492, 518, 526, 529

INDEX 577

Important Bird Area (IBA), 29, 71, 74, 77, 79, 386
Incubation periods, 18
Indigo Bunting. *See* Bunting, Indigo
Industry, 20, 21
Insecticides, and Peregrine Falcon, 160
International Panel on Climate Change (IPCC), 79
International Union for Conservation of Nature (IUCN), 70, 71, 73–74
Interspersion
    and American Coot, 170
    and Baltimore Oriole, 456
    and Bobolink, 442
    and Chuck-will's-widow, 218
    and Common Gallinule, 168
    and Gray Catbird, 332
    and Great Crested Flycatcher, 256
    and Indigo Bunting, 438
    and Long-eared Owl, 210
    and Red-tailed Hawk, 154
    and Rose-breasted Grosbeak, 434
    and Sora, 166
    and Yellow Warbler, 382
Introductions/stocking
    and Canada Goose, 88
    and European Starling, 338
    and House Finch, 460
    and House Sparrow, 468
    and Mute Swan, 90
    and Northern Bobwhite, 108
    and Ring-necked Pheasant, 110
    and Rock Pigeon, 192
Invasive species, 20, 21, 78, 90, 166, 192, 194, 338
Irruptive movement
    and Dickcissel, 440
    and Evening Grosbeak, 474
    and Pine Siskin, 464
    and Red Crossbill, 462
    and Red-breasted Nuthatch, 300
    and White-breasted Nuthatch, 302
    and White-winged Crossbill, 69
Ixobrychus exilis, 27, 30, 31, 55, 59, 74, 76, **122–123**, 168, 170, 190, 478, 523, 527

Jackdaw, Eurasian, 471, 528
Jay, Blue, ix, 56, 58, 63, **272–273**, 485, 502, 524, 528
Jewelweed, Spotted, 224
John Heinz National Wildlife Refuge, 79
Junco, Dark-eyed, 57, 59, 60, 61, 63, 68, 406, **426–427**, 444, 491, 515, 525, 529
Junco hyemalis, 57, 59, 60, 61, 63, 68, 406, **426–427**, 444, 491, 515, 525, 529

Kentucky Warbler. *See* Warbler, Kentucky
Kestrel, American, ix, 39, 55, 60, 64, **156–157**, 480, 496, 523, 527
Killdeer, 26, 27, 55, 64, **174–175**, 480, 496, 523, 527
King Rail. *See* Rail, King
Kingbird, Eastern, 56, 64, **258–259**, 484, 501, 524, 528

Kingfisher, Belted, 56, 64, 176, **226–227**, 286, 483, 498, 524, 528
Kinglet, Golden-crowned, 27, 40, 41, 57, 61, 68, 300, **318–319**, 486, 506, 524, 528

Lacustrine wetland, 20
Lake, 4, 20, 21, 88
    and Bald Eagle, 140
    and Bank Swallow, 288
    and Belted Kingfisher, 226
    and Black Tern, 190
    and Canada Goose, 88
    and Double-crested Cormorant, 118
    and Great Black-backed Gull, 188
    and Great Crested Flycatcher, 256
    and Herring Gull, 186
    and Hooded Merganser, 104
    and Killdeer, 174
    and Marsh Wren, 314
    and Merlin, 158
    and Osprey, 138
    and Prothonotary Warbler, 358
    and Ring-billed Gull, 184
    and Spotted Sandpiper, 176
    and Swamp Sparrow, 422
    and Trumpeter Swan, 92
    and White-throated Sparrow, 424
Lake Erie, 3, 4, 7, 20, 24
Land cover, 10, 38, 39, 42–45, 82, 521
Land form, 42–45, 521
Land use, 9
Landfill
    and Blue Grosbeak, 436
    and Great Black-backed Gull, 188
Lanius ludovicianus, 27, 49, 56, 61, 73, **260–261**, 484, 524, 528
Lark, Horned, 50, 56, 59, 60, 64, **280–281**, 412, 485, 503, 524, 528
Larus argentatus, 54, 55, 58, 184, **186–187**, 481, 523, 527
    delawarensis, 54, 55, 59, **184–185**, 481, 523, 527
    marinus, x, 54, 55, **188–189**, 481, 524
Larvae, food source, 80
    for Black-billed Cuckoo, 200
    for Pileated Woodpecker, 240
    for Worm-eating Warbler, 344
    for Yellow-billed Cuckoo, 198
    for Yellow-throated Vireo, 264
Laterallus jamaicensis, 27, 30, 31, 55, 527
Latitude, 4, 45, 46, 77, 232, 520
Latitudinal Shift, 45, 59–62, 96
    and American Black Duck, 96
    and Brown Creeper, 60, 304
    and Common Merganser, 60, 106
    and Northern Goshawk, 148
    and Red-breasted Nuthatch, 300
Laurel, Mountain, 14
    and Black-throated Blue Warbler, 388
    and Canada Warbler, 400
    and Chestnut-sided Warbler, 384
    and Dark-eyed Junco, 426
    and Northern Saw-whet Owl, 214

    and Swainson's Warbler, 360
    and Veery, 322
    and Worm-eating Warbler, 344
Laurel Highlands, 3
Lawrence's Warbler. *See* Warbler, Lawrence's
Leaf litter, 342
Leaf Roller, Oak, 436
Least Bittern. *See* Bittern, Least
Least Flycatcher. *See* Flycatcher, Least
Least Tern. *See* Tern, Least
Lichen, and Ruby-throated Hummingbird, 224
Limnothlypis swainsonii, 27, 57, 58, **360–361**, 448, 488, 528
Literature cited, 531–568
Livestock, 35
Loggerhead Shrike. *See* Shrike, Loggerhead
Logging, 9, 12, 15
    American Woodcock, 182
    Blackburnian Warbler, 380
    Black-throated Blue Warbler, 388
    Cerulean Warbler, 374
    Chestnut-sided Warbler, 384
    Eastern Whip-poor-will, 220
    Golden-winged Warbler, 350
    Great Blue Heron, 124
    Hooded Warbler, 370
    Mourning Warbler, 364
    Nashville Warbler, 362
    Northern Cardinal, 432
    Pine Warbler, 390
    Prairie Warbler, 396
    Prothonotary Warbler, 358
    Rose-breasted Grosbeak, 434
    Ruffed Grouse, 112
    Wild Turkey, 114
    Winter Wren, 310
Long-eared Owl. *See* Owl, Long-eared
Loon, Common, 474, 527
Loosestrife, Purple, 30, 31, 122
Lophodytes cucullatus, 27, 52, 55, **104–105**, 106, 478, 523, 527
Louisiana Waterthrush. *See* Waterthrush, Louisiana
Lowland and Intermediate Upland Physiographic section, 5, 12, 52, 477–493
Loxia curvirostra, 28, 50, 54, 58, 68, **462–463**, 464, 492, 526, 529
    L. leucoptera, 69

Magnolia, nesting substrate for Loggerhead Shrike, 260
Magnolia Warbler. *See* Warbler, Magnolia
Mallard, 55, 64, 94, 96, **98–99**, 106, 477, 495, 523, 527
Maple, 10, 13, 80
    and Purple Finch, 458
    and Yellow-bellied Sapsucker, 232
    and Yellow-throated Vireo, 264
Maple, Red, 10, 232
Maple, Sugar, 10, 12
Maps, interpretation of, 81–83
Marcellus Shale gas. *See* Energy development

Marsh bird, surveys, 1, 26, 29–30, 31, 61, 72, 162, 164, 166
Marsh Wren. *See* Wren, Marsh
Martin, Purple, ix, x, 56, 60, 64, 67, 77, **282–283**, 485, 503, 524, 528
Massasauga rattlesnake, 172
Meadowlark, Eastern, x, 18, 40, 41, 58, 64, 73, **446–447**, 448, 492, 517, 525, 529
Meadowlark, Western, 58, **448–449**, 492, 529
Megaceryle alcyon, 56, 64, 176, **226–227**, 286, 483, 498, 524, 528
Megascops asio, 30, 32, 56, 60, **204–205**, 528
Melanerpes carolinus, x, 56, 58, 59, 60, 62, 63, **230–231**, 483, 498, 524, 528
M. erythrocephalus, 27, 56, 60, 73, 77, **228–229**, 483, 498, 524, 528
Meleagris gallopavo, ix, 51, 53, 55, 58, 60, 62, 63, **114–215**, 124, 478, 495, 523, 527
Melospiza georgiana, ix, 57, 64, **422–423**, 491, 515, 525, 529
M. melodia, ix, 28, 41, 42, 57, 58, 62, 63, 270, **420–421**, 491, 515, 525, 529
Merganser
    Common, 55, 60, 61, **106–107**, 478, 523, 527
    Hooded, 27, 52, 55, **104–105**, 106, 478, 523, 527
    Red-breasted, 527
Mergus merganser, 55, 60, 61, **106–107**, 478, 523, 527
M. serrator, 527
Merlin, x, 54, 55, **158–159**, 471, 480, 523, 527
Microtine. *See* Small mammals, prey item for Eastern Screech-Owl
Migratory Bird Treat Act of 1918, 92, 188
Migratory Bird Treaty Act, 1972 amendment, 118
Migratory Bird Treaty Reform Act of 2004, 90
Mimus polyglottos, 57, 63, **334–335**, 336, 487, 508, 525, 528
Mine, 9, 20, 286
Mniotilta varia, 57, 63, **356–357**, 488, 510, 525, 528
Mockingbird, Northern, 57, 63, **334–335**, 336, 487, 508, 525, 528
Molothrus ater, 58, 59, 63, **452–453**, 492, 517, 526, 529
Monitoring, bird, 75–76
Monongahela River, 4
Moorhen, Common, 168
Moss, Beard, 376
    Spanish, 376
    Sphagnum, 400
Moth, food item for Eastern Whip-poor-will, 220
    Gypsy, 12, 198, 200
Mount Davis, 3
Mourning Dove. *See* Dove, Mourning
Mourning Warbler. *See* Warbler, Mourning
Mowing, 18
    Bobolink, 442
    Dickcissel, 440
    Eastern Meadowlark, 18, 446
    Field Sparrow, 410
    Ring-necked Pheasant, 110
    Savannah Sparrow, 414
    Sedge Wren, 212
    Upland Sandpiper, 178
    Vesper Sparrow, 412
Mudflat, 164, 166
Mute Swan. *See* Swan, Mute
Mycoplasma gallisepticum, 460
Myiarchus crinitus, 56, 59, 64, **256–257**, 484, 501, 524, 528

Nashville Warbler. *See* Warbler, Nashville
Natal fidelity, 100, 138
National Audubon Society, xi, xxii, 2, 71, 74, 75, 77
National Park Service, xi, xxii, 79
National Parks, 78
National Wildlife Refuge, 78
Native American, 9, 15
Natural Lands Trust, xxii
The Nature Conservancy, xxii, xxiii
NatureServe, 70, 71
Nest boxes, 35
    and American Kestrel, 156
    and Barn Owl, 202
    and Barred Owl, 208
    and Black-capped Chickadee, 296
    and Eastern Bluebird, 320
    and Great Crested Flycatcher, 256
    and Hooded Merganser, 94
    and Northern Rough-winged Swallow, 286
    and Prothonotary Warbler, 358
    and Tree Swallow, 284
    and Tufted Titmouse, 298
    and Wood Duck, 94
Nest predation. *See* Predation
New England Physiographic province, 5, 12, 52, 477–493
Newsletter, 2, 24
Nighthawk, Common, 27, 32, 56, 58, 59, 67, 73, 77, **216–217**, 482, 524, 528
Night-Heron, Black-crowned, 27, 55, 60, 61, 73, 76, 118, 126, **130–131**, 479, 523, 527
    Yellow-crowned, 27, 55, 73, **132–133**, 479, 523, 527
Nightjar, 30, 75
Nocturnal birds, surveying for, 1, 26, 29–30, 50, 32, 72, 182, 208, 214, 220
Nomenclature, 81
Nonnative species, 21, 90, 110, 192, 194
Nonnative vegetation, 18
North American Bird Conservation Initiative (NABCI), 70
North American Landbird Conservation Plan, 72
North American Ornithological Atlas Committee, 24
North American Waterfowl Management Plan, 70
North Atlantic Regional Shorebird Plan, 71, 73–74
Northern Bobwhite. *See* Bobwhite, Northern
Northern Cardinal. *See* Cardinal, Northern

Northern Flicker. *See* Flicker, Northern
Northern Goshawk. *See* Goshawk, Northern
Northern Hardwoods, 10, 46, 62, 65, 67, 68
    and Blackpoll Warbler, 386
    and Blue-gray Gnatcatcher, 316
    and Broad-winged Hawk, 152
    and Hermit Thrush, 326
    and Purple Finch, 458
    and White-throated Sparrow, 424
    and Yellow-bellied Sapsucker, 232
Northern Harrier. *See* Harrier, Northern
Northern Mockingbird. *See* Mockingbird, Northern
Northern Parula. *See* Parula, Northern
Northern Rough-winged Swallow. *See* Swallow, Northern Rough-winged
Northern Saw-whet Owl. *See* Owl, Northern Saw-whet
Northern Shoveler. *See* Shoveler, Northern
Northern Waterthrush. *See* Waterthrush, Northern
Northwestern Glaciated Plateau Physiographic section, 12
Nuthatch, Red-breasted, 52, 56, 64, 68, **300–301**, 486, 505, 524, 528
    White-breasted, 56, 63, **302–303**, 486, 505, 524, 528
Nyctanassa violacea, 27, 55, 73, **132–133**, 479, 523, 527
Nycticorax nycticorax, 27, 55, 60, 61, 73, 76, 118, 126, **130–131**, 479, 523, 527

Oak, 12, 13, 15, 80
    and Blackburnian Warbler, 380
    and Blue Jay, 272
    and Cerulean Warbler, 374
    and Prairie Warbler, 396
    and Red-headed Woodpecker, 228
    and Summer Tanager, 428
    and Yellow-throated Vireo, 264
Oak, Chestnut, 356
Oak-hickory forest
    and Blue-gray Gnatcatcher, 316
    and Cerulean Warbler, 374
    and Red-headed Woodpecker, 228
Oats, 17
Observer effects, 39, 40, 521
Ohio River, 4
Old field, 396, 408, 410, 420, 432
Olive-sided Flycatcher. *See* Flycatcher, Olive-sided
Open country, 43, 66
    and American Crow, 274
    and American Goldfinch, 466
    and Brown-headed Cowbird, 452
    and Eastern Kingbird, 258
    and Horned Lark, 280
    and Northern Bobwhite, 108
    and Northern Flicker, 238
    and Orchard Oriole, 454
    and Purple Martin, 282
    and Red-tailed Hawk, 154
    and Vesper Sparrow, 412
Opossum, 206

Orchard
- and Brown-headed Cowbird, 452
- and Cedar Waxwing, 340
- and Chipping Sparrow, 406
- and Eastern Bluebird, 320
- and Great Crested Flycatcher, 256
- and Least Flycatcher, 252
- and Rose-breasted Grosbeak, 434
- and Vesper Sparrow, 412
- and Warbling Vireo, 268

Orchard Oriole. *See* Oriole, Orchard
Oreothlypis ruficapilla, 52, 57, 59, 60, 68, **362–363**, 488, 525, 528
Organization, of 2nd Atlas, 24
Oriole, Baltimore, 58, 63, 454, **456–457**, 492, 518, 526, 529
- Northern. *See* Oriole, Baltimore
- Orchard, ix, 58, 60, 64, **454–455**, 492, 518, 526, 529

Osprey, 27, 37, 55, 74, **138–139**, 140, 479, 523, 527
Ovenbird, 57, 63, 79, **342–343**, 366, 368, 372, 382, 487, 509, 525, 528
Over-browsing. *See* Deer browsing
Owl, 75
- Barn, 27, 29, 30, 37, 49, 56, 73, **202–203**, 482, 524, 528
- Barred, 30, 32, 56, 59, 68, 206, **208–209**, 482, 524, 528
- Eastern Screech. *See* Screech-Owl, Eastern
- Great Horned, 30, 32, 56, **206–207**, 208, 482, 524, 528
- Long-eared, 27, 29, 30, 32, 56, 59, 60, 73, **210–211**, 482, 524, 528
- Northern Saw-whet, 27, 30, 32, 56, 68, **214–215**, 482, 524, 528
- Short-eared, 27, 29, 30, 32, 56, 73, **212–213**, 482, 524, 528

Oxyura jamaicensis, 472, 527

Palustrine wetlands, 20
Pandion haliaetus, 27, 37, 55, 74, **138–139**, 140, 479, 523, 527
Parasitism, by Brown-headed Cowbird
- on Blue Grosbeak, 436
- on Chipping Sparrow, 408, 452
- on Common Yellowthroat, 368
- on Golden-winged Warbler, 350
- on Hooded Warbler, 370
- on Kentucky Warbler, 366
- on Northern Cardinal, 432
- on Red-eyed Vireo, 270
- on Song Sparrow, 420, 452
- on Summer Tanager, 428
- on White-eyed Vireo, 262
- on Yellow Warbler, 382

Parkesia motacilla, 27, 57, 64, 68, 73, 77, 79, **346–347**, 348, 488, 509, 525, 528
P. noveboracensis, 27, 52, 57, 60, 61, 68, 77, **348–349**, 488, 525, 528
Partners in Flight (PIF), 70, 72, 73–74, 200, 374
Parula, Northern, 57, 58, 64, **376–377**, 489, 511, 525, 528

Passer domesticus, 21, 58, 63, 290, 292, 320, **468–469**, 493, 519, 526, 529
Passerculus sandwichensis, 57, 63, 412, 414–415, 490, 514, 525, 529
Passerina caerulea, 58, 60, **436–437**, 491, 516, 525, 529
P. cyanea, 58, 63, 436, **438–439**, 491, 516, 525, 529
Pasture, 17, 35
- and Barn Owl, 202
- and Brown-headed Cowbird, 452
- and Chipping Sparrow, 406
- and Eastern Bluebird, 320
- and Eastern Kingbird, 258
- and Eastern Meadowlark, 446
- and European Starling, 338
- and Field Sparrow, 410
- and Fish Crow, 276
- and Grasshopper Sparrow, 416
- and Indigo Bunting, 438
- and Prairie Warbler, 396
- and Red-headed Woodpecker, 228
- and Red-winged Blackbird, 444
- and Savannah Sparrow, 414
- and Wilson's Snipe, 180

PCB. *See* Pesticides
Pennsylvania Biological Survey, xxii, 29, 71, 72, 76
Pennsylvania Department of Conservation and Natural Resources, xi, xxii, xxiii, 12
Pennsylvania Game Commission (PGC), xi, xxi, xxii, 23, 29, 37, 48, 49, 50, 54, 71, 72, 76, 78, 88, 94, 96, 104, 108, 110, 112, 114, 118, 132, 140, 160, 202, 260, 446, 473
Pennsylvania Natural Diversity Inventory, 76
Pennsylvania Society for Ornithology (PSO), xxii, 23, 24, 74
The Pennsylvania State University (Penn State), xi, xxi, xxii, 4, 23, 24, 30, 34, 37
Peregrine Falcon. *See* Falcon, Peregrine
Persecution
- and Barred Owl, 208
- and Black-crowned Night-Heron, 130
- and Broad-winged Hawk, 152
- and Common Merganser, 106
- and Common Raven, 278
- and Cooper's Hawk, 146
- and Dickcissel, 440
- and Double-crested Cormorant, 118
- and Great Blue Heron, 124
- and Great Egret, 126
- and Killdeer, 174
- and Northern Mockingbird, 334
- and Pileated Woodpecker, 240
- and Red-tailed Hawk, 154
- and Upland Sandpiper, 178

Pesticides, 61
- and American Robin, 330
- and Black-billed Cuckoo, 200
- and Common Merganser, 61
- and Common Nighthawk, 216
- and Cooper's Hawk, 61, 146
- and Double-crested Cormorant, 118
- and Great Blue Heron, 61
- and Killdeer, 174
- and Northern Bobwhite, 108
- and Red-tailed Hawk, 154
- and Sedge Wren, 312
- and Sharp-shinned Hawk, 144
- and Short-eared Owl, 212
- and Upland Sandpiper, 178
- and Yellow-throated Vireo, 264

Pests, 12, 192, 194, 450, 468
- and Blue Jay, 272
- and Downy Woodpecker, 234
- and Swainson's Thrush, 324

Petrochelidon pyrrhonota, 56, 59, 64, 67, 288, **290–291**, 485, 504, 524, 528
Phalacrocorax auritus, 54, 55, 56, 58, 59, **118–119**, 126, 130, 478, 523, 527
Phasianus colchicus, x, 55, 58, 64, **110–111**, 478, 495, 523, 527
Pheasant, Ring-necked, x, 55, 58, 64, **110–111**, 478, 495, 523, 527
Pheucticus ludovicianus, 58, 63, **434–435**, 491, 516, 525, 529
Philadelphia, ix, 8, 9, 10
Philadelphia Academy of Natural Science, xxii
Phoebe, Eastern, 56, 63, **254–255**, 484, 500, 524, 528
Phragmites. *See* Reed, Common
Phsyiographic Provinces, 4, 5, 12, 52
Phsyiographic sections, 5, 12, 52
Picoides pubescens, 56, 59, 63, 230, **234–235**, 236, 248, 483, 499, 524, 528
P. villosus, 56, 59, 64, **236–237**, 483, 499, 524, 528
Pied-billed Grebe. *See* Grebe, Pied-billed
Piedmont Lowland Physiographic section, 5, 6, 12, 52, 477–493
Piedmont Physiographic province, 5, 12, 52, 477–493
Piedmont Upland Physiographic section, 5, 12, 52, 477–493
Pigeon, Passenger, 474
- Rock, 21, 55, 63, **192–193**, 481, 497, 524, 527
Pileated Woodpecker. *See* Woodpecker, Pileated
Pine, 10, 13, 15, 80
- and Blackburnian Warbler, 380
- and Chuck-will's-widow, 218
- and Cooper's Hawk, 146
- and Golden-crowned Kinglet, 318
- and Pine Warbler, 390
- and Prairie Warbler, 396
- and Purple Finch, 458
- and Red Crossbill, 462
- and Red-breasted Nuthatch, 300
- and Swainson's Thrush, 324
- and Yellow-rumped Warbler, 392

Pine, Austrian, 208
Pine, Eastern White. *See* Pine, White
Pine, Pitch, and Pine Warbler, 390

Pine, Red
	and Blackburnian Warbler, 380
	and Hermit Thrush, 326
	and Pine Warbler, 390
Pine, Scots, 208
Pine, Virginia, and Pine Warbler, 390
Pine, White
	and Blackpoll Warbler, 386
	and Brown Creeper, 304
	and Cooper's Hawk, 146
	and Magnolia Warbler, 378
	and Merlin, 158
	and Pine Siskin, 464
	and Red-breasted Nuthatch, 300
	and Ruffed Grouse, 112
	and Wild Turkey, 114
	and Yellow-rumped Warbler, 392
	and Yellow-throated Warbler, 394
Pine Barrens. *See* Barrens
Pine Siskin. *See* Siskin, Pine
Pine Warbler. *See* Warbler, Pine
Pintail, Northern, 474
Pipilo erythrophthalmus, 57, 63, **404–405**, 490, 514, 525, 529
Piping Plover. *See* Plover, Piping
Piranga olivacea, ix, 57, 59, 63, 71, 73, 78, **430–431**, 491, 516, 525, 529
	P. rubra, 28, 57, 58, 73, **428–429**, 491, 525, 529
Piscivore. *See* Fish: food item
Pittsburgh, 8, 10, 21
Pittsburgh Low Plateau Physiographic section, 5, 12, 52, 477–493
Planning, 2nd Atlas project, 23
Plantations, 10, 61
	and Barred Owl, 208
	and Blackburnian Warbler, 380
	and Golden-crowned Kinglet, 318
	and Pine Warbler, 390
	and Purple Finch, 458
	and Red Crossbill, 462
	and Red-breasted Nuthatch, 300
Plover, Piping, 21, 74, 186, **473**
	Upland. *See* Sandpiper, Upland
Pocono, 20
Podilymbus podiceps, 27, 30, 31, 55, 59, 74, **116–117**, 190, 478, 523, 527
Poecile atricapillus, 56, 63, 294, **296–297**, 486, 504, 524, 528
	P. carolinensis, 56, 59, 60, 64, **294–295**, 296, 485, 504, 524, 528
Point Count Surveys, 30, 33–35, 39–45
Poisoning. *See* Contaminants
Poleward shifts. *See* Latitudinal Shift
Polioptila caerulea, 57, 63, **316–317**, 486, 506, 524, 528
Pollution, 20
	and American Robin, 330
	and Common Merganser, 106
	and Common Nighthawk, 216
	and Killdeer, 174
	and Louisiana Waterthrush, 346
	and Northern Parula, 376
	and Spotted Sandpiper, 176

Pollution, light, x
Pollution, mine drainage, 20, 79
	and Belted Kingfisher, 226
	and Great Blue Heron, 124
	and Green Heron, 128
	and Killdeer, 176
	and Louisiana Waterthrush, 346
	and Spotted Sandpiper, 176
Pollution, noise, x
Pollution, runoff from agriculture/industry, 20, 176
Pond, 20
	and American Coot, 170
	and Belted Kingfisher, 226
	and Canada Goose, 88
	and Common Gallinule, 168
	and Double-crested Cormorant, 118
	and Great Egret, 126
	and Killdeer, 174
	and Prothonotary Warbler, 358
	and Ring-necked Duck, 472
	and Spotted Sandpiper, 176
	and Swamp Sparrow, 422
	and Trumpeter Swan, 92
	and White-throated Sparrow, 424
Pooecetes gramineus, x, 57, 60, 64, **412–413**, 490, 514, 525, 529
Poplar, 268
	Yellow, 360
Population control
	and Brown-headed Cowbird, 452
	and Canada Goose, 88
	and Common Grackle, 450
	and Double-crested Cormorant, 118
	and Eurasian Collared-Dove, 194
	and European Starling, 338
	and Herring Gull, 186
	and Mute Swan, 90
	and Red-winged Blackbird, 444
	and Ring-billed Gull, 184
	and Rock Pigeon, 192
Population size, 39–45, 62–64, 520–522
Porzana carolina, 27, 30, 31, 55, 59, 61, 74, **166–167**, 168, 170, 480, 523, 527
Possible breeding, definition, 28
Postglacial, 4, 20
Potomac River, 4
Powdermill Nature Reserve, xxii, 24, 35
Power Line Corridors. *See* Rights of way corridor
Prairie Warbler. *See* Warbler, Prairie
Prairie-Chicken, Greater, 474
Precipitation, 6–8, 61
Predation
	of American Kestrel by Cooper's Hawk, 156
	of Barn Swallow by Merlin, 158
	of Cedar Waxwing by Merlin, 158
	of Common Nighthawk nests, 216
	of Great Black-backed Gull nests, 188
	of Hairy Woodpecker nests by bear, 236
	of Herring Gull nests, 186
	of Northern Goshawks by Fishers, 148

	of Ovenbird nests, 342
	of Prothonotary Warbler nests, 358
	of Red-breasted Nuthatch nests, 300
	of Red-headed Woodpecker by Cooper's Hawk and Raccoon, 228
	of Ring-billed Gull nests, 184
	of Rock Pigeon by Peregrine Falcon, 192
	of terns nests by Ring-billed Gull, 184
	of Tree Swallow by Merlin, 158
	of wading bird nests by Ring-billed Gull, 184
	of White-eyed Vireo, 262
	of Wood Thrush nests, 328
	of Yellow-breasted Chat nests, 402
Presque Isle, 20–21
Probable breeding, definition, 28
Progne subis, ix, x, 56, 60, 64, 67, 77, **282–283**, 485, 503, 524, 528
Project coordinator, 36
Project director, 37
Project organization, 23–26
Prothonotary Warbler. *See* Warbler, Prothonotary
Protonotaria citrea, 27, 57, 59, 60, 74, **358–359**, 488, 525, 528
Purple Finch. *See* Finch, Purple
Purple Martin. *See* Martin, Purple
Pymatuning Reservoir, 20

Quarry, habitat for Bank Swallow, 288
	for Northern Rough-winged Swallow, 286
Quiscalus quiscula, 58, 63, **450–451**, 492, 517, 526, 529

Rabbit, 206
Raccoon, 228
Rail, 102, 168
	Black, 27, 30, 31, 55, 474, 527
	King, 27, 30, 31, 50, 55, 58, 59, 74, **162–163**, 480, 527
	Virginia, 27, 30, 31, 55, 61, 74, **164–165**, 166, 168, 170, 480, 523, 527
Rallus elegans, 27, 30, 31, 50, 55, 58, 59, 74, **162–163**, 480, 527
	R. limicola, 27, 30, 31, 55, 61, 74, **164–165**, 166, 168, 170, 480, 523, 527
Rat, 206
Raven, Common, ix, x, 56, 59, 64, 158, **278–279**, 485, 503, 524, 528
Redhead, 474
Reading Prong Physiographic section, 5, 12, 52, 477–493
Reclaimed surface mine, 10, 18–19, 66
	and Blue Grosbeak, 436
	and Bobolink, 442
	and Brown Thrasher, 336
	and Clay-colored Sparrow, 408
	and Common Nighthawk, 216
	and Common Yellowthroat, 368
	and Dickcissel, 440
	and Eastern Bluebird, 320
	and Eastern Meadowlark, 446

INDEX   581

Reclaimed surface mine (*continued*)
    and Field Sparrow, 410
    and Grasshopper Sparrow, 416
    and Henslow's Sparrow, 418
    and Horned Lark, 280
    and Northern Harrier, 142
    and Prairie Warbler, 396
    and Purple Finch, 458
    and Red-breasted Nuthatch, 300
    and Red-winged Blackbird, 444
    and Savannah Sparrow, 414
    and Short-eared Owl, 212
    and Upland Sandpiper, 178
    and Vesper Sparrow, 412
    and Yellow-breasted Chat, 402
Record review, 35
Red Crossbill. *See* Crossbill, Red
Red List, 70, 71, 73–74
Red-bellied Woodpecker. *See* Woodpecker, Red-bellied
Red-breasted Nuthatch. *See* Nuthatch, Red-breasted
Red-eyed Vireo. *See* Vireo, Red-eyed
Red-headed Woodpecker. *See* Woodpecker, Red-headed
Red-shouldered Hawk. *See* Hawk, Red-shouldered
Redstart, American, 57, 63, 80, **372–373**, 489, 511, 525, 528
Red-tailed Hawk. *See* Hawk, Red-tailed
Red-winged Blackbird. *See* Blackbird, Red-winged
Reed, Common, 30, 122, 314
Reforestation, 18
Region (2nd Atlas), 24, 25, 35, 36, 50–51
Regional Coordinators, xi–xii, 25, 26, 29, 30, 35, 36, 37
Regulus satrapa, 27, 40, 41, 57, 61, 68, 300, **318–319**, 486, 506, 524, 528
Reintroduction
    of Bald Eagle, x, 140
    of Osprey, 138
    of Peregrine Falcon, 160
    of Trumpeter Swan, 92
    of Wild Turkey, 114
Reptiles, food item, 158, 204
Reservoir, 20, 118, 138, 140, 162
Residential. *See* development
Results, summary, 48–69
Rhododendron, 14
    and Black-throated Blue Warbler, 388
    and Canada Warbler, 400
    and Northern Saw-whet Owl, 214
    and Northern Waterthrush, 348
    and Swainson's Warbler, 360
Ridge and Valley Physiographic province, 4, 5, 10, 12, 17, 21, 42, 52, 477–493
Rights of way corridor
    and Canada Warbler, 400
    and Common Yellowthroat, 368
    and Dark-eyed Junco, 426
    and Ovenbird, 342
    and Yellow-breasted Chat, 402
Ring-billed Gull. *See* Gull, Ring-billed

Ring-necked Duck. *See* Duck, Ring-necked
Ring-necked Pheasant. *See* Pheasant, Ring-necked
Riparia riparia, 56, 73, 77, 226, 286, **288–289**, 485, 504, 524, 528
Riparian. *See* River
Riparian buffer, 20
River, 4, 9, 20–21, 46, 66–67, 520
    and Bald Eagle, 140
    and Baltimore Oriole, 456
    and Bank Swallow, 288
    and Black-crowned Night Heron, 130
    and Carolina Wren, 306
    and Cerulean Warbler, 374
    and Chimney Swift, 222
    and Common Merganser, 106
    and Double-crested Cormorant, 118
    and Fish Crow, 276
    and Great Black-backed Gull, 188
    and Great Egret, 126
    and Great Heron, 128
    and Herring Gull, 186
    and Killdeer, 174
    and Marsh Wren, 314
    and Merlin, 158
    and Mute Swan, 90
    and Northern Parula, 376
    and Northern Rough-winged Swallow, 286
    and Orchard Oriole, 454
    and Osprey, 138
    and Peregrine Falcon, 162
    and Prothonotary Warbler, 358
    and Ring-billed Gull, 184
    and Ruby-throated Hummingbird, 224
    and Spotted Sandpiper, 176
    and Swainson's Warbler, 360
    and Veery, 322
    and Warbling Vireo, 268
    and White-eyed Vireo, 262
    and Worm-eating Warbler, 344
    and Yellow-crowned Night Heron, 132
    and Yellow-throated Vireo, 264
    and Yellow-throated Warbler, 394
Roadkill, 134, 136, 278
Robin, American, ix, 26, 27, 28, 57, 58, 62, 63, 75, **330–331**, 438, 444, 487, 507, 525, 528
Rock Pigeon. *See* Pigeon, Rock
Roof, nesting
    and Common Nighthawk, 216
    and Herring Gull, 186
    and Killdeer, 174
    and Ring-billed Gull, 184
Rose-breasted Grosbeak. *See* Grosbeak, Rose-breasted
Ruby-throated Hummingbird. *See* Hummingbird, Ruby-throated
Ruddy Duck. *See* Duck, Ruddy
Ruffed Grouse. *See* Grouse, Ruffed
Rye-grass, 440

Safe Dates, 26, 27, 36, 50, 527–529
Sandhill Crane. *See* Crane, Sandhill

Sandpiper, Spotted, 55, 100, **176–177**, 481, 523, 527
    Upland, 27, 50, 55, 73, 76, **178–179**, 481, 523, 527
Sapsucker, Yellow-bellied, ix, 27, 56, 59, 60, 62, 64, **232–233**, 300, 483, 498, 524, 528
Savannah Sparrow. *See* Sparrow, Savannah
Sayornis phoebe, 56, 63, **254–255**, 484, 500, 524, 528
Scarlet Tanager. *See* Tanager, Scarlet
Scolopax minor, 27, 37, 55, 59, 60, 73, 75, 180, **182–183**, 410, 481, 523, 527
Screech-Owl, Eastern, 30, 32, 56, 60, **204–205**, 481, 524, 528
Scrub. *See* Successional, early; Shrub
Sedge, 30, 120, 312, 314
Sedge Wren. *See* Wren, Sedge
Seiurus aurocapilla, 57, 63, 79, **342–343**, 366, 368, 372, 382, 487, 509, 525, 528
Serviceberry, and Yellow-bellied Sapsucker, 232
Set-aside, 110
Setophaga americana, 57, 58, 64, **376–377**, 489, 511, 525, 528
    S. caerulescens, x, 57, 61, 63, 68, 73, **388–389**, 489, 511, 525, 529
    S. cerulea, 27, 57, 59, 60, 64, 73, 77, **374–375**, 376, 489, 511, 525, 528
    S. citrina, x, 57, 58, 59, 60, 63, **370–371**, 489, 510, 525, 528
    S. coronata, ix, 57, 58, 59, 64, 68, 300, **392–393**, 489, 511, 525, 529
    S. discolor, 28, 57, 59, 64, 74, **396–397**, 490, 512, 525, 529
    S. dominica, 28, 57, **394–395**, 490, 512, 525, 529
    S. fusca, x, 27, 57, 59, 63, 68, 73, 310, 376, **380–381**, 489, 511, 525, 529
    S. magnolia, 57, 59, 61, 63, 68, 310, **378–379**, 489, 511, 525, 529
    S. pensylvanica, x, 57, 59, 63, **384–385**, 489, 511, 525, 529
    S. petechia, ix, 41, 57, 63, 372, **382–383**, 489, 511, 525, 529
    S. pinus, 57, 60, 64, **390–391**, 406, 489, 511, 525, 529
    S. ruticilla, 57, 63, 80, **372–373**, 489, 511, 525, 528
    S. striata, 27, 54, 57, 68, 73, **386–387**, 489, 525, 529
    S. virens, 28, 57, 59, 63, 68, 73, 310, 376, **398–399**, 490, 512, 525, 529
Shale gas. *See* Energy development
Sharp-shinned Hawk. *See* Hawk, Sharp-shinned
Shooting. *See* Harvesting; Persecution
Short-eared Owl. *See* Owl, Short-eared
Shoveler, Northern, 27, 55, **472**, 527
Shrike, Loggerhead, 27, 49, 56, 61, 73, **260–261**, 484, 524, 528
Shrub
    and Acadian Flycatcher, 246
    and American Goldfinch, 466
    and Black-throated Blue Warbler, 388

and Blue Grosbeak, 436
and Blue-headed Vireo, 266
and Blue-winged Warbler, 352
and Brewster's Warbler, 354
and Brown Thrasher, 336
and Canada Warbler, 400
and Chestnut-sided Warbler, 384
and Chipping Sparrow, 406
and Clay-colored Sparrow, 408
and Eastern Towhee, 404
and Field Sparrow, 410
and Golden-winged Warbler, 350
and Gray Catbird, 332
and Hooded Warbler, 370
and Indigo Bunting, 438
and Lawrence's Warbler, 354
and Loggerhead Shrike, 260
and Long-eared Owl, 210
and Mallard, 98
and Northern Bobwhite, 108
and Northern Cardinal, 432
and Northern Waterthrush, 348
and Orchard Oriole, 454
and Ring-necked Pheasant, 110
and Rose-breasted Grosbeak, 434
and Ruffed Grouse, 112
and Swamp Sparrow, 422
and Vesper Sparrow, 412
and White-eyed Vireo, 262
and White-throated Sparrow, 424
and Wild Turkey, 114
and Yellow Warbler, 382
and Yellow-bellied Flycatcher, 244
and Yellow-breasted Chat, 402
Shrub associated birds, 12, 73–74, 78
Shrubby wetland. *See* Wetland, forested
Shrublands/shrubby habitat, x, 14, 15
Sialia sialis, 7, 41, 42, 57, 63, **284, 320–321**, 487, 506, 525, 528
Silviculture. *See* Logging
Siskin, Pine, 28, 58, 59, 61, 73, **464–465**, 492, 526, 529
Sites of Special Concern, 72
Sitta canadensis, 52, 56, 64, 68, **300–301**, 486, 505, 524, 528
  S. carolinensis, 56, 63, **302–303**, 486, 505, 524, 528
Slope, 3, 10, 15, 43, 521
  and Black-and-white Warbler, 356
  and Great Crested Flycatcher, 256
  and Northern Parula, 376
  and Worm-eating Warbler, 344
  and Yellow-throated Vireo, 264
Small mammals, prey item for Eastern Screech-Owl, 204, 212
  for Merlin, 158
  for Short-eared Owl, 212
Snags, 11, 13
  and Brown Creeper
  and Downy Woodpecker, 234
  and Great Crested Flycatcher, 256
  and Red-bellied Woodpecker, 230
  and Red-breasted Nuthatch, 300
  and Red-headed Woodpecker, 228

and Tree Swallow, 286
and Winter Wren, 310
Snipe, Common. *See* Snipe, Wilson's
Snipe, Wilson's, 27, 55, 59, 74, **180–181**, 481, 523, 527
Snowy Egret. *See* Egret, Snowy
Soil, 4, 14, 15, 79, 182
Soil Bank Program, 108
Song Sparrow. *See* Sparrow, Song
Sora, 27, 30, 31, 55, 59, 61, 74, **166–167**, 168, 170, 480, 523, 527
South Mountain Physiographic section, 5, 12, 52, 477–493
Soybeans, 17, 280
Spanworm, 12
  Elm, 386, 474
Sparrow
  Bachman's, 28, 474, 529
  Chipping, ix, 41, 57, 58, 62, 63, 390, **406–407**, 490, 514, 522, 525, 529
  Clay-colored, 28, 54, 57, 58, **408–409**, 490, 525, 529
  Field, x, 40, 41, 57, 63, **410–411**, 490, 514, 525, 529
  Grasshopper, x, 57, 62, 64, 73, 412, **416–417**, 490, 515, 525, 529
  Henslow's, 28, 29, 57, 64, 73, 76, 78, 412, **418–419**, 491, 515, 525, 529
  House, 21, 58, 63, 290, 292, 320, **468–469**, 493, 519, 526, 529
  Lark, 28, 474, 529
  Savannah, 57, 63, 412, **414–415**, 490, 514, 525, 529
  Song, ix, 28, 41, 42, 57, 58, 62, 63, 270, **420–421**, 491, 515, 525, 529
  Swamp, ix, 57, 64, **422–423**, 491, 515, 525, 529
  Vesper, x, 57, 60, 64, **412–413**, 490, 514, 525, 529
  White-throated, 52, 57, 60, 61, 68, 348, **424–425**, 491, 525, 529
Sparrowhawk, Eurasian, 468
Spatterdock, 120
Sphyrapicus varius, ix, 27, 56, 59, 60, 62, 64, **232–233**, 300, 483, 498, 524, 528
Spinus pinus, 28, 58, 59, 61, 73, **464–465**, 492, 526, 529
  S. tristis, 58, 63, **466–467**, 493, 519, 526, 529
Spiza americana, 28, 58, 59, 61, 73, **440–441**, 491, 525, 529
Spizella pallida, 28, 54, 57, 58, **408–409**, 490, 525, 529
  S. passerina, ix, 41, 57, 58, 62, 63, 390, **406–407**, 490, 514, 522, 525, 529
  S. pusilla, x, 40, 41, 57, 63, **410–411**, 490, 514, 525, 529
Spotted Sandpiper. *See* Sandpiper, Spotted
Sprawl development, ix. *See also* Development
Sproul State Forest, 14
Spruce, 13, 35
Spruce
  and Blackburnian Warbler, 380
  and Golden-crowned Kinglet, 318

  and Magnolia Warbler, 378
  and Northern Parula, 376
  and Pine Siskin, 464
  and Pine Warbler, 390
  and Red Crossbill, 462
  and Red-breasted Nuthatch, 300
  and Swainson's Thrush, 324
  and Yellow-rumped Warbler, 392
Spruce, Norway
  and Golden-crowned Kinglet, 318
  and Hermit Thrush, 326
  and Loggerhead Shrike, 260
  and Merlin, 158
Spruce, Red
  and Blackpoll Warbler, 386
  and Red Crossbill, 462
  and Yellow-bellied Flycatcher, 244
Squirrel, 210
  Red, 148
Spruce Budwom
  and Blackpoll Warbler, 386
  and Sharp-shinned Hawk, 144
Starling, European, 21, 57, 63, 228, 238, 320, **338–339**, 487, 508, 525, 528
Start date, 2nd Atlas, 26
State Forest, 78
State Game Land, 78, 104
State Park, 78
Statistical methods, 38–47, 520–522
Stelgidopteryx serripennis, 56, 60, 64, 81, **286–287**, 485, 503, 524, 528
Sterna hirundo, 21, 27, 74, 186, **473**, 527
Sternula antillarum, 471, **473**, 527
Stilt, Black-necked, 55, 58, **473**, 527
Stream, 20–21, 29, 73, 79
  and Acadian Flycatcher, 246
  and Bald Eagle, 140
  and Bank Swallow, 288
  and Belted Kingfisher, 226
  and Cerulean Warbler, 374
  and Dark-eyed Junco, 426
  and Eastern Phoebe, 254
  and Green Heron, 128
  and Hooded Merganser, 104
  and Kentucky Warbler, 366
  and Louisiana Waterthrush, 79, 346
  and Magnolia Warbler, 378
  and Northern Parula, 376
  and Northern Rough-winged Swallow, 286
  and Northern Waterthrush, 348
  and Orchard Oriole, 454
  and Prothonotary Warbler, 358
  and Ruby-throated Hummingbird, 224
  and Spotted Sandpiper, 176
  and Swainson's Thrush, 324
  and Swainson's Warbler, 360
  and Veery, 322
  and Warbling Vireo, 268
  and Willow Flycatcher, 250
  and Winter Wren, 310
  and Worm-eating Warbler, 344
  and Yellow-breasted Chat, 402

INDEX 583

Stream (*continued*)
    and Yellow-crowned Night-Heron, 132
    and Yellow-throated Vireo, 264
    and Yellow-throated Warbler, 394
Streptopelia decaocto, x, 54, 56, **194–195**, 481, 524, 527
Strix varia, 30, 32, 56, 59, 68, 206, **208–209**, 482, 524, 528
Sturnella magna, x, 18, 40, 41, 58, 64, 73, **446–447**, 448, 492, 517, 525, 529
S. neglecta, 58, **448–449**, 492, 529
Sturnus vulgaris, 21, 57, 63, 228, 238, 320, **338–339**, 487, 508, 525, 528
Suburban, 15, 21. *See also* Development
Succession, 18
    and Black Tern, 190
    and Eastern Towhee, 404
    and Grasshopper Sparrow, 416
    and Henslow's Sparrow, 418
    and House Wren, 308
    and Least Flycatcher, 252
    and Rose-breasted Grosbeak, 434
    and Sedge Wren, 312
    and Vesper Sparrow, 412
    and Virginia Rail, 164
Successional, early, x, 11, 15, 18, 20, 58, 59, 61, 73–74, 78, 79
    and Alder Flycatcher, 248
    and American Goldfinch, 466
    and American Woodcock, 182
    and Blue-winged Warbler, 352
    and Brown Thrasher, 336
    and Canada Warbler, 400
    and Chipping Sparrow, 406
    and Field Sparrow, 410
    and Golden-winged Warbler, 350
    and Gray Catbird, 332
    and Nashville Warbler, 362
    and Northern Cardinal, 432
    and Orchard Oriole, 454
    and Prairie Warbler, 396
    and Ruffed Grouse, 112
    and White-eyed Vireo, 262
    and Yellow Warbler, 382
    and Yellow-breasted Chat, 402
Summer Tanager. *See* Tanager, Summer
Susquehanna Lowland Physiographic section, 12
Susquehanna River, 4
Swainson's Thrush. *See* Thrush, Swainson's
Swainson's Warbler. *See* Warbler, Swainson's
Swallow, Bank, 56, 73, 77, 226, 286, **288–289**, 485, 504, 524, 528
    Barn, ix, 56, 63, 158, 284, **292–293**, 485, 504, 524, 528
    Cliff, 56, 59, 64, 67, 288, **290–291**, 485, 504, 524, 528
    Northern Rough-winged, 56, 60, 64, 81, **286–287**, 485, 503, 524, 528
    Tree, 56, 60, 64, 158, **284–285**, 485, 503, 524, 528
Swamp, 20
    and Alder Flycatcher, 248
    and American Woodcock, 180
    and Blackpoll Warbler, 386
    and Common Yellowthroat, 368
    and Green Heron, 128
    and Green-winged Teal, 102
    and Nashville Warbler, 362
    and Northern Waterthrush, 348
    and Prothonotary Warbler, 358
    and Swainson's Warbler, 360
    and Swamp Sparrow, 422
Swamp Sparrow. *See* Sparrow, Swamp
Swan, Mute, 37, 55, **90–91**, 124, 477, 523, 527
    Trumpeter, x, 54, 55, **92–93**, 477, 527
Swift, Chimney, 27, 56, 63, 73, **222–223**, 482, 497, 524, 528
Sycamore, American
    and Black-crowned Night-Heron, 130
    and Cerulean Warbler, 374
    and Warbling Vireo, 268
    and Yellow-crowned Night-Heron, 132
    and Yellow-throated Warbler, 394

Tachycineta bicolor, 56, 60, 64, 158, **284–285**, 485, 503, 524, 528
Tallgrass prairies, 17, 312, 418
Tamarack, and Yellow-rumped Warbler, 392
Tanager, Scarlet, ix, 57, 59, 63, 71, 73, 78, **430–431**, 491, 516, 525, 529
    Summer, 28, 57, 58, 73, **428–429**, 491, 525, 529
Teal, 100–103, 170
    Blue-winged, 55, 58, **100–101**, 477, 523, 527
    Green-winged, 27, 55, 74, **102–103**, 477, 523, 527
Temperature, 4–8
Tern, Black, 27, 55, 58, 74, 168, **190–191**, 481, 524, 527
    Common, 21, 27, 74, 186, **473**, 527
    Least, 471, **473**, 527
Thistle, 466
Thrasher, Brown, 57, 64, 74, **336–337**, 487, 508, 525, 528
Threatened Species, 26, 29, 70, 73–74, 76, 178
Thrush, Hermit, 40, 41, 42, 43, 57, 59, 60, 61, 64, 68, 310, 322, **326–327**, 487, 507, 525, 528
    Swainson's, 27, 57, 68, 73, **324–325**, 487, 507, 525, 528
    Wood, 57, 59, 63, 73, 77, 78, 79, 82, 326, **328–329**, 487, 507, 525, 528
Thryomanes bewickii, 474, 528
Thryothorus ludovicianus, x, 7, 56, 58, 59, 60, 62, 63, **306–307**, 366, 486, 505, 524, 528
Tidal, 3, 20, 276
Tidal Marshes, 162, 314, 422
Timbering. *See* Logging
Timothy, 17, 18, 440
Titmouse, Tufted, 56, 60, 63, 296, **298–299**, 486, 505, 524, 528
Topography, 3, 7, 9, 14, 42–45, 82, 346. *See also* Slope
Tornado damage, 158, 364
Towhee, Eastern, 57, 63, **404–405**, 490, 514, 525, 529

Toxostoma rufum, 57, 64, 74, **336–337**, 487, 508, 525, 528
Trap and transfer. *See* Reintroduction
Tree diseases. *See* Adelgid, Hemlock Woolly; Cankerworm; Spanworm
Tree farm, 15, 458
Tree Swallow. *See* Swallow, Tree
Trees, large, 11
    and Yellow-throated Vireo, 264
    and Brown Creeper, 304
    and Common Merganser, 106
    and Cooper's Hawk, 146
    and Golden-crowned Kinglet, 318
    and Hooded Merganser, 104
    and House Wren, 308
    and Mourning Warbler, 364
    and Orchard Oriole, 454
    and Pileated Woodpecker, 240
    and Red-tailed Hawk, 154
    and Swainson's Warbler, 360
    and White-breasted Nuthatch, 302
    and Winter Wren, 310
    and Wood Duck, 94
    and Yellow-crowned Night-Heron, 132
Trees, mature. *See* Trees, large
Troglodytes aedon, ix, 56, 63, 294, 300, **308–309**, 358, 486, 506, 524, 528
    T. hiemalis, 57, 61, 64, 68, 73, 300, **310–311**, 486, 506, 524, 528
Trumpeter Swan. *See* Swan, Trumpeter
Tufted Titmouse. *See* Titmouse, Tufted
Turdus migratorius, ix, 26, 27, 28, 57, 58, 62, 63, 75, **330–331**, 438, 444, 487, 507, 525, 528
Turkey, Wild, ix, 51, 53, 55, 58, 60, 62, 63, **114–115**, 124, 478, 495, 523, 527
Turkey Vulture. *See* Vulture, Turkey
Tympanuchus cupido, 474
Tyrannus tyrannus, 56, 64, **258–259**, 484, 501, 524, 528
Tyto alba, 27, 29, 30, 37, 49, 56, 73, **202–203**, 482, 524, 528

Understory. *See* Forest understory
U.S. Department of Agriculture, 17, 20, 79, 100, 102, 110, 184, 186
U.S. Fish & Wildlife Service (USFWS), xi, xxii, 23, 37, 70, 71, 73–74, 96, 100, 102, 118, 180, 182
U.S. Geological Survey (USGS), 24, 25, 36, 47, 48, 75, 83
USGS, Patuxent Wildlife Research Center, xxii, xxiii
Upland Sandpiper. *See* Sandpiper, Upland
Upper Delaware Scenic and Recreational River, 79
Upper Mississippi Valley/Great Lakes Waterbird Conservation Plan, 71, 73–74
Urban. *See* development

Veery, 57, 63, **322–323**, 326, 487, 507, 525, 528
Vehicles, collisions with
    and Barred Owl, 208
    and Cooper's Hawk, 146

and Killdeer, 174
and Red-tailed Hawk, 154
and Sharp-shinned Hawk, 144
Vermivora chrysoptera, x, 27, 29, 57, 59, 60, 64, 74, 77, 78, **350–351**, 352, 354, 410, 488, 509, 525, 528
V. cyanoptera, x, 57, 59, 64, 74, 350, **352–353**, 354, 488, 509, 525, 528
Vesper Sparrow. *See* Sparrow, Vesper
Viburnum, 182
Vireo, Blue-headed, 56, 59, 60, 61, 63, 68, 73, **266–267**, 310, 484, 501, 524, 528
Red-eyed, ix, 40, 41, 56, 62, 63, 82, 264, **270–271**, 484, 502, 524, 528
Warbling, ix, 56, 60, 64, **268–269**, 484, 502, 524, 528
White-eyed, ix, 40, 41, 56, 59, 64, **262–263**, 484, 501, 524, 528
Yellow-throated, 56, 60, 64, 73, **264–265**, 484, 501, 524, 528
Vireo flavifrons, 56, 60, 64, 73, **264–265**, 484, 501, 524, 528
V. gilvus, ix, 56, 60, 64, **268–269**, 484, 502, 524, 528
V. griseus, ix, 40, 41, 56, 59, 64, **262–263**, 484, 501, 524, 528
V. olivaceus, ix, 40, 41, 56, 62, 63, 82, 264, **270–271**, 484, 502, 524, 528
V. solitarius, 56, 59, 60, 61, 63, 68, 73, **266–267**, 310, 484, 501, 524, 528
Virginia Rail. *See* Rail, Virginia
Volunteers, list of, xii–xxi
Vulture, Black, 55, 59, 60, **134–135**, 136, 479, 523, 527
Turkey, 55, 64, 134, **136–137**, 479, 523, 527

Warbler, Audubon's. *See* Warbler, Yellow-rumped
Black-and-white, 57, 63, **356–357**, 488, 510, 525, 528
Blackburnian, x, 27, 57, 59, 63, 68, 73, 310, 376, **380–381**, 489, 511, 525, 529
Blackpoll, 27, 54, 57, 68, 73, **386–387**, 489, 525, 529
Black-throated Blue, x, 57, 61, 63, 68, 73, **388–389**, 489, 511, 525, 529
Black-throated Green, 28, 57, 59, 63, 68, 73, 310, 376, **398–399**, 490, 512, 525, 529
Blue-winged, x, 57, 59, 64, 74, 350, **352–353**, 354, 488, 509, 525, 528
Brewster's, **354–355**, 525, 528
Canada, 57, 59, 61, 64, 68, **400–401**, 490, 512, 525, 529
Cerulean, 27, 57, 59, 60, 64, 73, 77, **374–375**, 376, 489, 511, 525, 528
Chestnut-sided, x, 57, 59, 63, **384–385**, 489, 511, 525, 529
Golden-winged, x, 27, 29, 57, 59, 60, 64, 74, 77, 78, **350–351**, 352, 354, 410, 488, 509, 525, 528
Hooded, x, 57, 58, 59, 60, 63, **370–371**, 489, 510, 525, 528
Kentucky, 27, 57, 64, 73, **366–367**, 488, 510, 525, 528
Lawrence's, **354–355**, 525, 528
Magnolia, 57, 59, 61, 63, 68, 310, **378–379**, 489, 511, 525, 529
Mourning, 57, 64, **364–365**, 488, 510, 525, 528
Myrtle. *See* Yellow-rumped
Nashville, 52, 57, 59, 60, 68, **362–363**, 488, 525, 528
Pine, 57, 60, 64, **390–391**, 406, 489, 511, 525, 529
Prairie, 28, 57, 59, 64, 74, **396–397**, 490, 512, 525, 529
Prothonotary, 27, 57, 59, 60, 74, **358–359**, 488, 525, 528
Swainson's, 27, 57, 58, **360–361**, 448, 488, 528
Worm-eating, 57, 64, 73, 78, **344–345**, 488, 509, 525, 528
Yellow, ix, 41, 57, 63, 372, **382–383**, 489, 511, 525, 529
Yellow-rumped, ix, 57, 58, 59, 64, 68, 300, **392–393**, 489, 511, 525, 529
Yellow-throated, 28, 57, **394–395**, 490, 512, 525, 529
Warbling Vireo. *See* Vireo, Warbling
WatchList (Audubon), 71, 73–74, 344, 360, 396
Water depth, 61
and Black Tern, 190
and Pied-billed Grebe, 116
and Prothonotary Warbler, 358
and Sora, 166
and Virginia Rail, 164
Water extraction, 20
Water levels. *See* Water depth
Water willow, 126, 190
Waterthrush
Louisiana, 27, 64, 57, 68, 73, 77, 79, **346–347**, 348, 488, 509, 525, 528
Northern, 27, 52, 57, 60, 61, 68, 77, **348–349**, 488, 525, 528
Waxwing, Cedar, 57, 59, 63, 158, **340–341**, 487, 508, 525, 528
Waynesburg Hills Physiographic section, 5, 12, 52, 477–493
Weather, 4–8, 12, 33, 35, 61, 77
and Carolina Wren, 306
and Eastern Bluebird, 320
and Northern Mockingbird, 334
and Purple Martin, 282
and Wild Turkey, 114
and Wilson's Snipe, 180
West Nile Virus
and American Crow, 274
and American Kestrel, 156
and Blue Jay, 272
and Common Raven, 278
and Cooper's Hawk, 146
and Eastern Bluebird, 320
and Eastern Screech-Owl, 204
and Fish Crow, 276
and Great Horned Owl, 206
and Northern Goshawk, 148
and Tufted Titmouse, 298
Western Meadowlark. *See* Meadowlark, Western
Wet meadows
and Common Yellowthroat, 368
and Sedge Wren, 312
and Sora, 166
and Willow Flycatcher, 250
and Wilson's Snipe, 180
Wetland, 15, 18–21, 66–67
and American Woodcock, 182
and Bald Eagle, 140
and Black-crowned Night-heron, 130
and Eastern Kingbird, 258
and Northern Harrier, 142
and Northern Shoveler, 472
and Red-shouldered Hawk, 150
and Red-winged Blackbird, 444
and Sandhill Crane, 172
and Short-eared Owl, 212
and Song Sparrow, 420
and Swamp Sparrow, 422
and Willow Flycatcher, 250
and Wilson's Snipe, 180
and Yellow Warbler, 382
and Yellow-breasted Chat, 402Wetland drainage. *See* Wetland loss
Wetland, emergent, 20
and American Bittern, 120
and Black Tern, 190
and Blue-winged Teal, 100
and Blue-winged Warbler, 352
and Chipping Sparrow, 406
and Green Heron, 128
and Green-winged Teal, 102
and King Rail, 162
and Least Bittern, 122
and Marsh Wren, 314
and Pied-billed Grebe, 116
and Red-winged Blackbird, 444
and Sora, 166
and Swamp Sparrow, 422
Wetland, forested, 15, 20
and Alder Flycatcher, 250
and American Redstart, 372
and Blue-winged Warbler, 352
and Hooded Merganser, 104
and Hooded Warbler, 370
and Northern Waterthrush, 348
and Swamp Sparrow, 422
and White-eyed Vireo, 262
and White-throated Sparrow, 424
and Winter Wren, 320
and Wood Duck, 94
and Worm-eating Warbler, 344
and Yellow Warbler, 382
and Yellow-bellied Flycatcher, 244
Wetland, headwater, and Northern Waterthrush, 348
Wetland, wood. *See* Wetland, forested

Wetland loss, 20
  and Alder Flycatcher, 248
  and American Bittern, 120
  and American Black Duck, 96
  and American Coot, 170
  and Common Gallinule, 168
  and Great Blue Heron, 124
  and Green Heron, 128
  and King Rail, 162
  and Least Bittern, 122
  and Marsh Wren, 314
  and Sora, 166
  and Swamp Sparrow, 422
  and Virginia Rail, 164
  and Wood Duck, 94
Wetland mitigation, and Virginia Rail, 164
Wetland Reserve Program, 100, 102, 120
Wheat, 17
Whip-poor-will, Eastern, 27, 32, 56, 73, 77, **220–221**, 482, 524, 528
  Mexican, 220
White-breasted Nuthatch. See Nuthatch, White-breasted
White-eyed Vireo. See Vireo, White-eyed
White-throated Sparrow. See Sparrow, White-throated
White-winged Crossbill. See Crossbill, White-winged
Wigeon, American, 27, 55, **471–472**, 527
Wild Turkey. See Turkey, Wild
Wildlife Action Plan (WAP), 71–72, 73–74, 108, 142, 148, 150, 164, 170, 202, 222, 228, 250, 310, 314, 336, 346, 374, 380, 388, 402, 414, 416, 418, 428, 430
Wildlife Habitat Inventive Program, 100, 102, 110
Willow, 382
Willow Flycatcher. See Flycatcher, Willow
Wilson's Snipe. See Snipe, Wilson's
Wind, 7, 15, 30, 33, 35, 40
Windfarm. See Energy development
Windows, collisions with, 22
  and Cooper's Hawk, 146
  and Ovenbird, 342
  and Scarlet Tanager, 430
  and Sharp-shinned Hawk, 144
Wind-throw, 362, 364, 400
Winter Raptor Survey
  and Black Vulture, 134
  and Northern Harrier, 142
  and Red-tailed Hawk, 154
Winter Wren. See Wren, Winter

Wintering grounds, issues, 80
  and American Kestrel, 156
  and Black-and-white Warbler, 356
  and Blackburnian Warbler, 380
  and Black-throated Blue Warbler, 388
  and Broad-winged Hawk, 152
  and Brown-headed Cowbird, 452
  and Dickcissel, 440
  and Eastern Wood-Pewee, 242
  and Great Blue Heron, 124
  and Indigo Bunting, 438
  and Louisiana Waterthrush, 346
  and Nashville Warbler, 362
  and Northern Waterthrush, 348
  and Ovenbird, 342
  and Prothonotray Warbler, 358
  and Red-winged Blackbird, 444
  and Sharp-shinned Hawk, 144
  and Sora, 166
  and Swainson's Warbler, 360
  and Upland Sandpiper, 178
  and Veery, 322
  and Wood Thrush, 328
  and Yellow-rumped Warbler, 392
  and Yellow-throated Vireo, 264
Wires, collisions with
  and Osprey, 138
  and Ovenbird, 342
  and Yellow-bellied Sapsucker, 232
Wood Duck. See Duck, Wood
Wood Thrush. See Thrush, Wood
Woodcock, American, 27, 37, 55, 59, 60, 73, 75, 180, **182–183**, 410, 481, 523, 527
Woodlot, 10
  and American Redstart, 372
  and Barred Owl, 208
  and Black-throated Blue Warbler, 388
  and Brown-headed Cowbird, 452
  and Cerulean Warbler, 374
  and Chuck-will's-widow, 218
  and Eastern Screech-Owl, 204
  and Golden-crowned Kinglet, 318
  and Great Crested Flycatcher, 256
  and Great Horned Owl, 206
  and Hooded Warbler, 370
  and House Wren, 308
  and Long-eared Owl, 210
  and Ovenbird, 342
  and Pileated Woodpecker, 240
  and Red-headed Woodpecker, 228
  and Song Sparrow, 420
  and Veery, 322

  and White-breasted Nuthatch, 302
  and Wood Thrush, 328
Woodpecker, Downy, 56, 59, 63, 230, **234–235**, 236, 248, 483, 499, 524, 528
  Hairy, 56, 59, 64, **236–237**, 483, 499, 524, 528
  Pileated, x, 56, 58, 64, 82, **240–241**, 483, 499, 524, 528
  Red-bellied, x, 56, 58, 59, 60, 62, 63, **230–231**, 483, 498, 524, 528
  Red-headed, 27, 56, 60, 73, 77, **228–229**, 483, 498, 524, 528
Wood-Pewee, Eastern, 56, 59, 63, **242–243**, 254, 483, 499, 524, 528
Worm-eating Warbler. See Warbler, Worm-eating
Wren
  Bewick's, 27, 474, 528
  Carolina, x, 7, 56, 58, 59, 60, 62, 63, **306–307**, 366, 486, 505, 524, 528
  House, ix, 56, 63, 294, 300, **308–309**, 358, 486, 506, 524, 528
  Marsh, 27, 57, 74, 77, 102, 168, 190, 312, **314–315**, 486, 524, 528
  Pacific, 310
  Sedge, 27, 57, 73, **312–313**, 486, 524, 528
  Winter, 57, 61, 64, 68, 73, 300, **310–311**, 486, 506, 524, 528

Yellow Warbler. See Warbler, Yellow
Yellow-bellied Flycatcher. See Flycatcher, Yellow-bellied
Yellow-bellied Sapsucker. See Sapsucker, Yellow-bellied
Yellow-billed Cuckoo. See Cuckoo, Yellow-billed
Yellow-breasted Chat. See Chat, Yellow-breasted
Yellow-crowned Night-Heron. See Night-Heron, Yellow-crowned
Yellow-rumped Warbler. See Warbler, Yellow-rumped
Yellowthroat, Common, 57, 58, 63, 368–369, 372, 382, 488, 510, 525, 528
Yellow-throated Vireo. See Vireo, Yellow-throated
Yellow-throated Warbler. See Warbler, Yellow-throated

Zenaida macroura, 56, 63, **196–197**, 481, 497, 524, 527
Zonotrichia albicollis, 52, 57, 60, 61, 68, 348, **424–425**, 491, 525, 529